The Physics of Plasmas

The Physics of Plasmas provides a comprehensive introduction to the subject suitable for adoption as a self-contained text for courses at advanced undergraduate and graduate level. The extensive coverage of basic theory is illustrated with examples drawn from fusion, space and astrophysical plasmas.

A particular strength of the book is its discussion of the various models used to describe plasma physics including particle orbit theory, fluid equations, ideal and resistive magnetohydrodynamics, wave equations and kinetic theory. The relationships between these distinct approaches are carefully explained giving the reader a firm grounding in the fundamentals, and developing this into an understanding of some of the more specialized topics. Throughout the text, there is an emphasis on the physical interpretation of plasma phenomena and exercises, designed to test the reader's understanding at a variety of levels, are provided.

Students of physics and astronomy, engineering and applied mathematics will find a clear and rigorous explanation of the fundamental properties of plasmas with minimal mathematical formality. This book will also serve as a reference source for physicists and engineers engaged in research on aspects of fusion and space plasmas.

Before retiring, T.J.M. BOYD was Professor of Physics at the University of Essex. He has taught graduate students on plasma physics courses in Europe and North America. His research interests have included atomic collision theory, computational physics and plasma physics. Professor Boyd has co-authored two previous books, *Plasma Dynamics* (1969) with J.J. Sanderson, and *Electricity* (1979) with C.A. Coulson.

JEFF SANDERSON is Professor Emeritus at the University of St Andrews. His research interests are in theoretical plasma physics and specifically plasma instabilities, collisionless shock waves and transport phenomena. Professor Sanderson has taught plasma physics for over 30 years, principally at St Andrews University, and the UKAEA Culham Summer School, but also by invitation in the USA, Europe and Pakistan. As well as co-authoring *Plasma Dynamics* (1969) he was a contributor to two Culham textbooks and co-editor with R.A. Cairns of *Laser Plasma Interactions* (1980).

The Physics of Plasmas

T.J.M. BOYD
University of Essex

J.J. SANDERSON
University of St Andrews

CAMBRIDGE UNIVERSITY PRESS
Cambridge, New York, Melbourne, Madrid, Cape Town, Singapore, São Paulo

Cambridge University Press
The Edinburgh Building, Cambridge CB2 8RU, UK

Published in the United States of America by Cambridge University Press, New York

www.cambridge.org
Information on this title: www.cambridge.org/9780521459129

First published 2003
Reprinted 2005

A catalogue record for this publication is available from the British Library

ISBN 978-0-521-45290-8 hardback
ISBN 978-0-521-45912-9 paperback

Transferred to digital printing (with amendments) 2007

The colour figure within this publication has been produced as a black and white image for this digital reprinting. At the time of going to press the original figure was available in colour for download from http://www.cambridge.org/9780521459129

Contents

Preface *page* xi

1 Introduction **1**
1.1 Introduction 1
1.2 Thermonuclear fusion 2
1.2.1 The Lawson criterion 3
1.2.2 Plasma containment 4
1.3 Plasmas in space 6
1.4 Plasma characteristics 7
1.4.1 Collisions and the plasma parameter 10

2 Particle orbit theory **12**
2.1 Introduction 12
2.2 Constant homogeneous magnetic field 14
2.2.1 Magnetic moment and plasma diamagnetism 16
2.3 Constant homogeneous electric and magnetic fields 16
2.3.1 Constant non-electromagnetic forces 18
2.4 Inhomogeneous magnetic field 19
2.4.1 Gradient drift 19
2.4.2 Curvature drift 21
2.5 Particle drifts and plasma currents 22
2.6 Time-varying magnetic field and adiabatic invariance 24
2.6.1 Invariance of the magnetic moment in an inhomogeneous field 25
2.7 Magnetic mirrors 26
2.8 The longitudinal adiabatic invariant 28
2.8.1 Mirror traps 30
2.9 Magnetic flux as an adiabatic invariant 31

2.10 Particle orbits in tokamaks 33
2.11 Adiabatic invariance and particle acceleration 35
2.12 Polarization drift 37
2.13 Particle motion at relativistic energies 38
2.13.1 Motion in a monochromatic plane-polarized electromagnetic wave 38
2.14 The ponderomotive force 40
2.15 The guiding centre approximation: a postscript 41
Exercises 43

3 Macroscopic equations **48**
3.1 Introduction 48
3.2 Fluid description of a plasma 49
3.3 The MHD equations 58
3.3.1 Resistive MHD 59
3.3.2 Ideal MHD 60
3.4 Applicability of the MHD equations 61
3.4.1 Anisotropic plasmas 67
3.4.2 Collisionless MHD 69
3.5 Plasma wave equations 71
3.5.1 Generalized Ohm's law 73
3.6 Boundary conditions 74
Exercises 76

4 Ideal magnetohydrodynamics **77**
4.1 Introduction 77
4.2 Conservation relations 78
4.3 Static equilibria 82
4.3.1 Cylindrical configurations 85
4.3.2 Toroidal configurations 89
4.3.3 Numerical solution of the Grad–Shafranov equation 100
4.3.4 Force-free fields and magnetic helicity 102
4.4 Solar MHD equilibria 105
4.4.1 Magnetic buoyancy 106
4.5 Stability of ideal MHD equilibria 108
4.5.1 Stability of a cylindrical plasma column 111
4.6 The energy principle 119
4.6.1 Finite element analysis of ideal MHD stability 123
4.7 Interchange instabilities 124
4.7.1 Rayleigh–Taylor instability 124
4.7.2 Pressure-driven instabilities 128

4.8 Ideal MHD waves 130
Exercises 133

5 Resistive magnetohydrodynamics 140
5.1 Introduction 140
5.2 Magnetic relaxation and reconnection 142
5.2.1 Driven reconnection 145
5.3 Resistive instabilities 148
5.3.1 Tearing instability 151
5.3.2 Driven resistive instabilities 155
5.3.3 Tokamak instabilities 157
5.4 Magnetic field generation 162
5.4.1 The kinematic dynamo 163
5.5 The solar wind 169
5.5.1 Interaction with the geomagnetic field 177
5.6 MHD shocks 179
5.6.1 Shock equations 182
5.6.2 Parallel shocks 186
5.6.3 Perpendicular shocks 188
5.6.4 Oblique shocks 189
5.6.5 Shock thickness 190
Exercises 193

6 Waves in unbounded homogeneous plasmas 197
6.1 Introduction 197
6.2 Some basic wave concepts 198
6.2.1 Energy flux 200
6.2.2 Dispersive media 200
6.3 Waves in cold plasmas 202
6.3.1 Field-free plasma ($\mathbf{B}_0 = 0$) 209
6.3.2 Parallel propagation ($\mathbf{k} \parallel \mathbf{B}_0$) 210
6.3.3 Perpendicular propagation ($\mathbf{k} \perp \mathbf{B}_0$) 214
6.3.4 Wave normal surfaces 217
6.3.5 Dispersion relations for oblique propagation 222
6.4 Waves in warm plasmas 227
6.4.1 Longitudinal waves 228
6.4.2 General dispersion relation 230
6.5 Instabilities in beam–plasma systems 238
6.5.1 Two-stream instability 240
6.5.2 Beam–plasma instability 241
6.6 Absolute and convective instabilities 244

6.6.1 Absolute and convective instabilities in systems with weakly coupled modes 245
Exercises 248

7 Collisionless kinetic theory **252**
7.1 Introduction 252
7.2 Vlasov equation 254
7.3 Landau damping 256
7.3.1 Experimental verification of Landau damping 263
7.3.2 Landau damping of ion acoustic waves 265
7.4 Micro-instabilities 268
7.4.1 Kinetic beam–plasma and bump-on-tail instabilities 273
7.4.2 Ion acoustic instability in a current-carrying plasma 274
7.5 Amplifying waves 276
7.6 The Bernstein modes 277
7.7 Inhomogeneous plasma 283
7.8 Test particle in a Vlasov plasma 287
7.8.1 Fluctuations in thermal equilibrium 288
Exercises 289

8 Collisional kinetic theory **296**
8.1 Introduction 296
8.2 Simple transport coefficients 297
8.2.1 Ambipolar diffusion 300
8.2.2 Diffusion in a magnetic field 301
8.3 Neoclassical transport 304
8.4 Fokker–Planck equation 307
8.5 Collisional parameters 313
8.6 Collisional relaxation 317
Exercises 321

9 Plasma radiation **324**
9.1 Introduction 324
9.2 Electrodynamics of radiation fields 325
9.2.1 Power radiated by an accelerated charge 326
9.2.2 Frequency spectrum of radiation from an accelerated charge 328
9.3 Radiation transport in a plasma 330
9.4 Plasma bremsstrahlung 334
9.4.1 Plasma bremsstrahlung spectrum: classical picture 336
9.4.2 Plasma bremsstrahlung spectrum: quantum mechanical picture 338

9.4.3 Recombination radiation 339
9.4.4 Inverse bremsstrahlung: free–free absorption 341
9.4.5 Plasma corrections to bremsstrahlung 342
9.4.6 Bremsstrahlung as plasma diagnostic 343
9.5 Electron cyclotron radiation 344
9.5.1 Plasma cyclotron emissivity 346
9.5.2 ECE as tokamak diagnostic 347
9.6 Synchrotron radiation 348
9.6.1 Synchrotron radiation from hot plasmas 348
9.6.2 Synchrotron emission by ultra-relativistic electrons 351
9.7 Scattering of radiation by plasmas 355
9.7.1 Incoherent Thomson scattering 355
9.7.2 Electron temperature measurements from Thomson scattering 358
9.7.3 Effect of a magnetic field on the spectrum of scattered light 360
9.8 Coherent Thomson scattering 361
9.8.1 Dressed test particle approach to collective scattering 361
9.9 Coherent Thomson scattering: experimental verification 365
9.9.1 Deviations from the Salpeter form factor for the ion feature: impurity ions 366
9.9.2 Deviations from the Salpeter form factor for the ion feature: collisions 369
Exercises 370

10 Non-linear plasma physics **376**
10.1 Introduction 376
10.2 Non-linear Landau theory 377
10.2.1 Quasi-linear theory 377
10.2.2 Particle trapping 382
10.2.3 Particle trapping in the beam–plasma instability 384
10.2.4 Plasma echoes 388
10.3 Wave–wave interactions 389
10.3.1 Parametric instabilities 392
10.4 Zakharov equations 397
10.4.1 Modulational instability 402
10.5 Collisionless shocks 405
10.5.1 Shock classification 408
10.5.2 Perpendicular, laminar shocks 411
10.5.3 Particle acceleration at shocks 421
Exercises 423

11 Aspects of inhomogeneous plasmas **425**
11.1 Introduction 425
11.2 WKBJ model of inhomogeneous plasma 426
11.2.1 Behaviour near a cut-off 429
11.2.2 Plasma reflectometry 432
11.3 Behaviour near a resonance 433
11.4 Linear mode conversion 435
11.4.1 Radiofrequency heating of tokamak plasma 439
11.5 Stimulated Raman scattering 441
11.5.1 SRS in homogeneous plasmas 441
11.5.2 SRS in inhomogeneous plasmas 442
11.5.3 Numerical solution of the SRS equations 447
11.6 Radiation from Langmuir waves 450
11.7 Effects in bounded plasmas 453
11.7.1 Plasma sheaths 453
11.7.2 Langmuir probe characteristics 456
Exercises 458

12 The classical theory of plasmas **464**
12.1 Introduction 464
12.2 Dynamics of a many-body system 465
12.2.1 Cluster expansion 469
12.3 Equilibrium pair correlation function 472
12.4 The Landau equation 476
12.5 Moment equations 480
12.5.1 One-fluid variables 485
12.6 Classical transport theory 487
12.6.1 Closure of the moment equations 488
12.6.2 Derivation of the transport equations 491
12.6.3 Classical transport coefficients 495
12.7 MHD equations 501
12.7.1 Resistive MHD 503
Exercises 505

Appendix 1 **Numerical values of physical constants and plasma parameters** **507**
Appendix 2 **List of symbols** **509**

References 517
Index 523

Preface

The present book has its origins in our earlier book *Plasma Dynamics* published in 1969. Many who used *Plasma Dynamics* took the trouble to send us comments, corrections and criticism, much of which we intended to incorporate in a new edition. In the event our separate preoccupations so delayed this that we came to the conclusion that we should instead write another book, that might better reflect changes of emphasis in the subject since the original publication. In writing we had two aims. The first was to describe topics that have a place in any core curriculum for plasma physics, regardless of subsequent specialization and to do this in a way that, while keeping physical understanding firmly in mind, did not compromise on a proper mathematical framework for developing the subject. At the same time we felt the need to go a step beyond this and illustrate and extend this basic theory with examples drawn from topics in fusion and space plasma physics.

In developing the subject we have followed the traditional approach that in our experience works best, beginning with particle orbit theory. This combines the relative simplicity of describing the dynamics of a single charged particle, using concepts familiar from classical electrodynamics, before proceeding to a variety of magnetohydrodynamic (MHD) models. Some of the intrinsic difficulties in getting to grips with magnetohydrodynamics stem from the persistent neglect of classical fluid dynamics in most undergraduate physics curricula. To counter this we have included in Chapter 3 a brief outline of some basic concepts of fluid dynamics before characterizing the different MHD regimes. This leads on to a detailed account of ideal MHD in Chapter 4 followed by a selection of topics illustrating different aspects of resistive MHD in Chapter 5. Plasmas support a bewildering variety of waves and instabilities and the next two chapters are given over to classifying the most important of these. Chapter 6 continues the MHD theme, dealing with waves which can be described macroscopically. In contrast to normal fluids, plasmas are characterized by modes which have to be described microscopically, i.e. in terms of kinetic theory, because only particular particles in the distribution interact with the modes in question. An introduction to plasma kinetic theory is included in Chapter 7 along with a full discussion of the basic modes, the physics of which is governed largely by wave–particle interactions. The development of kinetic theory is continued in Chapter 8 but with a change of emphasis. Whereas the effect of collisions between plasma particles is disregarded in Chapter 7, these move centre stage in Chapter 8 with an introduction to another key topic, plasma transport theory.

A thorough grounding in plasma physics is provided by a selection of topics from the first eight chapters, which make up a core syllabus irrespective of subsequent specialization. The remaining chapters develop the subject and provide a basis for more specialized courses, although arguably Chapter 9 on plasma radiation is properly part of any core syllabus. This chapter, which discusses the principal sources of plasma radiation, excepting bound–bound transitions, along with an outline of radiative transport and the scattering of radiation by laboratory plasmas, provides an introduction to a topic which underpins a number of key plasma diagnostics. Chapters 10 and 11 deal in turn and in different ways with aspects of non-linear plasma physics and with effects in inhomogeneous plasmas. Both subjects cover such a diversity of topics that we have been limited to a discussion of a number of examples, chosen to illustrate the methodology and physics involved. In Chapter 10 we mainly follow a tutorial approach, outlining a variety of important non-linear effects, whereas in Chapter 11 we describe in greater detail a few particular examples by way of demonstrating the effects of plasma inhomogeneity and physical boundaries. The book ends with a chapter on the classical theory of plasmas in which we outline the comprehensive mathematical structure underlying the various models used, highlighting how these relate to one another.

An essential part of getting to grips with any branch of physics is working through exercises at a variety of levels. Most chapters end with a selection of exercises ranging from simple quantitative applications of basic results on the one hand to others requiring numberical solution or reference to original papers.

We are indebted to many who have helped in a variety of ways during the long period it has taken to complete this work. For their several contributions, comments and criticism we thank Hugh Barr, Alan Cairns, Angela Dyson, Pat Edwin, Ignazio Fidone, Malcolm Haines, Alan Hood, Gordon Inverarity, David Montgomery, Ricardo Ondarza-Rovira, Sean Oughton, Eric Priest, Bernard Roberts, Steven Schwartz, Greg Tallents, Alexey Tatarinov and Andrew Wright. We are indebted to Dr J.M. Holt for permission to reproduce Fig. 9.16. Special thanks are due to Andrew Mackwood who prepared the figures and to Misha Sanderson who shared with Andrew the burden of producing much of the LaTeX copy. Finally, we thank Sally Thomas, our editor at CUP, for her ready help and advice in bringing the book to press.

T.J.M. Boyd, Dedham
J.J. Sanderson, St Andrews

1

Introduction

1.1 Introduction

The plasma state is often referred to as the *fourth* state of matter, an identification that resonates with the element of *fire*, which along with earth, water and air made up the elements of Greek cosmology according to Empedocles.† Fire may indeed result in a transition from the gaseous to the plasma state, in which a gas may be fully or, more likely, partially ionized. For the present we identify as *plasma* any state of matter that contains enough free charged particles for its dynamics to be dominated by electromagnetic forces. In practice quite modest degrees of ionization are sufficient for a gas to exhibit electromagnetic properties. Even at 0.1 per cent ionization a gas already has an electrical conductivity almost half the maximum possible, which is reached at about 1 per cent ionization.

The outer layers of the Sun and stars in general are made up of matter in an ionized state and from these regions winds blow through interstellar space contributing, along with stellar radiation, to the ionized state of the interstellar gas. Thus, much of the matter in the Universe exists in the plasma state. The Earth and its lower atmosphere is an exception, forming a plasma-free oasis in a plasma universe. The upper atmosphere on the other hand, stretching into the ionosphere and beyond to the magnetosphere, is rich in plasma effects.

Solar physics and in a wider sense cosmic electrodynamics make up one of the roots from which the physics of plasmas has grown; in particular, that part of the subject known as magnetohydrodynamics – MHD for short – was established largely through the work of Alfvén. A quite separate root developed from the physics of gas discharges, with glow discharges used as light sources and arcs as a means of cutting and welding metals. The word *plasma* was first used by Langmuir in 1928 to describe the ionized regions in gas discharges. These origins

† Empedocles, who lived in Sicily in the shadow of Mount Etna in the fifth century BC, was greatly exercised by fire. He died testing his theory of buoyancy by jumping into the volcano in 433BC.

are discernible even today though the emphasis has shifted. Much of the impetus for the development of plasma physics over the second half of the twentieth century came from research into controlled thermonuclear fusion on the one hand and astrophysical and space plasma phenomena on the other.

To a degree these links with 'big science' mask more bread-and-butter applications of plasma physics over a range of technologies. The use of plasmas as sources for energy-efficient lighting and for metal and waste recycling and their role in surface engineering through high-speed deposition and etching may seem prosaic by comparison with fusion and space science but these and other commercial applications have laid firm foundations for a new plasma technology. That said, our concern throughout this book will focus in the main on the physics of plasmas with illustrations drawn where appropriate from fusion and space applications.

1.2 Thermonuclear fusion

While thermonuclear fusion had been earlier indentified as the source of energy production in stars it was first discussed in detail by Bethe, and independently von Weizsäcker, in 1938. The chain of reactions proposed by Bethe, known as the carbon cycle, has the distinctive feature that after a sequence of thermonuclear burns involving nitrogen and oxygen, carbon is regenerated as an end product enabling the cycle to begin again. For stars with lower central temperatures the proton–proton cycle

$$\begin{aligned}
{}_1\mathrm{H}^1 + {}_1\mathrm{H}^1 &\rightarrow {}_1\mathrm{D}^2 + \mathrm{e}^+ + \nu && (1.44\,\mathrm{MeV})\\
{}_1\mathrm{D}^2 + {}_1\mathrm{H}^1 &\rightarrow {}_2\mathrm{He}^3 + \gamma && (5.49\,\mathrm{MeV})\\
{}_2\mathrm{He}^3 + {}_2\mathrm{He}^3 &\rightarrow {}_2\mathrm{He}^4 + 2\,{}_1\mathrm{H}^1 && (12.86\,\mathrm{MeV})
\end{aligned}$$

where e^+, ν and γ denote in turn a positron, neutrino and gamma-ray, is more important and is in fact the dominant reaction chain in lower main sequence stars (see Salpeter (1952)). Numbers in brackets denote the energy per reaction. In the first reaction in the cycle, the photon energy released following positron–electron annihilation (1.18 MeV) is included; the balance (0.26 MeV) carried by the neutrino escapes from the star. The third reaction in the cycle is only possible at temperatures above about 10^7 K but accounts for almost half of the total energy release of 26.2 MeV. The proton–proton cycle is dominant in the Sun, the transition to the carbon cycle taking place in stars of slightly higher mass. The energy produced not only ensures stellar stability against gravitational collapse but is the source of luminosity and indeed all aspects of the physics of the outer layers of stars.

The reaction that offers the best energetics for controlled thermonuclear fusion in the laboratory on the other hand is one in which nuclei of deuterium and tritium fuse to yield an alpha particle and a neutron:

$$ {}_1\mathrm{D}^2 + {}_1\mathrm{T}^3 \rightarrow {}_2\mathrm{He}^4 + {}_0\mathrm{n}^1 \quad (17.6\,\mathrm{MeV}) $$

The total energy output $\Delta E = 17.6\,\mathrm{MeV}$ is distributed between the alpha particle which has a kinetic energy of about 3.5 MeV and the neutron which carries the balance of the energy released. The alpha particle is confined by the magnetic field containing the plasma and used to heat the fuel, whereas the neutron escapes through the wall of the device and has to be contained by a neutron-absorbing blanket.

1.2.1 The Lawson criterion

Although the D–T reaction rate peaks at temperatures of the order of 100 keV it is not necessary for reacting nuclei to be as energetic as this, otherwise controlled thermonuclear fusion would be impracticable. Thanks to quantum tunnelling through the Coulomb barrier, the reaction rate for nuclei with energies of the order of 10 keV is sufficiently large for fusion to occur. A simple and widely used index of thermonuclear gain is provided by the Lawson criterion. For equal deuterium and tritium number densities, $n_\mathrm{D} = n_\mathrm{T} = n$, the thermonuclear power generated by a D–T reactor per unit volume is $P_\mathrm{fus} = \frac{1}{4}n^2\langle\sigma v\rangle\Delta E$, where $\langle\sigma v\rangle$ denotes the reaction rate, σ being the collisional cross-section and v the relative velocity of colliding particles. For a D–T plasma at a temperature of 10 keV, $\langle\sigma v\rangle \sim 1.1 \times 10^{-22}\,\mathrm{m^3\,s^{-1}}$ so that $P_\mathrm{fus} \sim 7.7 \times 10^{-35}n^2\,\mathrm{W\,m^{-3}}$. About 20% of this output is alpha particle kinetic energy which is available to sustain the fuel at thermonuclear reaction temperatures, the balance being carried by the neutrons which escape from the plasma. Thus the power absorbed by the plasma is $P_\alpha = \frac{1}{4}\langle\sigma v\rangle n^2 E_\alpha$ where $E_\alpha = 3.5\,\mathrm{MeV}$. This is the heat added to unit volume of plasma per unit time as a result of fusion.

We have to consider next the energy lost through radiation, in particular as bremsstrahlung from electron–ion collisions. We shall find in Chapter 9 that bremsstrahlung power loss from hot plasmas may be represented as $P_\mathrm{b} = \alpha n^2 T^{1/2}$, where α is a constant and T denotes the plasma temperature. Above some critical temperature the power absorbed through alpha particle heating outstrips the bremsstrahlung loss. Other energy losses besides bremsstrahlung have to be taken into consideration. In particular, heat will be lost to the wall surrounding the plasma at a rate $3nk_\mathrm{B}T/\tau$ where τ is the containment time and k_B is Boltzmann's constant. Balancing power gain against loss we arrive at a relation for $n\tau$. Lawson (1957) introduced an efficiency factor η to allow power available for heating to be expressed in terms of the total power leaving the plasma. The *Lawson criterion* for power

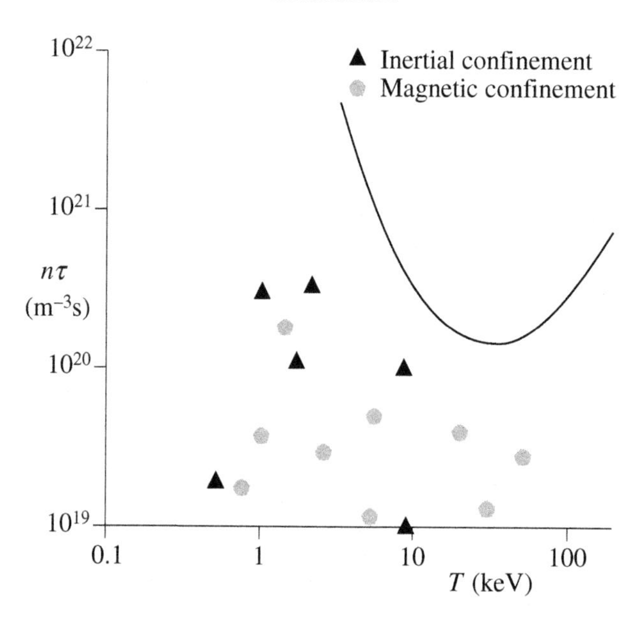

Fig. 1.1. The Lawson criterion for ignition of fusion reactions. Data points correspond to a range of magnetic and inertial confinement experiments showing a progression towards the Lawson curve.

gain is then

$$n\tau > \frac{3k_B T}{\left[\frac{\eta}{4(1-\eta)}\langle\sigma v\rangle \Delta E - \alpha T^{1/2}\right]} \tag{1.1}$$

This condition is represented in Fig. 1.1. Using Lawson's choice for $\eta = 1/3$ (which with hindsight is too optimistic), the power-gain condition reduces to $n\tau > 10^{20}\,\mathrm{m}^{-3}\,\mathrm{s}$. The data points shown in Fig. 1.1 are $n\tau$ values from a range of both magnetically and inertially contained plasmas over a period of about two decades, showing the advances made in both confinement schemes towards the Lawson curve.

1.2.2 Plasma containment

Hot plasmas have to be kept from contact with walls so that from the outset magnetic fields have been used to contain plasma in controlled thermonuclear fusion experiments. Early devices such as Z-pinches, while containing and pinching the plasma radially, suffered serious end losses. Other approaches trapped the plasma in a magnetic bottle or used a closed toroidal vessel. Of the latter the *tokamak*, a contraction of the Russian for *toroidal magnetic chamber*, has been the most successful. Its success compared with competing toroidal containment schemes is

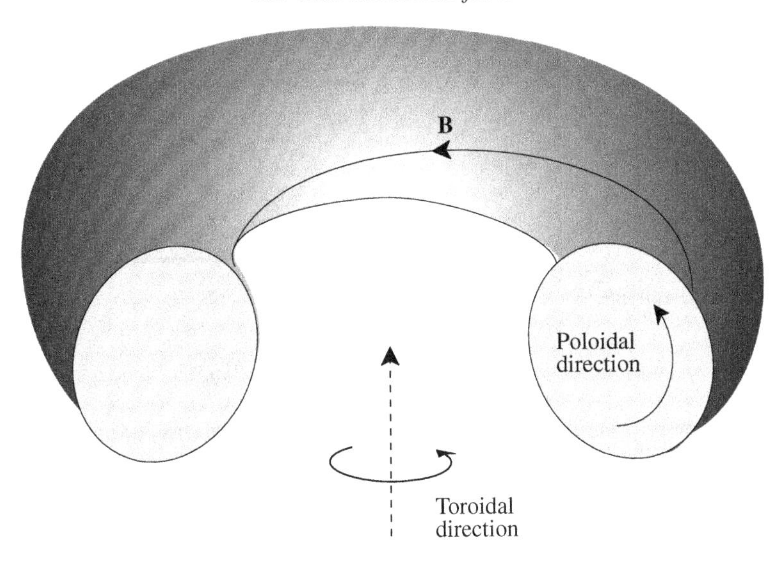

Fig. 1.2. Tokamak cross-section.

attributable in large part to the structure of the magnetic field used. Tokamak fields are made up of two components, one *toroidal*, the other *poloidal*, with the resultant field winding round the torus as illustrated in Fig. 1.2. The toroidal field produced by currents in external coils is typically an order of magnitude larger than the poloidal component and it is this aspect that endows tokamaks with their favourable stability characteristics. Whereas a plasma in a purely toroidal field drifts towards the outer wall, this drift may be countered by balancing the outward force with the magnetic pressure from a poloidal field, produced by currents in the plasma. Broadly speaking, the poloidal field maintains toroidal stability while the toroidal field provides radial stability. For a typical tokamak plasma density the Lawson criterion requires containment times of a few seconds.

Inertial confinement fusion (ICF) offers a distinct alternative to magnetic containment fusion (MCF). In ICF the plasma, formed by irradiating a target with high-power laser beams, is compressed to such high densities that the Lawson criterion can be met for confinement times many orders of magnitude smaller than those needed for MCF and short enough for the plasma to be confined inertially. The ideas behind inertial confinement are represented schematically in Fig. 1.3(a) showing a target, typically a few hundred micrometres in diameter filled with a D–T mixture, irradiated symmetrically with laser light. The ionization at the target surface results in electrons streaming away from the surface, dragging ions in their wake. The back reaction resulting from ion blow-off compresses the target and the aim of inertial confinement is to achieve compression around 1000 times

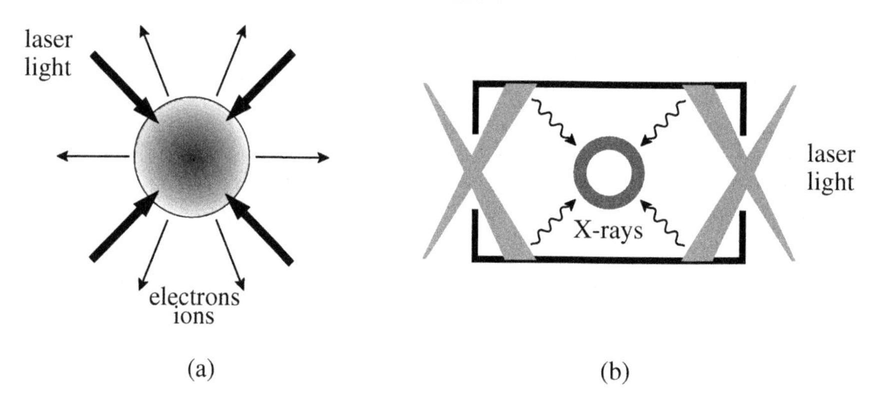

Fig. 1.3. Direct drive (a) and indirect drive (hohlraum) (b) irradiation of targets by intense laser light.

liquid density with minimal heating of the target until the final phase when the compressed fuel is heated to thermonuclear reaction temperatures. An alternative to the *direct drive* approach illustrated in Fig. 1.3(a) is shown in Fig. 1.3(b) in which the target is surrounded by a *hohlraum*. Light enters the hohlraum and produces X-rays which in turn provide target compression and *indirect drive* implosion.

1.3 Plasmas in space

Thermonuclear burn in stars is the source of plasmas in space. From stellar cores where thermonuclear fusion takes place, keV photons propagate outwards towards the surface, undergoing energy degradation through radiation–matter interactions on the way. In the case of the Sun the surface is a black body radiator with a temperature of 5800 K. Photons propagate outwards through the radiation zone across which the temperature drops from about 10^7 K in the core to around 5×10^5 K at the boundary with the convection zone. This boundary is marked by a drop in temperature so steep that radiative transfer becomes unstable and is supplanted as the dominant mode of energy transport by the onset of convection.

Just above the convection zone lies the photosphere, the visible 'surface' of the Sun, in the sense that photons in the visible spectrum escape from the photosphere. UV and X-ray surfaces appear at greater heights. Within the photosphere the Sun's temperature falls to about 4300 K and then unexpectedly begins to rise, a transition that marks the boundary between photosphere and chromosphere. At the top of the chromosphere temperatures reach around 20 000 K and heating then surges dramatically to give temperatures of more than a million degrees in the corona.

The surface of the Sun is characterized by magnetic structures anchored in the photosphere. Not all magnetic field lines form closed loops; some do not close in the photosphere with the result that plasma flowing along such field lines is not bound to the Sun. This outward flow of coronal plasma in regions of open magnetic field constitutes the solar wind. The interaction between this wind and the Earth's magnetic field is of great interest in the physics of the Sun–Earth plasma system. The Earth is surrounded by an enormous magnetic cavity known as the magnetosphere at which the solar wind is deflected by the geomagnetic field, with dramatic consequences for each. The outer boundary of the magnetosphere occurs at about $10R_E$, where R_E denotes the Earth's radius. The geomagnetic field is swept into space in the form of a huge cylinder many millions of kilometres in length, known as the magnetotail. Perhaps the most dramatic effect on the solar wind is the formation of a shock some $5R_E$ upstream of the magnetopause, known as the bow shock. We shall discuss a number of these effects later in the book by way of illustrating basic aspects of the physics of plasmas.

1.4 Plasma characteristics

We now introduce a number of concepts fundamental to the nature of any plasma whatever its origin. First we need to go a step beyond our statement in Section 1.1 and obtain a more formal identification of the plasma condition. Perhaps the most notable feature of a plasma is its ability to maintain a state of charge neutrality. The combination of low electron inertia and strong electrostatic field, which arises from even the slightest charge imbalance, results in a rapid flow of electrons to re-establish neutrality.

The first point to note concerns the nature of the electrostatic field. Although at first sight it might appear that the Coulomb force due to any given particle extends over the whole volume of the plasma, this is in fact not the case. Debye, in the context of electrolytic theory, was the first to point out that the field due to any charge imbalance is shielded so that its influence is effectively restricted to within a finite range. For example, we may suppose that an additional ion with charge Ze is introduced at a point P in an otherwise neutral plasma. The effect will be to attract electrons towards P and repel ions away from P so that the ion is surrounded by a neutralizing 'cloud'. Ignoring ion motion and assuming that the number density of the electron cloud n_c is given by the Boltzmann distribution, $n_c = n_e \exp(e\phi/k_B T_e)$, where T_e is the electron temperature, we solve Poisson's equation for the electrostatic potential $\phi(r)$ in the plasma.

Since $\phi(r) \to 0$ as $r \to \infty$, we may expand $\exp(e\phi/k_B T_e)$ and with $Zn_i = n_e$, Poisson's equation for large r and spherical symmetry about P becomes

$$\frac{1}{r^2}\frac{\mathrm{d}}{\mathrm{d}r}\left(r^2\frac{\mathrm{d}\phi}{\mathrm{d}r}\right) = \frac{n_e e^2}{\varepsilon_0 k_B T_e}\phi = \frac{\phi}{\lambda_D^2} \tag{1.2}$$

say, where ε_0 is the vacuum permittivity. Now matching the solution of (1.2), $\phi \sim \exp(-r/\lambda_D)/r$, with the potential $\phi = Ze/4\pi\varepsilon_0 r$ as $r \to 0$ we see that

$$\phi(r) = \frac{Ze}{4\pi\varepsilon_0 r}\exp(-r/\lambda_D) \tag{1.3}$$

where

$$\lambda_D = \left(\frac{\varepsilon_0 k_B T_e}{n_e e^2}\right)^{1/2} \simeq 7.43 \times 10^3 \left(\frac{T_e(\text{eV})}{n_e}\right)^{1/2} \text{ m} \tag{1.4}$$

is called the *Debye shielding length*. Beyond a *Debye sphere*, a sphere of radius λ_D, centred at P, the plasma remains effectively neutral. By the same argument λ_D is also a measure of the penetration depth of external electrostatic fields, i.e. of the thickness of the boundary sheath over which charge neutrality may not be maintained.

The plausibility of the argument used to establish (1.3) requires that a large number of electrons be present within the Debye sphere, i.e. $n_e\lambda_D^3 \gg 1$. The inverse of this number is proportional to the ratio of potential energy to kinetic energy in the plasma and may be expressed as

$$g = \frac{e^2}{\varepsilon_0 k_B T_e \lambda_D} = \frac{1}{n_e\lambda_D^3} \ll 1 \tag{1.5}$$

Since g plays a key role in the development of formal plasma theory it is known as the *plasma parameter*. Broadly speaking, the more particles there are in the Debye sphere the less likely it is that there will be a significant resultant force on any given particle due to 'collisions'. It is, therefore, a measure of the dominance of collective interactions over collisions.

The most fundamental of these collective interactions are the *plasma oscillations* set up in response to a charge imbalance. The strong electrostatic fields which drive the electrons to re-establish neutrality cause oscillations about the equilibrium position at a characteristic frequency, the *plasma frequency* ω_p. Since the imbalance occurs over a distance λ_D and the electron thermal speed V_e is typically $(k_B T_e/m_e)^{1/2}$ we may express the electron plasma frequency ω_{pe} by

$$\omega_{pe} = \frac{(k_B T_e/m_e)^{1/2}}{\lambda_D} = \left(\frac{n_e e^2}{m_e\varepsilon_0}\right)^{1/2} \tag{1.6}$$

which reduces to $\omega_{pe} \simeq 56.4 n_e^{1/2}\,\mathrm{s}^{-1}$. Note that any applied fields with frequencies less than the electron plasma frequency are prevented from penetrating the plasma by the more rapid electron response which neutralizes the field. Thus a plasma is not transparent to electromagnetic radiation of frequency $\omega < \omega_{pe}$. The corresponding frequency for ions, the *ion plasma frequency* ω_{pi}, is defined by

$$\omega_{pi} = \left(\frac{n_i (Ze)^2}{m_i \varepsilon_0}\right)^{1/2} \simeq 1.32 Z \left(\frac{n_i}{A}\right)^{1/2} \tag{1.7}$$

where Z denotes the charge state and A the atomic number.

1.4.1 Collisions and the plasma parameter

We have seen that the effective range of an electric field, and hence of a collision, is the Debye length λ_D. Thus any particle interacts at any instant with the large number of particles in its Debye sphere. Plasma collisions are therefore *many-body interactions* and since $g \ll 1$ collisions are predominantly weak, in sharp contrast with the strong, binary collisions that characterize a neutral gas. In gas kinetics a collision frequency ν_c is defined by $\nu_c = n V_{th} \sigma(\pi/2)$ where $\sigma(\pi/2)$ denotes the cross-section for scattering through $\pi/2$ and V_{th} is a thermal velocity. Such a deflection in a plasma would occur for particles 1 and 2 interacting over a distance b_0 for which $e_1 e_2 / 4\pi\varepsilon_0 b_0 \sim k_B T$ so that $\nu_c = (n V_{th} \pi b_0^2)$. However, the cumulative effect of the much more frequent weak interactions acts to increase this by a factor $\sim 8\ln(\lambda_D/b_0) \approx 8\ln(4\pi n \lambda_D^3)$. For electron collisions with ions of charge Ze it follows that the electron–ion collision time $\tau_{ei} \equiv \nu_{ei}^{-1}$ is given by

$$\tau_{ei} = \frac{2\pi \varepsilon_0^2 m_e^{1/2} (k_B T_e)^{3/2}}{Z^2 n_i e^4 \ln\Lambda} \tag{1.8}$$

where $\ln\Lambda = \ln 4\pi n \lambda_D^3$ is known as the *Coulomb logarithm*. For singly charged ions the electron–ion collision time is

$$\tau_{ei} = 3.44 \times 10^{11} \frac{T_e^{3/2}\,(\mathrm{eV})}{n_i \ln\Lambda}\,\mathrm{s}$$

in which we have replaced the factor 2π in (1.8) with the value found from a correct treatment of plasma transport in Chapter 12. The Coulomb logarithm is

$$\ln\Lambda = 6.6 - \frac{1}{2}\ln\left(\frac{n}{10^{20}}\right) + \frac{3}{2}\ln T_e\,(\mathrm{eV})$$

The *electron mean free path* $\lambda_e = V_e \tau_{ei}$ is

$$\lambda_e = 1.44 \times 10^{17} \frac{T_e^2(\mathrm{eV})}{n_i \ln\Lambda}$$

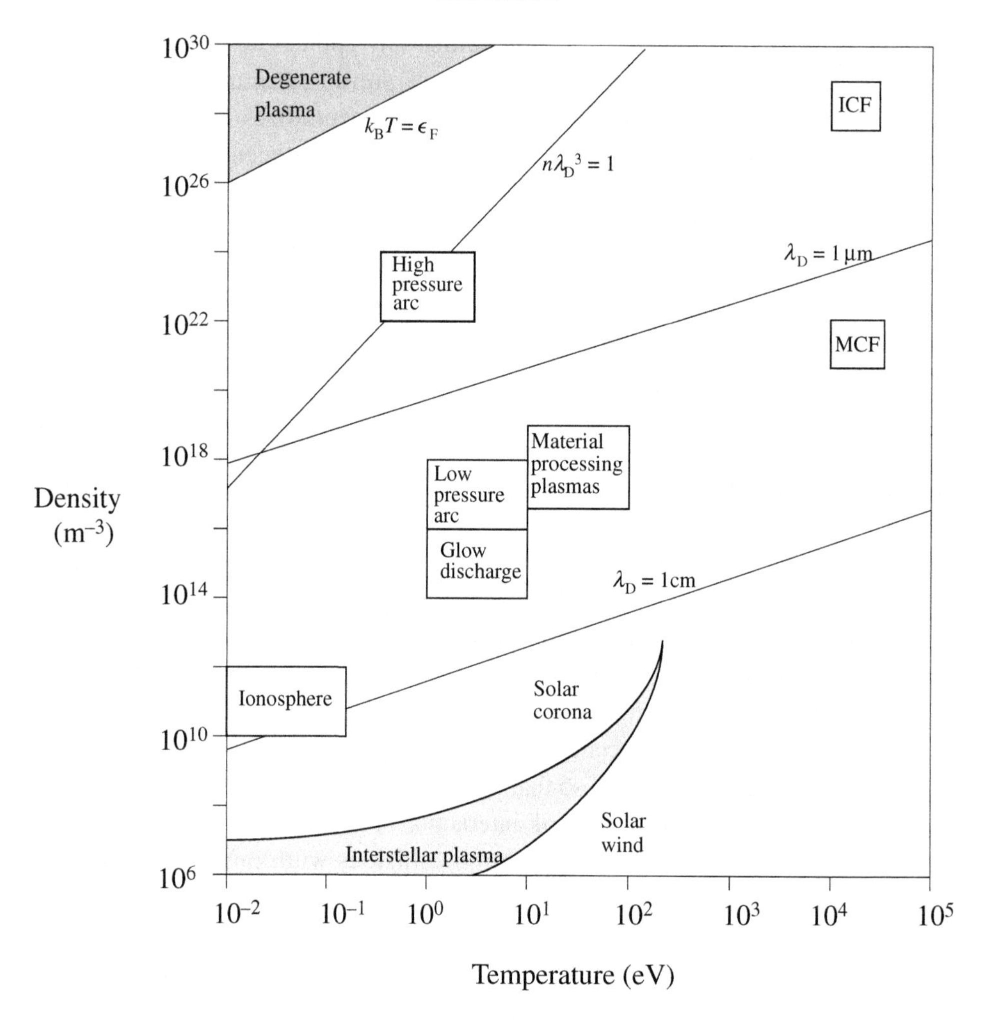

Fig. 1.4. Landmarks in the plasma universe.

Table 1.1 lists approximate values of various plasma parameters along with typical values of the magnetic field associated with each for a range of plasmas across the plasma universe. These and other representative plasmas are included in the diagram of parameter space in Fig. 1.4 which includes the parameter lines $\lambda_D = 1\ \mu m$, 1 cm and $n\lambda_D^3 = 1$ together with the line marking the boundary at which plasmas become degenerate $k_B T = \epsilon_F$, where ϵ_F denotes the Fermi energy.

Table 1.1. *Approximate values of parameters across the plasma universe.*

Plasma	n (m^{-3})	T (keV)	B (T)	ω_{pe} (s^{-1})	λ_D (m)	$n\lambda_D^3$	ν_{ei} (Hz)
Interstellar	10^6	10^{-5}	10^{-9}	$6 \cdot 10^4$	0.7	$3 \cdot 10^5$	$4 \cdot 10^8$
Solar wind (1 AU)	10^7	10^{-2}	10^{-8}	$2 \cdot 10^5$	7	$4 \cdot 10^9$	10^{-4}
Ionosphere	10^{12}	10^{-4}	10^{-5}	$6 \cdot 10^7$	$2 \cdot 10^{-3}$	10^4	10^4
Solar corona	10^{12}	0.1	10^{-3}	$6 \cdot 10^7$	0.07	$4 \cdot 10^8$	0.5
Arc discharge	10^{20}	10^{-3}	0.1	$6 \cdot 10^{11}$	$7 \cdot 10^{-7}$	40	10^{10}
Tokamak	10^{20}	10	10	$6 \cdot 10^{11}$	$7 \cdot 10^{-5}$	$3 \cdot 10^7$	$4 \cdot 10^4$
ICF	10^{28}	10	—	$6 \cdot 10^{15}$	$7 \cdot 10^{-9}$	$4 \cdot 10^3$	$4 \cdot 10^{11}$

2
Particle orbit theory

2.1 Introduction

On the face of it, solving an equation of motion to determine the orbit of a single charged particle in prescribed electric and magnetic fields may not seem like the best way of going about developing the physics of plasmas. Given the central role of collective interactions hinted at in Chapter 1 and the subtle interplay of currents and fields that will be explored in the chapters on MHD that follow, it is at least worth asking "Why bother with orbit theory?". One attraction is its relative simplicity. Beyond that, key concepts in orbit theory prove useful throughout plasma physics, sometimes shedding light on other plasma models.

Before developing particle orbit theory it is as well to be clear about conditions under which this description might be valid. Intuitively we expect orbit theory to be useful in describing the motion of high energy particles in low density plasmas where particle collisions are infrequent. More specifically, we need to make sure that the effect of self-consistent fields from neighbouring charges is small compared with applied fields. Then if we want to solve the equation of motion analytically the fields in question need to show a degree of symmetry. We shall find that scaling associated with an applied magnetic field is one reason – indeed the principal reason – for the success of orbit theory. Particle orbits in a magnetic field define both a natural length, r_{L}, the particle Larmor radius, and frequency, Ω, the cyclotron frequency. For many plasmas these are such that the scale length, L, and characteristic time, T, of the physics involved satisfy an ordering $r_{\mathrm{L}}/L \ll 1$ and $2\pi/\Omega T \ll 1$. This natural ordering lets us solve the dynamical equations in inhomogeneous and time-dependent fields by making perturbation expansions using r_{L}/L and $2\pi/\Omega T$ as small parameters. In this way Alfvén showed that one could filter out the rapid gyro-motion about magnetic field lines and focus on the dynamics of the centre of this motion, the so-called guiding centre. Alfvén's guiding centre model and the concept of adiabatic invariants (quantities that are not exact constants of the motion but, in certain circumstances, nearly so) play a key role in orbit theory. In large part, this chapter is taken up with the development and application of Alfvén's ideas.

Throughout this chapter we shall assume that radiative effects are negligible. For the present we suppose that particle energies are such that we need only solve the non-relativistic Lorentz equation for the motion of a particle of mass m_j and charge e_j at a position $\mathbf{r}_j(t)$ moving in an electric field $\mathbf{E}$ and a magnetic field $\mathbf{B}$

$$m_j\ddot{\mathbf{r}}_j = e_j\left[\mathbf{E}(\mathbf{r},t) + \dot{\mathbf{r}}_j \times \mathbf{B}(\mathbf{r},t)\right] \tag{2.1}$$

under prescribed initial conditions. This needs to be done for particles of each species. An important part of this procedure requires checking for self-consistency of the assumed fields. For the most part this means ensuring that fields induced by the motion of particles are negligible compared with the applied fields. For this we use Maxwell's equations

$$\nabla \times \mathbf{E} = -\frac{\partial \mathbf{B}}{\partial t} \tag{2.2}$$

$$\nabla \times \mathbf{B} = \varepsilon_0\mu_0\frac{\partial \mathbf{E}}{\partial t} + \mu_0\mathbf{j} \tag{2.3}$$

$$\nabla \cdot \mathbf{E} = q/\varepsilon_0 \tag{2.4}$$

$$\nabla \cdot \mathbf{B} = 0 \tag{2.5}$$

in which $\mathbf{j}(\mathbf{r},t)$ and $q(\mathbf{r},t)$ are current and charge densities defined by

$$\mathbf{j}(\mathbf{r},t) = \sum_{j=1}^{N} e_j\dot{\mathbf{r}}_j(t)\delta(\mathbf{r} - \mathbf{r}_j(t)) \tag{2.6}$$

$$q(\mathbf{r},t) = \sum_{j=1}^{N} e_j\delta(\mathbf{r} - \mathbf{r}_j(t)) \tag{2.7}$$

where δ denotes the Dirac delta function and sums are taken over *all* plasma particles. Checking for self-consistency, though not often stressed, is important since it may impose limits on the use of orbit theory and, in some cases, necessary conditions on the plasma or fields which would not otherwise be obvious. In the following applications of orbit theory we discuss self-consistency only when it gives rise to such limitations. In general whenever charge distributions or current densities are significant, orbit theory is no longer adequate and statistical or fluid descriptions are then essential.

2.2 Constant homogeneous magnetic field

The simplest problem in orbit theory is that of the non-relativistic motion of a charged particle in a constant, spatially uniform magnetic field, $\mathbf{B}$, with $\mathbf{E} = 0$. Moreover, we shall see that it is straightforward to deal with more general cases as perturbations of this basic motion. For simplicity of notation we discard the subscript j on e_j and m_j except where we wish specifically to distinguish between ions and electrons. Taking the direction of $\mathbf{B}$ to define the z-axis, that is $\mathbf{B} = B\hat{\mathbf{z}}$, the scalar product of (2.1) with $\hat{\mathbf{z}}$ gives,

$$\ddot{z} = 0 \tag{2.8}$$

so that $\dot{z} = v_\parallel =$ const. Also from (2.1),

$$m\ddot{\mathbf{r}} \cdot \dot{\mathbf{r}} = 0$$

so that

$$\tfrac{1}{2}m\dot{r}^2 = W = \text{const.}$$

Hence the magnitude of velocity components both perpendicular ($v_\perp$) and parallel ($v_\parallel$) to $\mathbf{B}$ are constant and the kinetic energy

$$W = W_\perp + W_\parallel = \tfrac{1}{2}m(v_\perp^2 + v_\parallel^2)$$

It is no surprise that kinetic energy is conserved since the force is always perpendicular to the velocity of the particle and, in consequence, does no work on it. Moreover, conservation of kinetic energy is not restricted to uniform magnetic fields.

The particle trajectory is determined by (2.8) together with the x and y components of (2.1):

$$\ddot{x} = \Omega\dot{y} \qquad \ddot{y} = -\Omega\dot{x}$$

where $\Omega = eB/m$. A convenient way of dealing with motion transverse to $\mathbf{B}$ starts by defining $\zeta = x + iy$ so that

$$\ddot{\zeta} + i\Omega\dot{\zeta} = 0$$

Integrating once with respect to time gives

$$\dot{\zeta}(t) = \dot{\zeta}(0)\exp(-i\Omega t)$$

and by defining $\dot{\zeta}(0) = v_\perp \exp(-i\alpha)$ it follows that

$$\dot{x} = v_\perp \cos(\Omega t + \alpha) \qquad \dot{y} = -v_\perp \sin(\Omega t + \alpha) \tag{2.9}$$

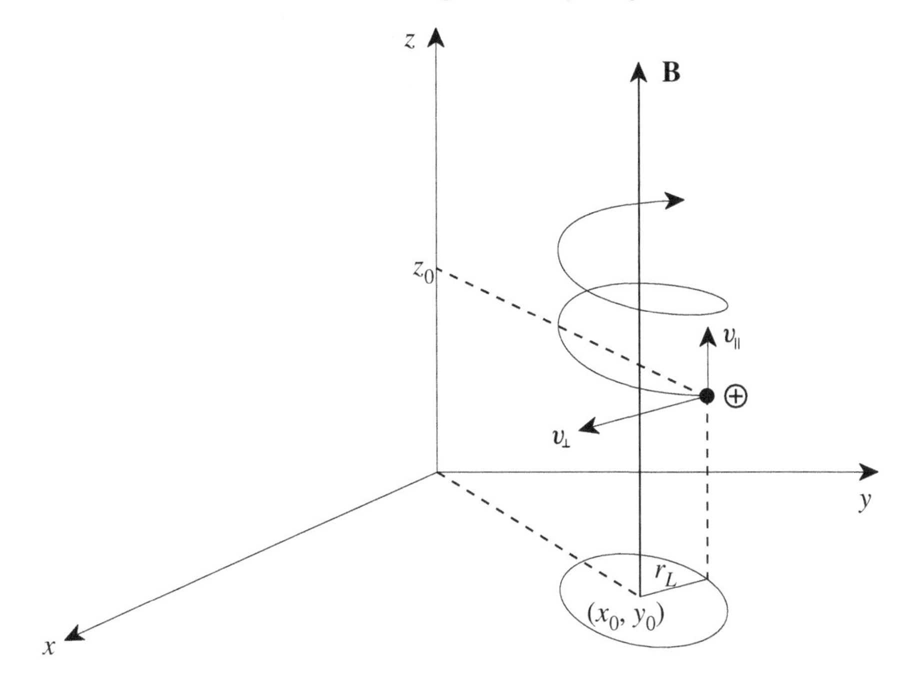

Fig. 2.1. Orbit of a positively charged particle in a uniform magnetic field.

Integrating a second time determines the particle orbit

$$\left.\begin{aligned} x &= \left(\frac{v_\perp}{\Omega}\right)\sin(\Omega t + \alpha) + x_0 \\ y &= \left(\frac{v_\perp}{\Omega}\right)\cos(\Omega t + \alpha) + y_0 \end{aligned}\right\} \tag{2.10}$$

and

$$z = v_\parallel t + z_0 \tag{2.11}$$

where α, x_0, y_0, and z_0, together with $v_\perp$ and $v_\parallel$, are determined by the initial conditions. The quantity $\phi(t) \equiv (\Omega t + \alpha)$ is sometimes referred to as the *gyro-phase*. The superposition of uniform motion in the direction of the magnetic field on the circular orbits in the plane normal to **B** defines a helix of constant pitch with axis parallel to **B** as shown in Fig. 2.1 for a positively charged particle. Referred to the moving plane $z = v_\parallel t + z_0$ the orbit projects as a circle with centre (x_0, y_0) and radius $r_\mathrm{L} = v_\perp/|\Omega|$. The centre of this circle, known as the *guiding centre*, describes the locus $\mathbf{r}_\mathrm{g} = (x_0, y_0, v_\parallel t + z_0)$. It is important to emphasize that the guiding centre is not the locus of a particle as such. The radius of the circle, r_L, is known as the *Larmor radius* and the frequency of rotation, Ω, as the *Larmor frequency*, *cyclotron frequency*, or *gyro-frequency*. The sense of rotation for a prescribed magnetic field is determined by Ω which depends on the sign of the charge. Viewed from $z = +\infty$, positive and negative particles rotate in clockwise and anticlockwise directions, respectively. For electrons $|\Omega_e| = 1.76 \times 10^{11}\ B\,\mathrm{s}^{-1}$, while for protons $\Omega_p = 9.58 \times 10^7\ B\,\mathrm{s}^{-1}$ where B is measured in teslas. In many applications of orbit theory we need concern ourselves only with guiding centre motion. However aspects of the Larmor motion are needed for discussion later in the chapter and we next look briefly at one of these.

2.2.1 Magnetic moment and plasma diamagnetism

We can formally associate a microscopic current I_L with the Larmor motion. The magnetic field from this microcurrent is determined by Ampère's law and is oppositely directed to the applied magnetic field. In this sense the response of the particle to the magnetic field is *diamagnetic*. In the same formal sense we may associate with this microcurrent a *magnetic moment* $\boldsymbol{\mu}_B$ given by

$$\boldsymbol{\mu}_B = -\pi r_L^2 I_L \mathbf{B}/|\mathbf{B}| = -\pi r_L^2 \left(\frac{e\Omega}{2\pi}\right)\frac{\mathbf{B}}{|\mathbf{B}|}$$

where $2\pi/|\Omega|$ is a Larmor period. From this it follows that

$$\boldsymbol{\mu}_B = -\frac{W_\perp}{B^2}\mathbf{B}$$

If we now extend this argument to all plasma particles we can find an expression for the magnetization per unit volume by summing individual moments over the distribution of particles. For n particles per unit volume the magnetization $\mathbf{M} = n\langle\boldsymbol{\mu}_B\rangle$ where the brackets denote an average. Then using $\mathbf{j}_M = \boldsymbol{\nabla} \times \mathbf{M} = \mu_0^{-1}\boldsymbol{\nabla} \times \mathbf{B}_{ind}$, the magnitude of the induced field B_{ind} relative to the applied magnetic field is

$$\frac{B_{ind}}{B} \sim \frac{\mu_0 n\langle W_\perp\rangle}{B^2}$$

where $\langle W_\perp\rangle$ denotes the average kinetic energy perpendicular to $\mathbf{B}$. Since we require the induced field to be small compared with the applied field this implies that the kinetic energy of the plasma must be much less than the magnetic energy.

2.3 Constant homogeneous electric and magnetic fields

We now introduce a constant, uniform electric field which may be resolved into components $\mathbf{E}_\parallel$ in the direction of $\mathbf{B}$ and $\mathbf{E}_\perp$, which is taken to define the direction of the y-axis. Thus $\mathbf{B} = (0, 0, B)$, $\mathbf{E} = (0, E_\perp, E_\parallel)$, and the components of (2.1) are

$$\ddot{x} = \Omega \dot{y} \tag{2.12}$$

$$\ddot{y} = \frac{eE_\perp}{m} - \Omega \dot{x} \tag{2.13}$$

$$\ddot{z} = \frac{eE_\parallel}{m} \tag{2.14}$$

Integrating (2.14) once gives

$$\dot{z} = v_\parallel + \frac{eE_\parallel t}{m}$$

from which it is clear that for sufficiently long times the non-relativistic approximation breaks down unless $E_\parallel = 0$. Further, since charges of opposite sign are accelerated in opposite directions, a non-zero $E_\parallel$ gives rise to arbitrarily large currents and charge separation. Therefore, from (2.3) and (2.4) significant fluctuating fields are induced contrary to our assumption of constant fields. Thus for consistency it is necessary to set $E_\parallel = 0$ in this approximation.

Equations (2.12) and (2.13) are solved as in Section 2.2. Now

$$\ddot{\zeta} + i\Omega\dot{\zeta} = \frac{ieE}{m} \tag{2.15}$$

where $E_\perp$ has been replaced by E. Integrating once gives

$$\dot{\zeta}(t) = \dot{\zeta}(0)e^{-i\Omega t} + v_{\mathrm{E}}(1 - e^{-i\Omega t})$$

where $v_{\mathrm{E}} = E/B$. Hence

$$\dot{x} = u\cos(\Omega t + \alpha) + v_{\mathrm{E}} \qquad \dot{y} = -u\sin(\Omega t + \alpha)$$

where u and α are constants defined by

$$\dot{\zeta}(0) - v_{\mathrm{E}} = ue^{-i\alpha}$$

The velocity of the guiding centre is now

$$\mathbf{v}_{\mathrm{g}} = (v_{\mathrm{E}}, 0, v_\parallel)$$

Thus the effect of an electric field perpendicular to the magnetic field is to produce a drift orthogonal to both. This means that the guiding centre is no longer tied to a particular field line but drifts across field lines. The drift velocity, which may be written

$$\mathbf{v}_{\mathrm{E}} = (\mathbf{E} \times \mathbf{B})/B^2 \tag{2.16}$$

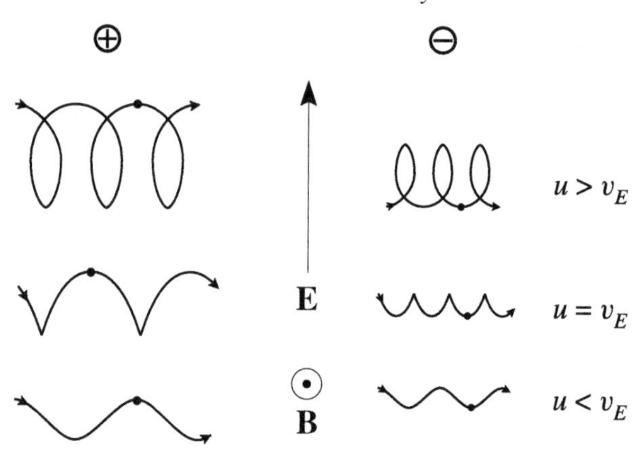

Fig. 2.2. Drift produced by a constant uniform electric field perpendicular to the magnetic field.

depends only on the fields; in particular, being independent of particle charge it cannot give rise to a current. The non-relativistic approximation implies a further restriction on the electric field; by (2.16), $E \ll cB$, where c is the speed of light. The trajectories of positive and negative particles in the plane defined by (2.11), found from a second integration of (2.15), are shown in Fig. 2.2. In its Larmor cycle a positive charge slows when moving in opposition to the electric field and accelerates when moving with it. Thus the Larmor orbit is continuously distorted as the instantaneous Larmor radius alternately becomes shorter in one half-cycle and longer in the next as in Fig. 2.2. The net effect is a drift to the right. The electrons also drift to the right since the opposite action of the field is compensated for by the anti-clockwise rotation. The mass difference between positively and negatively charged particles is represented schematically in Fig. 2.2 by the smaller electron drift per cycle which is fully compensated by the proportionately bigger Larmor frequency.

2.3.1 Constant non-electromagnetic forces

It may happen that the particles are subject to non-electromagnetic forces such as gravity. If such a force, $\mathbf{F}$, is constant it is equivalent to an electric field $\mathbf{E}' = \mathbf{F}/e$ and is subject to the same restrictions found for $\mathbf{E}$, i.e. its component parallel to $\mathbf{B}$ must be negligible and $F \ll ceB$. The drift velocity is then

$$\mathbf{v}_{\mathrm{F}} = (\mathbf{F} \times \mathbf{B})/eB^2$$

In contrast to $\mathbf{v}_{\mathrm{E}}$ this drift contributes to a current. There is a nice antithesis here in that a non-electromagnetic drift produces a current whereas the electromagnetic

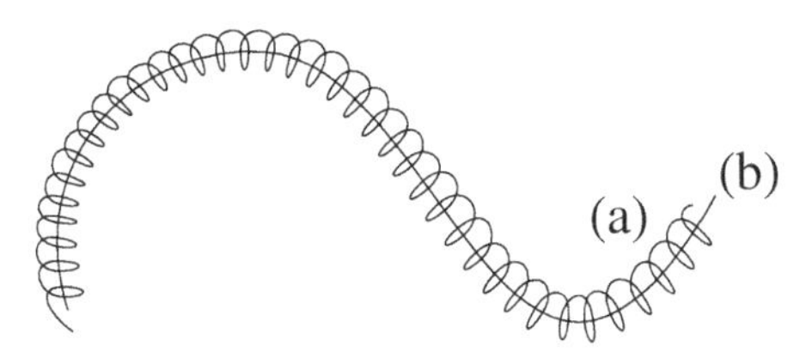

Fig. 2.3. Particle motion showing (a) exact and (b) guiding centre trajectories.

drift $\mathbf{v}_{\mathrm{E}}$ does not. Note that drifts from gravitational forces while usually insignificant in laboratory plasmas normally need to be taken into account when applying orbit theory in space plasmas.

2.4 Inhomogeneous magnetic field

In practice the fields we encounter are generally both space- and time-dependent. Restricting ourselves for the present to spatially inhomogeneous magnetic fields, $\mathbf{B}(\mathbf{r})$, it is necessary to solve (2.1) numerically in general. However, if the inhomogeneity is small – that is, the field experienced by the particle in traversing a Larmor orbit is almost constant – it is possible to determine the trajectory as a perturbation of the basic motion found in Section 2.2. With $\mathbf{B}(\mathbf{r}) \simeq \mathbf{B}(\mathbf{r}_0) + (\delta\mathbf{r} \cdot \boldsymbol{\nabla})\mathbf{B}|_{\mathbf{r}=\mathbf{r}_0}$, where $\mathbf{r}_0$ is the instantaneous position of the guiding centre and $\delta\mathbf{r} = \mathbf{r} - \mathbf{r}_0$, we require that $\delta\mathbf{B}$, the change in $\mathbf{B}$ over a distance r_{L}, be such that

$$|\delta\mathbf{B}| = |(\delta\mathbf{r} \cdot \boldsymbol{\nabla})\mathbf{B}| \ll |\mathbf{B}|$$

i.e. $r_{\mathrm{L}} \ll L$ where L is a distance over which the field changes significantly.

This perturbation approach was first applied systematically by Alfvén. It is known as the *guiding centre approximation* and has proved a robust tool in the application of orbit theory to cases of practical interest. Alfvén recognized that in many such applications one need not bother with the fast Larmor motion which can be averaged out to leave the slower guiding centre motion illustrated in Fig. 2.3.

The most general inhomogeneous field represented by the nine components of $[\partial B_i/\partial x_j]$ gives rise to distinct kinds of drift from the *gradient* and *curvature* terms. We look at each of these in turn.

2.4.1 Gradient drift

Taking $\mathbf{B} = (0, 0, B(y))$ and $\mathbf{E} = 0$, (2.1) gives

$$\ddot{x} = \Omega(y)\dot{y} \qquad \ddot{y} = -\Omega(y)\dot{x}$$

so that

$$\ddot{\zeta} = -i\Omega(y)\dot{\zeta} \tag{2.17}$$

and

$$\ddot{z} = 0 \tag{2.18}$$

With the assumption $\delta B \ll B$, Ω may be expanded about the initial position of the guiding centre to give

$$\Omega(y) \simeq \Omega(y_0) + (y - y_0) \left.\frac{\mathrm{d}\Omega}{\mathrm{d}y}\right|_{y_0} = \Omega_0 + (y - y_0)\Omega_0'$$

Hence

$$\ddot{\zeta} + i\Omega_0\dot{\zeta} = -i\Omega_0'(y - y_0)\dot{\zeta} \tag{2.19}$$

The terms on the left are zero-order while that on the right is first-order and, as such, y and $\dot{\zeta}$ may be replaced by their zero-order (that is, uniform **B**) values from (2.9) and (2.10) giving

$$\ddot{\zeta} + i\Omega_0\dot{\zeta} = -i\Omega_0' \frac{v_\perp^2}{\Omega_0} \cos(\Omega_0 t + \alpha) e^{-i(\Omega_0 t + \alpha)}$$

On integrating once

$$\dot{\zeta}(t) = \dot{\zeta}(0)e^{-i\Omega_0 t} - \frac{i\Omega_0' v_\perp^2}{\Omega_0^2} e^{-i(\Omega_0 t + \alpha)} [\sin(\Omega_0 t + \alpha) - \sin\alpha]$$

Then

$$\dot{x}(t) = v_\perp \cos(\Omega_0 t + \alpha) - \frac{\Omega_0' v_\perp^2}{2\Omega_0^2} [1 - \cos 2(\Omega_0 t + \alpha) - 2\sin(\Omega_0 t + \alpha)\sin\alpha] \tag{2.20}$$

$$\dot{y}(t) = -v_\perp \sin(\Omega_0 t + \alpha) - \frac{\Omega_0' v_\perp^2}{2\Omega_0^2} [\sin 2(\Omega_0 t + \alpha) - 2\cos(\Omega_0 t + \alpha)\sin\alpha] \tag{2.21}$$

These solutions satisfy identical initial conditions to the uniform **B** case. Also, from (2.18) $\dot{z} = v_\parallel$. From (2.20) and (2.21) we see that oscillations occur at $2\Omega_0$ in addition to those at Ω_0. In the present context, however, the term of interest is the non-oscillatory one in (2.20). The average of the velocity over one period ($T = 2\pi/\Omega_0$) is

$$\langle \mathbf{v} \rangle = (-v_\perp^2 \Omega_0'/2\Omega_0^2, 0, v_\parallel)$$

Thus a magnetic field in the z direction with a gradient in the y direction gives rise to a drift in the x direction. This *grad B drift velocity* may be written

$$\mathbf{v}_\mathrm{G} = [W_\perp(\mathbf{B} \times \boldsymbol{\nabla})B]/eB^3 \tag{2.22}$$

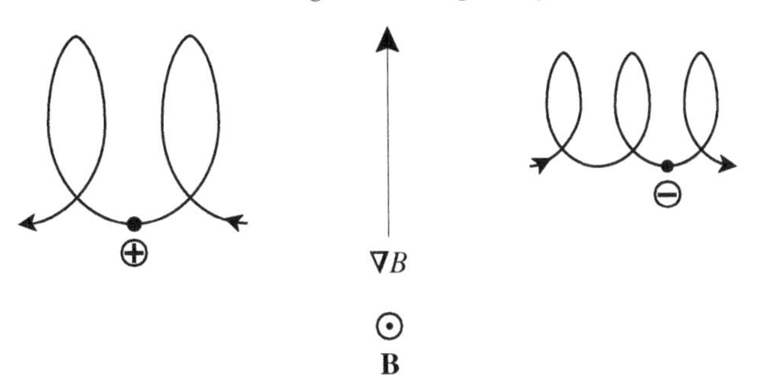

Fig. 2.4. Drift produced by an inhomogeneous magnetic field.

In this case, the drift depends on properties of the particle and, in particular, occurs in opposite directions for positive and negative charges. However, this of itself does not imply a flow of current as we shall see in Section 2.5. The physical source of the ∇B drift is clear from Fig. 2.4 which shows the Larmor radius bigger in regions of weaker **B**. Thus there will be a drift perpendicular both to **B** and ∇B but in this case the opposite rotation of positive and negative charges leads to drifts in opposite directions.

2.4.2 Curvature drift

In practice magnetic fields not only vary spatially but are generally curved and the curvature of field lines in turn gives rise to a drift. For example, if $\mathbf{B} = (0, B_y(z), B)$, where B_y and $\mathrm{d}B_y/\mathrm{d}z$ are taken as small quantities,† one has

$$\ddot{x} = \Omega \dot{y} - \Omega_y \dot{z} \qquad \ddot{y} = -\Omega \dot{x} \qquad \ddot{z} = \Omega_y \dot{x}$$

where $\Omega_y = (eB_y/m)$. Hence,

$$\ddot{\zeta} + i\Omega\dot{\zeta} = -\Omega_y(z)\dot{z} = -\Omega_y(z)v_\parallel$$

neglecting squares of small quantities. It is straightforward to show that there is a drift in the x-direction given by

$$v_C = \langle \dot{x} \rangle = -v_\parallel^2 \Omega_y' / \Omega^2 = -mv_\parallel^2 (\mathrm{d}B_y/\mathrm{d}z)/eB^2$$

This *curvature drift velocity* may be written more generally

$$\mathbf{v}_C = 2W_\parallel (\mathbf{B} \times (\mathbf{B} \cdot \nabla)\mathbf{B})/eB^4 \tag{2.23}$$

† It is convenient to keep B for the z component of **B**. Since $|\mathbf{B}|$ does not appear in the remainder of this section no confusion arises.

There is also a drift in the y-direction given by $(v_{\parallel}\Omega_y/\Omega)$. This merely keeps the guiding centre moving parallel to **B** in the Oyz-plane since, on average,

$$\frac{\langle\dot{y}\rangle}{\langle\dot{z}\rangle} = \frac{v_{\parallel}\Omega_y/\Omega}{v_{\parallel}} = \frac{B_y}{B}$$

To see the physical origin of the curvature drift, picture the particle moving along a curved field line with velocity $v_{\parallel}$. Because of the curvature the particle will feel a centrifugal force $\mathbf{F} = mv_{\parallel}^2\mathbf{R}_c/R_c^2$ where $\mathbf{R}_c$ is a vector from the local centre of curvature to the position of the charge. From Section 2.3.1, $\mathbf{v}_F = (\mathbf{F}\times\mathbf{B})/eB^2$ and so

$$\mathbf{v}_C = \frac{mv_{\parallel}^2}{eB^2}\frac{\mathbf{R}_c\times\mathbf{B}}{R_c^2}$$

Since the magnetic field lines are defined by $\mathrm{d}\mathbf{l}\times\mathbf{B} = 0$ so that $\mathrm{d}y/B_y = \mathrm{d}z/B$, it follows that $R_c^{-1} \simeq \mathrm{d}^2y/\mathrm{d}z^2 \simeq B^{-1}\,\mathrm{d}B_y/\mathrm{d}z$. Hence

$$v_C = -mv_{\parallel}^2(\mathrm{d}B_y/\mathrm{d}z)/eB^2$$

If the magnetic field is characterized by both gradient and curvature terms in $\nabla\mathbf{B}$ and there are no currents present so that $\nabla\times\mathbf{B} = 0$, then $\partial B_y/\partial z = \partial B/\partial y$ and the total drift velocity is given by

$$\mathbf{v}_B = [(W_{\perp} + 2W_{\parallel})(\mathbf{B}\times\nabla)B]/eB^3 \tag{2.24}$$

Taken together these drifts describe completely the lowest order motion of the guiding centre across an inhomogeneous time-independent magnetic field. The drift velocities are $O(r_L/L)$ times the particle velocities.

Of the other possible inhomogeneities in the magnetic field, the divergence terms are considered in Section 2.6.1. However neither these, nor either of the remaining components of $[\partial B_i/\partial x_j]$ describing shear, give rise to drift motion.

2.5 Particle drifts and plasma currents

We have already sounded a note of caution about too readily identifying particle guiding centre drifts with plasma current. A current density properly involves an average over a distribution of particles, a procedure that makes no direct appeal to the guiding centre motion of particles. Guiding centre drifts on the other hand are found from time-averaging the motion of a single particle. There are no grounds for supposing the two averages are identical and in general they are not.

Let us form a current density corresponding to the ∇B drift from (2.22) by summing over ions (i) and electrons (e) and using $\langle W_\perp \rangle$ to denote average values of $W_\perp$ (cf. Section 2.2.1). Then

$$\begin{aligned} \mathbf{j}_\mathrm{G} &= (n_\mathrm{i}\langle W_{\perp \mathrm{i}}\rangle + n_\mathrm{e}\langle W_{\perp \mathrm{e}}\rangle)\mathbf{B} \times \nabla B / B^3 \\ &= [n\langle W_\perp \rangle (\mathbf{B} \times \nabla B)]/B^3 \end{aligned} \tag{2.25}$$

To arrive at the total current density we must remember to include the contribution from the plasma diamagnetism. In Section 2.2.1 we saw that the magnetization per unit volume of plasma, $\mathbf{M}$, is given by

$$\mathbf{M} = -\frac{n\langle W_\perp \rangle}{B^2}\mathbf{B}$$

from which the magnetization current density is

$$\mathbf{j}_\mathrm{M} = -\nabla \times \left(\frac{n\langle W_\perp \rangle}{B^2}\mathbf{B}\right) \tag{2.26}$$

The total current density is then the sum of $\mathbf{j}_\mathrm{G}$ and $\mathbf{j}_M$; for simplicity we suppose there is no field curvature.

If we now turn to the configuration of Section 2.4.1 with $\mathbf{B} = (0, 0, B(y))$ we find that part of the magnetization current density cancels the contribution from the field gradient, so that

$$j_x = (\mathbf{j}_\mathrm{G} + \mathbf{j}_\mathrm{M}) \cdot \hat{\mathbf{x}} = -\frac{n\langle W_\perp \rangle}{B^2}\frac{\mathrm{d}B}{\mathrm{d}y} - \frac{\mathrm{d}}{\mathrm{d}y}\left[\frac{n\langle W_\perp \rangle}{B}\right] = -\frac{1}{B}\frac{\mathrm{d}}{\mathrm{d}y}(n\langle W_\perp \rangle) \tag{2.27}$$

The effects of plasma magnetization and guiding centre drift in an inhomogeneous magnetic field combine to produce a current perpendicular both to the magnetic field and to the direction in which the field varies, provided $n\langle W_\perp \rangle$ is spatially non-uniform. If we now substitute (2.27) into (2.3) we find, neglecting displacement current,

$$\left[\frac{\mathrm{d}B}{\mathrm{d}y} + \frac{\mu_0}{B}\frac{\mathrm{d}}{\mathrm{d}y}(n\langle W_\perp \rangle)\right] = 0 \tag{2.28}$$

so that $n\langle W_\perp \rangle + B^2/2\mu_0 = \text{const}$. Thus we see that our picture is consistent only if the increasing magnetic field is compensated by a corresponding decrease in $n\langle W_\perp \rangle$ (or as we shall see in Section 3.4.2, by decreasing pressure). The particular case of decreasing density is illustrated in Fig. 2.5.

In general if one keeps the displacement current term in Maxwell's equations there is then a time-dependent electric field which gives rise to plasma oscillations. Any appeal to orbit theory in conditions where charge separation is significant is of doubtful value. However, if the time dependence of the electric field is slow enough

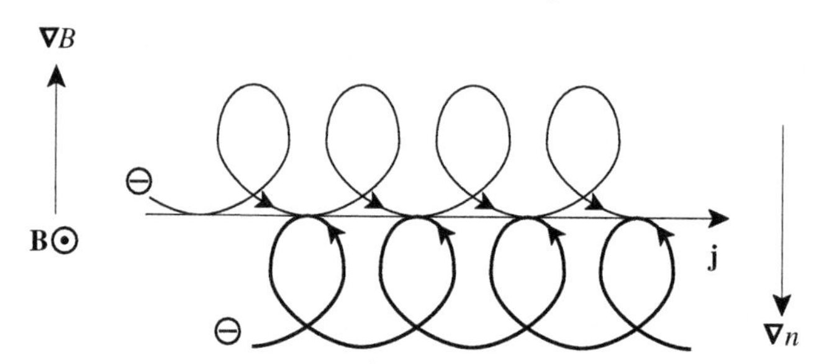

Fig. 2.5. Current density in an inhomogeneous magnetized plasma. In any current sheet perpendicular to the plane of the figure there are more electrons flowing to the left than to the right because of the density gradient.

so that plasma oscillations do not occur we may include a time-dependent electric field as we shall see in Section 2.12.

2.6 Time-varying magnetic field and adiabatic invariance

In Section 2.4 we introduced certain inhomogeneous magnetic fields. In practice, one often has to deal with fields that are time-dependent. In keeping with Section 2.4, which was restricted to weakly inhomogeneous fields, we shall consider only magnetic fields varying slowly in time ($|\dot{\mathbf{B}}|/|\mathbf{B}| \ll |\Omega|$). For simplicity, consider a magnetic field which varies in time but not in space. For particle motion in such a field, can one find conserved quantities that are counterparts to $W_\perp$, $W_\parallel$ for motion in a constant and uniform magnetic field? We shall demonstrate that such invariants do exist in particular circumstances.

A time-dependent axial magnetic field induces an azimuthal electric field, $\mathbf{E}$, so that, unlike $v_\parallel$, $v_\perp$ is no longer constant and taking the scalar product of (2.1) with $v_\perp$ gives

$$\frac{\mathrm{d}}{\mathrm{d}t}\left(\tfrac{1}{2}mv_\perp^2\right) = e\mathbf{E}\cdot\mathbf{v}_\perp$$

Thus in executing a Larmor orbit the particle energy changes by

$$\delta\left(\tfrac{1}{2}mv_\perp^2\right) = \oint \mathbf{E}\cdot\mathrm{d}\mathbf{r}_\perp = e\int(\nabla\times\mathbf{E})\cdot\mathrm{d}\mathbf{S}$$

where $\mathrm{d}\mathbf{r}_\perp = \mathbf{v}_\perp\,\mathrm{d}t$ and $\mathrm{d}\mathbf{S}$ is an element of the surface enclosed by the orbital path. Hence, from (2.2)

$$\delta\left(\tfrac{1}{2}mv_\perp^2\right) = -e\int\frac{\partial\mathbf{B}}{\partial t}\cdot\mathrm{d}\mathbf{S}$$

Since the field changes slowly

$$\delta\left(\tfrac{1}{2}mv_\perp^2\right) \simeq \pi {r_L}^2 |e| \dot{B} = \frac{mv_\perp^2}{2}\frac{2\pi}{|\Omega|}\frac{\dot{B}}{B}$$

Note in passing that the negative sign disappears since for positive (negative) charges $e > 0$ (< 0) and $\mathbf{B} \cdot \mathrm{d}\mathbf{S} < 0$ (> 0). If we denote by δB the change in magnitude of the magnetic field during one orbit it follows that

$$\delta W_\perp = W_\perp \frac{\delta B}{B}$$

i.e.

$$\delta(W_\perp / B) = 0$$

From Section 2.2.1 where we identified the magnetic moment of a charged particle as

$$\mu_\mathrm{B} = \frac{W_\perp}{B} \tag{2.29}$$

we see from this analysis that μ_B is an approximate constant of the motion, a property first recognized by Alfvén.

The magnetic moment is one of a number of entities which are approximate constants of the motion for particles in magnetic fields. In Hamiltonian dynamics such quantities are known as *adiabatic invariants*. In particular the *action* $\oint p\,\mathrm{d}q$, where p, q are conjugate canonical variables and the integral is taken over a period of the motion in q, is adiabatically invariant. The condition critical for adiabatic invariance is that the particle trajectory changes *slowly* on the time scale of the basic periodic motion. The number of invariants is determined by the periodicities that characterize the motion. We have established that the adiabatic invariance of μ_B is associated with Larmor precession in a magnetic field. We shall find that a charged particle may be trapped between magnetic mirror fields, as a result of which another periodicity appears and with it a second adiabatic invariant. If in addition we allow for curvature drift then for a suitably configured field a third invariant may be identified corresponding to the magnetic flux enclosed by the drift orbit of the guiding centre.

2.6.1 Invariance of the magnetic moment in an inhomogeneous field

The magnetic moment also turns out to be invariant for motion in spatially inhomogeneous magnetic fields for which the matrix $[\partial B_i/\partial x_j]$ has non-zero diagonal elements, the divergence terms. To demonstrate this, consider the axially symmetric magnetic field increasing slowly with z as in Fig. 2.6. Writing the divergence

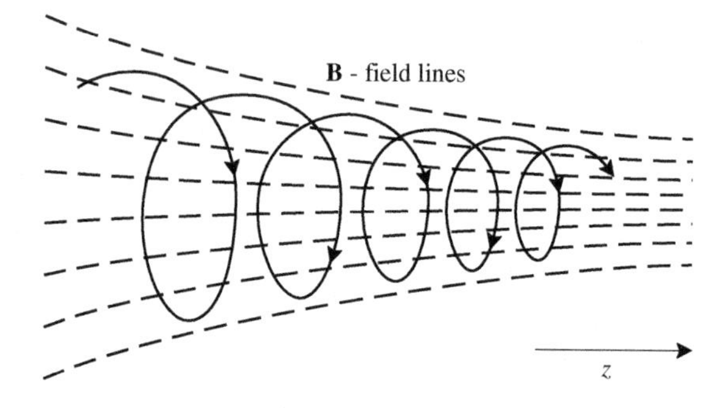

Fig. 2.6. Magnetic field increasing in the direction of the field.

property of **B** in cylindrical polar coordinates and integrating gives

$$rB_r = -\int_0^r r\frac{\partial B_z}{\partial z}\,\mathrm{d}r$$

Since the field is approximately constant over one Larmor orbit and $|B_r| \ll B_z$

$$B_r(r_\mathrm{L}) \simeq -\frac{r_\mathrm{L}}{2}\frac{\partial B_z}{\partial z} \simeq -\frac{r_\mathrm{L}}{2}\frac{\partial B}{\partial z}$$

With this approximation the z component of (2.1) gives

$$m\frac{\mathrm{d}v_\parallel}{\mathrm{d}t} = -\tfrac{1}{2}|e|r_\mathrm{L}v_\perp\frac{\partial B}{\partial z} = -\frac{W_\perp}{B}\frac{\partial B}{\partial z} = -\mu_\mathrm{B}\frac{\partial B}{\partial z}$$

Thus,

$$\frac{\mathrm{d}}{\mathrm{d}t}\left(\tfrac{1}{2}m{v_\parallel}^2\right) = -\mu_\mathrm{B}v_\parallel\frac{\partial B}{\partial z} = -\mu_\mathrm{B}\frac{\mathrm{d}B}{\mathrm{d}t} \tag{2.30}$$

From (2.29)

$$\frac{\mathrm{d}}{\mathrm{d}t}\left(\tfrac{1}{2}m{v_\perp}^2\right) \equiv \frac{\mathrm{d}}{\mathrm{d}t}(\mu_\mathrm{B}B) \tag{2.31}$$

Adding (2.30) and (2.31) and using energy conservation we find

$$\frac{\mathrm{d}\mu_\mathrm{B}}{\mathrm{d}t} = 0 \tag{2.32}$$

The invariance of μ_B in spatially varying magnetic fields has important implications which we explore in the next section.

2.7 Magnetic mirrors

Consider a particle moving in the inhomogeneous field introduced in Section 2.6 towards the region of increasing B. It follows from the invariance of $W_\perp/B$ that

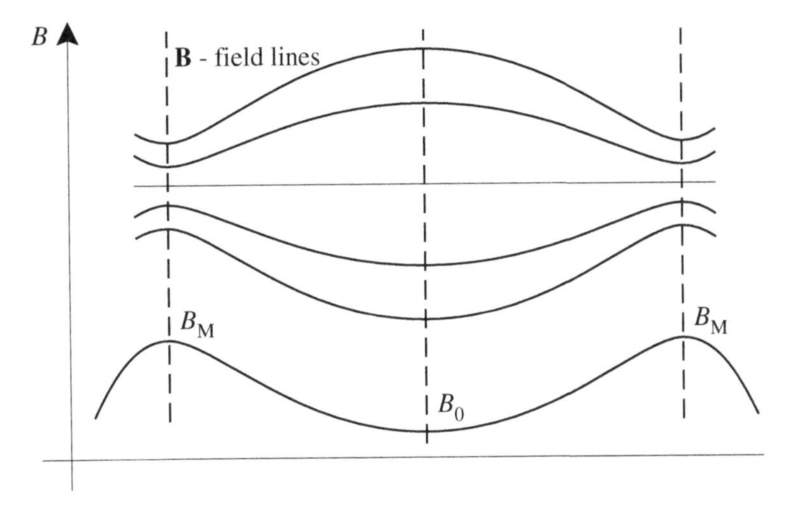

Fig. 2.7. Magnetic mirror field.

$W_\perp$ must increase. Since energy is conserved this increase must be at the expense of $W_\parallel$. Thus it may happen that for some value of B (B_R say) $W_\parallel = 0$, in which case the particle cannot penetrate further into the magnetic field and suffers reflection at this point (provided $(\mathbf{v} \times \mathbf{B}) \cdot \hat{\mathbf{z}} \neq 0$). Such a field configuration has the properties of a *magnetic mirror*.

It is convenient to define the *pitch angle*, θ, of the particle by

$$\tan\theta = \frac{v_\perp}{v_\parallel} \tag{2.33}$$

Then, from the invariance of $\mu_B = W_\perp/B$, it follows that $\sin^2\theta/B$ is constant. Defining the constant by B_R^{-1} we have

$$\sin\theta = (B/B_R)^{1/2} \tag{2.34}$$

For a particle which penetrates to the point where B reaches its maximum value B_M before being reflected, $B_R = B_M$. Hence particles with pitch angles such that $\sin\theta > (B/B_M)^{1/2}$ suffer reflection before reaching the region of maximum field; those having $\sin\theta \leq (B/B_M)^{1/2}$ are not reflected.

If one arranges two mirror fields in the configuration shown in Fig. 2.7 then particles with $\sin\theta > (B/B_M)^{1/2}$ will be reflected to and fro. This configuration constitutes a *magnetic bottle* or *adiabatic mirror trap*. Taking B_0 to be the value of the magnetic field in the mid-plane of the bottle, the *mirror ratio* is defined by

$$R = \frac{B_M}{B_0}$$

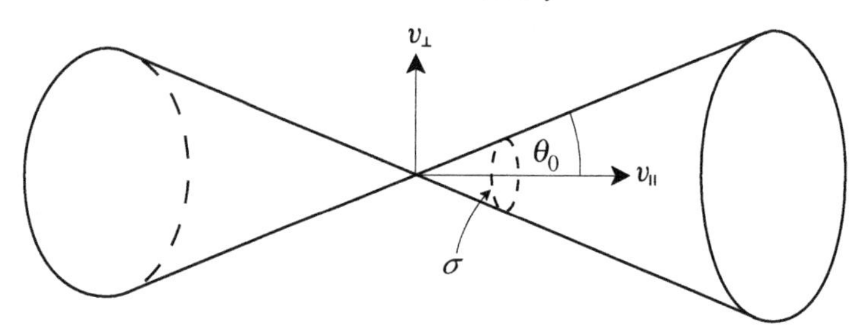

Fig. 2.8. Loss cone for a magnetic mirror. Particles with velocities within the cone will escape from the mirror.

and particles will be reflected if

$$\sin\theta_0 > R^{-1/2}$$

The particles which are lost from the magnetic bottle are those within the solid angle σ in velocity space shown in Fig. 2.8. This solid angle σ defines the *loss cone*. Denoting the probability of loss from the bottle by P where

$$P = \frac{\sigma}{2\pi} = \int_0^{\theta_0} \sin\theta \, \mathrm{d}\theta$$

we have

$$P = 1 - \left(\frac{R-1}{R}\right)^{1/2} \simeq \frac{1}{2R} \qquad \text{if } R \gg 1$$

Thus the higher the mirror ratio, the less likely it is that particles will escape.

Mirror traps have been used to contain laboratory plasmas but losses through the mirrors led to their being abandoned in favour of toroidal devices. However, the concept of magnetic trapping is fundamental to an understanding of many naturally occurring plasmas such as the Earth's radiation belts and the energetic particles associated with solar flares which move in closed loop fields associated with active regions of the Sun. Before discussing naturally occurring magnetic traps we first identify a second adiabatic invariant.

2.8 The longitudinal adiabatic invariant

Given that the number of adiabatic invariants reflects the distinct system periodicities, we expect a second invariant to arise in connection with the reflection of particles between the fields of a mirror trap. This invariant is associated with the guiding centre motion and is known as the *longitudinal invariant*, J, defined by

$$J = \oint v_\parallel \, \mathrm{d}s \tag{2.35}$$

where ds is an element of the guiding centre path and the integral is evaluated over one complete traverse of the guiding centre. The invariance of J is useful in situations in which the mirror points are no longer stationary; therefore B is taken to be slowly varying in both space and time. Since J is a function of $\dot{s}$ $(\equiv v_\parallel)$, s, t, using $W = \frac{1}{2}mv_\parallel^2 + \mu_B B$ we may write

$$J(W, s, t) = \int_{s_1}^{s} \left[\frac{2}{m}(W - \mu_B B)\right]^{1/2} \mathrm{d}s \tag{2.36}$$

Then

$$\begin{aligned}
\frac{\mathrm{d}J}{\mathrm{d}t} &= \left(\frac{\partial J}{\partial t}\right)_{W,s} + \left(\frac{\partial J}{\partial W}\right)_{s,t}\frac{\mathrm{d}W}{\mathrm{d}t} + \left(\frac{\partial J}{\partial s}\right)_{W,t}\frac{\mathrm{d}s}{\mathrm{d}t} \\
&= -\int_{s_1}^{s}\left[\frac{2}{m}(W-\mu_B B)\right]^{-1/2}\left(\frac{\mu_B}{m}\frac{\partial B}{\partial t}\right)\mathrm{d}s \\
&\quad + \left\{v_\parallel \dot{v}_\parallel + \frac{\mu_B}{m}\frac{\partial B}{\partial t} + \frac{\mu_B}{m}v_\parallel\frac{\partial B}{\partial s}\right\}\int_{s_1}^{s}\left[\frac{2}{m}(W-\mu_B B)\right]^{-1/2}\mathrm{d}s \\
&\quad + \left[\frac{2}{m}(W-\mu_B B)\right]^{1/2} v_\parallel - v_\parallel\int_{s_1}^{s}\left[\frac{2}{m}(W-\mu_B B)\right]^{-1/2}\left(\frac{\mu_B}{m}\frac{\partial B}{\partial s}\right)\mathrm{d}s
\end{aligned}$$

Now, at the turning point $s = s_1$, $v_\parallel = 0$; then

$$\begin{aligned}
\frac{\mathrm{d}J(W, s_1, t)}{\mathrm{d}t} &= -\oint\left[\frac{2}{m}(W-\mu_B B)\right]^{-1/2}\left(\frac{\mu_B}{m}\frac{\partial B}{\partial t}\right)\mathrm{d}s \\
&\quad + \frac{\mu_B}{m}\frac{\partial B}{\partial t}\oint\left[\frac{2}{m}(W-\mu_B B)\right]^{-1/2}\mathrm{d}s \\
&= -\oint\left(\frac{\mu_B}{m}\frac{\partial B}{\partial t}\right)\frac{\mathrm{d}s}{v_\parallel} + \frac{\mu_B}{m}\frac{\partial B}{\partial t}\oint\frac{\mathrm{d}s}{v_\parallel} \\
&= -\int_0^{\tau_\parallel}\frac{\mu_B}{m}\frac{\partial B}{\partial t}\mathrm{d}t + \frac{\mu_B}{m}\frac{\partial B}{\partial t}\int_0^{\tau_\parallel}\mathrm{d}t \\
&= -\frac{\mu_B}{m}\left[B(\tau_\parallel) - B(0) - \tau_\parallel\frac{\partial B}{\partial t}\right] \\
&\sim O(\tau_\parallel/t_F)^2
\end{aligned}$$

where $\tau_\parallel$ is the transit time and t_F the time scale for changes in B. Provided $t_F \gg \tau_\parallel$

$$\frac{\mathrm{d}J}{\mathrm{d}t} = 0 \tag{2.37}$$

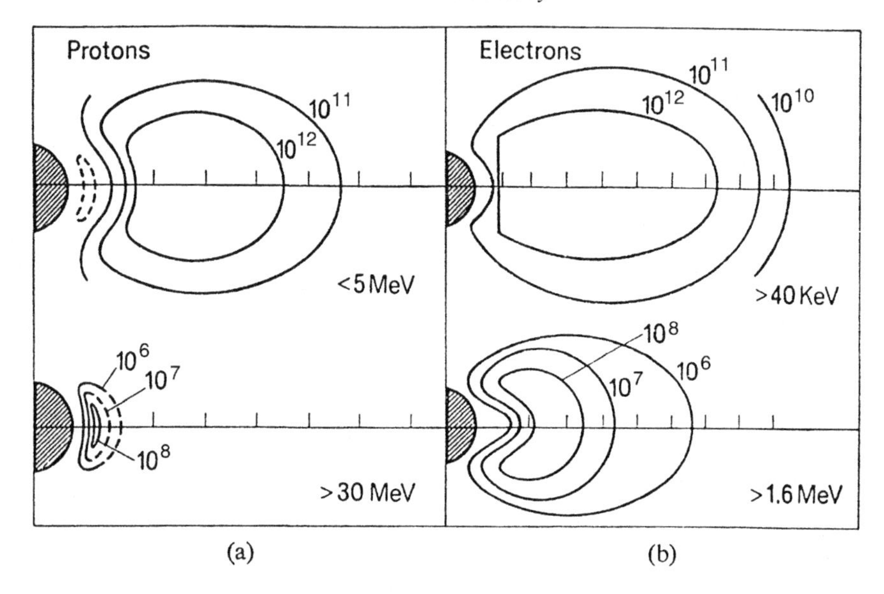

Fig. 2.9. Earth's radiation belts showing schematic spatial distributions of trapped protons (a) and electrons (b) across the energy ranges indicated. The contour label is the flux of charged particles in units (number $m^{-2} s^{-1}$).

Thus J is an adiabatic invariant. Since the transit time between mirror points is many Larmor periods, the condition on $\partial B/\partial t$ in order that J be invariant is clearly more stringent than that for μ_B. One can establish the adiabatic invariance of J more generally by including a slowly varying time-dependent electric field. A relativistic proof of J-invariance was given by Northrop and Teller (1960).

2.8.1 Mirror traps

Invariance of J is a particularly useful concept in determining particle trajectories in complex magnetic fields and is often used in practice in preference to integrating the guiding centre equations. One such application serves to characterize particles injected into, and trapped by, the geomagnetic field. Van Allen and co-workers first identified regions of energetic particles encircling the Earth in 1958. These structures are known as the *Van Allen radiation belts* and similar belts have since been identified in other planets where magnetospheres are present. The morphology of the Earth's radiation belts is complex, consisting of two regions, an inner and an outer belt, shown schematically in Fig. 2.9. Protons of energies from 30 to a few hundred MeV were observed extending out to about $2R_E$, where R_E is the Earth radius. Protons populating the outer belt are much less energetic. There is no comparable demarcation in the electron energy distribution across the two regions represented in Fig. 2.9. The distribution of electrons in the outer radiation belt stretches virtually as far as the magnetopause at $\sim 10R_E$.

The particle populations in the Van Allen belts are distinct from ionospheric distributions on the one hand and those of the solar wind on the other. Identifying the sources of the Van Allen populations provides a key to understanding the morphology of the radiation belts. The inner belt is characterized not only by its energy distribution but by the fact that it is more stable than the outer belts. Taken together these observations suggest distinct sources for inner and outer belt components. Cosmic rays colliding with atoms result in disintegrations into nuclear components. These include neutrons travelling outwards which decay to produce energetic protons and electrons. Observations have confirmed that cosmic rays are indeed a source for inner belt protons with energies of tens of MeV. The low energy component on the other hand is thought to derive from an ionospheric source while those at intermediate energies are variously solar wind particles that have been accelerated as well as particles injected from the plasma sheet during auroral events.

Although the detailed dynamics of the radiation belts is complicated it is clear that the geomagnetic field serves as a magnetic trap for charged particles. The geomagnetic field may be represented as a dipole (at any rate out to distances of about $5R_E$ beyond which the field is distorted by the solar wind) with the field lines bunching at the north and south poles. Energetic particles injected into the Earth's magnetic field will describe helical trajectories and undergo reflection in the stronger field regions around the magnetic poles, transit times for protons bouncing between mirror points being of the order of a second. The stability of the Van Allen belts is essentially a reflection of the invariance of J.

In addition to the bounce motion between mirror points, the results of Section 2.4 mean that particles drift azimuthally since field lines are curved and there exists a magnetic field gradient normal to the direction of **B**. Electrons drift from west to east and protons vice versa. For an electron with energy 40 keV, the time taken for the guiding centre to complete a circuit is of the order of an hour. The guiding centre of a particle generates a surface of rotation, which in some circumstances may be closed. The periodicity associated with this drift leads to a third adiabatic invariant in magnetic fields with suitable morphology.

2.9 Magnetic flux as an adiabatic invariant

The third adiabatic invariant is the flux Φ of the magnetic field through the surface of rotation. A formal proof of the adiabatic invariance of Φ was given by Northrop (1961). We present here a précis of Northrop's proof. It is convenient to use a set of curvilinear coordinates (α, β, s) where $\alpha(\mathbf{r}, t)$, $\beta(\mathbf{r}, t)$ are parameters characterizing a field line and therefore constant on it while s represents distance along the line. The parameters α and β, known as *Clebsch variables*, are chosen so that

$$\mathbf{A} = \alpha \nabla \beta \qquad \mathbf{B} = \nabla \alpha \times \nabla \beta \tag{2.38}$$

where $\mathbf{A}$ is the magnetic vector potential.

For time-dependent magnetic fields the particle energy $K = \frac{1}{2}mv_\parallel^2 + \mu_B B + e\alpha\partial\beta/\partial t$ is no longer a constant of motion. The flux Φ through the longitudinal invariant surface is a function of J, μ_B, K and t but since the first pair are known adiabatic invariants we suppress this dependence and represent the flux as $\Phi(K, t)$. Then

$$\Phi(K, t) = \oint \mathbf{A} \cdot \mathrm{d}\mathbf{l} = \oint \alpha \nabla \beta \cdot \mathrm{d}\mathbf{l} = \oint \alpha \, \mathrm{d}\beta$$

in which the contour is any simply connected curve lying on the surface and

$$\frac{\mathrm{d}\Phi}{\mathrm{d}t} = \frac{\partial \Phi(K, t)}{\partial K} \langle \dot{K} \rangle_\parallel + \frac{\partial \Phi(K, t)}{\partial t} \tag{2.39}$$

where $\langle \dot{K} \rangle_\parallel$ indicating an average of $\dot{K}$ over motion along the field lines has been substituted for $\dot{K}$ since changes over a longitudinal transit time are not of interest here. It is straightforward to show that (see Exercise 2.4)

$$\frac{\partial \Phi}{\partial K} = \oint \frac{\partial \alpha}{\partial K} \mathrm{d}\beta = \frac{1}{e} \oint \frac{\mathrm{d}\beta}{\langle \dot{\beta} \rangle} = \frac{\tau_P}{e}$$

$$\frac{\partial \Phi}{\partial t} = \oint \frac{\partial \alpha}{\partial t} \mathrm{d}\beta = -\frac{1}{e} \oint \frac{\langle \dot{K} \rangle_\parallel \mathrm{d}\beta}{\langle \dot{\beta} \rangle} = -\frac{\tau_P}{e} \langle \langle \dot{K} \rangle_\parallel \rangle_P$$

where τ_P denotes the period of precession of the guiding centre and $\langle \langle \dot{K} \rangle_\parallel \rangle_P$ is the average of $\langle \dot{K} \rangle_\parallel$ over a precession period. From these relations we see that

$$\frac{\mathrm{d}\Phi}{\mathrm{d}t} = \frac{\tau_P}{e} \left[\langle \dot{K} \rangle_\parallel - \langle \langle \dot{K} \rangle_\parallel \rangle_P \right] \tag{2.40}$$

Although $\mathrm{d}\Phi/\mathrm{d}t \neq 0$ it is evident that the average rate of change of the magnetic flux over a period of precession does vanish, i.e.

$$\left\langle \frac{\mathrm{d}\Phi}{\mathrm{d}t} \right\rangle_P = 0 \tag{2.41}$$

Hence although K is no longer invariant for time-dependent magnetic fields, a new adiabatic invariant Φ has been found in its place. A fuller discussion of the properties of invariant surfaces has been given by Northrop (1963).

A summary of the properties of the three adiabatic invariants μ_B, J, and Φ is set out in Table 2.1. It is perhaps worth noting that the stringent requirements demanded of Φ (associated with the loss of phase information in averaging over closed trajectories) make it a less useful invariant in practice than either μ_B or J.

Table 2.1. *Adiabatic invariants*

Invariant	Particle characteristic motion	Periodicity	Validity conditions
Magnetic moment $\mu_B = W_\perp/B$	Larmor orbit	$\tau_L = 2\pi/\lvert\Omega\rvert$ $v_\perp^2 \tau_L \simeq$ const.	$\tau_L \ll t_F$
Longitudinal invariant $J = \oint v_\parallel \, ds$	Longitudinal bounce between mirror fields	$\tau_\parallel$ $\langle v_\parallel^2 \rangle \tau_\parallel \simeq$ const.	$\tau_L \ll \tau_\parallel \ll t_F$ μ_B constant
Flux invariant $\Phi = \int \mathbf{B} \cdot d\mathbf{S}$	Azimuthal precession; drift velocity, v_P	τ_P $\langle v_P^2 \rangle \tau_P \simeq$ const.	$\tau_L \ll \tau_\parallel \ll \tau_P \ll t_F$ μ_B constant J constant

2.10 Particle orbits in tokamaks

Electron trapping in spatially inhomogeneous magnetic fields is important in the physics of tokamaks. In Chapter 1 we saw that a tokamak is characterized by a combination of toroidal and poloidal magnetic fields, B_t and B_p respectively, since a toroidal field on its own is not capable of containing a plasma in equilibrium. Anticipating our discussion of toroidal equilibria in Chapter 4, tokamaks are characterized by an ordering of these fields such that $B_t \gg B_p$. The magnetic field lines are helices wound on a toroidal surface. Particles whose guiding centres follow such helical field lines and whose velocity components along the field are high enough that they cycle round the torus make up the population of *passing* particles. In contrast particles with lower velocities parallel to the field contribute to the population of particles trapped on the outer side of the torus between magnetic mirrors created by the poloidal variation of the field. The tokamak magnetic field varies as $1/R$ where $R = R_0 + r\cos\theta$; R_0 is the major radius of the tokamak and r the minor radius of the surface on which the guiding centre of the particle lies, with θ the poloidal angle. The ratio $r/R_0 = \epsilon$ is known as the *inverse aspect ratio* and serves as an expansion parameter. The magnetic field $B(\theta)$ may then be expressed as

$$B(\theta) = B(0)(1 - \epsilon\cos\theta)/(1 - \epsilon) \tag{2.42}$$

From the adiabatic invariance of μ_{B} we find

$$\frac{v_{\|}^2}{v_0^2} = 1 - \frac{v_{\perp 0}^2}{v_0^2}\left(\frac{1-\epsilon\cos\theta}{1-\epsilon}\right)$$

It is clear from this expression that $v_{\|}$ will vanish for θ satisfying

$$\frac{v_{\|0}^2}{v_{\perp 0}^2} = \epsilon(1-\cos\theta) \tag{2.43}$$

where the right-hand side can assume values up to 2ϵ (at $\theta = \pi$). Clearly if $v_{\|0}^2/v_{\perp 0}^2 > 2\epsilon$, (2.43) cannot be satisfied and this condition defines the population of *passing* particles. On the other hand if $v_{\|0}^2/v_{\perp 0}^2 < 2\epsilon$ there will be some value of θ given by (2.43) for which $v_{\|} = 0$. This condition serves to define the population of *trapped* particles.

Consider in turn the characteristics of passing and trapped particles. Two effects contribute to the dynamics of passing particles. Superposed on the motion parallel to the magnetic field which gives rise to rotation in the poloidal direction with velocity $\mathbf{v}_{\mathrm{p}} = (\mathbf{B}_{\mathrm{p}}/B_{\mathrm{t}})v_{\|}$ is a combination of gradient and curvature drifts in the toroidal magnetic field. Given that B_{t} is approximately inversely proportional to the major radius R, it follows that the drift velocity $\mathbf{v}_{\mathrm{B}}$ defined by (2.24) is in the vertical direction, $\hat{\mathbf{z}}$. If we suppose that the cross-section is approximately circular, the motion of the guiding centre projected on to the poloidal plane may be represented as

$$\frac{\dot{r}}{r\dot{\theta}} = \frac{v_{\mathrm{B}}\sin\theta}{v_{\mathrm{p}} + v_{\mathrm{B}}\cos\theta} \tag{2.44}$$

The drift orbit is thus

$$\frac{r}{r_0} = \left[1 + \frac{v_{\mathrm{B}}}{v_{\mathrm{p}}}\cos\theta\right]^{-1} \tag{2.45}$$

where $r = r_0$ for $\theta = \pi/2$. The displacement of this distorted circle in the direction of the major radius is determined by

$$|\Delta_{\mathrm{pass}}| = r_0\left(\frac{v_{\mathrm{B}}}{v_{\|}}\right)\left(\frac{B_{\mathrm{t}}}{B_{\mathrm{p}}}\right) = \frac{q}{\Omega}\frac{(v_{\|}^2 + \frac{1}{2}v_{\perp}^2)}{v_{\|}} \sim \left(\frac{v_{\|}^2 + \frac{1}{2}v_{\perp}^2}{v_{\perp}v_{\|}}\right) q r_{\mathrm{L}} \tag{2.46}$$

where $q = r_0 B_{\mathrm{t}}/RB_{\mathrm{p}}$ is a quantity known as the *safety factor* (see Section 4.3.1) and r_L is the particle Larmor radius in the toroidal magnetic field. For tokamaks, q is typically about 3 near the plasma edge so that for passing particles the shift of the drift orbit, shown in Fig. 2.10, is significantly bigger than a Larmor radius.

Dealing with the trapped particles is more difficult but by making use of constants of the motion one can determine the size of the drift orbits for this population

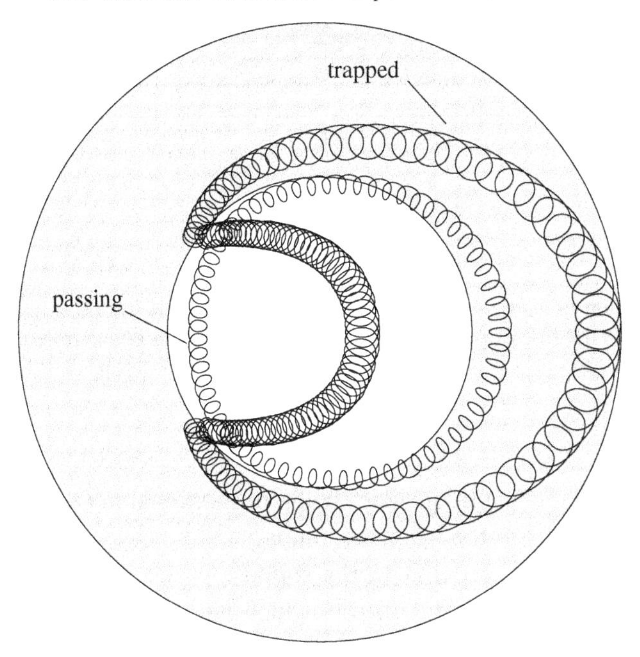

Fig. 2.10. Orbits of passing and trapped particles in a tokamak projected on the poloidal plane.

as well. The procedure is outlined in Exercise 2.5. Trapped particles describe the *banana* orbits shown in Fig. 2.10. The width of a banana orbit is approximately

$$|\Delta_{\mathrm{tr}}| \sim 2\left(\frac{2R}{r}\right)^{\frac{1}{2}} q r_{\mathrm{L}} \tag{2.47}$$

We see from this estimate that trapped particle orbits can be an order of magnitude bigger than a Larmor radius.

2.11 Adiabatic invariance and particle acceleration

Particle acceleration is of widespread interest in both laboratory and space plasmas. As an example we consider an idea originally put forward by Fermi (1949) to account for the very energetic particles ($O(10^{18}\,\mathrm{eV})$) in cosmic radiation. How such enormous energies are attained is obviously a key question in cosmic ray theory. Fermi postulated that there are regions of space in which clumps of magnetic field of higher than average intensity occur with charged particles trapped between them. He argued that these magnetic clumps would not be static and trapped particles could be accelerated if such regions were approaching one another. By the same token, particles would lose energy in mirror regions that were separating. Fermi

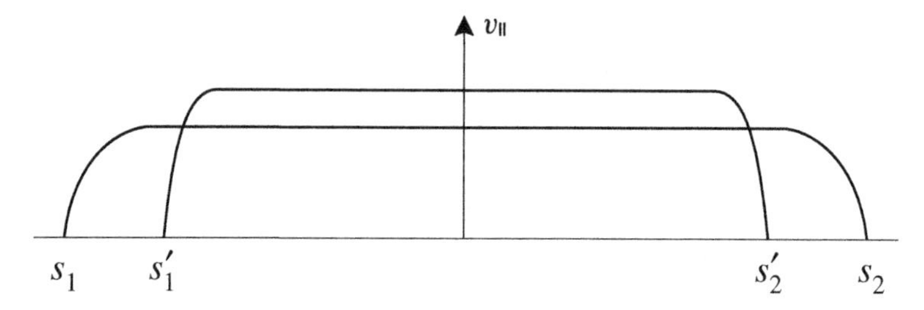

Fig. 2.11. Variation of $v_\parallel$ between mirror points s_1 and s_2. If the field maxima approach one another, so do the mirror points and at some later time the phase trajectory is given by the curve from s_1' to s_2'.

showed that the probability of head-on collisions was greater than that of overtaking collisions, their relative frequencies being proportional to $(v_\parallel + v_\mathrm{B})/(v_\parallel - v_\mathrm{B})$ where v_B is the velocity of the magnetic clump.

To see how Fermi acceleration works suppose a charged cosmic ray particle is trapped between two magnetic mirrors which move towards one another sufficiently slowly that J is a good adiabatic invariant. Suppose too that at $t = 0$ the coordinates of the mirror points in the phase space diagram, Fig. 2.11, are s_1 and s_2, while at some later time, t', they shift to s_1' and s_2' respectively. The invariance of J means that the area enclosed by the phase space orbit is constant. Denoting $s_1 - s_2$ by s_0, $s_1' - s_2'$ by s_0' gives

$$v_\parallel' \simeq v_\parallel \frac{s_0}{s_0'}$$

where $v_\parallel$, $v_\parallel'$ are the parallel components of velocity at the mid-plane at $t = 0$, $t = t'$, respectively. Then

$$W_\parallel' = \left(\frac{s_0}{s_0'}\right)^2 W_\parallel$$

and

$$W' = W_\perp' + W_\parallel' = \frac{m}{2}\left[v_\perp^2 + \left(\frac{s_0}{s_0'}\right)^2 v_\parallel^2\right]$$

by making use of the invariance of μ_B (assuming the field is constant at the mid-plane). Thus the energy of a particle trapped between slowly approaching magnetic mirrors increases. As proposed originally, Fermi acceleration suffers from a serious limitation. For increasing $v_\parallel$, the pitch angle defined by (2.33) decreases so that at some stage a particle being accelerated falls into the loss cone and escapes, thus limiting the gain in energy.

2.12 Polarization drift

Up to this point we have not allowed for any time-dependent electric fields, having set $\dot{E} = 0$ (in Section 2.5) to exclude plasma oscillations. We now want to relax this condition and allow for a *slow* time variation in an applied electric field. Doing so introduces yet another drift, the *polarization drift*, to add to those discussed earlier. To see how polarization drift comes about, we return to the configuration analysed in Section 2.3 with an electric field acting in a direction perpendicular to an applied magnetic field. By slow time dependence of the electric field we mean slow compared with the Larmor period of the particle. Starting with (2.12) and (2.13) which we used to identify the $\mathbf{E} \times \mathbf{B}$ drift, we introduce a transformation $\dot{X} = \dot{x} - eE_\perp/m\Omega$ to move to the drift frame. In this frame

$$\ddot{X} = \Omega\left(\dot{y} - \frac{e}{m\Omega^2}\frac{\mathrm{d}E_\perp}{\mathrm{d}t}\right) \qquad \ddot{y} = -\Omega\dot{X} \tag{2.48}$$

If we now apply a second transformation

$$\dot{Y} = \dot{y} - \frac{e}{m\Omega^2}\frac{\mathrm{d}E_\perp}{\mathrm{d}t}$$

and make use of the fact that the time scale of variation of the electric field is much longer than a Larmor period, the equations of motion in this new frame reduce to the Larmor equations for motion in a magnetic field alone. This establishes that in addition to the $\mathbf{E} \times \mathbf{B}$ drift there is now a polarization drift, with drift velocity $\mathbf{v}_\mathrm{P}$ given by

$$\mathbf{v}_\mathrm{P} = \frac{m}{eB^2}\frac{\mathrm{d}\mathbf{E}}{\mathrm{d}t} \tag{2.49}$$

The polarization drift has distinct properties compared with the $\mathbf{E} \times \mathbf{B}$ drift. Since $\mathbf{v}_\mathrm{P}$ is charge-dependent, electron and ion drifts are in opposite directions but the mass dependence means that the ion drift is dominant. Associated with the drift is a *polarization current density*

$$\mathbf{j}_\mathrm{P} \simeq \frac{n_\mathrm{i} m_\mathrm{i}}{B^2}\frac{\mathrm{d}\mathbf{E}}{\mathrm{d}t}$$

which contributes to the total current density in general (see Section 2.5). The name polarization drift comes from the fact that the electric field inside most plasmas derives not from an externally applied source but from the polarization of the plasma due to charge separation.

2.13 Particle motion at relativistic energies

To describe the motion of particles of relativistic energy we have to revert to the full Lorentz equation,

$$\frac{\mathrm{d}}{\mathrm{dt}}(\gamma m\mathbf{v}) = e(\mathbf{E} + \mathbf{v} \times \mathbf{B}) \tag{2.50}$$

where $\gamma = (1 - v^2/c^2)^{1/2}$. It is possible to set about solving the Lorentz equation for the various field configurations considered earlier. For example a particle moving with relativistic velocity in a constant, uniform magnetic field $\mathbf{B}$, with energy $\mathcal{E} = \gamma mc^2$ is now characterized by a Larmor frequency

$$\Omega = \frac{eB}{m\gamma} \tag{2.51}$$

In the case of motion in constant, uniform electric and magnetic fields we no longer need to insist that $E_\parallel = 0$ as in Section 2.3. In general one can repeat the analysis of slowly varying magnetic fields in the drift approximation using the full Lorentz equation and again establish the existence of adiabatic invariants. However, these exercises are of limited value and rather than go down this road we turn instead to a problem of greater practical interest, the relativistic motion of particles in an electromagnetic field.

2.13.1 Motion in a monochromatic plane-polarized electromagnetic wave

For a plane-polarized wave propagating in the x direction, $\mathbf{E} = (0, E, 0)$, and $\mathbf{B} = (0, 0, B)$. We can conveniently describe the fields by their vector potential, $\mathbf{A} = (0, a(\tau)\cos\omega\tau, 0)$, where a is the amplitude of the wave, ω the frequency, and $\tau = t - x/c$. For a truly monochromatic wave a is constant. In practice, however, a can never be constant since the amplitude must grow from zero when the wave is switched on. Moreover, treatments which assume a is constant lead to solutions depending critically on the initial phase and thereby predict electron drifts in arbitrary directions (see Exercise 2.7). For an almost monochromatic wave, $a(\tau)$ must be a slowly varying function and the only significant effect of including its variation is to ensure that the fields are initially zero; dependence on the initial phase does not then appear. $\mathbf{E}$ and $\mathbf{B}$ are given by the usual equations

$$\mathbf{E} = -\frac{\partial \mathbf{A}}{\partial t} \qquad \mathbf{B} = \nabla \times \mathbf{A}$$

from which it follows in the case of a plane-polarized wave

$$E = cB = -\frac{\mathrm{d}A}{\mathrm{d}\tau}$$

The relativistic Lorentz equation gives

$$\frac{\mathrm{d}}{\mathrm{d}t}(\gamma\dot{x}) = \frac{eB}{m}\dot{y} = -\frac{e}{mc}\dot{y}\frac{\mathrm{d}A}{\mathrm{d}\tau} \tag{2.52}$$

$$\frac{\mathrm{d}}{\mathrm{d}t}(\gamma\dot{y}) = \frac{e}{m}(E - B\dot{x}) = -\frac{e}{mc}(c - \dot{x})\frac{\mathrm{d}A}{\mathrm{d}\tau} \tag{2.53}$$

$$\frac{\mathrm{d}}{\mathrm{d}t}(\gamma\dot{z}) = 0 \tag{2.54}$$

$$\frac{\mathrm{d}}{\mathrm{d}t}(\gamma c) = \frac{eE}{mc}\dot{y} = -\frac{e}{mc}\dot{y}\frac{\mathrm{d}A}{\mathrm{d}\tau} \tag{2.55}$$

Subtracting (2.52) from (2.55) gives, on integrating,

$$1 - \dot{x}/c = (1 - v^2/c^2)^{1/2} \tag{2.56}$$

assuming that the electron is initially at rest at the origin. Since

$$\frac{\mathrm{d}\tau}{\mathrm{d}t} = 1 - \frac{\dot{x}}{c} \tag{2.57}$$

(2.53) may be integrated directly:

$$\frac{\dot{y}/c}{(1 - v^2/c^2)^{1/2}} = -\frac{eA}{mc} \tag{2.58}$$

Substituting for $\dot{y}$ from (2.58) and using (2.56) and (2.57), (2.52) becomes

$$\frac{\mathrm{d}}{\mathrm{d}t}(\gamma\dot{x}/c) = \left(\frac{e}{mc}\right)^2 A\frac{\mathrm{d}A}{\mathrm{d}\tau}\frac{\mathrm{d}\tau}{\mathrm{d}t}$$

Hence,

$$\gamma\dot{x}/c = \frac{1}{2}\left(\frac{eA}{mc}\right)^2$$

Finally, using (2.56) and (2.57) again,

$$\dot{x} = \frac{c}{2}\left(\frac{eA}{mc}\right)^2\frac{\mathrm{d}\tau}{\mathrm{d}t} = \frac{c}{2}a_0^2\frac{\mathrm{d}\tau}{\mathrm{d}t}\cos^2\omega\tau \tag{2.59}$$

$$\dot{y} = -c\left(\frac{eA}{mc}\right)\frac{\mathrm{d}\tau}{\mathrm{d}t} = -ca_0\frac{\mathrm{d}\tau}{\mathrm{d}t}\cos\omega\tau \tag{2.60}$$

and from (2.54)

$$\dot{z} = 0 \tag{2.61}$$

where $a_0 = (ea/mc)$. Averaging $\dot{x}$ over one period $(T = 2\pi/\omega)$, $a(\tau)$ may be treated as constant,

$$\langle \dot{x} \rangle = \frac{\int \dot{x}\, \mathrm{d}t}{\int \mathrm{d}t} = \frac{\frac{c}{2} a_0^2 \int \cos^2 \omega\tau\, \mathrm{d}\tau}{\int (\mathrm{d}\tau + \mathrm{d}x/c)}$$

$$= \frac{\frac{c}{2} a_0^2 \int \cos^2 \omega\tau\, \mathrm{d}\tau}{\int \mathrm{d}\tau \left(1 + \frac{1}{2} a_0^2 \cos^2 \omega\tau\right)} = \frac{c a_0^2}{4 + a_0^2} \tag{2.62}$$

Similarly,

$$\langle \dot{y} \rangle = 0$$

For $a_0 < 1$, the motion is effectively the quiver motion of the electron in the **E** field of the wave (i.e. along Oy) but in addition the electron drifts in the direction of propagation of the wave with drift velocity

$$\mathbf{v}_\mathrm{W} = \frac{e^2}{2m^2\omega^2} \langle \mathbf{E} \times \mathbf{B} \rangle \tag{2.63}$$

For highly relativistic electrons $a_0 \gg 1$ and $\langle \dot{x} \rangle \to c$. Electrons with relativistic energies appear as a consequence of the interaction of ultra-intense laser light (typically $\sim 10^{20}\ \mathrm{W\,cm^{-2}}$) with plasmas. Evidence of electrons drifting in the direction of propagation of the light has been found from both simulations and experiments. An interesting effect of this drift velocity is to predict a Doppler shift in the frequency of light scattered by free electrons (see Kibble (1964)).

It is a straightforward exercise to integrate the equations of motion to determine the electron trajectory, which has the form of a figure-of-eight. Experiments by Chen, Maksimchuk and Umstadter (1998) have confirmed the figure-of-eight trajectory.

2.14 The ponderomotive force

Spatial inhomogeneities give rise to another non-linear effect which plays a key role in the interaction of intense electromagnetic radiation with plasmas. For consistency we ought to use the full Lorentz equation but to keep the argument simple we revert to non-relativistic dynamics. Let us represent the spatially varying oscillating electric field as

$$\mathbf{E}(\mathbf{r}, t) = \mathbf{E}(\mathbf{r}) \cos \omega t \tag{2.64}$$

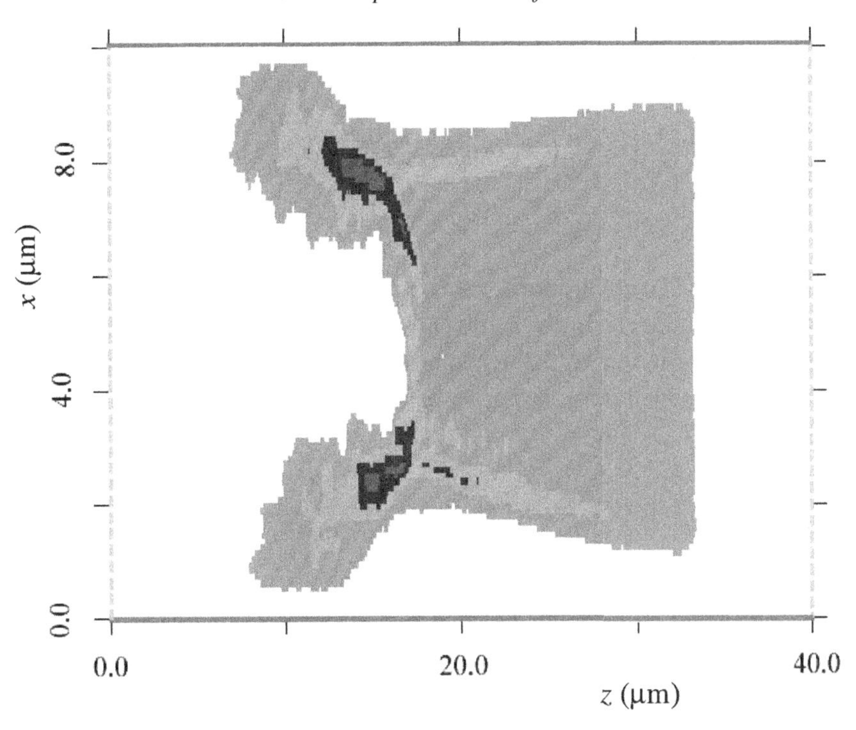

Fig. 2.12. Snapshot of the channel formation in the interaction of laser light of intensity 3×10^{19} W cm^{-2} with a plasma slab. The strong ponderomotive force from the intense laser beam incident from the left forms a channel approximately 4 μm across and 10 μm deep (at 850 fs). White denotes densities below $n_c = 10^{21}$ cm^{-3}, green $(1–4)n_c$, blue $(4–7)n_c$, red $(7–12)n_c$ and magenta over $12n_c$ (after Dyson (1998)).
At the time of going to press a colour version of this figure was available for download from http://www.cambridge.org/9780521459129

and write $\mathbf{E}(\mathbf{r})$ in terms of an expansion about the initial position of the particle $\mathbf{r}_0$

$$\mathbf{E}(\mathbf{r}) = \mathbf{E}(\mathbf{r}_0) + (\delta\mathbf{r}\cdot\boldsymbol{\nabla})\mathbf{E}|_{\mathbf{r}=\mathbf{r}_0} + \cdots \tag{2.65}$$

To lowest order we use the value $\mathbf{E}(\mathbf{r}_0)$ for the electric field so that the corresponding particle displacement $\delta\mathbf{r}$, is

$$\delta\mathbf{r} = -\frac{e\mathbf{E}}{m\omega^2}\cos\omega t \tag{2.66}$$

To next order we must keep the $\mathbf{v}\times\mathbf{B}$ contribution to the Lorentz equation giving

$$\ddot{\mathbf{r}}_1 = \frac{e}{m}\left[(\delta\mathbf{r}\cdot\boldsymbol{\nabla})\mathbf{E} + \dot{\mathbf{r}}\times\mathbf{B}\right] \tag{2.67}$$

Substituting for $\delta\mathbf{r}$ from (2.66) and averaging out the time dependence of the fields gives

$$\begin{aligned}\langle\ddot{\mathbf{r}}_1\rangle &= -\frac{1}{2}\left(\frac{e}{m\omega}\right)^2[(\mathbf{E}\cdot\nabla)\mathbf{E}+\mathbf{E}\times(\nabla\times\mathbf{E})]\\ &= -\left(\frac{e}{2m\omega}\right)^2\nabla E^2(\mathbf{r}_0) \qquad (2.68)\end{aligned}$$

Since the force is inversely proportional to particle mass it acts principally on electrons. The effect is that electrons are displaced from regions of high field intensity. If we suppose that the individual contributions from all electrons in unit volume of plasma may be added we arrive at an expression for the plasma *ponderomotive force*

$$\mathbf{F}_{\mathrm{PM}} = -\frac{\omega_{\mathrm{p}}^2}{\omega^2}\nabla\left\langle\frac{1}{2}\varepsilon_0 E^2\right\rangle \qquad (2.69)$$

Although the ponderomotive force displaces electrons from regions of high electric field to regions of weak field, the consequent charge separation creates a powerful electrostatic field which acts to pull the ions in the wake of the electrons. Simulations using large numbers of plasma particles provide a dramatic illustration of the ponderomotive force in the interaction of intense light with dense plasmas. High intensity laser light produces such a strong ponderomotive force that the plasma is pushed aside and a channel formed. Figure 2.12 shows channel formation when an intense laser beam is incident from the left on a plasma slab.

2.15 The guiding centre approximation: a postscript

Throughout this chapter we have made repeated use of Alfvén's idea of averaging out motion on the fast Larmor time scale to allow attention to focus on the motion of the guiding centre. To lowest order the particle gyrates about its guiding centre $\mathbf{r}_g$ where $\mathbf{r} = \mathbf{r}_g - (\dot{\mathbf{r}}\times\mathbf{b})/\Omega$ with $\mathbf{b} = \mathbf{B}/|\mathbf{B}|$. In this way we examined a number of guiding centre drifts in isolation by making assumptions about field configurations and imposing restrictions on the kind of inhomogeneity allowed in the fields. We found that gradient and curvature components of $[\partial B_i/\partial x_j]$ gave rise to drifts of the guiding centre, whereas divergence and shear components do not. There remains the question as to whether the drifts already identified for particular choices of field inhomogeneity provide a correct description of guiding centre motion in the case of a *general* static inhomogeneous magnetic field. To try to answer this one might represent the magnetic field making use of the expansion parameter $\epsilon = r_{\mathrm{L}}/L$; then to first order:

$$\mathbf{B} = \mathbf{B}_g + (\mathbf{r}-\mathbf{r}_g)\cdot\nabla\mathbf{B}\,|_{\mathbf{r}=\mathbf{r}_g}$$

One may then set about a general formulation of drifts to this order (Kruskal (1962), Northrop (1963), Morozov and Solov'ev (1966)). The answer turns out to be reassuring, but with one surprise. To order ϵ, the velocity of the guiding centre $\mathbf{v}_g$ is given by Balescu (1988)

$$\mathbf{v}_g = \left[v_\parallel + \frac{\epsilon v_\perp^2}{2\Omega}\mathbf{b}\cdot(\nabla\times\mathbf{b})\right]\mathbf{b} + \epsilon\frac{\mathbf{E}\times\mathbf{B}}{B^2} + \epsilon\frac{v_\perp^2}{2\Omega B}\mathbf{b}\times\nabla B + \frac{\epsilon v_\parallel^2}{\Omega}\mathbf{b}\times(\mathbf{b}\cdot\nabla)\mathbf{b} \qquad (2.70)$$

This result does indeed reproduce the $\mathbf{E}\times\mathbf{B}$, ∇B and curvature drifts. However, an unexpected contribution appears in the velocity parallel to the magnetic field, where intuitively all we expect is $v_\parallel\mathbf{b}$. Various attempts have been made at a physical interpretation of the $O(\epsilon)$ contribution to the parallel component of the guiding centre velocity. Morozov and Solov'ev (1966) argued that it ought to be removed by transforming to a new system of coordinates which ensures both conservation of energy as well as adiabatic invariance of μ_B. It turns out that the root of the problem lies in the averaging used by Alfvén to remove the Larmor motion from the dynamics, since this destroys the Hamiltonian structure of the system. A reappraisal by Littlejohn (1979, 1981) pointed the way round the difficulty. The issues at stake have been discussed in detail by Balescu (1988). The outcome is that by following Littlejohn's procedure, the physically spurious $O(\epsilon)$ term in the guiding centre velocity parallel to the magnetic field disappears.

Exercises

2.1 For non-relativistic motion in inhomogeneous magnetic fields:

(a) Verify (2.22) for $\mathbf{B} = (0, 0, B(\mathbf{r}))$ in which spatial inhomogeneities are small but otherwise arbitrary. [Hint: use (2.5) to show that $\mathbf{B}$ is not a function of z.]

(b) Verify that there is no mean drift velocity for a charged particle moving in the field $\mathbf{B} = (0, B_y(x), B)$. Assume B_y, $\mathrm{d}B_y/\mathrm{d}x$ are small.

(c) With $\mathbf{B} = (0, B_y(z), B)$, where both B_y, $\mathrm{d}B_y/\mathrm{d}z$ are small quantities, show that the curvature drift $\mathbf{v}_C = -\hat{\mathbf{x}}mv_\parallel^2(\mathrm{d}B_y/\mathrm{d}z)/eB^2$ and is given generally by (2.23). Show that there is in addition a drift in the y-direction given by $v_\parallel B_y/B$. Compare the relative magnitude of this drift with the curvature drift.

(d) A charged particle whose guiding centre lies initially on Oz moves in a converging, time-independent field

$$\mathbf{B} = B_0\left[-\frac{\alpha}{2}\mathbf{r} + (1+\alpha z)\hat{\mathbf{z}}\right]$$

where $\alpha > 0$ is a constant such that $\alpha r_L \ll 1$. Show that the particle motion is governed by the equations

$$\begin{aligned} \ddot{x} - \Omega\dot{y} &= \alpha\Omega(z\dot{y} + \tfrac{1}{2}y\dot{z}) \\ \ddot{y} - \Omega\dot{x} &= -\alpha\Omega(z\dot{x} + \tfrac{1}{2}x\dot{z}) \end{aligned} \qquad \ddot{z} = \tfrac{1}{2}\alpha\Omega(x\dot{y} - y\dot{x})$$

Show that to first order in αr_L, $\ddot{z} = \frac{1}{2} r_L v_\perp^2$ and interpret this result physically. Establish the adiabatic variance of μ_B.

(e) Near the equator the geomagnetic field may be approximated by $B(z) = B_0(1 + (z/z_0)^2)$ where z denotes the coordinate along the field and B_0, z_0 are positive constants. Show that particle motion near the equator is simple harmonic with period $\tau = 2\pi z_0(m/2\mu_B B_0)^{1/2}$ provided μ_B is a good adiabatic invariant.

(f) A particle with charge e and mass m moves in the bumpy field $\mathbf{B} = (b\sin kz, b\cos kz, B_0)$ where k, b and B_0 are constant and $b/B_0 \ll 1$. The particle is initially at the origin with velocity $\mathbf{v}(0) = (0, 0, \dot{z}_0)$. Assuming that $\Omega \neq k\dot{z}_0$, find $\dot{x}$, $\dot{y}$ to first order in b and show that

$$\dot{z}^2 = \dot{z}_0^2 + \frac{2\Omega^2\dot{z}_0^2}{(\Omega - k\dot{z}_0)^2}[\cos(\Omega - k\dot{z}_0)t - 1]$$

(g) The field in the geomagnetic tail (see Section 5.5.1) may be modelled in one dimension as

$$\begin{aligned} B_z &= B_0 & z &\geq L \\ &= B_0 z/L & -L &\leq z \leq L \\ &= -B_0 & z &\leq -L \end{aligned}$$

where $z = 0$ corresponds to the neutral sheet and $2L$ is the thickness of the plasma sheet. In such a representation μ_B is no longer adiabatically invariant. Write down the components of the Lorentz equation and integrate to find $\dot{y}$ and $\dot{z}$.

2.2 The geomagnetic field may be represented in terms of a dipole of moment $\mathbf{M} = -M\hat{\mathbf{z}}$, with components

$$B_R = -\frac{\mu_0 M}{2\pi}\frac{\sin\lambda}{r^3} \qquad B_\lambda = \frac{\mu_0 M}{4\pi}\frac{\cos\lambda}{r^3} \qquad B_\phi = 0$$

in which $\lambda = (\pi/2 - \theta)$ denotes latitude.

Show that the equation for a field line is $r = r_0\cos^2\lambda$ and write down an expression for the magnitude of the magnetic field $B(r, \lambda)$. Show that the ratio of the magnetic field to that at the equator B_0 is

$$\frac{B}{B_0} = \frac{(1 + 3\sin^2\lambda)^{1/2}}{\cos^6\lambda}$$

Using the representation for B/B_0, obtain an expression for the bounce time τ_B for trapped particles. Compute bounce times for magnetospheric electrons of energy 10 keV for a range of values of r_0/R_E where R_E is the Earth radius.

2.3 The combined gradient and curvature drift velocity $\mathbf{v}_B$ in (2.24) may be expressed in terms of the pitch angle α:

$$\mathbf{v}_B = \frac{mv^2}{2eB^3}(1 + \cos^2\alpha)\mathbf{B} \times \nabla B$$

(a) Use this result to show that at the equator $\mathbf{v}_B = -(3mv^2/2eBr)\hat{\boldsymbol{\phi}}$ where $\hat{\boldsymbol{\phi}}$ is the azimuthal unit vector.

(b) Show that the drift of electrons and ions contributes to a current, known as the Earth's ring current, given by

$$I_{\text{ring}} = -\frac{3\int nmv^2\,dV}{4\pi r^2 B}$$

In which direction does I_{ring} flow? For 1 MeV protons and 100 keV electrons, assuming $n \sim 10^7\,\text{m}^{-3}$, show that $I_{\text{ring}} \sim 1$ MA at $r \sim 4R_E$.

(c) For these parameters estimate the time needed for a proton to drift round the Earth.

(d) Find the extent to which the geomagnetic field is perturbed by the ring current. [To do this you need to include the diamagnetic contribution from the Larmor motion (see Section 2.2.1).]

2.4 Verify the results used in Section 2.9 to establish that Φ is an adiabatic invariant. [To show this you need to differentiate $K[\alpha(\beta, K, t), \beta, t]$ implicitly.]

2.5 Show that the drift orbit for passing particles in a tokamak is given by (2.45).

One way of dealing with trapped particles in a tokamak is to start from an integral of the motion in the guiding centre approximation. Conservation of toroidal momentum is expressed by $R(mv_\parallel + eA_\phi) = \text{constant}$, where A_ϕ is the toroidal component of the vector potential $\mathbf{A}$. If we make use of this integral of the motion at points at which the particle orbit intersects the plane $z = 0$ we have $-e(R_1 A_{\phi 1} - R_2 A_{\phi 2}) = m(R_1 v_{\parallel 1} - R_2 v_{\parallel 2})$. For trapped particles

$$R_1 A_{\phi 1} - R_2 A_{\phi 2} = \int_{R_2}^{R_1} \frac{\partial}{\partial R}(RA_\phi)\,dR = \int_{R_2}^{R_1} RB_z\,dR = \overline{RB_z}\Delta R$$

Show that at the points of intersection, taking $B_z = B_{p0}$ where B_{p0} denotes the poloidal field at the mid-plane where we set $|v_{\parallel 1}| = |v_{\parallel 2}|$, the width of the orbit of a trapped particle, the so-called *banana orbit* (see Fig 2.10), is given by (2.47).

2.6 Consider a particle of charge e and mass m moving in an electric field $E\sin(kx-\omega t)$. Particles whose velocities are close to the phase velocity of the wave are strongly affected by the wave field and exchange energy with the wave. The motion of a particle initially at x_0 moving with velocity v_0 is perturbed by the wave field in such a way that $x = x_0+v_0t+x_1+x_2$, $v = v_0+v_1+v_2$ where subscripts 1 and 2 denote corrections proportional to E and E^2 respectively.

Show that

$$v_1 = \frac{eE}{m\tilde{\omega}}[\cos(kx_0 - \tilde{\omega}t) - \cos kx_0]$$

$$\dot{v}_2 = -\frac{ke^2E^2}{m^2\tilde{\omega}^2}\cos(kx_0 - \tilde{\omega}t)[\sin(kx_0 - \tilde{\omega}t) - \sin kx_0 + \tilde{\omega}t\cos kx_0]$$

where $\tilde{\omega} = \omega - kv_0$.

Show that the rate of change of the energy of the particle averaged over random initial positions is given by

$$\langle\delta\dot{T}\rangle = \frac{e^2E^2}{2m}\left[\frac{\sin\tilde{\omega}t}{\tilde{\omega}} + \frac{kv_0}{\tilde{\omega}^2}(\sin\tilde{\omega}t - \tilde{\omega}t\cos\tilde{\omega}t)\right]$$

Using the representation of the delta function

$$\delta(\tilde{\omega}) = \lim_{t\to\infty}\frac{\sin\tilde{\omega}t}{\pi\tilde{\omega}}$$

show that

$$\langle\delta\dot{T}\rangle = \frac{\pi e^2E^2}{2m|k|}\frac{\partial}{\partial v_0}\left[v_0\delta\left(v_0 - \frac{\omega}{k}\right)\right]$$

2.7 Calculate the mean velocity of a charged particle in the electric field $\mathbf{E} = \mathbf{E}_0\cos(\omega t + \theta_0)$, where $\mathbf{E}_0$ and θ_0 are constants, assuming that relativistic effects are negligible. Repeat the calculation assuming that $\mathbf{E}_0$ is a slowly varying function (i.e. terms of order $(\mathrm{d}E_0/\mathrm{d}t)/\omega E_0$ may be neglected) with $\mathbf{E}_0(0) = 0$. Comment on the different results in the two cases.

2.8 Consider a relativistic charged particle moving in a constant, uniform magnetic field $\mathbf{B} = B_0\hat{\mathbf{k}}$. Show that the velocity of the particle is again given by (2.9) but with $\Omega = \Omega_0/\gamma$ in place of Ω_0.

2.9 Show that the Lorentz equation (2.50) may be rearranged to give

$$\dot{\boldsymbol{\beta}} = \frac{e}{m\gamma c}[\mathbf{E} - (\boldsymbol{\beta}\cdot\mathbf{E})\boldsymbol{\beta} + c\boldsymbol{\beta}\times\mathbf{B}]$$

where $\boldsymbol{\beta} = \mathbf{v}/c$. Note that for $\mathbf{E} = 0$, the result of the previous exercise follows by inspection.

2.10 Integrate (2.59) and (2.60), ignoring terms of order $(\mathrm{d}a/\mathrm{d}\tau)/a\omega$. Sketch the trajectory of the particle.

2.11 The vector potential for a monochromatic, plane wave of arbitrary polarization propagating in the x-direction is

$$\mathbf{A} = a(\tau)[0, \alpha \cos \omega\tau, (1-\alpha^2)^{1/2} \sin \omega\tau]$$

where $\tau = t - x/c$, $0 \le \alpha \le 1$ and the amplitude $a(\tau)$ is slowly varying.

(a) Integrate the relativistic Lorentz equation to obtain the velocity of an electron interacting with this wave, assuming that the electron starts from rest at the origin. Hence show that the drift velocity is the same as that for a plane-polarized wave ($\alpha = 1$).

(b) In the case in which the incident wave is circularly polarized show that the projection of the electron trajectory on the Oyz-plane is a circle of radius $(eE/\gamma m\omega^2)$.

(c) The gyration of an electron induces a magnetic field which is parallel (antiparallel) to the direction of wave propagation according to whether the wave field is left (right) circularly polarized. If one assumes (as in Section 2.2) that one may add the contributions from individual electrons, the resulting current is in turn the source of a magnetic field (the *inverse Faraday effect*). Show that the inverse Faraday field, B_{F}, may be expressed in terms of the Compton field, $B_{\mathrm{C}} = m\omega/e$, as

$$B_{\mathrm{F}} = \frac{\omega_{\mathrm{p}}^2}{2\omega^2} \frac{a_0^2}{(1+a_0^2)} B_{\mathrm{C}}$$

where $a_0 = eE/m\omega c$. Estimate B_{F} in the case of laser light with intensity $I = 10^{20}\,\mathrm{W\,cm^{-2}}$ interacting with a dense plasma.

2.12 Consider the case of a circularly polarized wave field propagating in the direction of a constant uniform magnetic field $\mathbf{B} = B_0\hat{\mathbf{x}}$. Show that in the plane perpendicular to the magnetic field the Lorentz equation may be written as

$$\ddot{\zeta} + i\Omega\dot{\zeta} = \left(\frac{eE}{\gamma m\omega^2}\right) e^{i\phi}$$

where $\zeta = y + iz$. Solve the Lorentz equation for the special case of *cyclotron resonance*, $\omega = \Omega$.

3

Macroscopic equations

3.1 Introduction

When the fields induced by the motion of the plasma particles are significant in determining that motion, particle orbit theory is no longer an apt description of plasma behaviour. The problem of solving the Lorentz equation self-consistently, where the fields are the result of the motion of many particles, is no longer practicable and a different approach is required. In this chapter, by treating the plasma as a fluid, we derive various sets of equations which describe both the dynamics of the plasma in electromagnetic fields and the generation of those fields by the plasma.

The fluid equations of neutral gases and liquids are usually derived by treating the fluid as a continuous medium and considering the dynamics of a small volume of the fluid. The aim is to develop a macroscopic model that, as far as possible, is independent of the detail of what happens at the molecular level. In this sense the approach is the opposite of that adopted in particle orbit theory where we seek information about a plasma by examining the motion of individual ions and electrons. In experiments one seldom makes measurements or observations at the microscopic level so we require a macroscopic description of a plasma similar to the fluid description of neutral gases and liquids. This is obtained here by an extension of the methods of fluid dynamics, an approach that conveniently skims over some fundamental difficulties inherent in plasmas. The chief of these is that a plasma is not really one fluid but at least two, one consisting of ions and the other electrons. The fact that these two fluids are comprised of particles with opposite charges and very unequal masses gives rise to phenomena that do not occur in neutral fluids, even those with more than one molecular component. Nevertheless, a single fluid description of a plasma is in many situations a useful and plausible model and one that is widely employed. Our first objective, therefore, is the derivation of the one-fluid, *magnetohydrodynamic (MHD) equations*.

The fundamental assumption of MHD is that fields and fluid fluctuate on the same time and length scales. Since the plasma is treated as a single fluid, these are necessarily determined by the slower rates of change of the heavy ions. However, in so far as electrons may behave independently of ions a plasma is able to support rapid wave fluctuations that are beyond the scope of the MHD equations. A second objective, therefore, is the derivation of the so-called *plasma wave equations*. Like the MHD equations, these are macroscopic fluid equations but the assumptions underlying them are quite different. In particular, since the rapid motion of the more mobile electrons must be distinguished from the slow response of the ions, this is necessarily a two-fluid description.

The final task of this chapter is a discussion of the boundary conditions applicable in the solution of the macroscopic equations we derive. These, of course, vary from problem to problem but, as in electromagnetic theory, there are certain general results which it is useful to establish once and for all.

3.2 Fluid description of a plasma

Before embarking on the actual derivation of the MHD equations it is helpful to discuss briefly some general concepts of fluid dynamics. First, as already mentioned, the fluid is treated as a continuous medium so that all macroscopic quantities are continuous functions of position $\mathbf{r}$ and time t. This assumption of continuity presupposes that one is interested in phenomena which vary on a hydrodynamic length scale L_{H} which, at the very least, is much greater than the average interparticle distance. This then leads on to the concept of a fluid element, a volume of fluid small enough that any macroscopic quantity has a negligible variation across its dimension but large enough to contain very many particles and so to be insensitive to particle fluctuations. To distinguish it from an element of volume δV, we denote the volume of a fluid element by $\delta\tau$.

Since any quantity F is a function of position and time its variation

$$\mathrm{d}F(\mathbf{r}, t) = \frac{\partial F}{\partial t}\mathrm{d}t + \frac{\partial F}{\partial r_i}\mathrm{d}r_i$$

and, in particular, its time rate of change is given by

$$\frac{\mathrm{d}F}{\mathrm{d}t} = \frac{\partial F}{\partial t} + \frac{\partial F}{\partial r_i}\frac{\mathrm{d}r_i}{\mathrm{d}t} = \frac{\partial F}{\partial t} + \mathbf{v}\cdot\boldsymbol{\nabla}F \tag{3.1}$$

where $\mathbf{v}$ is the velocity at the point $\mathbf{r}$ and time t. If, as will usually be the case, we are interested in the rate of change of F following a fluid element then $\mathbf{v} = \mathbf{u}(\mathbf{r}, t)$, the velocity of the fluid element or *flow velocity*, and for this special case it is customary to replace $\mathrm{d}/\mathrm{d}t$ by D/Dt to indicate this particular choice. Thus

$$\frac{DF}{Dt} \equiv \frac{\partial F}{\partial t} + \mathbf{u} \cdot \nabla F \tag{3.2}$$

is the time derivative of F as we follow the motion of a fluid element. This is known as the *material* or *substantive derivative* and the term $(\mathbf{u} \cdot \nabla)F$ is called the *convective derivative*.

Frequently in fluid theory, and particularly in deriving the fluid equations, we need the time derivatives of surface and volume integrals. Generalizing Leibnitz's theorem (see Fig. 3.1)

$$\frac{\mathrm{d}}{\mathrm{d}t}\int_{a(t)}^{b(t)} F(x,t)\mathrm{d}x = \int_a^b \frac{\partial F}{\partial t}\mathrm{d}x + \frac{\mathrm{d}b}{\mathrm{d}t}F(b,t) - \frac{\mathrm{d}a}{\mathrm{d}t}F(a,t)$$

to two and three dimensions, respectively, we have

$$\frac{\mathrm{d}}{\mathrm{d}t}\int_{A(t)} \mathbf{F}(\mathbf{r},t)\cdot \mathrm{d}\mathbf{A} = \int_{A(t)} \frac{\partial \mathbf{F}}{\partial t}\cdot \mathrm{d}\mathbf{A} + \oint_{C(t)} \mathbf{F}(\mathbf{r},t)\cdot \mathbf{v}_C \times \mathrm{d}\mathbf{l} \tag{3.3}$$

and

$$\frac{\mathrm{d}}{\mathrm{d}t}\int_{V(t)} F(\mathbf{r},t)\mathrm{d}V = \int_{V(t)} \frac{\partial F}{\partial t}\mathrm{d}V + \int_{A(t)} F(\mathbf{r},t)\mathbf{v}_A \cdot \mathrm{d}\mathbf{A} \tag{3.4}$$

In (3.3) the area of integration $A(t)$ is bounded by the closed curve $C(t)$ and $\mathbf{v}_C(\mathbf{r},t)$ is the velocity of the line element $\mathrm{d}\mathbf{l}$. In (3.4) the volume of integration $V(t)$ is bounded by the surface $A(t)$ and $\mathbf{v}_A(\mathbf{r},t)$ is the velocity of the surface element $\mathrm{d}\mathbf{A}$. These equations are quite general and may be applied to any surface or volume. Equation (3.3) is for future reference in Chapter 4. Here we are concerned with (3.4). Two cases of particular interest are:

(i) **Fixed volume** V

Here V is constant in time so its boundary is fixed and $\mathbf{v}_A \equiv 0$ giving

$$\frac{\mathrm{d}}{\mathrm{d}t}\int_V F(\mathbf{r},t)\mathrm{d}V = \int_V \frac{\partial F}{\partial t}\mathrm{d}V \tag{3.5}$$

(ii) **Fluid element** $\delta\tau$

Here $\mathbf{v}_A = \mathbf{u}$, the fluid velocity and, writing D/Dt for $\mathrm{d}/\mathrm{d}t$ to indicate this special choice, we get

$$\begin{aligned}\frac{D}{Dt}\int_{\delta\tau} F(\mathbf{r},t)\mathrm{d}\tau &= \int_{\delta\tau}\frac{\partial F}{\partial t}\mathrm{d}\tau + \int_{\delta S} F\mathbf{u}\cdot \mathrm{d}\mathbf{S} \\ &= \int_{\delta\tau}\left[\frac{\partial F}{\partial t} + \nabla\cdot(F\mathbf{u})\right]\mathrm{d}\tau \end{aligned} \tag{3.6}$$

where δS is the surface of the fluid element $\delta\tau$ and we have applied Gauss' divergence theorem.

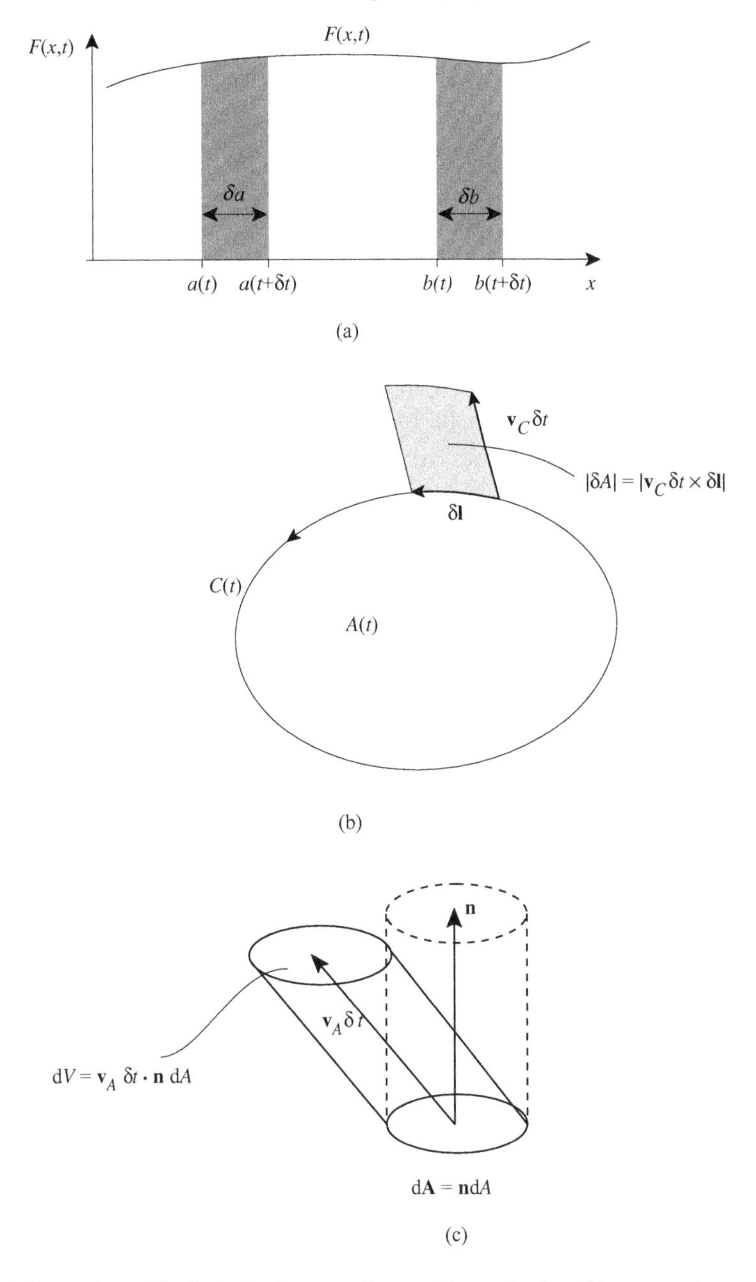

Fig. 3.1. Illustration of Leibnitz's theorem (in one dimension) and its extension to integrals over (b) two dimensions and (c) three dimensions.

The first of the macroscopic equations that we derive expresses conservation of fluid mass. Consider a fixed closed surface A lying entirely within the fluid and enclosing a volume V. If $\rho(\mathbf{r}, t)$ is the mass density of the fluid at position $\mathbf{r}$ and time t then the total mass of fluid enclosed by A at time t is $\int_V \rho \, \mathrm{d}V$ and the net rate at which mass is flowing outwards across the surface A is $\int_A \rho \mathbf{u} \cdot \mathrm{d}\mathbf{A}$. Thus

$$\frac{\mathrm{d}}{\mathrm{d}t} \int_V \rho \, \mathrm{d}V = -\int_A \rho \mathbf{u} \cdot \mathrm{d}\mathbf{A}$$

and using (3.5) with $F = \rho$ and Gauss' divergence theorem once more, this becomes

$$\int_V \left[\frac{\partial \rho}{\partial t} + \nabla \cdot (\rho \mathbf{u}) \right] \mathrm{d}V = 0$$

But this is valid for any volume V lying entirely within the fluid from which we conclude that the integrand must be identically zero at every point in the fluid. Hence, mass conservation is expressed by the *equation of continuity*

$$\frac{\partial \rho}{\partial t} + \nabla \cdot (\rho \mathbf{u}) = 0 \tag{3.7}$$

Using (3.7) and setting $F = \rho f$ in (3.6) we get

$$\frac{D}{Dt} \int_{\delta\tau} \rho f \, \mathrm{d}\tau = \int_{\delta\tau} \rho \frac{Df}{Dt} \, \mathrm{d}\tau \tag{3.8}$$

in which we note that on the right-hand side D/Dt acts only on f even though ρ is variable. This is a useful formula both in the derivation of the fluid equations and in applications. For example, it enables us to show quite generally that if $\phi(\mathbf{r}, t)$ represents the amount of any macroscopic quantity per unit mass, so that the total amount in a fluid element is $\int_{\delta\tau} \rho\phi \, \mathrm{d}\tau$, and if this changes under the action of influences represented by $Q(\mathbf{r}, t)$ at a rate given by $\int_{\delta\tau} Q \, \mathrm{d}\tau$ then

$$\frac{D}{Dt} \int_{\delta\tau} \rho\phi \, \mathrm{d}\tau = \int_{\delta\tau} \rho \frac{D\phi}{Dt} \, \mathrm{d}\tau = \int_{\delta\tau} Q \mathrm{d}\tau$$

and hence, since $\delta\tau$ is arbitrary,

$$\rho \frac{D\phi}{Dt} = Q \tag{3.9}$$

We now use (3.9) to obtain the equation of motion of a fluid element. Here $\phi = \mathbf{u}(\mathbf{r}, t)$, the fluid velocity or the momentum per unit mass. The forces which produce changes in the momentum of the fluid element can be long range or short range. Long range forces are approximately the same for all particles in the fluid element and can be treated as 'body' or volume forces represented by $\int_{\delta\tau} \rho \mathbf{F} \, \mathrm{d}\tau$, where $\mathbf{F}$ is the force per unit mass. Short range forces arising from particle interactions, although acting throughout the fluid element, produce net changes in its momentum only at its surface. The force per unit area (stress) is represented by the *stress tensor* whose elements Φ_{ij} specify the i-component of the force on unit area

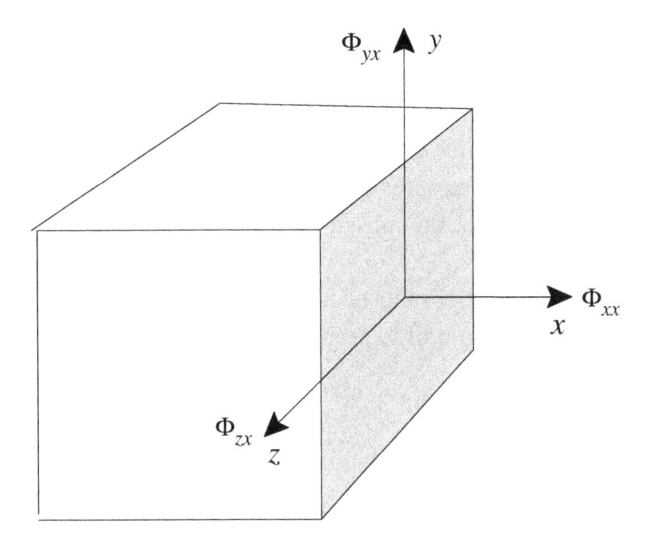

Fig. 3.2. Stress tensor.

normal to the j-direction; see, for example, Batchelor (1967). Figure 3.2 illustrates the stress tensor in Cartesian geometry. One normal and two tangential components of force act on each element of surface normal to the x, y and z directions; only the forces acting on the first of these are shown explicitly. It follows that the i-component of the total force on the fluid element is given by $\int_{\delta S} \Phi_{ij} n_j \, \mathrm{d}S$ where $\mathbf{n}$ is the unit vector normal to the surface element $\mathrm{d}S$. Thus, with $\phi = \mathbf{u}(\mathbf{r}, t)$ we have

$$\frac{D}{Dt} \int_{\delta\tau} \rho \mathbf{u} \, \mathrm{d}\tau = \int_{\delta\tau} \rho \mathbf{F} \, \mathrm{d}\tau + \int_{\delta S} \mathbf{\Phi} \cdot \mathbf{n} \, \mathrm{d}S$$

and, on using (3.8) and Gauss' theorem, this becomes

$$\int_{\delta\tau} \left(\rho \frac{D\mathbf{u}}{Dt} - \rho \mathbf{F} - \nabla \cdot \mathbf{\Phi} \right) \mathrm{d}\tau = 0$$

giving in differential form

$$\rho \frac{D\mathbf{u}}{Dt} = \rho \mathbf{F} + \nabla \cdot \mathbf{\Phi}$$

or

$$\rho \frac{Du_i}{Dt} = \rho F_i + \frac{\partial \Phi_{ij}}{\partial r_j} \tag{3.10}$$

In a neutral fluid at rest the stress tensor is isotropic (there is no preferred direction) and is written

$$\Phi_{ij} = -P \delta_{ij} \tag{3.11}$$

where P $(= -\Phi_{ii}/3)$ is the thermodynamic pressure and δ_{ij} is the Kronecker delta; the negative sign is introduced because by definition a positive normal component of Φ_{ij} represents a tension rather than a compression. For future reference we note that a magnetized plasma *does* have a preferred direction and may not be isotropic but have different pressures parallel and perpendicular to the direction of the magnetic field; we ignore this possibility for the moment and return to it in Section 3.4.1.

For a fluid in motion the stress tensor is, in general, no longer isotropic but it is customary to write it as the sum of isotropic and non-isotropic parts:

$$\Phi_{ij} = -P\delta_{ij} + d_{ij} \tag{3.12}$$

Here we define the pressure P to be the (negative) mean normal stress

$$P = -\Phi_{ii}/3 \tag{3.13}$$

This is an appropriate definition of the pressure of a fluid in motion since it is the quantity that would be measured in an experiment. However, it should be noted that it cannot be assumed to be the same as the thermodynamic pressure which is defined for a fluid in equilibrium. We shall ignore this difference since it gives rise to a correction which is important only for fluids with rotational and vibrational degrees of freedom and is therefore negligible for a fully ionized plasma. This is equivalent to the neglect of the bulk viscosity compared with the shear viscosity.

The non-isotropic part of the stress tensor, d_{ij}, is called the *viscous stress tensor* and by definition of P, $d_{ii} = 0$. The elements d_{ij} are related to the gradients of the components of the flow velocity, $\partial u_i/\partial r_j$, since it is the rate of change of momentum across the surface of the fluid element which produces the stress. Since the flow velocity changes very little on the scale of the fluid element it follows that the gradients are very small and a linear relationship may be assumed. It may then be shown (see Batchelor (1967)) that

$$d_{ij} = \mu\left(\frac{\partial u_i}{\partial r_j} + \frac{\partial u_j}{\partial r_i} - \frac{2}{3}\delta_{ij}\nabla\cdot\mathbf{u}\right) \tag{3.14}$$

where the coefficient of proportionality μ is called the *coefficient of (shear) viscosity*. It is found experimentally and can be shown theoretically (see Sections 8.2 and 12.6.3) that μ is a function of temperature and may, therefore, vary across the fluid.

Substituting (3.14) in (3.12) and (3.12) in (3.10) gives

$$\rho\frac{Du_i}{Dt} = \rho F_i - \frac{\partial P}{\partial r_i} + \frac{\partial}{\partial r_j}\left[\mu\left(\frac{\partial u_i}{\partial r_j} + \frac{\partial u_j}{\partial r_i} - \frac{2}{3}\delta_{ij}\frac{\partial u_k}{\partial r_k}\right)\right] \tag{3.15}$$

which is a general form of the *Navier–Stokes equation*. If temperature differences across the fluid are not too large, μ may be treated as constant giving

$$\rho \frac{D\mathbf{u}}{Dt} = \rho \mathbf{F} - \nabla P + \mu \left[\nabla^2 \mathbf{u} + \frac{1}{3} \nabla (\nabla \cdot \mathbf{u}) \right] \tag{3.16}$$

Assuming for the moment that $\mathbf{F}$ is given, it is clear that, whichever of these forms of the equation of motion may be appropriate, we need at least one more equation for P. We may anticipate that this will be provided by consideration of energy conservation and indeed it is. But this alone does not close the set of equations; closure is achieved by means of the relations of *classical thermodynamics*. There are two reasons why we need the thermodynamic relations. In the first place energy balance introduces the *internal energy* of the fluid element and this is a thermodynamic variable; it depends on the thermodynamic state of the fluid. Secondly, in addition to the coefficient of viscosity appearing in the momentum equation, energy balance brings in more transport coefficients and these, too, are functions of the state variables such as ρ and T.

There is any number of state variables, each of which has its particular use, but experiments have established empirically that for fluids in equilibrium all thermodynamic properties can be expressed in terms of any two state variables. We shall take P and ρ as the two independent variables so that every other state variable is then expressed as a function of these two by means of an *equation of state*. Thus our set of equations will be closed by the energy equation (for P) plus as many equations of state as there are state variables (other than ρ and P) appearing in the transport coefficients or elsewhere in the energy equation.

Fluids in motion are clearly not in equilibrium. Nevertheless, it has been found that classical thermodynamics may be applied to non-equilibrium states provided that the fluid passes through a series of quasi-static equilibrium states. Then if P and ρ, say, are given by their instantaneous values all the other state variables can be defined in terms of these two by their equations of state. In order that a quasi-static equilibrium be established we assume that changes in the macroscopic variables take place on a time scale long compared with the relaxation time for the attainment of local equilibrium.

The first law of thermodynamics is a statement of energy conservation in that it equates the change in the internal energy per unit mass $\mathcal{E}$ between two equilibrium states to the sum of the increase in heat energy per unit mass and the work done per unit mass on the system, that is

$$\mathrm{d}\mathcal{E} = \mathrm{d}Q + \mathrm{d}W \tag{3.17}$$

Note that $\mathcal{E}$ is a *state* variable so that $\mathrm{d}\mathcal{E}$ depends only on the initial and final states and not on the manner in which the change in internal energy is brought about. On the other hand Q and W are *path* variables and there is an infinite choice of values of $\mathrm{d}Q$ and $\mathrm{d}W$ for implementing any internal energy change $\mathrm{d}\mathcal{E}$. In particular, if a change of state is accomplished with no gain or loss of heat, i.e. $\mathrm{d}Q = 0$, we have an *adiabatic* change.

In our case the system is a fluid element which is not in equilibrium but in motion with flow velocity $\mathbf{u}$. We must be careful, therefore, to separate the work that goes into changing the kinetic energy of the flow from that which increases the internal energy. The former is obtained by taking the scalar product of $\mathbf{u}$ with the equation of motion (3.10) from which we get

$$\frac{D(u_i^2/2)}{Dt} = u_i F_i + \frac{u_i}{\rho}\frac{\partial \Phi_{ij}}{\partial r_j} \tag{3.18}$$

Now the total work done on the fluid element due to the body forces is $\int_{\delta\tau} \rho \mathbf{u} \cdot \mathbf{F}\, \mathrm{d}\tau$ and

$$\int_{\delta S} \mathbf{u} \cdot \mathbf{\Phi} \cdot \mathbf{n}\, \mathrm{d}S = \int_{\delta\tau} \nabla \cdot (\mathbf{u} \cdot \mathbf{\Phi})\, \mathrm{d}\tau \tag{3.19}$$

due to the surface forces. Hence the rate of work per unit mass is

$$u_i F_i + \frac{1}{\rho}\frac{\partial (u_i \Phi_{ij})}{\partial r_j} = u_i F_i + \frac{u_i}{\rho}\frac{\partial \Phi_{ij}}{\partial r_j} + \frac{\Phi_{ij}}{\rho}\frac{\partial u_i}{\partial r_j} \tag{3.20}$$

Comparing the total rate of work per unit mass (3.20) with the rate of change of kinetic energy per unit mass (3.18), we see that the rate of work expended on the internal energy per unit mass is just the last term in (3.20); that is

$$\frac{DW}{Dt} = \frac{\Phi_{ij}}{\rho}\frac{\partial u_i}{\partial r_j} = \frac{1}{\rho}\mathbf{\Phi} : \nabla \mathbf{u} \tag{3.21}$$

in dyadic notation.

The heat energy arises from two sources. There is both *Joule heating* $\int_{\delta\tau}(j^2/\sigma)\, \mathrm{d}\tau$, where $\mathbf{j}$ is the current density and σ the electrical conductivity, and heat conduction through the surface of the fluid element given by

$$\int_{\delta S} \kappa \nabla T \cdot \mathbf{n}\, \mathrm{d}S = \int_{\delta\tau} \nabla \cdot (\kappa \nabla T) \mathrm{d}\tau \tag{3.22}$$

where κ is the coefficient of heat conduction. Thus the rate of change of heat per unit mass is

$$\frac{DQ}{Dt} = \frac{1}{\rho\sigma} j^2 + \frac{1}{\rho}\nabla \cdot (\kappa \nabla T) \tag{3.23}$$

and from (3.17), (3.21) and (3.23) we get

$$\frac{D\mathcal{E}}{Dt} = \frac{1}{\rho}\mathbf{\Phi} : \nabla \mathbf{u} + \frac{1}{\rho}\nabla \cdot (\kappa \nabla T) + \frac{1}{\rho\sigma} j^2 \tag{3.24}$$

It is assumed that the transport coefficients μ, κ, σ are known functions of the state variables ρ and T so it remains only to specify the equations of state for T and $\mathcal{E}$. These follow from the assumption that the plasma behaves like a perfect gas.

In a perfect gas the total pressure and internal energy can be computed by adding up all the contributions from the individual particles as if they were independent of each other. In equilibrium it then follows that the contributions of the particles to the pressure and the internal energy are both proportional to their mean square velocities. In fact, the internal energy associated with each degree of freedom is $\frac{1}{2}k_B T$ where k_B is Boltzmann's constant and we find (see Section 12.5)

$$P = N k_B T \tag{3.25}$$

where N is the total number density of the particles. Since $\mathcal{E}$ is the internal energy per unit mass it follows that the internal energy per unit volume is

$$\rho\mathcal{E} = \frac{s}{2} N k_B T = \frac{s}{2} P \tag{3.26}$$

where s is the number of degrees of freedom per particle. For a plasma, where the particles are either ions or electrons, $s = 3$ corresponding to the three directions of translational motion, but in order to obtain the general result we shall not make this identification for the time being. Substitution of (3.26) in (3.24) gives, after some straightforward manipulation using (3.7), (3.12) and (3.14),

$$\begin{aligned} \frac{DP}{Dt} &= -\gamma P \nabla \cdot \mathbf{u} + (\gamma - 1) \nabla \cdot (\kappa \nabla T) \\ &\quad + \frac{\gamma - 1}{\sigma} j^2 + (\gamma - 1)\mu \left(\frac{\partial u_i}{\partial r_j} + \frac{\partial u_j}{\partial r_i} - \frac{2}{3}\delta_{ij} \nabla \cdot \mathbf{u} \right) \frac{\partial u_i}{\partial r_j} \end{aligned} \tag{3.27}$$

where $\gamma = (s + 2)/s$ is the ratio of the specific heats.

Since the total number density N is not one of our fluid variables we must re-write the equation of state (3.25) in terms of ρ. For a plasma consisting of ions and electrons with number densities, n_i and n_e, and charges, Ze and $-e$, respectively we have

$$N = n_i + n_e \approx n_i(1 + Z) \tag{3.28}$$

and

$$\rho = m_i n_i + m_e n_e \approx m_i n_i \tag{3.29}$$

where the first approximation follows from the quasi-neutrality condition $n_e \approx Z n_i$ and the second from the strong inequality of the masses $m_i \gg Z m_e$. Then using (3.28) and (3.29) we write (3.25) as

$$P = R_0 \rho T \tag{3.30}$$

where

$$R_0 = \frac{Nk_{\rm B}}{\rho} \approx \frac{(1+Z)k_{\rm B}}{m_{\rm i}} \tag{3.31}$$

is the gas constant.

Summarizing, the set of fluid equations comprises the continuity equation (3.7) for ρ, the equation of motion (3.15) for $\mathbf{u}$, the energy equation (3.27) for P and the equation of state (3.30) for T; it is assumed that the transport coefficients μ, κ and σ are given functions of ρ and T though in practice they are often treated as constants.

3.3 The MHD equations

So far we have applied the arguments of classical fluid dynamics to obtain a closed set of equations for the plasma fluid variables but, except for the introduction of Joule heating, we have taken almost no account of the fact that a plasma is a *conducting* fluid. This we do now by specifying the force per unit mass $\mathbf{F}$. Except in astrophysical contexts, where gravity is an important influence on the motion of the plasma, electromagnetic forces are dominant. For a fluid element with charge density q and current density $\mathbf{j}$ we then have

$$\rho\mathbf{F} = q\mathbf{E} + \mathbf{j} \times \mathbf{B} \tag{3.32}$$

where the fields $\mathbf{E}$ and $\mathbf{B}$ are determined by Maxwell's equations (2.2)–(2.5).

Equations (2.6) and (2.7) for q and $\mathbf{j}$ are not suitable in a fluid model. However, our first objective is to obtain a macroscopic description of the plasma in which the fields are those induced by the plasma motion. Thus, we now introduce the basic assumption of MHD that the fields vary on the same time and length scales as the plasma variables. If the frequency and wavenumber of the fields are ω and k respectively, we have $\omega\tau_{\rm H} \sim 1$ and $kL_{\rm H} \sim 1$, where $\tau_{\rm H}$ and $L_{\rm H}$ are the hydrodynamic time and length scales. A dimensional analysis then shows (see Exercise 3.5) that both the electrostatic force $q\mathbf{E}$ and displacement current $\varepsilon_0\mu_0\partial\mathbf{E}/\partial t$ may be neglected in the non-relativistic approximation $\omega/k \ll c$. Consequently, (3.32) becomes

$$\rho\mathbf{F} = \mathbf{j} \times \mathbf{B} \tag{3.33}$$

and (2.3) is replaced by Ampère's law

$$\mathbf{j} = \frac{1}{\mu_0}\boldsymbol{\nabla} \times \mathbf{B} \tag{3.34}$$

Now, Poisson's equation (2.4) is redundant (except for determining q) and just one further equation for $\mathbf{j}$ is required to close the set.

Here we run into the main problem with a one-fluid model. Clearly, a current exists only if the ions and electrons have distinct flow velocities and so, at least to this extent, we are forced to recognize that we have two fluids rather than one. For the moment we side-step this difficulty by following usual practice in MHD and adopting Ohm's law

$$\mathbf{j} = \sigma(\mathbf{E} + \mathbf{u} \times \mathbf{B}) \tag{3.35}$$

as the extra equation for $\mathbf{j}$. The usual argument for this particular form of Ohm's law is that in the non-relativistic approximation the electric field in the frame of a fluid element moving with velocity $\mathbf{u}$ is $(\mathbf{E} + \mathbf{u} \times \mathbf{B})$. However, this argument is over-simplified, unless $\mathbf{u}$ is constant so that the frame is inertial, and later, when we discuss the applicability of the MHD equations, we shall see that the assumption of a scalar conductivity in magnetized plasmas is rarely justified. The status of (3.35) should be regarded, therefore, as that of a 'model' equation, adopted for mathematical simplicity.

This closes the set of equations for the variables ρ, $\mathbf{u}$, P, T, $\mathbf{E}$, $\mathbf{B}$ and $\mathbf{j}$ but before listing them it is useful to reduce the set by eliminating some of the variables. Although in electrodynamics it is customary to think of the magnetic field being generated by the current, in MHD we regard Ampère's law (3.34) as determining $\mathbf{j}$ in terms of $\mathbf{B}$. Then Ohm's law (3.35) becomes

$$\mathbf{E} = \frac{1}{\sigma\mu_0}\nabla \times \mathbf{B} - \mathbf{u} \times \mathbf{B} \tag{3.36}$$

so determining $\mathbf{E}$. Finally, substituting (3.36) in (2.2), treating σ as a constant, and using (2.5), we get the induction equation for $\mathbf{B}$

$$\frac{\partial \mathbf{B}}{\partial t} = \frac{1}{\sigma\mu_0}\nabla^2\mathbf{B} + \nabla \times (\mathbf{u} \times \mathbf{B}) \tag{3.37}$$

Since we have eliminated $\mathbf{j}$ and $\mathbf{E}$, this is now the only equation we need add to the set derived at the end of the last section for the fluid variables.

3.3.1 Resistive MHD

Although we have a closed set of equations it is still too complicated for general application and some further reduction is essential. In fact, there is a natural reduction consistent with the assumptions already made. In a collision-dominated plasma the electron and ion distribution functions remain close to local Maxwellian distributions. These are the 'quasi-static equilibrium states' through which the plasma was assumed to pass when we invoked thermodynamics in Section 3.2. A plasma with local Maxwellian distribution functions has zero viscosity and heat conduction so it follows that these terms in the momentum and energy equations

Table 3.1. *Resistive MHD equations*

Evolution equations:

$$\begin{aligned} D\rho/Dt &= -\rho\nabla\cdot\mathbf{u} \\ \rho D\mathbf{u}/Dt &= (\nabla\times\mathbf{B})\times\mathbf{B}/\mu_0 - \nabla P \\ DP/Dt &= -\gamma P\nabla\cdot\mathbf{u} + (\gamma-1)(\nabla\times\mathbf{B})^2/\sigma\mu_0^2 \\ \partial\mathbf{B}/\partial t &= \nabla^2\mathbf{B}/\sigma\mu_0 + \nabla\times(\mathbf{u}\times\mathbf{B}) \end{aligned}$$

Equation of state:

$T = P/R_0\rho$

Equation of constraint:

$\nabla\cdot\mathbf{B} = 0$

Definitions:

$$\begin{aligned} \mathbf{E} &= (\nabla\times\mathbf{B})/\sigma\mu_0 - \mathbf{u}\times\mathbf{B} \\ \mathbf{j} &= (\nabla\times\mathbf{B})/\mu_0 \\ q &= -\varepsilon_0\nabla\cdot(\mathbf{u}\times\mathbf{B}) \end{aligned}$$

Approximations:

Strong collisions:	$\tau_i \ll (m_e/m_i)^{1/2}\tau_H$	$\lambda_c \ll L_H$
Non-relativistic:	$\omega/k \sim L_H/\tau_H \sim u \ll c$	
Quasi-neutrality:	$\omega\lvert\Omega_e\rvert/\omega_p^2 \ll 1$	
Small Larmor radius:	$r_{L_i} \ll L_H$	
Scalar conductivity:	$\lvert\Omega_e\rvert \ll \nu_c$	

are small in some sense yet to be clarified. Neglecting these terms leaves electrical resistivity as the only remaining dissipative mechanism and so the reduced set of equations describes *resistive MHD*. These equations are listed in Table 3.1. The approximations on which the resistive MHD equations are based are discussed below in Section 3.4.

3.3.2 Ideal MHD

Going one step further and neglecting all dissipation leads to *ideal MHD*. This is sometimes referred to as the infinite conductivity limit, but since that is the same as the limit of no collisions we would be in danger of employing contradictory arguments in our derivation. The proper approach comes from a dimensional analysis of the two terms on the right-hand side of (3.37). We see that the ratio of the convective term to the diffusion term is

$$\frac{\sigma\mu_0|\nabla\times(\mathbf{u}\times\mathbf{B})|}{|\nabla^2\mathbf{B}|} \sim \mu_0\sigma u L_H \equiv R_M \tag{3.38}$$

where R_M is called the *magnetic Reynolds number* by analogy with the hydrodynamic Reynolds number R which measures the relative magnitude of the inertial, convective term to the diffusion term in the Navier–Stokes equation (3.16):

$$\frac{\rho|D\mathbf{u}/Dt|}{\mu|\nabla^2\mathbf{u}|} \sim \frac{\rho L_H^2}{\mu\tau_H} \sim \frac{\rho u L_H}{\mu} = R \tag{3.39}$$

In R_M the resistivity σ^{-1} plays the role of the kinematic viscosity (μ/ρ) in R. We see from (3.38) and (3.39) that ideal MHD corresponds to the limit of infinite Reynolds numbers and that both these limits can be achieved consistently by letting $L_H \to \infty$; ideal MHD is therefore properly regarded as the limit of large scale length.

Comparing the two terms on the right-hand side of the pressure evolution equation (see Table 3.1) we have

$$\frac{|\nabla \times \mathbf{B}|^2}{\mu_0^2 \sigma P |\nabla \cdot \mathbf{u}|} \sim \frac{1}{\beta R_M} \tag{3.40}$$

where $\beta = \mu_0 P/B^2$ is the ratio of plasma pressure to magnetic pressure. This confirms that the $R_M \to \infty$ limit removes the dissipative term from this equation as well so that it reduces to

$$\frac{DP}{Dt} = -\gamma P \nabla \cdot \mathbf{u} \tag{3.41}$$

On substituting for $\nabla \cdot \mathbf{u}$ from the continuity equation, (3.41) becomes

$$\frac{DP}{Dt} - \frac{\gamma P}{\rho}\frac{D\rho}{Dt} = \frac{1}{\rho^\gamma}\frac{D}{Dt}(P\rho^{-\gamma}) = 0 \tag{3.42}$$

and hence, for any fluid element,

$$P\rho^{-\gamma} = \text{const.} \tag{3.43}$$

which is the *adiabatic gas law*. The ideal MHD equations and the approximations governing their validity (discussed in the following section) are listed in Table 3.2.

3.4 Applicability of the MHD equations

Magnetohydrodynamics, especially ideal MHD, is widely employed throughout plasma physics, on occasions it has to be said, with scant regard to its range of validity. A mathematically rigorous discussion of validity requires the two-fluid approach of Chapter 12 but we can gain considerable insight into the applicability and the likely limitations of the MHD equations by plausible physical arguments such as we have used in their derivation. We do this by taking in turn each of the

Table 3.2. *Ideal MHD equations*

Evolution equations:

$$\begin{aligned} D\rho/Dt &= -\rho\boldsymbol{\nabla}\cdot\mathbf{u} \\ \rho D\mathbf{u}/Dt &= (\boldsymbol{\nabla}\times\mathbf{B})\times\mathbf{B}/\mu_0 - \boldsymbol{\nabla}P \\ D(P\rho^{-\gamma})/Dt &= 0 \\ \partial\mathbf{B}/\partial t &= \boldsymbol{\nabla}\times(\mathbf{u}\times\mathbf{B}) \end{aligned}$$

Equation of constraint:

$\boldsymbol{\nabla}\cdot\mathbf{B} = 0$

Definitions:

$$\begin{aligned} \mathbf{E} &= -\mathbf{u}\times\mathbf{B} \\ \mathbf{j} &= (\boldsymbol{\nabla}\times\mathbf{B})/\mu_0 \\ q &= -\varepsilon_0\boldsymbol{\nabla}\cdot(\mathbf{u}\times\mathbf{B}) \end{aligned}$$

Approximations:

Strong collisions:	$\tau_i \ll (m_e/m_i)^{1/2}\,\tau_H$	$\lambda_c \ll L_H$
Non-relativistic:	$\omega/k \sim L_H/\tau_H \sim u \ll c$	
Quasi-neutrality:	$\omega\lvert\Omega_e\rvert/\omega_p^2 \ll 1$	
Small Larmor radius:	$r_{L_i}/L_H \ll \beta^{1/2}$	
Large R_M:	$\left(r_{L_i}/L_H\right)^2 \ll \beta\,(m_i/m_e)^{1/2}\,(\tau_i/\tau_H)$	

assumptions made in the derivation of the equations, identifying the underlying approximation, and writing it in terms of a dimensionless parameter.

Since classical thermodynamics assumes the establishment of quasi-static equilibrium states (local Maxwellians) and these are brought about by collisions we require the collision time $\tau_c \ll \tau_H$. Let us now be more precise about what we mean by this strong inequality. For τ_H we take the minimum hydrodynamic time scale of interest, i.e. the time for significant change in the most rapidly fluctuating of the macroscopic variables. The ion and electron collision times, τ_i and τ_e, are defined as the times for significant particle deflection (momentum change). Since ions are effective in deflecting both other ions and electrons we may, for order of magnitude arguments, consider only the scattering off ions. For low Z and $T_i \approx T_e$ it then follows that the collision time for each species is inversely proportional to its thermal speed and hence $\tau_i \sim (m_i/m_e)^{1/2}\tau_e$. Thus, both ions and electrons will be in local equilibrium states provided

$$\tau_i \ll \tau_H \tag{3.44}$$

However, a one-fluid model naturally assumes a single temperature and this imposes an even stronger collisionality condition. Temperature equilibration depends on energy exchange between ions and electrons and since the energy exchange per collision is proportional to m_e/m_i it follows that initially different temperatures will be approximately equal after a time $(m_i/m_e)\tau_e$. We must, therefore, replace (3.44) by the stronger collisionality condition

$$\tau_i \ll (m_e/m_i)^{1/2}\tau_H \tag{3.45}$$

Here it is worth noting that although, in the two-fluid model of Chapter 12, temperature equilibration is assumed to take place on the hydrodynamic time scale so that (3.44) suffices, one then has separate ion and electron energy equations. There is no inconsistency however because contact between the two-fluid and one-fluid models is established with the derivation of the resistive MHD equations and, as we shall see, (3.45) reappears as the condition for the neglect of heat conduction. This confirms that it is in order to impose the stronger collisionality condition (3.45) for the one-fluid model.

In terms of scale lengths we require the mean free paths of the ions and electrons to be very small compared with L_H. Here there is no ambiguity because the mean free path λ_c is the product of the thermal speed and the collision time and so has the same order of magnitude for ions and electrons. The inequality

$$\lambda_c \ll L_H \tag{3.46}$$

then enables us to define a fluid element of dimension $(\delta\tau)^{1/3}$ such that

$$N^{-1/3} \ll \lambda_c < (\delta\tau)^{1/3} \ll L_H \tag{3.47}$$

where $N^{-1/3}$ is the mean interparticle separation. For plasmas with many particles in the Debye sphere, i.e. $N\lambda_D^3 \gg 1$, the first inequality in (3.47) is satisfied since $N^{1/3}\lambda_c \sim (N\lambda_D^3)^{4/3}/\ln(N\lambda_D^3) \gg 1$. Thus, a fluid element with dimensions of several mean free paths will retain its identity on account of collisions; it will not be sensitive to microscopic fluctuations because it contains many particles, and the variation of macroscopic quantities within it will be negligible. Approximations (3.45) and (3.47) underlie the derivation of the fluid equations.

The neglect of electrostatic forces and the displacement current in the electromagnetic equations was a consequence of the basic assumption of MHD that the fields and flow are strongly correlated and therefore change on the same scales; furthermore the flow, being dominated by ion inertia is non-relativistic:

$$\frac{\omega}{k} \sim \frac{L_H}{\tau_H} \sim u \ll c \tag{3.48}$$

Here, for future reference, we note that the adoption of this approximation marks a fundamental difference between MHD and the plasma wave equations where wave and electron speeds may be relativistic.

So far all of these approximations were mentioned, if not spelled out precisely, in the course of the derivation of the MHD equations. Those we consider next were not identified on account of adopting an empirical Ohm's law (3.35). The statement that the electric field in the frame of the fluid element is $(\mathbf{E} + \mathbf{u} \times \mathbf{B})$ skips over the fact that $\mathbf{j}$ must also be transformed and the further complication that in general the fluid element is not an inertial frame. A rigorous derivation of (3.35) for a moving deformable conductor can be found in Jeffrey (1966) assuming the approximation†

$$\frac{\varepsilon_0}{\sigma} \ll \tau_{\mathrm{H}}$$

which is easily satisfied in almost all plasmas. However, we shall not reproduce this derivation since it starts from the scalar relationship $\mathbf{j} = \sigma \mathbf{E}$ and for a magnetized plasma we *cannot* assume a scalar conductivity. The correct relationship between current and fields is given by the *generalized Ohm's law*, the derivation of which requires the two-fluid approach of Chapter 12 rather than the one-fluid model used here. Nevertheless, there is a simple argument leading to the generalized Ohm's law which is appropriate in the MHD approximation and sufficient to enable us to identify the further approximations required to obtain (3.35).

The argument is based on the strong inequality of the masses, $Zm_{\mathrm{e}} \ll m_{\mathrm{i}}$, and quasi-neutrality, $Zn_{\mathrm{i}} \approx n_{\mathrm{e}}$, the condition for the latter being

$$\frac{|q|}{en_{\mathrm{e}}} \sim \frac{\varepsilon_0 |\nabla \cdot \mathbf{E}|}{en_{\mathrm{e}}} \sim \frac{\omega |\Omega_{\mathrm{e}}|}{\omega_{\mathrm{p}}^2} \ll 1 \tag{3.49}$$

where Ω_{e} is the electron Larmor frequency, ω_{p} is the plasma frequency and we have used (2.4) to estimate $|q|$ and (2.2) to estimate $|\mathbf{E}|$. A consequence of the mass inequality is that the flow velocity is determined by the much heavier ions but within the fluid element the forces acting on the more mobile electrons may produce an electron flow velocity which is different from that of the ions, so giving rise to a current. Thus,

$$\mathbf{u} \approx \mathbf{u}_{\mathrm{i}} \qquad \mathbf{j} = Zen_{\mathrm{i}}\mathbf{u}_{\mathrm{i}} - en_{\mathrm{e}}\mathbf{u}_{\mathrm{e}} \approx \frac{Ze\rho}{m_{\mathrm{i}}}(\mathbf{u} - \mathbf{u}_{\mathrm{e}}) \tag{3.50}$$

from which we see that fluctuations in $\mathbf{u}_{\mathrm{e}}$ occur on the same scales as those in $\rho, \mathbf{u}$ and $\mathbf{j}$, i.e. the hydrodynamic scales τ_{H} and L_{H}. In writing an equation of motion for $\mathbf{u}_{\mathrm{e}}$, therefore, we may ignore electron inertia and viscosity, since they involve derivatives of $\mathbf{u}_{\mathrm{e}}$, but we must include a term to account for the collisional interaction with the ions; we take this to be proportional to the product of the collision frequency $\nu_{\mathrm{c}} = \tau_{\mathrm{e}}^{-1}$ and the electron momentum per unit volume relative

† This ensures that any net space charge decays away in a time much shorter than τ_{H} so that the only electric fields present are those generated by the action of the magnetic field.

to the ions. The balance of forces on the electrons is then given by

$$en_e(\mathbf{E} + \mathbf{u}_e \times \mathbf{B}) + m_e n_e \nu_c(\mathbf{u}_e - \mathbf{u}_i) + \nabla p_e = 0 \tag{3.51}$$

where p_e is the electron pressure. Substituting for $\mathbf{u}_e$ and $\mathbf{u}_i$ from (3.50) we get

$$\mathbf{E} + \mathbf{u} \times \mathbf{B} - \mathbf{j}/\sigma = \frac{m_i}{Ze\rho}(\mathbf{j} \times \mathbf{B} - \nabla p_e) \tag{3.52}$$

where we have identified $Ze^2\rho/m_i m_e \nu_c \approx n_e e^2/m_e \nu_c = \sigma$ as the coefficient of electrical conductivity. Equation (3.52) is the MHD form of the *generalized Ohm's law* which is discussed further in Section 3.5.1. It reduces to (3.35) if we ignore the terms on the right-hand side. To understand the significance of neglecting these terms, it is instructive to re-write (3.52) as

$$\mathbf{j} + \frac{|\Omega_e|}{\nu_c}(\mathbf{j} \times \mathbf{b}) = \sigma \tilde{\mathbf{E}} \tag{3.53}$$

where $\mathbf{b}$ is a unit vector parallel to $\mathbf{B}$ and

$$\tilde{\mathbf{E}} = \mathbf{E} + \mathbf{u} \times \mathbf{B} + \frac{m_i}{Ze\rho}\nabla p_e$$

is the 'effective' electric field. This representation quite clearly shows the distinctive roles of the ∇p_e and $\mathbf{j} \times \mathbf{B}$ terms.

The ∇p_e term may be treated as an extra component in the effective electric field and, comparing it with the Lorentz force $\mathbf{u} \times \mathbf{B}$ we have

$$\frac{m_i|\nabla p_e|}{Ze\rho|\mathbf{u} \times \mathbf{B}|} \sim \left(\frac{r_{L_i}}{L_H}\right)\left(\frac{c_s}{u}\right)\left(\frac{T_e}{T_i}\right)^{1/2}$$

where r_{L_i} is the ion Larmor radius and $c_s = (k_B T_e/m_i)^{1/2}$ is the ion acoustic speed. Since we have taken $T_e \approx T_i$, we see that neglect of the ∇p_e term requires $(r_{L_i}/L_H) \ll (u/c_s)$. From the momentum equation $|\mathbf{u}| \sim c_s$ if $\beta \gtrsim 1$ and $|\mathbf{u}| \sim c_s/\beta^{1/2}$ if $\beta \ll 1$ so the small Larmor radius condition

$$r_{L_i} \ll L_H \tag{3.54}$$

covers both cases.

In contrast, the role of the $\mathbf{j} \times \mathbf{B}$ term is far more fundamental since its presence means that there is no scalar relationship between $\mathbf{j}$ and $\tilde{\mathbf{E}}$; there must be a component of $\tilde{\mathbf{E}}$ which is perpendicular to $\mathbf{j}$ (and $\mathbf{B}$) to balance the $\mathbf{j} \times \mathbf{b}$ term. This is the *Hall effect*. The condition for the recovery of a scalar conductivity is clearly

$$|\Omega_e| \ll \nu_c \tag{3.55}$$

This condition may be satisfied for certain conducting materials but is seldom true for magnetized plasmas, particularly fusion and space plasmas; hence the statement following (3.35) that the scalar Ohm's law is a model equation adopted for mathematical simplicity.

In summary, the conditions for the applicability of the dissipative MHD equations are (3.45)–(3.49) and (3.54)–(3.55). As well as the last of these, the first is often not satisfied and we return to this point below but, for the moment, we continue the analysis by showing that these approximations are also the basis for the resistive MHD equations.

Clearly, the condition for the neglect of viscosity in the momentum equation, as in ideal MHD, is $R \gg 1$. Noting that the viscosity coefficient $\mu \sim P\tau_i$, this condition may be expressed as $R^{-1} \sim (\mu/\rho u L_H) \sim (P\tau_i/\rho u L_H) \sim (c_s/u)(\lambda_c/L_H) \ll 1$ which is satisfied provided (3.46) holds. The same approximation justifies neglecting viscosity in the energy equation (3.27), as may be seen by comparison with the pressure terms $(\mu u^2 \tau_H/L_H^2 P \sim \lambda_c/L_H)$.

Since the coefficient of thermal conductivity $\kappa \sim n k_B^2 T \tau_e/m_e$, comparison of the thermal conduction term in (3.27) with the pressure term gives

$$\frac{|\nabla \cdot (\kappa \nabla T)|}{|P \nabla \cdot \mathbf{u}|} \sim \left(\frac{m_i}{m_e}\right)^{1/2} \frac{\tau_i}{\tau_H} \left(\frac{c_s}{u}\right)^2 \sim \left(\frac{m_i}{m_e}\right)^{1/2} \frac{\tau_i}{\tau_H} \beta \tag{3.56}$$

showing that thermal conductivity may be neglected if (3.45) is satisfied. Thus, no new approximations are required for resistive MHD. By contrast, ideal MHD does require the additional approximation $R_M \gg 1$ (or $R_M \gg \beta^{-1}$ if $\beta \ll 1$; see (3.40)). However, the condition for the neglect of the Hall current is no longer (3.55) since this arose from a comparison of the two current terms in (3.53) both of which are now neglected. A suitable approximation is provided by a comparison of the $\mathbf{j} \times \mathbf{B}$ and $\mathbf{u} \times \mathbf{B}$ terms so that we now require

$$\frac{m_i |\mathbf{j} \times \mathbf{B}|}{Ze\rho |\mathbf{u} \times \mathbf{B}|} \sim \frac{1}{\beta} \left(\frac{r_{L_i}}{L_H}\right) \left(\frac{c_s}{u}\right) \left(\frac{T_e}{T_i}\right)^{1/2} \ll 1$$

For $\beta \sim 1$ this is the same condition (3.54) as for the neglect of the ∇p_e term. For low β plasmas the somewhat stronger small Larmor radius approximation

$$r_{L_i}/L_H \ll \beta^{1/2} \qquad (\beta \ll 1) \tag{3.57}$$

is required.

For completeness let us write the large magnetic Reynolds number approximation in terms of τ_i and τ_H; using $\sigma = n_e e^2/m_e \nu_c$ we have

$$R_M^{-1} = (\mu_0 \sigma u L_H)^{-1} \sim \frac{1}{\beta} \left(\frac{r_{L_i}}{L_H}\right)^2 \left(\frac{m_e}{m_i}\right)^{1/2} \left(\frac{\tau_H}{\tau_i}\right) \ll 1$$

Combining this with (3.45) we have

$$\frac{1}{\beta}\left(\frac{r_{L_i}}{L_H}\right)^2 \ll \left(\frac{m_i}{m_e}\right)^{1/2}\left(\frac{\tau_i}{\tau_H}\right) \ll 1 \tag{3.58}$$

Apart from (3.47) to (3.49), which are fundamental to MHD and may therefore be taken for granted, (3.58) summarizes the approximations of ideal MHD; the strong inequalities involving (r_{L_i}/L_H) represent the large scale length (high conductivity) and small ion Larmor radius approximations, while the final inequality represents the high collisionality approximation. It is this last approximation that is least likely to be valid for fusion and space plasmas which are usually better described as collisionless rather than collision-dominated. Despite this, a fluid description of the plasma in these circumstances may still be tenable though some re-examination of the model, particularly of the energy equation, is required. It turns out that the magnetic field, by acting as a localizing agent, is able to compensate in part for insufficient collisionality. However, discussion of this requires a proper recognition of the anisotropic nature of a magnetized plasma.

3.4.1 Anisotropic plasmas

Since a magnetized plasma has a natural preferred direction the assumption of isotropic pressure cannot strictly be justified. This point is relevant when the magnetic field is sufficiently strong (or collisions sufficiently weak) so that

$$r_{L_i} \ll \lambda_c \tag{3.59}$$

Consequently, in the plane perpendicular to the magnetic field it is the Larmor orbits rather than collisions that restrict the free flow of the particles. It then may happen that, instead of (3.11), the equilibrium stress tensor takes the form

$$\mathbf{\Phi} = -\begin{pmatrix} P_\perp & 0 & 0 \\ 0 & P_\perp & 0 \\ 0 & 0 & P_\parallel \end{pmatrix} \tag{3.60}$$

with different components parallel and perpendicular to the field. In ideal MHD one then obtains, instead of the adiabatic gas law (3.42), two separate adiabatic conditions called the *double adiabatic approximation*.

We can see roughly how this comes about by returning to (3.24) and (3.26) and splitting them into two pairs of equations for separate internal energy components

$\mathcal{E}_\parallel$ and $\mathcal{E}_\perp$. Dropping all the dissipative terms and using (3.60) we write (3.24) as

$$\frac{D\mathcal{E}_\parallel}{Dt} = -\frac{P_\parallel}{\rho}\frac{\partial u_3}{\partial r_3} \tag{3.61}$$

$$\frac{D\mathcal{E}_\perp}{Dt} = -\frac{P_\perp}{\rho}\left(\nabla\cdot\mathbf{u} - \frac{\partial u_3}{\partial r_3}\right) \tag{3.62}$$

and from (3.26) we substitute

$$\rho\mathcal{E}_\parallel = P_\parallel/2 \tag{3.63}$$

$$\rho\mathcal{E}_\perp = P_\perp \tag{3.64}$$

An equation for $\partial u_3/\partial r_3$ is obtained from the parallel component of the magnetic convection equation on expanding the right-hand side and using (2.5) to get

$$\frac{\partial B}{\partial t} + (\mathbf{u}\cdot\nabla)B = B\frac{\partial u_3}{\partial r_3} - B\nabla\cdot\mathbf{u} \tag{3.65}$$

Substituting for $\partial u_3/\partial r_3$ from (3.65) and, as before, for $\nabla\cdot\mathbf{u}$ from the continuity equation, it is easily verified that (3.61) and (3.62) become

$$\frac{D}{Dt}\left(\frac{P_\parallel B^2}{\rho^3}\right) = 0 \tag{3.66}$$

and

$$\frac{D}{Dt}\left(\frac{P_\perp}{\rho B}\right) = 0 \tag{3.67}$$

which are the double adiabatic conditions replacing (3.42) when the pressure is anisotropic.

An interesting parallel may be drawn between the double adiabatic conditions and the adiabatic invariants μ_B and J of particle orbit theory. Since $\mu_B = mv_\perp^2/2B$ and $P_\perp/\rho \propto \langle v_\perp^2/2\rangle$, where the bracket denotes an average over all the particles in the fluid element, we may regard (3.67) as a macroscopic representation of the invariance of μ_B. Likewise, treating the fluid element as a flux tube and using a well-known result of ideal MHD that the length of a flux tube $l \propto B/\rho$ (see Section 4.2), we have

$$\frac{P_\parallel B^2}{\rho^3} \propto \langle v_\parallel^2\rangle l^2 \propto J^2$$

so that (3.66) becomes a statement of the invariance of J. These insights go some way towards explaining why ideal MHD can be applied with success even in the collisionless regime. However, they are not rigorous and uncritical use of MHD beyond the validity of its approximations can lead to erroneous results, as illustrated by Kulsrud (1983).

The role of collisions in MHD is twofold; they not only establish the quasi-static local equilibrium state but serve to define the dimensions of the fluid element. In the collisionless limit the adiabatic invariants describe the plasma locally and the ion Larmor radius defines the perpendicular dimension of the fluid element provided (3.54) is valid, a condition well satisfied in almost all magnetized plasmas. The outstanding problem is the definition of the parallel dimension of the fluid element since, in the collisionless limit, particles may flow freely along the field lines. In particular, there must be negligible heat flow from one fluid element to another and since the heat conduction coefficient parallel to **B** increases with the collision time a negligible parallel temperature gradient is required. This imposes a severe restriction on the applicability of double adiabatic theory.

3.4.2 Collisionless MHD

One way to get around the problem of establishing a fluid description for the parallel flow in the collisionless limit is to use a one-dimensional kinetic equation to describe $v_\parallel$-dependent plasma behaviour. This is done in the guiding centre plasma model (Grad, 1967) but it is considerably more complicated than either ideal MHD or double adiabatic theory and we shall not discuss it here.

With the objective of finding a simpler theory applicable to fusion plasmas, Freidberg (1987) proposed an alternative fluid model which he called *collisionless MHD*. This model is worthy of consideration not only because it confronts the problem that the strong collisions condition (3.45) is satisfied in neither fusion nor space plasmas but also because it provides a link between the fluid description and particle orbit theory as developed in Chapter 2. Indeed, it is the relationship between particle drifts and plasma currents that is the key to establishing the collisionless MHD equations in the plane perpendicular to the magnetic field.

The first step is to write the velocity $\mathbf{v}$ of a particle as

$$\mathbf{v} = v_\parallel \mathbf{b} + \mathbf{v}_\perp + \mathbf{v}_\mathrm{g}$$

where the perpendicular component has been separated into its rapidly changing gyration around the field line and its slowly changing guiding centre velocity. Next we express $\mathbf{v}_\mathrm{g}$ in terms of all its possible components, evaluated in Chapter 2,

$$\mathbf{v}_\mathrm{g} = \mathbf{v}_\mathrm{E} + \mathbf{v}_\mathrm{G} + \mathbf{v}_\mathrm{C} + \mathbf{v}_\mathrm{P}$$

where $\mathbf{v}_\mathrm{E}$, $\mathbf{v}_\mathrm{G}$, $\mathbf{v}_\mathrm{C}$ and $\mathbf{v}_\mathrm{P}$ are given by (2.16), (2.22), (2.23) and (2.49), respectively. We note that, in the MHD approximation (3.48), $\mathbf{v}_\mathrm{E}$ is of higher order than each of the other terms and since it is the same for both ions and electrons and is independent of $\mathbf{v}$ we see that to a first approximation the plasma flow velocity is

$$\mathbf{u} = \mathbf{E} \times \mathbf{B}/B^2$$

On taking the cross product of this equation with $\mathbf{B}$ we get

$$\mathbf{E}_\perp + \mathbf{u} \times \mathbf{B} = 0$$

the perpendicular component of Ohm's law in ideal MHD.

Now, calculating the guiding centre current $\mathbf{j}_g$ by summing over all the individual particle contributions as in Chapter 2 we find

$$\mathbf{j}_g = \frac{1}{B}\mathbf{b} \times \left(P_\perp \frac{\nabla B}{B} + P_\parallel (\mathbf{b} \cdot \nabla)\mathbf{b}) + \rho \frac{d\mathbf{u}_\perp}{dt} \right)$$

where, as appropriate in a fluid model, we have identified the partial pressures $P_\perp$ and $P_\parallel$ with $n\bar{W}_\perp$ and $2n\bar{W}_\parallel$, respectively. This correspondence has already been noted in the preceding section.

To this guiding centre current we must add the magnetization current $\mathbf{j}_M = \nabla \times \mathbf{M}$, where $\mathbf{M} = -(P_\perp/B)\mathbf{b}$ so that the current density perpendicular to the magnetic field is

$$\mathbf{j}_\perp = \mathbf{j}_g + \mathbf{j}_M$$

The cross product of this equation with $\mathbf{B}$ then gives the perpendicular momentum equation

$$\rho \left(\mathbf{b} \times \frac{d\mathbf{u}_\perp}{dt} \right) \times \mathbf{b} = \mathbf{j} \times \mathbf{B} - \nabla_\perp P_\perp - (P_\parallel - P_\perp)(\mathbf{b} \cdot \nabla)\mathbf{b} \qquad (3.68)$$

This is the perpendicular momentum equation found from both the guiding centre plasma model and the double adiabatic theory of Chew, Goldberger and Low (1956).

Freidberg replaced the parallel momentum equation by the heuristic assumption of incompressibility, $\nabla \cdot \mathbf{u} = 0$, which implies that the density and pressures are convected with the plasma. The main role of the 'parallel' equations is to describe the propagation of sound waves along the field lines. But these waves do not couple strongly with most ideal MHD instabilities which involve incompressible wave motion. Thus, as Freidberg points out, for the most part the model is inaccurate only where it does not matter and collisionless MHD should at least provide a credible basis for the discussion of ideal MHD stability. If it is assumed that $P_\perp = P_\parallel$ then the equations and predictions of collisionless MHD and incompressible, ideal MHD are virtually identical. Furthermore, in those situations where the two models produce different predictions, neither model can be considered reliable. The evolution equations and conditions governing collisionless MHD for $P_\perp = P_\parallel = P$ are set out in Table 3.3.

Table 3.3. *Collisionless MHD equations*

Evolution equations:

$$
\begin{aligned}
D\rho/Dt &= 0 \\
\rho(D\mathbf{u}_\perp/Dt)_\perp &= (\nabla \times \mathbf{B}) \times \mathbf{B}/\mu_0 - \nabla_\perp P \\
DP/Dt &= 0 \\
\partial\mathbf{B}/\partial t &= \nabla \times (\mathbf{u} \times \mathbf{B})
\end{aligned}
$$

Equations of constraint:

$$\nabla \cdot \mathbf{B} = 0 \qquad \nabla \cdot \mathbf{u} = 0$$

Definitions:

$$
\begin{aligned}
\mathbf{E} &= -\mathbf{u} \times \mathbf{B} \\
\mathbf{j} &= (\nabla \times \mathbf{B})/\mu_0 \\
q &= -\varepsilon_0 \nabla \cdot (\mathbf{u} \times \mathbf{B})
\end{aligned}
$$

Approximations:

Collisionless:	$\tau_H \ll \tau_i \qquad L_H \ll \lambda_c$
Non-relativistic:	$\omega/k \sim L_H/\tau_H \sim u \ll c$
Quasi-neutrality:	$\omega\lvert\Omega_e\rvert/\omega_p^2 \ll 1$
Small Larmor radius:	$r_{L_i}/L_H \ll \beta^{1/2}$

3.5 Plasma wave equations

The interaction of plasma and electromagnetic fields generates a very wide spectrum of wave phenomena of which only the low frequency limit is described by MHD. A fluid description of plasma wave propagation is feasible but cannot be derived from a collision-dominated model since most wave frequencies are greater than collision frequencies. Also, such a description must be two-fluid since much of the physics is related to the differences in ion and electron motion.

Neighbouring particles of a given species will tend to move coherently in response to the fields but disperse on account of their random, thermal velocities. The persistence of a fluid element depends, therefore, on the dominance of the first effect over the second and we can express this in terms of a strong inequality by requiring the distance moved by a particle on account of its thermal speed V_{th} in one wave period to be much less than the wavelength of the field fluctuations

$$\frac{2\pi}{\omega} V_{th} \ll \lambda$$

or

$$V_{th} \ll \frac{\omega}{k} = v_p \tag{3.69}$$

Table 3.4. *Cold plasma wave equations*

Wave equations:

$$\begin{aligned} \partial n_\alpha/\partial t + \boldsymbol{\nabla}\cdot(n_\alpha \mathbf{u}_\alpha) &= 0 \\ m_\alpha n_\alpha(\partial/\partial t + \mathbf{u}_\alpha\cdot\boldsymbol{\nabla})\mathbf{u}_\alpha &= e_\alpha n_\alpha(\mathbf{E} + \mathbf{u}_\alpha \times \mathbf{B}) \end{aligned}$$

Maxwell equations:

$$\begin{aligned} \boldsymbol{\nabla}\times\mathbf{E} &= -\partial\mathbf{B}/\partial t \\ \boldsymbol{\nabla}\times\mathbf{B} &= \varepsilon_0\mu_0\partial\mathbf{E}/\partial t + \mu_0\mathbf{j} \\ \boldsymbol{\nabla}\cdot\mathbf{E} &= q/\varepsilon_0 \\ \boldsymbol{\nabla}\cdot\mathbf{B} &= 0 \end{aligned}$$

Approximations:

Cold plasma: $V_{\mathrm{th}} \ll \omega/k = v_{\mathrm{p}}$

Collisionless: $\nu_{\mathrm{c}} \ll \omega$

where v_{p} is the phase velocity of the wave. This is the *cold plasma approximation* and in the limit $V_{\mathrm{th}}/v_{\mathrm{p}} \to 0$ the fluid description is exact for a collisionless plasma since, at any given point in the plasma, the velocity of all the particles of a given species is uniquely determined by the species flow velocity $\mathbf{u}_\alpha$ and the forces acting on the particles are given by the fields at that point. Thus, the *cold plasma wave equations* are simply the (separate) ion and electron continuity equations and equations of motion. In the latter the electric field is retained because the non-relativistic approximation ($v_{\mathrm{p}} \ll c$) is *not* assumed and the fields $\mathbf{E}$ and $\mathbf{B}$ are determined by the full Maxwell equations, i.e. the displacement current is retained. The equations are listed in Table 3.4.

The cold plasma wave equations provide a very good description of wave phenomena in collisionless plasmas, especially at the high frequency end of the spectrum. However, they inevitably become invalid at wave resonances where $k \to \infty$. The effects of finite temperature may be investigated by introducing pressure gradients into the equations of motion and adding the adiabatic equations of state to determine the isotropic (or anisotropic) pressures. These are the *warm plasma wave equations* which, for the case of isotropic pressures, are listed in Table 3.5. In the adiabatic equations $\rho_\alpha = m_\alpha n_\alpha$ and we have allowed for distinct ratios of the specific heats. The warm plasma wave equations are model equations for they have no rigorous derivation and, as discussed later, the fluid model omits important kinetic effects like Landau damping. Nevertheless, they provide a simple description of finite temperature modifications of cold plasma waves and of the further fluid modes which propagate in a warm plasma but disappear in the cold plasma limit.

Table 3.5. *Warm plasma wave equations*

Wave equations:

$$\begin{aligned}\partial n_\alpha/\partial t + \nabla\cdot(n_\alpha\mathbf{u}_\alpha) &= 0\\ m_\alpha n_\alpha(\partial/\partial t + \mathbf{u}_\alpha\cdot\nabla)\mathbf{u}_\alpha &= e_\alpha n_\alpha(\mathbf{E} + \mathbf{u}_\alpha\times\mathbf{B}) - \nabla p_\alpha\\ D(p_\alpha\rho_\alpha^{-\gamma_\alpha})/Dt &= 0\end{aligned}$$

Maxwell equations:

$$\begin{aligned}\nabla\times\mathbf{E} &= -\partial\mathbf{B}/\partial t\\ \nabla\times\mathbf{B} &= \varepsilon_0\mu_0\partial\mathbf{E}/\partial t + \mu_0\mathbf{j}\\ \nabla\cdot\mathbf{E} &= q/\varepsilon_0\\ \nabla\cdot\mathbf{B} &= 0\end{aligned}$$

Collisionless approximation: $\nu_c \ll \omega$

3.5.1 Generalized Ohm's law

For warm plasma waves it is sometimes necessary to include in the equations of motion a term to represent the exchange of momentum between species. As in Section 3.4 (see (3.51)), we write the rate of flow of momentum from electrons to ions as $m_e n_e \nu_c(\mathbf{u}_e - \mathbf{u}_i)$. Using the approximations (3.50) and ignoring quadratic terms $(\mathbf{u}_\alpha\cdot\nabla)\mathbf{u}_\alpha$ it is then straightforward (see Exercise 3.6) to combine the equations of motion to obtain a generalized Ohm's law in the form

$$\mathbf{E} + \mathbf{u}\times\mathbf{B} - \mathbf{j}/\sigma - \frac{1}{\sigma\nu_c}\frac{\partial\mathbf{j}}{\partial t} = \frac{m_i}{Ze\rho}(\mathbf{j}\times\mathbf{B} - \nabla p_e) \tag{3.70}$$

This differs from (3.52) through the additional $\partial\mathbf{j}/\partial t$ term which is negligible in the MHD approximation because $\mathbf{j}$ is assumed to vary on the hydrodynamic time scale and $\nu_c\tau_H \gg 1$. In the warm plasma wave equations, $\mathbf{j}$ and $\mathbf{u}_e$ may vary on a collision time scale and it is easily verified that inserting the inertial term $m_e n_e \partial\mathbf{u}_e/\partial t$ in (3.51) and using (3.50) yields (3.70).

Neither derivation of the generalized Ohm's law presented in this chapter is mathematically rigorous. However, as discussed in Section 12.6.2, Balescu (1988) has shown that in the MHD approximation the effect of the inertial terms is transient and dies out after a few collision times so that the MHD version of the generalized Ohm's law (3.52) can be rigorously derived. On the other hand, there is no rigorous derivation of (3.70) and this form of the generalized Ohm's law, like the warm plasma wave equations and the scalar Ohm's law (3.35), has the status of a model equation.

3.6 Boundary conditions

In problems where the plasma may be treated as infinite the boundary conditions take the simple form of prescribed values at infinity and perhaps at certain internal points. More realistically, they are conditions to be satisfied by the solutions obtained in different regions on the boundary between them. Typically, a plasma may be surrounded by a vacuum and the boundary conditions, applied at the plasma–vacuum interface, relate the solution of the fluid and field equations in the plasma to the solution of the field equations in the vacuum; the vacuum may extend to infinity or be surrounded by a wall and further appropriate boundary conditions are applied to the vacuum fields.

Although in reality all variables change continuously across boundaries they often do so very rapidly and it is convenient to treat the boundary as an infinitesimally thin surface across which discontinuous changes take place. Differential equations become invalid when the variables or their derivatives are discontinuous but by integrating the equations over an infinitesimal volume or surface which straddles the boundary we derive conditions which relate the values of the variables on either side of the boundary in terms of some surface quantity. The electromagnetic boundary conditions are a familiar example of this procedure. Provided that there are only volume distributions of current and charge the field variables are continuous across the boundary between two media. However, if either medium is a conductor containing a surface current or charge, then the tangential component of the magnetic field and the normal component of the electric field suffer discontinuities determined by the surface current and charge respectively.

In ideal MHD there is no space charge and therefore no surface charge. On the other hand, the thickness of the skin current in a good conductor decreases as the conductivity increases and, in the ideal MHD limit, such currents become surface currents flowing in a skin of infinitesimal thickness. Here, since $\mathbf{E}$ is determined by Ohm's law, we are concerned with the boundary conditions on $\mathbf{B}$. As in electromagnetism these are obtained by integrating $\nabla \cdot \mathbf{B} = 0$ and Ampère's law over a small cylindrical volume and rectangular surface, respectively, leading to the well-known results (see Fig. 3.3 and Exercise 3.7)

$$[\mathbf{n} \cdot \mathbf{B}]_1^2 = 0 \tag{3.71}$$

$$[\mathbf{n} \times \mathbf{B}]_1^2 = \mu_0 \mathbf{J}_s \tag{3.72}$$

where $\mathbf{n}$ is the unit vector normal to the boundary surface from side 1 to side 2, $[X]_1^2 = X_2 - X_1$ is the change in X across the surface, and $\mathbf{J}_s$ is the surface current.

Another important boundary condition at a plasma–vacuum surface in ideal MHD is obtained by applying the same procedure used to obtain (3.71) to the

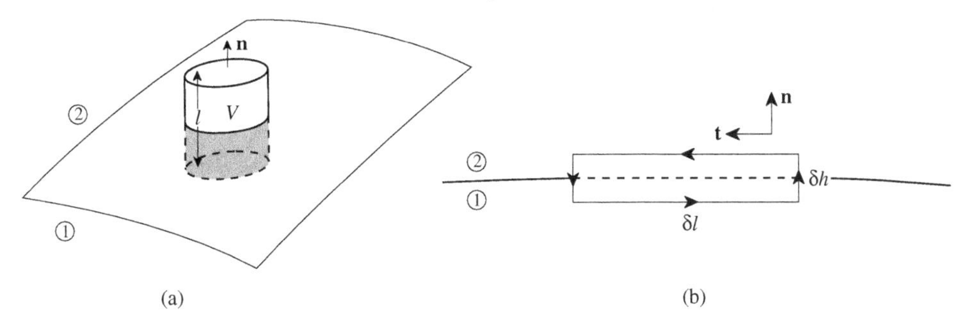

Fig. 3.3. Boundary volume and surface integrals.

momentum equation. In the next chapter we show that this equation may be written

$$\rho \frac{D\mathbf{u}}{Dt} = -\nabla \cdot \left[\left(P + \frac{B^2}{2\mu_0} \right) \mathbf{I} - \frac{\mathbf{BB}}{\mu_0} \right]$$

where $\mathbf{I}$ is the unit dyadic. Integrating this equation over the volume of the infinitesimal cylinder whose ends are either side of the boundary surface as shown in Fig. 3.3(a) and using Gauss' theorem gives

$$\int_V \rho \frac{D\mathbf{u}}{Dt} \, \mathrm{d}V = \int_S \left[\left(P + \frac{B^2}{2\mu_0} \right) \mathbf{n} - (\mathbf{B} \cdot \mathbf{n})\mathbf{B} \right] \mathrm{d}S$$

where V is the volume of the cylinder and S is its surface. As the length, l, of the cylinder tends to zero the contribution to the surface integral from the curved surface vanishes leaving only the normal contribution from either side of the boundary. Further, since the plasma is a perfect conductor, there is no normal component of $\mathbf{B}$ at the surface and so the second term in the square bracket also vanishes as $l \to 0$. Finally, since the acceleration of the plasma must remain finite, the volume integral vanishes as l, and hence V, tends to zero. Thus,

$$[P + B^2/2\mu_0]_1^2 = 0 \tag{3.73}$$

i.e. the total pressure (plasma plus magnetic) is continuous at the boundary. Of course, if the plasma density $\rho \to 0$ at the boundary then $P \to 0$ and (3.73) requires that the magnetic pressure be continuous. Since in this case there can be no surface current, this is consistent with (3.71) and (3.72) expressing the continuity of both normal and tangential components of $\mathbf{B}$.

The vanishing of the normal component of $\mathbf{B}$ at the surface of a perfect conductor also means that in ideal MHD, at a plasma–vacuum interface, (3.71) is replaced by the stronger conditions

$$\mathbf{B} \cdot \mathbf{n} = 0 = \tilde{\mathbf{B}} \cdot \mathbf{n} \tag{3.74}$$

where $\mathbf{B}$ and $\tilde{\mathbf{B}}$ are the plasma and vacuum fields, respectively.

Exercises

3.1 In general the motion of a plasma is determined by the action of both applied and induced fields. Under what circumstances is it necessary to investigate plasma dynamics using fluid equations rather than particle orbit theory?

What are the main features distinguishing the MHD and plasma wave descriptions?

What is the essential difference between the hot and cold plasma wave equations? State this in terms of a strong inequality.

3.2 What are the essential properties of a plasma fluid element and how may they be expressed in terms of dimensionless parameters?

Why do short range inter-particle forces produce a net change in the momentum of a fluid element only at its surface?

3.3 What is the assumption that allows us to use the thermodynamic equations of state in MHD? Express this in terms of a strong inequality.

The first law of thermodynamics (3.17) relates the change in internal energy of a plasma to the work done and heat exchange in effecting the transition from one equilibrium state of the plasma to another. What is the fundamental distinction between the *state* variable $\mathcal{E}$ and the *path* variables W and Q? What defines an adiabatic change of state?

3.4 Carry out the steps indicated in the text to obtain (3.27) describing the evolution of the plasma pressure. What is the physical significance of each of the terms on the right-hand side of this equation?

3.5 By a dimensional analysis of the Maxwell equations, show that the displacement current and the electrostatic force may be neglected in the non-relativistic approximation. What does this mean physically in terms of the reaction of the plasma to the electrostatic field compared with its reaction to the electromagnetic fields in MHD?

Verify your answer by showing, from the equation of charge conservation and $\mathbf{j} = \sigma \mathbf{E}$, that any net space charge will decay away in a time of order ε_0/σ.

Why, in general, is the non-relativistic approximation not adequate in plasma wave theory?

3.6 Obtain (3.70) by combining the ion and electron equations of motion. Show, also, that it may be derived by retaining the neglected electron inertial term $m_e n_e \partial \mathbf{u}_e/\partial t$ in (3.51) and substituting for $\mathbf{u}_i$ and $\mathbf{u}_e$ from (3.50).

3.7 By integrating $\boldsymbol{\nabla} \cdot \mathbf{B} = 0$ and Ampère's law across the boundaries illustrated in Fig. 3.3 derive the boundary conditions (3.71) and (3.72).

4
Ideal magnetohydrodynamics

4.1 Introduction

Ideal MHD is used to describe macroscopic behaviour across a wide range of plasmas and in this chapter we consider some of the most important applications. Being dissipationless the ideal MHD equations are conservative and this leads to some powerful theorems and simple physical properties. We begin our discussion by proving the most important theorem, due to Alfvén (1951), that the magnetic field is 'frozen' into the plasma so that one carries the other along with it as it moves. This kinematic effect arises entirely from the evolution equation for the magnetic field and represents the conservation of magnetic flux through a fluid element. Of course, any finite resistivity allows some slippage between plasma and field lines but discussion of these effects entails non-ideal behaviour and is postponed until the next chapter.

The concept of field lines frozen into the plasma leads to very useful analogies which aid our understanding of the physics of ideal MHD. It also suggests that one might be able to contain a thermonuclear plasma by suitably configured magnetic fields, although research has shown that this is no easily attainable goal. Further, since the ideal MHD equations are so much more amenable to mathematical analysis they can be used to investigate realistic geometries. The theory has thereby provided a useful and surprisingly accurate description of the macroscopic behaviour of fusion plasmas showing why certain field configurations are more favourable to containment than others.

Notwithstanding the wide applicability of ideal MHD in space and laboratory plasma physics a note of caution needs to be sounded over results derived from it. Since the neglected dissipative terms are of higher differential order than the non-dissipative terms, even a very small amount of dissipation can lead to solutions which are significantly different from those of ideal MHD. Mathematically, higher differential order means that singular perturbation theory must be used to examine the effects of dissipative terms and the 'ideal' solution cannot be recovered by taking the dissipative solution to an appropriate limit. This is illustrated in Exercise 4.1 where it is shown in a particular example that as one approaches the ideal limit of no dissipation the effect of the dissipative terms is restricted to a narrow sheath in which steepening gradients compensate for vanishing dissipative coefficients. Generally, whilst an ideal MHD solution may be valid over most of a plasma volume it cannot be entirely divorced from, and must be matched to, the non-ideal solution at the boundary of such sheaths. Because of the steep gradients, often the most interesting physics takes place in the sheath with dramatic consequences for the whole plasma, as we shall discover in the next chapter.

4.2 Conservation relations

The basic physical properties of ideal MHD are related to the conservation of mass, momentum, energy and magnetic flux, so it is helpful to write the fluid equations in the form of conservation relations. The continuity equation

$$\frac{\partial \rho}{\partial t} = -\nabla \cdot (\rho \mathbf{u}) \tag{4.1}$$

is already in the required form, which is not surprising since it was derived by expressing the conservation of mass in an arbitrary fixed volume within the plasma.

The equation expressing the conservation of momentum is obtained by taking the partial time derivative of $\rho\mathbf{u}$ and using the continuity and momentum equations for $\partial\rho/\partial t$ and $\partial\mathbf{u}/\partial t$ to get

$$\frac{\partial(\rho\mathbf{u})}{\partial t} = -\nabla \cdot (\rho\mathbf{u}\mathbf{u}) - \nabla P + \frac{1}{\mu_0}(\nabla \times \mathbf{B}) \times \mathbf{B}$$

Now, using the vector identity $(\nabla \times \mathbf{B}) \times \mathbf{B} = (\mathbf{B}\cdot\nabla)\mathbf{B} - \nabla(B^2/2)$ and $\nabla\cdot(\mathbf{BB}) = \mathbf{B}\cdot\nabla\mathbf{B}$ this becomes

$$\frac{\partial(\rho\mathbf{u})}{\partial t} = -\nabla \cdot (\rho\mathbf{u}\mathbf{u} + P\mathbf{I} - \mathbf{T}) = \nabla \cdot \mathbf{\Pi} \tag{4.2}$$

say. In (4.2) $\mathbf{I}$ is the unit dyadic and

$$\mathbf{T} = \frac{1}{\mu_0}\mathbf{BB} - \frac{B^2}{2\mu_0}\mathbf{I}$$

is the *Maxwell stress tensor*, $\{\varepsilon_0\mathbf{EE} + \mathbf{BB}/\mu_0 - \frac{1}{2}(\varepsilon_0 E^2 + B^2/\mu_0)\mathbf{I}\}$, in the non-relativistic limit, $\varepsilon_0\mu_0 E^2/B^2 \ll 1$.

Similarly the equation of energy conservation follows from the partial time derivative of the total energy density

$$U = \frac{1}{2}\rho u^2 + \frac{P}{\gamma - 1} + \frac{B^2}{2\mu_0}$$

comprising kinetic, internal, and magnetic energies, respectively. On using the ideal MHD equations (Table 3.2) to evaluate the derivatives the result is

$$\frac{\partial U}{\partial t} = -\nabla \cdot \mathbf{S} \tag{4.3}$$

where

$$\mathbf{S} = \left(\frac{1}{2}\rho u^2 + \frac{\gamma P}{\gamma - 1}\right)\mathbf{u} + \frac{1}{\mu_0}\mathbf{B} \times (\mathbf{u} \times \mathbf{B})$$

is the total energy flux (see Exercise 4.2). Each of these conservation equations expresses the time rate of change of the conserved quantity, at any given point in the fluid, as the (negative) divergence of the corresponding flux at that point. Integrating these equations over an arbitrary volume and using Gauss' theorem gives the rate of change of the conserved quantity within the volume in terms of the flux through its surface.

It is less obvious that the evolution equation for the magnetic field is a conservation equation but, in what has become known as the *frozen flux theorem*, Alfvén showed that a consequence of

$$\frac{\partial \mathbf{B}}{\partial t} = \nabla \times (\mathbf{u} \times \mathbf{B}) \tag{4.4}$$

is that the magnetic flux, through any surface S bounded by a closed contour C moving with the fluid, is constant. From the two-dimensional extension of Leibnitz's theorem (see (3.3) in Section 3.2) we have

$$\frac{\mathrm{d}}{\mathrm{d}t}\int_S \mathbf{B} \cdot \mathrm{d}\mathbf{S} = \int_S \frac{\partial \mathbf{B}}{\partial t} \cdot \mathrm{d}\mathbf{S} + \oint_C \mathbf{B} \cdot \mathbf{v}_C \times \mathrm{d}\mathbf{l} \tag{4.5}$$

so that in the case of a surface S whose boundary C is moving with the local flow velocity $\mathbf{u}$, as illustrated in Fig. 4.1, we have

$$\frac{D}{Dt}\int_S \mathbf{B} \cdot \mathrm{d}\mathbf{S} = \int_S \frac{\partial \mathbf{B}}{\partial t} \cdot \mathrm{d}\mathbf{S} + \oint_C \mathbf{B} \cdot \mathbf{u} \times \mathrm{d}\mathbf{l} \tag{4.6}$$

where the first term on the right-hand side of (4.6) represents the change of flux due to the time rate of change of $\mathbf{B}$ and the second term represents the change in the surface area due to the movement of the bounding contour C (see Fig. 3.1(b)). Now interchanging dot and cross in the triple scalar product and using Stokes' theorem to convert the line integral to a surface integral we see that

$$\frac{D}{Dt}\int_S \mathbf{B} \cdot \mathrm{d}\mathbf{S} = \int_S \left(\frac{\partial \mathbf{B}}{\partial t} - \nabla \times (\mathbf{u} \times \mathbf{B})\right) \cdot \mathrm{d}\mathbf{S} = 0 \tag{4.7}$$

on account of (4.4). Thus, the flux through a surface S moving with the fluid is constant. Representing the flux through the surface S by the totality of the field

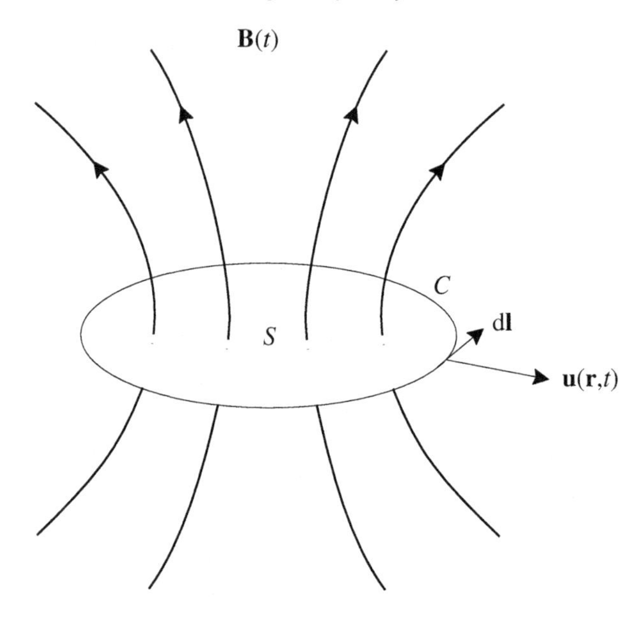

Fig. 4.1. Magnetic flux motion.

lines threading the loop C and bearing in mind that the theorem is true for arbitrary C we see that the field lines are constrained to move with C, i.e. with the fluid.

Although not tangible physical quantities, 'field lines' and 'flux tubes' are helpful concepts for understanding the properties of magnetic fields. At every point, whether in the fluid or in the vacuum, field lines follow the direction of the magnetic field and are, therefore, defined by the equations

$$\frac{\mathrm{d}x}{B_x} = \frac{\mathrm{d}y}{B_y} = \frac{\mathrm{d}z}{B_z}$$

A *magnetic* (or *flux) surface* is one that is everywhere tangential to the field, i.e. the normal to the surface is everywhere perpendicular to **B**, as shown in Fig. 4.2(a). Figure 4.2(b) shows an open-ended cylindrical magnetic surface which defines a *flux tube* and it is sometimes helpful to picture the density of the field lines through the tube as a representation of the field strength; in other words, the number of field lines threading the tube is proportional to the flux.

Clearly, the flux through a magnetic surface is zero and by the frozen flux theorem must remain so if the surface moves with the fluid. Let us now imagine a long thin fluid element which at any given moment lies along a flux tube. As it moves, its cylindrical surface remains a magnetic surface and the flux through its ends remains constant; hence the notion that the field lines are 'frozen' into the fluid (see Fig. 4.2(c)).

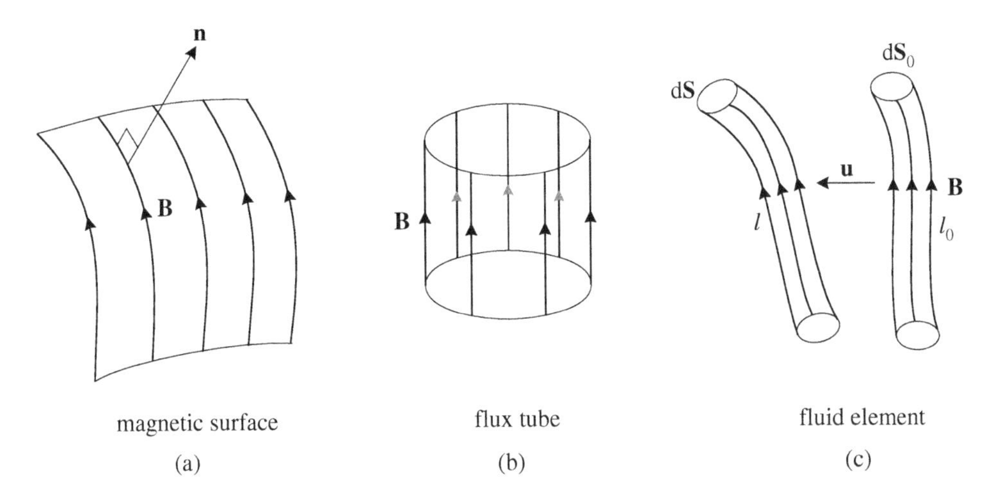

Fig. 4.2. Field lines and flux tubes.

Next we consider the distortion of a flux tube of initial cross-section dS_0 and length l_0 as it moves with the fluid. Since the flux through the tube is constant a motion which stretches the tube to length $l > l_0$ and narrows its cross-section to $dS < dS_0$ will result in an increase in the field strength since $B_0\,dS_0 = B\,dS$. As the mass of the fluid is also conserved, i.e. $\rho_0\,dS_0\,l_0 = \rho\,dS\,l$, we may divide these equations to get the result

$$\frac{B_0}{\rho_0 l_0} = \frac{B}{\rho l} \tag{4.8}$$

For an incompressible fluid (ρ = const.) this becomes

$$B = (B_0/l_0)l \tag{4.9}$$

i.e. the field strength is proportional to the length of the flux tube; these results were first noted by Walén (1946).

Flux conservation has a profound effect on the structure of the field. A fluid element may change its shape but it does not break into separate pieces. The same is true, therefore, of a flux tube with the result that the topology of the field cannot change. This is a severe constraint since field configurations in which a lower energy state could be arrived at by line breaking and reconnection are both easy to imagine and to realize in practice. Such changes, which are forbidden in ideal MHD by flux conservation, become possible with dissipation, however small, and the consequences for plasma stability are dramatic, as we shall see in Chapter 5.

4.3 Static equilibria

The fact that the field lines are frozen into a perfectly conducting fluid leads naturally to the notion that by controlling the magnetic field configuration one might be able to contain the fluid. For the high temperature plasmas needed for fusion reactions this is a vital matter since contact between plasma and wall is likely to be deleterious to both. We turn, therefore, to the momentum equation to examine the effect of the field on the motion of the fluid but, because we wish to examine possible equilibrium configurations, we begin with the specially simple case in which the fluid is at rest.

For mathematical consistency this poses a problem for if $\mathbf{u} = 0$ then $R_{\mathrm{M}} = 0$ and ideal MHD is invalid! However, a dimensional analysis of the resistive MHD equations in Table 3.1 in this case shows that the pressure increases and the magnetic field diffuses on a time scale that is proportional to the plasma conductivity σ. The requirement, therefore, is that σ should be sufficiently large that this diffusion time, $\tau_{\mathrm{D}} \sim \mu_0 \sigma L_{\mathrm{H}}^2$, is greater than any other time of interest. We assume that such times as will arise, e.g. the time for the growth of instabilities (see Sections 4.5–4.7) or the period of plasma waves (see Section 4.8), are less than τ_{D}. With this proviso we may set $\mathbf{u}$ and all time derivatives equal to zero in the ideal MHD equations of Table 3.2 to get

$$\mathbf{j} \times \mathbf{B} = \nabla P \tag{4.10}$$

$$\nabla \cdot \mathbf{B} = 0 \tag{4.11}$$

$$\mathbf{j} = \frac{1}{\mu_0} \nabla \times \mathbf{B} \tag{4.12}$$

If an equilibrium state is to be established the $\mathbf{j} \times \mathbf{B}$ force must balance the pressure gradient and to investigate this it is convenient to use the static momentum equation (4.10) in conservative form.

Defining the *total stress tensor*

$$\mathcal{T}_{ik} = [(P + B^2/2\mu_0)\delta_{ik} - B_i B_k/\mu_0] \tag{4.13}$$

(4.2) becomes

$$\frac{\partial \mathcal{T}_{ik}}{\partial r_i} = 0 \tag{4.14}$$

The total stress tensor may be reduced to diagonal form by transformation to the principal axes. The eigenvalues may be obtained from the secular equation

$$|\mathcal{T}_{ik} - \delta_{ik}\lambda| = 0$$

the solution being

$$\lambda_1 = P + B^2/2\mu_0 = \lambda_2, \qquad \lambda_3 = P - B^2/2\mu_0$$

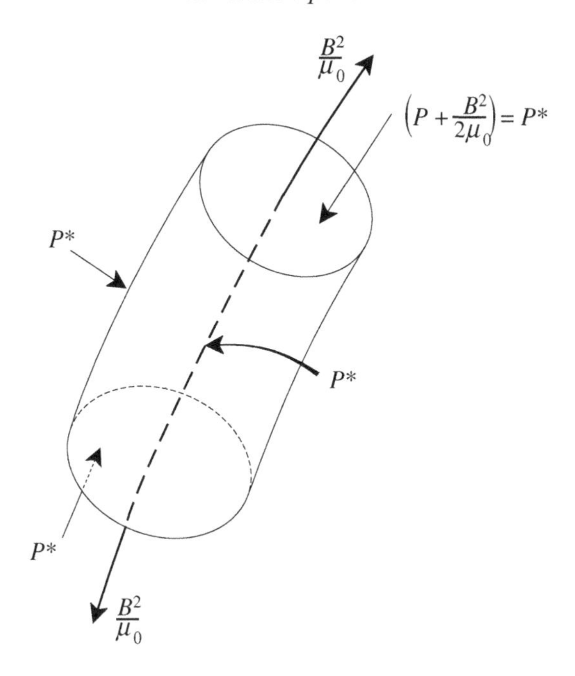

Fig. 4.3. Total stress tensor relative to principal axes.

Thus, referred to the principal axes, $\mathcal{T}_{ik}$ takes the form

$$\begin{bmatrix} P + B^2/2\mu_0 & 0 & 0 \\ 0 & P + B^2/2\mu_0 & 0 \\ 0 & 0 & P - B^2/2\mu_0 \end{bmatrix}$$

The principal axes are oriented so that the axis corresponding to λ_3 is parallel to **B** and the axes corresponding to λ_1,λ_2 are perpendicular to **B**. From this we see that the stress caused by the magnetic field amounts to a pressure $B^2/2\mu_0$ in directions transverse to the field and a tension $B^2/2\mu_0$ along the lines of force. In other words, the total stress amounts to an isotropic pressure which is the sum of the fluid pressure and the *magnetic pressure* $B^2/2\mu_0$, and a *tension* B^2/μ_0 along the lines of force. This is illustrated in Fig. 4.3. The ratio of fluid pressure to magnetic pressure, $2\mu_0 P/B^2$, is an important parameter commonly denoted by β. In MHD, it is often convenient to picture a tube of force behaving like an elastic string under tension. Thus stretching the tube of force increases the tension, which means that the field is increased as explained in the previous section.

From the equilibrium condition (4.10) it follows that

$$\mathbf{B} \cdot \nabla P = 0 \tag{4.15}$$

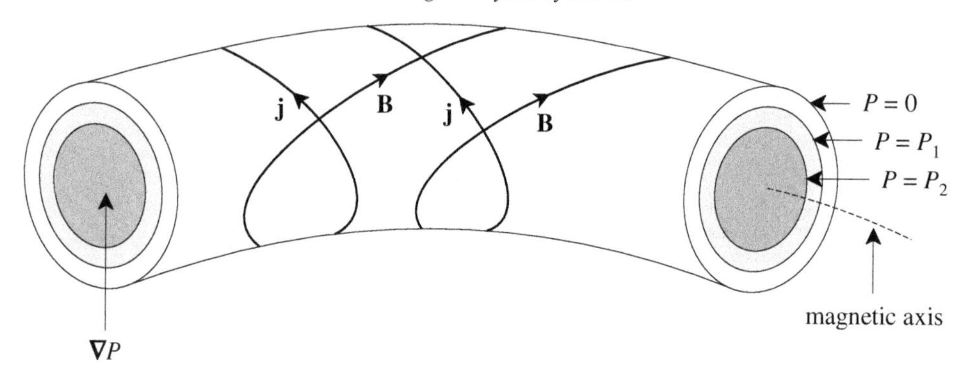

Fig. 4.4. Nested isobaric surfaces.

and

$$\mathbf{j} \cdot \boldsymbol{\nabla} P = 0 \tag{4.16}$$

This means that both **B** and **j** lie on surfaces of constant pressure so that the current flows between magnetic surfaces. Supposing the constant pressure (isobaric) surfaces to be closed, then, since (4.15) states that no magnetic field line passes through the surface, one may picture the surface as made up from a winding of field lines, i.e. the isobaric surfaces are also magnetic surfaces. Likewise, from (4.16) the isobaric surfaces are made up from lines of current density; these lines will, in general, intersect the field lines. The cross-section in Fig. 4.4 shows a set of nested surfaces on which the pressure increases in passing from the outside towards the axis; the currents are such that $\mathbf{j} \times \mathbf{B}$ points towards the axis. The implication here is that a plasma may be contained entirely by the magnetic force, an arrangement referred to as *magnetic confinement.*

An important integral relationship, known as the *virial theorem*, follows from (4.14) and shows that such magnetic confinement cannot be achieved without the aid of external currents. To demonstrate this we integrate the identity

$$\frac{\partial}{\partial r_k}(r_i \mathcal{T}_{ik}) = r_k \frac{\partial \mathcal{T}_{ik}}{\partial r_i} + \mathcal{T}_{ii}$$

over an arbitrary volume V, bounded by a surface S. Then, since $\partial \mathcal{T}_{ik}/\partial r_i = 0$ for equilibrium, on using Gauss' theorem we get

$$\int (3P + B^2/2\mu_0)\mathrm{d}V = \int [(P + B^2/2\mu_0)\mathbf{n} \cdot \mathbf{r} - (B^2/\mu_0)(\mathbf{r} \cdot \mathbf{b})(\mathbf{n} \cdot \mathbf{b})]\mathrm{d}S$$

Now if the fields are due entirely to the plasma they must decrease as $S \to \infty$ at least as fast as a dipole field ($\propto r^{-3}$) so that the surface integral vanishes. The volume integral, on the other hand, is positive definite and does not vanish, from which we conclude that we cannot set $\partial \mathcal{T}_{ik}/\partial r_i = 0$ and there is no equilibrium without an externally applied field.

4.3.1 Cylindrical configurations

For simplicity let us first consider the radial equilibrium of cylindrical plasmas. We assume cylindrical symmetry so that all variables are independent of θ and z and the lines of **j** and **B** lie on surfaces of constant r. Then

$$\mu_0 \mathbf{j} = \left[0, -\frac{\mathrm{d}B_z(r)}{\mathrm{d}r}, \frac{1}{r}\frac{\mathrm{d}}{\mathrm{d}r}(r B_\theta(r))\right] \tag{4.17}$$

and the radial component of (4.10) gives

$$\frac{\mathrm{d}}{\mathrm{d}r}[P + (B_\theta^2 + B_z^2)/2\mu_0] = -B_\theta^2/\mu_0 r \tag{4.18}$$

which expresses the (radial force) balance between the total (gas and magnetic) pressure gradient and the magnetic tension due to the curvature (if any) of the magnetic field. Of course, it is possible to remove magnetic curvature by choosing $B_\theta = 0$ and then the gas and magnetic pressure gradients must be oppositely directed and in balance. In this case, setting $B_\theta = 0$ in (4.18) and integrating, we have

$$P^* \equiv P + B^2/2\mu_0 = \text{const.} \tag{4.19}$$

Such a field may be produced by currents flowing azimuthally; early devices designed to contain plasma in this configuration were known as *theta-pinches* (since θ is used to denote the azimuthal coordinate). Azimuthal currents are induced by discharging a current suddenly in a metal conductor enclosing the discharge tube. The induced currents flow in the opposite direction and an axial magnetic field is generated in the region between. The $\mathbf{j}\times\mathbf{B}$ force acts to push the plasma towards the axis until the external magnetic pressure is balanced by the (total) internal pressure; from (4.19)

$$P(r) + B^2(r)/2\mu_0 = B_0^2/2\mu_0 \tag{4.20}$$

where B_0 is the external magnetic field. Note that this means that the plasma acts as a diamagnetic medium, $B(r) < B_0$ (see Section 2.2.1). In the absence of any initial magnetic field there is only the induced field which remains entirely outside the plasma since it cannot penetrate in ideal MHD and (4.20) reduces to

$$P = B_0^2/2\mu_0 \tag{4.21}$$

i.e. the plasma is radially contained by the magnetic field generated by the azimuthal current flowing in its outer surface. If there is an internal magnetic field

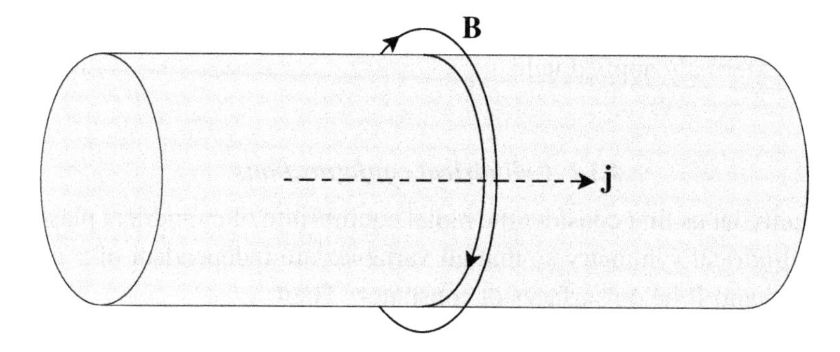

Fig. 4.5. Z-pinch configuration of axial current and azimuthal magnetic field.

$B(r)$, the current penetrates the plasma and (4.20) applies. It is customary to define the plasma β with respect to the external magnetic field, i.e. $\beta(r) = 2\mu_0 P(r)/B_0^2$, so that from (4.20)

$$\beta(r) = 1 - (B(r)/B_0)^2$$

which may take any value in the range $0 < \beta < 1$.

Another configuration which has been important for studying plasma containment is the *Z-pinch*. Here the **j**, **B** lines of the theta-pinch are interchanged so that with **j** now axial and **B** azimuthal, the $\mathbf{j} \times \mathbf{B}$ force is again directed towards the axis. Consider a fully ionized plasma contained in a cylindrical discharge tube with a current flowing parallel to the axis of the tube, as shown in Fig. 4.5. Under the action of the $\mathbf{j} \times \mathbf{B}$ force the plasma is squeezed or 'pinched' into a filament along the axis of the tube. Since $B_z = 0$ the static condition (4.18) may be written

$$\frac{\mathrm{d}P}{\mathrm{d}r} = -\frac{B}{\mu_0 r}\frac{\mathrm{d}}{\mathrm{d}r}(rB) \tag{4.22}$$

If the radius of the pinch is a, then multiplying (4.22) by r^2 and integrating gives

$$\int_0^a r^2 \frac{\mathrm{d}P}{\mathrm{d}r}\,\mathrm{d}r = -\frac{1}{\mu_0}\int_0^a (rB)\,\mathrm{d}(rB)$$

i.e.

$$(r^2 P)_{r=a} - 2\int_0^a rP\,\mathrm{d}r = -\frac{1}{2\mu_0}(rB)^2_{r=a}$$

If we suppose that the plasma pressure vanishes at $r = a$ the first term vanishes altogether and we get

$$2\int_0^a rP\,\mathrm{d}r = \frac{1}{2\mu_0}(rB)^2_{r=a} \tag{4.23}$$

Now from (4.17)

$$\mu_0 j = \frac{1}{r}\frac{\mathrm{d}}{\mathrm{d}r}(rB) \tag{4.24}$$

which integrates to give

$$(rB)_{r=a} = \mu_0 \int_0^a jr\,\mathrm{d}r$$

and, on defining the total current I flowing in the plasma column by

$$I = \int_0^a 2\pi rj\,\mathrm{d}r \tag{4.25}$$

we get

$$2\int_0^a rP\,\mathrm{d}r = \frac{\mu_0 I^2}{8\pi^2}$$

Assuming quasi-neutrality, $Zn_\mathrm{i}(r) = n_\mathrm{e}(r)$, and uniform ion and electron temperatures, we may substitute

$$P(r) = n_\mathrm{i}(r)k_\mathrm{B}T_\mathrm{i} + n_\mathrm{e}(r)k_\mathrm{B}T_\mathrm{e} = n_\mathrm{e}(r)k_\mathrm{B}(T_\mathrm{e} + T_\mathrm{i}/Z) \tag{4.26}$$

to obtain

$$I^2 = 8\pi k_\mathrm{B}(T_\mathrm{e} + T_\mathrm{i}/Z)N_\mathrm{e}/\mu_0 \tag{4.27}$$

where

$$N_\mathrm{e} = \int_0^a 2\pi r n_\mathrm{e}(r)\mathrm{d}r \tag{4.28}$$

is the number of electrons per unit length (electron line density) of the plasma column. Equation (4.27), known as *Bennett's relation*, determines the total current required for containment of a plasma of specified temperature and line density. (Note, however, that this analysis assumes a stable configuration; in fact, the Z-pinch, as we shall see in Section 4.5.1, is highly unstable.)

A device in which the field lines wind around the axis in a helical path is called a *screw pinch*. The rate at which the field lines rotate about the axis of the cylinder is an important parameter for equilibrium and stability. Referring to Fig. 4.6, this can be measured by $\mathrm{d}\theta/\mathrm{d}z$ and this in turn is determined from the equation for the magnetic field lines

$$\frac{r\,\mathrm{d}\theta}{\mathrm{d}z} = \frac{B_\theta(r)}{B_z(r)}$$

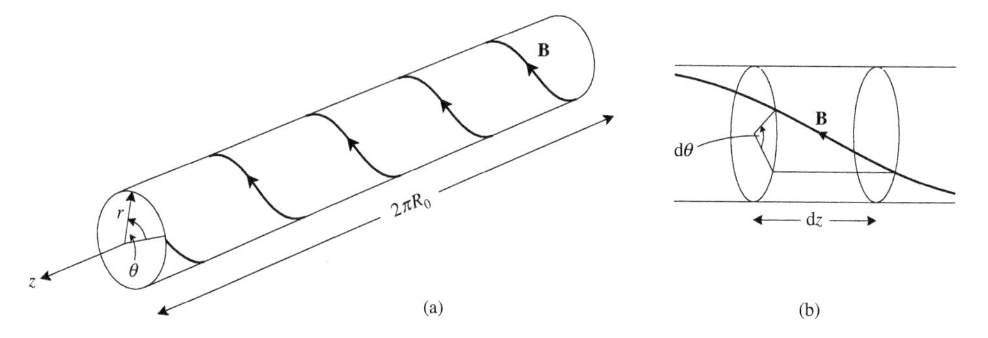

Fig. 4.6. Screw pinch geometry.

The total angle of rotation for a cylinder of length $2\pi R_0$ (the one-dimensional equivalent of a torus of major radius R_0) is called the *rotational transform* and is denoted by

$$\iota(r) = \int_0^{2\pi R_0} \left(\frac{\mathrm{d}\theta}{\mathrm{d}z}\right) \mathrm{d}z = \frac{2\pi R_0 B_\theta(r)}{r B_z(r)}$$

A related parameter, the significance of which will be seen when we consider stability in the next section, is the *MHD safety factor*

$$q(r) = 2\pi/\iota(r) = r B_z(r)/R_0 B_\theta(r) \tag{4.29}$$

The *pitch* of a field line is defined as the length of its projection on the axis in one complete rotation about the axis so that it is given by

$$\int_0^{2\pi} \left(\frac{\mathrm{d}z}{\mathrm{d}\theta}\right) \mathrm{d}\theta = \frac{2\pi r B_z(r)}{B_\theta(r)} = \frac{4\pi^2 R_0}{\iota(r)}$$

In the screw pinch the ability to manipulate two profiles, $B_\theta(r)$ and $B_z(r)$, instead of only one gives much greater flexibility in controlling the various physical parameters which influence stability and other conditions necessary for containment. One such parameter is the safety factor, while another is the plasma β for which one needs an optimum value between high β to achieve large $n\tau$ (to satisfy the Lawson criterion) and low β because of instability thresholds. Multiplying (4.18) by πr^2 and integrating from 0 to a gives

$$\int_0^a 2\pi r P \,\mathrm{d}r + \int_0^a 2\pi r \left(\frac{B_z^2}{2\mu_0}\right) \mathrm{d}r - \pi a^2 \left(\frac{B_z^2(a) + B_\theta^2(a)}{2\mu_0}\right) = 0$$

which may be written as

$$\langle P\rangle + \langle B_z^2/2\mu_0\rangle - \frac{B_z^2(a)}{2\mu_0} - \frac{B_\theta^2(a)}{2\mu_0} = 0 \tag{4.30}$$

where

$$\langle X\rangle = \frac{1}{\pi a^2}\int_0^a 2\pi r X(r)\,\mathrm{d}r$$

is the average value of $X(r)$ over a cross-section of radius a. Defining

$$\langle\beta\rangle = \frac{2\mu_0\langle P\rangle}{B^2(a)} \tag{4.31}$$

and similarly the corresponding 'poloidal' and 'toroidal' parameters as

$$\beta_\mathrm{p} = \frac{2\mu_0\langle P\rangle}{B_\theta^2(a)} \qquad \beta_\mathrm{t} = \frac{2\mu_0\langle P\rangle}{B_z^2(a)} \tag{4.32}$$

we may write (4.30) as

$$\frac{1}{\langle\beta\rangle} = \frac{1}{\beta_\mathrm{p}} + \frac{1}{\beta_\mathrm{t}} = 1 + \frac{\langle B_z^2\rangle}{2\mu_0\langle P\rangle} \tag{4.33}$$

This equation shows the flexibility of the screw pinch. Clearly, $\langle\beta\rangle$ can take values in the range $0 \le \langle\beta\rangle \le 1$. Furthermore, any particular value can be achieved given the range of choices for β_p and β_t. This contrasts sharply with theta- and Z-pinches where $\langle\beta\rangle$ is given by β_t and β_p respectively and, in the latter case, $\langle\beta\rangle = \beta_\mathrm{p} = 1$. Note from the first equality in (4.33) that whenever β_p and β_t differ in magnitude, $\langle\beta\rangle$ is given approximately by the smaller of these parameters.

4.3.2 Toroidal configurations

So far the discussion of plasma confinement has concentrated entirely on radial containment. There is nothing to prevent the plasma from flowing freely along the field lines and in cylindrical discharges plasma will be lost through the ends of the device unless something is done to prevent this. The obvious answer is to bend the cylinder round into a torus so that, rather than flowing out of the ends, the plasma flows round and round the device. This, however, introduces a second equilibrium constraint, known as the *toroidal force balance*.

Qualitatively, we can see how the problem arises by picturing what happens to the magnetic surface surrounding the initially cylindrical plasma as it is bent into a torus. There results a net outward force on the toroidal plasma which has two components, one due to the plasma pressure and the other due to the magnetic pressure.

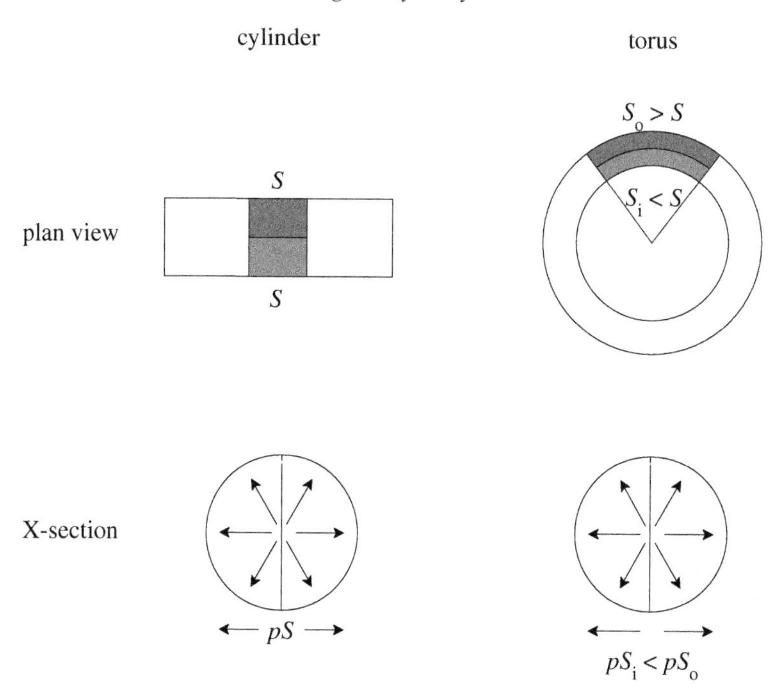

Fig. 4.7. Net outward force in a torus due to plasma pressure.

The first of these is akin to the force in an inflated tyre and is simply due to the fact that the total surface area of the outer half of the tube, or in our case magnetic surface, is greater than that of the inner half, while the pressure is constant over the (isobaric) surface so that the force outwards is greater than the force inwards. This is demonstrated schematically in Fig. 4.7 by plan and cross-sectional views of cylindrical and toroidal configurations. Segments of equal area (S) in the cylinder are stretched on the outer surface $S_o(> S)$ and compressed on the inner surface $S_i(< S)$ of the torus.

The other force involving the magnetic field is best understood by considering poloidal and toroidal fields separately. When there is only a poloidal field (the case, illustrated in Fig. 4.8, of bending a Z-pinch into a torus) this force is similar to the outward force experienced by a current-carrying circular loop. Conservation of magnetic flux generated by the current means that field lines are more densely packed inside the loop than outside so that the field strength is greater inside giving a net $\mathbf{j} \times \mathbf{B}$ force radially outwards.

For the purely toroidal field consider the simple case, shown in Fig. 4.9, of no initial internal field. Then the current flows in a thin skin at the plasma surface and the external field, generated by the current I_c in the toroidal field coils, is

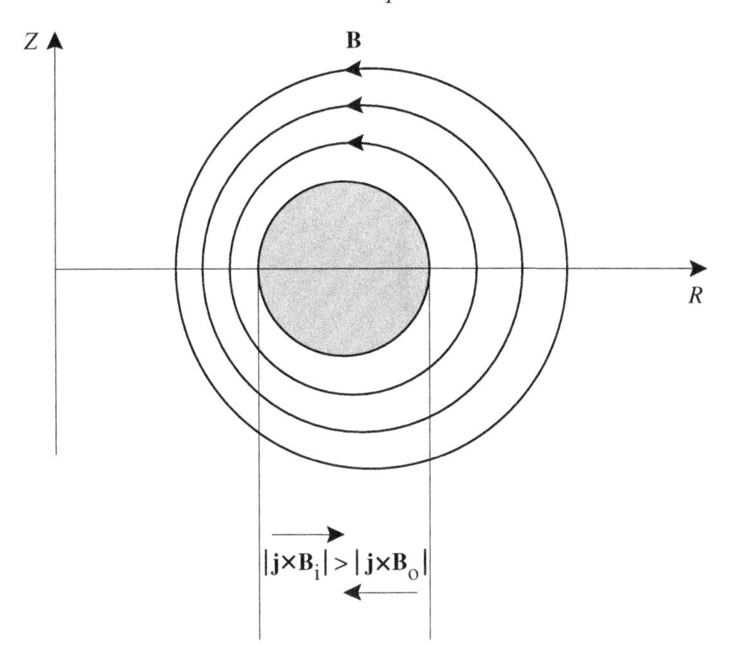

Fig. 4.8. Net outward $\mathbf{j} \times \mathbf{B}$ force in a torus due to the poloidal field.

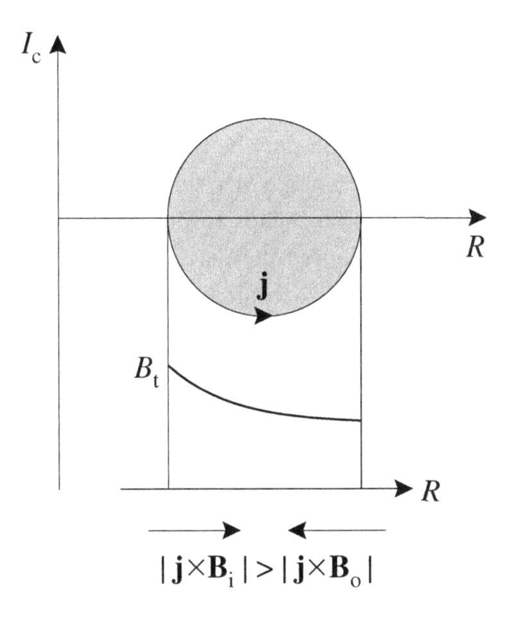

Fig. 4.9. Net outward $\mathbf{j} \times \mathbf{B}$ force in a torus due to the toroidal field.

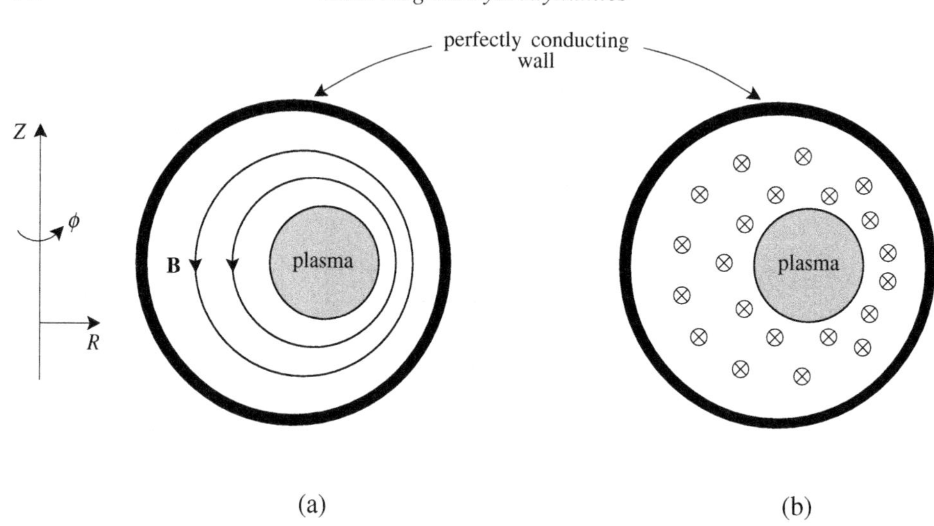

Fig. 4.10. A perfectly conducting wall can provide toroidal force balance for (a) poloidal fields but not for (b) toroidal fields.

$B_t = \mu_0 I_c / 2\pi R$. The R^{-1} dependence means that B_t is again stronger on the inner side of the torus than on the outer side with the same result as before.

The net toroidal force is usually quite small compared with the forces involved in radial pressure balance but the plasma and its self-generated 'containing' magnetic field cannot provide compensation which must, therefore, be applied externally. Provided there is a poloidal component of the magnetic field outside the plasma, as shown in Fig. 4.10(a), compensation may be provided by surrounding the plasma with a perfectly conducting wall. Then, since field lines cannot penetrate the wall they are compressed as the plasma moves towards the wall and the increase in magnetic pressure eventually balances the net toroidal force. Note that this does not work if the external field is purely toroidal since the field lines are not trapped between plasma and wall but can slip around the plasma as can be seen from Fig. 4.10(b). Since walls are not perfect conductors the compressed field leaks away in a finite time. This may be shorter than other times of interest which is a drawback to this method.

Another approach is to impose a vertical magnetic field by means of external coils as shown in Fig. 4.11. By suitable choice of current direction the vertical field reinforces the poloidal field on the outer side of the plasma and opposes it on the inner side providing the desired compensation. Here again, note that this does not work for a purely toroidal field since the vertical field is everywhere at right angles to the toroidal field and therefore has no effect. It follows that a poloidal component of magnetic field is essential for toroidal force balance.

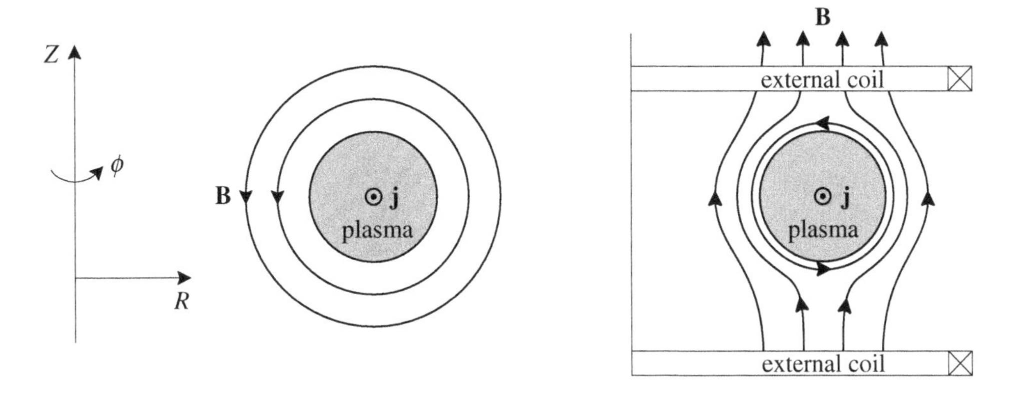

Fig. 4.11. Toroidal force balance provided by external coils.

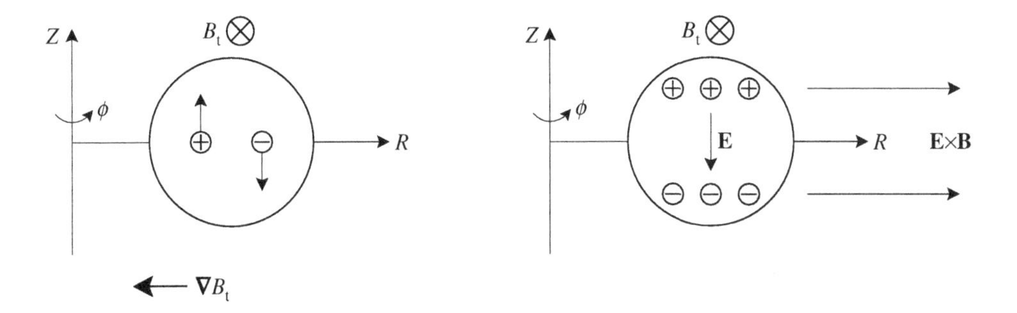

Fig. 4.12. $\mathbf{E} \times \mathbf{B}$ drift due to toroidal field.

Herein lies the dilemma of plasma containment. A theta-pinch has good radial stability (the plasma sits in a magnetic well) but if bent into a torus, with no poloidal field, there is no toroidal equilibrium. On the other hand, a closed toroidal system with a magnetic field that is predominantly poloidal will have good toroidal equilibrium but poor radial stability. The challenge is to find the optimal mix of poloidal and toroidal fields which can provide toroidal equilibrium without sacrificing radial stability.

The problem of maintaining toroidal equilibrium can be described in terms of particle orbits and was mentioned briefly in Section 2.10. As noted there and above, the toroidal field generated by external coils decreases with major radius R across the plasma. Consequently, there is a drift of the particle guiding centre relative to the lines of force which is a combination of grad B and curvature drifts. Such drifts are in opposite directions for ions and electrons so that an electric field is created and the $\mathbf{E} \times \mathbf{B}$ drift (the same for both species) is radially outward as shown in Fig. 4.12. Thus, if there is only a toroidal field there is no toroidal equilibrium.

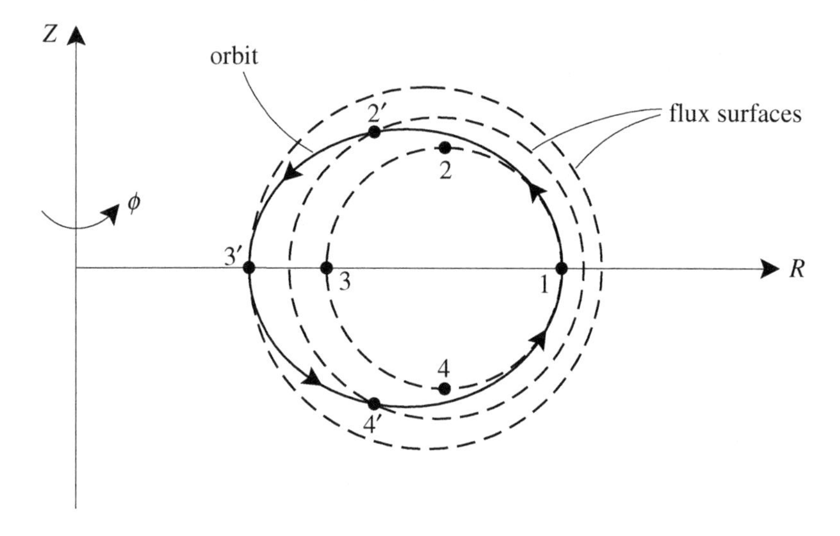

Fig. 4.13. Illustration of poloidal field compensation for particle drift in a torus.

Now the introduction of a poloidal field B_p can compensate for the particle drift as illustrated in Fig. 4.13. For simplicity, consider a flux surface on which the field lines rotate once in the poloidal direction during one circuit in the toroidal direction. With no drift a particle would simply gyrate about a field line, which for example starts at point 1 on the surface, reaches point 2 a quarter of the way round, point 3 halfway round and so on, returning to point 1 after one toroidal revolution. An upward drift (the case illustrated) causes the particle to leave this particular field line and move continuously across magnetic surfaces arriving at points 2′ and 3′ instead of 2 and 3. Thereafter, an upward drift means that the particle moves back towards the original magnetic surface arriving back at point 1 via point 4′.

The example we have considered is a particularly simple one. In general, neither the particle nor the field line will arrive back at the same point after one revolution but will be displaced through some angle in the poloidal plane as illustrated in Fig. 4.14. In an equivalent cylinder of length $2\pi R_0$ this angle is the rotational transform. In a torus, in general, the change in poloidal angle per toroidal revolution depends on the starting point, so the rotational transform is defined as the average change over a large number of revolutions

$$\iota = \lim_{N\to\infty} \frac{1}{N} \sum_{n=1}^{N} \Delta\theta_n$$

If ι is a rational fraction of 2π the line will eventually return to its starting position (i.e. the field lines are closed); if not, it is said to be *ergodic*.

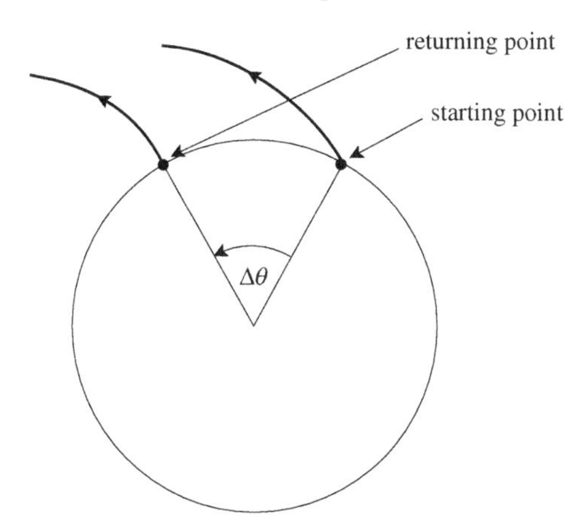

Fig. 4.14. Rotation of field lines in a torus.

The safety factor $q = 2\pi/\iota$ and is therefore equal to the average number of toroidal revolutions required to complete one poloidal revolution or

$$q = \lim_{N\to\infty} \frac{N_\mathrm{t}}{N_\mathrm{p}}$$

where N_t, N_p are the toroidal and poloidal winding numbers, respectively, and the limit is taken over an infinite number of revolutions. If $\Delta\phi$ is the average change in toroidal angle for one poloidal revolution $q = \Delta\phi/2\pi$ and since

$$\frac{R\mathrm{d}\phi}{\mathrm{d}s} = \frac{B_\mathrm{t}}{B_\mathrm{p}}$$

where ds is the distance moved in the poloidal plane, it follows that

$$q = \frac{1}{2\pi}\oint \frac{B_\mathrm{t}}{RB_\mathrm{p}}\,\mathrm{d}s \tag{4.34}$$

where the integral is over a single poloidal circuit around the flux surface. For a torus of circular cross-section and large aspect ratio $R_0 \gg a$, where R_0, a are the major and minor axes, respectively, we may put $R = R_0$ and $B_\mathrm{t} \approx B_\mathrm{t}(R_0)$ so that

$$q(r) = \frac{rB_\mathrm{t}}{R_0 B_\mathrm{p}} \tag{4.35}$$

in agreement with (4.29) for the cylindrical case.

Another representation for q can be obtained by considering the magnetic flux through an annulus between two neighbouring flux surfaces, as illustrated in Fig. 4.15. The poloidal flux through the annulus is

$$\mathrm{d}\psi = 2\pi RB_\mathrm{p}\,\mathrm{d}x \tag{4.36}$$

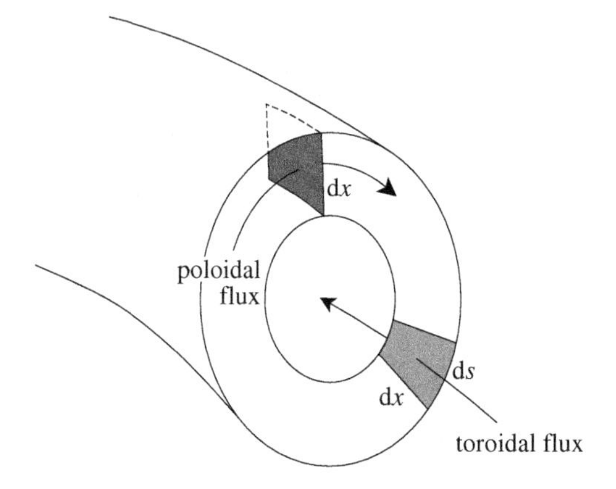

Fig. 4.15. Magnetic flux through an annulus between flux surfaces.

where $\mathrm{d}x$ is the separation at R between the flux surfaces. The toroidal flux is

$$\mathrm{d}\Phi = \oint \mathrm{d}s(B_{\mathrm{t}}\,\mathrm{d}x) \tag{4.37}$$

Substituting (4.36) in (4.34) and using (4.37) then gives

$$q = \frac{\mathrm{d}\Phi}{\mathrm{d}\psi}$$

expressing q as the rate of change of toroidal flux with poloidal flux.

To describe the equilibrium configuration of a plasma torus we use the cylindrical coordinates (R, ϕ, Z) shown in Fig. 4.16(a) and assume toroidal axisymmetry, i.e. there is no dependence on the azimuthal coordinate ϕ. It follows from $\mathbf{B} = \nabla \times \mathbf{A}$ that we may write

$$B_R = -\frac{1}{R}\frac{\partial \psi}{\partial Z} \qquad B_Z = \frac{1}{R}\frac{\partial \psi}{\partial R} \tag{4.38}$$

where $\psi = RA_\phi$ is called the *flux function*. We note that $(\mathbf{B} \cdot \nabla)\psi = 0$, on using (4.38), so that the magnetic surfaces are surfaces of constant ψ which may, therefore, be used to label them. In fact, $2\pi\psi$ is the total poloidal flux through a circle of radius R centred at the origin in the $Z = 0$ plane. Referring to Fig. 4.16(b) we have

$$\int \mathbf{B} \cdot \mathrm{d}\mathbf{S} = \int \nabla \times \mathbf{A} \cdot \mathrm{d}\mathbf{S} = \oint_{C(R)} \mathbf{A} \cdot \mathrm{d}\mathbf{l} = 2\pi R A_\phi = 2\pi\psi(R, 0)$$

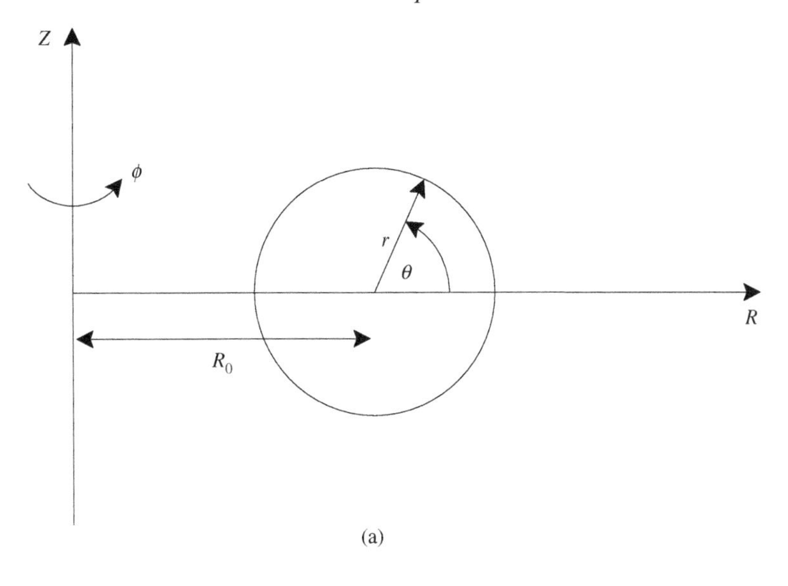

(a)

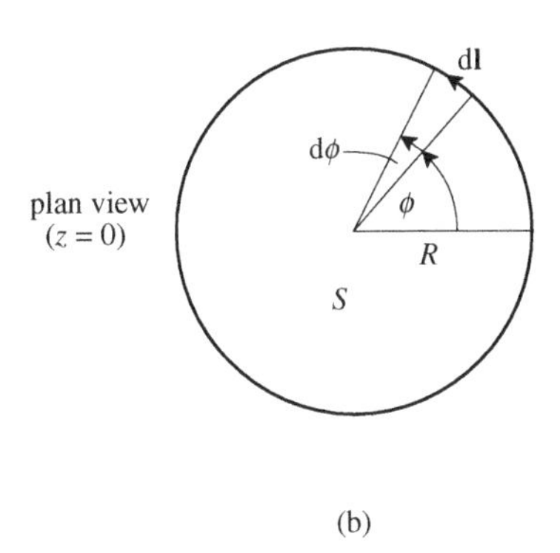

(b)

Fig. 4.16. Coordinate configuration of a torus in (a) side and (b) plan view.

where we have used Stokes' theorem to convert the surface integral to a line integral. Now substituting (4.12) in (4.10) and using (4.38) for B_R and B_Z we get

$$\mu_0 \frac{\partial P}{\partial R} + \frac{B_t}{R}\frac{\partial}{\partial R}(RB_t) + \frac{1}{R^2}\frac{\partial \psi}{\partial R}\Delta^* \psi = 0 \quad (4.39)$$

$$\frac{\partial \psi}{\partial R}\frac{\partial}{\partial Z}(RB_t) - \frac{\partial \psi}{\partial Z}\frac{\partial}{\partial R}(RB_t) = 0 \quad (4.40)$$

$$\mu_0 \frac{\partial P}{\partial Z} + B_t \frac{\partial B_t}{\partial Z} + \frac{1}{R^2}\frac{\partial \psi}{\partial Z}\Delta^* \psi = 0 \quad (4.41)$$

where

$$\Delta^*\psi = \frac{\partial^2\psi}{\partial R^2} - \frac{1}{R}\frac{\partial\psi}{\partial R} + \frac{\partial^2\psi}{\partial Z^2} \tag{4.42}$$

and we have identified B_ϕ as the toroidal magnetic field B_t.

Equation (4.40) may be written as $\nabla\psi \times \nabla(RB_t) = 0$ showing that surfaces of constant RB_t are also surfaces of constant ψ, i.e. $RB_t = F(\psi)$, for some function F. In fact, $F(\psi)$ is related to the total poloidal current I_p through the disc of radius R shown in Fig. 4.16(b). We have

$$I_p = \int \mathbf{j}\cdot d\mathbf{S} = \frac{1}{\mu_0}\int \nabla\times\mathbf{B}\cdot d\mathbf{S} = \frac{1}{\mu_0}\oint_{C(R)} \mathbf{B}\cdot d\mathbf{l} = \frac{2\pi RB_t}{\mu_0} \tag{4.43}$$

Thus, $F(\psi) = \mu_0 I_p(\psi)/2\pi$. Substituting this result in (4.39) and (4.41), multiplying by $\partial\psi/\partial Z$ and $\partial\psi/\partial R$ respectively, and subtracting gives

$$\frac{\partial P}{\partial R}\frac{\partial\psi}{\partial Z} - \frac{\partial P}{\partial Z}\frac{\partial\psi}{\partial R} = 0$$

i.e. $\nabla P\times\nabla\psi = 0$. Hence $P = P(\psi)$, a result we already know since the magnetic surfaces are isobaric surfaces.

Finally, since P and RB_t are functions of ψ we may write (4.39) as

$$\Delta^*\psi + FF' + \mu_0 R^2 P' = 0 \tag{4.44}$$

where the prime denotes differentiation with respect to ψ. This is the general equation for axisymmetric, toroidal equilibria and is known as the *Grad–Shafranov equation*. It is a non-linear partial differential equation, derived from the ideal MHD equations for static, toroidal equilibria with azimuthal symmetry ($\partial/\partial\phi \equiv 0$), for the flux function ψ which determines the poloidal magnetic field. It expresses the balance between plasma pressure gradient (third term) and the $\mathbf{j}\times\mathbf{B}$ contributions (first and second terms). Of the latter, the first term represents the toroidal current and poloidal field, which we identified as essential for toroidal stability, and the second term comes from the poloidal current and toroidal field determining radial stability. Indeed, it is easy to see that if we drop the first term in (4.44) we can integrate and by using (4.43) recover (4.19), the equation for radial equilibrium in a theta-pinch.

Solutions to the Grad–Shafranov equation provide a complete characterization of axisymmetric ideal MHD equilibria. The nature of the equilibrium configuration is determined by the choice of the two arbitrary functions $P(\psi)$ and $F(\psi)$, for the pressure and current profiles, together with the boundary conditions. Given P and F, (4.44) is solved as a boundary value problem to find the flux function $\psi(R, Z)$. The main difficulty lies in the fact that P and F are themselves functions of ψ, which is not known until (4.44) is solved. Although some progress

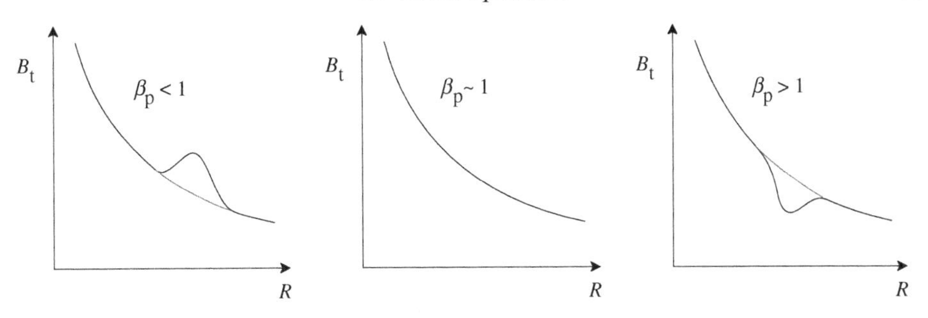

Fig. 4.17. Toroidal field variation with major radius in a tokamak.

can be made analytically either in closed form under certain restrictive conditions (see Exercise 4.4) or by expansion in terms of the inverse aspect ratio, a/R_0, the Grad–Shafranov equation generally has to be solved numerically (see Section 4.3.3).

Before discussing numerical solutions it is instructive to consider briefly three limiting cases within the general equation. By setting $FF'(\psi) = 0$ we discard the poloidal current in the plasma. The toroidal magnetic field behaves as R^{-1} and pressure balance is maintained solely by $\mathbf{j}_t \times \mathbf{B}_p$. For this case $\beta_p = 1$. On the other hand, $P'(\psi) = 0$ corresponds to the special case in which the magnetic field is force-free, in the sense to be described in Section 4.3.4, and there is no containment. When pressure balance is maintained predominantly by $\mathbf{j}_p \times \mathbf{B}_t$, the poloidal current term in the Grad–Shafranov equation $FF'(\psi) \gg |\Delta^* \psi|$ and the configuration is characterized by $\beta_p \gg 1$ (high beta tokamak). The behaviour of B_t as a function of the major radius of the tokamak is shown in Fig. 4.17 for different magnitudes of β_p.

Three examples of toroidal configurations are illustrated in Fig. 4.18. Since the compensating force required for toroidal equilibrium is usually much smaller than those involved in radial force balance one would expect $|B_p| \ll |B_t|$ as in the tokamak and screw pinch. In fact, for tokamaks $|B_t| > |B_p| R_0/a$ which means that a field line makes several transits around the torus before completing one spiral of the minor axis. The toroidal current flows mainly in the plasma column.

In the screw pinch, on the other hand, the current flows mainly in a sheath surrounding the plasma column. Likewise, the poloidal field does not penetrate the plasma and in the vacuum $|B_t| \sim |B_p| R_0/a$.

The reversed field pinch differs in that $|B_t| \sim |B_p|$ so that the field lines spiral many times around the magnetic axis in going once round the torus. One would not expect containment from such a high poloidal field because the associated large toroidal current is inimical to radial stability. However, as we shall discuss in Section 4.7.1, strong magnetic shear can act as a stabilizing mechanism to provide

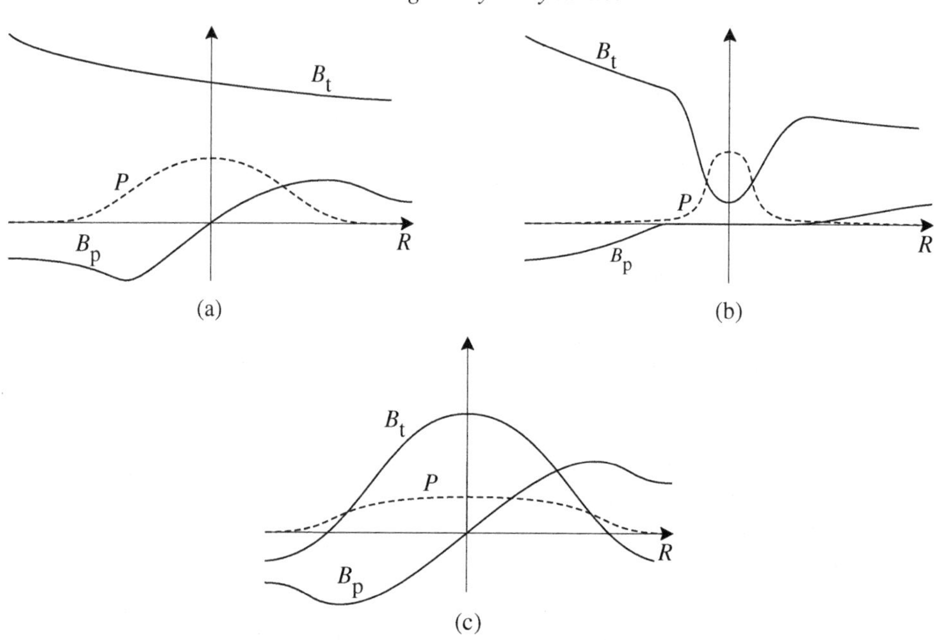

Fig. 4.18. Toroidal and poloidal fields in (a) tokamak, (b) screw pinch, (c) reversed field pinch.

a stable configuration in which the toroidal field reverses its direction in the outer part of the plasma column, known as the *reversed field pinch.*

4.3.3 Numerical solution of the Grad–Shafranov equation

Structurally, the Grad–Shafranov equation in the form $\Delta^* \psi = S(r, \psi)$, where S is a non-linear functional of ψ, is a second-order, non-linear elliptic partial differential equation. Apart from a number of special cases, two-dimensional axisymmetric equilibria have to be determined numerically. Equilibria are determined by the choice of $S(r, \psi)$, i.e. $P(\psi)$ and $F(\psi)$, along with specified boundary conditions. In the simplest case the plasma might be contained within a conducting shell at which ψ is specified; this corresponds to a fixed-boundary problem. If, on the other hand, a region of vacuum is present between the plasma and the conducting shell we have a free–boundary problem since neither the position nor the shape of the boundary are known. In practice both the position and shape of the plasma are determined by external coils carrying known currents. Alternatively, the inverse problem prescribes the boundary and then leaves the current distributions needed to provide this to be determined. Codes are used to solve the Grad–Shafranov equation for tokamak equilibria. Here we limit ourselves to solving a simple model

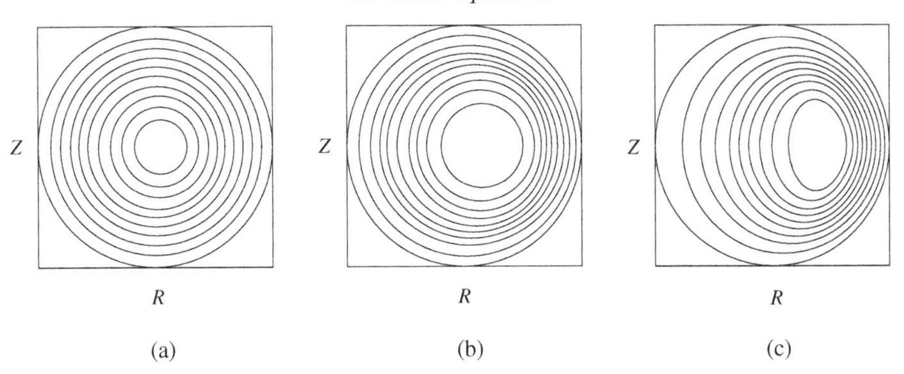

Fig. 4.19. Flux surfaces for tokamak equilibria for (a) $\beta_p < 1$, (b) $\beta_p \sim 1$, (c) $\beta_p > 1$.

problem of a plasma bounded by a perfectly conducting boundary with circular cross-section. By and large, techniques for solving elliptic equations use direct or iterative methods. The matrix equation resulting from a finite difference discretization in the (R, Z) plane, though large, is sparse and is readily solved iteratively (Press *et al.* (1989)).

To characterize particular tokamak equilibria means prescribing functional forms for $P(\psi)$ and $F(\psi)$. Sections of a cosine function were used to compute the flux surfaces shown in Fig. 4.19 for values of β_p corresponding to A^{-1}, A^0 and A where A is the aspect ratio R_0/a. For $\beta_p \sim A^{-1}$ the current and magnetic field are approximately collinear so that the entire plasma is very nearly force-free. The poloidal current $j_p \sim (B_p/B_t)j_t$ gives rise to an enhancement of B_t; $FF' < 0$ and the flux surfaces for this case are approximately concentric circles. At a later stage of the discharge the rise in plasma pressure means that $\beta_p \sim 1$ and a diamagnetic current now tends to annul the poloidal current with the result that the magnetic surfaces are displaced toward the outer wall of the tokamak as shown in Fig. 4.19(b).

Increasing β_p to values of the order of A and above results in flux surfaces that are displaced yet further outwards. This shift is known as the *Shafranov shift*. Flux surfaces are now compressed on the outside with a corresponding expansion on the inside. To obtain high-beta equilibria FF' must be large and positive. The poloidal current is reversed from the $\beta_p < 1$ case and is comparable in magnitude to the toroidal current. The B_t profile (Fig. 4.17) shows that a diamagnetic well is now created and this feature is largely responsible for radial pressure balance. While radial pressure balance for high-beta tokamaks is achieved largely by the poloidal diamagnetic currents induced in the plasma (as in a theta-pinch), the toroidal current provides toroidal force balance but is limited on account of the stability requirement $q \geq 1$ (see Section 4.5).

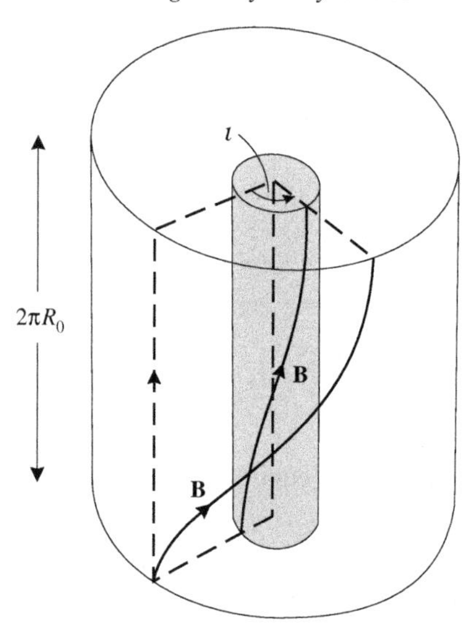

Fig. 4.20. Cylindrical flux tube with uniform twist.

4.3.4 Force-free fields and magnetic helicity

Although in the outer regions of stars both gas pressure and gravity may be negligible, the plasma can carry significant currents. If there is to be equilibrium in such regions it follows that the Lorentz force $\mathbf{j} \times \mathbf{B}$ must vanish also, i.e. the field must be *force-free*. This means that $\mathbf{j}$ and $\mathbf{B}$ are parallel so that we may write

$$\nabla \times \mathbf{B} = \alpha \mathbf{B} \tag{4.45}$$

where, in general, α will be spatially dependent. However, taking the divergence of (4.45) and using $\nabla \cdot \mathbf{B} = 0$, we see that α must obey

$$\mathbf{B} \cdot \nabla \alpha = 0$$

i.e. α is constant along a field line. The implication of (4.45) is that as one follows a particular field line the neighbouring field lines rotate in a constant manner about it.

A simple example of a force-free field can be obtained by starting with a cylindrical flux tube in which the field lines are initially all parallel to the axis, $\mathbf{B} = B_z(r)\hat{\mathbf{z}}$, and rotating one end of it through an angle $\iota = 2\pi R_0 \epsilon$, keeping the other end fixed, as shown in Fig. 4.20. Note that ι is the same for all r so that everywhere

$$\frac{B_\theta(r)}{B_z(r)} = r\frac{\mathrm{d}\theta}{\mathrm{d}z} = \frac{r\iota}{2\pi R_0} = \epsilon r \tag{4.46}$$

where ϵ is the uniform twist (rotation per unit length) of the field. Putting $P = 0$ in (4.18) and substituting for $B_\theta(r)$ from (4.46) gives

$$\frac{2\epsilon^2 r}{1+\epsilon^2 r^2} + \frac{1}{B_z}\frac{\mathrm{d}B_z}{\mathrm{d}r} = 0$$

which integrates to yield the result

$$B_z(r) = \frac{B_0}{1+\epsilon^2 r^2} \qquad B_\theta(r) = \frac{\epsilon r B_0}{1+\epsilon^2 r^2}$$

where B_0 is the value of B_z on the axis. It is easily verified from (4.45) that $\alpha(r) = 2\epsilon/(1+\epsilon^2 r^2)$ in this example.

An important theorem about force-free fields is that an isolated conducting fluid mass cannot have a field that is force-free everywhere. This follows from the virial theorem (see p. 84) on putting $\nabla P = 0$ and means that a force-free field must be anchored on a bounding surface; it cannot arise entirely from currents within a finite volume.

Further theorems relating to force-free fields in closed systems were proved by Woltjer (1958). He showed the invariance of the quantity

$$K = \int_V \mathbf{A}\cdot\mathbf{B}\,\mathrm{d}\tau$$

where $\mathbf{A}$ is the vector potential ($\mathbf{B} = \nabla\times\mathbf{A}$) with the integral taken over the whole volume of the closed system, and then used this to prove that force-free fields with constant α represent the state of minimum energy in a closed system. K, known as the *magnetic helicity*, provides a measure of the complexity of the magnetic field topology since it represents the interlinking of magnetic field lines. In ideal MHD a closed field line interlinking another with given connectivity maintains this connectivity through any plasma motion with the consequence that the topological properties of the field are preserved.

To establish the invariance of K we use (3.5) to get

$$\frac{\mathrm{d}K}{\mathrm{d}t} = \int_V \mathbf{A}\cdot\frac{\partial\mathbf{B}}{\partial t}\,\mathrm{d}\tau + \int_V \frac{\partial\mathbf{A}}{\partial t}\cdot\mathbf{B}\,\mathrm{d}\tau$$

Then substituting $\mathbf{B} = \nabla\times\mathbf{A}$ and using the expansion of $\nabla\cdot(\partial\mathbf{A}/\partial t\times\mathbf{A})$ we obtain

$$\frac{\mathrm{d}K}{\mathrm{d}t} = \int_V \left[\nabla\cdot\left(\frac{\partial\mathbf{A}}{\partial t}\times\mathbf{A}\right) + 2\frac{\partial\mathbf{A}}{\partial t}\cdot\nabla\times\mathbf{A}\right]\mathrm{d}\tau$$

From (4.4) we have

$$\frac{\partial\mathbf{A}}{\partial t} = \mathbf{u}\times\nabla\times\mathbf{A}$$

so the second term in the integral vanishes. Using Gauss' theorem the first term may be converted to the integral of $(\partial \mathbf{A}/\partial t \times \mathbf{A})$ over the surface of the closed system. However, $\partial \mathbf{A}/\partial t$ must vanish on this surface since $\mathbf{A}$ is continuous across the surface and the motions inside a closed system cannot affect the vector potential outside the system. This establishes the invariance of K.

Next we examine the stationary values of the magnetic energy

$$W_{\mathrm{B}} = \int_V (B^2/2\mu_0)\mathrm{d}\tau$$

subject to the condition that K is constant, i.e. we take the variation of $W_{\mathrm{B}} - \frac{1}{2}\alpha_0 K$, obtaining

$$\delta\left(W_{\mathrm{B}} - \frac{1}{2}\alpha_0 K\right) = \frac{1}{\mu_0}\int_V \left[\mathbf{B}\cdot\delta\mathbf{B} - \frac{\alpha_0}{2}(\delta\mathbf{A}\cdot\mathbf{B} + \mathbf{A}\cdot\delta\mathbf{B})\right]\mathrm{d}\tau$$

Substituting $\delta\mathbf{B} = \boldsymbol{\nabla}\times\delta\mathbf{A}$ and again using the expansion of the divergence of a vector product this becomes

$$\begin{aligned}
&\delta\left(W_{\mathrm{B}} - \frac{1}{2}\alpha_0 K\right) \\
&\quad = \frac{1}{\mu_0}\int_V \left[\boldsymbol{\nabla}\cdot\left(\delta\mathbf{A}\times\mathbf{B} + \frac{\alpha_0}{2}\mathbf{A}\times\delta\mathbf{A}\right) + (\boldsymbol{\nabla}\times\mathbf{B} - \alpha_0\mathbf{B})\cdot\delta\mathbf{A}\right]\mathrm{d}\tau \\
&\quad = \frac{1}{\mu_0}\int_V (\boldsymbol{\nabla}\times\mathbf{B} - \alpha_0\mathbf{B})\cdot\delta\mathbf{A}\,\mathrm{d}\tau
\end{aligned}$$

on using Gauss' theorem to convert the first term to a surface integral which again vanishes since $\delta\mathbf{A}$ is zero on the boundary. Thus, for arbitrary choice of $\delta\mathbf{A}$ there is an extremum if and only if

$$\boldsymbol{\nabla}\times\mathbf{B} = \alpha_0\mathbf{B} \tag{4.47}$$

in which α_0 is a constant.

Because of their importance in astrophysics much effort has been put into finding such fields. Taking the curl of (4.47) gives the *Helmholtz equation*

$$(\nabla^2 + \alpha_0^2)\mathbf{B} = 0 \tag{4.48}$$

the solutions of which are well-known. However, it must be remembered that this increases the differential order of the equation and not all solutions of (4.48) satisfy (4.47).

Another approach to the investigation of force-free fields uses the Clebsch variable representation

$$\mathbf{B} = \boldsymbol{\nabla}\alpha\times\boldsymbol{\nabla}\beta \tag{4.49}$$

introduced in Section 2.9. It is clear from this representation that (4.49) satisfies $\nabla \cdot \mathbf{B} = 0$. Also, since

$$\mathbf{B} \cdot \nabla\alpha = 0 = \mathbf{B} \cdot \nabla\beta \tag{4.50}$$

it follows that α and β are constant on each field line and, therefore, may be used to label each field line. Substitution of (4.45) in (4.50) and using (4.49) for $\mathbf{B}$ gives

$$[\nabla \times (\nabla\alpha \times \nabla\beta)] \cdot \nabla\alpha = 0 = [\nabla \times (\nabla\alpha \times \nabla\beta)] \cdot \nabla\beta \tag{4.51}$$

as coupled differential equations to be solved for α and β.

Only a limited range of solutions of such non-linear equations may be found by analytical methods. However, numerical solutions may be generated using variational techniques. This is discussed in Sturrock (1994) in which the Clebsch variable representation is used to show that $\delta W = 0$ leads to force-free field configurations, provided α and β are constant on the bounding surface (see Exercise 4.5).

4.4 Solar MHD equilibria

Magnetic fields play a key role in solar physics ranging from their creation through dynamo action to their role in sunspot formation and in dramatic, if transient, phenomena such as solar flares. As a consequence, many aspects of solar physics are governed by magnetohydrodynamics. The plasma beta serves as an index of the relative importance of magnetic effects. In this section we make use of a simple flux tube model, developed by Parker (1955), to gain insights into aspects of solar MHD equilibria. Parker's flux tube model is particularly useful in view of the subsequent realization that virtually all the magnetic flux extruding from the surface of the Sun is concentrated into isolated flux tubes or bundles of these.

The long-held view that the background magnetic field at the Sun's surface was weak was undermined by high resolution observations that uncovered a hierarchy of magnetic structures. While the mean field over large regions of the surface of the Sun is indeed no more than ~ 0.5 mT, these observations showed that the magnetic flux through the surface, far from being uniform, is concentrated into flux tubes with intensities typically a few hundred times the mean field, over diameters of a few hundred kilometres. This localization of flux is not what one might expect intuitively. The region across which the flux tube bursts through the surface is known as a *magnetic knot* and was first identified by Beckers and Schröter (1968) from high resolution H_α pictures. They estimated that around 90% of the flux in active regions is accounted for by flux tubes appearing at magnetic knots. Beyond the appearance of knots the picture is yet more complicated, with knots attracting one another to form aggregates of flux tubes which in turn break up or, more rarely, go on to develop into sunspots, extending across regions with scale lengths typically hundreds of times that of a magnetic knot. These observations led to a reappraisal of the solar magnetic field. The picture now appears to be one in which magnetic flux penetrating the surface of the Sun is concentrated into intense flux tubes distributed over the surface.

Of the many questions raised by these observations perhaps the most puzzling is why the Sun's magnetic field should appear as intense isolated flux tubes surrounded by field-free zones, a configuration that is to say the least counter-intuitive. Conventionally one might picture the magnetic field spreading to fill the whole of space allowed by the tension B^2/μ_0 along the lines of force. However Parker (1955) pointed out that allowing for the solar gravitational field results in *magnetic buoyancy* and this buoyancy is responsible for isolating the magnetic field into individual flux tubes.

4.4.1 Magnetic buoyancy

Consider the magnetohydrostatic equilibrium of a flux tube deep in the photosphere (see Fig. 4.21). Following Parker, we assume that the flux tube is slender, in the sense that its diameter is small compared with scale lengths characterizing variations along the tube, as for example the radius of curvature. Allowing for the Sun's gravitational field, the magnetohydrostatic condition is now expressed as

$$\frac{1}{\mu_0}(\nabla \times \mathbf{B}) \times \mathbf{B} - \nabla P - \rho \nabla \psi = 0 \tag{4.52}$$

where ψ is the gravitational potential. For simplicity we assume that initially the flux tube is horizontal and aligned in the x-direction. The fluid pressure within the flux tube, $P_{\text{int}}(x, z)$, is governed by a purely hydrostatic equilibrium. If we neglect the field from any neighbouring flux tubes pressure balance across the flux tube is simply expressed by equating the external pressure P_{ext} to the total internal pressure

$$P_{\text{ext}}(x, z) = P_{\text{int}}(x, z) + \frac{B^2(x, z)}{2\mu_0} \tag{4.53}$$

Thus the field strength is determined by the ambient fluid pressure which in turn is a function of the gravitational potential ψ, so that $B = B(\psi)$. Assuming the fluid is governed by the equation of state for an ideal gas $P = \rho k_{\text{B}} T/m$, where ρ is the mass density and m the particle mass, and the temperature is uniform, the external density ρ_e must exceed the internal density ρ_i. There is, therefore, a buoyancy force $A(\rho_e - \rho_i)\nabla\psi$ per unit length where A denotes the cross-section of the flux tube. The magnitude of this force $\mathcal{F}$ per unit length is then

$$\mathcal{F} = \frac{B^2}{2\mu_0 \Lambda} \tag{4.54}$$

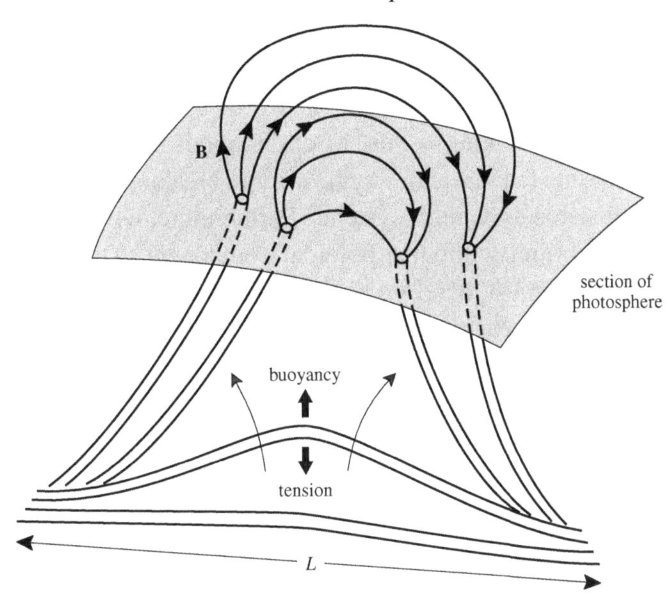

Fig. 4.21. Buoyancy of flux tubes from the convective zone to the surface of the Sun.

where $\Lambda \equiv k_{\mathrm{B}}T/mg$ is the local scale height, and the flux tube will rise through the photosphere in response to it. In general a flux tube rising in this way will not remain horizontal (see Fig. 4.21) and once some curvature develops the buoyancy will be countered by magnetic tension along the field lines. Provided the length of the flux tube $L > 2\Lambda$ magnetic buoyancy forces continue to act. Parker also showed that raising a section of a long flux tube will result in a flow of plasma from that section and so serve to enhance buoyancy.

The simplicity of the magnetic buoyancy concept set alongside evidence that the background magnetic field of the Sun appears as a distribution of isolated flux tubes is appealing though the model needs modification to allow for realities such as turbulence in the convective zone and the fact that solar rotation distorts flux tubes. At a deeper level one might question where flux tubes come from in the first place. There are no grounds for supposing that the magnetic fields generated by dynamo action would produce long flux tubes. It would seem that some mechanism, possibly an instability, is needed to cause the fields generated deep in the convective zone to fragment and concentrate. Further questions are posed by observations of the apparent mutual attraction of magnetic knots and their coalescence leading to the formation of sunspots. Many issues affecting magnetic buoyancy are discussed in detail by Parker (1979), Priest (1987) and Hughes (1991).

4.5 Stability of ideal MHD equilibria

Having discussed various plasma equilibria we now turn to a consideration of their stability. The most striking feature of observations of Z-pinch dynamics is a tendency for the plasma to twist and wriggle prior to breaking up. Z-pinches appear to be inherently unstable dynamical systems. Furthermore, we have seen that there is no toroidal equilibrium without a poloidal component of the magnetic field. Toroidal configurations must therefore have a non-zero toroidal current which may then act as a source of free energy for the development of instabilities. The question of stability is of vital importance for plasma containment and for explaining natural phenomena like solar flares, sunspots and prominences.

The terms *stable, unstable* describe the behaviour of dynamical systems in equilibrium towards small perturbations of the system. If a perturbation causes forces to act on the system tending to restore it to its equilibrium configuration, the system is said to be in *stable equilibrium* (with respect to the class of perturbations considered). If, on the other hand, the system tends to depart further and further from the equilibrium configuration as a result of the perturbation, it is in *unstable equilibrium.*

In general, plasma instabilities may be broadly categorized as macroscopic or microscopic. The first class involves the physical (spatial) displacement of plasma and may be discussed within the framework of the MHD equations. Microscopic instabilities need to be described on the basis of kinetic theory since they arise from changes in the velocity distribution functions and this information is lost in the MHD description. Although microscopic instabilities can be very important, usually they are less catastrophic than MHD instabilities and so the latter are normally one's first concern. Likewise, *ideal* MHD stability may be regarded as a first step towards MHD stability because the introduction of dissipation allows slippage between fluid and field which usually facilitates instability; if this is so we may expect that an ideal MHD stability condition is necessary but not always sufficient for maintaining the equilibrium.

In this section the ideal MHD stability of some static configurations is discussed. The investigation of ideal MHD stability can be approached in a number of ways. We emphasize at the outset that we shall confine our attention to *linear* stability analyses. Within a linear framework stability considerations can be approached from the point of view of an initial value problem or alternatively, from a normal mode perspective. The first determines the evolution in time of a prescribed initial perturbation and in so doing provides more information than is needed to answer the question of stability. The normal mode approach leads to an eigenvalue equation. Since in practice most stability problems can only be resolved numerically, the normal mode route generally offers advantages over solving the initial value problem. Here we shall describe both methods, starting with the initial value problem.

Since we are concerned with small departures from equilibrium we can apply perturbation theory to the ideal MHD equations in Table 3.2. We write

$$\left.\begin{aligned} \rho(\mathbf{r},t) &= \rho_0(\mathbf{r}) + \rho_1(\mathbf{r},t) \\ \mathbf{u}(\mathbf{r},t) &= \mathbf{u}_0(\mathbf{r}) + \mathbf{u}_1(\mathbf{r},t) = \mathbf{u}_1(\mathbf{r},t) \\ P(\mathbf{r},t) &= P_0(\mathbf{r}) + P_1(\mathbf{r},t) \\ \mathbf{B}(\mathbf{r},t) &= \mathbf{B}_0(\mathbf{r}) + \mathbf{B}_1(\mathbf{r},t) \end{aligned}\right\} \tag{4.55}$$

where the subscripts 0 and 1 denote equilibrium and perturbation values, respectively, and we have put $\mathbf{u}_0 = 0$ since the equilibrium is static. Ignoring products of the perturbations gives, to zero order, the equilibrium equation

$$\mu_0 \nabla P_0 = (\nabla \times \mathbf{B}_0) \times \mathbf{B}_0 \tag{4.56}$$

and, to first order,

$$\frac{\partial \rho_1}{\partial t} = -\mathbf{u}_1 \cdot \nabla \rho_0 - \rho_0 \nabla \cdot \mathbf{u}_1 \tag{4.57}$$

$$\rho_0 \frac{\partial \mathbf{u}_1}{\partial t} = -\nabla P_1 + \frac{1}{\mu_0}(\nabla \times \mathbf{B}_0) \times \mathbf{B}_1 + \frac{1}{\mu_0}(\nabla \times \mathbf{B}_1) \times \mathbf{B}_0 \tag{4.58}$$

$$\frac{\partial \mathbf{B}_1}{\partial t} = \nabla \times (\mathbf{u}_1 \times \mathbf{B}_0) \tag{4.59}$$

$$\frac{\partial P_1}{\partial t} = -\mathbf{u}_1 \cdot \nabla P_0 - \gamma P_0 \nabla \cdot \mathbf{u}_1 \tag{4.60}$$

In deriving (4.60) we have used (4.57) to simplify the final expression.

Now since $\mathbf{u}_1(\mathbf{r}, t)$ is the only time-dependent variable on the right-hand sides of (4.57), (4.59) and (4.60) we may integrate them partially with respect to time. It is convenient to choose initial conditions such that the constants of integration $\rho_1(\mathbf{r}, 0)$, $\mathbf{B}_1(\mathbf{r}, 0)$ and $P_1(\mathbf{r}, 0)$ are all zero; this simply means that we start at the equilibrium configuration and disturb it dynamically by means of a non-zero $\mathbf{u}_1(\mathbf{r}, 0)$. The integrated equations are

$$\rho_1(\mathbf{r},t) = -\boldsymbol{\xi}(\mathbf{r},t) \cdot \nabla \rho_0 - \rho_0 \nabla \cdot \boldsymbol{\xi}(\mathbf{r},t) \tag{4.61}$$

$$\mathbf{B}_1(\mathbf{r},t) = \nabla \times (\boldsymbol{\xi}(\mathbf{r},t) \times \mathbf{B}_0(\mathbf{r})) \tag{4.62}$$

$$P_1(\mathbf{r},t) = -\boldsymbol{\xi}(\mathbf{r},t) \cdot \nabla P_0(\mathbf{r}) - \gamma P_0(\mathbf{r}) \nabla \cdot \boldsymbol{\xi}(\mathbf{r},t) \tag{4.63}$$

where the *displacement vector*

$$\boldsymbol{\xi}(\mathbf{r},t) \equiv \int_0^t \mathbf{u}_1(\mathbf{r},t')\mathrm{d}t' \tag{4.64}$$

From (4.64)

$$\frac{\partial \boldsymbol{\xi}}{\partial t} = \mathbf{u}_1(\mathbf{r}, t) \tag{4.65}$$

so that the initial conditions on $\boldsymbol{\xi}$ are

$$\boldsymbol{\xi}(\mathbf{r}, 0) = 0 \qquad \partial \boldsymbol{\xi}(\mathbf{r}, 0)/\partial t = \mathbf{u}(\mathbf{r}, 0) \neq 0 \tag{4.66}$$

We have written the integral of (4.57) for future reference but it may be noted that ρ_1 no longer appears in the remaining equations (4.58)–(4.60) which form a closed set. Thus, substituting (4.62) and (4.63) in (4.58) gives

$$\rho_0 \frac{\partial^2 \boldsymbol{\xi}}{\partial t^2} = \mathbf{F}(\boldsymbol{\xi}(\mathbf{r}, t)) \tag{4.67}$$

where

$$\begin{aligned} \mathbf{F}(\boldsymbol{\xi}) \quad = \quad & \nabla(\boldsymbol{\xi} \cdot \nabla P_0 + \gamma P_0 \nabla \cdot \boldsymbol{\xi}) + \frac{1}{\mu_0}(\nabla \times \mathbf{B}_0) \times [\nabla \times (\boldsymbol{\xi} \times \mathbf{B}_0)] \\ & + \frac{1}{\mu_0}\{[\nabla \times \nabla \times (\boldsymbol{\xi} \times \mathbf{B}_0)] \times \mathbf{B}_0\} \end{aligned} \tag{4.68}$$

The equilibrium configuration defines ρ_0, P_0, and $\mathbf{B}_0$ so that (4.67), together with appropriate boundary conditions and the initial values (4.66), determines the displacement vector $\boldsymbol{\xi}$ and hence the time evolution of ρ_1, $\mathbf{B}_1$, P_1 and $\mathbf{u}_1$. This is the initial value solution of the linear stability problem.

Finding the normal mode solution is less onerous. We now assume that we may separate the space and time dependence of the displacement, $\boldsymbol{\xi}(\mathbf{r}, t) = \boldsymbol{\xi}(\mathbf{r})T(t)$, so that (4.67) becomes

$$\begin{aligned} \ddot{T} &= -\omega^2 T \\ -\omega^2 \rho_0 \boldsymbol{\xi}(\mathbf{r}) &= \mathbf{F}[\boldsymbol{\xi}(\mathbf{r})] \end{aligned} \tag{4.69}$$

where the separation constant is chosen as $-\omega^2$ so that $T(t) = e^{i\omega t}$ and $\boldsymbol{\xi}(\mathbf{r}, t) = \boldsymbol{\xi}(\mathbf{r})e^{i\omega t}$. Since $\mathbf{F}(\boldsymbol{\xi})$ is linear in $\boldsymbol{\xi}$, (4.69) represents an eigenvalue problem in which the boundary conditions determine the possible values of ω^2. If these are discrete and labelled with suffix n, the general solution of (4.67) is

$$\boldsymbol{\xi}(\mathbf{r}, t) = \sum_n \boldsymbol{\xi}_n(\mathbf{r})e^{i\omega_n t} \tag{4.70}$$

where $\boldsymbol{\xi}_n(\mathbf{r})$ is the *normal mode* corresponding to the *normal frequency* ω_n. One can show from the properties of $\mathbf{F}(\boldsymbol{\xi})$ that for any discrete normal mode the eigenvalue ω_n^2 is real (see Exercise 4.7). It is then clear from (4.70) that if, for all the normal frequencies, $\omega_n^2 > 0$ then all the modes are periodic. This means that the system oscillates about the equilibrium position. On the other hand, if at least one of the normal frequencies is such that $\omega_n^2 < 0$ the corresponding normal mode will grow exponentially and the equilibrium configuration is unstable.

This property of real eigenvalues is directly related to the conservation of energy in ideal MHD. In principle, by suitable choice of initial perturbation, one could excite each normal mode in turn. The mode can either oscillate about the equilibrium position or the perturbation can grow continuously as the potential energy of an unstable equilibrium position is converted to kinetic energy. Damped or growing oscillations are not possible since these would require an energy sink or source. Thus, ω_n is either real or pure imaginary and ω_n^2 is real. This leads ultimately to the most elegant and efficient method of investigating stability, the *energy principle*. It is also important for the analysis of *neutral* or *marginal stability*. In general the transition from stability to instability takes place at $\Im\omega = 0$ and $\Re\omega$ must be calculated. In ideal MHD, however, stability boundaries are defined by $\omega = 0$ which makes their determination much easier.

4.5.1 Stability of a cylindrical plasma column

By way of illustration we apply the normal mode analysis to determine the stability of a cylindrical plasma of length L and circular cross-section of radius a. We assume that the equilibrium fields in the plasma and the surrounding vacuum are

$$\mathbf{B}_0 = (0, 0, B_0) \tag{4.71}$$

and

$$\tilde{\mathbf{B}}_0 = (0, B_\theta(r), B_z) \tag{4.72}$$

respectively, where B_0 and B_z are constants and $B_\theta(r)$ is the azimuthal field due to the current flowing along the plasma column. Substituting (4.71) in (4.12) gives $\mathbf{j} = 0$ everywhere except at the edge of the plasma column and, if the total 'skin' current is I, it follows that

$$B_\theta(r) = \frac{\mu_0 I}{2\pi r} \tag{4.73}$$

Also, since $\mathbf{j}(r) = 0$ $(r < a)$, from (4.10) we see that P_0 is constant so that (4.69) reduces to

$$-\rho_0\omega^2\boldsymbol{\xi}(\mathbf{r}) = \gamma P_0\boldsymbol{\nabla}(\boldsymbol{\nabla}\cdot\boldsymbol{\xi}) + \frac{1}{\mu_0}(\boldsymbol{\nabla}\times\mathbf{B}_1)\times\mathbf{B}_0 \tag{4.74}$$

where, for computational convenience, we have re-introduced $\mathbf{B}_1$ using (4.62).

We are interested in perturbations with poloidal and axial periodicity so we set

$$\boldsymbol{\xi}(\mathbf{r}) = [\xi_r(r), \xi_\theta(r), \xi_z(r)]e^{i(m\theta+kz)} \tag{4.75}$$

where m and $(kL/2\pi)$ take integer values. Also, to lighten the algebra we assume that

$$\nabla \cdot \boldsymbol{\xi} = 0 \tag{4.76}$$

which infers that the plasma is incompressible. It is shown in the next section that allowing for compressibility makes the plasma more rather than less stable so that for considerations of stability our assumption errs on the side of caution.

The procedure is to solve (4.74) within the plasma column, together with the field equations in the vacuum surrounding the plasma and to apply the boundary conditions across the plasma–vacuum interface. It is this last step that determines the set of values of ω (the normal frequencies ω_n) for which (4.70) is an acceptable solution.

Starting with the plasma interior, it is easily seen that, on using (4.75) and (4.76), (4.62) becomes

$$\mathbf{B}_1 = ikB_0\boldsymbol{\xi} \tag{4.77}$$

and, since P_0 is constant, (4.63) gives

$$P_1 = 0 \tag{4.78}$$

On taking the divergence of (4.74) only the last term contributes giving

$$\nabla \cdot [(\nabla \times \mathbf{B}_1) \times \mathbf{B}_0] = \mathbf{B}_0 \cdot (\nabla \times \nabla \times \mathbf{B}_1) = -\mathbf{B}_0 \cdot \nabla^2\mathbf{B}_1 = 0$$

which, on using (4.77), becomes

$$\nabla^2 \xi_z = 0$$

In cylindrical polar coordinates this is the Bessel equation

$$\left[\frac{\mathrm{d}^2}{\mathrm{d}r^2} + \frac{1}{r}\frac{\mathrm{d}}{\mathrm{d}r} - \left(k^2 + \frac{m^2}{r^2}\right)\right]\xi_z(r) = 0$$

for which the solution having no singularity at $r = 0$ is

$$\xi_z(r) = \xi_z(a)\frac{I_m(kr)}{I_m(ka)} \tag{4.79}$$

where I_m is the modified Bessel function of the first kind of order m.

Next, we use (4.77) and the radial component of (4.74) to obtain

$$\xi_r(r) = -\frac{ikB_0^2}{(k^2B_0^2 - \mu_0\rho\omega^2)}\frac{\mathrm{d}\xi_z}{\mathrm{d}r} = -\frac{ik^2B_0^2\xi_z(a)}{(k^2B_0^2 - \mu_0\rho\omega^2)}\frac{I'_m(kr)}{I_m(ka)} \tag{4.80}$$

and, although we shall not require it explicitly, we could likewise find $\xi_\theta(r)$. The field perturbations in the plasma column are then given by (4.77).

In the vacuum, since $\mathbf{j} = 0$, we may represent the field perturbation by a scalar potential

$$\tilde{\mathbf{B}}_1 = \nabla\phi \tag{4.81}$$

where ϕ takes the form $\phi(r)e^{i(m\theta+kz)}$. Then $\nabla\cdot\tilde{\mathbf{B}}_1 = 0$ shows that $\phi(r)$ satisfies the same Bessel equation as $\xi_z(r)$ but here we must choose the solution which vanishes at infinity giving

$$\phi(r) = \phi(a)\frac{K_m(kr)}{K_m(ka)} \tag{4.82}$$

where K_m is the modified Bessel function of the second kind of order m.

Finally, we apply the ideal MHD boundary conditions (3.73) and (3.74) at the plasma–vacuum interface. The conditions are valid, of course, in both the equilibrium and the perturbed configurations so that zero-order terms cancel each other out and only terms linear in perturbed quantities need be retained. Nevertheless, this procedure requires some care because linear terms arise from the displacement of the interface as well as directly from the perturbations. Denoting the equilibrium position of a point on the interface by $\mathbf{r}_0$, its displacement is

$$\mathbf{r} - \mathbf{r}_0 = \int_0^t \mathbf{u}_1(\mathbf{r}_0, t')\mathrm{d}t' \tag{4.83}$$

Comparing this with (4.64) and assuming that $\mathbf{r} - \mathbf{r}_0$ remains a small quantity it is easily seen that to first order, $\mathbf{r} - \mathbf{r}_0 = \boldsymbol{\xi}(\mathbf{r}_0, t) \simeq \boldsymbol{\xi}(\mathbf{r}, t)$. The subtle difference between $\boldsymbol{\xi}(\mathbf{r}, t)$ and $\boldsymbol{\xi}(\mathbf{r}_0, t)$, which is of second order, is that (4.83) describes the displacement of a fluid element (labelled by $\mathbf{r}_0$) in a Lagrangian coordinate system moving with the fluid, whereas (4.64) defines a displacement vector from a fixed point $\mathbf{r}$ in an Eulerian coordinate system relative to a fixed inertial frame. Thus, we use the expansion

$$\begin{aligned} f(\mathbf{r}, t) &= f_0(\mathbf{r}, t) + f_1(\mathbf{r}, t) \\ &\simeq f_0(\mathbf{r}_0) + \boldsymbol{\xi}\cdot\nabla f_0 + f_1(\mathbf{r}, t) \end{aligned}$$

for each quantity appearing in the boundary conditions.

Keeping only first-order terms (3.73) gives

$$P_1 + \frac{\mathbf{B}_0\cdot\mathbf{B}_1}{\mu_0} = \frac{\tilde{\mathbf{B}}_0\cdot\tilde{\mathbf{B}}_1}{\mu_0} + (\boldsymbol{\xi}\cdot\nabla)\frac{\tilde{B}_0^2}{2\mu_0}$$

which, on using (4.71)–(4.73), (4.77), (4.78) and (4.81), becomes

$$ikB_0^2\xi_z(a) = \left[ikB_z + \frac{im}{a}B_\theta(a)\right]\phi(a) - \frac{B_\theta^2(a)}{a}\xi_r(a) \tag{4.84}$$

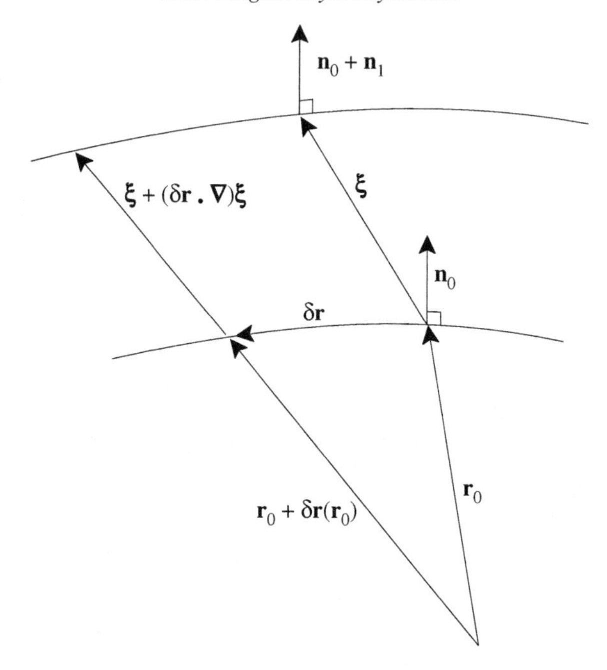

Fig. 4.22. Interface perturbations.

Linearizing (3.74) we have

$$\mathbf{B}_1 \cdot \mathbf{n}_0 + \mathbf{B}_0 \cdot \mathbf{n}_1 = 0 \tag{4.85}$$

and

$$[\tilde{\mathbf{B}}_1 + (\boldsymbol{\xi} \cdot \boldsymbol{\nabla})\tilde{\mathbf{B}}_0] \cdot \mathbf{n}_0 + \tilde{\mathbf{B}}_0 \cdot \mathbf{n}_1 = 0 \tag{4.86}$$

where $\mathbf{n}_0 = \hat{\mathbf{r}}$, the radial unit vector, and $\mathbf{n}_1$ is the perturbation in $\mathbf{n}_0$ due to the migration of the interface. To find $\mathbf{n}_1$ we note that any infinitesimal displacement $\delta\mathbf{r}(\mathbf{r}_0)$ from a point $\mathbf{r}_0$ on the interface will remain on the interface provided $\delta\mathbf{r} \cdot \mathbf{n}_0 = 0$. Applying this condition on the perturbed interface we have, with reference to Fig. 4.22,

$$[\delta\mathbf{r} + (\delta\mathbf{r} \cdot \boldsymbol{\nabla})\boldsymbol{\xi}] \cdot (\mathbf{n}_0 + \mathbf{n}_1) = (\delta\mathbf{r} \cdot \boldsymbol{\nabla})\xi_r + \delta\mathbf{r} \cdot \mathbf{n}_1 = 0$$

and, since $\delta\mathbf{r}$ may be chosen arbitrarily, this gives

$$\mathbf{n}_1 = -\boldsymbol{\nabla}\xi_r$$

Substituting this in the boundary conditions, (4.85) is satisfied identically and (4.86) becomes

$$\left(\frac{\mathrm{d}\phi}{\mathrm{d}r}\right)_{r=a} - ikB_z\xi_r(a) - \frac{im}{a}B_\theta(a)\xi_r(a) = 0$$

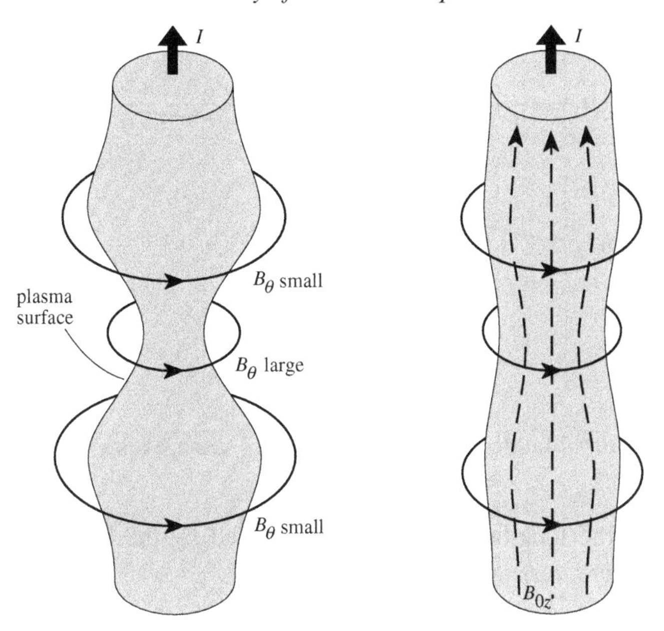

Fig. 4.23. Sausage instability.

Now we use (4.80) and (4.82) to substitute for $\xi_r(a)$ and $\phi'(a)$ and (4.84) to eliminate the constant $\phi(a)$ obtaining the dispersion relation

$$\frac{\omega^2}{k^2} = \frac{B_0^2}{\mu_0 \rho} - \frac{(B_z + mB_\theta(a)/ka)^2}{\mu_0 \rho} \frac{I_m'(ka)K_m(ka)}{I_m(ka)K_m'(ka)} - \frac{B_\theta^2(a)}{\mu_0 \rho} \frac{I_m'(ka)}{kaI_m(ka)} \tag{4.87}$$

Each term on the right-hand side of this equation is real confirming that ω can be either real or pure imaginary, i.e. there are no solutions corresponding to damped or growing oscillations. Instability occurs if $\omega^2 < 0$ and, since $K_m' < 0$ while I_m, K_m and I_m' are all positive, this requires the third term to be larger in magnitude than the sum of the first two.

The $m = 0$ case illustrates the roles of the equilibrium magnetic fields. Both the internal and external longitudinal fields, B_0 and B_z respectively, enhance stability while B_θ has the opposite effect. This may be explained physically with reference to Fig. 4.23, which shows an axially symmetric perturbation of the equilibrium configuration of the Z-pinch. Since $B_\theta \propto r^{-1}$, the external magnetic pressure on the plasma surface is increased where the perturbation squeezes the plasma into a neck and is decreased where the perturbation fattens the plasma into a bulge. This gradient in the external magnetic field causes the perturbation to grow and, without an internal field, the necks contract to the axis, giving a sausage-like appearance to the plasma; hence the name *sausage instability*. On the other hand, the magnetic

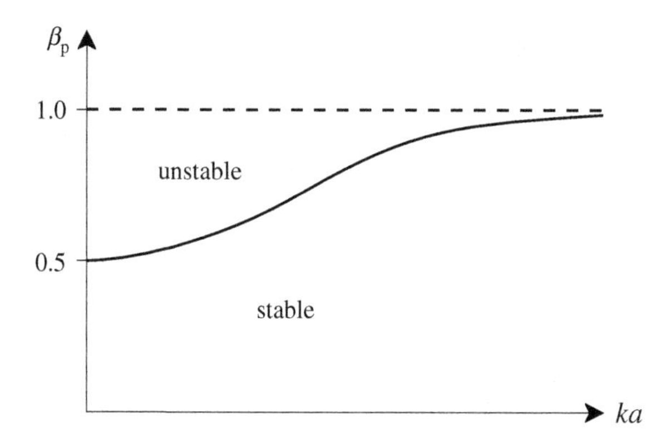

Fig. 4.24. Marginal stability curve for sausage instability.

pressure due to the longitudinal fields is increased by the plasma perturbations which are, therefore, resisted.

In the case that $B_z = 0$ the condition for stability is

$$B_0^2 > B_\theta^2(a)\frac{I_0'(ka)}{kaI_0(ka)} \tag{4.88}$$

and since $I_0'(x)/xI_0(x) < 1/2$ for all x there is stability for all k provided $B_0^2 > B_\theta^2(a)/2$. However, for given plasma pressure P_0 and current I, B_0 is bounded above by the equilibrium pressure condition

$$P_0 + B_0^2/2\mu_0 = B_\theta^2/2\mu_0 \tag{4.89}$$

It is, therefore, somewhat more useful to write the stability condition in terms of the 'poloidal' $\beta_p = 2\mu_0 P_0/B_\theta^2(a)$; see (4.32). Combining (4.88) and (4.89) we have

$$\beta_p = \frac{2\mu_0 P_0}{B_\theta^2} < 1 - \frac{I_0'(ka)}{kaI_0(ka)} \sim \begin{cases} 1/2 & ka \to 0 \\ 1 - 1/ka & ka \to \infty \end{cases}$$

as the marginal stability condition illustrated in Fig. 4.24.

For $m > 0$ it is convenient to assume $|ka| \ll 1$ so that we may use the approximations

$$I_m(x) \approx \frac{(x/2)^m}{m!} \qquad K_m \approx \frac{(m-1)!}{2}\left(\frac{x}{2}\right)^{-m} \qquad (|x| \ll 1)$$

to simplify the dispersion relation (4.87) which then reduces to

$$\mu_0\rho\omega^2 = k^2B_0^2 + \left[kB_z + \frac{m}{a}B_\theta(a)\right]^2 - \frac{m}{a^2}B_\theta^2(a) \tag{4.90}$$

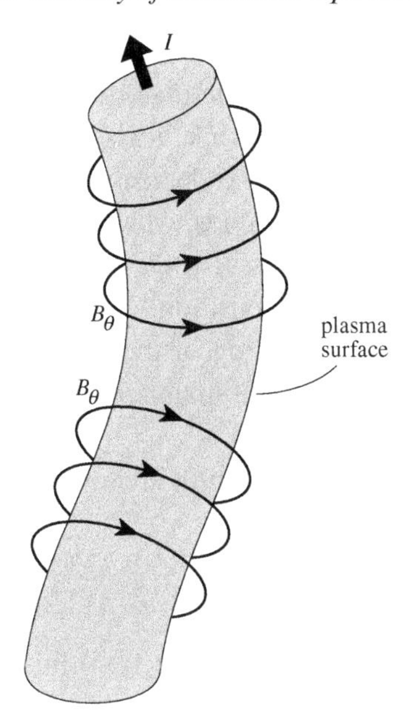

Fig. 4.25. Kink instability.

Differentiating ω^2 with respect to k it is easily verified that the minimum value of ω^2 occurs at

$$k = -\frac{m B_z B_\theta(a)}{a(B_0^2 + B_z^2)} \tag{4.91}$$

and is given by

$$\omega_{\min}^2 = \frac{B_\theta^2(a)}{\mu_0 \rho a^2}\left(\frac{m^2 B_0^2}{B_0^2 + B_z^2} - m\right) \tag{4.92}$$

Using the equation of equilibrium pressure balance

$$P_0 + \frac{B_0^2}{2\mu_0} = \frac{B_z^2}{2\mu_0} + \frac{B_\theta^2(a)}{2\mu_0} \tag{4.93}$$

and (4.32) this may be written in terms of the ratio of the external field components $B_\theta(a)/B_z$ and the 'toroidal' $\beta_t = 2\mu_0 P_0/B_z^2$ as

$$\omega_{\min}^2 = \frac{m B_\theta^2(a)}{\mu_0 \rho a^2}\left[\frac{(m-1)(1 + B_\theta^2(a)/B_z^2 - \beta_t) - 1}{(2 + B_\theta^2(a)/B_z^2 - \beta_t)}\right]$$

showing that for low β_t plasmas in devices with $|B_\theta/B_z| \ll 1$ only the $m = 1$ and $m = 2$ modes can become unstable.

The unstable $m = 1$ mode is known as the *kink instability* since it arises from the perturbation shown in Fig. 4.25. The distortion grows because the magnetic pressure on the concave side of the kink is increased (the B_θ lines are closer together), while that on the convex side is decreased (the B_θ lines are further apart). Again, the action of the longitudinal fields, whether internal or external, enhances stability since the tension in the lines of force caused by their stretching tries to restore the pinch to its equilibrium position. The shorter the wavelength of the perturbation the greater is the stretching of the field lines. The balance between the stabilizing z-components and destabilizing θ-component of the magnetic field leads to an important stability condition.

To investigate this we put $m = 1$ in (4.90) and use (4.93) to write it in the form

$$\omega^2 = \frac{k^2 B_z^2}{\mu_0 \rho} \left[2 \left(1 + \frac{B_\theta(a)}{kaB_z} \right) + \frac{B_\theta^2(a)}{B_z^2} - \beta_t \right]$$

For $|B_\theta(a)/B_z| \ll 1$ and $\beta_t \ll 1$, it follows from this equation that the $m = 1$ mode is stable for $|B_\theta/B_z| < |ka|$ and since $|k|$ cannot be less than $2\pi/L$, where L is the length of the plasma column, the stability condition is

$$|B_\theta/B_z| < 2\pi a/L$$

In a toroidal device, where $L = 2\pi R_0$, this may be written in terms of the safety factor (4.29) as

$$q(a) = aB_z(a)/R_0 B_\theta(a) > 1 \tag{4.94}$$

a result known as the *Kruskal–Shafranov stability criterion*. It says that the ratio of toroidal to poloidal magnetic field must exceed the aspect ratio R_0/a. In (4.94) we have implicitly extrapolated the Kruskal–Shafranov condition to the case of a diffuse pinch with variable B_z rather than the sharp-edged plasma which was the subject of our calculation. In fact, the same result can be obtained by a simple physical argument due to Johnson *et al.* (1958). We consider an $m = 1$ helical perturbation of the plasma such as that shown in Fig. 4.25 and we examine the displacement of a field line by inspecting two cross-sections one-quarter wavelength apart as shown in Fig. 4.26. If the displacement vector $\boldsymbol{\xi}$ is horizontal at the first cross-section it will be vertical at the second and if the pitch of $\mathbf{B}_0$ is such that the angle of rotation $\theta_0 > \pi/2$ then the vertical (downwards) displacement of the field line due to the perturbation means that the angle of rotation has increased. This in turn implies that the perturbation has increased B_θ relative to B_z, thereby perturbing the magnetic pressure balance such as to accelerate the downward displacement of the plasma. It is easy to see that if $\theta_0 < \pi/2$ the angle of rotation is decreased by the perturbation, so B_θ is decreased relative to B_z and growth of the perturbation

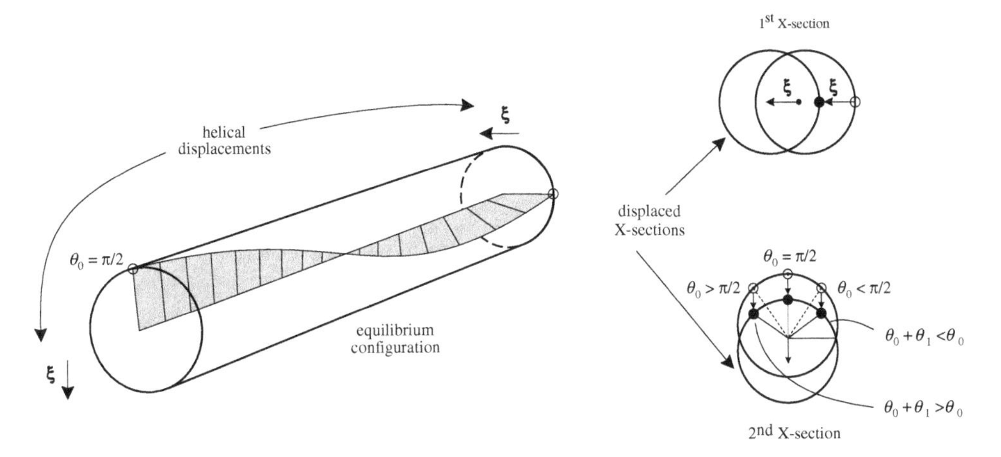

Fig. 4.26. Illustration of Kruskal–Shafranov stability criterion for the kink instability. The equilibrium configuration on the left shows the $\pi/2$ rotation of a magnetic field line on traversing a quarter wavelength. The cross-sections on the right show the result of imposing a kink displacement $\boldsymbol{\xi}$ which is initially horizontal but subsequently vertical due to the $\pi/2$ rotation.

is resisted. The stability condition is therefore $\iota/4 < \pi/2$ at the boundary of the plasma and from (4.29) this gives (4.94).

The Kruskal–Shafranov criterion, by restricting $B_\theta(a)$, sets a limit on the toroidal current that may be safely driven through the plasma. It is for this reason that a tokamak has a small ratio of poloidal to toroidal field component.

Tokamaks are also low β devices. This second restriction is related to the unfavourable curvature of the poloidal field B_θ with respect to the plasma. Whenever the external field (the field containing the plasma) is concave towards the plasma, any ripple on the plasma will tend to grow for the same reasons that the sausage and kink perturbations grow. Containment of these so-called *ballooning instabilities* restricts β, as discussed later in Section 4.7.2. Conversely, containing fields which are convex towards the plasma tend to smooth out ripple perturbations.

4.6 The energy principle

So far, in discussing the stability of equilibria, we have progressed from an initial value problem to a normal mode analysis, this being made possible by the linearization which led to (4.67) in which only $\boldsymbol{\xi}(\mathbf{r}, t)$ is time dependent. The role of the initial values is merely to determine the mixture of the normal modes in a particular solution and this is of secondary importance when one's main interest is stability; in these circumstances the 'convenient' choice of initial conditions in the integrals (4.61)–(4.63) is not a serious loss of generality. Nevertheless, a normal mode analysis still involves significant effort as we have seen from our relatively simple example of a sharp-edged, cylindrical plasma. Investigating the stability of more realistic plasmas and geometries normally is only possible by numerical analysis. If, however, one merely wants to answer the question 'Is the equilibrium configuration stable?' there is an even more direct approach based on energy considerations.

Since the ideal MHD equations are dissipationless they conserve energy and a stable equilibrium configuration must correspond to a minimum in the potential energy W. This is the physical basis of the *energy principle* which states that if there exists a displacement $\boldsymbol{\xi}$ for which the change in potential energy $\delta W < 0$ the equilibrium is unstable.

To find an expression for δW we note that since the equilibrium is static the (change in) kinetic energy K is $\frac{1}{2}\rho_0\dot{\xi}^2$ integrated over the whole plasma volume

$$\begin{aligned} K(\dot{\boldsymbol{\xi}}, \dot{\boldsymbol{\xi}}) &= \frac{1}{2}\int \rho_0 \dot{\boldsymbol{\xi}} \cdot \dot{\boldsymbol{\xi}}\,\mathrm{d}\mathbf{r} = -\frac{\omega^2}{2}\int \rho_0 \boldsymbol{\xi} \cdot \boldsymbol{\xi}\,\mathrm{d}\mathbf{r} \\ &= \frac{1}{2}\int \boldsymbol{\xi} \cdot \mathbf{F}(\boldsymbol{\xi})\,\mathrm{d}\mathbf{r} \end{aligned} \tag{4.95}$$

by (4.69). Thus, by conservation of energy

$$\delta W(\boldsymbol{\xi}, \boldsymbol{\xi}) = -\frac{1}{2}\int \boldsymbol{\xi} \cdot \mathbf{F}(\boldsymbol{\xi})\,\mathrm{d}\mathbf{r} \tag{4.96}$$

From (4.95) and (4.96) we may write

$$\omega^2 = \delta W(\boldsymbol{\xi}, \boldsymbol{\xi})/K(\boldsymbol{\xi}, \boldsymbol{\xi}) \tag{4.97}$$

which is the variational formulation of the linear stability problem. Given that the operator $\mathbf{F}(\boldsymbol{\xi})$ is self-adjoint†, that is

$$\int \boldsymbol{\eta} \cdot \mathbf{F}(\boldsymbol{\xi})\,\mathrm{d}\mathbf{r} = \int \boldsymbol{\xi} \cdot \mathbf{F}(\boldsymbol{\eta})\,\mathrm{d}\mathbf{r} \tag{4.98}$$

for any allowable displacement vectors $\boldsymbol{\xi}$ and $\boldsymbol{\eta}$, it is easy to show (see Exercise 4.7) that any $\boldsymbol{\xi}$ for which ω^2 is an extremum is an eigenfunction of (4.69) with eigenvalue ω^2. This establishes the equivalence of the variational principle and the normal mode analysis. But in practice it is analytically and computationally much easier to investigate stability via the variational principle. One chooses a trial function $\boldsymbol{\xi} = \sum_n a_n \psi_n$, where the ψ_n are a suitable set of basis functions, and minimizes δW with respect to the coefficients a_n subject to the normalization condition

$$K(\boldsymbol{\xi}, \boldsymbol{\xi}) = \text{const.} \tag{4.99}$$

† The direct proof of (4.98) is rather lengthy (see Freidberg (1987)).

If $\delta W < 0$ the equilibrium is unstable and the variational principle guarantees that a lower bound for the growth rate γ of the instability is $(-\delta W/K)^{1/2}$.

The energy principle goes a stage further in that one is not restricted to the normalization condition (4.99). Often, great analytical simplification is achieved by choosing some other normalization condition. Minimization of δW with the result $\delta W < 0$ then indicates instability but information on the growth rate is lost since K is unknown. Each step from initial value problem through normal mode analysis and variational principle to energy principle brings analytical and computational simplification at the expense of detailed knowledge, from full solution of the evolution of a linear perturbation to mere determination of the stability of the equilibrium. Since MHD instabilities tend to be the fastest growing and the most catastrophic (bulk movement of the plasma) the stability question is usually all one needs answer. Furthermore, δW may be written in a form that gives good physical insight into the cause of instability. The energy principle is, therefore, within the limits of ideal MHD, very effective and widely used.

Returning to (4.96), with $\mathbf{F}(\boldsymbol{\xi})$ given by (4.68), it is straightforward, if tedious, to recast δW in a more useful and illuminating form; the details are left as an exercise (see Exercise 4.8). The objective is to express δW as the sum of three terms representing the changes in potential energy within the plasma (δW_{P}), the surface (δW_{S}) and the vacuum (δW_{V}). Using vector identities one expresses the integrand $\boldsymbol{\xi} \cdot \mathbf{F}(\boldsymbol{\xi})$ as a sum of divergence terms and scalar functions. Then, using Gauss' theorem, the integral of the divergence terms is converted to an integral over the surface of the plasma. In the surface integral the boundary condition (3.73) is used, thereby introducing the vacuum magnetic field. From a practical point of view this is a most important step because the boundary condition is now incorporated in the energy principle and there is no need to find trial functions obeying the boundary condition, which is a cumbersome constraint on the use of the energy principle in its original form. Next, boundary condition (3.73) is used to eliminate the tangential component of $\nabla(P + B^2/2\mu_0)$ in the surface integral; since (3.73) holds all over the surface the tangential component of the gradient must be continuous and hence

$$[\mathbf{n}_0 \times \nabla(P + B^2/2\mu_0)]_1^2 = 0 \tag{4.100}$$

Finally, using Gauss' theorem again, part of the surface integral is converted to a volume integral over the vacuum. The result is

$$\delta W = \delta W_{\mathrm{P}} + \delta W_{\mathrm{S}} + \delta W_{\mathrm{V}} \tag{4.101}$$

where

$$\delta W_{\mathrm{P}} = \frac{1}{2}\int [B_1^2/\mu_0 - \boldsymbol{\xi}\cdot(\mathbf{j}_0\times\mathbf{B}_1) - P_1(\boldsymbol{\nabla}\cdot\boldsymbol{\xi})]\,\mathrm{d}\mathbf{r} \tag{4.102}$$

$$\delta W_{\mathrm{S}} = \frac{1}{2}\int (\boldsymbol{\xi}\cdot\mathbf{n}_0)^2[\boldsymbol{\nabla}(P_0 + B_0^2/2\mu_0)]_1^2\cdot\mathrm{d}\mathbf{S} \tag{4.103}$$

$$\delta W_{\mathrm{V}} = \int (\tilde{B}_1^2/2\mu_0)\,\mathrm{d}\mathbf{r} \tag{4.104}$$

and the integrals are taken over the plasma, surface, and vacuum, respectively.

The three terms in δW_{P} are, respectively, the increase in magnetic energy, the work done against the perturbed $\mathbf{j}\times\mathbf{B}$ force and the change in internal energy due to the compression (or expansion) of the plasma. The surface energy δW_{S} is the work done by displacing the boundary. If there is no surface current the total pressure gradient is continuous across the boundary and $\delta W_{\mathrm{S}} = 0$; likewise, $\delta W_{\mathrm{S}} = 0$ if the boundary is fixed ($\boldsymbol{\xi}\cdot\mathbf{n}_0 = 0$). The vacuum contribution δW_{V} is simply the increase in the energy of the vacuum field. This also vanishes if $\boldsymbol{\xi}\cdot\mathbf{n}_0 = 0$ since there is no perturbation of the vacuum field. For fixed boundary problems, therefore, $\delta W = \delta W_{\mathrm{P}}$. Instabilities may be classified as *internal* (fixed boundary) or *external* (free boundary) modes.

After further manipulation (see Exercise 4.8) δW_{P} can be expressed in the form

$$\begin{aligned}\delta W_{\mathrm{P}} = {} & \frac{1}{2}\int [B_{1\perp}^2/\mu_0 + (B_0^2/\mu_0)(\boldsymbol{\nabla}\cdot\boldsymbol{\xi}_\perp + 2\boldsymbol{\xi}_\perp\cdot\boldsymbol{\kappa})^2 + \gamma P_0(\boldsymbol{\nabla}\cdot\boldsymbol{\xi})^2 \\ & - 2(\boldsymbol{\xi}_\perp\cdot\boldsymbol{\nabla}P_0)(\boldsymbol{\kappa}\cdot\boldsymbol{\xi}_\perp) - j_\parallel(\boldsymbol{\xi}_\perp\times\mathbf{b})\cdot\mathbf{B}_{1\perp}]\mathrm{d}\mathbf{r}\end{aligned} \tag{4.105}$$

where $\boldsymbol{\kappa} = (\mathbf{b}\cdot\boldsymbol{\nabla})\mathbf{b}$ is the curvature of the equilibrium magnetic field $\mathbf{B}_0 = B_0\mathbf{b}$ and vector quantities have been separated into parallel and perpendicular components relative to $\mathbf{b}$, i.e. $\mathbf{X} = X_\parallel\mathbf{b} + \mathbf{X}_\perp$. The first three (positive) terms in the integral represent the potential energy associated with the shear Alfvén wave, the compressional Alfvén wave, and the sound wave, respectively. We show in Section 4.8 that these are the three natural wave modes supported by an ideal MHD plasma. It is also clear from this form of δW_{P} that compressibility ($\boldsymbol{\nabla}\cdot\boldsymbol{\xi} \neq 0$) is stabilizing, so it is often assumed for simplicity that the plasma is incompressible on the understanding that this is a 'worst case' assumption and any necessary correction is favourable to stability. The only possible destabilizing terms are the last two. The instabilities arising from these are said to be *pressure-driven* and *current-driven*, respectively, although, since $\boldsymbol{\nabla}P_0 = \mathbf{j}_{0\perp}\times\mathbf{B}_0$, both types are driven by the energy in the current but by different components. This distinction is also used to classify ideal MHD instabilities. The kink instability is an example of the (parallel) current-driven kind so instabilities in this class are known as *kink instabilities*. The pressure-driven modes are called *interchange instabilities* for reasons that will become clear when we discuss them in Section 4.7.

4.6.1 Finite element analysis of ideal MHD stability

Practical determination of ideal MHD stability on the basis of the energy principle has to be done computationally. Whereas codes that use finite difference methods have to ensure that the energy-conserving properties of MHD equilibria reflected in the self-adjointness of the operator $\mathbf{F}(\boldsymbol{\xi})$ are preserved by the difference scheme, an alternative approach, the *finite element method* (FEM), has an advantage in that it appeals directly to $\mathbf{F}(\boldsymbol{\xi})$, thus ensuring that energy conservation is built into the numerics. The FEM method was developed for the stress analysis of structures and was first applied to problems of MHD stability independently by Takeda *et al.* (1972) and by Boyd, Gardner and Gardner (1973), who analysed the stability of a cylindrical tokamak. Subsequently, the approach was generalized and FEM codes were developed to describe toroidal configurations.

Normalized growth rates for the helical $m = 2$ mode in a plama column with free boundary were determined and compared with values obtained by Shafranov (1970) from an analysis valid for small values of the axial wavenumber k. The vacuum region is defined by $a < r < b$. Figure 4.27 shows the computed normalized growth rate as a function of $nq(a)$ for $k = 0.2$ and three values of the ratio a/b. The

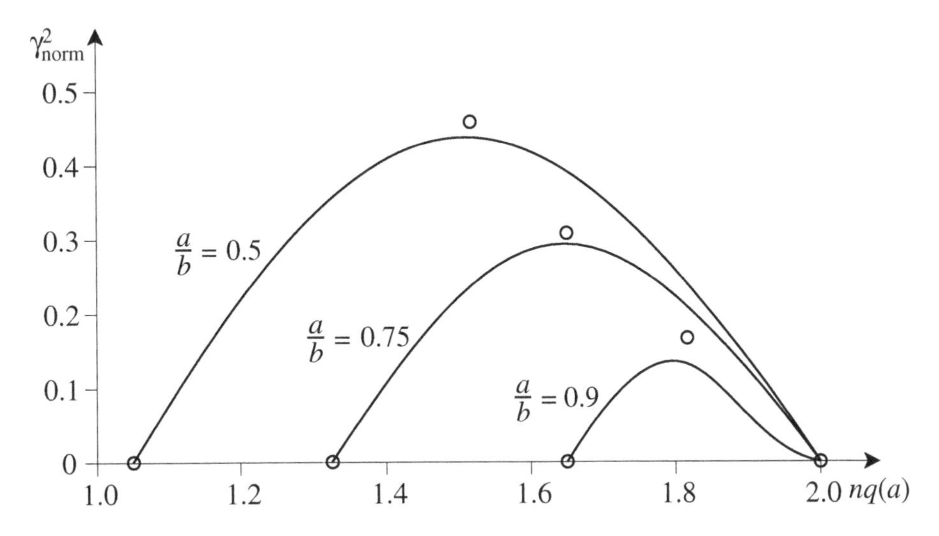

Fig. 4.27. Square of normalized growth rates versus $nq(a)$ for helical $m = 2$ mode in plasma column with free boundary. The circles indicate Shafranov's analytical results (after Boyd, Gardner and Gardner (1973)).

range of the instability coincides exactly with the Shafranov range (see Biskamp (1993))

$$m - 1 + \left(\frac{a}{b}\right)^{2m} < nq(a) < m \tag{4.106}$$

The behaviour of growth rates as a function of $nq(a)$ shows good agreement with Shafranov's theoretical result.

4.7 Interchange instabilities

Since interchange instabilities are pressure-driven they are essentially hydrodynamic. As an example consider a case where a fluid of density ρ_1 lies in a horizontal layer over one of density ρ_2 as shown in Fig. 4.28(a). Now let us perturb the equilibrium by rippling the boundary layer. We may think of the ripple arising from the interchange of neighbouring fluid elements as suggested by Fig. 4.28(b). The fluid element of density ρ_1 has moved downwards with consequent loss of gravitational potential energy while the opposite is true of the fluid element of density ρ_2. Clearly, the net change in potential energy δW has the same sign as $(\rho_2 - \rho_1)$ and it follows that the equilibrium is stable if and only if $\rho_2 \geq \rho_1$. This comes as no surprise; it is intuitively obvious that the only stable equilibrium is to have the denser fluid supporting the less dense. The instability that arises when $\rho_2 < \rho_1$ is called the *Rayleigh–Taylor instability*. In a non-ideal fluid the increase in surface tension due to the stretching of the boundary provides a stabilizing effect and prevents the growth of the instability for perturbations with wavelengths below a certain critical value.

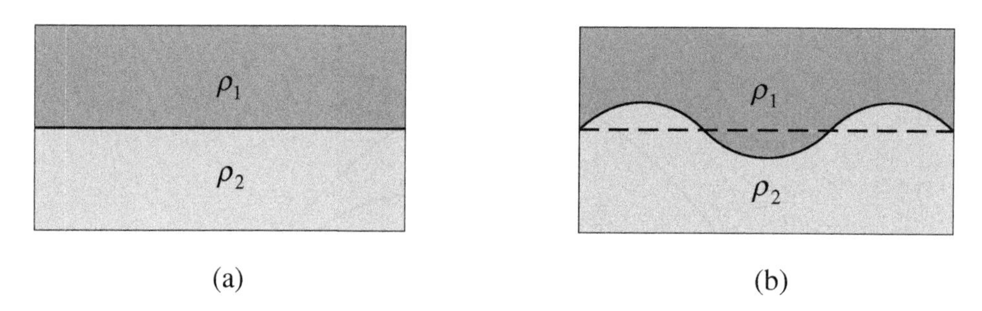

Fig. 4.28. Rippled boundary layer between fluids of differing densities.

4.7.1 Rayleigh–Taylor instability

Kruskal and Schwarzschild (1954) showed that an MHD analogue of the Rayleigh–Taylor instability arises when a plasma is supported against gravity by a magnetic

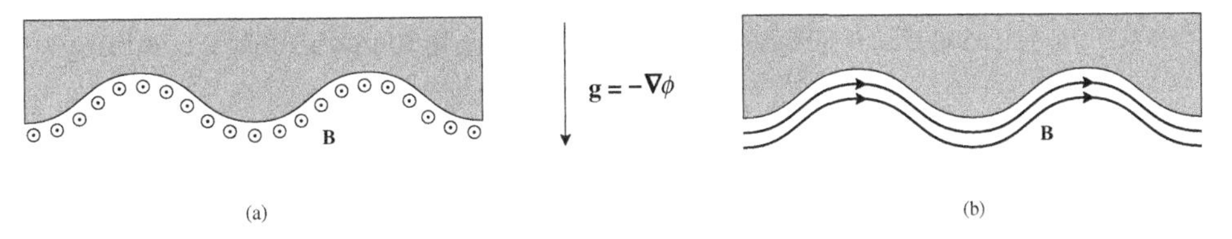

Fig. 4.29. Boundary perturbations for field lines (a) perpendicular and (b) parallel to the perturbation.

field. A simple illustration of the origin of the instability can be constructed by supposing that the field lines are straight and perpendicular to the perturbation, as in Fig. 4.29(a). We may think of the perturbation as brought about by the interchange of flux tubes and elongated fluid elements with no change in magnetic energy since there is no bending of the field lines. The fluid elements, on the other hand, have lost potential energy; hence, the perturbations will grow and the equilibrium is unstable.

Although gravity is of no significance in laboratory plasmas any other acceleration of the plasma may take on the role of gravity. For example we saw in Section 2.4.2 that particles moving in curved magnetic fields feel a centrifugal force which acts like an equivalent gravitational force.

To discuss the instability quantitatively we consider the gravitational case in plane geometry. This minimizes the algebra without losing the essential physics. The first point to note is that had we taken the field lines parallel to the perturbation in our simple illustration they would have been bent by the perturbation as indicated in Fig. 4.29(b), thereby increasing the magnetic energy and providing a stabilizing effect akin to that of surface tension in the hydrodynamic case. Clearly, perturbations parallel to the field maximize this stabilizing effect but one cannot assume perturbations will not arise in a direction perpendicular to the field for which there is no stabilizing effect. However, by introducing magnetic shear we can make sure that for any given mode with propagation vector $\mathbf{k}$ this least stable condition, $\mathbf{k}\cdot\mathbf{B} = 0$, is restricted to specific layers and does not occur throughout the plasma. Thus, in our analysis we want to allow for arbitrary orientation of $\mathbf{k}$ to $\mathbf{B}$ and vertical variation of the direction of $\mathbf{B}$. Without loss of generality, we may choose coordinates such that the y axis is vertical and the z axis is parallel to $\mathbf{k}$, the direction of propagation of the mode under investigation. Then the equilibrium magnetic field $\mathbf{B}_0(y) = [B_x(y), 0, B_z(y)]$ and the gravitational acceleration $\mathbf{g} = (0, -g, 0)$.

With these preliminaries we now proceed to a normal mode analysis of the MHD Rayleigh–Taylor instability in which the perturbation

$$\boldsymbol{\xi}(\mathbf{r}, t) = [0, \xi_y(y), \xi_z(y)]e^{i(kz-\omega t)}$$

and, for the reasons discussed earlier, we impose the incompressibility condition

$$\nabla \cdot \boldsymbol{\xi} = \frac{\mathrm{d}\xi_y}{\mathrm{d}y} + ik\xi_z = 0 \tag{4.107}$$

The dispersion relation is obtained from (4.67) but the inclusion of the gravitational force $\rho\mathbf{g}$ in the equation of motion leads to an extra term $\rho_1\mathbf{g}$ in $\mathbf{F}(\boldsymbol{\xi})$ which, on using (4.61) and (4.107), becomes

$$\begin{aligned} \mathbf{F}(\boldsymbol{\xi}) &= \nabla(\boldsymbol{\xi} \cdot \nabla P_0) - \boldsymbol{\xi} \cdot \nabla \rho_0 \mathbf{g} + \frac{1}{\mu_0}(\nabla \times \mathbf{B}_0) \times [\nabla \times (\boldsymbol{\xi} \times \mathbf{B}_0)] \\ &\quad + \frac{1}{\mu_0}\{[\nabla \times \nabla \times (\boldsymbol{\xi} \times \mathbf{B}_0)] \times \mathbf{B}_0\} \end{aligned} \tag{4.108}$$

Remembering that equilibrium variables vary only with y, we find for the y and z components of (4.67)

$$\begin{aligned} \rho_0\omega^2\xi_y &= \frac{\mathrm{d}}{\mathrm{d}y}(\xi_y P_0') - g\xi_y\rho_0' + \frac{k^2 B_z^2}{\mu_0}\xi_y + \frac{ikB_z'}{\mu_0}(\mathbf{B}_0 \cdot \boldsymbol{\xi}) \\ &\quad + \frac{1}{\mu_0}(ikB_z\mathbf{B}_0 \cdot \boldsymbol{\xi}' - \mathbf{B}_0 \cdot \mathbf{B}_0'\xi_y' - \mathbf{B}_0 \cdot \mathbf{B}_0''\xi_y \\ &\quad + ikB_z\mathbf{B}_0' \cdot \boldsymbol{\xi} - \mathbf{B}_0' \cdot \mathbf{B}_0'\xi_y) \end{aligned} \tag{4.109}$$

$$\rho_0\omega^2\xi_z = ik\xi_y P_0' + \frac{k^2 B_z^2}{\mu_0}\xi_z - \frac{k^2 B_z}{\mu_0}(\mathbf{B}_0 \cdot \boldsymbol{\xi}) - \frac{ik\xi_y}{\mu_0}(\mathbf{B}_0 \cdot \mathbf{B}_0') \tag{4.110}$$

Substituting for $\xi_z P_0'$ from (4.110) and for ξ_z from (4.107) then gives

$$\frac{\mathrm{d}}{\mathrm{d}y}\left[\left(\rho_0\omega^2 - \frac{(\mathbf{k} \cdot \mathbf{B}_0)^2}{\mu_0}\right)\frac{\mathrm{d}\xi_y}{\mathrm{d}y}\right] - k^2\left(\rho_0\omega^2 - \frac{(\mathbf{k} \cdot \mathbf{B}_0)^2}{\mu_0}\right)\xi_y - k^2 g\frac{\mathrm{d}\rho_0}{\mathrm{d}y}\xi_y = 0 \tag{4.111}$$

This differential equation contains all the information we need to discuss the Rayleigh–Taylor instability. For example, the hydrodynamic case is obtained by putting $\mathbf{B}_0 = 0$ and assuming $\rho_0' = 0$ except at $y = 0$, this being the boundary between the two fluids, labelled 1 for $y > 0$ and 2 for $y < 0$. Then in both fluids (4.111) is

$$\xi_y'' - k^2\xi_y = 0$$

with solutions

$$\xi_y(y) = \begin{cases} \xi_y(0)e^{-ky} & y > 0 \\ \xi_y(0)e^{ky} & y < 0 \end{cases}$$

Now, integrating (4.111) over the interval $-\epsilon < y < +\epsilon$ and letting $\epsilon \to 0$ we get

$$-(\rho_1 + \rho_2)k\omega^2\xi_y(0) - k^2 g(\rho_1 - \rho_2)\xi_y(0) = 0$$

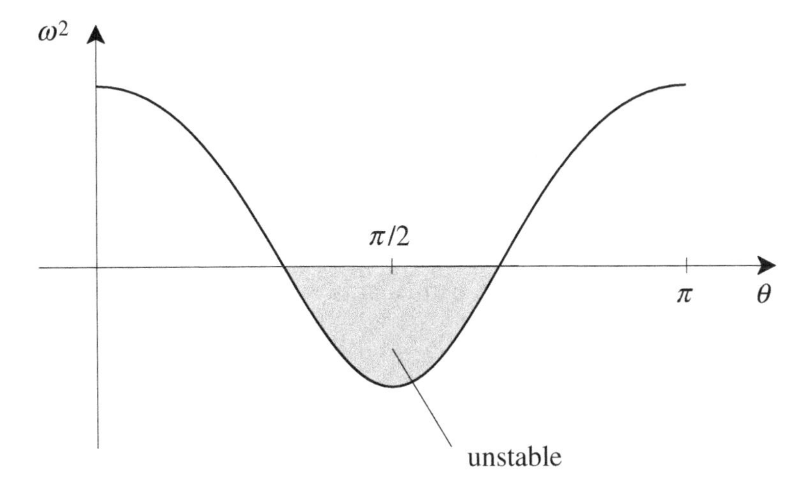

Fig. 4.30. Schematic illustration of Rayleigh–Taylor instability zone for uniform magnetic field.

giving

$$\omega^2 = -\frac{kg(\rho_1 - \rho_2)}{\rho_1 + \rho_2}$$

and confirming that the equilibrium is unstable if the denser fluid is above the less dense, i.e. $\rho_1 > \rho_2$. Clearly, the fastest growth rate $(kg)^{1/2}$ occurs when $\rho_1 \gg \rho_2$.

We next demonstrate the stabilizing effect of a magnetic field by supposing that the fluids are plasmas and that there is a uniform $\mathbf{B}_0$ throughout. From (4.111) we see that this simply replaces $\rho_0\omega^2$ by $\rho_0\omega^2 - (\mathbf{k}\cdot\mathbf{B}_0)^2/\mu_0$ to give

$$\omega^2 = -\frac{kg(\rho_1 - \rho_2)}{\rho_1 + \rho_2} + \frac{2(\mathbf{k}\cdot\mathbf{B}_0)^2}{\mu_0(\rho_1 + \rho_2)} \tag{4.112}$$

showing that shorter wavelength modes such that

$$k \geq \frac{\mu_0 g(\rho_1 - \rho_2)}{2B_0^2\cos^2\theta}$$

where θ is the angle between $\mathbf{k}$ and $\mathbf{B}_0$, are stabilized.

The same procedure (see Exercise 4.9) may be used to find the dispersion relation

$$\omega^2 = -kg + (\mathbf{k}\cdot\mathbf{B}_0)^2/\mu_0\rho_0 \tag{4.113}$$

for the case of an unmagnetized plasma of constant density ρ_0 supported by a uniform magnetic field $\mathbf{B}_0$. The qualitative features of both this result and (4.112) are sketched in Fig. 4.30 for a given wavenumber k, showing that there is always an interval around $\theta = \pi/2$ for which instability occurs. If, however, instead of a uniform $\mathbf{B}_0$ we have a rotating $\mathbf{B}_0(y)$ then $\theta = \theta(y)$ and the instability of a given mode is restricted to a layer about the surface $y = y_s$ where $\mathbf{k} \cdot \mathbf{B}_0(y_s) = 0$. Thus, the magnetic field has two stabilizing roles. Any bending of the field lines resists the growth of the perturbation and magnetic shear limits the region of instability when it occurs.

Surfaces where $\mathbf{k} \cdot \mathbf{B}_0 = 0$ play a crucial role in stability analysis quite generally and are called *resonant surfaces*. In particular, we note that (4.111) has a singular point wherever

$$\rho_0 \omega^2 - (\mathbf{k} \cdot \mathbf{B}_0)^2/\mu_0 = 0$$

Since $(\mathbf{k} \cdot \mathbf{B}_0)^2 \geq 0$, such singularities occur only for stable configurations but at marginal stability ($\omega = 0$) they occur at the resonant surfaces $\mathbf{k} \cdot \mathbf{B}_0 = 0$. As might be expected, the instability tends to be localized around the resonant surfaces.

4.7.2 Pressure-driven instabilities

Another interchange instability, discussed by Kruskal and Schwarzschild (1954), may arise in plasma contained by a magnetic field. Here the pressure gradient plays a role akin to gravity in the Rayleigh–Taylor instability. It is clear from (4.105) that if the pressure gradient ∇P_0 and magnetic curvature $\boldsymbol{\kappa}$ act in the same direction relative to $\boldsymbol{\xi}_\perp$ a pressure-driven instability may arise. On the other hand, if $(\boldsymbol{\xi}_\perp \cdot \nabla P_0)(\boldsymbol{\xi}_\perp \cdot \boldsymbol{\kappa}) < 0$ this term is stabilizing. Since ∇P_0 is generally directed towards the centre of the plasma, the stability condition requires that magnetic curvature be directed outwards (i.e. away from the plasma centre). This confirms our earlier observation that cusp fields have favourable curvature whilst those which are concave towards the plasma have unfavourable curvature. The sausage instability is an example of an unstable interchange mode resulting from unfavourable curvature. A simple argument similar to that used for the Rayleigh–Taylor instability allows us to find a stability condition for interchange perturbations. We consider a cross-section of the plasma perpendicular to the field lines (again assumed straight) and perturb it by interchanging two flux tubes of equal strength. This leaves the total magnetic energy unchanged but, in general, alters the internal energy of the plasma because both the pressure and the volume may change. The plasma initially in flux tube 1 with pressure P_1 and volume V_1 goes to the position of flux tube 2 with volume V_2 and, by the adiabatic gas law, pressure $P_1(V_1/V_2)^\gamma$. Thus, the change in internal energy for flux tube 1 is

$$[P_1(V_1/V_2)^\gamma V_2 - P_1 V_1]/(\gamma - 1)$$

Similarly, the change for flux tube 2 is

$$[P_2(V_2/V_1)^\gamma V_1 - P_2V_2]/(\gamma - 1)$$

and letting $P_2 = P_1 + \delta P$, $V_2 = V_1 + \delta V$ the total change is, to lowest order in $\delta P, \delta V$,

$$\delta W = \gamma P_1(\delta V)^2/V_1 + \delta P \delta V$$

A sufficient condition for stability is, therefore,

$$\delta P \delta V > 0 \tag{4.114}$$

Now if $S(l)$ is the cross-sectional area of a flux tube its volume is $\int S\,\mathrm{d}l$, where $\mathrm{d}l$ is the line element and the integral is taken along the length of the tube. Also, since the flux is constant along the tube, i.e. $B(l)S(l) = \Phi$, we may write

$$V = \Phi \int \frac{\mathrm{d}l}{B(l)}$$

Thus, if $\delta P < 0$ as we move away from the centre of the plasma we require

$$\delta \int \frac{\mathrm{d}l}{B(l)} < 0 \tag{4.115}$$

which says that the magnetic field, averaged along a flux tube, must increase as we move away from the centre of the plasma. This is the *minimum B stability condition*, so called since it requires the plasma to occupy the region where the average field is minimum.

In two- and three-dimensional configurations there is usually a mixture of favourable and unfavourable curvature; toroidal confinement devices, for example, tend to have favourable curvature on the inside of the torus and unfavourable curvature on the outside. A flux tube may, therefore, experience good and bad curvature as it winds around the torus. Perturbations will tend to grow in regions of unfavourable curvature causing the plasma in such regions to balloon out, so giving rise to the name *ballooning instabilities*. Since, by definition one might say, ballooning instabilities are not controlled by a minimum B (or magnetic well) configuration they must be avoided by keeping the plasma pressure down, i.e. stability thresholds for ballooning instabilities set maximum values for β.

Of course, magnetic shear may also be used to stabilize pressure-driven instabilities. The simple argument leading to (4.114) no longer applies when the field is sheared and, as with the Rayleigh–Taylor instability, the instability is localized near the resonant surfaces. For cylindrical plasmas, where perturbations take the form (4.75), the resonant surfaces occur at $r = r_s$ where

$$\frac{m}{r_s} B_\theta(r_s) + kB_z(r_s) = 0$$

and linear stability analysis involves the minimization of δW around $r = r_s$ for any given m and k. From such an analysis (see Wesson (1981)) one can show that a cylindrical pinch is stable provided $(q'/q)^2 > -(8\mu_0 P'/rB_z^2)$ where q is the safety factor defined by (4.29). This is *Suydam's criterion* which says that the magnetic shear must be large enough to overcome the destabilizing effect of the pressure gradient.

4.8 Ideal MHD waves

One of the most interesting aspects of a plasma in a magnetic field is the great variety of waves which it can support. A more complete treatment of waves in plasmas is deferred to Chapter 6. However, a discussion of ideal MHD at this point would be incomplete without some account of the natural waves which may propagate through the plasma. We may expect such waves to be widespread in space plasmas, where ideal MHD is in general a valid model, and it was in this context that Alfvén first discovered and described the nature and properties of these waves.

In order to concentrate on the basic properties we avoid the complications of boundary conditions by assuming an infinite plasma. Likewise, for simplicity, we assume that the unperturbed plasma is static and homogeneous. Thus, our starting point is a plasma with

$$\rho = \rho_0 \qquad P = P_0 \qquad \mathbf{u} = 0 \qquad \mathbf{j} = 0 \qquad \mathbf{B} = B_0\hat{\mathbf{z}} \tag{4.116}$$

where ρ_0, P_0, and B_0 are constants.

As in our stability investigations we assume a small perturbation of the system so that in the linear approximation we arrive, as before, at (4.67) but with the simplification that now, in (4.68), $\nabla P_0 = 0$ and $\nabla \times \mathbf{B}_0 = 0$. Also, since the plasma is infinite we may carry out a Fourier analysis in space as well as time, i.e. we assume $\boldsymbol{\xi}(\mathbf{r}, t) = \sum_{\mathbf{k},\omega} \boldsymbol{\xi}(\mathbf{k}, \omega)e^{-i(\mathbf{k}\cdot\mathbf{r}-\omega t)}$. Thus (4.67) reads

$$\rho_0\omega^2\boldsymbol{\xi} = \mathbf{k}\gamma P_0(\mathbf{k}\cdot\boldsymbol{\xi}) + \frac{1}{\mu_0}\{[\mathbf{k}\times(\mathbf{k}\times(\boldsymbol{\xi}\times\mathbf{B}_0))]\times\mathbf{B}_0\} \tag{4.117}$$

Without loss of generality we can choose Cartesian axes such that $\mathbf{k} = k_\perp\hat{\mathbf{y}} + k_\parallel\hat{\mathbf{z}}$ and then after expanding the vector products the three components of (4.117) are

$$(\omega^2 - k_\parallel^2 v_A^2)\xi_x = 0 \tag{4.118}$$

$$(\omega^2 - k_\perp^2 c_s^2 - k^2 v_A^2)\xi_y - k_\perp k_\parallel c_s^2 \xi_z = 0 \tag{4.119}$$

$$-k_\perp k_\parallel c_s^2 \xi_y + (\omega^2 - k_\parallel^2 c_s^2)\xi_z = 0 \tag{4.120}$$

where $c_s = (\gamma P_0/\rho_0)^{1/2}$ is the sound speed and $v_A = (B_0^2/\mu_0\rho_0)^{1/2}$ is the Alfvén

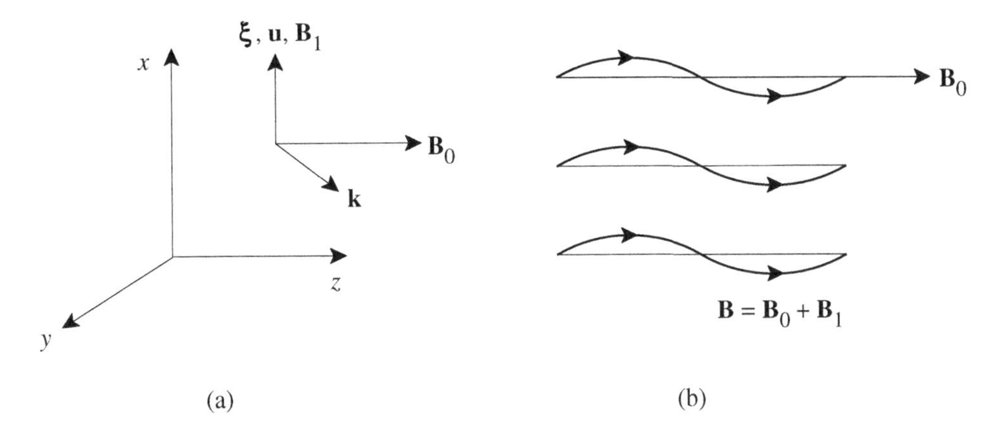

Fig. 4.31. Shear Alfvén wave.

speed. The condition for a non-trivial solution ($\boldsymbol{\xi} \neq 0$) is that the determinant of the coefficients should be zero and this gives the *dispersion relation*

$$\begin{vmatrix} (\omega^2 - k_\parallel^2 v_A^2) & 0 & 0 \\ 0 & (\omega^2 - k_\perp^2 c_s^2 - k^2 v_A^2) & -k_\perp k_\parallel c_s^2 \\ 0 & -k_\perp k_\parallel c_s^2 & (\omega^2 - k_\parallel^2 c_s^2) \end{vmatrix} = 0$$

i.e.

$$(\omega^2 - k_\parallel^2 v_A^2)(\omega^4 - k^2(c_s^2 + v_A^2)\omega^2 + k^2 k_\parallel^2 c_s^2 v_A^2) = 0 \tag{4.121}$$

with solutions

$$\omega^2 = k_\parallel^2 v_A^2 \tag{4.122}$$

$$\omega^2 = \frac{1}{2} k^2 (c_s^2 + v_A^2)[1 \pm (1 - \delta)^{1/2}] \tag{4.123}$$

where

$$\delta = 4 \frac{k_\parallel^2}{k^2} \frac{c_s^2 v_A^2}{(c_s^2 + v_A^2)^2} \tag{4.124}$$

Since $0 \leq \delta \leq 1$, all three solutions are real and the waves propagate without growth or decay. There is neither dissipation to cause decay nor free energy (currents) to drive instabilities.

Taking each mode in turn, (4.122) is the dispersion relation for the *shear Alfvén wave*. As is clear from (4.118)–(4.120), this mode is decoupled from the other two and its displacement vector $\xi_x \hat{\mathbf{x}}$ is perpendicular to both $\mathbf{B}_0$ and $\mathbf{k}$, i.e. the wave, illustrated in Fig. 4.31(a), is transverse. Note that, from (4.62) and (4.64), $\mathbf{B}_1$ and $\mathbf{u} = \mathbf{u}_1$ are in the same direction as $\boldsymbol{\xi}$. Since $\mathbf{k} \cdot \boldsymbol{\xi} = 0$, we see that ρ_1 and P_1 are both zero, i.e. the wave is incompressible. It propagates, as shown in Fig. 4.31(b), like waves along plucked strings under tension, the strings being the magnetic field lines. In fact Alfvén, using the analogy with elastic strings, pointed out that the phase velocity obtained is exactly what one would expect if one substitutes the magnetic tension B_0^2/μ_0 for T in the expression $\omega/k = (T/\rho_0)^{1/2}$ for the phase velocity of transverse waves along strings with line density ρ_0 and tension T. The energy in the wave oscillates between plasma kinetic energy $\frac{1}{2}\rho_0 u_1^2$ and perturbed magnetic energy $B_1^2/2\mu_0$. This confirms the statement about the first term in the integrand of δW_P in (4.105).

The two remaining modes have $\xi_x = 0 = u_x$. Observing that the minimum (maximum) value of ω^2, when we take the plus (minus) sign in (4.123), is given by $\delta = 1$, it follows that $\omega_F \geq k_\parallel v_A \geq \omega_S$, where $\omega_{F,S}$ are the *fast* and *slow* wave frequencies corresponding to the plus and minus signs, respectively. Since both the magnetic (Alfvén) and acoustic wave speeds appear in the dispersion relation for these waves and they are compressional they are known as the *fast* and *slow magnetoacoustic waves*. To discuss these modes we note, from (4.119) and (4.120), that they decouple when propagation is either parallel or perpendicular to $\mathbf{B}_0$.

For perpendicular propagation ($k_\parallel = 0$) the fast wave has

$$\omega^2 = k^2(c_s^2 + v_A^2)$$

and the displacement vector $\boldsymbol{\xi} = \xi_y \hat{\mathbf{y}}$ is parallel to $\mathbf{k} = k_\perp \hat{\mathbf{y}}$. From (4.62), $\mathbf{B}_1$ is parallel to $\mathbf{B}_0$ so the compression of the magnetic field combines with that of the plasma ($P_1 \propto \mathbf{k} \cdot \boldsymbol{\xi}$) to drive the wave. The slow wave does not propagate in this direction ($\omega = 0$).

In the case of parallel propagation ($k_\perp = 0$), one mode has $\omega = k v_A$ and $\xi_y \neq 0$ and the other, $\omega = k c_s$ and $\xi_z \neq 0$. In this limit the magnetoacoustic waves have separated into a *compressional Alfvén wave* and an *acoustic wave*. Which mode is fast and which slow depends on the relative magnitudes of v_A and c_s but usually $\beta < 1$ in which case the acoustic wave is the slow mode. In the acoustic wave the displacement vector is along $\mathbf{B}_0$ so the field plays no role; the wave is driven by the fluctuations in gas pressure. On the other hand, in the compressional Alfvén wave $\mathbf{k} \cdot \boldsymbol{\xi} \sim \mathbf{k} \cdot \mathbf{u} \sim \boldsymbol{\nabla} \cdot \mathbf{u} = 0$ so that the compressibility of the plasma has no effect.

For propagation at arbitrary angles to the magnetic field, these modes are coupled. In the low β limit $c_s \ll v_A$ so that $\omega_F \approx k v_A$ and $\omega_S \approx k_\parallel c_s$. Thus, for the fast mode, from (4.120) we see that $|\xi_z|/|\xi_y| \sim \beta \ll 1$, i.e. the plasma motion is almost perpendicular to the field lines. The oscillation in energy is between plasma kinetic energy and field energy (compression and tension). Likewise, for the slow mode in the low β limit, from (4.119) we see that $|\xi_y|/|\xi_z| \sim \beta \ll 1$ so the plasma motion is almost parallel to $\mathbf{B}_0$. Here energy oscillates within the plasma between kinetic and internal energy.

Exercises

4.1 Consider the steady flow $\mathbf{u} = u(z)\hat{\mathbf{x}}$ of a viscous conducting fluid between infinite horizontal planes at $z = 0$ and $z = 2d$ driven by the motion of the upper plane with velocity $V_0\hat{\mathbf{x}}$ relative to the fixed plane at $z = 0$. Given that the flow, known as *Couette flow*, is subject to a constant applied magnetic field $B_0\hat{\mathbf{z}}$ but that there is no electric field (short-circuit condition), verify that the solution of the equation set

$$\begin{aligned} \rho\frac{D\mathbf{u}}{Dt} &= \mathbf{j}\times\mathbf{B} - \nabla P + \mu\left[\nabla^2\mathbf{u} + \frac{1}{3}\nabla(\nabla\cdot\mathbf{u})\right] \\ \mathbf{j} &= \sigma(\mathbf{u}\times\mathbf{B}) \\ &= \nabla\times\mathbf{B}/\mu_0 \end{aligned}$$

is $u(z) = V_0\sinh(Hz/d)/\sinh 2H$ where $H = B_0 d(\sigma/\mu)^{1/2}$ is the *Hartmann number*. (Assume that all variables depend only on z).

Consider the limits $H \to 0$ and $H \to \infty$ to show that as $B_0 \to 0$ the hydrodynamic flow $u(z) = V_0 z/2d$ is recovered, whilst as $\mu \to 0$ the inviscid solution ($u(z) \equiv 0$) is not retrieved but the flow is effectively restricted to a boundary layer at the moving plane, the thickness of which tends to zero as H^{-1}.

4.2 Derive the equation of energy conservation (4.3) from the ideal MHD equations in Table 3.2.

4.3 Show that the Bennett electron density profile $n_e(r) = n_e(0)[1+(r/r_0)^2]^{-2}$ in a Z-pinch corresponds to uniform electron flow velocity across the plasma column. Evaluate the scale length r_0.

Using (4.23)–(4.28) show that the pressure, magnetic field and current profiles are given by

$$\begin{aligned} P(r) &= \frac{\mu_0 I^2}{8\pi^2}\frac{r_0^2}{(r^2+r_0^2)^2} \\ B(r) &= \frac{\mu_0 I}{2\pi}\frac{r}{r^2+r_0^2} \\ j(r) &= \frac{I}{\pi}\frac{r_0^2}{(r^2+r_0^2)^2} \end{aligned} \tag{E4.1}$$

Sketch these profiles and show that for $r > r_0$ both the plasma and magnetic pressure gradients are negative. It is the magnetic tension $-B^2/\mu_0 r$, acting against this combined outward pressure, which constrains the plasma in the equilibrium configuration.

4.4 Solov'ev (1968) found an exact axisymmetric solution to the Grad–Shafranov equation (4.44) with

$$\begin{aligned} P &= -|P'|\psi + P_0 \\ F^2 &= 2\gamma(1+\alpha^2)^{-1}|P'|\psi + F_0^2 \end{aligned}$$

where P_0 and (F_0/R_0) denote the pressure and magnetic field on the magnetic axis $R = R_0$, $Z = 0$, $\psi = 0$. Check that with this choice a solution to the Grad–Shafranov equation (setting $\mu_0 = 1$) is

$$\psi = \frac{|P'|}{2(1+\alpha^2)}\left[(R^2-\gamma)Z^2 + \frac{\alpha^2}{4}(R^2 - R_0^2)^2\right]$$

Setting $R - R_0 = x \ll R_0$ show that the magnetic surfaces have approximately elliptical cross-sections

$$[x + \delta(x, Z)]^2 + \left(\frac{R_0^2 - \gamma}{\alpha^2 R_0^2}\right) Z^2 = x_0^2$$

Determine the Shafranov shift $\delta(x, Z)$ which has the effect of shifting the magnetic axis, compressing the magnetic surfaces on the outside and relaxing their separation on the inside (see Fig. 4.19(c)).

4.5 Starting from the expression

$$W_B = \int_V (B^2/2\mu_0)\mathrm{d}\tau$$

for the magnetic energy in a volume V, show that $\delta W = 0$ leads to

$$\int \mathrm{d}\tau \mathbf{B} \cdot (\nabla\delta\alpha \times \nabla\beta + \nabla\alpha \times \nabla\delta\beta) = 0$$

if $\mathbf{B} = \nabla\alpha \times \nabla\beta$, where the Clebsch variables α and β are constant on the bounding surface.

Integrate this equation by parts to obtain

$$\int \mathrm{d}\tau \, \{[\nabla\beta \cdot (\nabla \times \mathbf{B})]\delta\alpha - [\nabla\alpha \cdot (\nabla \times \mathbf{B})]\delta\beta\} = 0$$

and hence deduce the coupled differential equations (4.51) for α and β.

4.6 A simple model represents a sunspot as a flux cylinder in which $\mathbf{B} = B(r)\hat{\mathbf{z}}$, $\mathbf{g} = -g\hat{\mathbf{z}}$ and with radially decreasing magnetic pressure and increasing plasma pressure. Starting from the magnetohydrostatic condition (4.52) and assuming that the field vanishes at the edge of the sunspot, show that the pressure balance condition is

$$P_{\mathrm{int}}(r, z) + \frac{B^2(r)}{2\mu_0} = P_{\mathrm{ext}}(z)$$

where $P_{\rm int}$ and $P_{\rm ext}$ are the internal and external plasma pressures, respectively. Show also that the density

$$\rho = -\frac{1}{g}\frac{\mathrm{d}P_{\rm ext}}{\mathrm{d}z}$$

is a function of z only. Hence, deduce the temperature ratio

$$\frac{T_{\rm int}(r,z)}{T_{\rm ext}(z)} = \frac{P_{\rm int}(r,z)}{P_{\rm ext}(z)} = 1 - \frac{B^2(r)}{2\mu_0 P_{\rm ext}(z)}$$

showing that a vertical magnetic field may produce a temperature deficit inside the sunspot.

4.7 By taking the scalar product of (4.67) with $\boldsymbol{\xi}^*$ and subtracting its complex conjugate show that

$$[\omega^2 - (\omega^*)^2]\int \rho_0|\boldsymbol{\xi}|^2\,\mathrm{d}\mathbf{r} = \int [\boldsymbol{\xi}^*\cdot\mathbf{F}(\boldsymbol{\xi}) - \boldsymbol{\xi}\cdot\mathbf{F}(\boldsymbol{\xi}^*)]\mathrm{d}\mathbf{r}$$

Deduce that the self-adjointness of $\mathbf{F}(\boldsymbol{\xi})$ implies that the eigenvalues ω^2 are real.

From the equation

$$\omega^2 = \delta W(\boldsymbol{\xi}^*,\boldsymbol{\xi})/K(\boldsymbol{\xi}^*,\boldsymbol{\xi})$$

where $\delta W(\boldsymbol{\xi}^*,\boldsymbol{\xi}) = -\frac{1}{2}\int \boldsymbol{\xi}^*\cdot\mathbf{F}(\boldsymbol{\xi})\mathrm{d}\mathbf{r}$ and $K(\boldsymbol{\xi}^*,\boldsymbol{\xi}) = \frac{1}{2}\int \rho_0|\boldsymbol{\xi}|^2\,\mathrm{d}\mathbf{r}$, show that the varation $\boldsymbol{\xi}\to\boldsymbol{\xi}+\delta\boldsymbol{\xi}$, $\omega^2\to\omega^2+\delta\omega^2$, for small $\delta\boldsymbol{\xi}$ and $\delta\omega^2$, leads to

$$\delta\omega^2 = \delta W(\delta\boldsymbol{\xi}^*,\boldsymbol{\xi}) + \delta W(\boldsymbol{\xi}^*,\delta\boldsymbol{\xi}) - \omega^2[K(\delta\boldsymbol{\xi}^*,\boldsymbol{\xi}) + K(\boldsymbol{\xi}^*,\delta\boldsymbol{\xi})]/K(\boldsymbol{\xi}^*,\boldsymbol{\xi})$$

Hence, using the self-adjoint property of $\mathbf{F}$ and setting $\delta\omega^2 = 0$, corresponding to ω^2 being an extremum, show that

$$\int \mathrm{d}\mathbf{r}\{\delta\boldsymbol{\xi}^*\cdot[\mathbf{F}(\boldsymbol{\xi}) + \omega^2\rho_0\boldsymbol{\xi}] + \delta\boldsymbol{\xi}\cdot[\mathbf{F}(\boldsymbol{\xi}^*) + \omega^2\rho_0\boldsymbol{\xi}^*]\} = 0$$

and from this deduce the equivalence of the variational principle discussed in Section 4.6 and the normal mode eigenvalue equation (4.67).

4.8 By following the steps indicated in the text derive (4.101) from (4.96).

Show further, by separating vector quantities into parallel and perpendicular components, $\mathbf{X} = X_{\parallel}\mathbf{b} + \mathbf{X}_{\perp}$, where $\mathbf{b}$ is a unit vector in the direction of the magnetic field, that the expression (4.102) for δW_p may be written in the form (4.105).

4.9 Show that the same procedure used to derive (4.112) leads to the dispersion relation (4.113) for the case of an unmagnetized plasma of constant density ρ_0 supported by a constant magnetic field $\mathbf{B}_0$.

4.10 Consider a plasma with a perturbed rippled boundary (see Fig. 4.29(a)) containing a uniform magnetic field $\mathbf{B}_0 = B_0\hat{\mathbf{z}}$ in a gravitational field. According to Section 2.3.1 a positive ion is then subject to a gravitational drift along $-\hat{\mathbf{x}}$. Sketch the effect of this drift on the charge on adjacent sides of the ripple, including the direction of the induced electric field $\delta\mathbf{E}$. Show that the resultant $\delta\mathbf{E} \times \mathbf{B}_0$ drift acts in such a way as to amplify the original rippled perturbation.

4.11 Magnetic buoyancy can lead to instability. By considering an isolated flux tube rising adiabatically along the z-axis in a conducting fluid containing a magnetic field $\mathbf{B} = B_0(z)\hat{\mathbf{x}}$, show that this rise becomes unstable if the field decays with z faster than the density ρ, i.e. if $\mathrm{d}(B_0/\rho)/\mathrm{d}z < 0$.

The magnetic buoyancy instability is a special case of the Rayleigh–Taylor instability. How is the instability condition modified when curvature of the field lines is taken into account (see Parker (1979))?

4.12 Show that if the radius of curvature $\mathbf{R}_\mathrm{c}$ (see Section 2.4.2.) and $\mathbf{g}$ are in the same direction, the drift $\mathbf{v}_\mathrm{B}$, given by (2.24), is equivalent to a gravitational drift with $g = (v_\perp^2 + 2v_\parallel^2)/2R_\mathrm{c}$.

By averaging over a velocity distribution for a thermal plasma show that this orbit theory result leads to the condition $g = 2P/\rho R_\mathrm{c}$ where P and ρ denote the ion pressure and density respectively.

By analogy with the result from Exercise 4.10, a plasma in a curved magnetic field should show a similar tendency for charge to build up and hence become unstable. Deduce that instability occurs when the plasma is confined by a magnetic field that is concave towards the plasma. By analogy with the Rayleigh–Taylor growth rate show that the growth rate of this *flute instability* is $\gamma = (2\nabla P/\rho R_\mathrm{c})^{1/2}$.

4.13 Show that Suydam's criterion in Section 4.7.2. can be expressed in the form $s^2 > (8r^2/B_\theta^2)\kappa_\mathrm{c}\nabla P$, where $s = \mathrm{d}\ln q/\mathrm{d}\ln r$ is the shear parameter and $\boldsymbol{\kappa}_\mathrm{c} = -B_\theta^2\hat{\mathbf{r}}/B^2 r$ is the field line curvature vector in cylindrical geometry.

For toroidal geometry substitute $\boldsymbol{\kappa} = \boldsymbol{\kappa}_\mathrm{c} + \boldsymbol{\kappa}_\mathrm{t}$ where $\boldsymbol{\kappa}_\mathrm{t} = -\hat{\mathbf{R}}/R$ is the curvature of the toroidal magnetic field. Show that although $\kappa_\mathrm{t} \sim \kappa_\mathrm{c}(R/r)$, $\boldsymbol{\kappa}_\mathrm{t}$ is along $\boldsymbol{\nabla}P$ on the outside of the torus but in the opposite direction on the inside and hence, following a field line, destabilizing and stabilizing contributions alternate so that the effect of toroidal curvature averages out to lowest order.

Show that by going to next order and averaging over a field line, assuming concentric circular flux surfaces, one finds an approximate toroidal curvature of $(1 - q^2/2)\kappa_\mathrm{c}$. A rigorous calculation gives in fact $(1 - q^2)\kappa_\mathrm{c}$.

Using this result write down the toroidal analogue of Suydam's criterion (see Biskamp (1993)).

4.14 In practice since most problems in MHD have to be solved numerically it is often preferable to integrate the MHD equations numerically from the start. A simple introduction to computational MHD is provided by a one-dimensional Lagrangian code. In a Lagrangian finite difference scheme the grid points of the finite difference scheme move with the fluid. This has the advantage that the advective term in the MHD equations is replaced by the Lagrangian time derivative which means that the complication inherent in these terms is transferred to the equation of motion for the mesh $\mathrm{d}x/\mathrm{d}t = v$. In other words in a Lagrangian scheme one labels a differential mass of fluid by its position at $t = 0$, say $x_0(t = 0)$ and then determines x as well as the velocity, pressure, density etc. of this same differential mass of fluid as time evolves.

Defining a space mesh with mesh points j moving with the fluid velocity, i.e. $x_j^{n+1} = x_j^n + v_j^{n+\frac{1}{2}}\Delta t$ and the cell width by $\Delta_{j+\frac{1}{2}}^{n+1} = x_{j+1}^{n+1} - x_j^{n+1}$ the MHD equations integrated on this mesh have the form:

$$\frac{\mathrm{d}}{\mathrm{d}t}(X\Delta) = 0 \qquad \frac{\mathrm{d}}{\mathrm{d}t}(P\rho^{-\gamma}) = 0 \qquad \frac{\mathrm{d}v}{\mathrm{d}t} = -\frac{1}{\rho}\frac{\partial}{\partial x}\left(P + \frac{B^2}{2\mu_0}\right)$$

where $X = (\rho, B)$.

The fluid properties are expressed as cell quantities defined at the centre of each cell, i.e.

$$X_{j+\frac{1}{2}}^{n+1} = X_{j+\frac{1}{2}}^n \frac{\left(x_{j+1}^n - x_j^n\right)}{\left(x_{j+1}^{n+1} - x_j^{n+1}\right)} \qquad P_{j+\frac{1}{2}}^{n+1} = \left(\frac{\rho_{j+\frac{1}{2}}^{n+1}}{\rho_{j+\frac{1}{2}}^n}\right)^{\gamma} P_{j+\frac{1}{2}}^n$$

The pressure is then used to recalculate the velocity of the boundary for each cell:

$$v_j^{n+\frac{1}{2}} = v_j^{n-\frac{1}{2}} - \frac{2\Delta t}{\left(\rho_{j+\frac{1}{2}}^n + \rho_{j-\frac{1}{2}}^n\right)} \frac{\left(P_{j+\frac{1}{2}}^{*n} - P_{j-\frac{1}{2}}^{*n}\right)}{\left(x_{j+\frac{1}{2}}^n - x_{j-\frac{1}{2}}^n\right)}$$

where $P^* = P + B^2/2\mu_0$. Not only is a one-dimensional Lagrangian mesh simple conceptually, but the mesh itself reflects fluid behaviour through successive bunching and spreading of cell boundaries giving rise to sound waves on the mesh. As wave profiles steepen to form shocks the Lagrangian mesh automatically accumulates mesh points in the region of the shock front which is beneficial for spatial resolution.

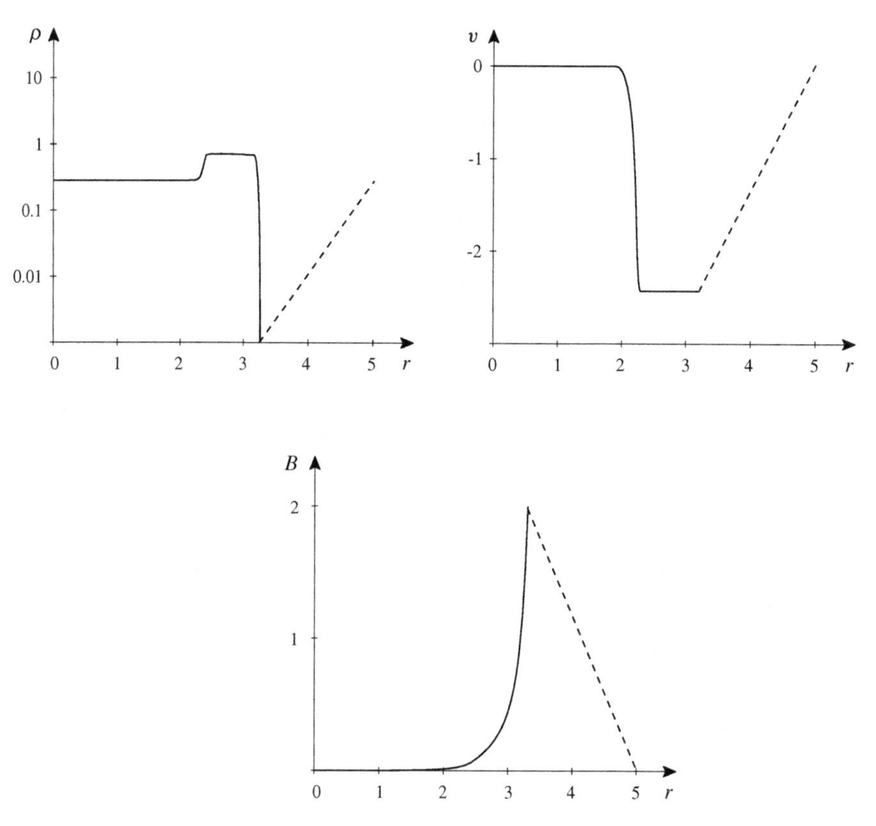

Fig. 4.32. Time-frame from a 1D Lagrangian code output showing density ρ, velocity v, and magnetic field B profiles as functions of radius, representing a shock imploding on a plasma slab from the right (arbitrary units).

The project involves using the Lagrangian scheme outlined to model shock implosion leading to the formation of a pinch. A shock is applied at the right-hand boundary of a plasma slab and propagates inwards. Choose suitable initial values for plasma parameters and integrate the equations of motion to determine ρ, v and B as the shock propagates inwards. The profiles in Fig. 4.32 are typical.

However, your output is likely to show small scale oscillations at the shock front which are computational rather than physical. Representing a shock on a difference mesh presents problems since by definition a shock front is steep and short wavelengths cannot be represented on the mesh. We shall find in Chapter 5 that shock thickness is determined by viscosity. Within the framework of ideal MHD we have to resort to introducing an *artificial viscosity* which has the effect of suppressing discontinuities with a wavelength less than the step size of the space mesh, while leaving longer wavelengths unaffected. This artifice serves to remove oscillations that otherwise appear in the region of shock compression. To make this change we replace $P^{n+\frac{1}{2}}_{j+\frac{1}{2}}$ by $\left(P^{n+\frac{1}{2}}_{j+\frac{1}{2}} + r^{n+\frac{1}{2}}_{j+\frac{1}{2}}\right)$ where r denotes the numerical viscosity. The particular representation adopted for the artificial viscosity is not critical. Numerical viscosity was included in the scheme used to produce the snapshots of the implosion in Fig. 4.32.

5

Resistive magnetohydrodynamics

5.1 Introduction

Although ideal MHD is often a good model for astrophysical and space plasmas and is widely employed in fusion research it is never universally valid, for the reasons discussed in Section 4.1. In this chapter we consider some of the most important effects which arise when allowance is made for finite resistivity and, in the case of shock waves, other dissipative mechanisms. Even though the dissipation may be very weak the changes it introduces are fundamental. For example, finite resistivity enables the plasma to move across field lines, a motion forbidden in ideal MHD. Usually, the effects of this diffusion are concentrated in a boundary layer so that mathematically the problem is one of matching solutions, of the non-ideal equations in the boundary layer and ideal MHD elsewhere. On the length scale of the plasma the boundary layer may be treated as a discontinuity in plasma and field variables and, depending on the strength of the flow velocity, this discontinuity may appear as a shock wave.

A comparison of Tables 3.1 and 3.2 reveals that the difference between resistive and ideal MHD is the appearance of extra terms proportional to the plasma resistivity, $\eta \equiv \sigma^{-1}$, in the evolution equations for P and $\mathbf{B}$. Although one is tempted, therefore, to think of ideal MHD as the zero resistivity limit we know from the discussion in Section 3.3.2 that it is properly regarded as the infinite magnetic Reynolds number or large length scale limit, an observation that arises from a comparison of the diffusion and convection† terms in the induction equation

$$\frac{\partial \mathbf{B}}{\partial t} = \frac{\eta}{\mu_0}\nabla^2 \mathbf{B} + \nabla \times (\mathbf{u} \times \mathbf{B}) \tag{5.1}$$

The characteristic times for resistive diffusion and convection are, respectively, $\tau_R \sim \mu_0 L^2/\eta$ and the Alfvén transit time $\tau_A \sim L/v_A$, where, anticipating that we shall be interested mainly in flows dominated by the magnetic field, we have

† The convection term is known also as the *advection* term.

Table 5.1. *Characteristic lengths, times and Lundquist numbers*

	L_H (m)	τ_R (s)	τ_A (s)	S
Arc discharge	10^{-1}	10^{-3}	10^{-3}	1
Tokamak	1	1	10^{-8}	10^8
Earth's core	10^6	10^{12}	10^5	10^7
Sunspot	10^7	10^{14}	10^5	10^9
Solar corona	10^9	10^{18}	10^6	10^{12}

approximated $|\mathbf{u}|$ by the Alfvén speed v_A and L is an appropriate length scale. With this choice and $L = L_H$, the hydrodynamic length scale, the magnetic Reynolds number is usually denoted by $S = \tau_R/\tau_A$ and referred to as the *Lundquist number*. For high temperature laboratory plasmas S is typically 10^6–10^8 and several orders of magnitude greater still for astrophysical plasmas. Table 5.1 shows characteristic values for various plasmas.

Some of these time scales at first sight look rather surprising. For example, they indicate that the diffusion time for a sunspot is millions of years when we know that sunspots seldom last longer than a few months. By contrast, they suggest that the Earth's magnetic field should have diffused away relatively early in its lifetime. The fallacy comes from equating diffusion time with lifetime. The Earth's field persists because some regenerative process is at work compensating for diffusive decay and sunspots disappear on a time scale governed, not by the slow diffusion of their fields through the photosphere, but by some much faster mechanism. How do these other physical processes come into effect when the very large values of S in Table 5.1 suggest that ideal MHD is a more than adequate approximation for fusion and space plasmas?

The answer to this question is twofold. First, we note that $S = \mu_0 v_A L/\eta$ and we have used $L = L_H$ in Table 5.1. Then we must remember that the dimensional comparison of diffusion and convection terms is a crude argument. If, somewhere in the plasma, the convection term vanishes there will be a local region in which the diffusion term, however small, will come into play. Thus, the significance of large S is not that resistivity is entirely negligible but rather that, compared with L_H, the length scale of the region in which it need be considered is very small. In other words, although ideal MHD may be valid for most of the plasma, there can be narrow boundary layers such as current sheets, in which we must apply resistive MHD. We shall see that within such regions plasma relaxation involves the reconnection of magnetic field lines, generally reducing a complex field topology to one with simpler connectivity, thereby enabling the system to arrive at a lower energy state. These topological changes in the magnetic field take place on a time

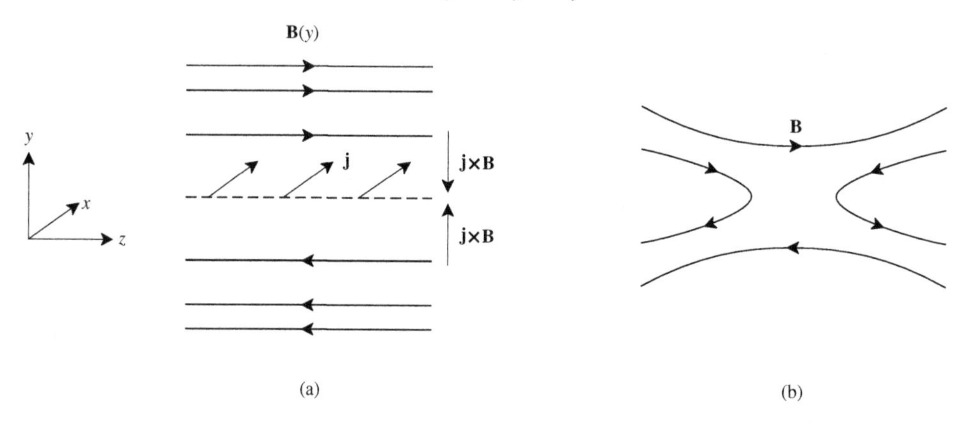

Fig. 5.1. Magnetic reconnection in slab plasma.

scale intermediate between τ_A and τ_R. Such fast reconnections taking place at current sheets are vital for violent events like solar flares on the one hand and major disruptions in tokamaks on the other. The concept of relaxation of stressed magnetic fields and energy release in one form or other underpins much of the discussion in this chapter.

5.2 Magnetic relaxation and reconnection

To understand how magnetic field changes occur in a real plasma with small but finite resistivity let us consider the simplest model of a slab plasma, as in Fig. 5.1(a), in which the field is slowly varying with y, decreasing in magnitude, reversing sign, and then increasing again. The plane ($y = 0$) in which $\mathbf{B} = 0$ is called the *neutral sheet*. If the field lines define the z-axis, the current $\mathbf{j}$ is parallel to the x-axis and the Lorentz force $\mathbf{j}\times\mathbf{B}$ acts downwards for $y > 0$ and upwards for $y < 0$. In ideal MHD, either these forces are opposed by a plasma pressure gradient maintaining equilibrium or plasma and field lines will move together towards the $y = 0$ plane until these forces are in balance. However, with the introduction of finite resistivity, no matter how small, the field is no longer frozen into the plasma and slippage of field lines across the plasma allows breaking of the field lines with reconnection to lines of opposite polarity as shown in Fig. 5.1(b). This may happen at various points along the neutral lines, as depicted in Fig. 5.2, giving rise to so-called *magnetic islands*, i.e. sets of nested magnetic surfaces each with its own magnetic axis. The dashed line in Fig. 5.2 is the separatrix marking the boundary between the regions of different field topology. The topological change takes place because the magnetic energy associated with the magnetic islands is less than that in the original, MHD equilibrium configuration. We can readily imagine this if we

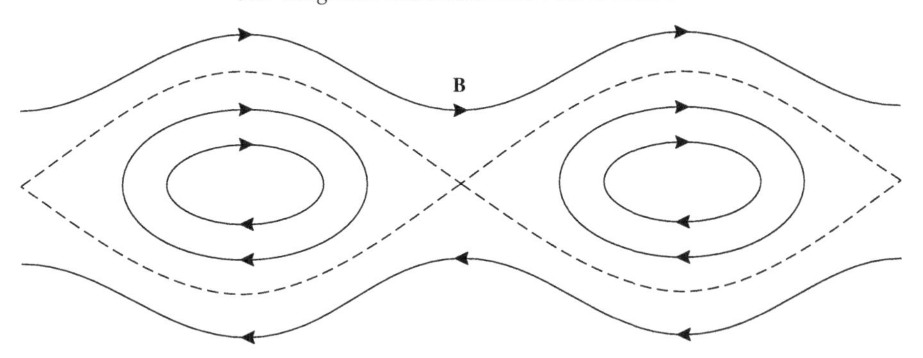

Fig. 5.2. Magnetic islands.

think of the field lines as stretched strings; the tension in them has been reduced because breaking and reconnecting allows them to contract around the island axes. The stored (potential) energy in the final configuration is less than in the original configuration. The null-points of the magnetic field define *O-points*, at the axes of the magnetic islands, and *X-points*, at the intersections of the separatrix.

Some dissipation is essential for any system to attain a lower energy state from its initial state by a relaxation process and Taylor (1974) provided a mathematical basis for this by applying a modification of Woltjer's theorem to plasmas with small but finite resistivity. As discussed in Section 4.3.4, Woltjer showed that the helicity, $K = \int_V \mathbf{A} \cdot \mathbf{B} \, d\tau$, of an ideal plasma is invariant when the integral is taken over the volume V of a closed system. It follows that K is conserved for every volume enclosed by a flux surface, i.e. every infinitesimal flux tube. This amounts to an infinite set of integral constraints ensuring a one-to-one correspondence between initial and final flux surfaces. Clearly, this no longer holds in a plasma with finite resistivity since the continual breaking and reconnecting of field lines destroys the identity of infinitesimal flux tubes. *Taylor's hypothesis* states that only the helicity associated with the *total* volume of the plasma is conserved. This replaces an infinite set of constraints by a single constraint and allows the system access to lower energy states which in ideal MHD are forbidden. It means, also, that the final state of the plasma is largely independent of its initial conditions. Indeed a feature of certain toroidal discharges is that after an initial, violently unstable phase, the discharge relaxes to a grossly stable, quiescent state which depends only on a few external parameters and not on the history of the discharge. The characteristics of reversed field pinches, in particular, may be interpreted on the basis of Taylor's hypothesis. By contrast, relaxation does not play such a prominent role in tokamaks on account of the strong toroidal magnetic field.

Assuming that the plasma is contained by perfectly conducting walls the only flux surface that retains its identity is the plasma boundary. Taylor argued, therefore, that the energy should be minimized subject to the single constraint of constant total magnetic helicity, i.e.

$$K_0 = \int_{V_0} \mathbf{A} \cdot \mathbf{B}\, \mathrm{d}\tau = \text{const.}$$

where V_0 is the total plasma volume. If the plasma is almost ideal and its kinetic energy is negligible compared with the magnetic energy ($\beta \approx 0$) one arrives, as in Woltjer's theorem, at the condition for force-free fields

$$\nabla \times \mathbf{B} = \alpha \mathbf{B} \tag{5.2}$$

but with the fundamental difference that here α is a constant, related to K_0, with the same value on *all* field lines. In view of the assumed boundary conditions (see Section 4.3.4), a second constant determining the solution of (5.2) is the total toroidal flux. For a linear discharge with cylindrical symmetry ($\partial/\partial\theta \equiv \partial/\partial z \equiv 0$)) we have from (5.2)

$$B_r(r) = 0 \qquad \alpha B_\theta = -\frac{\mathrm{d}B_z}{\mathrm{d}r} \qquad \alpha B_z = \frac{1}{r}\frac{\mathrm{d}}{\mathrm{d}r}(r B_\theta)$$

giving

$$\frac{\mathrm{d}^2 B_z}{\mathrm{d}r^2} + \frac{1}{r}\frac{\mathrm{d}B_z}{\mathrm{d}r} + \alpha^2 B_z = 0$$

This is Bessel's equation of order zero with the result that

$$B_z(r) = B_0 J_0(\alpha r) \qquad\qquad B_\theta(r) = B_0 J_1(\alpha r) \tag{5.3}$$

where J_0 and J_1 are Bessel functions of the first kind of order zero and one, respectively, and B_0 is the value of the magnetic field on the axis. Since $J_0(x)$ changes sign at $x = 2.4$ it follows that field reversal will occur if, in the relaxed state, the plasma radius a is such that $\alpha a > 2.4$.

To appreciate the implications of this condition in practice it is helpful to introduce two measurable quantities, the *field reversal parameter* F, which is the normalized toroidal field at $r = a$, i.e.

$$F = B_{z0}(a)/\langle B_z \rangle$$

where $\langle B_z \rangle$ denotes the average toroidal magnetic field and the *pinch parameter* θ, which represents the normalized toroidal current,

$$\theta = \frac{B_{\theta 0}(a)}{\langle B_z \rangle} = \frac{aI}{2\psi_z} = \frac{\alpha a}{2}$$

where ψ_z is the toroidal flux and we have set $\mu_0 = 1$. It follows using (5.3) that

$$F = \frac{\alpha a}{2}\frac{J_0(\alpha a)}{J_1(\alpha a)}$$

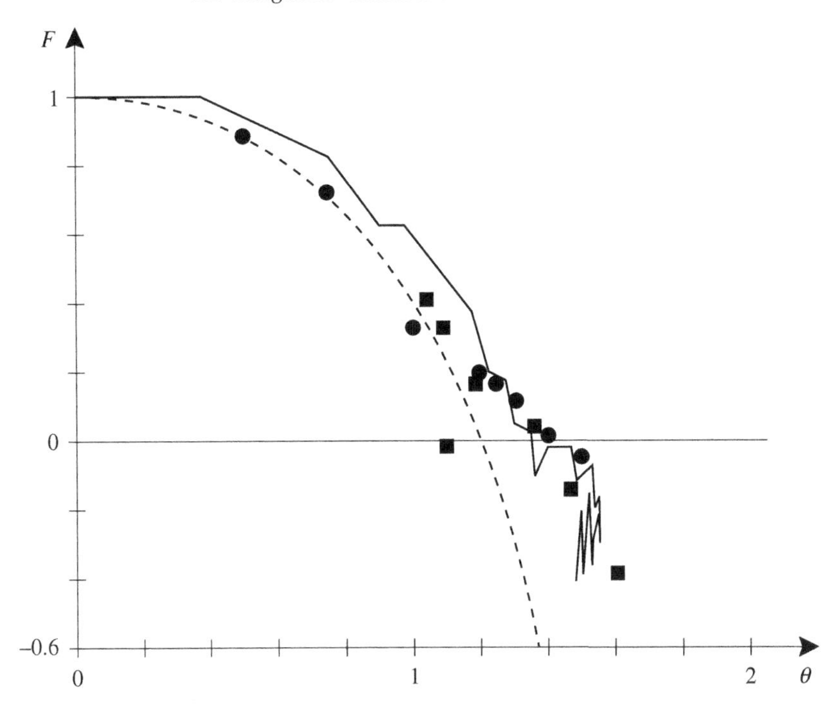

Fig. 5.3. Measured $(F\text{–}\theta)$ characteristics: ■, Zeta; ●, HBTX; ——, Di Marco; and the theoretical curve (dashed line).

Figure 5.3 plots F as a function of θ. We see that as the current increases, $B_z(a)$ decreases and changes sign at $\theta = 1.2$, corresponding to field reversal. Figure 5.3 also shows a measured $(F\text{–}\theta)$ characteristic (Di Marco (1983)) which, while broadly in agreement, lies above the theoretical curve. The discrepancy between the characteristics showing less pronounced field reversal in practice than predicted, is generally attributed to the behaviour of the current density in the boundary region. A relaxed state with $\alpha = \mathbf{j} \cdot \mathbf{B}/B^2$ constant is inconsistent with the physical boundary condition $\mathbf{j} = 0$ at the plasma edge. From Fig. 5.3 we see also that the current does not increase beyond $\theta \sim 1.5$ (i.e. $\alpha a \sim 3$) which is in agreement with the predicted value ($\alpha a = 3.1$). Further increasing the voltage does not result in higher current. Data from a high-beta toroidal pinch experiment (HBTX) and from Zeta, analysed by Bodin and Newton (1980), provide additional support for the relaxation hypothesis.

5.2.1 Driven reconnection

A distinction may be drawn between *spontaneous* and *driven* reconnection. What has been considered so far is the relaxation of plasma in reversed field pinches to a state of lower energy by magnetic field reconnection. Examples of spontaneous reconnection arising in other magnetically contained plasmas crop up later in this chapter, as for instance in Section 5.3.1, where we shall see that reconnection is responsible for the tearing mode instability. However a less benign form of reconnection, sometimes referred to as driven reconnection, may take place. The most dramatic events of this kind occur in nature where plasmas collide, as happens in solar flares and when the solar wind strikes the Earth's magnetosphere. It is possible to distinguish the two types of magnetic reconnection by a qualitative argument. In the spontaneous case if we suppose that a current sheet is present initially then referring to (5.1) we may use the Lundquist number as an index of the *spontaneous* reconnection rate M_{sp} where

$$M_{\mathrm{sp}} = S^{-1} \ll 1$$

The need for fast reconnection first became evident in attempts to explain the explosive onset of solar flares. Through fast reconnection the magnetic field can change its morphology and so release energy. The importance of neutral points at which the field vanishes was first realized by Giovanelli (1947) and stressed by Dungey (1953) who showed that rapid dissipation of the magnetic field was possible at X-type neutral points. Sweet (1958) noted that energy release from a magnetic field requires the field to be stressed in some way and the model used to represent this was one in which oppositely directed fields collide. Figure 5.4 shows schematically magnetic fields being pushed together by flows into a narrow region. In the flow regions the resistivity is low and hence the magnetic field is frozen in the flow. The two regions are separated by a *current sheet* since the reversal of the magnetic field **B** requires a current to flow in the thin layer separating them. Within this layer resistive diffusion plays a key role. As the two regions come together the plasma is squeezed out along the field lines allowing the fields to get closer and closer to the neutral sheet. At some stage the field lines break and reconnect in a new configuration at a magnetic null-point, X. The large stresses in the acutely bent field lines in the vicinity of the null-point result in a double-action magnetic catapult that ejects plasma in both directions, with velocity of $O(v_{\mathrm{A}})$. This in turn allows plasma to flow into the reconnection zone from the sides.

This model was later developed by Parker (1963) and is generally referred to as the *Sweet–Parker model*. A simple quasi-static argument using momentum balance shows that the plasma is ejected at the Alfvén velocity, v_{A}. Denoting the length of the current sheet by $2L$ and the thickness by $2d$, mass conservation dictates that $uL = v_{\mathrm{A}} d$ where u is the plasma flow speed in the direction normal to **B**. Under steady state conditions the rate at which magnetic flux is convected towards the

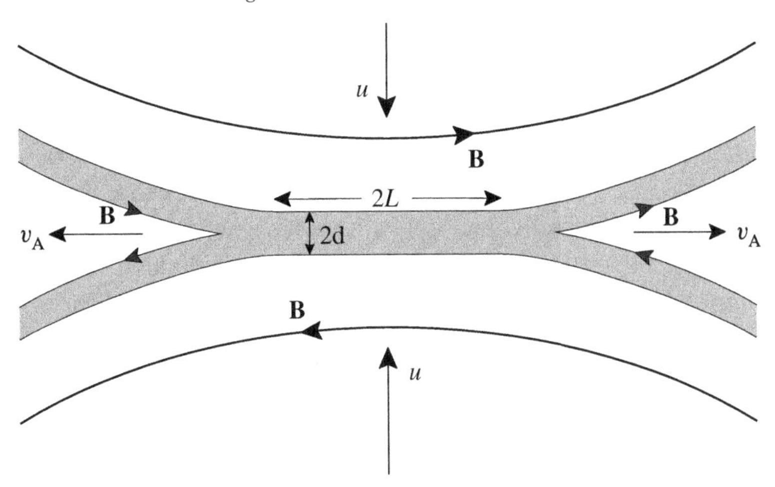

Fig. 5.4. Driven reconnection at X-type neutral point.

current sheet by the plasma flow is balanced by the rate of ohmic dissipation so that

$$u = \eta/2\mu_0 d \tag{5.4}$$

This equation combined with the expression of mass conservation gives

$$u = v_A/S^{1/2} \qquad d = L/S^{1/2} \tag{5.5}$$

Identifying $M_{dr} = u/v_A = M_A$ (the Alfvén Mach number) as a dimensionless index for the driven reconnection rate, it follows from (5.5) that $M_{dr} = S^{-1/2}$. From Table 5.1 we find reconnection that is many orders of magnitude too slow to characterize the evolution of solar flares. By the same token an inverse aspect ratio $d/L \sim 10^{-6}$ implies a current sheet only metres thick for typical values of L, which is unrealistic.

The realization that the Sweet–Parker reconnection rate was much too slow pointed to the need for incorporating some faster mechanism into the current sheet model. Petschek (1964) attempted to do this by means of a slow MHD shock model which again involves a current sheet but leads to a reconnection rate now effectively independent of resistivity. Petschek reasoned that the magnetic fields would meet at a relatively narrow apex, rather than across the entire region envisaged in the Sweet–Parker model and found a maximum reconnection rate

$$M_{dr} \sim (\ln S)^{-1} \tag{5.6}$$

This weak dependence on resistivity seemed to fit the bill and led to the model being widely adopted despite a longstanding debate about its validity, over issues such as precisely how to define a Petschek regime and the boundary conditions governing it. These misgivings were strengthened by insights gained from numerical experiments on reconnection by Biskamp (1986). Biskamp (1993) has given a detailed critique of the Petschek slow-shock model. On general grounds Petschek's model has been seen as counter-intuitive in that two plane regions of highly conducting plasma with oppositely directed magnetic fields pushed together might be expected to generate a flat current sheet configuration rather than the cone required by the Petschek model. However Biskamp's criticism centres on the treatment of the diffusion region where a boundary layer solution, matching the ideal MHD solution outside the diffusion zone to the resistive MHD solution within, is required. Biskamp's numerical experiments of driven reconnection do not show a Petschek-like configuration in the small η limit. Although features characteristic of slow shocks are confirmed by the simulations, Biskamp found that as the reconnection rate increases, both the length and width of the diffusion region increase, counter to Petschek's predictions.

Whatever doubts persist over models for magnetic reconnection in solar flares, observations by Innes *et al.* (1997) have provided the first direct evidence for reconnection. They report ultraviolet observations of explosive events in the solar chromosphere which point to the presence of oppositely directed plasma jets ejected from small sites above the solar surface. Observations of these jets show signs of some anisotropy in that jets directed away from the solar surface may stream freely up to the corona while downward jets should suffer attenuation on account of the increasing density of the chromosphere. The stream exhibiting a blue shift, indicative of plasma flowing away from the solar surface, is of very much greater extent than the red stream.

5.3 Resistive instabilities

The ability of a plasma, through magnetic reconnection, to reach lower energy states means that ideal MHD stability theory needs re-examination. Modification of the theory by the introduction of a small but finite resistivity leads to the discovery of new instabilities. These *resistive instabilities* were first derived in the seminal paper of Furth, Killeen and Rosenbluth (1963). In this paper the resistive MHD equations were solved in the boundary layer in which $S \lesssim 1$ and field line diffusion takes place; the ideal MHD equations were solved outside this region and the solutions matched at the boundary. Three instabilities were discovered with growth times much smaller than τ_R but much greater than τ_A. One of these, the *tearing instability*, arises spontaneously while the others are driven instabilities.

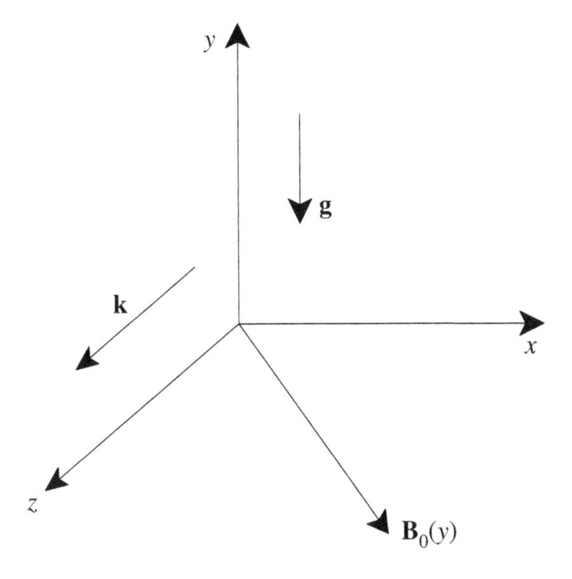

Fig. 5.5. Coordinate axes for calculation of resistive instabilities.

A discussion of the original calculation may be found in Miyamoto (1989) and accounts of both the linear and non-linear theory of resistive instabilities have been given by Bateman (1978) and by White (1983). In what follows we make no attempt to repeat details of the full calculation in the original paper but derive instead the basic equations from which the linear dispersion relation is obtained and, in the spirit of Wesson (1981), use heuristic arguments to determine the parametric dependence of the linear growth rates.

The procedure followed is close to that used in Chapter 4, in discussing linear stability in ideal MHD, first obtaining (4.67) and then (4.111) for the Rayleigh–Taylor instability. We use the geometry illustrated in Fig. 5.5 and again assume that equilibrium quantities vary only in the y direction. Incompressibility is also assumed so that (4.107) holds. This is justified since the growth rates of the resistive instabilities are very small on the hydromagnetic time scale τ_A and this means that fluid and magnetic pressure changes tend to be compensating, having a negligible effect on the dynamics of the instabilities. Two generalizations of the assumptions used to derive (4.111) are required. However, one of these, the replacement of the ideal by the resistive equation for P, turns out to be of no consequence since we shall eliminate the ∇P term.

The other generalization is important since it introduces the driving term for one of the resistive instabilities. We want to allow for variable resistivity in which case, following the usual derivation of the induction equation from the Maxwell equation and Ohm's law, we get

$$\begin{aligned}\frac{\partial \mathbf{B}}{\partial t} &= -\nabla \times \mathbf{E} \\ &= \nabla \times (\mathbf{u} \times \mathbf{B}) - \nabla \times \left(\frac{\eta}{\mu_0} \nabla \times \mathbf{B}\right) \\ &= \nabla \times (\mathbf{u} \times \mathbf{B}) + \frac{\eta}{\mu_0} \nabla^2 \mathbf{B} - \nabla \eta \times \frac{(\nabla \times \mathbf{B})}{\mu_0} \end{aligned} \tag{5.7}$$

Since we are now treating η as a variable we need an extra equation to determine it and for this we assume that it does not change within the fluid element but only on account of its advection so that

$$\frac{D\eta}{Dt} = \frac{\partial \eta}{\partial t} + \mathbf{u} \cdot \nabla \eta = 0 \tag{5.8}$$

Linearizing (5.7) and (5.8) we have

$$\begin{aligned}\frac{\partial \mathbf{B}_1}{\partial t} &= \nabla \times (\mathbf{u}_1 \times \mathbf{B}_0) + \frac{\eta_0}{\mu_0} \nabla^2 \mathbf{B}_1 \\ &\quad - \nabla \eta_1 \times \frac{(\nabla \times \mathbf{B}_0)}{\mu_0} - \nabla \eta_0 \times \frac{(\nabla \times \mathbf{B}_1)}{\mu_0} \end{aligned} \tag{5.9}$$

and

$$\frac{\partial \eta_1}{\partial t} = -\mathbf{u}_1 \cdot \nabla \eta_0 \tag{5.10}$$

Integrating (5.10) gives

$$\eta_1 = -\xi_y \eta_0'(y) \tag{5.11}$$

where we have used (4.64) and the prime denotes differentiation with respect to y. Now substituting (5.11) in (5.9), the y component is

$$\frac{\partial B_y}{\partial t} = i(\mathbf{k} \cdot \mathbf{B}_0) u_y + \frac{\eta_0}{\mu_0} \nabla^2 B_y - i \frac{(\mathbf{k} \cdot \mathbf{B}_0)'}{\mu_0} \eta_0' \xi_y$$

where we have dropped the subscript 1 on first-order variables. Assuming, as in our discussion of the Rayleigh–Taylor instability, that all variables are of the form $A(y)e^{ikz+\gamma t}$ this equation may be integrated to obtain

$$B_y = i(\mathbf{k} \cdot \mathbf{B}_0)\xi_y + \frac{\eta_0}{\mu_0 \gamma}(B_y'' - k^2 B_y) - i \frac{(\mathbf{k} \cdot \mathbf{B}_0)'}{\mu_0 \gamma} \eta_0' \xi_y \tag{5.12}$$

This is one of the basic equations from which the dispersion relation is obtained.

The second equation comes from the linearized equation of motion which is

$$\rho_0 \frac{\partial \mathbf{u}_1}{\partial t} = -\nabla P_1 + \rho_1 \mathbf{g} + \frac{1}{\mu_0}(\nabla \times \mathbf{B}_1) \times \mathbf{B}_0 + \frac{1}{\mu_0}(\nabla \times \mathbf{B}_0) \times \mathbf{B}_1 \tag{5.13}$$

and the linearized continuity equation

$$\frac{\partial \rho_1}{\partial t} = -\mathbf{u}_1 \cdot \nabla \rho_0 \tag{5.14}$$

where, since incompressibility is assumed, we have used $\nabla \cdot \mathbf{u}_1 = 0$. Integrating (5.14) gives

$$\rho_1 = -\xi_y \rho_0'$$

and substituting this in (5.13) the y and z components are

$$\begin{aligned} \rho_0 \gamma^2 \xi_y &= -\frac{\mathrm{d}P_1}{\mathrm{d}y} + \xi_y \rho_0' g - \frac{1}{\mu_0}[B_x B_{0x}' \\ &\quad + B_z B_{0z}' + B_{0x} B_x' + B_{0z}(B_z' - ikB_y)] \qquad (5.15) \\ \rho_0 \gamma^2 \xi_z &= -ikP_1 + \frac{1}{\mu_0}(B_y B_{0z}' - ikB_x B_{0x}) \qquad (5.16) \end{aligned}$$

Now we use (4.107) for ξ_z in (5.16) and substitute the resulting expression for P_1 in (5.15) to get

$$\begin{aligned} &\mu_0 \gamma^2[(\rho_0 \xi_y')' - k^2 \rho_0 \xi_y] + \mu_0 k^2 g \rho_0' \xi_y = \\ &\quad k^2(B_x B_{0x}' + B_z B_{0z}' + B_{0x} B_x' + B_{0z} B_z' - ikB_y B_{0z}) - ik(B_y B_{0z}' - ikB_x B_{0x})' \end{aligned}$$

Also, since

$$\nabla \cdot \mathbf{B}_1 = B_y' + ikB_z = 0$$

we may eliminate B_z to obtain

$$\begin{aligned} &\mu_0 \gamma^2[(\rho_0 \xi_y')' - k^2 \rho_0 \xi_y] + \mu_0 k^2 g \rho_0' \xi_y \\ &\quad = ikB_{0z}\left(B_y'' - k^2 B_y - \frac{B_0'' B_y}{B_{0z}}\right) \\ &\quad = i(\mathbf{k} \cdot \mathbf{B}_0)\left[B_y'' - k^2 B_y - \frac{(\mathbf{k} \cdot \mathbf{B}_0)''}{(\mathbf{k} \cdot \mathbf{B}_0)} B_y\right] \qquad (5.17) \end{aligned}$$

which is the second basic equation relating ξ_y and B_y. It is easily verified that putting $\eta_0 \equiv 0$ in (5.12) and substituting $B_y = i(\mathbf{k} \cdot \mathbf{B}_0)\xi_y$ in (5.17) reproduces (4.111) with γ replacing $i\omega$.

5.3.1 Tearing instability

We begin our heuristic analysis of (5.12) and (5.17) by concentrating on the instability which arises by spontaneous reconnection of antiparallel field lines. The role of the η_0' term in (5.12), like that of the g term in (5.17), is to provide a driving force for a resistive instability. For the moment let us drop both of these terms. Now, in the limit of vanishing resistivity we may ignore the diffusion term in (5.12) except near the resonant surfaces where $\mathbf{k}\cdot\mathbf{B}_0 \approx 0$ and we define the width ϵL of the resistive boundary layer by equating the magnitudes of the convection and diffusion terms in (5.12)

$$(\mathbf{k}\cdot\mathbf{B}_0)\xi_y \sim \frac{\eta_0}{\mu_0\gamma}\nabla^2 B_y \sim \frac{\eta_0\gamma\rho_0}{(\mathbf{k}\cdot\mathbf{B}_0)}\xi_y''$$

where the second approximation arises from (5.17). In this we have ignored the variation (on scale length L) of equilibrium variables compared with the variation (on scale length ϵL) of first-order variables and we have assumed $k\epsilon L \ll 1$, i.e. the wavelength of the perturbation is much greater than the width of the boundary layer. Then replacing ξ_y'' by $\xi_y/(\epsilon L)^2$ and $(\mathbf{k}\cdot\mathbf{B}_0)$ by $\epsilon L\mathbf{k}\cdot\mathbf{B}_0'(0)$, since $(\mathbf{k}\cdot\mathbf{B}_0) = 0$ at $y = 0$ but $(\mathbf{k}\cdot\mathbf{B}_0)' \neq 0$, we get

$$\epsilon L \sim \left(\frac{\gamma\eta_0\rho_0}{k^2(B_0')^2}\right)^{1/4} \tag{5.18}$$

Using (5.12) to eliminate ξ_y from (5.17) it is clear that in the boundary layer we have a fourth-order differential equation for B_y whereas outside this region, where the diffusion term is negligible, the equation reduces to second order. This is an eigenvalue problem in which the eigenvalues, some of which lead to positive γ and hence instability, are determined by matching the solutions at the boundaries of the resistive region. Figure 5.6 shows schematically the variation of the key quantities $\mathbf{k}\cdot\mathbf{B}_0$, $(\mathbf{k}\cdot\mathbf{B}_0)'$, B_y and B_y'. The point to note is the rapid change in B_y' across the boundary layer which, on the scale L of the whole plasma, appears as a discontinous change at the resonant surface ($y = 0$) and hence a very large B_y''. The actual change is determined by the eigenvalue but we shall take it as given in terms of a dimensionless quantity, usually denoted by Δ', and defined by

$$\frac{\Delta'}{L} = \lim_{\epsilon\to 0}\frac{B_y'(\epsilon L/2) - B_y'(-\epsilon L/2)}{B_y(0)} \tag{5.19}$$

The same is true for ξ_y' so the dominant terms in (5.17) are those in ξ_y'' and B_y'' and we may write

$$\begin{aligned}\frac{\mu_0\gamma^2\rho_0\xi_y}{(\epsilon L)^2} &\sim (\mathbf{k}\cdot\mathbf{B}_0)B_y'' \\ &\sim (\mathbf{k}\cdot\mathbf{B}_0)\frac{[B_y'(\epsilon L/2) - B_y'(-\epsilon L/2)]}{\epsilon L} \\ &\sim (\mathbf{k}\cdot\mathbf{B}_0)\frac{B_y(0)\Delta'}{\epsilon L^2}\end{aligned} \tag{5.20}$$

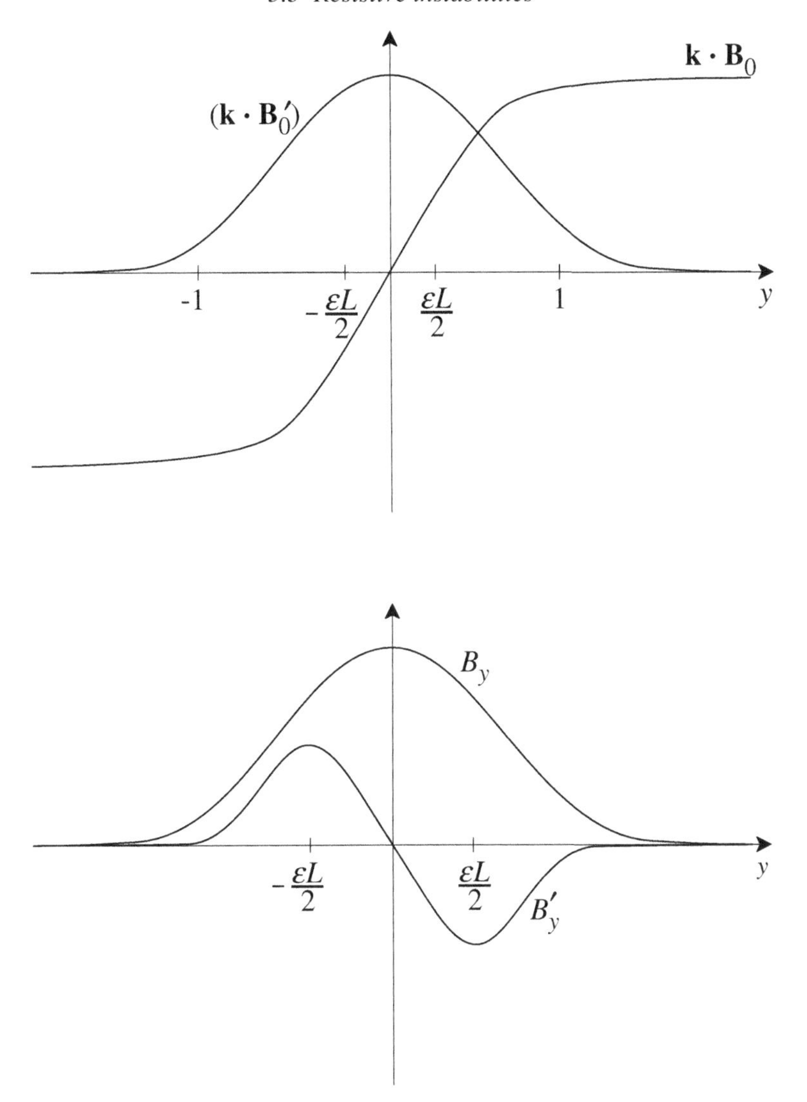

Fig. 5.6. Characteristic variation in tearing instability.

in the limit $\epsilon \to 0$ on using (5.19). Also, from (5.12), $B_y \sim (\mathbf{k} \cdot \mathbf{B}_0)\xi_y$ so that (5.20) becomes

$$\frac{\mu_0 \gamma^2 \rho_0}{(\epsilon L)} \sim \frac{(\mathbf{k} \cdot \mathbf{B}_0)^2 \Delta'}{L} \sim k^2 (B_0')^2 \epsilon^2 L \Delta'$$

and, substituting for ϵL from (5.18), we find the parametric dependence of the

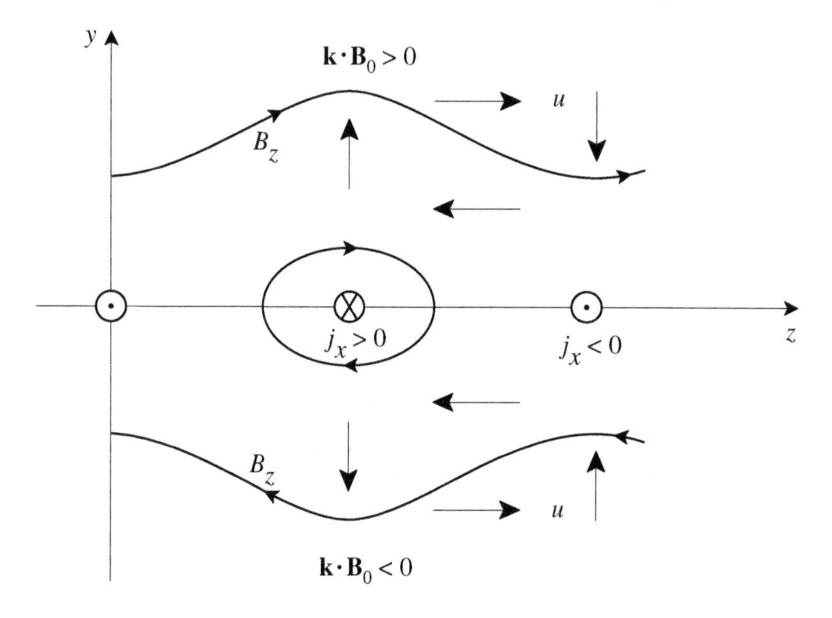

Fig. 5.7. Directions of current, field and velocity variations in the tearing instability.

growth rate γ to be

$$\gamma \sim \left[\frac{k^2(B_0')^2(\Delta')^4\eta_0^3}{\mu_0^4\rho_0 L^4}\right]^{1/5} \tag{5.21}$$

In terms of the resistive diffusion and convection times τ_R and τ_A, this may be written as

$$\gamma \sim \left[\frac{(kL)^2(\Delta')^4}{\tau_A^2\tau_R^3}\right]^{1/5} \sim (kL)^{2/5}S^{2/5}(\Delta')^{4/5}\tau_R^{-1} \tag{5.22}$$

showing that the time scale for the development of this instability is two-fifths Alfvénic $(S/\tau_R)^{2/5}$ and three-fifths diffusive $(1/\tau_R)^{3/5}$ and thus intermediate between these widely differing time scales. In a tokamak τ_R would be measured typically in seconds, τ_A in tens of nanoseconds and γ^{-1} in milliseconds. This wide separation of time scales justifies the simple order of magnitude analysis we have used.

The growth rate increases with resistivity, η, and with shear, $(\mathbf{k}\cdot\mathbf{B}_0)'$. It is driven by the Lorentz force which acts towards the resonant surface. Finite resistivity allows field lines to break and reconnect at the nodal X points and then contract towards the axes of the magnetic islands passing through the O points. Figure 5.7 shows the directions of current, field and velocity perturbations. Note that the equilibrium current $\mathbf{j}_0$, maintaining the variation in $\mathbf{B}_0(y)$, is in the x direction so the growth of the instability increases the current at the O points and decreases it at the X points causing the current to break up into filaments. The characteristics of this instability, namely the breaking of field lines and filamentation of the current, are reflected in the name given to it, i.e. the *tearing mode*.

Finally, an important feature of this instability is its dependence on Δ', the discontinuity in B_y' at the resonant surface. This means that the growth rate γ is determined by the global state of the plasma and not by the local equilibrium conditions at the resonant surface.

5.3.2 Driven resistive instabilities

Reintroduction of the η_0' term in (5.12) and the g term in (5.17) allows additional possibilities for instability since these terms appear in the eigenvalues. To find the parametric dependence of the growth rates of these driven instabilities we need only replace (5.20) by the corresponding approximations when the gravitational or variable resistivity terms dominate the dynamics in the boundary layer.

In the case that the g term dominates we have

$$\frac{\gamma^2 \rho_0 \xi_y}{(\epsilon L)^2} \sim k^2 g \rho_0' \xi_y$$

which, on substitution from (5.18), leads to

$$\gamma \sim \left(\frac{\eta_0}{\rho_0}\right)^{1/3} \left(\frac{k g \rho_0'}{B_0'}\right)^{2/3} \sim \left[\frac{(kL)^2 \tau_A^2}{\tau_R \tau_G^4}\right]^{1/3} \sim (kL)^{2/3} S^{2/3} \left(\frac{\tau_A}{\tau_G}\right)^{4/3} \tau_R^{-1} \quad (5.23)$$

where $\tau_G = (\rho_0/g\rho_0')^{1/2}$ is a gravitational time scale. Keeping only the dominant inertial and gravitational terms in (5.17), it is easily seen that ξ_y grows when $\rho_0' > 0$ or $\mathbf{g} \cdot \nabla \rho_0 < 0$, i.e. when there is an inverted density gradient as for the Rayleigh–Taylor instability. This is the *gravitational interchange mode*. From (5.23) we see that, unlike the tearing mode, this instability growth rate is reduced by increased magnetic shear, and, since it depends on $\rho_0'(0)$, it is a local instability. A similar resistive interchange instability may occur in a curved magnetic field when there is a pressure gradient aligned with the curvature (see the discussion in Section 4.7.2).

Turning now to the third resistive instability, we first substitute for $(B_y'' - k^2 B_y)$ from (5.12) in (5.17) to bring the η_0' term into the equation of motion and then balance this term with the inertia term to obtain

$$\frac{\mu_0 \gamma^2 \rho_0}{(\epsilon L)^2} \xi_y \sim \frac{(\mathbf{k} \cdot \mathbf{B}_0)(\mathbf{k} \cdot \mathbf{B}_0)' \eta_0'}{\eta_0} \xi_y$$

Table 5.2. *Resistive instability characteristics*

Mode	Range	$\gamma\tau_R$	ϵ
Tearing	$kL < 1$	$(kL)^{2/5}S^{2/5}(\Delta')^{4/5}$	$(kL)^{-2/5}S^{-2/5}(\Delta')^{1/5}$
Gravitational interchange	$\mathbf{g}\cdot\nabla\rho_0 < 0$	$(kL)^{2/3}S^{2/3}(\tau_A/\tau_G)^{4/3}$	$(kL)^{-1/3}S^{-1/3}(\tau_A/\tau_G)^{1/3}$
Rippling	$\eta_0' \neq 0$	$(kL)^{2/5}S^{2/5}(L\eta_0'/\eta_0)^{4/5}$	$(kL)^{-2/5}S^{-2/5}(L\eta_0'/\eta_0)^{1/5}$

Hence,

$$\gamma \sim \left[\frac{k^2(B_0')^2(\eta_0')^4}{\mu_0^4\rho_0\eta_0}\right]^{1/5} \sim \left[\left(\frac{L\eta_0'}{\eta_0}\right)^4 \frac{(kL)^2}{\tau_R^3\tau_A^2}\right]^{1/5} \sim (kL)^{2/5}S^{2/5}(L\eta_0'/\eta_0)^{4/5}\tau_R^{-1} \tag{5.24}$$

As for the other driven instability, this is local. To understand its physical origin we linearize Ohm's law, assuming $\mathbf{E} = 0$, to get

$$\eta_0\mathbf{j}_1 + \eta_1\mathbf{j}_0 = \mathbf{u}_1 \times \mathbf{B}_0$$

and substitute for η_1 from (5.11) so that

$$\eta_0\mathbf{j}_1 = \xi_y\eta_0'\mathbf{j}_0 + \mathbf{u}_1 \times \mathbf{B}_0$$

Thus, there is an additional (driving) force $\mathbf{F}_d$ due to the η_0' term in $\mathbf{j}_1$, given by

$$\mathbf{F}_d = \xi_y\frac{\eta_0'}{\eta_0}\mathbf{j}_0 \times \mathbf{B}_0 = (\boldsymbol{\xi}\cdot\nabla\eta_0)(\mathbf{j}_0 \times \mathbf{B}_0/\eta_0)$$

Assuming the variation of $\eta_0(y)$ is monotonic across the resonant surface whilst $\mathbf{B}_0$ changes sign, it follows that $\mathbf{F}_d$ is stabilizing on one side and destabilizing on the other. Physically, it is clear that $\mathbf{F}_d$ is destabilizing on the side of lower resistivity since this is where the current is increased. Thus, the fluid motion is amplified only on the lower resistivity side and this creates a rippling effect which gives the mode its name. The motion of the plasma in relation to the field lines for both driven instabilities is illustrated in Fig. 5.8.

Table 5.2 lists the main parametric properties of the three resistive instabilities. The tearing mode, because it is endemic and not dependent on an imposed driving force and since its growth rate increases with magnetic shear, is usually the most dangerous. Also, the other two are local instabilities with the resistive interchange mode stabilized by shear and the rippling mode stabilized by high temperature which increases the heat conductivity and invalidates the 'adiabatic' assumption (5.8).

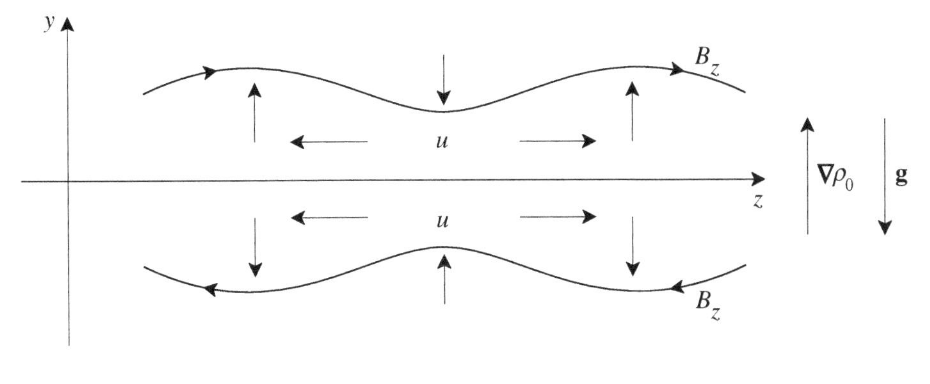

(a) Resistive **g** mode

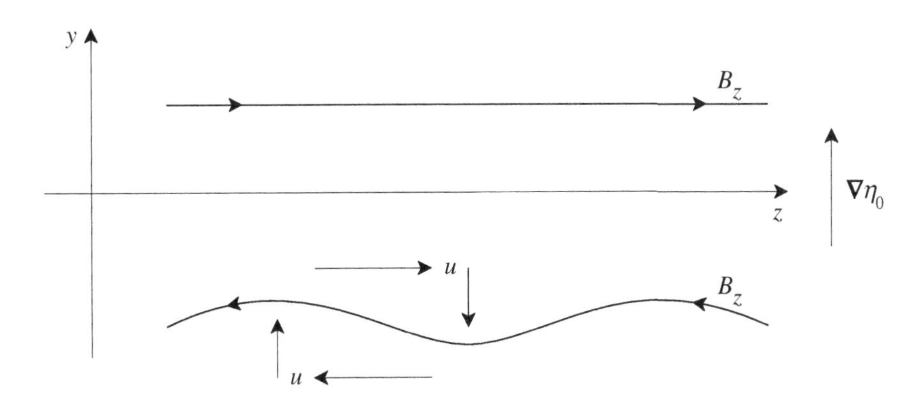

(b) Rippling mode

Fig. 5.8. Plasma motion relative to field lines in (a) gravitational interchange mode and (b) rippling mode.

5.3.3 Tokamak instabilities

Although we have considered only the simplest geometry of a slab plasma, the tearing mode makes its appearance in cylindrical and toroidal plasmas in which the resonant surfaces occur at the mode rational surfaces. In toroidal geometry in which perturbations vary as $A(r)e^{i(m\theta-n\phi)}$, θ and ϕ being the poloidal and toroidal angles and r the minor radius, the condition $\mathbf{k}\cdot\mathbf{B}_0 = 0$ becomes

$$\frac{m}{r}B_\theta - \frac{n}{R_0}B_\phi = 0$$

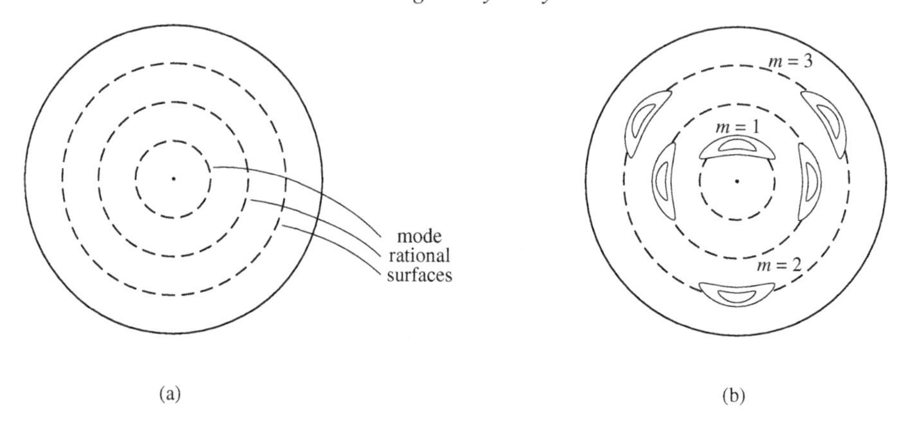

Fig. 5.9. Formation of magnetic islands at resonant surfaces in a tokamak.

For a circular torus of large aspect ratio we may substitute (4.35) to write this as

$$\left(\frac{m}{n} - q(r)\right) B_\theta(r) = 0$$

showing that resonant surfaces occur at $r = r_{mn}$ where

$$q(r_{mn}) = m/n$$

These are the mode rational surfaces. Figure 5.9 shows the formation of magnetic islands at the resonant surfaces corresponding to $m = 1, 2$ and 3. The mode rational surfaces on which these islands appear are shown in Fig. 5.9(a). With magnetic shear, $q(r)$ increases with r and the magnetic field lines on any surface $r = r_{mn}$ close on themselves after n circuits around the torus. (For all other values of r the field lines continue to wind around the torus without ever closing and eventually cover all of the surface. Such field lines and surfaces are said to be ergodic, as mentioned in the discussion of the rotational transform in Section 4.3.2.) The magnetic islands that form at the mode rational surfaces also twist around the torus closing on themselves after n times. Since heat flows rapidly along field lines one of the consequences of this structure is an increase in transport across the plasma.

Tearing mode instabilities are believed to be the source of *Mirnov oscillations* which are magnetic fluctuations, first detected by Mirnov and Semenov (1971), occurring during the current rise in tokamaks. The azimuthal variation of the fluctuations shows them to be associated with a succession of decreasing m numbers as the current rises and the q value at the plasma surface decreases according to

$$q(a) = \frac{aB_\phi}{RB_\theta(a)} = \frac{2\pi a^2 B_\phi}{\mu_0 RI}$$

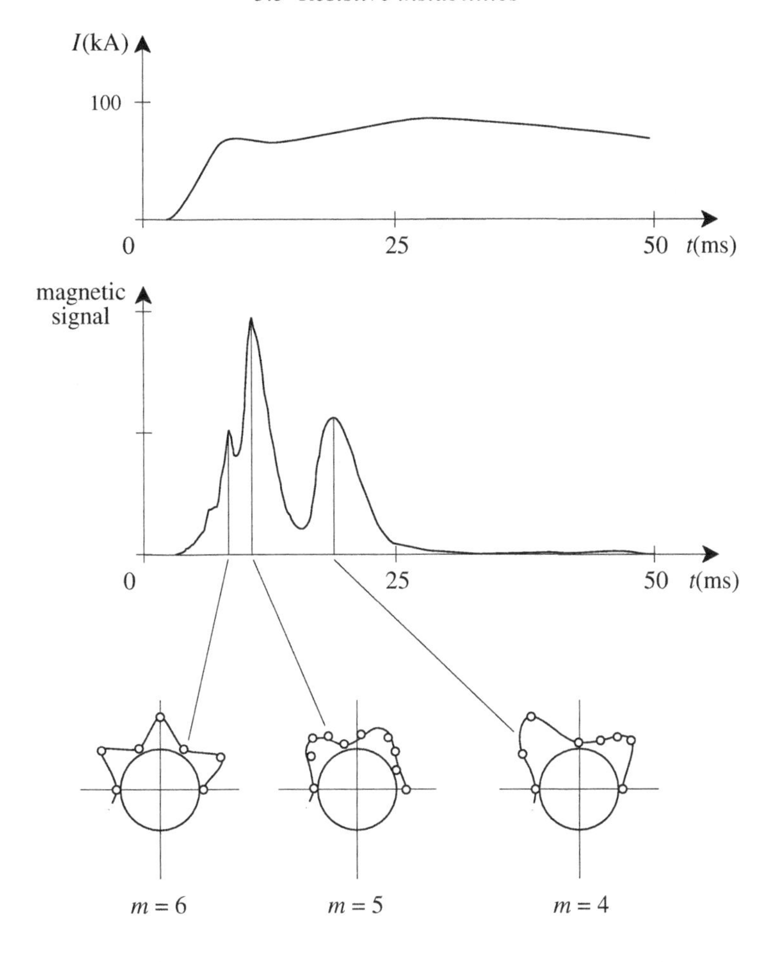

Fig. 5.10. Mirnov oscillations (after Mirnov and Semenov (1971)).

where from (4.73) we have substituted $\mu_0 I = 2\pi a B_\theta(a)$. Time variations are illustrated in Fig. 5.10. Note that these oscillations persist beyond the current rise.

Sawtooth oscillations, observed in the soft X-ray emission from tokamaks, are evidence of another instability, in this case thought to be the $m = 1$ kink mode occurring in the centre of the plasma. The name arises from the shape of the oscillations which show a slow rise over a period of about 1 ms followed by a sudden fall; X-rays from the outer region of the plasma show the inverse pattern of a slow decay followed by a rapid rise, indicating that temperature changes in the outer plasma compensate those occurring in the central plasma due to the instability. These are illustrated in Fig. 5.11.

The slow build-up in the inner region arises because the plasma is hotter there and, since conductivity increases with temperature, any increase in axial current

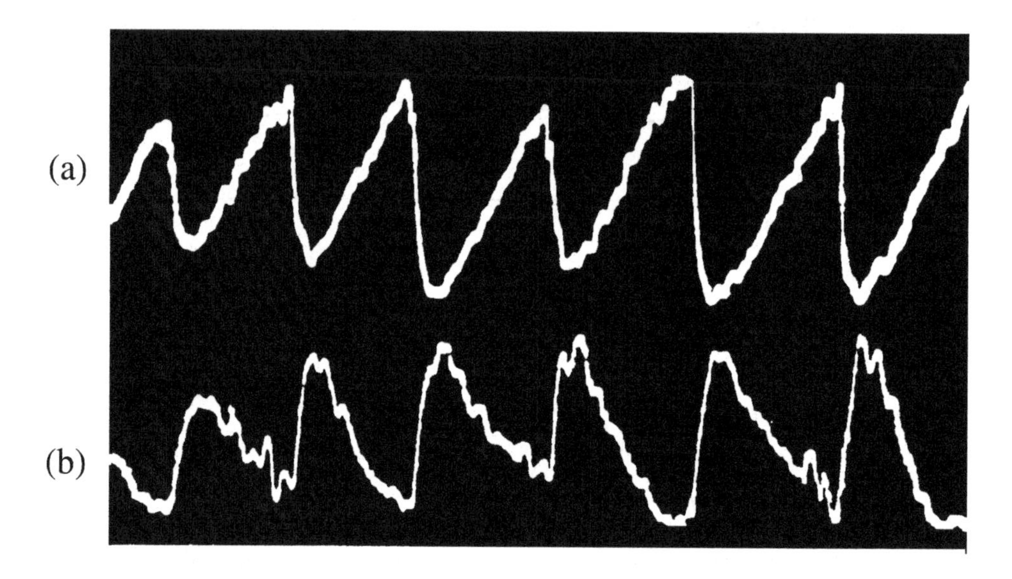

Fig. 5.11. Sawtooth oscillations in X-ray emission from (a) the central region and (b) the outer region of a tokamak plasma (after Wesson (1987)).

means increased ohmic heating leading to an unstable concentration of current in the centre of the plasma. Now as the axial current increases so does the poloidal magnetic field and for sufficiently large current the safety factor q may fall below unity allowing a $q = 1$ surface to appear near the axis. This is illustrated in Fig. 5.12 which shows the subsequent development of the $m = 1$ instability at the $q = 1$ surface. The shaded region is the hot $q < 1$ island. The growth of the $q > 1$ island displaces it to the outer region where it slowly decays and disappears by thermal conduction to the cooler plasma. The equilibrium field structure is restored and the procedure is repeated. The rapid reconnection phase, (iii)–(v), produces the sudden fall (rise) in temperature in the inner (outer) region of the plasma and the corresponding patterns in soft X-ray emission.

Neither Mirnov nor sawtooth instabilities prevent the satisfactory operation of tokamaks. On the other hand, a third resistive instability leads to the collapse of the plasma current and it is therefore known as the *disruptive instability*. The principal characteristics of the disruptive instability are a rapid broadening of the current profile with consequent decrease in the poloidal field, followed by a loss of thermal energy from the plasma to a degree that quenches the discharge. In general the instability is interpreted in terms of a model in which magnetic surfaces are destroyed by tearing modes with different helicities at different resonant surfaces.

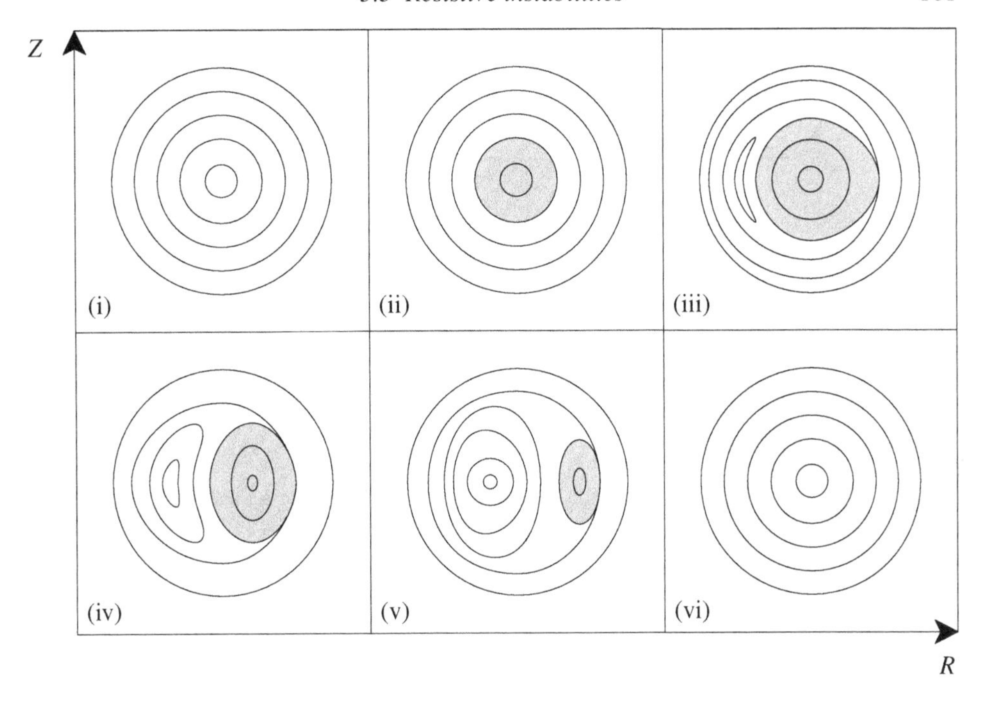

Fig. 5.12. Development of $m = 1$ instability at $q = 1$ surface (after Wesson (1987)).

Plasma containment breaks down with loss of energy from the tokamak by means of heat transport along the field.

The disruptive instability is the least well understood of the tokamak instabilities but observations indicate that the $m = 2$ tearing mode is crucially involved. In the earliest phase the $m = 2$ instability is saturated at a low level but slowly increasing density or current triggers the precursor phase in which the unstable oscillations reach much higher amplitudes. Since other low m modes are observed in this phase it is possible that a non-linear interaction of the $m = 2$ mode with the $m = 1$ sawtooth or $m = 3$ ($n = 1$ or 2) modes is involved, though other interactions with the outer region of the plasma or the limiter are possible. Whatever the mechanism, the growth in amplitude over a period of about 10 ms triggers the fast phase in which the central temperature collapses and the radial current profile flattens in a time of the order of 1 ms. This is followed by the quench phase in which the plasma current decays to zero. Fuller discussions of the disruptive instability are to be found in Wesson (1987) and Biskamp (1993). The phases of a disruptive instability driven by increasing density are illustrated schematically in Fig. 5.13 which sketches the time development of the $m = 2$ magnetic field fluctuations, central temperature, and plasma current.

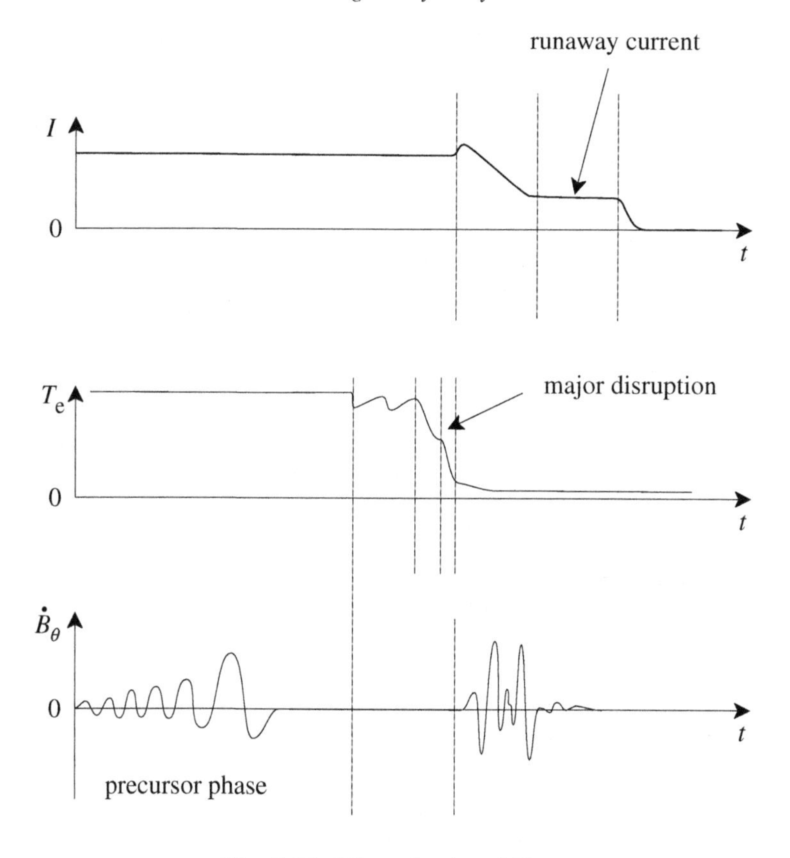

Fig. 5.13. Disruptive instability.

5.4 Magnetic field generation

Magnetic fields pervade the Universe ranging in magnitude from background levels of about 10^{-10} T to as high as 10^8 T in neutron stars. Understanding why large scale magnetic fields occur in stars and galaxies remains a key concern in astrophysics (Field (1995)). A detailed discussion of the problems of field generation has been given by Parker (1979). Here we limit our discussion to a brief outline of some basic ideas. The contraction of gas leading to formation of galaxies carried with it some part of the primordial magnetic field so that each galaxy, when formed, had within it a magnetic field. In the same way diffuse interstellar gas clouds condensing to form stars had remnants of the primordial field trapped within. However, in due course magnetic buoyancy and diffusion might be expected to disperse such fields except possibly for some fragment preserved within the (stable) inner core of stars. Consequently, one of the most long-standing and challenging problems in MHD has been to explain just how the observed magnetic fields of stars and planets are sustained.

There are two possible mechanisms by which magnetic fields may be regenerated, by means of a dynamo or a battery. A dynamo mechanism necessarily requires some initial field on which the fluid motion can act. This is not a requirement for the battery mechanism suggested by Biermann (1950) which in the event proved to be incapable of generating sufficiently strong fields. However, as Parker (1979) remarks, the real significance of the Biermann battery is that it guarantees, if all else fails, a seed field for stars and galaxies. We turn now to an outline of some aspects of dynamo action.

5.4.1 The kinematic dynamo

One obvious source of energy for the regeneration of magnetic fields is the kinetic energy in flow fields. Dynamo action amounts to the systematic conversion of the kinetic energy of the flow field into magnetic field energy. The full dynamo problem is formidable since the regeneration of the field must come via the convection term in (5.1) and so what is required is the simultaneous solution of this equation for **B** and the equation of motion for **u**. The difficulties of this task are such that work has mostly been concentrated on the *kinematic dynamo problem* in which one tries to devise a flow field which will maintain a magnetic field against resistive decay, i.e. (5.1) is to be solved for **B** when $\mathbf{u}(\mathbf{r}, t)$ is given.

Paradoxically, a major advance in kinematic dynamo theory was made in 1934 by Cowling's proof of an *anti-dynamo theorem* (Cowling, 1976). Using a simple argument he showed that a steady, axisymmetric magnetic field cannot be maintained. In the case that he considered both the flow and the field lines are in a meridional plane through the axis of symmetry. In any such plane the field lines must be closed curves enclosing at least one neutral line as shown in Fig. 5.14. Then if we integrate Ohm's law around this line we get

$$\begin{aligned}\oint \mathbf{j}\cdot d\mathbf{l} &= \sigma\oint \mathbf{E}\cdot d\mathbf{l} + \sigma\oint \mathbf{u}\times\mathbf{B}\cdot d\mathbf{l} \\ &= \sigma\int \nabla\times\mathbf{E}\cdot d\mathbf{S} = -\sigma\int \frac{\partial \mathbf{B}}{\partial t}\cdot d\mathbf{S} = 0 \qquad (5.25)\end{aligned}$$

where the integral of the convection term vanishes on account of the fact that **B** is zero along the neutral line and the final integral vanishes because, by assumption, $\partial\mathbf{B}/\partial t = 0$. But (5.25) implies that $j_\phi = 0$ which is clearly incompatible with Ampère's law, $\mu_0\mathbf{j} = \nabla\times\mathbf{B}$, and the contradiction proves the theorem. The physical interpretation of the theorem is that while the convection term can transport the field lines in the meridional plane it cannot create new field lines to replace those that diffuse through the plasma and disappear at the neutral point.

The proof of Cowling's theorem can be extended to include an azimuthal component of the magnetic field. In this case $\mathbf{u}\times\mathbf{B}\cdot d\mathbf{l} = 0$ on the neutral line since

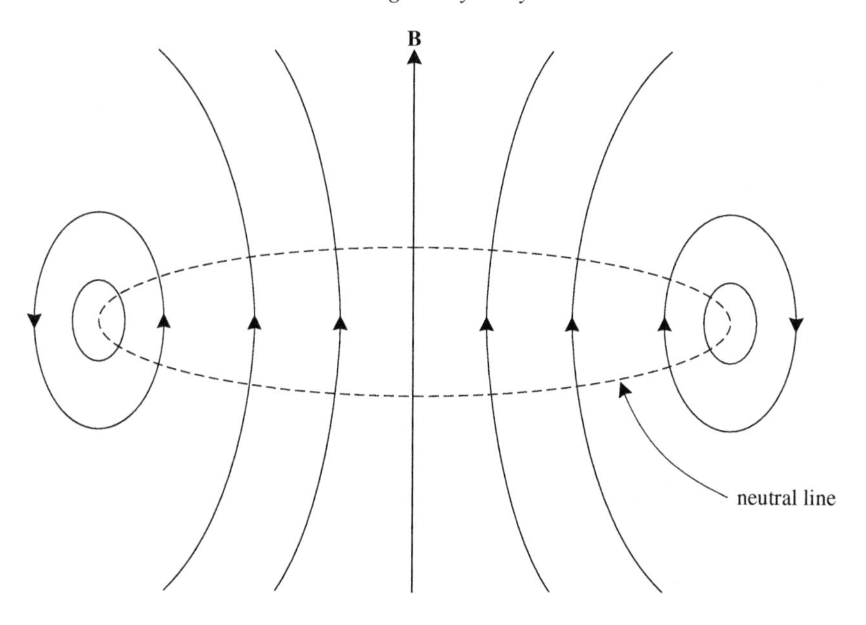

Fig. 5.14. Geometry of field lines in Cowling's theorem.

B and d**l** are parallel. Further generalizations of the anti-dynamo theorem exist for field configurations which are topologically similar to the axisymmetric case. The conclusion is that although a spherical body like a star or planet may have a dominant axially symmetric dipole field at its surface, the magnetic field within the body must be considerably more complicated if it is to be maintained by dynamo action.

Cowling's theorem and extensions of it are important in ruling out fields and fluid motions with certain simple structures. The question remains as to what properties are required of the motion for magnetic fields to be generated? In the most general terms the answer appears to be that differential (non-uniform) rotation and turbulent convection are required.

Following Cowling's theorem it took more than twenty years before Herzenberg (1958) and Backus (1958) proved existence theorems for possible dynamo mechanisms for steady and oscillating fields, respectively. This led to the development of other dynamo models and the emphasis of research was able to turn from mere existence to possible relevance. Subsequently, progress has been more rapid so that the kinematic problem is now broadly understood and, despite its mathematical complexity, considerable advances have been made towards the ultimate goal of a self-consistent solution of the dynamic problem. The brief qualitative account of the essential physics of the dynamo action presented here follows closely the decriptions given by Moffat (1993) and Field (1995).

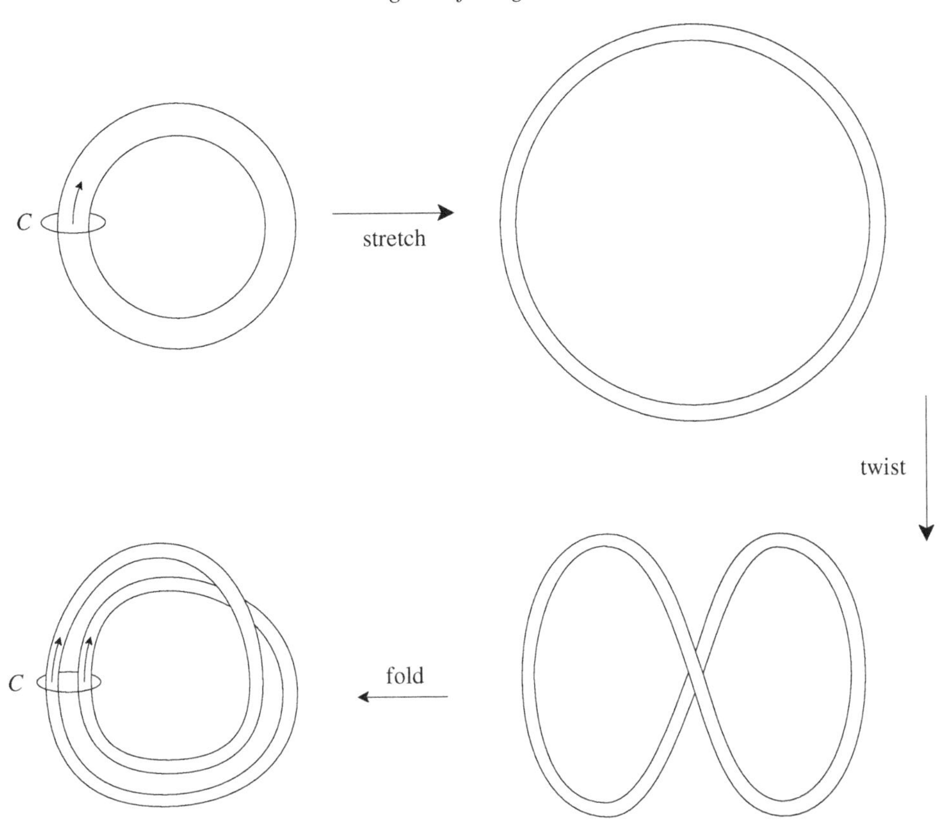

Fig. 5.15. Increase in field strength by stretch, twist and fold sequence (after Moffat (1993)).

We have seen already in Section 4.2 how a flow which stretches a magnetic flux tube increases the field strength. Continuing the analogy, we may suppose that the flux tube is stretched to twice its original length and then twisted and folded, as illustrated in Fig. 5.15, to obtain a field of twice the original strength through the fixed loop C. Of course, such a stretch–twist–fold cycle, suggested originally by Zel'dovich in 1972, merely illustrates how the field strength *may* be increased from an arbitrarily low level and it is necessary to explain what kind of a flow might realistically produce this effect. The answer lies in the combination of two physical processes known as the α- and Ω-effects.

It is assumed that **u** and **B** consist of slowly varying axisymmetric mean components and weak small scale fluctuating components:

$$\begin{aligned}\mathbf{u} &= \langle\mathbf{u}\rangle + \mathbf{u}' \\ \mathbf{B} &= \langle\mathbf{B}\rangle + \mathbf{B}'\end{aligned}$$

Averaging the induction equation over the small scale fluctuations gives

$$\frac{\partial \langle \mathbf{B} \rangle}{\partial t} = \nabla \times [\langle \mathbf{u} \rangle \times \langle \mathbf{B} \rangle] + \nabla \times \langle \mathbf{u}' \times \mathbf{B}' \rangle + \eta \nabla^2 \langle \mathbf{B} \rangle \tag{5.26}$$

Subtracting this equation from (5.1) then gives the evolution equation for $\mathbf{B}'$, namely,

$$\frac{\partial \mathbf{B}'}{\partial t} = \nabla \times [\langle \mathbf{u} \rangle \times \mathbf{B}' + \mathbf{u}' \times \langle \mathbf{B} \rangle] + \nabla \times [\mathbf{u}' \times \mathbf{B}' - \langle \mathbf{u}' \times \mathbf{B}' \rangle] + \eta \nabla^2 \mathbf{B}' \tag{5.27}$$

Since we may transform to a coordinate system in which $\langle \mathbf{u} \rangle = 0$ the basic task of kinematic dynamo theory is to solve (5.27) for $\mathbf{B}'$ in terms of $\mathbf{u}'$ and $\langle \mathbf{B} \rangle$ and substitute in the second term on the right-hand side of (5.26). This is the crucial 'extra' term (compared with the induction equation containing only the mean axisymmetric flow and field) which permits dynamo action to take place. It does so by means of a regenerative cycle in which a toroidal field $\mathbf{B}_t$ is created from a poloidal field $\mathbf{B}_p$ by means of the Ω-effect while the combination of diffusion and the α-effect acts on $\mathbf{B}_t$ to regenerate $\mathbf{B}_p$. The Ω-effect arises from the first term on the right-hand side of (5.26) and is relatively simple to understand as explained below. The α-effect comes from the second term $\nabla \times \langle \mathbf{u}' \times \mathbf{B}' \rangle$ and is much more subtle. Specifically, under certain simplifying assumptions one can show that

$$\langle \mathbf{u}' \times \mathbf{B}' \rangle = -\alpha \langle \mathbf{B} \rangle + \beta \nabla \times \langle \mathbf{B} \rangle$$

where $\alpha = \langle \mathbf{u}' \cdot \nabla \times \mathbf{u}' \rangle \tau / 3$, $\beta = \langle \mathbf{u}' \cdot \mathbf{u}' \rangle \tau / 3$ and τ is the velocity correlation time. The β term enhances the diffusion whilst the α term, provided the helicity is non-zero, regenerates the mean poloidal field.

The Ω-effect takes its name from the symbol used to denote the rate of angular rotation of a conducting sphere, which is assumed to vary with distance from the axis of rotation. Such a differential rotation rate arises when convection currents are subject to a combination of buoyancy and Coriolis forces. The essential point is that, by conservation of angular momentum, descending fluid elements increase their rate of rotation. Consequently, a field line passing through the rotating sphere is wound around the axis of rotation rather than simply being carried around, as would be the case if the rotation rate was uniform. As Fig. 5.16 shows, this creates a toroidal field component B_t from what was originally a purely poloidal field B_p; note, also, that B_t is antisymmetric about the equatorial plane.

The α-effect does the opposite, creating a poloidal component B_p from a purely toroidal field B_t. Consider the flow field

$$\mathbf{u} = (0, u_0 \cos(kx - \omega t), u_0 \sin(kx - \omega t))$$

which represents a circularly polarized wave travelling along the x-axis as shown in Fig. 5.17(a). It is easily verified that the vorticity $\boldsymbol{\omega} \equiv \nabla \times \mathbf{u} = -k\mathbf{u}$ and hence

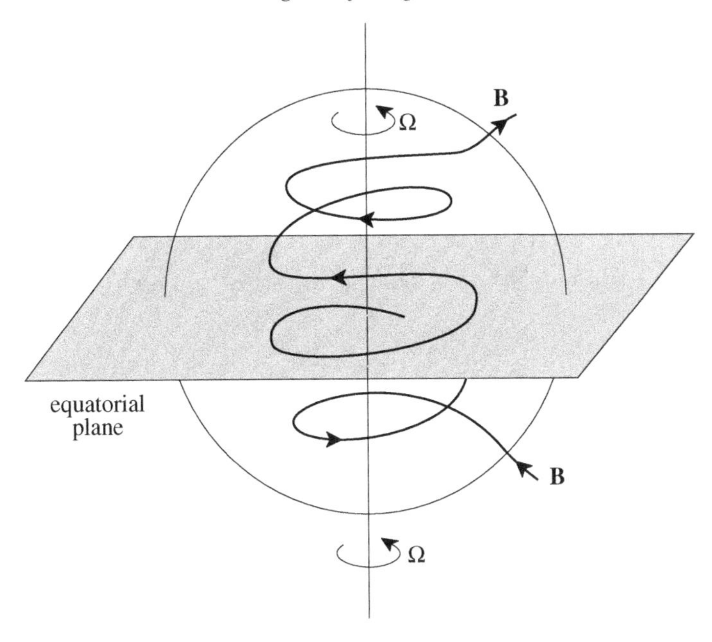

Fig. 5.16. Illustration of Ω-effect (after Moffat (1993)).

the kinetic helicity $\boldsymbol{\omega} \cdot \mathbf{u} = -ku_0^2$ is constant. Now it turns out that such a flow can deform a straight magnetic field line into a helix, as indicated in Fig. 5.17(b), though the way in which this is done is quite subtle and requires finite resistivity.

We have seen already in Section 4.3.4 that a current density parallel to the magnetic field produces a helical field. However, the electromotive force $\mathbf{u} \times \mathbf{B}$, resulting from the interaction of flow and field, gives rise to a current component perpendicular to $\mathbf{B}$ in ideal MHD and it is only when the effect of finite resistivity shifts the phase of the magnetic field perturbation $\mathbf{b}$ relative to $\mathbf{u}$ that a space-averaged $\langle \mathbf{u} \times \mathbf{b} \rangle$ leads to a current component parallel to $\mathbf{B}$, i.e. $\mathbf{j} = \alpha \mathbf{B}$; the constant α is proportional to the phase shift which, in turn, is proportional to the resistivity. Steenbeck, Krause and Rädler (1966) showed that the α-effect occurs in any turbulent flow field provided that the mean helicity $\langle \boldsymbol{\omega} \cdot \mathbf{u} \rangle \neq 0$.

In toroidal geometry the x-axis transforms to the toroidal direction and we see that the α-effect produces a poloidal component B_p from a toroidal field B_t. Thus, the combined $\alpha\Omega$-mechanism comprises a regenerative cycle. It is now widely accepted that it is by this cycle that the magnetic fields of the Earth and Sun are maintained and there is a nice irony in that the very effect (finite resistivity) that makes a regenerative mechanism necessary is also essential to its operation.

Although the kinematic problem is generally well understood the dynamic problem is still wide open. In both the Earth and Sun the $\alpha\Omega$-mechanism operates in

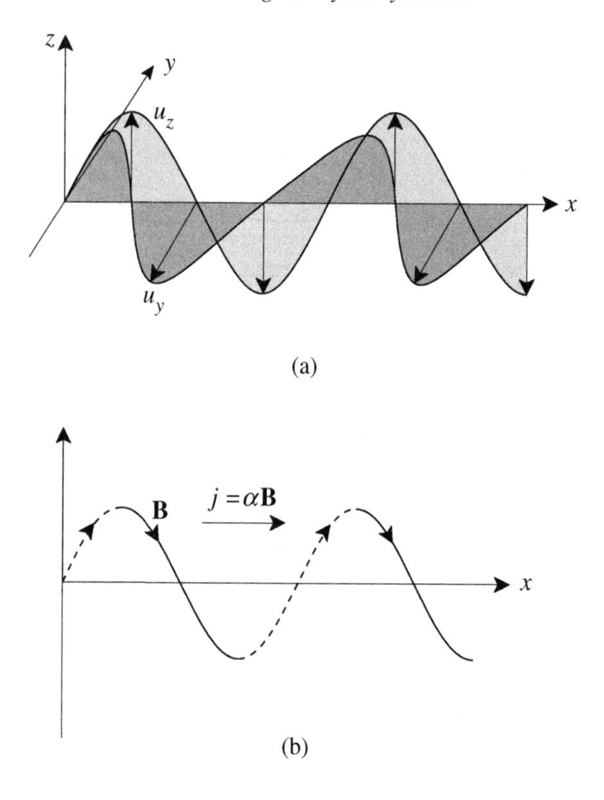

Fig. 5.17. Illustration of α-effect (after Moffat (1993)).

a convection zone which is a spherical annulus. For the Earth this is the liquid outer core which lies between the solid inner core and the Earth's mantle. The most widely accepted model is one originally proposed by Braginsky (1991) who suggested that the slow solidification of the liquid at the inner core boundary (ICB) releases an excess of lighter elements, such as sulphur, in the liquid alloy of the outer core. The buoyancy of the lighter fluid causes it to rise towards the core–mantle boundary (CMB) generating convection currents as it does so. Beyond that, however, essential questions regarding the length scale of the rising elements, the degree of turbulence that they generate and the rate at which they mix with the heavier liquid, which determines the diffusion rate, remain unanswered. The thrust of research is therefore, on the experimental side, a detailed study of the variation with time of the Earth's field so that, by the application of inverse theory, the field at the CMB may be reconstructed and, on the theoretical side, the development of self-consistent models.

A computer simulation by Glatzmaier and Roberts (1995) which simulated the Earth's field over a period of 40 000 years produced the first convincing evidence that fluid motions in the Earth's core could sustain the geomagnetic field. The three-dimensional model has not only generated a stable, dipole field over a period roughly three times the magnetic diffusion time ($\sim$13 000 years) but one which reproduced other geomagnetic features such as magnetic axis displacement (from rotational axis) and field reversal.

5.5 The solar wind

The continual streaming of plasma from the Sun's surface is known as the *solar wind* and we have noted already that there are important consequences for the geomagnetic field arising from reconnection in the region where the solar wind strikes the magnetosphere. The solar wind extends the solar plasma to the Earth and well beyond; in fact, the solar wind is thought to continue to a boundary with the interstellar medium at 50–100 AU. Correlation between variations of the Earth's magnetic field and activity on the Sun had been observed since the nineteenth century but attempts to explain the connection on the basis of static models of the solar corona failed to provide satisfactory solutions and, although there had been earlier suggestions that plasma ejected from the Sun was the cause of geomagnetic storms and the shape of comet tails, it was not until 1958 that Parker established the theoretical basis for the solar wind by solving the steady flow problem. Soon afterwards satellite observations confirmed its existence and began to compile its physical properties in ever increasing detail.

Parker (1958) showed that no hydrostatic equilibrium between the solar corona and interstellar space was possible and that non-equilibrium is what gives rise to a supersonic low density flow which is the solar wind. Although both the high temperature of the corona and the outflowing solar wind have been long observed, the precise mechanism that leads to plasma flowing out from the Sun's surface is not well understood. The slowest flows can be attributed to a thermally driven wind but non-thermal additions are needed in the supersonic region. Energy and momentum transfer from hydromagnetic waves, such as Alfvén waves, are likely sources and Alfvén-like fluctuations have been observed in the solar wind. Parameters characteristic of the solar wind are given in Table 5.3 at distances of a solar radius ($R_\odot = 6.96 \times 10^8$ m) and at 1 AU $= 215 R_\odot$.

The solar wind is composed largely of protons and is permeated by a magnetic field. The magnetic field is frozen in the radial flow outwards from the surface. However because of the Sun's rotation the magnetic field is twisted into a spiral. It exhibits a complex structure attributable in part to the admixture of open and closed magnetic structures at the Sun's surface. In a general sense open magnetic structures are favourable to the generation of the solar wind while closed structures oppose it.

Table 5.3. *Solar wind characteristics*

$r/R_\odot$	1	215
Composition	H^+, He^{++}	H^+, He^{++}
Number density n_i (m^{-3})	$2 \cdot 10^{14}$	$7 \cdot 10^6$
Ion temperature T_i (K)	10^6	10^5
Plasma flow velocity u ($km\,s^{-1}$)	1	400
Magnetic field B (nT)	10^5	10

The flow problem that Parker solved was hydrodynamic rather than hydromagnetic in that he assumed that the dominant forces in the equation of motion are the pressure gradient and gravity. Thus, for the steady, spherically symmetric flow of an isothermal plasma the equation of motion is

$$\rho u \frac{\mathrm{d}u}{\mathrm{d}r} = -\frac{\mathrm{d}P}{\mathrm{d}r} - \frac{GM_\odot \rho}{r^2} \tag{5.28}$$

where G is the gravitational constant and $M_\odot$ is the mass of the Sun. Mass conservation requires

$$4\pi r^2 \rho u = \text{const.} \tag{5.29}$$

and the assumption of constant temperature means that we can define a constant *isothermal sound speed* by

$$u_c^2 = P/\rho. \tag{5.30}$$

Differentiating (5.29) and (5.30) and eliminating $\mathrm{d}\rho/\mathrm{d}r$ gives $\mathrm{d}P/\mathrm{d}r$ in terms of $\mathrm{d}u/\mathrm{d}r$ which may be substituted in (5.28) to give

$$\left(u - \frac{u_c^2}{u}\right)\frac{\mathrm{d}u}{\mathrm{d}r} = \frac{2u_c^2}{r} - \frac{GM_\odot}{r^2} \tag{5.31}$$

This equation has a critical point at $r = r_c \equiv GM_\odot/2u_c^2$, $u = u_c$, where $\mathrm{d}u/\mathrm{d}r$ is undefined. Its analytic solution is

$$\left(\frac{u}{u_c}\right)^2 - \log\left(\frac{u}{u_c}\right)^2 = 4\log\left(\frac{r}{r_c}\right) + \frac{4r_c}{r} + C \tag{5.32}$$

where C is a constant of integration. The solution curves are sketched in Fig. 5.18; the critical point A at (r_c, u_c) is a saddle point with $\mathrm{d}u/\mathrm{d}r$ becoming infinite as $u \to u_c$ $(r \neq r_c)$ and vanishing as $r \to r_c$ $(u \neq u_c)$, in accordance with (5.31). The trajectories through the critical point separate the various classes of solution. Of these I and III are double-valued and, therefore, physically unacceptable while II and IV have solar wind speeds which are entirely supersonic (II) or subsonic (IV),

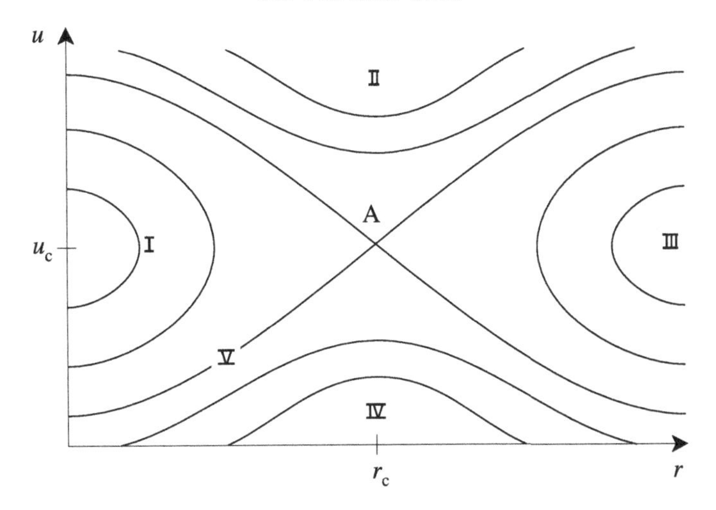

Fig. 5.18. Solution curves for Parker's solar wind model.

neither of which accords with observation. Since the solar wind has a flow speed which is subsonic at the Sun and supersonic at the Earth the only acceptable solution is the (positive slope) trajectory (V) through the critical point (r_c, u_c). It is apparent from (5.32) that $C = -3$ for this curve.

For $r \to \infty$, $u \gg u_c$ so that $u \sim (\log r)^{1/2}$ and, from (5.29) and (5.30), $P \sim \rho \sim r^{-2}(\log r)^{-1/2} \to 0$, as one would expect. The model predicts a flow speed of about $10\,\mathrm{km\,s^{-1}}$ at the Sun and about $100\,\mathrm{km\,s^{-1}}$ at the Earth, both of which are of the right order of magnitude. Unfortunately, its prediction for the density at the Earth is two orders of magnitude too high. Consequently, there have been many further developments of the model. In particular, the assumption of an isothermal plasma is known to be an over-simplification so the model has been extended to include heat conduction in an energy equation and, in a further refinement, a two-fluid model with separate ion and electron energy equations has been investigated since energy exchange between the species is negligible in the solar wind except for the region very close to the Sun. We shall not pursue these developments but an account of them may be found in Priest (1987).

Of course, a major simplification of Parker's model is the omission of the magnetic field. If the magnetic energy in the solar wind is negligible compared with its kinetic energy, as Parker assumed, then the field does not significantly affect the flow but, on the other hand, the flow drags the field lines out from the solar surface as indicated in Fig. 5.19. In plan view, looking down the polar axis, the field lines are spirals due to the combined effect of the solar wind and solar rotation. We can calculate the angle $\psi(r)$ between the flow velocity $u(r)\hat{\mathbf{r}}$ and the magnetic field $\mathbf{B}(r)$ by considering the motion of a fluid element which leaves a point P_0 on the

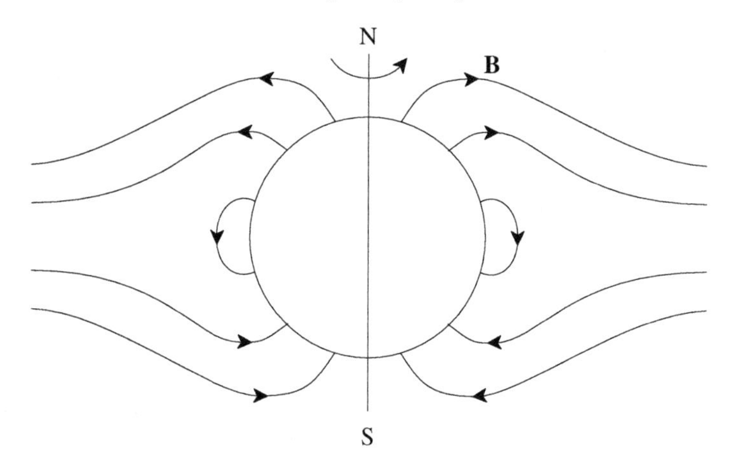

Fig. 5.19. Field lines at solar surface.

Sun's surface at time $t = 0$. To an observer at $P_0(t)$ rotating with the Sun, the fluid element traces out the spiral path $P_0(t) \rightarrow P(t)$ shown in Fig. 5.20 and, in the ideal MHD approximation, this must be a field line since there can be no motion perpendicular to the field lines. On the other hand, to an observer not rotating with the Sun but remaining at $P_0(0)$ the path of the fluid element is straight and radial, a motion that is made possible by the rotation of the field line about P_0. In plane polar coordinates with origin at $P_0(0)$ the velocities of fluid element and field line at $P(t)$ are $\mathbf{u} = [u(r), 0]$ and $\mathbf{v}_B = [0, \Omega(r - R_\odot)]$, respectively. Then the condition that there should be no motion of the fluid element perpendicular to the field line is

$$u \sin \psi(r) = v_B \cos \psi(r)$$

from which we get

$$\tan \psi(r) = \frac{\Omega(r - R_\odot)}{u(r)} \tag{5.33}$$

At the Earth this angle is about $\pi/4$.

In fact, the assumption that the kinetic energy dominates the magnetic energy, so that the field does not affect the flow, does not hold very close to the Sun's surface and the field lines wind up into a very tight spiral slowing the radial flow drastically at low latitudes. The radius r_A at which the kinetic and magnetic energies are equal, i.e. the flow speed is the Alfvén speed v_A, is called the *Alfvén radius*. For $r \ll r_A$, the field keeps the solar wind rotating with the Sun thereby increasing its angular momentum (at the expense of the Sun! – this effect is thought to have slowed the Sun's rotation significantly during its lifetime); well beyond $r = r_A$ the effect of the field on the wind becomes negligible and it continues its outward flow conserving

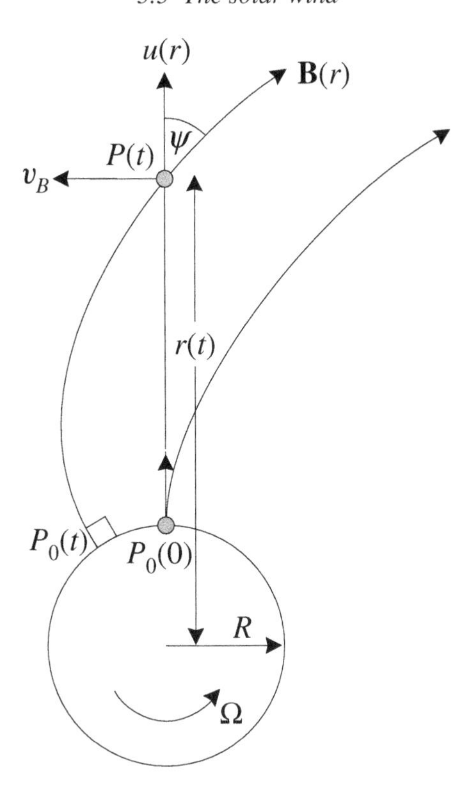

Fig. 5.20. Plasma motion in solar wind.

its angular momentum. In a more accurate model, therefore, the solar wind has a flow velocity with both radial and azimuthal components.

The azimuthal component of the flow velocity of the solar wind in the equatorial plane of the Sun was calculated by Weber and Davis (1967). They examined a steady state model with axial symmetry in which the field lines are completely drawn out by the solar wind in the equatorial plane so that there is no component of the field perpendicular to the plane. Thus, in the equatorial plane it is assumed that, in spherical polar coordinates, $\mathbf{B} = [B_r(r), 0, B_\phi(r)]$ and $\mathbf{u} = [u_r(r), 0, u_\phi(r)]$. Assuming that the field is radial at the solar surface ($r = R_\odot$), i.e. $\mathbf{B}(R_\odot) = (B_0, 0, 0)$ say, it follows from $\nabla \cdot \mathbf{B} = 0$ that

$$B_r(r) = B_0 R_\odot^2 / r^2 \tag{5.34}$$

Mass conservation requires

$$\rho u_r r^2 = \text{const.} \tag{5.35}$$

and the steady state induction equation, $\nabla \times (\mathbf{u} \times \mathbf{B}) = 0$, gives

$$\frac{1}{r}\frac{\mathrm{d}}{\mathrm{d}r}\left[r(u_r B_\phi - u_\phi B_r)\right] = 0$$

which may be integrated to obtain

$$u_r B_\phi - u_\phi B_r = -\Omega R_\odot^2 B_0/r \tag{5.36}$$

assuming $u_\phi(R_\odot) = \Omega R_\odot$. It then follows from (5.34) and (5.36) that

$$B_\phi(r) = \frac{u_\phi(r) - r\Omega}{u_r(r)} B_r(r) \tag{5.37}$$

The equation of motion is

$$\rho(\mathbf{u}\cdot\nabla)\mathbf{u} = -\nabla P + (\nabla \times \mathbf{B}) \times \mathbf{B}/\mu_0 - (GM_\odot \rho/r^3)\mathbf{r} \tag{5.38}$$

the ϕ-component of which is simply

$$\rho\frac{u_r}{r}\frac{\mathrm{d}}{\mathrm{d}r}(ru_\phi) = \frac{B_r}{\mu_0 r}\frac{\mathrm{d}}{\mathrm{d}r}(rB_\phi) \tag{5.39}$$

But from (5.34) and (5.35), $\rho u_r r^2$ and $B_r r^2$ are both constant so multiplying (5.39) by r^3 we may integrate to get

$$ru_\phi - \frac{B_r}{\mu_0 \rho u_r} r B_\phi = \text{const.} = L \tag{5.40}$$

say. Introducing the radial *Alfvén Mach number*

$$M_\mathrm{A} \equiv \frac{u_r}{B_r/(\mu_0\rho)^{1/2}} \tag{5.41}$$

we may substitute (5.37) in (5.40) to obtain

$$u_\phi(r) = \Omega r(M_\mathrm{A}^2 L/(\Omega r^2) - 1)/(M_\mathrm{A}^2 - 1) \tag{5.42}$$

This equation determines u_ϕ as a function of r and $M_\mathrm{A}(r)$. From observations it is known that $M_\mathrm{A} \ll 1$ near the surface of the Sun and that $M_\mathrm{A} \simeq 10$ at the Earth. The point $r = r_\mathrm{A}$, between the Sun and the Earth, at which $M_\mathrm{A} = 1$ is called the *Alfvén critical point*. At this point the numerator in (5.42) must vanish to keep $u_\phi(r_\mathrm{A})$ finite so that $L = \Omega r_\mathrm{A}^2$.

From (5.34), (5.35) and (5.41) we deduce that $M_\mathrm{A}^2/u_r r^2$ is a constant, which we evaluate at the Alfvén critical point to get

$$\frac{M_\mathrm{A}^2}{u_r r^2} = \frac{1}{v_\mathrm{A}(r_\mathrm{A}) r_\mathrm{A}^2} \tag{5.43}$$

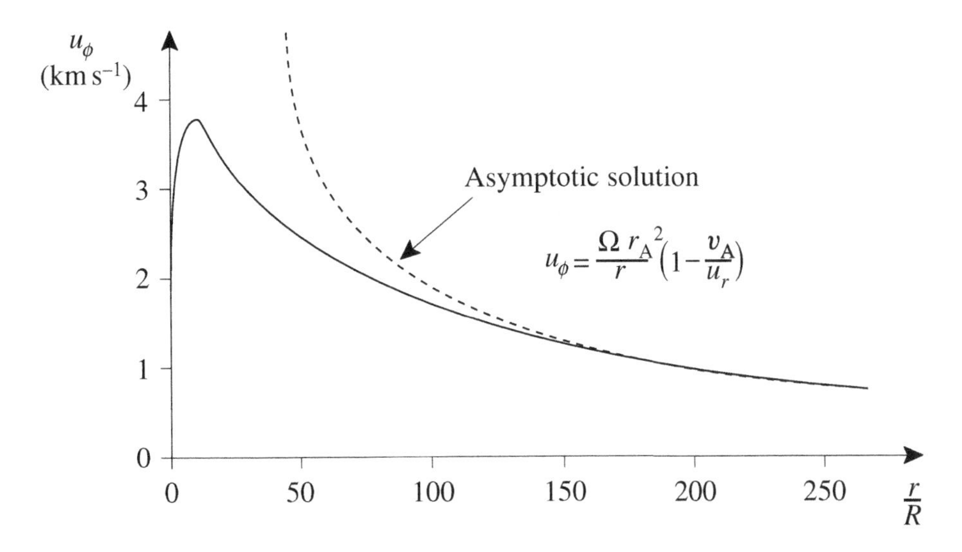

Fig. 5.21. Azimuthal flow velocity of solar wind (after Weber and Davies (1967)).

so that we may write (5.42) as

$$u_\phi(r) = \Omega r \frac{(1 - u_r/v_A)}{(1 - M_A^2)} \tag{5.44}$$

and (5.37) as

$$B_\phi(r) = -B_r \frac{\Omega r}{v_A} \frac{(1 - (r/r_A)^2)}{(1 - M_A^2)} = -\frac{B_0 R_\odot^2 \Omega}{v_A r} \frac{(1 - (r/r_A)^2)}{(1 - M_A^2)} \tag{5.45}$$

where we have also used (5.34). From these expressions we may obtain the asymptotic behaviour of the azimuthal components of **u** and **B**. First of all, for $r \to \infty$ the effect of the magnetic field is negligible and from Parker's solution we know that $u_r \sim (\log r)^{1/2}$ so that from (5.43) we see that $M_A \sim r(\log r)^{1/4}$ and hence, $u_\phi \sim r^{-1}$ and $B_\phi \sim (r(\log r)^{1/2})^{-1}$. For $r \ll r_A$, we obtain from (5.43)–(5.45), to lowest order in u_r/v_A and r/r_A,

$$\begin{aligned} u_\phi &= \Omega r \left[1 - \frac{u_r}{v_A}\left(1 - \left(\frac{r}{r_A}\right)^2\right)\right] \\ B_\phi &= -B_r \frac{r\Omega}{v_A}\left[1 - \left(\frac{r}{r_A}\right)^2 \left(1 - \frac{u_r}{v_A}\right)\right] \end{aligned}$$

The solution for u_ϕ obtained by Weber and Davis is shown in Fig. 5.21. The dashed line is the asymptotic solution

$$u_\phi \to \frac{\Omega r_A^2}{r}\left(1 - \frac{v_A}{u_r}\right)$$

which follows from (5.44). If, in the simple model discussed above, the plasma, constrained by the Sun's magnetic field, were to rotate with the angular velocity of the Sun out to $r = r_A$ and then experience no effect of the field for $r > r_A$, conservation of angular momentum would give $u_\phi \to \omega r_A^2/r$. The factor $(1 - v_A/u_r)$ represents a correction to this oversimplified picture on account of the angular momentum retained by the magnetic field at large r.

Weber and Davis go on to calculate u_r from the radial component of (5.38) using the adiabatic gas law for p

$$p\rho^{-\gamma} = \text{const.} = p_A \rho_A^{-\gamma}$$

where p_A and ρ_A are the solar wind pressure and density at $r = r_A$. The equation requires numerical solution and we shall not pursue the details here. However, it is of interest to note that the (u_r, r) phase plane now has three critical points occurring in succession at the slow magnetoacoustic, shear Alfvén, and fast magnetoacoustic wave speeds, i.e. at the characteristic wave speeds for an ideal plasma (see Section 4.8). The first of these is the equivalent of Parker's critical point, occurring at slightly below the sound speed c_s. The second is, of course, the Alfvén critical point already mentioned, and the third follows it almost immediately because in the solar wind $\beta \ll 1$ so that the fast wave speed is only slightly greater than v_A. The only acceptable solutions are ones passing through all three critical points and, of these, only one gives results of the right order of magnitude both at the Sun and at the Earth; this solution gives results for u_r and ρ which are essentially the same as Parker's solution. At the Earth the azimuthal speed is typically two orders of magnitude smaller than the radial speed.

The most serious criticism of these calculations is that they are based on a one-fluid model. Since the average electron–ion mean free path in the solar wind is of the order of 1 AU, only the fields bind electrons and ions together and at the very least a two-fluid model seems essential. Satellite observations have provided very detailed information about ion and electron velocity distributions in the solar wind and whereas ion distributions may, to first order, be represented as drifting Maxwellians, in which the drift velocity is much greater than the thermal speed, this is not the case for the electrons. For electrons the drift velocity is very much less than the thermal speed so the distribution is approximately isotropic and close to a power law. As Bryant (1993) has pointed out, such a distribution has no characteristic energy and therefore no meaningful temperature. A kinetic treatment may therefore be essential for a satisfactory description of electron properties.

5.5.1 Interaction with the geomagnetic field

One of the main aims of satellite observations has been to investigate the interaction of the solar wind with the magnetic fields of the planets and with that of the Earth in particular. We know from (5.33) that the angle ψ between the solar wind magnetic field and flow direction increases with distance from the Sun and decreases with flow speed. At a mean speed of 430 km s^{-1}, particles from the Sun take about four days to reach the Earth (1 AU $= 1.5 \times 10^8$ km) during which time the Sun has executed slightly more than one seventh of its 27 day rotation. A faster stream, making a smaller angle, will cause turbulence and interplanetary shock formation as it overtakes a slower stream. In this way events on the Sun, such as solar flares, lead to major perturbations in the planetary interaction.

The main effect of the solar wind on a planetary magnetic field is to create an asymmetry in the noon–midnight meridian plane. In ideal MHD there can be no interpenetration of the fields so the solar wind flows around the planet enclosing its field in a cavity called the *magnetosphere*. This is compressed on the dayside by the pressure of the solar wind and stretches out on the nightside in the *magnetotail*. The boundary of the magnetosphere is called the *magnetopause*.

There is, however, another important boundary beyond the magnetopause due to the fact that the solar wind speed is greater than the fast magnetosonic wave speed. As we shall see in the next section, in this situation, which is analogous to supersonic flow around a stationary object, a shock wave is created – the *bow shock*. The region between the bow shock and the magnetopause is known as the *magnetosheath*. At the bow shock the plasma in the solar wind is slowed, compressed and heated and it then flows through the magnetosheath and around the magnetosphere. We shall return to the transition at the bow shock later but for the moment our interest is in the inner boundary of the magnetosheath, namely, the magnetopause.

The model described so far, proposed by Chapman and Ferraro in the early 1930s, is illustrated in Fig. 5.22. The magnetopause, being a narrow boundary layer between oppositely directed magnetic fields, carries a strong current (the *Chapman–Ferraro current*) and, as discussed in Section 5.2, magnetic reconnection may take place; Dungey (1961) was the first to point this out. Reconnection takes place both at the 'nose' of the magnetopause between the northwards magnetopause field and southwards solar wind field and in the equatorial plane between the Earth's polar field lines which are dragged out into the magnetotail by the action of the solar wind. The effect of reconnection is fundamental because field lines now cross the magnetopause at the nose and in the tail and the magnetosphere is no longer enclosed. Since particles travel easily along field lines this means that interchange between the solar wind and magnetospheric plasma is possible.

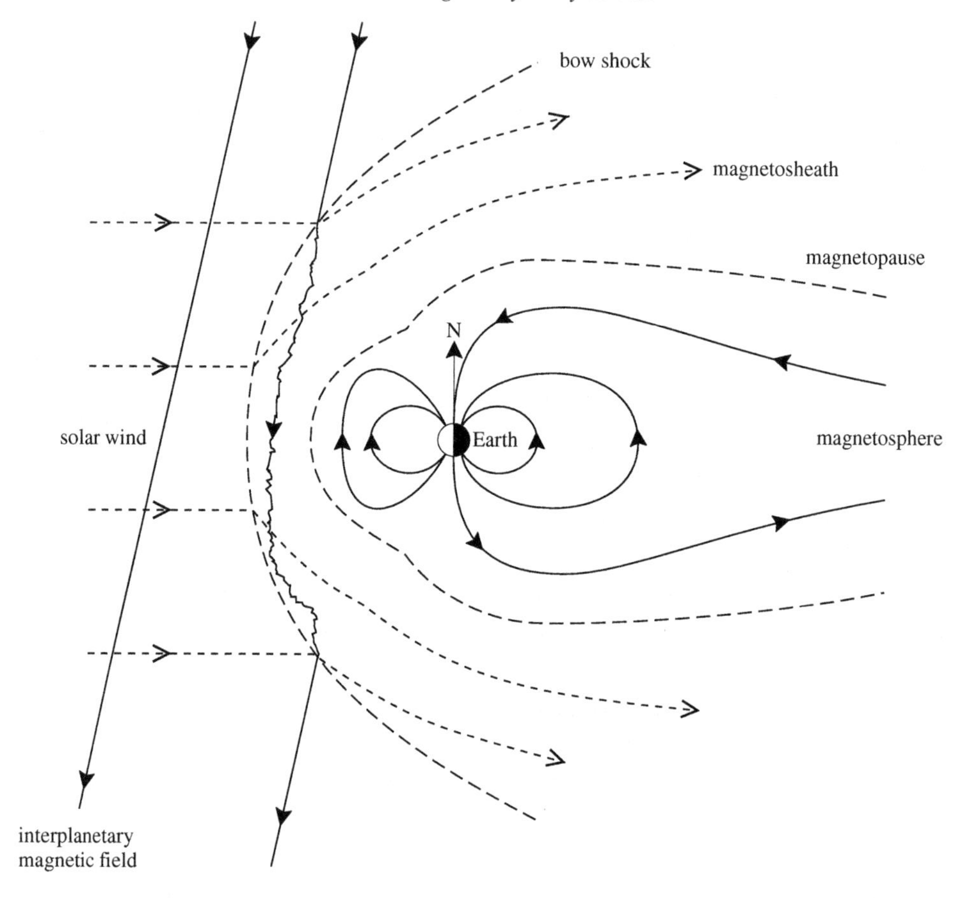

Fig. 5.22. Interaction of solar wind with Earth's geomagnetic field; arrows indicate direction of magnetic field (solid lines) and plasma flow (dashed lines) (after Cowley (1995)).

Detection of the reconnection predicted by Dungey has been discussed by Cowley (1995). One possibility is to detect plasma flow away from the X point (see Fig. 5.4 and the discussion in Section 5.2.1) although the motion of the magnetopause makes detection difficult. Since the current sheet between the oppositely directed magnetosheath and magnetosphere fields is only a few hundred kilometres thick and the speed of the transverse motion is typically several tens of kilometres per second, the magnetopause passes across the spacecraft in about ten seconds thereby requiring resolution of plasma data of just a few seconds. Observations were made by the Explorer satellites in the late 1970s and, with higher resolution, by the AMPTE-IRM spacecraft in the mid-1980s. Data show that reconnection takes place in about half of all magnetopause crossings where the angle between the fields is greater than 90°. When reconnection takes place it often does so in a pulsed manner on a time scale of about ten minutes, creating so-called *flux transfer events* (FTEs) which travel over the magnetopause. Neither the pulsed nature of FTEs nor the factors that determine when and where reconnection takes place are well understood. The proximity of oppositely directed fields is a necessary but not a sufficient condition.

The clearest evidence for the validity of Dungey's model has been obtained by an analysis of the ion velocity distributions on either side of the magnetopause confirming the pattern predicted by Cowley in 1982. The acceleration (by the contracting field lines), transmission and reflection of ions entering the current sheet produces characteristic D-shaped velocity distributions in the plane of the magnetopause on open field lines. A spherical distribution of incident ions should produce a D-shaped distribution of transmitted ions on open lines in the magnetosphere and a double D-shaped distribution of incident and reflected ions on open lines in the magnetosheath. Such D-shaped velocity distributions have been obtained by Smith and Rodgers (1991) from AMPTE-UKS spacecraft data.

Further support for Dungey's model comes from observations of magnetospheric flow correlations. Given the sensitivity to solar activity there is a wide variation of field-flow angle at the nose of the magnetopause and conditions are most favourable for reconnection when the field points south and least favourable when it points north. Following dayside reconnection there is excitation of magnetospheric flow via the open field lines from dayside to tail and the subsequent return of closed field lines through the magnetosphere. Experiments carried out by Cowley and collaborators have shown that dayside flows are excited within about five minutes of a switch of the interplanetary field from north to south and this has been correlated with nightside activity, including intense auroral displays associated with the consequent change in field structure which occur after a period of about 30–45 minutes.

5.6 MHD shocks

Supersonic flow gives rise to the generation of shock waves, a well-known illustration of this principle being the audible 'sonic bang' emanating from an aircraft in supersonic flight. This is no less true for flows in conducting media than for neutral fluid flows but, as we shall see, it is considerably more complex. In a neutral fluid any disturbance, such as that produced by a moving aircraft or a piston at one end of a tube, causes a wave to propagate through the fluid at the speed of sound c_s; such a wave is called a *compression* or *sound wave*. So long as the cause of the compression wave, the aircraft or piston, is itself moving more slowly than the speed of sound the wave will propagate ahead of the disturbance and adiabatic changes may take place in response to it. But if the disturbing agent increases its speed to c_s or greater it begins to catch up with and then overtake the compression

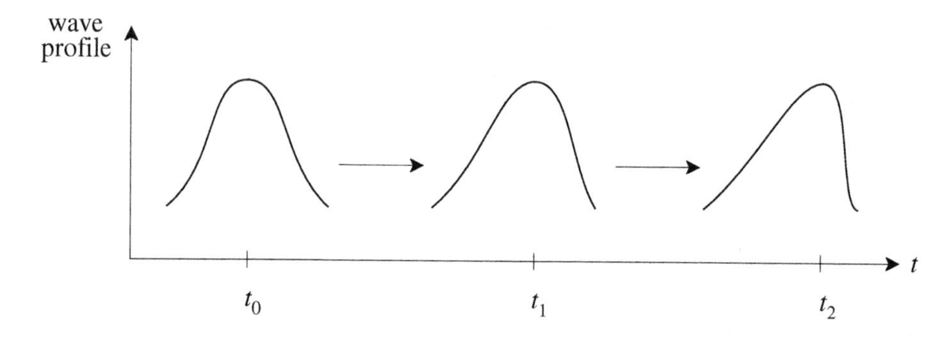

Fig. 5.23. Illustration of wave-front steepening in propagation of compression wave.

wave-front with the result that the fluid experiences a sudden, non-adiabatic change of state; this is what we mean by a *shock*.

The profile of the shock wave, travelling through the fluid and creating the change of state, is the result of a balance between convective and dissipative effects. Since the sound speed $c_s \propto \rho^{\gamma-1}$ is greatest at the peak of a finite amplitude wave the wave-front steepens, as illustrated in Fig. 5.23. However, as the wave-front steepens, dissipative effects, which are proportional to gradients in the fluid variables, become stronger and a steady profile is achieved when the convective steepening is counter-balanced by the dissipative flattening of the wave-front. It is this steady wave-front, propagating at supersonic speed through the undisturbed fluid, which constitutes the *shock wave*. The smaller the coefficients of dissipation the more nearly the shock wave aproaches a vertical discontinuity.

Fluid properties may change considerably across a shock wave and the shock is thus a steady transition region between the undisturbed (unshocked) fluid and the fluid through which the shock has passed. In Fig. 5.24, region 1 is the unshocked fluid and is said to be *in front of* the shock or *upstream*; region 2 is the shocked fluid, said to be *behind* the shock or *downstream*. Usually one regards a shock as a transition region between two uniform states as in Fig. 5.24(a) although in practice this is difficult to realize and the situation depicted in Fig. 5.24(b) is more likely. Here, the shocked fluid is not in a uniform state but is subject to a relief or expansion wave. This means that the state of the fluid behind the shock does not persist but changes with time after the shock has passed on. Nevertheless, it is convenient to take both region 1 and region 2 as uniform and this will be assumed unless otherwise stated. (The validity of this assumption depends on the time of relaxation from the state represented by the point A in Fig. 5.24(b) being longer than other times of interest.) Since the establishment of a new equilibrium state in a non-conducting fluid can only be achieved by collisions, the width or thickness of the shock is of the order of a few mean free paths.

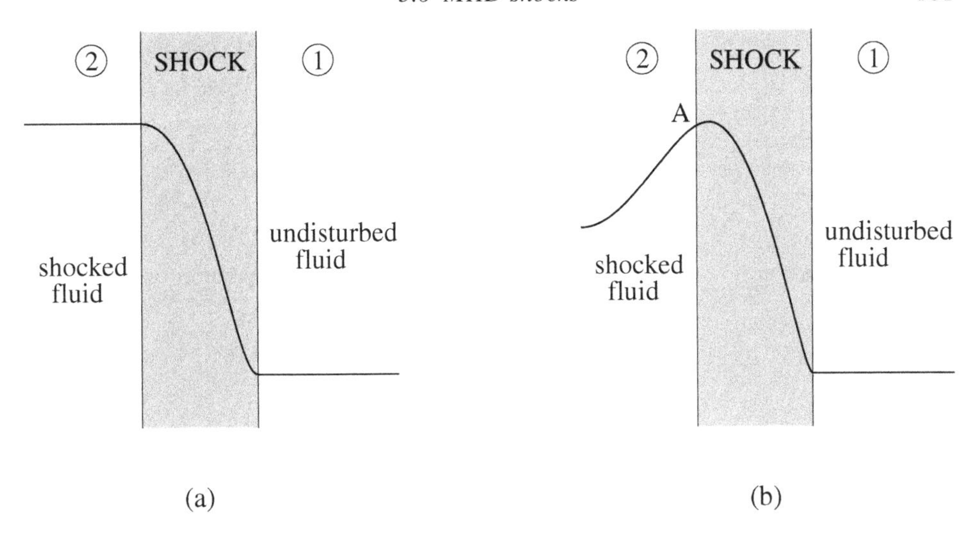

Fig. 5.24. Illustration of shock wave transition between (a) uniform states and (b) uniform initial state and final state subject to expansion wave.

The theory of hydrodynamic shocks is reasonably well understood; see Lighthill (1956). As usual, the hydromagnetic case is considerably more complicated. For a start, conducting fluids in a magnetic field can support two further modes of wave propagation. Returning to the piston analogy, transverse movements of the piston in a non-conducting fluid have no effect (beyond a boundary layer where viscous forces maintain a velocity gradient). This means that there is effectively only one (longitudinal) degree of freedom and, therefore, only one mode of propagation, with the velocity c_{s}. However, the transverse movement of a conducting piston in an infinitely conducting fluid carries any longitudinal component of a magnetic field with it, thereby producing a wave. Thus a conducting fluid in the presence of a magnetic field has three degrees of freedom (one longitudinal and two transverse) and hence three modes of wave propagation. There are three propagation speeds, therefore, generally known as *fast, intermediate*, and *slow* (see Section 6.4.2). Intermediate waves are purely transverse and do not steepen to form shock waves. The fast and slow modes in general contain both transverse and longitudinal components and these modes give rise to shocks.

More fundamental differences between shocks in neutral fluids and in plasmas arise when we come to consider shock structure. Particularly striking is the existence of shocks with thicknesses much less than the collisional mean free path. These *collisionless shocks* cannot be MHD shocks (though they may in certain limits be described by fluid equations) and we postpone their discussion until Chapter 10. However, for *collision-dominated shocks* two factors greatly facilitate discussion of the effects of a shock wave on a fluid. First, the shock transition region

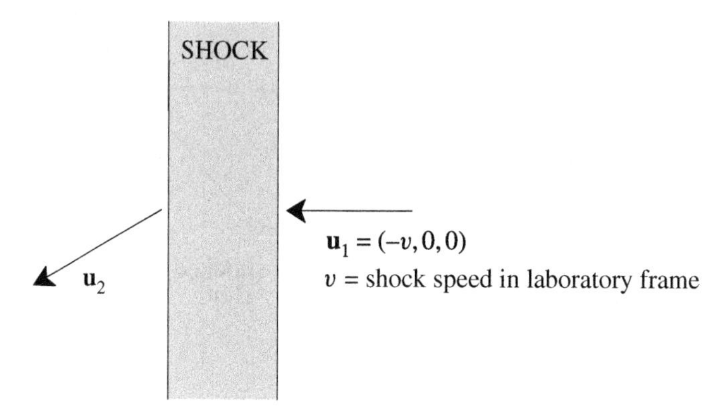

Fig. 5.25. Shock rest frame.

may for most purposes be approximated by a discontinuity in fluid properties. Second, the macroscopic conservation equations and the Maxwell equations may be integrated across the shock to give a set of equations which are independent of shock structure and relate fluid properties on either side of the shock. This most useful and straightforward aspect of shock theory is developed first.

5.6.1 Shock equations

For simplicity we restrict discussion to plane shocks moving in a direction normal to the plane of the shock. Let this be the x direction; then all variables are functions of x only inside the shock and are constant outside the shock (in regions 1 and 2). It is convenient to use a frame of reference in which the shock is at rest, In this frame, depicted in Fig. 5.25, the plasma enters the shock with velocity $\mathbf{u}_1$ and emerges with velocity $\mathbf{u}_2$. Steady state conditions apply (i.e. all variables are time-independent) and the Maxwell equation for $\mathbf{j}$ is

$$\mu_0 \mathbf{j} = \left(0, -\frac{\mathrm{d}B_z}{\mathrm{d}x}, \frac{\mathrm{d}B_y}{\mathrm{d}x}\right) \tag{5.46}$$

The MHD equations do not apply in the shock region since dissipative processes take place there; however they do apply on either side of the shock, i.e. in regions 1 and 2. Thus from Ohm's law

$$\mathbf{E}_1 = \frac{\mathbf{j}_1}{\sigma_1} - \mathbf{u}_1 \times \mathbf{B}_1 \quad \mathbf{E}_2 = \frac{\mathbf{j}_2}{\sigma_2} - \mathbf{u}_2 \times \mathbf{B}_2 \tag{5.47}$$

Now from $\nabla \cdot \mathbf{B} = 0$ and $\nabla \times \mathbf{E} = -\partial \mathbf{B}/\partial t = 0$ we find

$$\frac{\mathrm{d}B_x}{\mathrm{d}x} = 0 \tag{5.48}$$

$$\frac{\mathrm{d}E_y}{\mathrm{d}x} = 0 \tag{5.49}$$

$$\frac{\mathrm{d}E_z}{\mathrm{d}x} = 0 \tag{5.50}$$

The other equations to be integrated across the shock are the equations of conservation of mass, momentum and energy. These are discussed in their most general form in Section 12.5 but since we shall integrate them across the shock and evaluate them in the upstream and downstream plasmas, where the MHD approximation is assumed, the result is the same if we use the conservation equations derived in Section 4.2. Since variables depend on x only, we have from (4.1)–(4.3)

$$\frac{\mathrm{d}(\rho u_x)}{\mathrm{d}x} = 0 \tag{5.51}$$

$$\frac{\mathrm{d}\Pi_{xx}}{\mathrm{d}x} = 0 \qquad \frac{\mathrm{d}\Pi_{xy}}{\mathrm{d}x} = 0 \qquad \frac{\mathrm{d}\Pi_{xz}}{\mathrm{d}x} = 0 \tag{5.52}$$

$$\frac{\mathrm{d}S_x}{\mathrm{d}x} = 0 \tag{5.53}$$

Integrating (5.48)–(5.50) and using (5.46) and (5.47) gives, on observing that gradients are zero in regions 1 and 2,

$$[B_x]_1^2 = 0 \tag{5.54}$$

$$\left[u_x B_y - u_y B_x\right]_1^2 = 0 \tag{5.55}$$

$$[u_x B_z - u_z B_x]_1^2 = 0 \tag{5.56}$$

where $[\phi]_1^2 = (\phi_2 - \phi_1)$ in the usual notation. The integration of (5.51)–(5.53) is trivial and one finds

$$[\rho u_x]_1^2 = 0 \tag{5.57}$$

$$\left[\rho u_x^2 + P + (B_y^2 + B_z^2)/2\mu_0\right]_1^2 = 0 \tag{5.58}$$

$$\left[\rho u_x u_y - B_x B_y/\mu_0\right]_1^2 = 0 \tag{5.59}$$

$$[\rho u_x u_z - B_x B_z/\mu_0]_1^2 = 0 \tag{5.60}$$

$$\begin{aligned}\big[\rho u_x I + P u_x + \rho u_x u^2/2 + u_x(B_y^2 + B_z^2)/\mu_0 \\ - B_x(B_y u_y + B_z u_z)/\mu_0\big]_1^2 = 0\end{aligned} \tag{5.61}$$

where the internal energy

$$I = P/(\gamma - 1)\rho \tag{5.62}$$

Equations (5.54)–(5.61) relate fluid variables on one side of the shock to those on the other, and these equations are sometimes called the *jump conditions* across the shock. Defining the unit vector $\hat{\mathbf{n}}$ in the direction of shock propagation, the jump conditions may be written in general vector form as

$$[\rho \mathbf{u}\cdot\hat{\mathbf{n}}]_1^2 = 0 \tag{5.63}$$

$$\left[\rho\mathbf{u}(\mathbf{u}\cdot\hat{\mathbf{n}}) + (P + B^2/2\mu_0)\hat{\mathbf{n}} - (\mathbf{B}\cdot\hat{\mathbf{n}})\mathbf{B}/\mu_0\right]_1^2 = 0 \tag{5.64}$$

$$\begin{aligned}\big[\mathbf{u}\cdot\hat{\mathbf{n}}\{(\rho I + \rho u^2/2 + B^2/2\mu_0) + (P + B^2/2\mu_0)\}\\ - (\mathbf{B}\cdot\hat{\mathbf{n}})(\mathbf{B}\cdot\mathbf{u})/\mu_0\big]_1^2 &= 0\end{aligned} \tag{5.65}$$

$$\left[\mathbf{B}\cdot\hat{\mathbf{n}}\right]_1^2 = 0 \qquad \left[\hat{\mathbf{n}}\times(\mathbf{u}\times\mathbf{B})\right]_1^2 = 0 \tag{5.66}$$

The first three equations represent the conservation of mass, momentum, and energy, respectively, for the flow of plasma through the shock. The last pair of equations gives the jump conditions for the magnetic field expressing the continuity of the normal component of $\mathbf{B}$ and the tangential component of $\mathbf{E} = -\mathbf{u}\times\mathbf{B}$. When $\mathbf{B} = 0$, (5.63)–(5.65) reduce to the corresponding hydrodynamic equations known as the *Rankine–Hugoniot equations*.

After considerable manipulation (see Exercise 5.6), the velocity variables $\mathbf{u}_1$ and $\mathbf{u}_2$ may be eliminated from the energy equation (5.65) and the result is

$$[I]_1^2 + \frac{1}{2}(P_1 + P_2)[1/\rho]_1^2 + \frac{1}{4\mu_0}\{[B]_1^2\}^2[1/\rho]_1^2 = 0 \tag{5.67}$$

The hydrodynamic equivalent of this equation,

$$[I]_1^2 + \frac{1}{2}(P_1 + P_2)[1/\rho]_1^2 = 0 \tag{5.68}$$

relates the pressure and density on either side of the shock and is known as the *Hugoniot relation*. It assumes the role played by the law $P\rho^{-\gamma}$ = const. in adiabatic changes of state. Note that the *hydromagnetic Hugoniot* (5.67) reduces to (5.68) not only when the magnetic field is zero but also when $\mathbf{B}_1$ and $\mathbf{B}_2$ are both parallel to the direction of shock propagation, since (5.66) then gives $\mathbf{B}_1 = \mathbf{B}_2$.

Before discussing particular solutions of the shock equations, the compressive nature of shocks (i.e. $P_2 > P_1$) will be proved, assuming the plasma is a perfect gas. The proof follows as a consequence of the second law of thermodynamics – the law of increase of entropy. The entropy S of a perfect gas is given by

$$S = C_\text{v}\log(P/\rho^\gamma) + \text{const.} \tag{5.69}$$

where C_v is the specific heat at constant volume. Thus,

$$\frac{\mathrm{d}S_2}{\mathrm{d}\rho_2} = \frac{C_\text{v}}{P_2}\frac{\mathrm{d}P_2}{\mathrm{d}\rho_2} - \frac{\gamma C_\text{v}}{\rho_2}$$

which, in terms of the ratios $r = \rho_2/\rho_1$, $R = P_2/P_1$, may be written

$$\frac{\rho_1}{C_v}\frac{dS_2}{d\rho_2} = \frac{1}{R}\frac{dR}{dr} - \frac{\gamma}{r} \tag{5.70}$$

In this equation, we regard ρ_1 and P_1 as constants (the given values of ρ and P in region 1). A straightforward rearrangement of (5.67) gives

$$R = \frac{(\gamma+1)r - (\gamma-1) + (\gamma-1)(r-1)b^2}{(\gamma+1) - (\gamma-1)r} \tag{5.71}$$

where $b^2 = (\mathbf{B}_2 - \mathbf{B}_1)^2/2\mu_0 P_1$. Now differentiating (5.71) with respect to r and substituting in (5.70), we get

$$\frac{\rho_1}{C_v}\frac{dS_2}{d\rho_2} = \frac{\gamma(\gamma^2-1)(r-1)^2 + (\gamma-1)[\gamma(\gamma-1)r^2 - 2(\gamma^2-1)r + \gamma(\gamma+1)]b^2}{r[(\gamma+1)-(\gamma-1)r][(\gamma+1)r - (\gamma-1) + (\gamma-1)(r-1)b^2]} \tag{5.72}$$

The next step is to show that $dS_2/d\rho_2$ is positive and we do this by proving that both numerator and denominator of the expression in (5.72) are positive. Since $\gamma > 1$, it is easy to verify that this statement is true for $r = 1$. Writing (5.71) as

$$R = (A + Cb^2)/D$$

it follows for $r < 1$ that $C < 0$ and $D > 0$. Since the pressure ratio R must be positive, A must also be positive, and

$$r > (\gamma-1)/(\gamma+1) \tag{5.73}$$

Likewise, if $r > 1$ then $C > 0$, $A > 0$ and, therefore, $R > 0$ implies $D > 0$, i.e.

$$r < (\gamma+1)/(\gamma-1) \tag{5.74}$$

Combining (5.73) and (5.74),

$$\frac{\gamma-1}{\gamma+1} < r < \frac{\gamma+1}{\gamma-1} \tag{5.75}$$

Thus, in the expression for R, the denominator D is positive. Therefore, $R > 0$ implies that the numerator $(A + Cb^2)$ is also positive. Since the denominator of the right-hand side of (5.72) is simply $rD(A + Cb^2)$, it is therefore positive.

Turning now to the numerator in (5.72), it is clear, since $\gamma > 1$, that the first term is positive. The remaining term is quadratic in r and positive for $r = 1$. If this term is to be negative for some r it must pass through zero. However, equating it to zero one finds imaginary roots for r, which proves that the term is positive for all real r. The proof that $\mathrm{d}S_2/\mathrm{d}\rho_2 > 0$ is thus complete, showing that S_2 and ρ_2 increase or decrease together.

If $\rho_2 = \rho_1$, it follows from (5.67) that $I_2 = I_1$ and hence $P_2 = P_1$. Then, from (5.69), $S_2 = S_1$. This is the limiting case in which no shock is present. Now since the second law of thermodynamics requires $S_2 \geq S_1$, and S_2 and ρ_2 change in the same sense, it follows that $\rho_2 \geq \rho_1$ and (5.75) must be replaced by

$$1 \leq r < (\gamma + 1)/(\gamma - 1) \tag{5.76}$$

Then, from (5.71), discounting $r = 1$ (no shock)

$$\frac{P_2}{P_1} = R > 1 + \frac{(r-1)(\gamma-1)b^2}{(\gamma+1) - (\gamma-1)r}$$

i.e. shocks are compressive, confirming the qualitative arguments used earlier. Although this proof applies only to a perfect gas it seems that for all gases shocks are compressive. This may be proved quite generally for weak shocks (Landau and Lifshitz (1960)) and with some further assumptions (Ericson and Bazar (1960)) for shocks of arbitrary strength.

We shall now discuss some particular shocks but before we do this some general observations are helpful. From the jump conditions, provided there is non-zero mass flux through the shock ($\rho\mathbf{u} \cdot \hat{\mathbf{n}} \neq 0$) it is easy to show that $\hat{\mathbf{n}}$, $\mathbf{B}_1$ and $\mathbf{B}_2$ are coplanar (see Exercise 5.7). It then follows from (5.66) that

$$\mathbf{B} \cdot \hat{\mathbf{n}}[\mathbf{u}]_1^2 = [(\mathbf{u} \cdot \hat{\mathbf{n}})\mathbf{B}]_1^2$$

Thus, if $\mathbf{B} \cdot \hat{\mathbf{n}} \neq 0$ and $\mathbf{u}$ has a component perpendicular to the plane of $\hat{\mathbf{n}}$ and $\mathbf{B}$ it must be the same on both sides of the shock. If $\mathbf{B} \cdot \hat{\mathbf{n}} = 0$ the same result is obtained from (5.63) and (5.64). This means that we may choose a frame of reference in which $\hat{\mathbf{n}}$, $\mathbf{B}$ and $\mathbf{u}$ are coplanar. If this is the (x, y) plane all z components in the jump conditions (5.54)–(5.61) are zero and (5.56) and (5.60) are satisfied trivially.

The angle θ between $\mathbf{B}_1$ and $\hat{\mathbf{n}}$ is used to classify shocks as *parallel* ($\theta = 0$), *perpendicular* ($\theta = \pi/2$) and *oblique* ($0 < \theta < \pi/2$). We shall begin with the simple cases of parallel and perpendicular shocks for which, without loss of generality, we may set $\mathbf{u}_1 = (u_1, 0, 0)$, i.e. the unshocked fluid flow is normal to the stationary shock front. This means that we have chosen a frame of reference moving with the shock speed in the x direction and the tangential flow speed u_{1y} of the unshocked fluid in the y direction.

5.6.2 Parallel shocks

Here both $\mathbf{u}_1$ and $\mathbf{B}_1$ are parallel to $\hat{\mathbf{n}}$ and we have noted already that one possibility is that $\mathbf{B}_2 = \mathbf{B}_1$. In this case it follows from (5.66) that $\mathbf{u}_2$ is also parallel to $\hat{\mathbf{n}}$ and it is easily seen that the magnetic field drops out of the jump conditions leaving

$$\rho_1 u_1 = \rho_2 u_2 \tag{5.77}$$

$$\rho_1 u_1^2 + P_1 = \rho_2 u_2^2 + P_2 \tag{5.78}$$

$$\rho_1 u_1 I_1 + P_1 u_1 + \rho_1 u_1^3/2 = \rho_2 u_2 I_2 + P_2 u_2 + \rho_2 u_2^3/2 \tag{5.79}$$

where the last equation may also be written, using (5.62) and (5.77), as

$$\frac{\gamma P_1}{(\gamma - 1)\rho_1} + \frac{u_1^2}{2} = \frac{\gamma P_2}{(\gamma - 1)\rho_2} + \frac{u_2^2}{2} \tag{5.80}$$

This solution, therefore, corresponds to a hydrodynamic shock and it may be shown (see Exercise 5.8) that one must have u_1 supersonic (relative to the sound speed in region 1) while u_2 is subsonic (relative to the sound speed in region 2).

Strong shocks are defined as those for which the pressure ratio $R \gg 1$. In this case, it follows (see Exercise 5.8) that the *Mach number* in region 1, $M = (u_1/c_s(1)) \gg 1$, and the temperature ratio

$$\frac{T_2}{T_1} \approx \frac{2\gamma(\gamma - 1)}{(\gamma + 1)^2} M^2 \gg 1 \tag{5.81}$$

Since M may be as large as 100, it is clear from (5.81) that strong shocks may be used to generate high-temperature plasmas or to obtain a plasma from a neutral gas by creating a temperature T_2 behind the shock sufficiently high to cause ionization. The conversion of flow energy into thermal energy in this situation is easily demonstrated from (5.80). In view of (5.81), the initial thermal energy may be neglected compared with the final thermal energy. Also (see Exercise 5.8)

$$\frac{u_1}{u_2} = \frac{\rho_2}{\rho_1} \approx \frac{(\gamma + 1)}{(\gamma - 1)} \tag{5.82}$$

This ratio is 4 for $\gamma = 5/3$ so (5.80) may be approximated by

$$\frac{\gamma P_2}{(\gamma + 1)} = \frac{1}{2}\rho_1 u_1^2 \tag{5.83}$$

Thus, the flow energy in region 1 is converted by the shock into thermal energy in region 2.

The hydrodynamic shock is not, however, the only possible solution for propagation parallel to the initial magnetic field. It may happen that some of the flow energy is converted into magnetic energy so that $|\mathbf{B}_2| > |\mathbf{B}_1|$. Since the normal component of $\mathbf{B}$ is conserved this means that the passage of the shock creates a tangential component and it is said to be a *switch-on shock*. We shall return to this possibility in our discussion of oblique shocks.

5.6.3 Perpendicular shocks

Here we have

$$\mathbf{B}_1 = (0, B_1, 0) \tag{5.84}$$

and so, using (5.54), $\mathbf{B}_2$ may be written

$$\mathbf{B}_2 = (0, B_2, 0) \tag{5.85}$$

Since $\rho u_x \neq 0$, it follows from (5.55), (5.57) and (5.59) that

$$\begin{aligned} \mathbf{u}_2 &= (u_1/r, 0, 0) \\ \mathbf{B}_2 &= (0, rB_1, 0) \end{aligned}$$

Thus the magnetic field is constant in direction and increased in magnitude by the same ratio as the density. Finally, (5.58) and (5.61) are now

$$\rho_1 u_1^2 + P_1 + B_1^2/2\mu_0 = \rho_2 u_2^2 + P_2 + B_2^2/2\mu_0$$

and

$$\frac{\gamma P_1}{(\gamma - 1)\rho_1} + \frac{u_1^2}{2} + \frac{B_1^2}{\mu_0 \rho_1} = \frac{\gamma P_2}{(\gamma - 1)\rho_2} + \frac{u_2^2}{2} + \frac{B_2^2}{\mu_0 \rho_2}$$

These may be written

$$\gamma M^2(1 - 1/r) = (R - 1) + (r^2 - 1)/\beta \tag{5.86}$$

and

$$\gamma M^2 \left(1 - \frac{1}{r^2}\right) = \frac{2\gamma}{(\gamma - 1)}\left(\frac{R}{r} - 1\right) + 4(r - 1)/\beta$$

respectively, where $\beta = 2\mu_0 P_1/B_1^2$ is the plasma β ahead of the shock and the shock Mach number $M = u_1/c_s$, where $c_s = (\gamma P_1/\rho_1)^{1/2}$ is the sound speed in the upstream plasma. Eliminating R and excluding the solution $r = R = 1$, which corresponds to no shock, we get

$$2(2 - \gamma)r^2 + [2\gamma(\beta + 1) + \beta\gamma(\gamma - 1)M^2]r - \beta\gamma(\gamma + 1)M^2 = 0$$

If r_1 and r_2 are the roots of this equation, then

$$r_1 r_2 = -\beta\gamma(\gamma + 1)M^2/2(2 - \gamma)$$

and for $\gamma < 2$ one root is negative and, therefore, non-physical. Consequently, there is only one solution corresponding to a shock in this case. Since $r > 1$

$$\beta\gamma(\gamma + 1)M^2 > 2(2 - \gamma) + 2\gamma(\beta + 1) + \beta\gamma(\gamma - 1)M^2$$

which reduces to

$$\gamma M^2 > \gamma + 2/\beta$$

and, hence,

$$u_1^2 > B_1^2/\mu_0\rho_1 + \gamma P_1/\rho_1 = v_A^2 + c_s^2 = (c_s^*)^2$$

where the second equality defines c_s^*. Thus, for shocks to propagate perpendicular to a magnetic field the shock speed must be greater than c_s^*. The speed c_s^* assumes the role played by c_s in hydrodynamic shocks (see Exercise 5.9); this is not a surprising result since c_s^* is the speed of the fast compressional wave propagating perpendicular to a magnetic field (see Section 4.8). (Note that there is no shock corresponding to the slow wave since it does not propagate perpendicular to the magnetic field.) We see that the effect of the magnetic field is to increase the effective pressure by a factor $(1 + 2/\gamma\beta)$,

The shock strength R is reduced by the introduction of the magnetic field (see Exercise 5.9) since flow energy is now converted into magnetic energy as well as heat. However, since

$$B_2/B_1 = r < (\gamma + 1)/(\gamma - 1) \tag{5.87}$$

the increase in magnetic energy is limited while from (5.86)

$$P_2/P_1 = R = 1 + \gamma M^2(1 - 1/r) - (r^2 - 1)/\beta \tag{5.88}$$

so that for large Mach number, relative to a fixed value of β, the temperature ratio is approximately the same as for the hydrodynamic case (see Exercise 5.9).

5.6.4 Oblique shocks

In the case of oblique propagation where in general $\mathbf{u}$ and $\mathbf{B}$ have both x and y components it is convenient to choose a frame of reference, known as the *de Hoffmann–Teller frame*, in which $\mathbf{u}_1 \times \mathbf{B}_1 = 0$, that is

$$u_{1y} = u_{1x}B_{1y}/B_x$$

where $B_x = B_{1x} = B_{2x}$ by (5.54). Note that this is consistent with our choice for parallel shocks but not for perpendicular shocks for which $B_x = 0$. From (5.55) it follows that

$$u_{2y} = u_{2x}B_{2y}/B_x$$

and hence $\mathbf{u}_2 \times \mathbf{B}_2 = 0$, i.e. $\mathbf{u}$ and $\mathbf{B}$ are parallel on both sides of the shock; physically, this simply says $\mathbf{E} = 0$ on both sides of the shock. Now

$$\frac{u_{2y}}{u_{1y}} = \frac{u_{2x}}{u_{1x}}\frac{B_{2y}}{B_{1y}} = \frac{\rho_1}{\rho_2}\frac{B_{2y}}{B_{1y}} = \frac{1}{r}\frac{B_{2y}}{B_{1y}} \tag{5.89}$$

Also, from (5.59), we get

$$\frac{u_{2y}}{u_{1y}} - 1 = \frac{B_x B_{1y}}{\mu_0 \rho_1 u_{1x} u_{1y}} \left(\frac{B_{2y}}{B_{1y}} - 1 \right) = \frac{B_1^2}{\mu_0 \rho_1 u_1^2} \left(\frac{B_{2y}}{B_{1y}} - 1 \right)$$

which may be combined with (5.89) to give

$$\frac{u_{2y}}{u_{1y}} = \frac{u_1^2 - v_A^2}{u_1^2 - r v_A^2} = \frac{1}{r} \frac{B_{2y}}{B_{1y}} \tag{5.90}$$

Furthermore, with this choice of reference frame the magnetic terms in (5.61) are identically zero leaving only the hydrodynamic terms from which we get

$$\begin{aligned} \frac{P_2}{P_1} &= r + \frac{(\gamma - 1) r u_1^2}{2 c_s^2} \left(1 - \frac{u_2^2}{u_1^2} \right) \\ &= r + \frac{(\gamma - 1) r u_1^2}{2 c_s^2} \left[1 - \frac{\cos^2 \theta}{r^2} - \sin^2 \theta \left(\frac{u_1^2 - v_A^2}{u_1^2 - r v_A^2} \right)^2 \right] \end{aligned} \tag{5.91}$$

It is clear from (5.90) that $B_{2y} > B_{1y}$ for $u_1^2 \geq r v_A^2 > v_A^2$ and, conversely, $B_{2y} < B_{1y}$ for $u_1^2 \leq v_A^2 < r v_A^2$. The first case corresponds to the fast shock and the second to the slow shock and these are illustrated in Fig. 5.26 showing refraction (a) away from and (b) towards the normal, respectively. It is when the equalities hold in these relationships, i.e. $u_1^2 = r v_A^2$ so that $B_{1y} = 0$ for the fast shock, or $u_1^2 = v_A^2$ so that $B_{2y} = 0$ for the slow shock, that we get (c) *switch-on* and (d) *switch-off shocks*, respectively, corresponding to the tangential component of the magnetic field being switched on or off.

The switch-on shock is one of the possible solutions (the fast shock) when we let $\theta \to 0$ (i.e. parallel propagation). The slow wave in this limit has $u_1 = v_A$ so that $B_{2y} = 0$. In this case both $\mathbf{B}_1$ and $\mathbf{B}_2$ are parallel to the shock normal and $\mathbf{B}_1 = \mathbf{B}_2$. As noted earlier when discussing (5.67) this yields the hydrodynamic Hugoniot; in other words, the slow shock becomes a hydrodynamic shock at parallel propagation.

5.6.5 Shock thickness

We now wish to consider the structure of a shock, i.e. the variation of pressure, density, magnetic field etc., within the shock itself. Even for collision-dominated shock waves a quantitative calculation of shock structure involves considerable effort. The procedure is to solve the appropriate *transport equations* inside the shock region using either region 1 or region 2 as a set of boundary conditions. Just what form the appropriate transport equations take is not easily decided in general. Often the important dissipative mechanisms are viscosity, heat conductivity, and electrical conductivity (Joule heating). Of these, only Joule heating was retained in

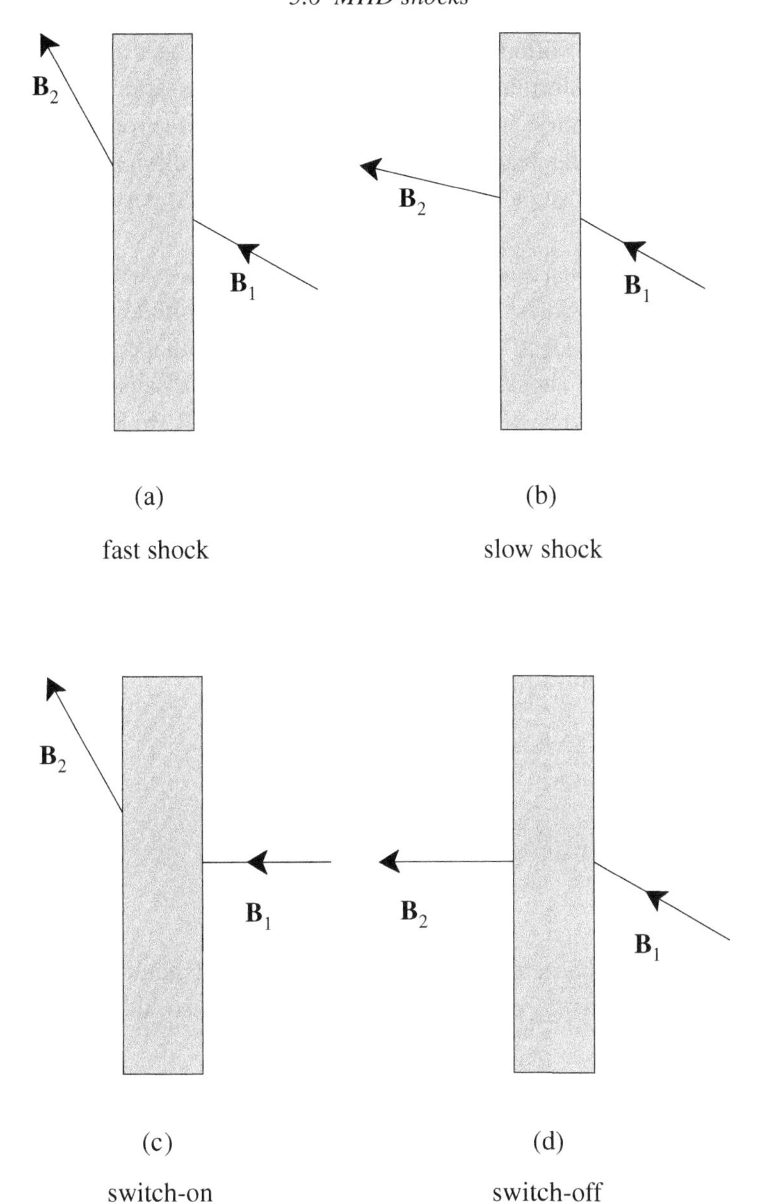

Fig. 5.26. Magnetic field refraction in oblique shocks.

the resistive MHD equations. The heat conduction term must be reintroduced into the energy equation; similarly, viscosity terms must be brought into the momentum and energy equations. However, there is considerable doubt as to whether such a one-fluid, hydromagnetic description is appropriate for a discussion of shock structure. In particular, it often happens that the electrons heat up first in a shock and then reach an equilibrium temperature with the ions after a longer period of time. Thus, a description involving separate ion and electron temperatures is usually necessary. Also, depending on the conditions, other dissipative mechanisms may be important – for instance, ionization if the unshocked gas has a zero or low degree of ionization.

However, leaving aside a quantitative discussion of shock structure, one may obtain estimates for the thickness of a shock by order of magnitude arguments. Since the conditions on either side of the shock are given by the initial conditions together with the solutions of the shock equations, the total rate of dissipation of energy is known and this must occur within the shock thickness, δ. If we know (or assume) that the dissipation is due principally to one particular mechanism, we can write an order of magnitude relationship. For example, suppose the appropriate dissipative process is viscosity. In the energy transport equation, the term involving viscosity is proportional to the square of the velocity gradient. Then, if other dissipative processes are negligible, the rate of dissipation of energy, $\Delta E/\Delta t$, is proportional to $\rho\nu(\Delta\mathbf{u}/\delta)^2$, where $\nu(=\mu/\rho)$ is the kinematic viscosity. Since $\Delta t \sim \delta/u_1$, the order of magnitude relationship is

$$\frac{\Delta E}{\Delta t} \sim \frac{u_1 \Delta E}{\delta} \sim \rho\nu\left(\frac{\mathbf{u}_1 - \mathbf{u}_2}{\delta}\right)^2$$

i.e.

$$\delta \sim \frac{\rho\nu(\mathbf{u}_1 - \mathbf{u}_2)^2}{u_1 \Delta E} \tag{5.92}$$

Applying this to the particular case of the strong hydrodynamic shock discussed in Section 5.6.2, the energy dissipated $\Delta E \approx \frac{1}{2}\rho_1 u_1^2$. Also, using (5.82), (5.92) implies

$$\delta \sim \nu/u_1 \tag{5.93}$$

or, in other words, the Reynolds number, $R = u_1\delta/\nu$, is of order unity. From kinetic theory (see Section 8.2), one can show approximately that $\nu \sim P\tau_c/\rho$, where τ_c is the ion–ion collision time. Using $P \sim \frac{1}{2}(P_1 + P_2) \approx \frac{1}{2}P_2$, it follows from (5.82), (5.83), and (5.93) that

$$\delta \sim \tau_c\left(\frac{P_2}{\rho_2}\right)^{1/2} \sim \tau_c\left(\frac{k_B T_2}{m_i}\right)^{1/2}$$

Thus the shock thickness is of the order of the ion collision mean free path.

To give a further example, consider dissipation by Joule heating. Here, the dissipation of energy occurs at a rate proportional to j^2/σ. Since $\mu_0\mathbf{j} = \nabla \times \mathbf{B}$, the order of magnitude relationship is

$$\frac{u_1 \Delta E}{\delta} \sim \frac{1}{\sigma}\left(\frac{\mathbf{B}_1 - \mathbf{B}_2}{\mu_0 \delta}\right)^2$$

i.e.

$$\delta \sim \frac{(\mathbf{B}_1 - \mathbf{B}_2)^2}{\sigma \mu_0^2 u_1 \Delta E} \tag{5.94}$$

From (3.38) this may be written in terms of the magnetic Reynolds number as

$$R_{\mathrm{M}} = \mu_0 \sigma \delta u_1 \sim \frac{(\mathbf{B}_1 - \mathbf{B}_2)^2}{\mu_0 \Delta E} \tag{5.95}$$

Now suppose this is applied to a strong shock propagating perpendicular to a magnetic field (see Section 5.6.3). With $\Delta E \approx \frac{1}{2}\rho_1 u_1^2$ and $B_1 \approx B_2/4$,

$$R_{\mathrm{M}} \sim \frac{B_2^2/2\mu_0}{\frac{1}{2}\rho_1 u_1^2}$$

which gives the order of magnitude of the magnetic Reynolds number required if Joule heating is to be an adequate dissipative mechanism for this shock.

As one might expect, it is found that a given dissipative mechanism can produce the required change of state up to a certain limiting shock strength. Beyond that other mechanisms must come into play with the result that the shock may show a more complicated structure with more than one characteristic thickness. We return to this question of shock structure with a specific calculation when discussing collisionless shocks in Section 10.5.2.

Exercises

5.1 With reference to the induction equation (5.1) explain the significance of the magnetic Reynolds number R_{M}. What is the relationship of the Lundquist number S to R_{M}?

(a) Why is it that even in plasmas for which $S \gg 1$ resistive diffusion cannot be completely ignored? What does a dimensional analysis of (5.1) tell us about the relative length scales for magnetic field changes due to diffusion and convection?

(b) Establish from (5.4) and the conservation of mass that the Sweet–Parker reconnection rate is $S^{-1/2}$ where S is the Lundquist number.

(c) Follow the discussion given by Parker (1979), (Section 15.6), summarizing Petschek's model and variations, notably that due to Sonnerup (1970).

(d) Refer to Biskamp (1993), (Section 6.2), for a contrasting critical summary of Petschek's model. Biskamp argues that not only is Petschek's concept difficult to accept intuitively but it is seriously flawed in the inappropriate treatment of the diffusion region. Numerical simulations of driven reconnection do not produce a Petschek-like configuration for small η. In particular Biskamp (1986) found that the sheet width L increased as η decreased in contrast to Petschek's scaling of $L \sim O(\eta)$.

5.2 Explain Fig. 5.6 with reference to the heuristic arguments used in Section 5.3.1 to obtain the parametric dependence of the growth rate (5.22) of the tearing mode. What are the physical characteristics of the tearing mode that make it much more of a threat to plasma stability than the gravitational and rippling modes?

5.3 Biermann (1950) showed that stellar magnetic fields could be generated at the expense of the thermal energy of the star. In the event the time needed for Biermann's mechanism to establish fields typical of those on the Sun for example proved to be rather too long. From the generalized Ohm's law (3.52) if no magnetic field is present initially show that this reduces to $\mathbf{E} = -(m_i/Ze\rho)\nabla p_e$. Then from Faraday's law, a magnetic field is generated provided $\nabla\rho \times \nabla p_e \neq 0$. For a spherically symmetric pressure gradient, no magnetic field is generated.

(a) Consider next the case of a rotating star for which

$$\nabla p = \rho\left(\mathbf{g} + \Omega^2\bar{\mathbf{r}}\right) \tag{E5.1}$$

in which Ω denotes the angular velocity of the star and $\bar{\mathbf{r}}$ is the displacement from the axis of rotation. Show that in the case of a rotating star a toroidal magnetic field is generated.

(b) Generally, both ohmic and Hall terms in the generalized Ohm's law act to limit the growth of the field. In the case of the Sun, by balancing energy input from the battery against ohmic dissipation with appropriate choices for parameters from Table 5.1 and using $\bar{r}\Omega \sim 2\times10^3\ \mathrm{m\,s^{-1}}$, show that the field $B \sim 0.01\ \mathrm{T}$ and that the time needed for the field to evolve to this magnitude is about 10^9 years, the order of the age of the Universe. However, it needs to be borne in mind that such rough estimates are changed by allowing for convection.

(c) The battery mechanism has proved to be important as a source of magnetic fields generated in targets irradiated by intense laser light where fields as large as $O(100\ \mathrm{T})$ have been detected over scale lengths of $O(100\ \mu\mathrm{m})$. Check the estimate given for the strength of the magnetic field generated.

5.4 Verify that (5.32) is a solution of (5.31) and that this corresponds to the solution curves sketched in Fig. 5.18. Show that the constant $C = -3$ for the only acceptable solution for the solar wind represented by trajectory V. What are possible explanations of why this solution does not predict the correct density at the Earth?

5.5 Using the results in Section 5.5 for the Weber and Davis model of the solar wind, obtain an expression for the magnitude of the interplanetary magnetic field $B_{\mathrm{IMF}} = (B_r^2 + B_\phi^2)^{1/2}$ and show that (i) $B_{\mathrm{IMF}} \sim 1/r^2$ near the Alfvén critical point $r = r_{\mathrm{A}}$ and (ii) $B_{\mathrm{IMF}} \sim 1/r(\log r)^{1/2}$ as $r \to \infty$.

5.6 Derive the hydromagnetic Hugoniot relation (5.67) from (5.54)–(5.61) by eliminating the velocity variables.

[Hint: First use (5.57), in the form $\rho_1 u_{x1} = \rho_2 u_{x2} = k$, say, to eliminate u_x in favour of k and then use the resulting (5.59) and (5.60) to eliminate u_y and u_z. Next, write (5.58) in the form $k^2[1/\rho]_1^2 = -[P + B^2/2\mu_0]_1^2]$ and multiply by $(\rho_1^{-1} + \rho_2^{-1})/2$ to obtain

$$\frac{k^2}{2}\left[\frac{1}{\rho^2}\right]_1^2 = -\frac{1}{2}\left(\frac{1}{\rho_1} + \frac{1}{\rho_2}\right)\left[P + \frac{B^2}{2\mu_0}\right]_1^2$$

Finally, substitute this equation, together with the converted (5.55) and (5.56) in (5.61) to eliminate k and obtain (5.67).]

5.7 In a plane shock show that the unit vector normal to the shock, $\hat{\mathbf{n}}$, and the magnetic fields on either side of the shock, $\mathbf{B}_1$ and $\mathbf{B}_2$, are coplanar provided that the mass flux through the shock is non-zero ($\rho\mathbf{u} \cdot \hat{\mathbf{n}} \neq 0$).

5.8 Use (5.77) to write (5.78) and (5.80) as

$$\begin{aligned} \gamma M_1^2(1 - 1/r) &= R - 1 \\ \gamma M_1^2(1 - 1/r^2) &= \frac{2\gamma}{\gamma - 1}\left(\frac{R}{r} - 1\right) \end{aligned}$$

By solving these equations for R, r and applying the condition $R > 1$ show that u_1 is supersonic. Rewriting the equations in terms of M_2 show that u_2 is subsonic.

Deduce that $R \gg 1$ implies $M_1 \gg 1$.

Verify (5.82).

5.9 For the perpendicular shock discussed in Section 5.6.3 it was shown that $u_1 > c_s^*(1)$. By rewriting the equations in terms of the downstream variables M_2 and β_2 show that $u_2 < c_s^*(2)$.

Show also that for $\gamma < 2$, $r < r_0$, where r_0 is the solution for zero magnetic field. Hence, deduce from (5.86) that the introduction of a magnetic field reduces the shock strength.

From (5.87) and (5.88) show that for fixed β the temperature ratio in the large Mach number limit is approximately the same as for the hydrodynamic shock.

6

Waves in unbounded homogeneous plasmas

6.1 Introduction

Historically studies of wave propagation in plasmas have provided one of the keystones in the development of plasma physics and they remain a focus in contemporary research. Much was already known about plasma waves long before the subject itself had any standing, early studies being prompted by practical concerns. The need to allow for the effect of the geomagnetic field in determining propagation characteristics of radio waves led to the development, by Hartree in 1931, of what has become known as Appleton–Hartree theory. About the same time another basic plasma mode, electron plasma oscillations, had been identified. In 1926 Penning suggested that oscillations of electrons in a gas discharge could account for the anomalously rapid scattering of electron beams, observed over distances much shorter than a collisional mean free path. These oscillations were studied in detail by Langmuir and were identified theoretically by Tonks and Langmuir in 1928.

Alfvén's pioneering work in the development of magnetohydrodynamics led him to the realization in 1942 that magnetic field lines, pictured as elastic strings under tension, should support a class of magnetohydrodynamic waves. The shear Alfvén wave, identified in Section 4.8, first appeared in Alfvén's work on cosmical electrodynamics. Following the development of space physics we now know that Alfvén (and other) waves pervade the whole range of plasmas in space from the Earth's ionosphere and magnetosphere to the solar wind and the Earth's bow shock and beyond.

There is a bewildering collection of plasma waves and schemes for classifying the various modes are called for. Plasma waves whether in laboratory plasmas or in space are in general non-linear features. Moreover, real plasmas are at the same time inhomogeneous and anisotropic, dissipative and dispersive. To avoid being overwhelmed by detail at the outset some radical simplifications are needed and so we begin by assuming that the medium is unbounded and consider only small disturbances so that a *linear* theory of wave propagation is adequate. Even this is a tall order and to begin with we make a further approximation and ignore the effects of plasma pressure. This allows us to discuss a number of electromagnetic modes in some detail since thermal effects play only a minor role in their dispersion characteristics. To make matters even more straightforward we move towards a general dispersion relation in stages, first identifying modes that propagate along, and transverse to, the magnetic field before dealing with oblique propagation. In all of this we are helped by the natural ordering of the electron and ion masses in separating modes into high and low frequency regimes. This ordering underpins a classification of dispersion characteristics in terms of wave normal surfaces which is discussed in outline for a cold plasma.

Dropping the cold plasma approximation and allowing for plasma pressure enables us to identify other waves, in particular electrostatic modes. Thermal effects bring dissipation, not usually via inter-particle collisions, though these may contribute particularly in partially ionized plasmas. In most plasmas of interest, interactions between plasma electrons and ions and the waves themselves are more important. Moreover, since these *wave–particle* interactions generally involve only those particles with thermal velocities close to the phase velocity of the wave they cannot be dealt with using a fluid model. Thus the discussion of the most important of these interactions, *Landau damping*, has to await the development of kinetic theory in Chapter 7.

6.2 Some basic wave concepts

Before embarking on a description of the propagation characteristics of small amplitude waves in plasmas we review briefly some basic wave concepts, familiar from the theory of electromagnetic wave propagation. We restrict our discussion to plane wave solutions of the wave equation, a plane wave being one for which the wave disturbance is constant over all points of a plane normal to the direction of propagation of the wave. For the plane wave solutions

$$\mathbf{E}(\mathbf{r}, t) = \mathbf{E}_0 \exp i(\mathbf{k} \cdot \mathbf{r} - \omega t) \qquad \mathbf{B}(\mathbf{r}, t) = \mathbf{B}_0 \exp i(\mathbf{k} \cdot \mathbf{r} - \omega t)$$

the vacuum divergence equations demand that $\mathbf{k} \cdot \mathbf{E}_0 = 0 = \mathbf{k} \cdot \mathbf{B}_0$ so that $(\mathbf{E}, \mathbf{B}, \mathbf{k})$ form a triad of orthogonal vectors.

The electric field in a plane wave is expressed in general by a superposition of two linearly independent solutions of the wave equation. Choosing the z-axis along the wave vector $\mathbf{k}$ gives

$$\mathbf{E}(z, t) = (E_x \hat{\mathbf{x}} + E_y \hat{\mathbf{y}}) \exp i(kz - \omega t) \tag{6.1}$$

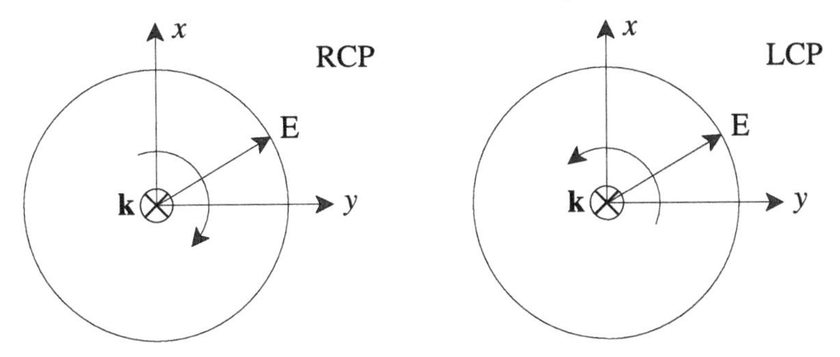

Fig. 6.1. Circularly polarized plane waves.

in which E_x, E_y are complex amplitudes

$$E_x = E_{x0}\exp(i\alpha) \qquad E_y = E_{y0}\exp(i\beta)$$

where E_{x0}, E_{y0} are real. With $\delta = \beta - \alpha$, (6.1) becomes

$$\mathbf{E}(z,t) = \left[E_{x0}\hat{\mathbf{x}} + E_{y0}e^{i\delta}\hat{\mathbf{y}}\right]\exp i(kz - \omega t + \alpha) \tag{6.2}$$

At each point in space the electric vector rotates in a plane normal to $\hat{\mathbf{z}}$ and as time evolves its tip describes an ellipse. This is most easily seen by setting $\delta = \pm\pi/2$ so that

$$\mathbf{E}(z,t) = (E_{x0}\hat{\mathbf{x}} \pm iE_{y0}\hat{\mathbf{y}})\exp i(kz - \omega t + \alpha)$$

from which

$$\left.\begin{aligned} E_x(z,t) &= E_{x0}\cos(kz - \omega t + \alpha) \\ E_y(z,t) &= \mp E_{y0}\sin(kz - \omega t + \alpha) \end{aligned}\right\} \tag{6.3}$$

Thus in general an electromagnetic wave is *elliptically polarized*. In the special case when E_{x0} or $E_{y0} = 0$ the electric field is *linearly* (or *plane*) *polarized*, while if $E_{x0} = E_{y0}$ the field is *circularly polarized*.

To an observer looking along the direction of propagation the negative sign in (6.3) corresponds to an electric field vector, at any point z, rotating in a *clockwise* direction. In this case, for $E_{x0} = E_{y0}$ the wave is said to be *right-circularly polarized* (RCP). For the positive sign, rotation is anticlockwise and the wave is *left-circularly polarized* (LCP). Both polarizations are illustrated in Fig. 6.1.

With $\delta = \pm\pi/2$ and defining the wave polarization in terms of the complex amplitudes in (6.2) by

$$\mathcal{P} = iE_x/E_y$$

we see that $\mathcal{P} > 0(< 0)$ represents clockwise (anticlockwise) rotation and $\mathcal{P} = +1(-1)$ indicates RCP(LCP).

6.2.1 Energy flux

Defining the Poynting vector $\mathbf{S} = \mathbf{E} \times \mathbf{H}$ allows us to describe the flux of energy associated with electromagnetic fields. Poynting's theorem is an expression of the electromagnetic energy flux as a balance between the rate of change of energy in the elecromagnetic field, with energy density $W = \frac{1}{2}(\mathbf{E} \cdot \mathbf{D} + \mathbf{B} \cdot \mathbf{H})$, and power dissipated ohmically in the system, $\mathbf{j} \cdot \mathbf{E}$. Then at any instant

$$\frac{\partial W}{\partial t} + \nabla \cdot \mathbf{S} = -\mathbf{j} \cdot \mathbf{E} \tag{6.4}$$

With harmonic time dependence, the time-averaged energy flux becomes

$$\langle \mathbf{S} \rangle = \frac{1}{2} \Re(\mathbf{E} \times \mathbf{H}^*) \tag{6.5}$$

The time-average of $\partial W / \partial t$ vanishes, leaving, for sources contained in a volume V bounded by a closed surface σ with unit normal vector $\mathbf{n}$,

$$\int_\sigma \langle \mathbf{S} \rangle \cdot \mathbf{n} \, \mathrm{d}\sigma = -\frac{1}{2} \Re \int_V \langle \mathbf{E} \cdot \mathbf{j}^* \rangle \mathrm{d}V \tag{6.6}$$

For a dissipation-free system $\Re \langle \mathbf{E} \cdot \mathbf{j} \rangle$ vanishes and so there is no net energy flux averaged over a cycle.

6.2.2 Dispersive media

So far we have considered only monochromatic waves. In practice even with such a monochromatic source as a laser there will be a spread in frequency ω and wavenumber k. Moreover in general $\omega = \omega(k)$ so that a wave-form that is not monochromatic will change as it propagates, exhibiting *dispersion*. Consider, for example, scalar waves propagating along the z-axis; using a Fourier representation

$$E(z, t) = \frac{1}{\surd(2\pi)} \int_{-\infty}^{\infty} a(k) \exp[i(kz - \omega(k)t)] \, \mathrm{d}k \tag{6.7}$$

and $\omega = \omega(k)$ is known as the *dispersion relation*. Equation (6.7) and

$$a(k) = \frac{1}{\surd(2\pi)} \int_{-\infty}^{\infty} E(z, 0) e^{-ikz} \, \mathrm{d}z$$

define a *wave packet*. Assume, for convenience, that $a(k)$ is peaked about some wavenumber k_0. The central question is this: 'Given a particular wave packet at $t = 0$ (the *pulse shape*), what does it look like at some later time?' Provided the medium is not too dispersive, $\omega(k)$ may be expanded about k_0:

$$\omega(k) = \omega_0 + \left(\frac{\mathrm{d}\omega}{\mathrm{d}k} \right)_{k=k_0} (k - k_0) + \cdots$$

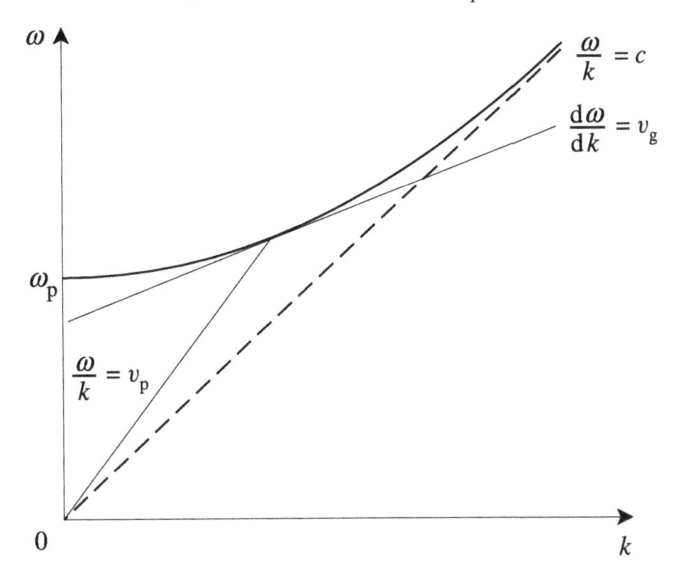

Fig. 6.2. Dispersion curve for electromagnetic wave.

where ω_0 stands for $\omega(k_0)$ and so

$$E(z,t) = \frac{1}{\sqrt{(2\pi)}}\int_{-\infty}^{\infty} a(k)\exp\{i[kz - \omega_0 t - (d\omega/dk)_{k_0}(k-k_0)t]\}\,dk$$

Then

$$E(z,t) \simeq E(z - v_g t, 0)\exp[i(k_0 v_g - \omega_0)t] \tag{6.8}$$

which represents a pulse travelling without distortion with a velocity $v_g = (d\omega/dk)_{k=k_0}$. This is the *group velocity*. The group velocity appears in this context as the propagation velocity of a wave packet, a concept first introduced by Hamilton.

To relate v_g to the phase velocity $v_p = \omega/k$ is straightforward:

$$v_g = \frac{d\omega}{dk} = \frac{d}{dk}(kv_p) = v_p + k\frac{dv_p}{dk}$$

or equivalently in terms of the wavelength

$$v_g = v_p - \lambda\frac{dv_p}{d\lambda}$$

Clearly, when the phase velocity is independent of wavelength there is no dispersion. Such is the case for the shear Alfvén wave introduced in Section 4.8. For $dv_p/d\lambda > 0$, $v_g < v_p$ and the wave is said to exhibit *normal dispersion*. An electromagnetic wave propagating in a plasma provides an example of normal dispersion since its dispersion relation (obtained in Section 6.3.1) is $\omega^2 = \omega_p^2 + k^2c^2$, which means that $d\omega/dk = kc^2/\omega = c^2/v_p$. Since $v_p = \omega/k > c$, it follows that $v_g < c$ as shown in Fig. 6.2.

6.3 Waves in cold plasmas

As discussed in Section 3.5, a cold plasma is one in which the thermal speeds of the particles are much smaller than the phase speeds of the waves and the cold plasma wave equations, given in Table 3.4, are simply the ion and electron equations of continuity and motion in the electromagnetic fields, which are governed by Maxwell's equations.

Since we shall discuss only small amplitude waves we shall be concerned with the linearized version of the cold plasma equations, namely

$$\frac{\partial n_1}{\partial t} + \nabla \cdot (n_0 \mathbf{u}_1) = 0 \tag{6.9}$$

$$\frac{\partial \mathbf{u}_1}{\partial t} = \frac{e}{m}(\mathbf{E}_1 + \mathbf{u}_1 \times \mathbf{B}_0) \tag{6.10}$$

$$\nabla \times \mathbf{E}_1 = -\frac{\partial \mathbf{B}_1}{\partial t} \tag{6.11}$$

$$\nabla \times \mathbf{B}_1 - \frac{1}{c^2}\frac{\partial \mathbf{E}_1}{\partial t} = \mu_0 \mathbf{j} = \mu_0 \sum e n_0 \mathbf{u}_1 \tag{6.12}$$

$$\nabla \cdot \mathbf{E}_1 = \frac{q}{\varepsilon_0} = \frac{1}{\varepsilon_0}\sum e n_1 \tag{6.13}$$

$$\nabla \cdot \mathbf{B}_1 = 0 \tag{6.14}$$

where the species label has been suppressed but the sums in (6.12) and (6.13) are over species and

$$\left.\begin{aligned} n &= n_0 + n_1 \\ \mathbf{u} &= \mathbf{u}_1 \\ \mathbf{E} &= \mathbf{E}_1 \\ \mathbf{B} &= \mathbf{B}_0 + \mathbf{B}_1 \end{aligned}\right\} \tag{6.15}$$

with the quantities n_0 and $\mathbf{B}_0$ being constant in time and space. Thus, the linearization describes a small departure from a plasma in equilibrium. The closed set of two-fluid wave equations is actually (6.9)–(6.12) (remembering that (6.9) and (6.10) must be written for ions and electrons), since (6.13) and (6.14) are essentially *initial conditions*; if they are satisfied at some time t_0, we can show that they must be satisfied at all other times.

In the usual way we eliminate $\mathbf{B}_1$ from (6.11) and (6.12) to get

$$\nabla \times \nabla \times \mathbf{E}_1 = -\frac{1}{c^2}\frac{\partial^2 \mathbf{E}_1}{\partial t^2} - \mu_0 \frac{\partial \mathbf{j}}{\partial t} \tag{6.16}$$

Next, using the second equality in (6.12) and solving (6.10) for $\mathbf{u}_1$, we obtain $\mathbf{j}$ in terms of $\mathbf{E}_1$ which we may express formally as

$$\mathbf{j} = \boldsymbol{\sigma} \cdot \mathbf{E}_1 \tag{6.17}$$

where $\boldsymbol{\sigma}$ is the *conductivity tensor*. Then, assuming all variables vary like $\exp i(\mathbf{k} \cdot \mathbf{r} - \omega t)$, (6.16) becomes

$$\mathbf{n} \times (\mathbf{n} \times \mathbf{E}_1) = -\mathbf{E}_1 - \frac{i}{\varepsilon_0 \omega} \boldsymbol{\sigma} \cdot \mathbf{E}_1 = -\boldsymbol{\varepsilon} \cdot \mathbf{E}_1 \tag{6.18}$$

where $\mathbf{n} = c\mathbf{k}/\omega$ is a dimensionless wave propagation vector and $\boldsymbol{\varepsilon}$ is the *cold plasma dielectric tensor*. The requirement that this equation should have a non-trivial solution yields the dispersion relation containing all the information about linear wave propagation in a cold plasma.

To find the elements of $\boldsymbol{\sigma}$ (and hence $\boldsymbol{\varepsilon}$) we must solve (6.10), the components of which, dropping the subscript 1 and writing $\Omega = eB_0/m$, are

$$-i\omega u_x - \Omega u_y = eE_x/m \tag{6.19}$$

$$-i\omega u_y + \Omega u_x = eE_y/m \tag{6.20}$$

$$-i\omega u_z = eE_z/m \tag{6.21}$$

and then substitute the results in the expression for $\mathbf{j}$ in (6.12). This is straightforward but we can minimize the computation involved by first carrying out the calculation for the variables, $\tilde{\mathbf{u}}$, $\tilde{\mathbf{E}}$, $\tilde{\mathbf{j}}$ with components

$$\left.\begin{array}{lll} u^{\pm} & = u_x \pm i u_y, & u_z \\ E^{\pm} & = E_x \pm i E_y, & E_z \\ j^{\pm} & = j_x \pm i j_y, & j_z \end{array}\right\} \tag{6.22}$$

since the conductivity tensor $\tilde{\boldsymbol{\sigma}}$, defined by

$$\tilde{\mathbf{j}} = \tilde{\boldsymbol{\sigma}} \cdot \tilde{\mathbf{E}} \tag{6.23}$$

is diagonal. By combining (6.19) and (6.20) in obvious ways we get the solutions

$$u^{\pm} = \frac{ieE^{\pm}}{m(\omega \mp \Omega)}, \qquad u_z = \frac{ieE_z}{m\omega} \tag{6.24}$$

and substituting in

$$\tilde{\mathbf{j}} = \sum_{\alpha} e_\alpha n_{0\alpha} \tilde{\mathbf{u}} \tag{6.25}$$

we see, by comparison with (6.23), that

$$\tilde{\sigma} = i\varepsilon_0 \begin{pmatrix} \sum_\alpha \frac{\omega_{p\alpha}^2}{\omega - \Omega_\alpha} & 0 & 0 \\ 0 & \sum_\alpha \frac{\omega_{p\alpha}^2}{\omega + \Omega_\alpha} & 0 \\ 0 & 0 & \sum_\alpha \frac{\omega_{p\alpha}^2}{\omega} \end{pmatrix} \tag{6.26}$$

Now from (6.22) it follows that the matrix that transforms $\mathbf{u}$, $\mathbf{E}$ and $\mathbf{j}$ to $\tilde{\mathbf{u}}$, $\tilde{\mathbf{E}}$, $\tilde{\mathbf{j}}$ is

$$\mathbf{T} = \begin{pmatrix} 1 & i & 0 \\ 1 & -i & 0 \\ 0 & 0 & 1 \end{pmatrix}$$

and its inverse

$$\mathbf{T}^{-1} = \begin{pmatrix} 1/2 & 1/2 & 0 \\ -i/2 & i/2 & 0 \\ 0 & 0 & 1 \end{pmatrix}$$

Thus, from (6.23) we obtain

$$\mathbf{j} = \mathbf{T}^{-1} \cdot \tilde{\mathbf{j}} = \mathbf{T}^{-1} \cdot \tilde{\boldsymbol{\sigma}} \cdot \mathbf{T} \cdot \mathbf{E}$$

and, comparing with (6.17), we see that

$$\boldsymbol{\sigma} = \mathbf{T}^{-1} \cdot \tilde{\boldsymbol{\sigma}} \cdot \mathbf{T}$$

giving

$$\boldsymbol{\sigma} = \begin{pmatrix} (\tilde{\sigma}_{11} + \tilde{\sigma}_{22})/2 & i(\tilde{\sigma}_{11} - \tilde{\sigma}_{22})/2 & 0 \\ -i(\tilde{\sigma}_{11} - \tilde{\sigma}_{22})/2 & (\tilde{\sigma}_{11} + \tilde{\sigma}_{22})/2 & 0 \\ 0 & 0 & \tilde{\sigma}_{33} \end{pmatrix} \tag{6.27}$$

where the components of $\tilde{\boldsymbol{\sigma}}$ are as in (6.26).

Now, returning to (6.18), we may write the dielectric tensor components $\varepsilon_{ij} = \delta_{ij} + (i/\varepsilon_0\omega)\sigma_{ij}$, that is

$$\boldsymbol{\varepsilon} = \begin{pmatrix} S & -iD & 0 \\ iD & S & 0 \\ 0 & 0 & P \end{pmatrix} \tag{6.28}$$

where

$$\left.\begin{aligned}
S &= \tfrac{1}{2}(R+L) = 1 - \frac{\omega_{\mathrm{p}}^2(\omega^2+\Omega_{\mathrm{i}}\Omega_{\mathrm{e}})}{(\omega^2-\Omega_{\mathrm{i}}^2)(\omega^2-\Omega_{\mathrm{e}}^2)} \\
D &= \tfrac{1}{2}(R-L) = \frac{\omega_{\mathrm{p}}^2\omega(\Omega_{\mathrm{i}}+\Omega_{\mathrm{e}})}{(\omega^2-\Omega_{\mathrm{i}}^2)(\omega^2-\Omega_{\mathrm{e}}^2)} \\
R &= 1 - \frac{\omega_{\mathrm{p}}^2}{(\omega+\Omega_{\mathrm{i}})(\omega+\Omega_{\mathrm{e}})} \\
L &= 1 - \frac{\omega_{\mathrm{p}}^2}{(\omega-\Omega_{\mathrm{i}})(\omega-\Omega_{\mathrm{e}})} \\
P &= 1 - \frac{\omega_{\mathrm{p}}^2}{\omega^2}
\end{aligned}\right\} \tag{6.29}$$

and $\omega_{\mathrm{p}}^2 = \omega_{\mathrm{pi}}^2 + \omega_{\mathrm{pe}}^2$ is the square of the plasma frequency. Note that, in combining the elements $\tilde{\sigma}_{11} \pm \tilde{\sigma}_{22}$, we have used the fact that $\omega_{\mathrm{pe}}^2\Omega_{\mathrm{i}} + \omega_{\mathrm{pi}}^2\Omega_{\mathrm{e}} = Ze^3B_0(n_{\mathrm{e}0} - Zn_{\mathrm{i}0})/m_{\mathrm{e}}m_{\mathrm{i}}\varepsilon_0 = 0$ because of equilibrium charge neutrality.

Finally, without loss of generality, we may choose axes such that $\mathbf{n} = (n\sin\theta, 0, n\cos\theta)$, as shown in Fig. 6.3, so that (6.18) may be written

$$(\mathbf{n}\cdot\mathbf{E})\mathbf{n} - n^2\mathbf{E} + \boldsymbol{\varepsilon}\cdot\mathbf{E} = 0$$

and hence

$$\begin{bmatrix} S - n^2\cos^2\theta & -iD & n^2\cos\theta\sin\theta \\ iD & S - n^2 & 0 \\ n^2\cos\theta\sin\theta & 0 & P - n^2\sin^2\theta \end{bmatrix}\begin{bmatrix} E_x \\ E_y \\ E_z \end{bmatrix} = 0 \tag{6.30}$$

Thus, taking the determinant of the coefficients, the *general dispersion relation for cold plasma waves* is

$$An^4 - Bn^2 + C = 0 \tag{6.31}$$

where

$$\left.\begin{aligned}
A &= S\sin^2\theta + P\cos^2\theta \\
B &= RL\sin^2\theta + PS(1+\cos^2\theta) \\
C &= PRL
\end{aligned}\right\} \tag{6.32}$$

We treat this as an equation to be solved for n^2 as a function of θ, the angle of propagation relative to the magnetic field $\mathbf{B}_0$; the dimensionless quantities $\omega_{\mathrm{p}}/\omega$, $\Omega_{\mathrm{i}}/\omega$, $\Omega_{\mathrm{e}}/\omega$, occurring in the coefficients (see (6.29)), are to be regarded as parameters which vary according to choice of wave frequency and equilibrium plasma.

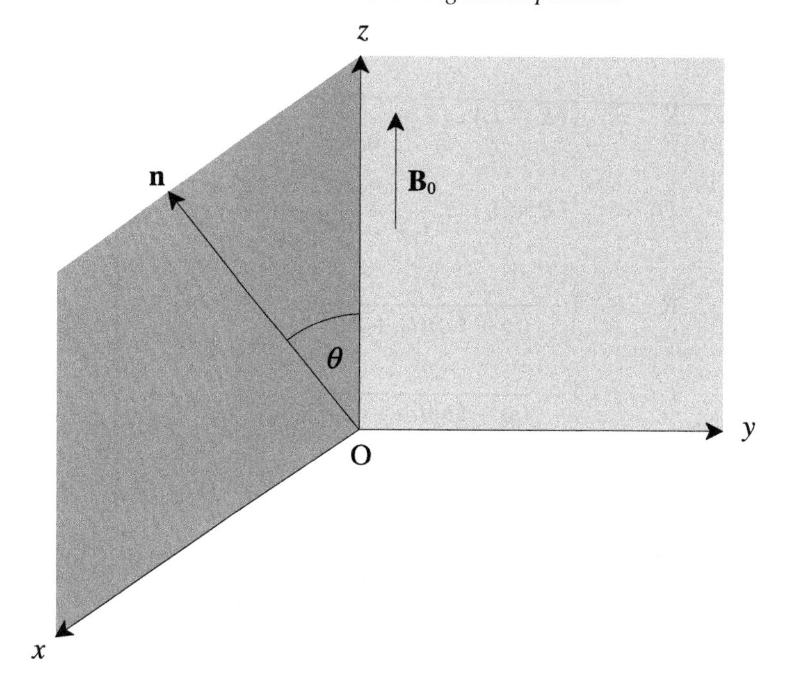

Fig. 6.3. Orientation of wave propagation vector relative to magnetic field.

Since a general discussion of the solutions of (6.31) is algebraically challenging, our approach will be to look initially at waves propagating parallel ($\theta = 0$) and perpendicular ($\theta = \pi/2$) to the magnetic field. To this end a useful alternative expression of (6.31) is obtained by solving it for $\tan^2\theta$ as a function of n^2 with the result

$$\tan^2\theta = -\frac{P(n^2 - R)(n^2 - L)}{(Sn^2 - RL)(n^2 - P)} \tag{6.33}$$

There can be only real solutions of (6.31) for n^2 since in the cold, non-streaming, plasma equations there are no sources of free energy to drive instabilities and no dissipation terms to produce decaying waves and it is a simple matter to prove this formally by showing that the discriminant of the bi-quadratic equation may be written in the form

$$B^2 - 4AC = (RL - PS)^2\sin^4\theta + 4P^2D^2\cos^2\theta \geq 0 \tag{6.34}$$

Thus, n is either pure real or pure imaginary corresponding to wave propagation or evanescence, respectively. The changeover from propagation to evanescence (or vice versa) takes place whenever n^2 passes through zero or infinity. From (6.31) and (6.32), it is clear that the first of these possibilities occurs whenever $C = 0$, that is

$$P = 0 \quad \text{or} \quad R = 0 \quad \text{or} \quad L = 0 \tag{6.35}$$

These are called *cut-offs* because, for given equilibrium conditions, they define frequencies above or below which the wave ceases to propagate at any angle ($k \to 0$ for finite ω, i.e. $v_p \to \infty$). From (6.29) the cut-off frequencies are:

$$\left.\begin{array}{ll} P = 0: & \omega = \omega_p \\ R = 0: & \omega = [\omega_p^2 + (\Omega_i - \Omega_e)^2/4]^{1/2} - (\Omega_i + \Omega_e)/2 \equiv \omega_R \\ L = 0: & \omega = [\omega_p^2 + (\Omega_i - \Omega_e)^2/4]^{1/2} + (\Omega_i + \Omega_e)/2 \equiv \omega_L \end{array}\right\} \tag{6.36}$$

Note that we have chosen the positive square root in order to get $\omega > 0$; we consider only positive ω since solutions with $\omega < 0$ merely correspond to waves travelling in the opposite direction. Note, also, that $\omega_R > \omega_L$ since $\Omega_e < 0$ and $|\Omega_e| \gg \Omega_i$.

At a resonance $v_p \to 0$ ($k \to \infty$ for finite ω) but this does not in general mean, as at a cut-off, that the wave ceases to propagate altogether; rather, it defines a cone of propagation. Letting $n^2 \to \infty$ in (6.31) shows that we require $A = 0$, i.e. $\tan^2\theta = -P/S$. This equation defines for given parameters the *resonant angle*, θ_{res}, above or below which the wave does not propagate; indeed, directly from (6.33) we get

$$\tan^2\theta_{res} = -P/S \tag{6.37}$$

Here we note that θ_{res}, if it exists, lies between 0 and $\pi/2$ because the dispersion relation, being a function only of $\sin^2\theta$ and $\cos^2\theta$, is symmetric about $\theta = 0$ and $\theta = \pi/2$. Physically, these are manifestations of the azimuthal symmetry about the direction of the magnetic field $\mathbf{B}_0$ and the symmetry with respect to the direction of wave propagation $\mathbf{k}$. Thus, a wave that experiences a *resonance* propagates either (a) for $0 \le \theta < \theta_{res}$ but not $\theta_{res} < \theta \le \pi/2$ or (b) for $\theta_{res} < \theta \le \pi/2$ but not $0 \le \theta < \theta_{res}$, as indicated in Fig. 6.4. From this we can see that when $\theta_{res} \to 0$ in case (a) or $\theta_{res} \to \pi/2$ in case (b) the wave does disappear altogether. These are called the *principal resonances* and like the cut-offs they define, again for given equilibrium conditions, frequencies above or below which a particular wave does not propagate.

From (6.37) the principal resonances occur at

$$\theta_{res} = 0: \quad P = 0 \quad \text{or} \quad S = \frac{1}{2}(R + L) \to \infty \tag{6.38}$$

$$\theta_{res} = \pi/2: \quad S = 0 \tag{6.39}$$

The first possibility in (6.38) is a degenerate case because when $P = 0$ and $\theta = 0$ all the coefficients A, B, and C vanish; indeed, we have seen already that $P = 0$ is also a cut-off where $n^2 = 0$. Exactly what occurs here depends on the order in

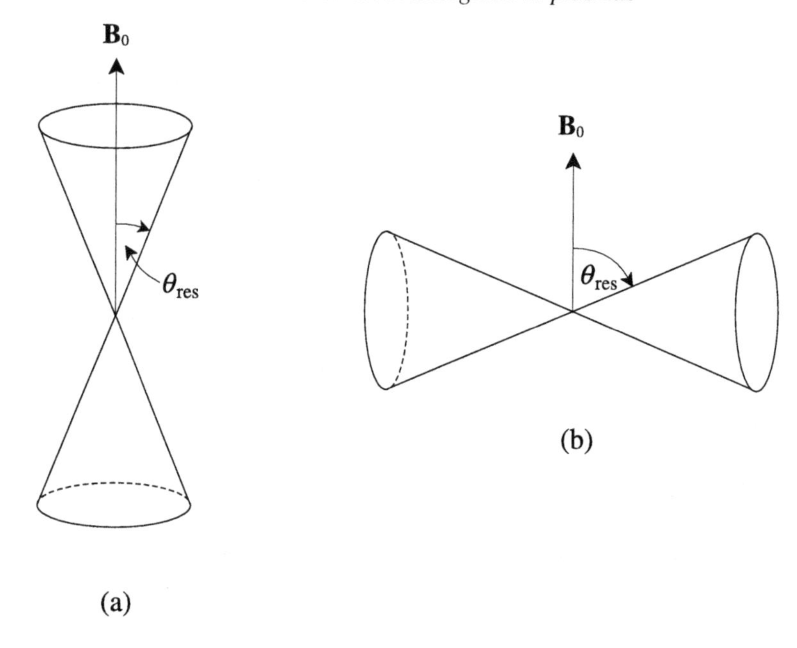

Fig. 6.4. Wave propagation cones.

which one takes the limits $\theta \to 0$ and $n^2 \to 0$ and nothing is gained by pursuing a general discussion of this case. The second possibility provides the interesting cases because either

$$R \to \infty \quad \text{as} \quad \omega \to -\Omega_e = |\Omega_e| \tag{6.40}$$

which is the *electron cyclotron resonance*, or

$$L \to \infty \quad \text{as} \quad \omega \to \Omega_i \tag{6.41}$$

which is the *ion cyclotron resonance*.

From (6.39) and (6.29) we see that the principal resonances at $\theta = \pi/2$ occur when

$$\omega^4 - \omega^2(\omega_p^2 + \Omega_i^2 + \Omega_e^2) - \Omega_i\Omega_e(\omega_p^2 - \Omega_i\Omega_e) = 0$$

which has the solutions

$$\omega^2 = \left(\frac{\omega_p^2 + \Omega_i^2 + \Omega_e^2}{2}\right)\left[1 \pm \left(1 + \frac{4\Omega_i\Omega_e(\omega_p^2 - \Omega_i\Omega_e)}{(\omega_p^2 + \Omega_i^2 + \Omega_e^2)^2}\right)^{1/2}\right] \tag{6.42}$$

Unlike the cyclotron resonances at $\theta = 0$ which involve either the ions or the electrons, these perpendicular resonances involve both ions and electrons together and are known, therefore, as the *hybrid resonances*. Since the second term in the square root in (6.42) is always much less than unity we may expand the square root to obtain the approximate solutions

$$\omega_{\mathrm{UH}}^2 \simeq (\omega_{\mathrm{p}}^2 + \Omega_{\mathrm{i}}^2 + \Omega_{\mathrm{e}}^2) \simeq \omega_{\mathrm{pe}}^2 + \Omega_{\mathrm{e}}^2 \tag{6.43}$$

$$\omega_{\mathrm{LH}}^2 \simeq -\frac{\Omega_{\mathrm{i}}\Omega_{\mathrm{e}}(\omega_{\mathrm{p}}^2 - \Omega_{\mathrm{i}}\Omega_{\mathrm{e}})}{\omega_{\mathrm{p}}^2 + \Omega_{\mathrm{i}}^2 + \Omega_{\mathrm{e}}^2} \simeq \begin{cases} |\Omega_{\mathrm{i}}\Omega_{\mathrm{e}}| & (\omega_{\mathrm{p}}^2 \gg \Omega_{\mathrm{e}}^2) \\ \omega_{\mathrm{pi}}^2 + \Omega_{\mathrm{i}}^2 & (\omega_{\mathrm{p}}^2 \ll \Omega_{\mathrm{e}}^2) \end{cases} \tag{6.44}$$

where the subscripts UH and LH denote the *upper hybrid* and *lower hybrid resonances*, respectively.

We shall discuss the physics of cut-offs and principal resonances as we meet the waves affected by them. We can now set out to investigate various special cases. By doing this systematically we shall find that the final picture that emerges enables us to construct a comprehensive picture of cold plasma wave propagation.

6.3.1 *Field-free plasma* ($\mathbf{B}_0 = 0$)

When there is no magnetic field there is no preferred direction so that without loss of generality we may take $\mathbf{n}$ to be in the z-direction, i.e. $\theta = 0$. Also, from (6.29), $S = P$ and $D = 0$ so that (6.30) takes the particularly simple diagonal form

$$\begin{bmatrix} 1 - \dfrac{\omega_{\mathrm{p}}^2}{\omega^2} - n^2 & 0 & 0 \\ 0 & 1 - \dfrac{\omega_{\mathrm{p}}^2}{\omega^2} - n^2 & 0 \\ 0 & 0 & 1 - \dfrac{\omega_{\mathrm{p}}^2}{\omega^2} \end{bmatrix} \begin{bmatrix} E_x \\ E_y \\ E_z \end{bmatrix} = 0 \tag{6.45}$$

Clearly, there are two types of wave in this case. Either $\mathbf{E} = (0, 0, E_z)$ and

$$\omega^2 = \omega_{\mathrm{p}}^2 \tag{6.46}$$

or $E_z = 0$ and

$$\omega^2 = \omega_{\mathrm{p}}^2 + k^2c^2 \tag{6.47}$$

The first of these solutions corresponds to the well-known, *longitudinal plasma oscillations*. Note that the terms *longitudinal* ($\mathbf{k} \parallel \mathbf{E}$) and *transverse* ($\mathbf{k} \perp \mathbf{E}$) indicate the direction of wave propagation relative to the electric field, $\mathbf{E}$, while the terms *parallel* and *perpendicular* indicate the direction of $\mathbf{k}$ relative to $\mathbf{B}_0$.

In this cold plasma limit the group velocity $v_{\mathrm{g}} = \mathrm{d}\omega/\mathrm{d}k = 0$, i.e. this wave does not propagate; if the disturbance producing the wave is local it remains so. It is an electrostatic wave as we can see from (6.11) that $\mathbf{B}_1 = 0$.

The second solution (6.47) has $\mathbf{k} \perp \mathbf{E}$ so this is a *transverse* wave. Since $k^2 < 0$ for $\omega^2 < \omega_p^2$ we see that $0 < \omega < \omega_p$ is a *stop-band* for transverse waves in a magnetic field-free plasma. The physical reason for this is simply that ω_p is the natural frequency with which the plasma responds to any imposed electric field. If the frequency of such a field is less than ω_p the plasma particles are able to respond quickly enough to neutralize it and it is damped out over a distance of about $|k|^{-1}$. This will be recognized as the first of the cut-offs, $P = 0$ in (6.36). The dispersion curve is sketched in Fig. 6.2, showing the characteristic behaviour of a cut-off, $\omega \to \omega_p$ (in this case) as $k \to 0$. As the frequency increases, the influence of the plasma decreases and the dispersion curve approaches the asymptote for propagation in vacuum, $\omega = kc$.

6.3.2 *Parallel propagation* ($\mathbf{k} \parallel \mathbf{B}_0$)

When wave propagation is along the magnetic field, $\theta = 0$ and (6.30) becomes

$$\begin{bmatrix} S - n^2 & -iD & 0 \\ iD & S - n^2 & 0 \\ 0 & 0 & P \end{bmatrix} \begin{bmatrix} E_x \\ E_y \\ E_z \end{bmatrix} = 0 \tag{6.48}$$

This shows, as in the field-free case, that the longitudinal $[\mathbf{E} = (0, 0, E_z)]$ and transverse $[\mathbf{E} = (E_x, E_y, 0)]$ waves are decoupled and that the dispersion relation, $P = 0$, i.e. $\omega^2 = \omega_p^2$, for the former is unchanged. This is only to be expected for the applied field $\mathbf{B}_0$ lies in the direction of the plasma oscillations so that there is no Lorentz force and therefore no effect on this mode.

The dispersion relation for the transverse waves can be obtained from (6.48) but we can get it and its solution directly from (6.33) on putting $\theta = 0$; eliminating the longitudinal wave ($P = 0$), the solutions are

$$n^2 = R = 1 - \frac{\omega_p^2}{(\omega + \Omega_i)(\omega + \Omega_e)} \tag{6.49}$$

$$n^2 = L = 1 - \frac{\omega_p^2}{(\omega - \Omega_i)(\omega - \Omega_e)} \tag{6.50}$$

The R and L modes, as we may call them, have cut-offs at ω_R and ω_L (see (6.36)) and principal resonances at $|\Omega_e|$ and Ω_i (see (6.40) and (6.41)). Remembering that $\Omega_e < 0$, it is clear from (6.49) and (6.50) that $n^2 > 0$ at the very lowest frequencies ($\omega \to 0$) and as $\omega \to \infty$ for both of these modes. Thus, the stop-bands lie between $|\Omega_e|$ and ω_R, and Ω_i and ω_L, for the R and L modes, respectively.

In order to sketch the dispersion curves for the propagating frequencies we take the high and low frequency limits of (6.49) and (6.50). The high frequency limit is easily dealt with for, as $\omega \to \infty$, both equations give the dispersion relation,

$\omega = kc$, for transverse waves in vacuo. However, as we reduce ω the terms in (6.49) and (6.50) containing the natural frequencies come into play and we get the approximate dispersion relations

$$R: \quad \omega^2 = k^2c^2 + \frac{\omega\omega_p^2}{\omega - |\Omega_e|} \tag{6.51}$$

$$L: \quad \omega^2 = k^2c^2 + \frac{\omega\omega_p^2}{\omega + |\Omega_e|} \tag{6.52}$$

From these equations it is clear that the phase velocity of the R mode is greater than that of the L mode so that we may label them *fast* and *slow*, respectively. Also the R mode cut-off ($k \to 0$) occurs above ω_p, whilst that of the L mode lies below ω_p; it is easily verified that ω_R and ω_L in (6.36) agree with the $k \to 0$ limit of (6.51) and (6.52) on neglecting terms in $\Omega_i/|\Omega_e|$.

Turning now to the low frequency limit of (6.49) and (6.50) and noting that $\omega_p^2/\Omega_i\Omega_e = (c/v_A)^2$ we obtain in both cases

$$\omega^2 = \frac{k^2v_A^2}{1 + (v_A/c)^2} \tag{6.53}$$

We can compare this result with the low frequency Alfvén waves discussed in Section 4.8. There we found three modes, the fast and slow magnetoacoustic waves and the (intermediate) shear Alfvén wave. For parallel propagation the magnetoacoustic waves decoupled into the compressional Alfvén wave and an acoustic wave with $\omega = kc_s$. In a cold plasma $c_s \to 0$ so the acoustic wave is the slow wave and disappears in this limit. Thus, the R and L modes may be identified with the two Alfvén waves. The slight discrepancy between (6.53) and the result $\omega = kv_A$ obtained from the ideal MHD equations may be traced directly to the retention of the displacement current in the cold plasma equations; it disappears in the non-relativistic limit ($v_p \approx v_A \ll c$).

To discover which of our two cold plasma modes is the fast, compressional wave and which the intermediate, shear wave we must resolve the degeneracy in (6.53) by keeping the next most significant term in ω. This means keeping the ω in ($\omega \pm \Omega_i$) whilst still ignoring it in ($\omega \pm \Omega_e$). Then in the non-relativistic limit (6.49) and (6.50) give

$$\left.\begin{aligned} R: \quad \omega^2 &= k^2v_A^2(1 + \omega/\Omega_i) \\ L: \quad \omega^2 &= k^2v_A^2(1 - \omega/\Omega_i) \end{aligned}\right\} \tag{6.54}$$

Thus, the R mode is the fast, compressional Alfvén wave with phase velocity $v_p > v_A$ and the L mode is the intermediate, shear Alfvén wave with $v_p < v_A$.

Collecting all this information about the R and L modes we can sketch their dispersion relations as shown in Figs. 6.5 and 6.6. Both modes have dispersion

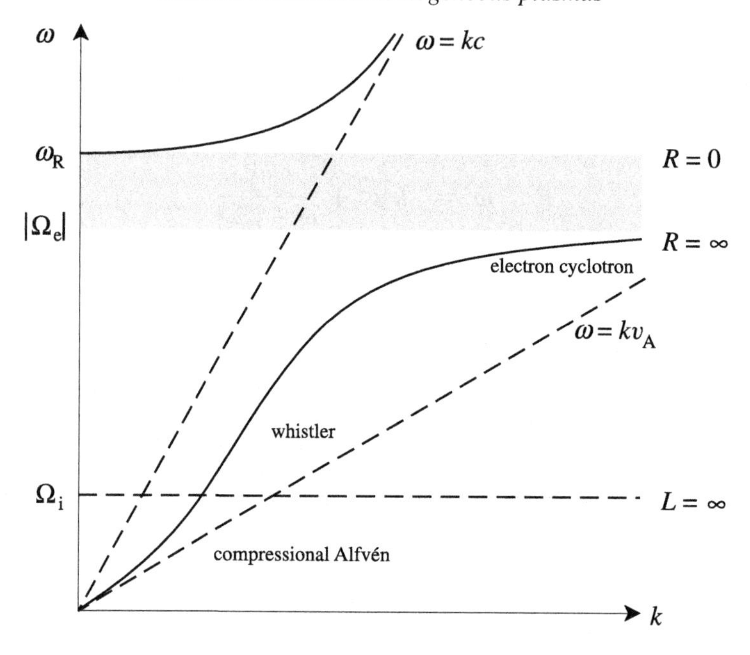

Fig. 6.5. Dispersion curves for R mode.

curves which are asymptotic to $\omega = kv_A$ and $\omega = kc$ at low and high frequencies, respectively. The horizontal asymptotes in both figures are at cut-offs ($k \to 0$) or principal resonances ($k \to \infty$).

Note that only the relevant cut-off and principal resonance affects a given wave so that the R mode continues to propagate above $\omega = \Omega_i$ but as it does so its phase velocity departs further and further from the Alfvén speed v_A. Choosing a value of ω such that $\Omega_i \ll \omega \ll |\Omega_e|$ and using the non-relativistic condition $v_p \ll c$, we may write (6.49) as

$$\omega \simeq k^2 c^2 |\Omega_e| / \omega_p^2 \tag{6.55}$$

which is the dispersion relation for *whistler waves*, so-called because they propagate in the ionosphere at audio-frequencies and can be heard as a whistle of descending pitch. They are triggered by lightning flashes and travel along the Earth's dipole field. From (6.55) we see that $\omega \propto k^2$ so both the phase velocity (ω/k) and the group velocity $(\mathrm{d}\omega/\mathrm{d}k)$ increase with k. This is what gives rise to the whistle; from a pulse initially containing a spread of frequencies the higher frequency waves travel faster arriving earlier at the detection point than the lower frequency waves and so a whistle of descending pitch is heard.

Near the principal resonances it is easy to show from (6.49) and (6.50) that the dispersion relations for the *electron cyclotron* and *ion cyclotron waves* are given

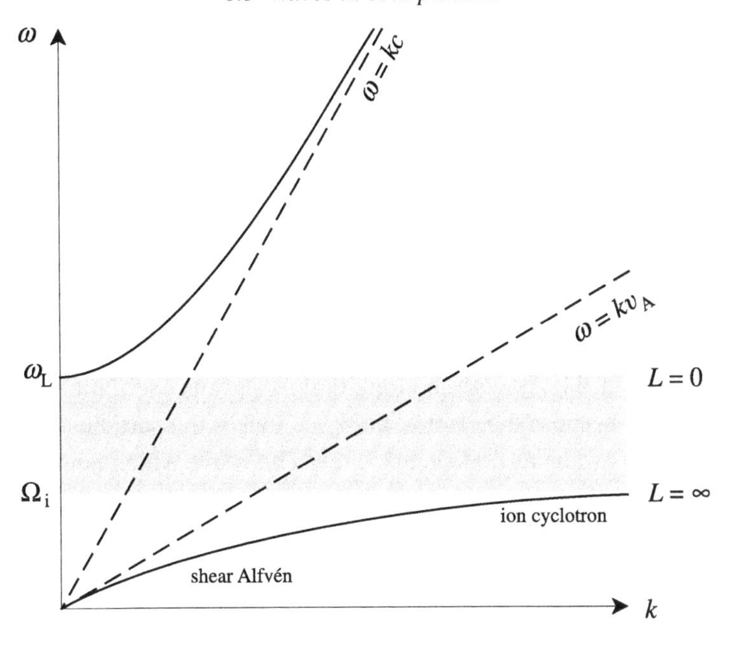

Fig. 6.6. Dispersion curves for L mode.

approximately by

$$R: \quad \omega = |\Omega_e|(1 + \omega_p^2/k^2c^2)^{-1} \simeq |\Omega_e|(1 + \omega_{pe}^2/k^2c^2)^{-1} \tag{6.56}$$

$$L: \quad \omega = |\Omega_i|(1 + \omega_{pi}^2/k^2c^2)^{-1} \tag{6.57}$$

respectively. To understand the physical origin of these resonances we note that (6.48) gives

$$\mathcal{P} = \frac{iE_x}{E_y} = \frac{n^2 - S}{D} = \frac{2n^2 - (R + L)}{R - L} = \begin{cases} +1 & (n^2 = R) \\ -1 & (n^2 = L) \end{cases}$$

showing that the R wave is RCP and the L wave is LCP. In Section 2.2 we saw that the electrons (ions) rotate about the magnetic field in a right (left) circular motion. Thus, the electric field of each wave rotates in the same sense as one of the particle species. So long as the wave frequency ω is less than the cyclotron frequency no resonance occurs but as the frequency of the $R(L)$ wave approaches $|\Omega_e|(\Omega_i)$ the electrons (ions) experience a near constant field and are continuously accelerated resulting in the absorption of the wave energy by the particles. The group velocity of both waves, $v_g \sim k^{-3} \to 0$ as $k \to \infty$.

6.3.3 Perpendicular propagation ($\mathbf{k} \perp \mathbf{B}_0$)

Putting $\theta = \pi/2$ in (6.30) gives

$$\begin{bmatrix} S & -iD & 0 \\ iD & S - n^2 & 0 \\ 0 & 0 & P - n^2 \end{bmatrix} \begin{bmatrix} E_x \\ E_y \\ E_z \end{bmatrix} = 0 \tag{6.58}$$

and we see that one of the solutions is the transverse wave with $\mathbf{E} \perp \mathbf{k}$, i.e. $\mathbf{E} = (0, 0, E_z)$, and dispersion relation $n^2 = P$. This is the same wave found in the field-free case (see (6.47) in Section 6.3.1†) which is unaffected by the introduction of the magnetic field $\mathbf{B}_0$. As with the longitudinal plasma oscillations for parallel propagation, this is because the electric field, E_z, makes the particles move parallel to $\mathbf{B}_0$ and therefore produces no Lorentz force. This wave which is independent of the magnetic field, is known as the *ordinary (O) mode.*

The dispersion relation for the other wave, called the *extraordinary (X) mode*, is most easily obtained from (6.33) and is given by

$$n^2 = \frac{RL}{S} \tag{6.59}$$

Thus, the X mode has cut-offs ($k \to 0$) at ω_R ($R = 0$) and ω_L ($L = 0$) and resonances ($k \to \infty$) at the upper and lower hybrid frequencies ($S = 0$). By careful examination of (6.36), (6.43) and (6.44) we may show that $\omega_R \geq \omega_{UH} \geq \omega_L \geq \omega_{LH}$ (with equality only for either n_0 or $B_0 = 0$) and hence deduce that the stop-bands for the X mode lie in the frequency intervals ω_{LH} to ω_L and ω_{UH} to ω_R. Also, we may write (6.59) as

$$\frac{\omega^2}{k^2c^2} = \frac{S}{RL} = \frac{1}{2}\left(\frac{1}{R} + \frac{1}{L}\right) \tag{6.60}$$

and by inspection of (6.29), we see that $R, L \to 1$ as $\omega \to \infty$ so that the X mode dispersion relation is asymptotic to $\omega = kc$ in this limit. Then as ω decreases the first cut-off occurs at ω_R and the mode is evanescent until we reach the first resonance (ω_{UH}) at $S = 0$. The X mode then propagates again until ω_L is reached where the $L = 0$ cut-off occurs. There is then another stop-band until the lower hybrid frequency ω_{LH} is reached at which propagation recommences down to $\omega = 0$. As $\omega \to 0$, $R, L \to c^2/v_A^2$ so the dispersion curve is asymptotic to $\omega = kv_A$. These observations are summarized in Fig. 6.7.

The dispersion curve for the O mode is shown in Fig. 6.2 and, as discussed earlier, the stop-band extends from $\omega = 0$ to $\omega = \omega_p$. Below ω_p there is, therefore, at most only the X mode propagating perpendicular to the magnetic field. At the very lowest frequencies ($\omega \to 0$) this is clearly the compressional Alfvén

† Note that in Section 6.3.1 the choice of axes was different with $\mathbf{k} = (0, 0, k)$ and $\mathbf{E} = (E_x, E_y, 0)$.

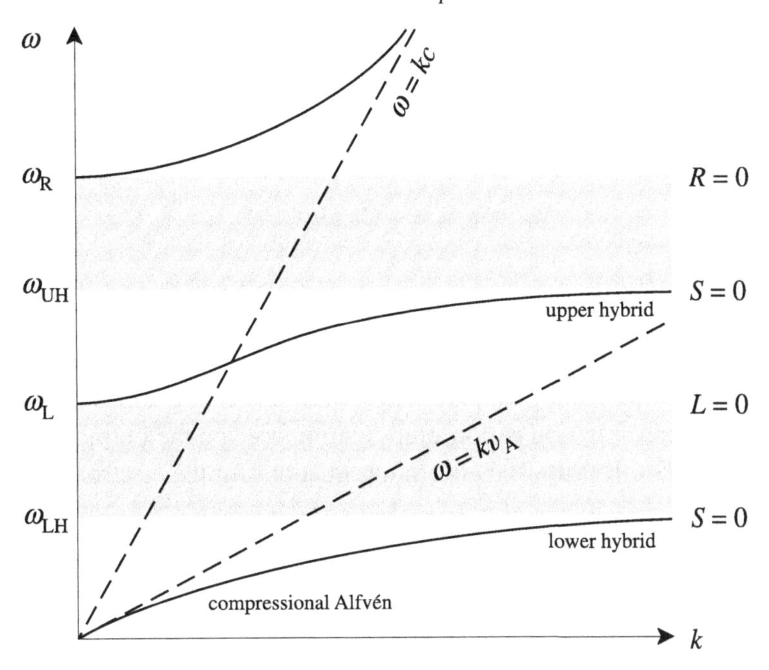

Fig. 6.7. Dispersion curves for X mode.

wave since the shear Alfvén wave propagates along $\mathbf{B}_0$ but not perpendicular to it. Whereas at parallel propagation the compressional Alfvén wave becomes the whistler and then the electron cyclotron wave as it approaches resonance, at perpendicular propagation it becomes the lower hybrid wave as resonance is approached. Note, also, that resonance is reached at a lower frequency ($\omega_{\mathrm{LH}} < |\Omega_{\mathrm{e}}|$) for perpendicular propagation. Between ω_{LH} and $|\Omega_{\mathrm{e}}|$ the resonant angle, given by (6.37), decreases from $\pi/2$ to 0 so that the cone of propagation (see Fig. 6.4(b)) narrows as the frequency increases until the wave is suppressed completely at $\omega = |\Omega_{\mathrm{e}}|$.

The physical mechanism of the lower hybrid resonance is more complicated than the simple cyclotron resonances because both types of particle are involved. From (6.44) we see that the lower hybrid frequency is proportional to the geometric mean of the cyclotron frequencies and for sufficiently high density ($\omega_{\mathrm{p}}^2 \gg \Omega_{\mathrm{e}}^2$) we have $\Omega_{\mathrm{i}} \ll \omega_{\mathrm{LH}} = (\Omega_{\mathrm{i}}|\Omega_{\mathrm{e}}|)^{1/2} \ll |\Omega_{\mathrm{e}}|$. Thus, on a time scale of the lower hybrid period the ions are effectively unmagnetized and they oscillate back and forth in response to the electric field. From (6.58) we see that as $\omega \to \omega_{\mathrm{LH}}$, i.e. $S \to 0$, $E_y \to 0$ and so the equation of motion of the ions to lowest order is

$$m_{\mathrm{i}}\ddot{x} = ZeE$$

giving an ion displacement in the x direction of magnitude

$$(\Delta x)_{\mathrm{i}} \sim ZeE/m_{\mathrm{i}}\omega^2$$

The ion displacement in the y direction is of first order and given by

$$m_{\mathrm{i}}\ddot{y} = -Ze\dot{x}B_0$$

from which we get

$$(\Delta y)_{\mathrm{i}} \sim \frac{\Omega_{\mathrm{i}}}{\omega}(\Delta x)_{\mathrm{i}} \tag{6.61}$$

The electrons, on the other hand, are magnetized and rotate about the field lines many times in a lower hybrid period. But superimposed on the Larmor orbits there is an oscillating $\mathbf{E} \times \mathbf{B}$ drift. The governing equations for the electrons are

$$m_{\mathrm{e}}\ddot{x} = -eE - e\dot{y}B_0 \tag{6.62}$$

$$m_{\mathrm{e}}\ddot{y} = e\dot{x}B_0 \tag{6.63}$$

From (6.63) we get

$$(\Delta y)_{\mathrm{e}} \sim \frac{|\Omega_{\mathrm{e}}|}{\omega}(\Delta x)_{\mathrm{e}} \tag{6.64}$$

and substituting this in (6.62) gives

$$(\Delta x)_{\mathrm{e}} \sim \frac{eE}{m_{\mathrm{e}}(\Omega_{\mathrm{e}}^2 + \omega^2)} \approx \frac{eE}{m_{\mathrm{e}}\Omega_{\mathrm{e}}^2}$$

From (6.61) and (6.64) we see that the x displacement of the ions is much greater than their y displacement while the opposite is true for the electrons and, for $\omega < (\Omega_{\mathrm{i}}|\Omega_{\mathrm{e}}|)^{1/2}$, we have $|(\Delta x)_{\mathrm{i}}| > |(\Delta x)_{\mathrm{e}}|$ so that the average motion of ions and electrons is as shown in Fig. 6.8(a). However, as $\omega \to \omega_{\mathrm{LH}} = (\Omega_{\mathrm{i}}|\Omega_{\mathrm{e}}|)^{1/2}$, $(\Delta x)_{\mathrm{i}} \to (\Delta x)_{\mathrm{e}}$ and the picture is as shown in Fig. 6.8(b). Now the ions and electrons not only oscillate in phase but maintain charge neutrality so the field cannot be maintained and the wave ceases to propagate.

For lower densities ($\omega_{\mathrm{p}}^2 \lesssim \Omega_{\mathrm{e}}^2$) the lower hybrid frequency decreases towards Ω_{i} with the result that the ion motion becomes more circular (see (6.61)) while the average electron motion becomes more elongated and $|(\Delta x)_{\mathrm{e}}| \ll |(\Delta x)_{\mathrm{i}}|$. Consequently, the role of the electrons in maintaining the space charge responsible for the electric field is diminished and the resonance becomes predominantly an ion affair with $\omega_{\mathrm{LH}} \simeq (\omega_{\mathrm{pi}}^2 + \Omega_{\mathrm{i}}^2)^{1/2}$. The resonance occurs when the ion motion in the x direction, which is a resultant of direct response to the electric field and Larmor oscillation about $\mathbf{B}_0$, is in phase with the electric field.

Similarly, at the upper hybrid resonance, although nominally both types of particle are involved, the motion of the ions is insignificant at this very high frequency,

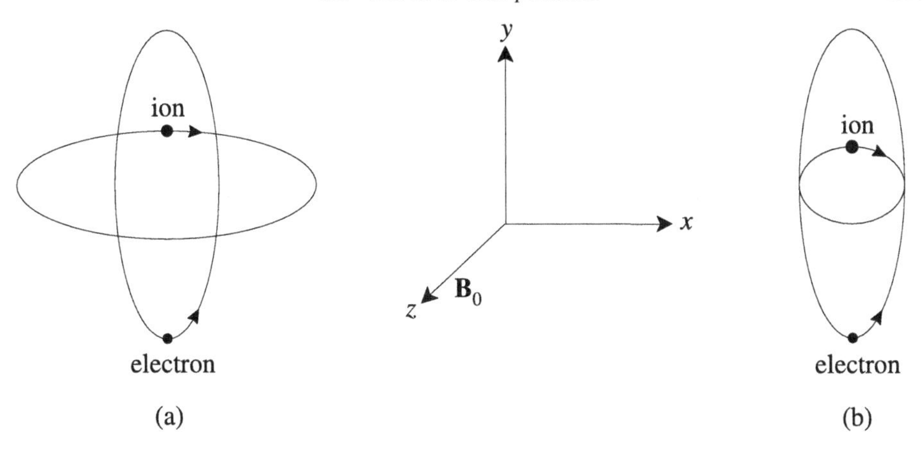

Fig. 6.8. Average particle orbits in lower hybrid wave for (a) $\omega < \omega_{LH}$, (b) $\omega \to \omega_{LH}$.

$\omega_{UH} = (\omega_p^2 + \Omega_e^2 + \Omega_i^2)^{1/2} \approx (\omega_{pe}^2 + \Omega_e^2)^{1/2}$, and the resonance is between the electron motion and the electric field as can be seen from (6.62). Poisson's equation provides an order of magnitude for the electric field, $E \sim n_e e(\Delta x)_e/\varepsilon_0$, so that, using (6.64) for $(\Delta y)_e$, (6.62) gives

$$\omega^2(\Delta x)_e \sim (\omega_{pe}^2 + \Omega_e^2)(\Delta x)_e$$

This is a second example of the symmetries found in the cold plasma wave theory between electron properties at high frequency and ion properties at low frequency, the first being the simple cyclotron resonances.

A final observation from (6.58) is that

$$\mathcal{P} = \frac{iE_x}{E_y} = \frac{-D}{S} = \frac{L-R}{L+R}$$

so that in general the X mode is elliptically polarized although this becomes linear at the resonances ($S \to 0$), as already noted, and circular at the cut-offs; the wave is RCP at $\omega = \omega_R$ ($R = 0$) and LCP at $\omega = \omega_L$ ($L = 0$).

6.3.4 Wave normal surfaces

Much information about cold plasma waves has been obtained by examining the special cases of parallel and perpendicular propagation. We shall now show that by combining the results of the last two sub-sections with some of the properties of the general dispersion relation we can make deductions about the waves propagating at oblique angles ($0 < \theta < \pi/2$) to the magnetic field $\mathbf{B}_0$.

First, let us summarize some of the properties of the solutions of the general dispersion relation (6.31):

(i) There are two solutions which are distinct except where the discriminant (6.34) vanishes. Except for the discrete points in parameter space where the surfaces $RL = PS$ and $PD = 0$ intersect, the discriminant can vanish only at $\theta = 0$ or $\pi/2$. For oblique propagation, therefore, we can use this distinction to label one of the solutions the fast (F) wave and the other the slow (S) wave. By extrapolation this labelling can be used at $\theta = 0$ and $\pi/2$, also, even when the discriminant vanishes at these angles. Since $n^2 = c^2/v_{\mathrm{p}}^2$ we have $n_{\mathrm{F}}^2 < n_{\mathrm{S}}^2$.

(ii) The phase velocity of a propagating wave may remain finite at all angles or may tend to zero ($k \to \infty$) as $\theta \to \theta_{\mathrm{res}}$. In the latter case the wave propagates in only one of the cones shown in Fig. 6.4. If both waves propagate and one of them suffers a resonance this must be the S wave. It can be shown (see Stix (1992)) that, if both waves propagate, at most one of them can suffer a resonance.

(iii) If the waves propagate at $\theta = 0$ one of them is the R wave and the other the L wave. These, also, are useful identifying labels but it should be remembered that the dispersion relations $n^2 = R, L$ apply *only* at $\theta = 0$ and the properties of RCP and LCP, likewise, do not apply at oblique propagation.

(iv) Similarly, the O and X labels may be used if the waves propagate at $\theta = \pi/2$ but, here again, one cannot extrapolate the dispersion relations $n^2 = P$, $n^2 = RL/S$ nor is it true that the dispersion relation of the O wave remains independent of the magnetic field for $\theta < \pi/2$.

All this information may be neatly summarized by drawing the *wave normal surfaces* at any given point in parameter space. The wave normal surface is a plot of the phase velocity in spherical polar coordinates but since there is no dependence on the azimuthal coordinate, ϕ, this reduces to a plane polar plot of v_{p} versus θ, the surface being generated by rotation of the figure about the polar ($\hat{\mathbf{z}}$) axis. In view of properties (i) and (ii) the only possible surfaces are the spheroid and lemniscoids shown in Fig. 6.9; the lemniscoid with propagation at $\theta = 0$ is called a dumb-bell and that with propagation at $\theta = \pi/2$ is called a wheel (imagine the polar plots rotated about the polar axis).

If both waves propagate, the permissible combinations of wave normal surfaces are two spheroids or a spheroid and a lemniscoid as illustrated in Fig. 6.10. Except for the discrete points of parameter space mentioned in (i) the wave normal surfaces may be tangential only at $\theta = 0$ or $\pi/2$. Clearly the outer surface is the F wave and we may add, as appropriate, the labels R or L at $\theta = 0$ and O or X at $\theta = \pi/2$. For example, the wave normal surfaces for the compressional and shear Alfvén waves in the low frequency regime ($\omega < \Omega_{\mathrm{i}}, \omega_{\mathrm{p}}$) correspond to a spheroid and dumb-bell

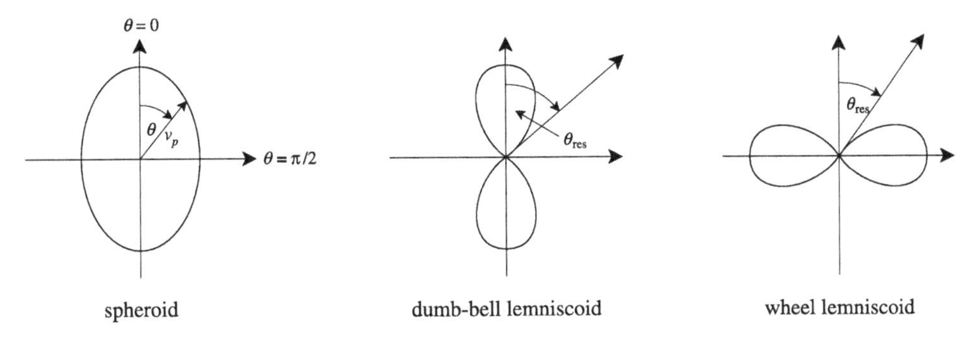

Fig. 6.9. Wave normal surfaces.

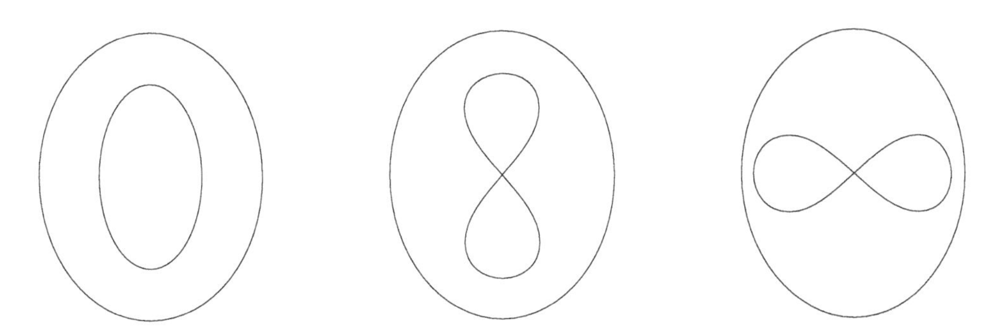

Fig. 6.10. Possible wave normal surfaces when both modes propagate.

lemniscoid, respectively. Ideal MHD suggests that these surfaces are tangential at $\theta = 0$ but (6.54) shows that this is true only in the limit $\omega \to 0$.

Now we come to the most important statement about the wave normal surfaces for the compilation of a general classification scheme. *The topology of the surfaces can change only at the cut-offs and principal resonances.* For example, at a cut-off $v_{\mathrm{p}} \to \infty$ so the F wave solution changes sign at infinity, i.e. a spheroid disappears. At a principal resonance $v_{\mathrm{p}} \to 0$ so a spheroid may become a lemniscoid. The converse of this occurs when $\theta_{\mathrm{res}} \to 0(\pi/2)$ for a wheel (dumb-bell) lemniscoid. Finally, lemniscoids disappear when $\theta_{\mathrm{res}} \to 0(\pi/2)$ for the dumb-bell (wheel). This means that the cut-offs and principal resonances are the natural classification boundaries in parameter space.

For a two-component plasma, parameter space is two dimensional and can be represented by a diagram with $\alpha^2 = \omega_{\mathrm{p}}^2/\omega^2$ as abscissa and $\beta^2 = |\Omega_{\mathrm{i}}\Omega_{\mathrm{e}}|/\omega^2$ as ordinate; thus, the horizontal axis is the direction of increasing density or decreasing frequency and the vertical axis is the direction of increasing magnetic field or decreasing frequency. The cut-offs and principal resonances divide this space

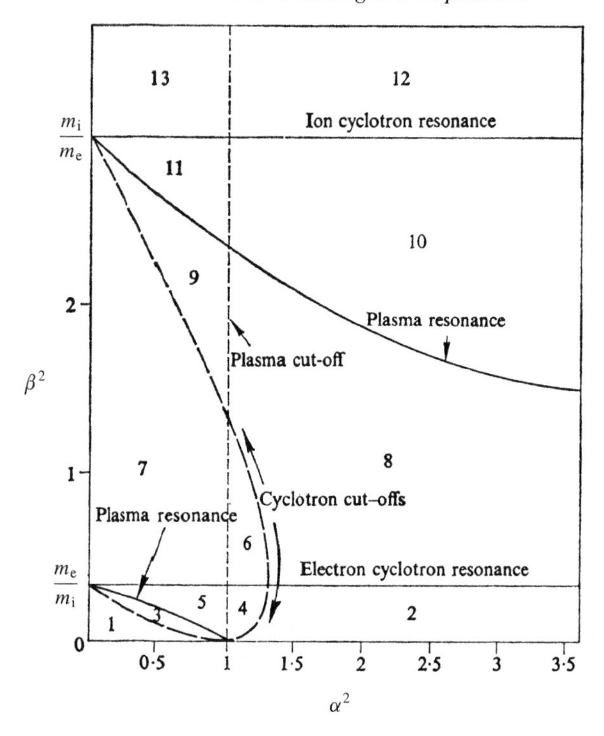

Fig. 6.11. Subdivision of parameter space by principal resonance and cut-off curves.

into thirteen regions which are numbered† alternately left and right of the plasma cut-off $P = 0$ (i.e. $\alpha^2 = 1$) and with increasing β^2, as shown in Fig. 6.11; the figure is illustrative and is not drawn to any realistic scale of mass ratio.

Since all the boundaries of parameter space represent a specific frequency we can deduce all the wave normal surfaces for the thirteen regions from the dispersion diagrams for the R, L, O and X waves (see Figs. 6.2 and 6.5–6.7). These tell us which of these waves propagate at $\theta = 0$ and $\pi/2$ and, if both propagate, by comparing the asymptotic behaviour, which is the F and which the S wave. For example, Fig. 6.2 shows that the O mode propagates only for $\omega > \omega_p$, i.e. in the odd numbered regions to the left of $P = 0$. Also, by comparing Figs. 6.2 and 6.7 we see that the X mode is the F wave for $\omega > \omega_R$ (as $k \to 0$, $\omega_X \to \omega_R$ and $\omega_O \to \omega_p$ with $\omega_R > \omega_p$). Similarly, from Figs. 6.5 and 6.6, both R and L modes propagate for $\omega > \omega_R$ and the R mode is clearly the F wave. Thus, in region 1 ($\omega > \omega_R$) both wave normal surfaces are spheroids, the F wave having the labels RX and the S wave LO. In crossing the $R = 0$ ($\omega = \omega_R$) boundary the RX mode is cut-off ($v_p \to \infty$) and only the LO mode propagates in

† This numbering system is not universal; our choice follows Allis *et al.* (1963).

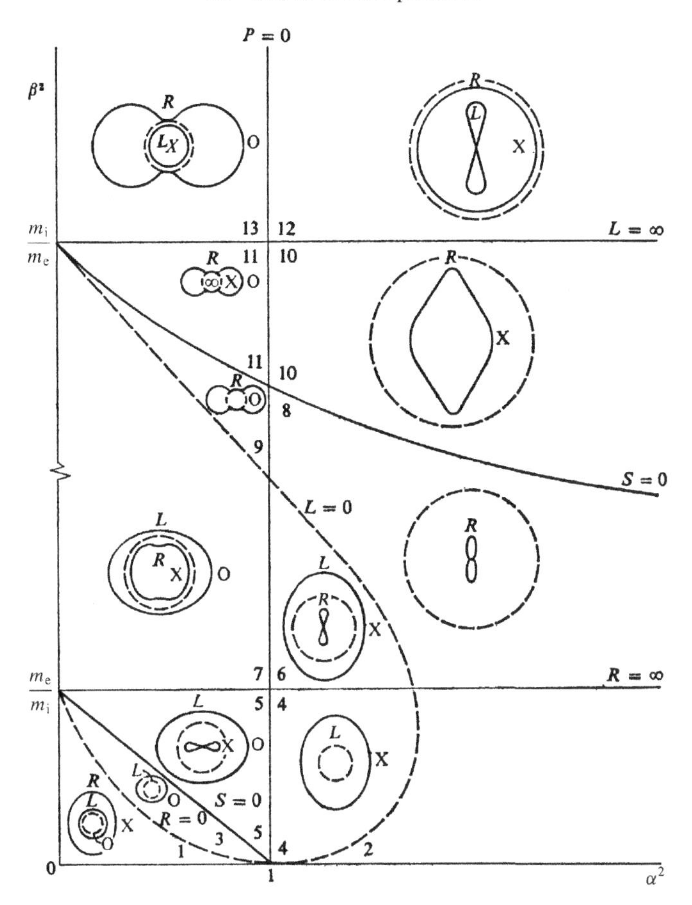

Fig. 6.12. CMA diagram showing the wave normal surfaces for a cold plasma. The surfaces are not drawn to scale but the dashed circle represents the velocity of light in each region (after Allis *et al.* (1963)).

region 3. In this manner one can traverse the whole of parameter space identifying the wave normal surfaces† (see Exercise 6.8) and obtain Fig. 6.12, which is called the *Clemmow–Mullaly–Allis (CMA) diagram.* The sketches of the wave normal surfaces in Fig. 6.12 are schematic, merely indicating type. The actual shape varies across the region; for example, the spheroidal wave normal surface of the SLX mode in region 13 pinches in ($v_p \to 0$) at $\theta = 0$ as one approaches the $\omega = \Omega_i$ boundary in anticipation of the disappearance of the L mode on crossing this boundary into region 11. Note that the S wave does not disappear completely at this boundary but its wave normal surface changes from a spheroid to a wheel lemniscoid; such a change is called a *re-shaping transition.* On the other hand, in

† There is one slight complication with the O and X labels in region 7; they switch waves across the surface $RL = PS$ because $n_X^2 - n_O^2 = (RL - PS)/S$.

crossing the same boundary from region 12 to region 10, the cut-off of the L mode does mean the complete disappearance of the S wave; this is called a *destructive transition.* In contrast, the F wave experiences no significant change on crossing this boundary and is said to undergo an *intact transition.*

6.3.5 Dispersion relations for oblique propagation

Since the boundaries of parameter space are all specified by particular frequencies we can use this to find approximate dispersion relations by comparing ω in a given region with its bounding frequencies and either expanding the coefficients in (6.32) in the small parameter ω/ω_B or ω_B/ω (where ω_B is the frequency at a boundary of the region) or approximating the coefficients by letting $\omega \to \omega_B$. We have used this technique already in the low frequency regime (region 12) by assuming $\omega/\Omega_i \ll 1$ and $\omega/\omega_p \ll 1$ and then later letting $\omega \to \Omega_i$. It is particularly useful when comparing ω with the plasma and cyclotron frequencies for, as one can see from (6.29), the parameters R, L, S and P can all be expressed in terms of $\alpha^2 = \omega_p^2/\omega^2$, $\beta^2 = |\Omega_i\Omega_e|/\omega^2$, $\beta_i = \Omega_i/\omega$ and $\beta_e = |\Omega_e|/\omega$:

$$\left.\begin{aligned} S &= 1 - \frac{\alpha^2(1-\beta^2)}{(1-\beta_i^2)(1-\beta_e^2)} \\ R &= 1 - \frac{\alpha^2}{(1+\beta_i)(1-\beta_e)} \\ L &= 1 - \frac{\alpha^2}{(1-\beta_i)(1+\beta_e)} \\ P &= 1 - \alpha^2 \end{aligned}\right\} \tag{6.65}$$

Low frequency regime ($\omega < \Omega_i$)

Let us use this method to recover the dispersion relations in region 12 for arbitrary angles of propagation. We may choose $\omega \ll \Omega_i$ and $\omega \ll \omega_p$ giving

$$S, R, L \approx 1 + \alpha^2/\beta^2 = 1 + \gamma \qquad (\gamma = c^2/v_A^2)$$

and

$$P \approx -\alpha^2$$

Substituting these approximations in (6.32) we get

$$\begin{aligned} A &\approx 1 + \gamma - (1 + \gamma + \alpha^2)\cos^2\theta \\ B &\approx (1+\gamma)[(1+\gamma-\alpha^2) - (1+\gamma+\alpha^2)\cos^2\theta] \\ C &\approx -\alpha^2(1+\gamma)^2 \end{aligned}$$

and (6.31) factorizes to give the solutions

$$n_1^2 = 1 + \gamma \tag{6.66}$$

$$n_2^2 = \frac{\alpha^2(1+\gamma)}{(1+\gamma+\alpha^2)\cos^2\theta - (1+\gamma)} \tag{6.67}$$

The first solution is independent of θ so the wave normal surface is a sphere. The second solution gives real values of n only for $0 \leq \theta \leq \theta_{\mathrm{res}}$, where $\cos^2\theta_{\mathrm{res}} = (1+\gamma)/(1+\gamma+\alpha^2)$ so this wave normal surface is a dumb-bell lemniscoid confirming the statement made in Section 6.3.4. The ideal MHD solutions, (4.122) and (4.123) with $c_{\mathrm{s}} = 0$, are recovered in the non-relativistic and low frequency limits $\gamma, \alpha^2 \to \infty$.

Near the ion cyclotron resonance ($\omega \to \Omega_{\mathrm{i}}$)

Next let us find the dispersion relations for oblique propagation as we approach the ion cyclotron resonance at $\omega = \Omega_{\mathrm{i}}$. Here we let $\omega = \Omega_{\mathrm{i}}(1-\epsilon)$, where $0 < \epsilon \ll 1$, giving $S \approx \gamma/(2\epsilon)$, $R \approx 1 + \gamma/2$, $L \approx \gamma/\epsilon$ and $P \approx -\gamma/\mu$, with $\mu \equiv \Omega_{\mathrm{i}}/|\Omega_{\mathrm{e}}| = Zm_{\mathrm{e}}/m_{\mathrm{i}} \ll 1$. Substituting these approximations in (6.32) we get

$$A \approx \frac{\gamma}{2\epsilon\mu}(\mu\sin^2\theta - 2\epsilon\cos^2\theta)$$

$$B \approx -\frac{\gamma^2}{2\epsilon\mu}(1+\cos^2\theta)$$

$$C \approx -\frac{\gamma^2}{2\epsilon\mu}(2+\gamma)$$

Since both μ and ϵ are small quantities it is clear that $B^2 \gg |4AC|$ so that, expanding the discriminant, the approximate solutions of (6.31) are

$$n^2 = \frac{C}{B} = \frac{2+\gamma}{1+\cos^2\theta} \tag{6.68}$$

and

$$n^2 = \frac{B}{A} = \frac{\gamma(1+\cos^2\theta)}{(-\mu\sin^2\theta + 2\epsilon\cos^2\theta)} \tag{6.69}$$

The first of these solutions, which may be written

$$\omega^2 = \frac{k^2 v_{\mathrm{A}}^2(1+\cos^2\theta)}{1+2v_{\mathrm{A}}^2/c^2} \approx k^2 v_{\mathrm{A}}^2(1+\cos^2\theta) \tag{6.70}$$

is the generalization of (6.53) for compressional Alfvén waves propagating at arbitrary angles as $\omega \to \Omega_{\mathrm{i}}$. It shows the increase in phase velocity ($v_{\mathrm{p}} \approx v_{\mathrm{A}}\sqrt{2}$) at

$\theta = 0$ as $\omega \to \Omega_{\mathrm{i}}$ (see Fig. 6.5); the wave normal surface is still a spheroid but no longer a sphere.

The more interesting solution is (6.69) which is the dispersion relation for ion cyclotron waves. It has a resonance at $\tan^2\theta_{\mathrm{res}} = 2\epsilon/\mu$, confirming that $\theta_{\mathrm{res}} \to 0$ as $\epsilon \to 0$ (i.e. $\omega \to \Omega_{\mathrm{i}}$). Dropping the term in μ and rewriting (6.69) as $n^2 \approx S(1+\cos^2\theta)/\cos^2\theta$, where $S \approx -\omega_{\mathrm{p}}^2\Omega_{\mathrm{i}}/|\Omega_{\mathrm{e}}|(\omega^2 - \Omega_{\mathrm{i}}^2)$, we get,

$$\omega^2 \approx \frac{k^2 v_{\mathrm{A}}^2 \cos^2\theta}{1+\cos^2\theta}\left(\frac{\Omega_{\mathrm{i}}^2 - \omega^2}{\Omega_{\mathrm{i}}^2}\right) \tag{6.71}$$

as the generalization of (6.57) for ion cyclotron waves.

High frequency regime

Simple dispersion relations are obtainable whenever the discriminant $(B^2 - 4AC)$ is a perfect square or can be expanded. Although the first of these possibilities (which yielded the solutions (6.66) and (6.67)) occurs rarely, it is clear from (6.34) that one can always find an expansion by letting $\theta \to 0$ or $\pi/2$; these denote the *quasi-parallel* ($Q_{\parallel}$) and *quasi-perpendicular* ($Q_{\perp}$) *approximations*, respectively. This method of approximation is particularly appropriate in the high frequency regime which we consider next.

If $\omega \gg |\Omega_{\mathrm{i}}\Omega_{\mathrm{e}}|^{1/2}$ it follows from (6.65) that

$$\left.\begin{array}{lcl} S & \approx & 1-\alpha^2/(1-\beta_{\mathrm{e}}^2) \\ R & \approx & 1-\alpha^2/(1-\beta_{\mathrm{e}}) \\ L & \approx & 1-\alpha^2/(1+\beta_{\mathrm{e}}) \\ P & = & 1-\alpha^2 \end{array}\right\} \tag{6.72}$$

and, since $\alpha^2 \approx \omega_{\mathrm{pe}}^2/\omega^2$, it is clear that the effect of the ions on wave propagation is negligible. This regime, which embraces all of regions 1–8 in the CMA diagram (provided we are not too close to the $S = 0$ ($\omega = \omega_{\mathrm{LH}}$) boundary in region 8), has been studied extensively in the context of waves in the ionosphere and gives rise to *magneto-ionic theory*.

To establish contact with this theory it is convenient to cast the solution of (6.31) in the form (see Exercise 6.11)

$$n^2 = 1 - \frac{2(A-B+C)}{2A - B \mp (B^2-4AC)^{1/2}} \tag{6.73}$$

Using the approximations (6.72) this becomes

$$n^2 = 1 - \frac{2\alpha^2(1-\alpha^2)}{2(1-\alpha^2) - \beta_{\mathrm{e}}^2\sin^2\theta \mp \Gamma} \tag{6.74}$$

where

$$\Gamma = [\beta_e^4 \sin^4\theta + 4\beta_e^2(1-\alpha^2)^2\cos^2\theta]^{1/2} \tag{6.75}$$

Equation (6.74) is the *collisionless Appleton–Hartree dispersion relation*. We shall consider it in the limits:

$$Q_\parallel : \quad \beta_e^2 \sin^4\theta \ll 4(1-\alpha^2)^2\cos^2\theta \tag{6.76}$$

$$Q_\perp : \quad \beta_e^2 \sin^4\theta \gg 4(1-\alpha^2)^2\cos^2\theta \tag{6.77}$$

Quasi-parallel ($\theta \to 0$)

The $Q_\parallel$ solutions are given by

$$n^2 \approx 1 - \frac{\alpha^2}{1 \mp \beta_e \cos\theta} \tag{6.78}$$

and comparison with (6.72) shows that the plus and minus signs correspond, in the limit $\theta \to 0$, to the L and R waves, respectively. Thus, (6.78) is the generalization, in the $Q_\parallel$ limit, of (6.51) and (6.52), giving

$$\omega^2 = k^2c^2 + \frac{\omega\omega_p^2}{\omega \mp |\Omega_e|\cos\theta}$$

and showing that the high frequency dispersion relations for oblique propagation are obtained from (6.51) and (6.52) by replacing $|\Omega_e|$ by $|\Omega_e|\cos\theta$, i.e. the component of the field along the direction of wave propagation.

For $\beta_e \geq 1$ the R wave has a resonance at $\cos\theta = \beta_e^{-1}$. According to (6.78) this occurs for any value of α^2 but near a resonance we need to make a more careful examination of the approximation. This is best done directly from (6.37) which gives

$$\tan^2\theta_{res} = \frac{(\alpha^2-1)(\beta_e^2-1)}{(\alpha^2+\beta_e^2-1)} = \frac{\beta_e^2-1}{1+\beta_e^2/(\alpha^2-1)} \tag{6.79}$$

showing that there is no real solution and, therefore, no resonance for $\alpha^2 < 1$ and confirming that the resonance occurs at $\cos\theta = \beta_e^{-1}$ provided $\beta_e^2 \ll \alpha^2 - 1$, which is consistent with the $Q_\parallel$ approximation (6.76).

Quasi-perpendicular ($\theta \to \pi/2$)

Turning to the $Q_\perp$ approximation we find from (6.74) the solutions

$$n^2 = \frac{1-\alpha^2}{1-\alpha^2\cos^2\theta} \tag{6.80}$$

and

$$n^2 = \frac{(1-\alpha^2)^2 - \beta_e^2 \sin^2\theta}{(1-\alpha^2) - \beta_e^2 \sin^2\theta} \tag{6.81}$$

For $\alpha^2 < 1$ the first of these has $n^2 > 0$ for all θ and reduces to (6.47) as $\theta \to \pi/2$ so this is the O mode. Rearranging (6.80) we get

$$\omega^4 - \omega^2(\omega_p^2 + k^2c^2) + k^2c^2\omega_p^2 \cos^2\theta = 0$$

with the approximate solutions

$$\omega^2 \approx (\omega_p^2 + k^2c^2)\left[1 - \frac{k^2c^2\omega_p^2\cos^2\theta}{(\omega_p^2 + k^2c^2)^2}\right] \tag{6.82}$$

and

$$\omega^2 \approx \frac{\omega_p^2 \cos^2\theta}{1 + \omega_p^2/k^2c^2} \tag{6.83}$$

Only the first solution (6.82) has $\alpha^2 < 1$ so this is the generalization of (6.47) showing a marginal decrease in phase velocity for propagation of the O mode away from the perpendicular direction. The second solution (6.83) has $\alpha^2 > 1$ and, from (6.80), we see that it propagates only for $0 \leq \theta < \cos^{-1}(\alpha^{-1})$. This is, in fact, the $Q_\perp$ approximation of the R mode dispersion relation for α^2, $\beta^2 > 1$, as we can see from (6.79). The symmetry of this equation with respect to α and β shows that for $1 < \alpha^2 \ll \beta_e^2 - 1$ the resonance occurs at $\cos\theta = \alpha^{-1}$. Given that region 6 has $1 < \alpha^2 < 2$ and $1 < \beta_e^2 < |\Omega_e|/\Omega_i - 1$, (6.83) is appropriate in this region except near the electron cyclotron resonance or $\theta = 0$.

The second solution (6.81) in the $Q_\perp$ approximation is the dispersion relation for the X mode. For perpendicular propagation this was given by (6.59) and if we substitute for R, L and S from (6.72) in this equation we get

$$n^2 = \frac{(1-\alpha^2)^2 - \beta_e^2}{1 - \alpha^2 - \beta_e^2}$$

Comparing this with (6.81) we see that the high frequency dispersion relation for the X mode, in the $Q_\perp$ approximation, is obtained simply by replacing $|\Omega_e|$ by $|\Omega_e|\sin\theta$, i.e. the component of the field perpendicular to the direction of wave propagation. The resonance occurs at

$$\sin^2\theta_{res} = \frac{1-\alpha^2}{\beta_e^2}$$

which has real solutions only for $0 < 1 - \alpha^2 < \beta_e^2$ and, checking against the exact (6.79), we see that we must add to this the condition $\beta_e^2 < 1$. Rewriting these limits as $\alpha^2 < 1$, $\beta_e^2 < 1$, $\alpha^2 + \beta_e^2 > 1$, we see that this is region 5 and putting $\omega = (1 - \epsilon)(\omega_p^2 + \Omega_e^2 \sin^2\theta)^{1/2}$, where again $0 < \epsilon \ll 1$, we find

$$\omega \approx (\omega_p^2 + \Omega_e^2 \sin^2\theta)^{1/2}[1 - \omega_p^2\Omega_e^2 \sin^2\theta/2k^2c^2(\omega_p^2 + \Omega_e^2 \sin^2\theta)]$$

for the dispersion relation near the resonance.

6.4 Waves in warm plasmas

Cold plasma theory has shown clearly the existence of a large number of waves in an anisotropic, loss-free plasma. The theory is valid provided the plasma is cold, i.e. the thermal velocity is much smaller than v_p. This approximation obviously breaks down near a resonance where the phase velocity $v_p \to 0$. We shall now consider some finite temperature modifications of the theory, still within the confines of a fluid description. This we may do by adding pressure terms to the fluid equations although we underline a fundamental difference between cold and warm plasma theory. Whereas cold plasma theory is properly a fluid theory, to describe warm plasma behaviour fully we need to make use of kinetic theory. In part this is because pressure is due to particle collisions which may lead to wave damping. However, even in a dissipation-free plasma the fluid equations give an incomplete picture of warm plasma wave motion.

A prime example of the shortcomings of the fluid approach appears in the description of electron plasma waves. In the cold plasma limit, these are simply oscillations at $\omega = \omega_p$, i.e. they do not propagate. In a finite temperature plasma, on the other hand, the dispersion relation is $\omega^2 = \omega_p^2 + k^2V^2$ where the thermal velocity V is given by

$$V^2 = (\gamma_i k_B T_{i0}/m_i + \gamma_e k_B T_{e0}/m_e) \tag{6.84}$$

Moreover, this result is obtained (for sufficiently small k) regardless of whether we use the fluid equations or kinetic theory. However, from a kinetic theory treatment, additional information is retrieved that is lost in fluid theory; in particular, we find that electron plasma waves in an equilibrium plasma are damped even though inter-particle collisions are negligible. This phenomenon, known as Landau damping, comes about because those electrons which have thermal velocities approximately equal to the wave phase velocity interact strongly with the wave. The physical consequences of such an interaction (wave damping in this example) are lost to a fluid analysis because of the averaging over individual particle velocities.

These shortcomings notwithstanding, a fluid description provides a simpler introduction than kinetic theory to wave characteristics in warm plasmas and we use it to give an indication of what new modes may arise and to see what modification of cold plasma modes may occur. We shall assume isotropic pressure and no heat flow; although it is simple enough, in the presence of a strong magnetic field, to justify a diagonal pressure tensor and no heat flow perpendicular to the magnetic field, the assumptions of equal parallel and perpendicular pressures and zero parallel heat flow are no more than mathematical expediencies in a collisionless theory. Thus we add pressure gradients to the equations of motion and use the adiabatic gas law

$$p_\alpha n_\alpha^{-\gamma_\alpha} = \text{const.} \tag{6.85}$$

to close the set of equations as in Table 3.5. The linearized equations are now

$$\frac{\partial n_\alpha}{\partial t} + n_{\alpha 0}\nabla\cdot\mathbf{u}_\alpha = 0 \tag{6.86}$$

$$n_{\alpha 0}m_\alpha\frac{\partial \mathbf{u}_\alpha}{\partial t} + \nabla p_\alpha - n_{\alpha 0}e_\alpha(\mathbf{E} + \mathbf{u}_\alpha\times\mathbf{B}_0) = 0 \tag{6.87}$$

$$\frac{p_\alpha}{p_{\alpha 0}} - \frac{\gamma_\alpha n_\alpha}{n_{\alpha 0}} = 0 \tag{6.88}$$

and the Maxwell equations (6.11)–(6.14); as before, variables with subscript zero are equilibrium values and those without subscript are the perturbations. Assuming plane wave variation $\sim \exp i(\mathbf{k}\cdot\mathbf{r} - \omega t)$ and eliminating all variables but $\mathbf{u}_\mathrm{i}$ and $\mathbf{u}_\mathrm{e}$ we arrive, after some tedious but straightforward algebra, at the equations

$$-\omega^2\mathbf{u}_\mathrm{i} + V_\mathrm{i}^2(\mathbf{k}\cdot\mathbf{u}_\mathrm{i})\mathbf{k} + \frac{\omega_\mathrm{pi}^2}{(k^2-\omega^2/c^2)}\left[\mathbf{k}\cdot(\mathbf{u}_\mathrm{i}-\mathbf{u}_\mathrm{e})\mathbf{k} - \frac{\omega^2}{c^2}(\mathbf{u}_\mathrm{i}-\mathbf{u}_\mathrm{e})\right] + i\omega\Omega_\mathrm{i}(\mathbf{u}_\mathrm{i}\times\mathbf{b}_0) = 0 \tag{6.89}$$

$$-\omega^2\mathbf{u}_\mathrm{e} + V_\mathrm{e}^2(\mathbf{k}\cdot\mathbf{u}_\mathrm{e})\mathbf{k} + \frac{\omega_\mathrm{pe}^2}{(k^2-\omega^2/c^2)}\left[\mathbf{k}\cdot(\mathbf{u}_\mathrm{e}-\mathbf{u}_\mathrm{i})\mathbf{k} - \frac{\omega^2}{c^2}(\mathbf{u}_\mathrm{e}-\mathbf{u}_\mathrm{i})\right] + i\omega\Omega_\mathrm{e}(\mathbf{u}_\mathrm{e}\times\mathbf{b}_0) = 0 \tag{6.90}$$

where

$$\left.\begin{aligned} V_\mathrm{i}^2 &= \frac{\gamma_\mathrm{i} p_\mathrm{i0}}{n_\mathrm{i0}m_\mathrm{i}} = \frac{\gamma_\mathrm{i} k_\mathrm{B} T_\mathrm{i0}}{m_\mathrm{i}} \\ V_\mathrm{e}^2 &= \frac{\gamma_\mathrm{e} p_\mathrm{e0}}{n_\mathrm{e0}m_\mathrm{e}} = \frac{\gamma_\mathrm{e} k_\mathrm{B} T_\mathrm{e0}}{m_\mathrm{e}} \end{aligned}\right\} \tag{6.91}$$

and $\mathbf{b}_0$ is the unit vector in the direction of $\mathbf{B}_0$.

6.4.1 Longitudinal waves

A simple case which illustrates both finite temperature modification of earlier results and the emergence of a new warm plasma mode arises when propagation and motion are parallel to $\mathbf{B}_0$; from (6.87) it follows that $\mathbf{E}$ is parallel to $\mathbf{k}$ so these are longitudinal waves. Thus, (6.89) and (6.90) become

$$(\omega^2 - k^2V_i^2 - \omega_{pi}^2)u_i + \omega_{pi}^2 u_e = 0 \quad (6.92)$$

$$\omega_{pe}^2 u_i + (\omega^2 - k^2V_e^2 - \omega_{pe}^2)u_e = 0 \quad (6.93)$$

and the dispersion relation is

$$(\omega^2 - k^2V_i^2 - \omega_{pi}^2)(\omega^2 - k^2V_e^2 - \omega_{pe}^2) - \omega_{pi}^2\omega_{pe}^2 = 0$$

with solution

$$\omega^2 = \frac{1}{2}(\omega_p^2 + k^2V^2)\left\{1 \pm \left[1 - \frac{4(k^4V_i^2V_e^2 + k^2V_i^2\omega_{pe}^2 + k^2V_e^2\omega_{pi}^2)}{(\omega_p^2 + k^2V^2)^2}\right]^{1/2}\right\} \quad (6.94)$$

Usually $V_i^2 \ll V_e^2$, so that the second term in the square root is small and the solutions are

$$\omega^2 \approx \omega_p^2 + k^2V^2 \quad (6.95)$$

$$\omega^2 \approx \frac{k^4V_i^2V_e^2 + k^2V_i^2\omega_{pe}^2 + k^2V_e^2\omega_{pi}^2}{\omega_p^2 + k^2V^2} \quad (6.96)$$

The first of these solutions, (6.95), may be further approximated to

$$\omega^2 = \omega_{pe}^2 + k^2\gamma_e k_B T_e/m_e \quad (6.97)$$

which is the dispersion relation for *electron plasma waves* or *Langmuir waves* and shows an important change from the cold plasma result. Instead of electron plasma oscillations we now have longitudinal waves which propagate with group velocity

$$v_g = \frac{d\omega}{dk} = \frac{kV^2}{\omega}$$

The ion terms in (6.95) are negligible compared with the electron terms and likewise, from (6.93), the ion flow velocity $|u_i| \ll |u_e|$. Essentially, the ions provide a static neutralizing background for the electron plasma waves.

We now turn to the second solution (6.96) which vanishes in the cold plasma limit and is, therefore, a new dispersion relation for ion waves. Using (6.91) it may be written

$$\frac{\omega^2}{k^2} \approx \frac{\gamma_i k_B T_{i0}}{m_i} + \frac{Z\gamma_e k_B T_{e0}}{m_i(1 + k^2\lambda_D^2)} \quad (6.98)$$

This is the dispersion relation for the *ion acoustic wave*. However, there is a fundamental distinction between this mode and a sound wave in a neutral gas which propagates on account of collisions. The potential energy to drive the ion acoustic wave is electrostatic in origin and is due to the difference in amplitudes of the electron and ion oscillations. In ion acoustic waves ions provide the inertia while the more mobile electrons neutralize the charge separation.

Finally consider ion waves in the limit $\omega_{\mathrm{p}}^2 \ll k^2 V_{\mathrm{e}}^2$. In this case (6.96) becomes

$$\omega^2 \approx \omega_{\mathrm{pi}}^2 + k^2 V_{\mathrm{i}}^2 \tag{6.99}$$

the ion counterpart to electron plasma waves. Comparison with (6.95) shows the symmetry between ion and electron waves which one would expect from the basic equations. Note that the retention of the ω_{pi}^2 term in (6.99) implies $T_{\mathrm{i0}} \ll T_{\mathrm{e0}}$. Also, from (6.92), we now have $|u_{\mathrm{e}}| \ll |u_{\mathrm{i}}|$ and the electrons provide a neutralizing background for the ion plasma oscillations; however, because of their high thermal velocities they play a dynamic rather than a static role. In fact these observations are academic since Landau damping restricts the propagation of these waves to a narrow band of wavelengths such that $(T_{\mathrm{i0}}/T_{\mathrm{e0}})^{1/2}\lambda_{\mathrm{D}} \ll \lambda \ll \lambda_{\mathrm{D}}$.

6.4.2 General dispersion relation

The general dispersion relation (which may be obtained from (6.89) and (6.90)) gives six roots for ω^2 corresponding to each wavenumber; these fall naturally into a high frequency group and a low frequency group. For propagation along $\mathbf{B}_0$, the high frequency group consists of RCP and LCP electromagnetic waves and the longitudinal electron plasma wave.

We shall not draw the CMA diagram for warm plasma waves, as the wave normal surfaces are now considerably more complicated than in the cold plasma limit. Instead, a typical (ω, k) dispersion plot is shown in Fig. 6.13 for a low β plasma with $|\Omega_{\mathrm{e}}| < \omega_{\mathrm{pe}}$. The high frequency curves come from the Appleton–Hartree dispersion relation (6.74) in which ion motion and pressure terms are ignored; since v_{p} is large the cold plasma approximation is good. The low frequency curves are due to Stringer (1963) and refer to a plasma having $\beta = 10^{-2}$, $v_{\mathrm{A}}/c = 10^{-3}$, $c_{\mathrm{s}}/v_{\mathrm{A}} = 10^{-1}$, $V_{\mathrm{i}}/c_{\mathrm{s}} = 0.33$ and $\theta = 45°$. The value chosen for β ensures that the high and low frequency parts of the (ω, k) diagram are well separated.

Stringer obtained the dispersion relation for the three low frequency modes from the linearized two-fluid equations (6.86)–(6.88) and the Maxwell equations by combining the ion and electron momentum equations into a one-fluid equation of motion

$$\rho_0 \frac{\partial \mathbf{u}}{\partial t} = -\nabla P + \mathbf{j} \times \mathbf{B}_0 \tag{6.100}$$

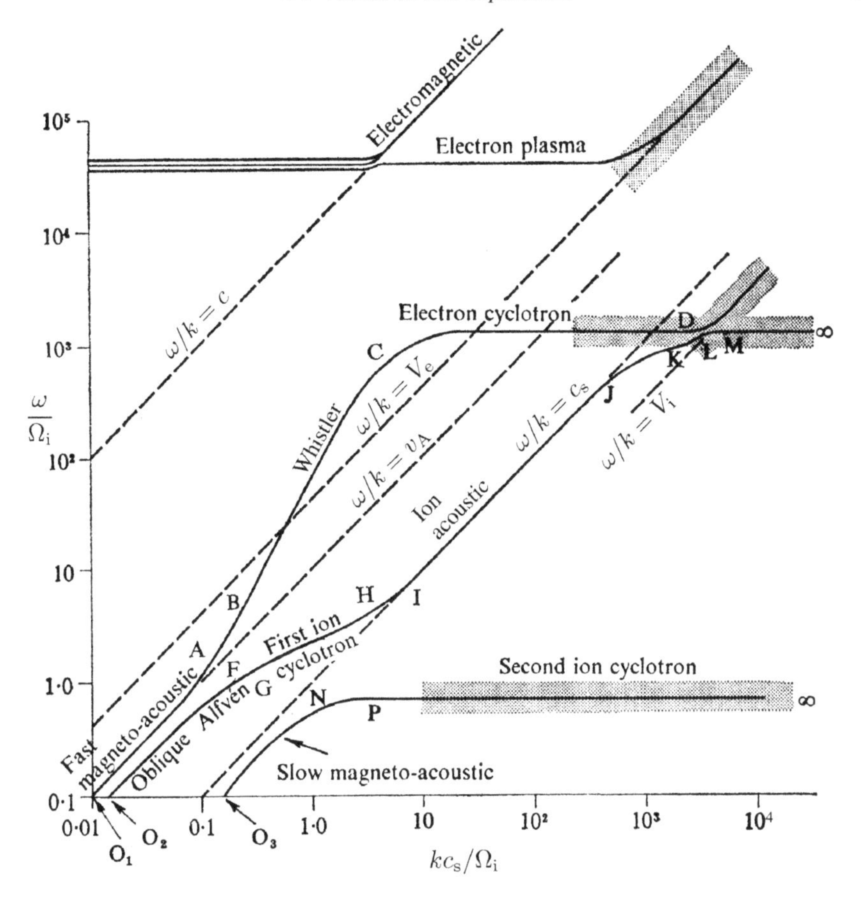

Fig. 6.13. Dispersion curves for oblique waves in low β plasma with $|\Omega_e| < \omega_{pe}$ (after Stringer (1963)).

where $P = p_e + p_i$, and a generalized Ohm's law

$$\frac{\partial \mathbf{j}}{\partial t} = \frac{n_0 e^2}{m_e}(\mathbf{E} + \mathbf{u} \times \mathbf{B}_0) - \frac{e}{m_e}\mathbf{j} \times \mathbf{B}_0 + \frac{e}{m_e}\nabla p_e \tag{6.101}$$

Then from the curl of the induction equation

$$\nabla \times \nabla \times \mathbf{E} = \mu_0 \frac{\partial \mathbf{j}}{\partial t} \tag{6.102}$$

on neglect of the displacement current. In the derivation of (6.101) terms of order Zm_e/m_i have been ignored but, in fact, this equation may be obtained directly from (3.70) by taking the $\nu_c \to 0$ limit; note that $\sigma = n_e e^2/m_e \nu_c$.

Now replacing ∇ by $i\mathbf{k}$ and $\partial/\partial t$ by $-i\omega$, (6.102) becomes

$$i\omega\mu_0 \mathbf{j} = k^2\mathbf{E} - (\mathbf{k} \cdot \mathbf{E})\mathbf{k} \tag{6.103}$$

and substituting this in the left-hand side of (6.101) gives

$$Q\mathbf{E} - \frac{c^2}{\omega_{\rm pe}^2}(\mathbf{k}\cdot\mathbf{E})\mathbf{k} + \mathbf{u}\times\mathbf{B}_0 + \frac{ip_{\rm e}}{n_0 e}\mathbf{k} - \frac{1}{n_0 e}\mathbf{j}\times\mathbf{B}_0 = 0 \tag{6.104}$$

where $Q = (1 + k^2c^2/\omega_{\rm pe}^2)$.

Next, without loss of generality, we may choose $\mathbf{k} = (k, 0, 0)$ and $\mathbf{B}_0 = B_0(\cos\theta, 0, \sin\theta)$, so that (6.103) gives

$$\left.\begin{array}{rcl} j_x &=& 0 \\ E_y &=& i\omega\mu_0 j_y/k^2 \\ E_z &=& i\omega\mu_0 j_z/k^2 \end{array}\right\} \tag{6.105}$$

and (6.100) gives

$$\left.\begin{array}{rcl} u_x &=& iB_0 j_y \sin\theta/\rho_0\omega(1 - k^2c_{\rm s}^2/\omega^2) \\ u_y &=& iB_0 j_z \cos\theta/\rho_0\omega \\ u_z &=& -iB_0 j_y \cos\theta/\rho_0\omega \end{array}\right\} \tag{6.106}$$

where $c_{\rm s}^2 = (\gamma_{\rm e} p_{\rm e0} + \gamma_{\rm i} p_{\rm i0})/\rho_0$ and we have used (6.86) and (6.88) to replace P by

$$\begin{aligned} P &= (\gamma_{\rm e} p_{\rm e0}\mathbf{k}\cdot\mathbf{u}_{\rm e} + \gamma_{\rm i} p_{\rm i0}\mathbf{k}\cdot\mathbf{u}_{\rm i})/\omega \\ &= [\gamma_{\rm e} p_{\rm e0}(\mathbf{k}\cdot\mathbf{u} - \mathbf{k}\cdot\mathbf{j}/n_0 e) + \gamma_{\rm i} p_{\rm i0}\mathbf{k}\cdot\mathbf{u}]/\omega \\ &= (\gamma_{\rm e} p_{\rm e0} + \gamma_{\rm i} p_{\rm i0})\mathbf{k}\cdot\mathbf{u}/\omega \end{aligned}$$

since $j_x = 0$. Finally, substituting (6.105) and (6.106) in (6.104) yields a vector equation, involving $\mathbf{j}$ as the only unknown, the y and z components of which are

$$\begin{aligned} \left[\frac{Q\mu_0\omega}{k^2} - \frac{B_0^2}{\rho_0\omega}\left(\cos^2\theta + \frac{\sin^2\theta}{(1 - k^2c_{\rm s}^2/\omega^2)}\right)\right] j_y + \frac{im_{\rm i}B_0\cos\theta}{e\rho_0} j_z &= 0 \\ -\frac{im_{\rm i}B_0\cos\theta}{e\rho_0} j_y + \left(\frac{Q\mu_0\omega}{k^2} - \frac{B_0^2\cos^2\theta}{\rho_0\omega}\right) j_z &= 0 \end{aligned}$$

Equating the determinant of the coefficients of this equation to zero gives Stringer's dispersion relation

$$\begin{aligned} &\left[\left(\frac{\omega}{k}\right)^4 - \left(\frac{\omega}{k}\right)^2 (c_{\rm s}^2 + v_{\rm A}^2/Q) + c_{\rm s}^2(v_{\rm A}^2/Q)\cos^2\theta\right] \\ &\quad\times\left[\left(\frac{\omega}{k}\right)^2 - (v_{\rm A}^2/Q)\cos^2\theta\right] - \left(\frac{\omega v_{\rm A}^2}{\Omega_{\rm i} Q}\right)^2 \left(\frac{\omega^2}{k^2} - c_{\rm s}^2\right)\cos^2\theta = 0 \end{aligned} \tag{6.107}$$

Table 6.1. *Dispersion curves (Fig. 6.13): slow branch*

Section	*Mode*	*Dispersion relation*	*Physical characteristics*
O_3N	slow magneto-acoustic	$\omega = kc_s \cos\theta$ $kc_s \ll \Omega_i$	**E** almost longitudinal coupling electron and ion fluids
$P\infty$	second ion cyclotron	$\omega = \Omega_i \cos\theta$ $kc_s \gg \Omega_i$	longitudinal wave $\mathbf{E} \parallel \mathbf{k}$

Table 6.2. *Dispersion curves (Fig. 6.13): intermediate branch*

Section	*Mode*	*Dispersion relation*	*Physical characteristics*
O_2F	oblique Alfvén	$\omega = kv_A \cos\theta$ $\omega \ll \Omega_i$	$\mathcal{P} < 0$
GH	first ion cyclotron	$\omega = \Omega_i[1 + (k^2c_s^2/\Omega_i^2)\sin^2\theta$ $-(\Omega_i^2/k^2v_A^2)(1 + \sec^2\theta)]^{1/2}$ $c_s\Omega_i/(v_A\cos\theta) \ll kc_s \sim \Omega_i$	$\mathcal{P} = -1$: Below Ω_i magnetic energy drives the wave; above Ω_i plasma pressure dominates
IJ	ion acoustic	$\omega = kc_s$ $\Omega_i < kc_s < \omega_{pi}$	longitudinal wave $\mathbf{E} \parallel \mathbf{k}$

The waves and approximate dispersion relations corresponding to the three low frequency branches (labelled slow, intermediate and fast) are given in Tables 6.1–6.3. Conditions which apply throughout the tables are $c_s^2 \ll v_A^2$ and $m_e/m_i \ll \cos^2\theta$.

The first thing to note in Fig. 6.13 is the appearance of the additional, slow branch ($O_3NP\infty$), corresponding to the slow magnetoacoustic wave discussed in Section 4.8. We have seen already that the cold plasma modes at low frequencies correspond to the fast magnetoacoustic and intermediate, shear Alfvén waves. In fact, we can recover the results of Section 4.8 from (6.107) by taking the low frequency limit $\omega \ll \Omega_i$. In this case the $(\omega/\Omega_i)^2$ term may be neglected leaving a dispersion relation equivalent to (4.121) but with v_A^2/Q replacing v_A^2. However, for $c_s^2 \ll v_A^2$ the slow mode has $\omega \approx kc_s\cos\theta$, independent of v_A, while the other two modes have $\omega \sim kv_A$, so that $c^2k^2/\omega_{pe}^2 \ll Zm_e/m_i$ giving $Q \approx 1$, and the solutions (4.122) and (4.123) are recovered. The wave normal surfaces for these waves in the ideal MHD approximation and for $c_s < v_A$ are sketched in Fig. 6.14.

Table 6.3. *Dispersion curves (Fig. 6.13): fast branch*

Section	*Mode*	*Dispersion relation*	*Physical characteristics*
O_1A	fast magneto-acoustic	$\omega = k(v_A^2 + c_s^2 \sin^2\theta)^{1/2}$ $\omega \ll \Omega_i$	$\mathcal{P} > 0$
BC	whistler	$\omega \simeq (k^2 v_A^2/\Omega_i)\cos\theta$ $\Omega_i \ll k v_A \cos\theta$	$\mathcal{P} \approx +1$; for $\omega > \Omega_i$ ion role decreases on account of inertia
CD	electron cyclotron	$\omega \simeq \lvert\Omega_e\rvert \cos\theta$ $\lvert\Omega_e\rvert < \omega_{pe}$	$\mathcal{P} = +1$; electron velocity increase $\perp \mathbf{B}_0$ is limited by increase $\parallel \mathbf{B}_0$ to maintain charge neutrality

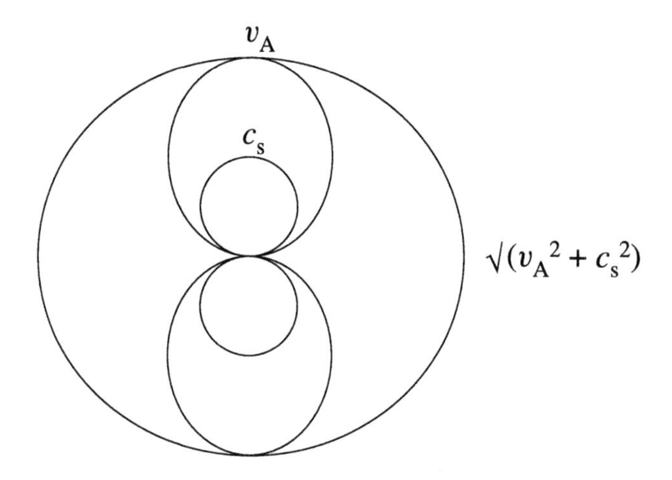

Fig. 6.14. Phase velocity surfaces of MHD waves for $v_A > c_s$.

Let us now examine what happens as we approach the ion cyclotron frequency. For the cold plasma we know that the intermediate wave disappears as the resonance, which is at $\theta = \pi/2$ for $\omega \ll \Omega_i$, approaches $\theta = 0$ as $\omega \to \Omega_i$. To find the resonances of the low frequency warm plasma waves we let $k \to \infty$ in (6.107) giving

$$\omega^2 = \Omega_i^2 \cos^2\theta (1 - c^2\omega^2/v_A^2\omega_{pe}^2 \cos^2\theta)^2$$

which, ignoring terms of order m_e/m_i, has the solutions

$$\omega = \Omega_i \cos\theta, \qquad \omega = |\Omega_e| \cos\theta \tag{6.108}$$

Now, writing (6.107) as

$$\left(\frac{\omega}{k}\right)^6 - \left(\frac{\omega}{k}\right)^4 \left[c_s^2 + (v_A^2/Q)(1+\cos^2\theta)\right] + \left(\frac{\omega}{k}\right)^2 \left\{(v_A^2/Q)\cos^2\theta \left[2c_s^2 + (v_A^2/Q)\left(1-\frac{\omega^2}{\Omega_i^2}\right)\right]\right\} + \frac{v_A^4 c_s^2 \cos^2\theta}{Q^2\Omega_i^2}(\omega^2 - \Omega_i^2\cos^2\theta) = 0$$

we may neglect the last term in the neighbourhood of the resonance at $\omega = \Omega_i \cos\theta$ and solve the resulting bi-quadratic for the other two modes, obtaining

$$\frac{\omega^2}{k^2} \approx c_s^2 + v_A^2(1+\cos^2\theta) \tag{6.109}$$

and

$$\frac{\omega^2}{k^2} \approx \frac{v_A^2\cos^2\theta[2c_s^2 + v_A^2(1-\omega^2/\Omega_i^2)]}{c_s^2 + v_A^2(1+\cos^2\theta)} \tag{6.110}$$

Here we have again put $Q = 1$ since $c^2k^2/\omega_{pe}^2 \sim Z^2 m_e/m_i \ll 1$. We can identify these modes by letting $c_s \to 0$ in which case (6.109) reduces to (6.70) for the compressional Alfvén wave and (6.110) to (6.71) for the ion cyclotron wave. Thus, (6.109) is the dispersion relation for the fast magnetoacoustic wave and (6.110) is for the shear Alfvén–ion cyclotron mode at $\omega \approx \Omega_i \cos\theta$. The interesting point about (6.110) is that the resonance stays at $\theta = \pi/2$ as $\omega \to \Omega_i$ and does not migrate towards $\theta = 0$. Consequently, there is no destructive transition at Ω_i and the mode persists for $\omega > \Omega_i$. The sharp reduction in v_p due to the vanishing of the second term in the square bracket in (6.110) at $\omega = \Omega_i$ gives rise to a so-called *pseudo-resonance*, indicated by the GH section of the intermediate wave dispersion curve in Fig. 6.13. The mode is called the *first* ion cyclotron wave in this region ($\omega \approx \Omega_i$) to distinguish it from the slow wave, which has a true resonance ($\omega \approx \Omega_i \cos\theta$ for all k so $v_p \to 0$ as $k \to \infty$ for all θ) at $\omega = \Omega_i$, and is called the *second* ion cyclotron wave. Thus, consideration of finite temperature has (i) introduced the slow magnetoacoustic wave which then disappears at the ion cyclotron resonance and (ii) demonstrated the continuation of the shear Alfvén–first ion cyclotron mode to frequencies above Ω_i.

To find approximate dispersion relations for the fast and intermediate modes between the ion and electron resonances we may take $\Omega_i \ll \omega \ll |\Omega_e|$. Since $\omega/k \sim c_s$ or v_A all terms in (6.107) are of similar magnitude except for the last one which has the factor $(\omega/\Omega_i)^2$. Thus, dropping all terms but this, one solution is

$$\omega^2 = k^2 c_s^2 \tag{6.111}$$

i.e. the ion acoustic wave. Then rewriting (6.107) in the form

$$\left(\frac{\omega}{k}\right)^6 - \left(\frac{\omega}{k}\right)^4 [c_s^2 + (v_A^2/Q)(1+\cos^2\theta)] +$$
$$\left(\frac{\omega}{k}\right)^2 (2c_s^2 + v_A^2/Q)(v_A^2/Q)\cos^2\theta - c_s^2(v_A^4/Q^2)\cos^4\theta +$$
$$\left(\frac{\omega}{\Omega_i}\right)^2 \left(c_s^2 - \left(\frac{\omega}{k}\right)^2\right)(v_A^4/Q^2)\cos^2\theta = 0$$

we may neglect the third and fourth terms compared with the final term and, anticipating that the second solution for $c_s^2 \ll v_A^2$ has $\omega/k \gtrsim v_A$, we may also drop the second term and the c_s^2 in the final term leading to the result

$$\omega \approx \frac{k^2 v_A^2 \cos\theta}{\Omega_i} = \frac{k^2 c^2 |\Omega_e| \cos\theta}{\omega_{pe}^2} \tag{6.112}$$

where we have again put $Q = 1$ since $k^2c^2/\omega_{pe}^2 \sim \omega/|\Omega_e| \ll 1$. This is the generalization for non-zero θ of (6.55), the dispersion relation for the whistler wave. Thus, between the resonances the intermediate wave emerges from the pseudo-resonance and propagates at a reduced phase velocity as an ion acoustic wave while the fast wave follows its cold plasma behaviour becoming a whistler.

As the electron cyclotron resonance at $\omega = |\Omega_e|\cos\theta$ is approached the pattern of behaviour seen at the ion cyclotron resonance is repeated. The slower (intermediate) wave suffers the destructive transition, which in the cold plasma was the fate of the fast wave, while the fast wave undergoes a pseudo-resonance and survives to continue propagation above $\omega = |\Omega_e|$, but at the reduced phase velocity $\omega/k = V_i$. Both of these occurrences can be attributed to the appearance of the new, longitudinal, warm plasma mode discussed in Section 6.4.1; see (6.96). The coupling of transverse and longitudinal waves that occurs for $\theta \neq 0$ enables the first ion cyclotron wave to emerge from the ion cyclotron resonance as the ion acoustic wave (6.98). Likewise, the electron cyclotron wave emerges from the electron cyclotron resonance as the ion plasma wave, the mode described by (6.99).

Longitudinal modes are not well described by (6.107) for the neglect of the displacement current in its derivation implied $\mathbf{k}\cdot\mathbf{j} = 0$, i.e. zero space charge. Stringer, therefore, derived an electrostatic dispersion relation from which the approximate results (shown in the tables) in the neighbourhood of these transitions are found.

For sufficiently low β plasmas $\omega_{pe} < |\Omega_e|$ and the high and low frequency branches overlap. For such cases (6.107) becomes invalid at the overlap, i.e. for $\omega > \omega_L \sim \omega_{pe}^2/|\Omega_e|$. An example is shown in Fig. 6.15 in which Stringer used the Appleton–Hartree dispersion relation (6.74) to calculate the curve for the fast wave for $\omega > \omega_{pe}^2/|\Omega_e|$.

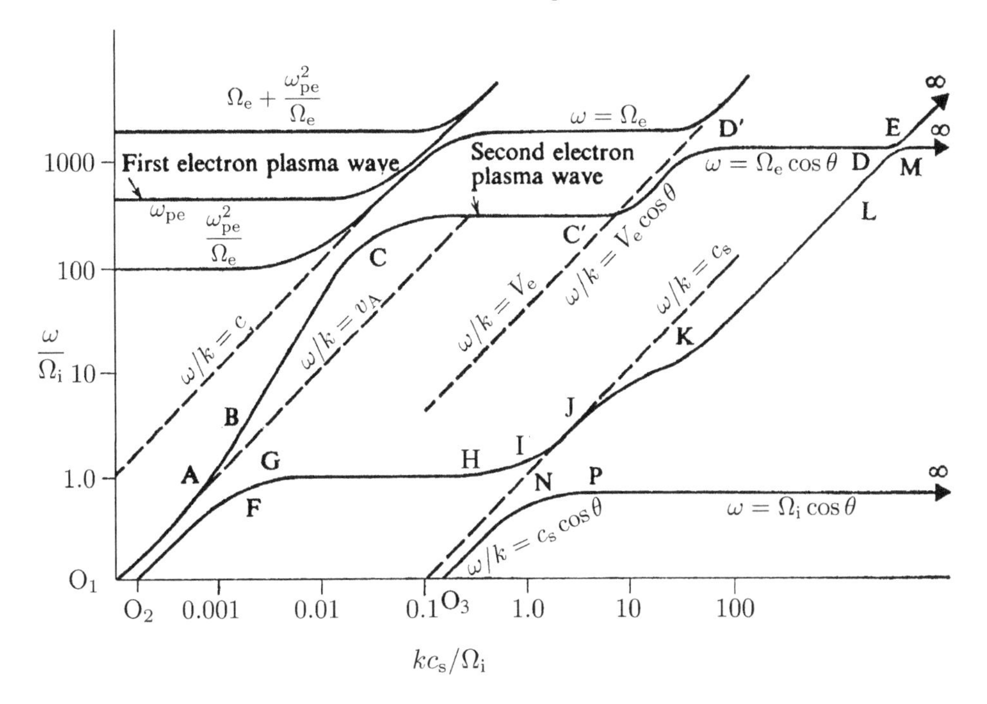

Fig. 6.15. Dispersion curves for oblique waves in very low β plasma with $\omega_{pe} < |\Omega_e|$ (after Stringer (1963)).

In general, the dispersion curves do not change appreciably as θ is varied provided the values 0 and $\pi/2$ are avoided. As $\theta \to 0$, for example, the gap between the first ion cyclotron–acoustic wave transition (HI in Fig. 6.13) and the slow magnetoacoustic–second ion cyclotron wave transition (NP in Fig. 6.13) shrinks. In the limit, the points (H, P) and (N, I) become coincident, that is, the curves $O_2G\infty$ and O_3J now intersect. The transition from finite θ to 0 is shown in Fig. 6.16; the presence of a transverse magnetic field couples longitudinal and transverse wave components so that the transverse Alfvén wave passes into the longitudinal acoustic wave while on the lower frequency branch a longitudinal mode passes into a transverse mode. At $\theta = 0$, however, no such coupling occurs and the transverse shear Alfvén wave now becomes a transverse ion cyclotron wave, as in the cold plasma limit, while the other branch O_3J is now entirely longitudinal. A similar transition occurs between the electron cyclotron wave and the ion acoustic wave.

The situation as $\theta \to \pi/2$ is more complicated and will not be discussed; a typical dispersion plot for the three low frequency branches is shown in Fig. 6.17. Observe that only the O_1CE branch survives in the limit $\theta = \pi/2$ and that the lower hybrid frequency (at which a resonance appeared for $\theta = \pi/2$ propagation in the cold plasma limit) now reappears as a pseudo-resonance.

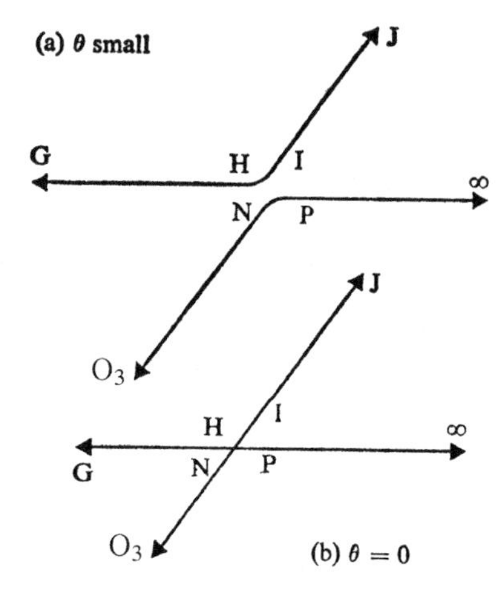

Fig. 6.16. Coupling of longitudinal and transverse wave components in (a) for small θ disappears in (b) for $\theta = 0$.

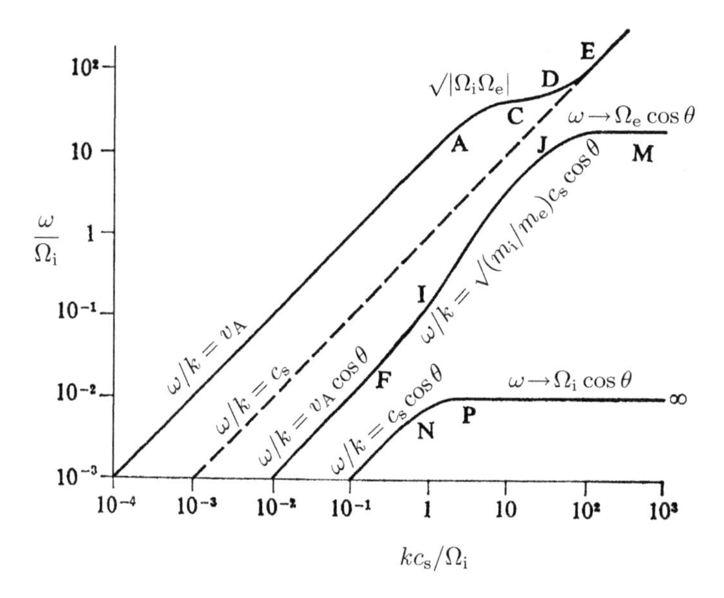

Fig. 6.17. Dispersion curves for low frequency waves in low β plasma as $\theta \to \pi/2$.

6.5 Instabilities in beam–plasma systems

The waves that we have considered so far are those that may arise when we perturb a plasma that is initially in equilibrium. In this section we take one step further to investigate the perturbation of a steady state plasma; in particular, we allow for non-zero flow velocities $\mathbf{u}_{0\alpha}$. Interstreaming or beam-carrying plasmas are of widespread interest so this is an important generalization. The most significant result of this extension of wave theory is the appearance of instabilities driven by the plasma streams.

To keep the analysis as simple as possible we consider a cold, unmagnetized plasma which, in the steady state, has interstreaming components. These may be ions or electrons (or both) and there may also be a stationary background plasma. Thus, the species label α now denotes the various components and we linearize the cold plasma equations using, instead of (6.15),

$$\left.\begin{aligned} n &= n_0 + n_1 \\ \mathbf{u} &= \mathbf{u}_0 + \mathbf{u}_1 \\ \mathbf{E} &= \mathbf{E}_1 \\ \mathbf{B} &= \mathbf{B}_1 \end{aligned}\right\} \tag{6.113}$$

where n_0 and $\mathbf{u}_0$ are constants, obtaining for the equations of continuity and motion

$$\frac{\partial n_1}{\partial t} + \nabla \cdot (n_0 \mathbf{u}_1 + n_1 \mathbf{u}_0) = 0 \tag{6.114}$$

$$\frac{\partial \mathbf{u}_1}{\partial t} + (\mathbf{u}_0 \cdot \nabla)\mathbf{u}_1 = \frac{e}{m}(\mathbf{E}_1 + \mathbf{u}_0 \times \mathbf{B}_1) \tag{6.115}$$

Note that we still have $\mathbf{E}_0 = 0$, otherwise there would be no steady state. For longitudinal waves ($\nabla \times \mathbf{E}_1 = 0$) there is no magnetic field perturbation so, again for simplicity, we consider this case. Then we need only Poisson's equation

$$\nabla \cdot \mathbf{E}_1 = \frac{1}{\varepsilon_0} \sum e n_1 \tag{6.116}$$

to close the set.

Assuming that all perturbed quantities vary as $\exp i(\mathbf{k} \cdot \mathbf{r} - \omega t)$, it is a simple matter to obtain

$$\mathbf{u}_1 = \frac{ie\mathbf{E}_1}{m(\omega - \mathbf{k} \cdot \mathbf{u}_0)}$$

from (6.115) and substitute it in (6.114) to find

$$n_1 = \frac{ien_0 k\mathbf{E}_1}{m(\omega - \mathbf{k} \cdot \mathbf{u}_0)^2}$$

Then from (6.116) we see that the condition for a non-trivial solution, $\mathbf{E}_1 \neq 0$, is

$$\sum_\alpha \frac{\omega_{\mathrm{p}\alpha}^2}{(\omega - \mathbf{k} \cdot \mathbf{u}_\alpha)^2} = 1 \tag{6.117}$$

where $\omega_{\mathrm{p}\alpha}$ and $\mathbf{u}_\alpha$ are, respectively, the plasma frequency and steady state streaming velocity for species α. This is the dispersion relation for longitudinal waves in a plasma containing particle streams. Note that if all the stream velocities are zero we recover the dispersion relation for longitudinal plasma oscillations.

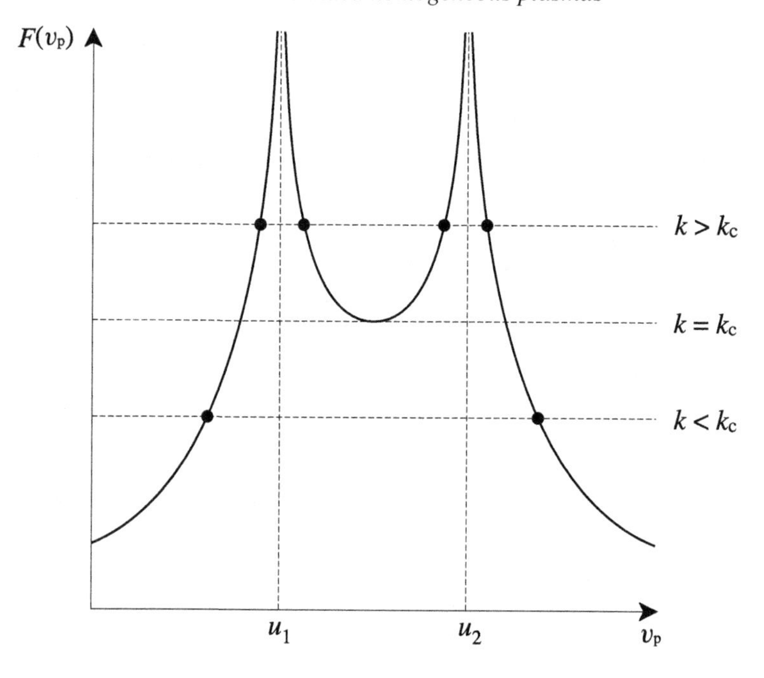

Fig. 6.18. Schematic plot of $F(v_\mathrm{p})$.

6.5.1 Two-stream instability

To demonstrate the onset of instability let us simplify further to the case of just two streams with velocities $\mathbf{u}_1$ and $\mathbf{u}_2$ which are parallel (if they are not, we can transform to a frame in which they are) and consider waves propagating in the same direction. Then we may re-write (6.117) as

$$F(v_\mathrm{p}) = \frac{\omega_{\mathrm{p}1}^2}{(v_\mathrm{p} - u_1)^2} + \frac{\omega_{\mathrm{p}2}^2}{(v_\mathrm{p} - u_2)^2} = k^2 \tag{6.118}$$

where $v_\mathrm{p} = \omega/k$. The function $F(v_\mathrm{p})$ is sketched in Fig. 6.18 and we see that for large enough k^2 there are four real solutions of (6.118). However, for $k < k_\mathrm{c}$ there are only two. Since (6.118) is a quartic equation in v_p with real coefficients there must be four roots and for $k < k_\mathrm{c}$ two of these form a complex conjugate pair $v_\mathrm{p} = (\omega_r \pm i\gamma)/k$, representing exponentially growing and damped waves. The growing wave solution is identified with the *two-stream instability*. The critical value k_c can be found by setting $\mathrm{d}F/\mathrm{d}v_\mathrm{p} = 0$ and is given by

$$k_\mathrm{c}^2 = \left(\omega_{\mathrm{p}1}^{2/3} + \omega_{\mathrm{p}2}^{2/3}\right)^3 / (u_1 - u_2)^2 \tag{6.119}$$

On the basis of this analysis it appears that there will always be some waves, of

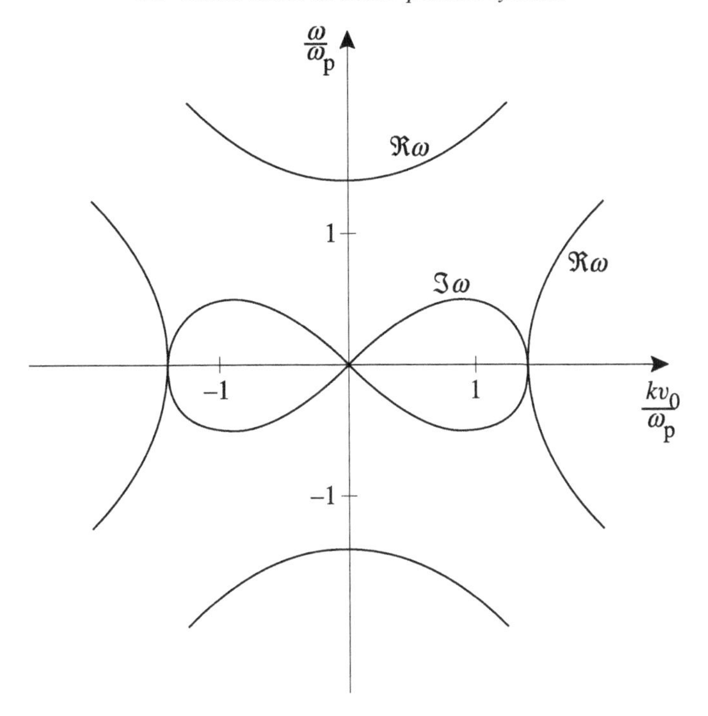

Fig. 6.19. Two-stream instability dispersion relation showing the real and imaginary parts of the frequency as functions of wavenumber.

long enough wavelength, which are unstable. This is another instance of the fluid description proving to be misleading. We shall see in the next chapter that when we allow for the thermal spread in particle velocities and analyse the problem using kinetic theory there appears a threshold relative velocity between the streams below which there is no instability for any value of k.

For counterstreaming beams penetrating one wavelength in a plasma period, a perturbation in density δn_1 on stream 1 will be amplified by particles bunching in stream 2. And since $\delta n_1 \propto n_1$, the perturbation grows exponentially in time. The phase condition for this to occur is

$$|u_1 - u_2|(2\pi/\omega_p) \sim (2\pi/k)$$

For $u_1 - u_2 = 2v_0$ this gives the condition for growth of the perturbation, i.e. $k \sim \omega_p/2v_0$.

In the case of opposing streams of equal strength we may put $\omega_{p1} = \omega_{p2} = \omega_p$ and $u_1 = -u_2 = v_0$ and from (6.118) the dispersion relation is

$$\frac{\omega_p^2}{(\omega - kv_0)^2} + \frac{\omega_p^2}{(\omega + kv_0)^2} = 1$$

with solution

$$\omega^2 = k^2 v_0^2 + \omega_p^2 \pm \omega_p [\omega_p^2 + 4k^2 v_0^2]^{1/2}$$

From (6.119) we see that instability occurs in the range $0 < k < \sqrt{2}\omega_p/v_0$. The maximum growth rate occurs at $k = \omega_p\sqrt{3}/2v_0$, obtained by setting $d\omega/dk = 0$, and is given by $\Im\omega = \omega_p/2$. The dispersion curves for real and imaginary ω are sketched in Fig. 6.19. The density perturbations grow until the electric fields which they create become large enough to scatter the electrons causing dispersion in the stream velocity which eventually extinguishes the instability.

6.5.2 Beam–plasma instability

We can also use (6.118) to discuss the instability which arises when a single electron beam with number density n_b and plasma frequency ω_{pb} flows with speed v_b through a stationary cold plasma. The dispersion relation is

$$\frac{\omega_p^2}{\omega^2} + \frac{\omega_{pb}^2}{(\omega - kv_b)^2} = 1 \tag{6.120}$$

which may be written as

$$(\omega^2 - \omega_p^2)[(\omega - kv_b)^2 - \omega_{pb}^2] = \omega_p^2 \omega_{pb}^2$$

where the left-hand side shows the four linear waves (two normal modes, a Langmuir wave and an electron beam mode), while the term on the right-hand side acts as a coupling term for these modes.

From (6.119), instability occurs for $k < k_c$ where

$$k_c = \frac{\omega_p}{v_b}\left[1 + \left(\frac{\omega_{pb}}{\omega_p}\right)^{2/3}\right]^{3/2}$$

which, in the weak-beam limit, $\omega_{pb} \ll \omega_p$, becomes $k_c = \omega_p/v_b$. For this value the beam modes have $\omega = \omega_p \pm \omega_{pb}$ so that the interaction is three-wave with $\omega = -\omega_p$ well separated. By letting $\omega = \omega_p + \Delta\omega$, $k = \omega_p/v_b + \Delta k$ and keeping only terms of lowest order in ω_{pb}/ω_p, (6.120) becomes

$$\Delta\omega(\Delta\omega - v_b \Delta k)^2 = \omega_p \omega_{pb}^2/2$$

The maximum growth rate is then

$$\gamma_{\max} = \sqrt{3}(\omega_p \omega_{pb}^2)^{1/3}/2^{4/3} \tag{6.121}$$

Dispersion curves in the weak-beam limit are sketched in Fig. 6.20.

There is an interesting formal similarity between (6.120) and the dispersion relation for an instability that appears when electrons drift through a neutralizing

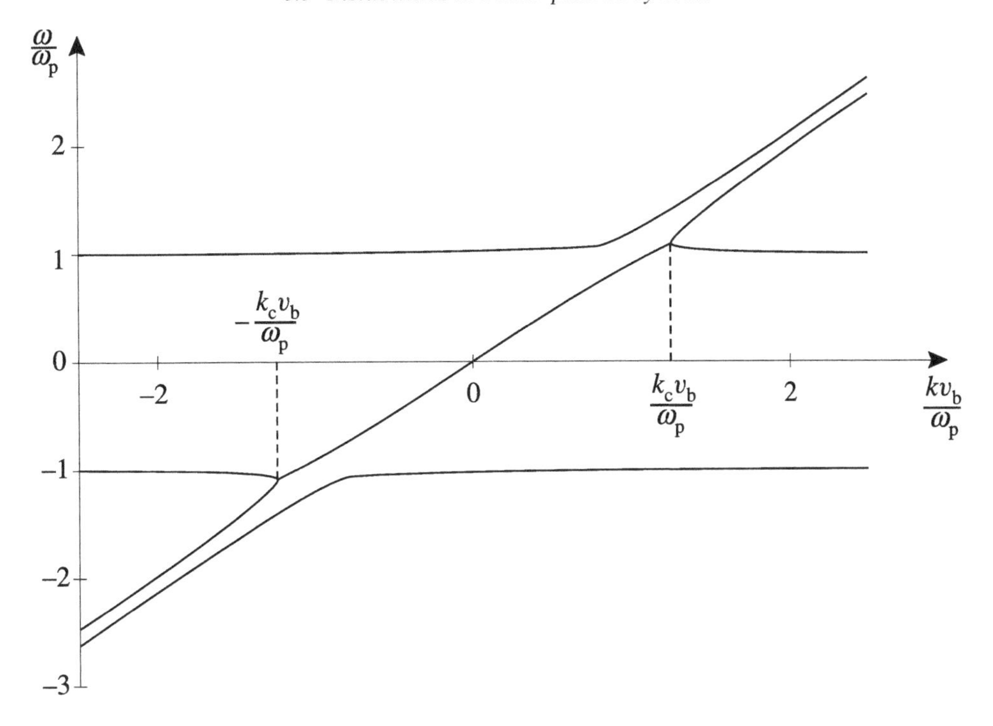

Fig. 6.20. Dispersion curves for weak-beam–plasma system with $n_b/n_0 = 2 \times 10^{-3}$.

background of stationary ions. This instability, first identified by Buneman (1959), will be discussed briefly in the next chapter since it is properly set in the context of instabilities in warm plasmas. Leaving that aside, if we simply identify the electron–ion drift velocity v_d with v_b in the weak-beam case, we may write the dispersion relation in the cold plasma limit (in the rest frame of the electrons) in one dimension as

$$\frac{\omega_{pe}^2}{\omega^2} + \frac{\omega_{pi}^2}{(\omega - k v_d)^2} = 1 \tag{6.122}$$

Formally, ω_{pi}^2 plays the role of ω_{pb}^2 in (6.120) and so by analogy with (6.121) the maximum growth rate for the Buneman instability is

$$\gamma_{max} = \frac{\sqrt{3}}{2\sqrt{2}} \left(\frac{Z m_e}{m_i} \right)^{1/3} \omega_{pe} \approx 0.05 \omega_{pe}$$

for $Z = 1$.

Two-stream and beam–plasma instabilities are widespread in both laboratory and space plasmas. Large electric field fluctuations have been measured in space plasmas and streaming instabilities have been detected at the boundary of the plasma sheet. Enhanced fluctuations near the plasma frequency have been observed upstream from the Earth's bow shock and correlated with fluxes of energetic electrons.

6.6 Absolute and convective instabilities

In this section we return to consider in more detail the interpretation of complex solutions to the dispersion relations, examples of which appeared in Sections 6.5.1 and 6.5.2. In these examples we supposed that the wavenumber was real and found pairs of complex roots in the dispersion relation, corresponding to modes that were either damped or growing in time. In practice it is often more convenient to look for complex roots of the wavenumber k, for real frequencies. Then the complex conjugate pair correspond to modes that are *evanescent*, i.e. the amplitude of a disturbance decays with distance from its source, or spatially *amplifying*. Beam–plasma systems have some parallels with electron beam–circuit systems. For example, in travelling wave tubes an input signal is amplified by interacting with beam electrons travelling down the tube synchronously with the electromagnetic wave. Twiss (1950, 1952) first drew attention to the distinct ways in which a pulsed perturbation at some point in a physical system can evolve and emphasized the need for a criterion to identify amplifying waves. Sturrock (1958) postulated that the distinction between amplifying and evanescent waves is not dynamical but kinematical and deciding which is which should be possible from a scrutiny of the dispersion relation alone. However, to draw this distinction one has to consider not a single mode but analyse instead the evolution of a wave packet.

A related problem appears when solving the dispersion relation for complex ω roots in terms of real k. A wave packet may evolve in time in either of two distinct ways. Considering for simplicity an unbounded system, a pulse that is localized initially at some point may propagate away from its source, growing in amplitude as it propagates, as represented in Fig. 6.21(a). Given a sufficiently long time the disturbance decays with time at any fixed point in space. Instabilities with these characteristics are classed as *convective* and the mode is said to be *C-unstable*. An alternative outcome in Fig. 6.21(b) shows the initial pulsed perturbation spreading across the entire region, with the amplitude of the disturbance growing in time everywhere. Such instabilities are said to be *absolute*, the mode in question being *A-unstable*. It is important to distinguish between these two possibilities. Clearly one distinction can be drawn depending on the frame of the observer. An observer in a frame moving faster than the speed at which an absolute instability spreads

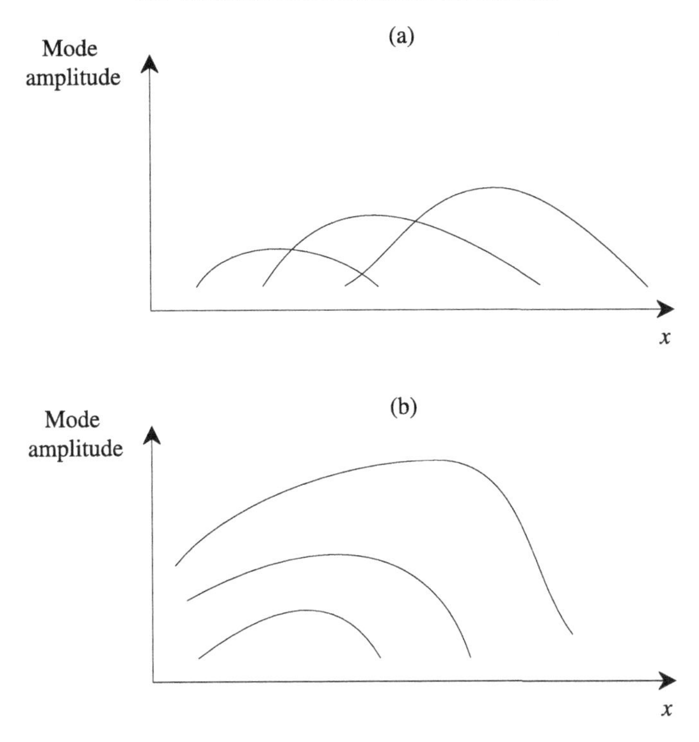

Fig. 6.21. Pulse amplification due to (a) convective and (b) absolute instability.

would classify the plasma as C-unstable. By contrast an observer in the frame moving with the peak of the disturbance in Fig. 6.21(a) would see the mode as A-unstable. Nevertheless, in practice there will usually be a preferred frame of reference and hence a real physical distinction between convective amplification and absolute instability. This distinction takes on particular significance in inhomogeneous plasmas where a mode may be unstable only over some localized region. Then a convectively unstable mode can grow only as long as it is contained within the unstable region. We shall return to this point in Chapter 11. While the terms 'amplifying' and 'evanescent' apply to the behaviour of modes with real ω, an amplifying wave has essentially the same character as one that is C-unstable (real k, complex ω).

6.6.1 Absolute and convective instabilities in systems with weakly coupled modes

As an example of the classification of instabilities as absolute or convective we consider a dissipation-free system in which two branches of the dispersion relation correspond to distinct linear modes. In the absence of any interaction between the modes the dispersion relation simply factors into two branches, i.e.

$$(\omega - \omega_1(k))\,(\omega - \omega_2(k)) = 0 \tag{6.123}$$

In the neighbourhood of a crossing point P at (ω_0, k_0) between the two branches

$$\left.\begin{array}{l} \omega_1(k) \simeq \omega_0 + (k - k_0)v_1 \\ \omega_2(k) \simeq \omega_0 + (k - k_0)v_2 \end{array}\right\} \tag{6.124}$$

and v_1 and v_2 are constant *group* velocities.

However, in general in the neighbourhood of such a point P the modes exhibit coupling. If we suppose that this is weak then the dispersion relation for the coupled modes in the neighbourhood of P may be represented by

$$[\omega - \omega_0 - (k - k_0)v_1]\,[\omega - \omega_0 - (k - k_0)v_2] = \epsilon \tag{6.125}$$

where ϵ is a small quantity. Equation (6.125) serves as a paradigm for mode-coupling leading to instability in many physical systems including plasmas. Solving for $\omega(k)$ and for $k(\omega)$ gives in turn

$$\omega(k) - \omega_0 = \frac{1}{2}\left[(k - k_0)(v_1 + v_2) \pm \left[(k - k_0)^2(v_1 - v_2)^2 + 4\epsilon^2\right]^{1/2}\right] \tag{6.126}$$

$$k(\omega) - k_0 = \frac{1}{2v_1 v_2}\left[(\omega - \omega_0)(v_1 + v_2) \pm \left[(\omega - \omega_0)^2(v_1 - v_2)^2 + 4\epsilon v_1 v_2\right]^{1/2}\right] \tag{6.127}$$

Admitting mode-coupling has the effect of shifting P into the complex plane. Representing $(\omega - \omega_0)$ as a function of $(k - k_0)$ in Fig. 6.22 throws up four distinct cases:

$$\left.\begin{array}{llllll} \text{(a)} & \epsilon > 0; & v_1 v_2 > 0 & \text{(c)} & \epsilon < 0; & v_1 v_2 > 0 \\ \text{(b)} & \epsilon > 0; & v_1 v_2 < 0 & \text{(d)} & \epsilon < 0; & v_1 v_2 < 0 \end{array}\right\} \tag{6.128}$$

(a) The functions $\omega(k)$ are real for all real k and the system is stable. Moreover the functions $k(w)$ are real for all real ω and so the modes propagate without amplification.

(b) Here $\omega(k)$ is real for all k and so the system is stable. However $k(\omega)$ is complex across the range of ω given by

$$(\omega - \omega_0)^2 < 4|\epsilon v_1 v_2|/(v_1 - v_2)^2 \tag{6.129}$$

There is no propagation over this range, i.e. the modes are evanescent.

(c) In this case there are complex roots of $\omega(k)$ for real k and of $k(\omega)$ for real ω. For

$$(k - k_0)^2 < 4|\epsilon|/(v_1 - v_2)^2 \tag{6.130}$$

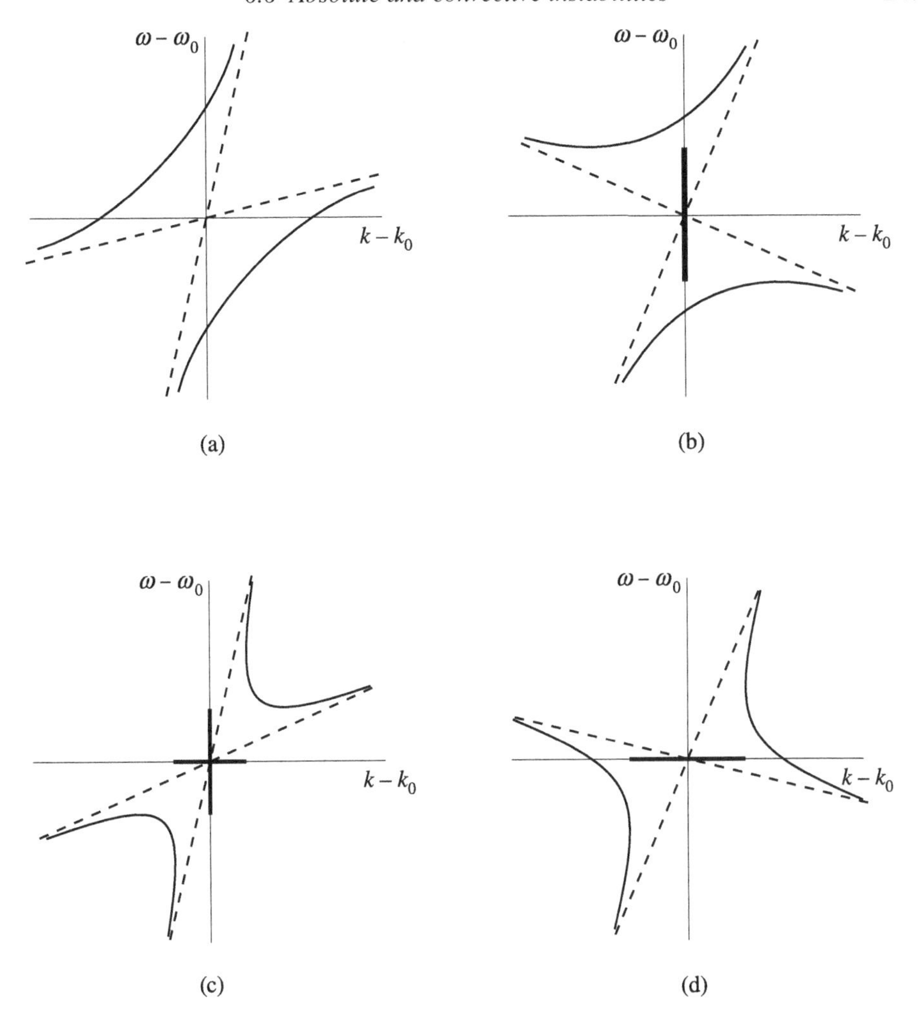

Fig. 6.22. Dispersion curves for weakly coupled modes.

the $\omega(k)$ are complex and one of the pair has $\Im\omega = \omega_i(k) > 0$ and so is unstable. The instability is *convective* since for $|\omega| \to \infty$ the roots $k(\omega)$ are approximately ω/v_1 and ω/v_2 and when $\omega_i \to \infty$, they fall in the same half k-plane. For v_1, $v_2 > 0$ they lie in the upper half k-plane. For real ω in the range (6.129) the roots $k(\omega)$ form a complex conjugate pair. The root with $k_i(\omega) < 0$ has crossed to the lower half-plane. Across the frequency range defined by (6.129) waves propagating in the positive x-direction will amplify.

(d) Here $k(\omega)$ is real for all real ω but $\omega(k)$ is complex across the range (6.130). The system is therefore unstable. Since $v_1 v_2 < 0$ as $\omega \to \infty$ it follows that the roots $k(\omega)$ fall in opposite half-planes. The roots coalesce at a point in the upper half ω-plane for which

$$\omega = \omega_c = \omega_0 + 2i\left(\frac{\sqrt{(\epsilon v_1 v_2)}}{|v_1 - v_2|}\right) \tag{6.131}$$

This corresponds to an *absolute instability* with growth rate ω_{ci}.

Exercises

6.1 Obtain expressions for the phase and group velocities of the following modes:

(i) Alfvén: $\omega^2 = k^2c^2/(1 + c^2/v_A^2)$

(ii) whistler: $\omega = -k^2c^2\Omega_e/\omega_p^2$

(iii) electron cyclotron ($\omega \to |\Omega_e|$): $\omega^2 - \omega\omega_p^2/(\omega + \Omega_e) = k^2c^2$

6.2 Using the data in Table 1.1, compute Alfvén wave speeds for plasmas in (i) interstellar space, (ii) solar corona, (iii) ionosphere and (iv) tokamak. Assume that the positive charges are protons in (i), (ii) and (iv) and oxygen (O^+) ions in (iii).

6.3 The energy density of a wave propagating in a plasma is the sum of contributions from the oscillating electric and magnetic fields and from the coherent particle motion induced by these fields. Suppose that the fields show a small degree of exponential growth so that their behaviour with time is described by $\exp(-i\omega_R + \gamma)t$ where $\gamma \ll \omega_R$. The rate of change of the energy density averaged over a period is given by

$$\frac{dW}{dt} = \frac{1}{2}\Re(\mathbf{E}^* \cdot \mathbf{j}) + \frac{1}{4}\frac{\partial}{\partial t}\left[\varepsilon_0|E|^2 + \frac{|B|^2}{\mu_0}\right]$$

where $*$ denotes the complex conjugate. Writing

$$j_i = i\omega\varepsilon_0[\delta_{ij} - \varepsilon_{ij}(\omega)]E_j$$

where ω denotes the complex frequency $\omega = \omega_R + i\gamma$, show, using a Taylor expansion along with the relation $(\partial/\partial t)|E|^2 = 2\gamma|E|^2$, that the wave energy density may be expressed in the form

$$W = \frac{1}{4}\left[\varepsilon_0 E_i^\star \frac{\partial}{\partial\omega}(\omega\varepsilon_{ij})E_j + \frac{|B|^2}{\mu_0}\right]$$

Apply this result to an electromagnetic wave propagating in an isotropic plasma ($\mathbf{B} = 0$), identifying the contribution to the energy density from coherent particle motion.

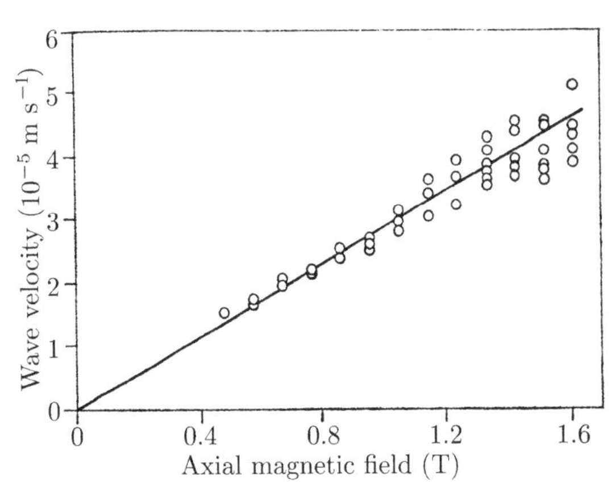

Fig. 6.23. Measurement of the dependence of Alfvén wave phase velocity on axial magnetic field in a hydrogen plasma compared with theory (after Wilcox *et al.* (1960)).

6.4 By introducing a term into (6.10) to allow phenomenologically for the effects of electron–ion collisions through a collision frequency ν_{ei}, show that the dispersion relation for electromagnetic waves in an isotropic plasma becomes

$$\frac{c^2k^2}{\omega^2} = 1 - \frac{\omega_{\mathrm{p}}^2}{\omega(\omega + i\nu_{\mathrm{ei}})}$$

Hence show that electromagnetic waves are damped as a result of electron–ion collisions, with damping coefficient $\gamma = \nu_{\mathrm{ei}}(\omega_{\mathrm{p}}^2/2\omega^2)$.

6.5 Figure 5 shows the Alfvén wave velocity, as a function of the axial magnetic field, measured in a hydrogen plasma by Wilcox, Boley and De Silva (1960). The plasma temperature was 1 eV and the proton density $5 \times 10^{21}\ \mathrm{m}^{-3}$. Using the cold plasma dispersion relation plot the Alfvén velocity as a function of magnetic field. Verify that the cold plasma approximation is valid in this parameter range. How might the discrepancy between the measured phase velocities and those from the simple Alfvén dispersion relation be explained?

6.6 Show that the dispersion relation for a wave propagating orthogonally to a magnetic field $\mathbf{B}_0$ with its electric vector aligned with $\mathbf{B}_0$ is $\omega^2 = \omega_{\mathrm{p}}^2 + k^2c^2$. Explain the physical significance of this result.

6.7 Show that the points of intersection of the plasma cut-off $P = 0$ with the plasma resonance $S = 0$ and cyclotron cut-offs $L = 0$ in a two-component, cold plasma occur at $\alpha^2 = 1$, $\beta_{\mathrm{i}} = 1 - m_{\mathrm{e}}/m_{\mathrm{i}}$ and $\alpha^2 = 1$, $\beta_{\mathrm{i}}^2 = 1 - m_{\mathrm{e}}/m_{\mathrm{i}} + (m_{\mathrm{e}}/m_{\mathrm{i}})^2$, respectively.

6.8 Verify the topological representation of the wave normal surfaces in the CMA diagram, Fig. 6.12.

6.9 The wave normal surfaces corresponding to fast and slow modes coincide when the discriminant (6.34) vanishes. Find the modes propagating along the magnetic field for which this is possible.

For propagation orthogonal to the magnetic field show that there is a curve in the CMA diagram on which coincidence is possible and obtain its equation.

6.10 Determine the group velocity v_g of the ion cyclotron wave satisfying the dispersion relation (6.71) and show that $v_g \to 0$ as $\omega \to \Omega_i$. Show that the wave is elliptically polarized, becoming LCP at the resonance frequency.

6.11 Derive (6.73) from (6.31). [Hint: Write (6.31) as $n^2 = (An^2 - C)/(An^2 + A - B)$ and then substitute the biquadratic solution of (6.31) for n^2 on the right-hand side.]

6.12 Determine the conditions under which the second term in the discriminant in (6.94) is *not* small compared with unity.

Show that in ion acoustic waves, electron and ion velocities are of comparable magnitude.

6.13 Show that in a homogeneous isotropic plasma the conduction current and displacement current in Langmuir waves cancel exactly.

6.14 Show that the dispersion relation for obliquely propagating ion cyclotron waves in a warm plasma is

$$\frac{k^2c_s^2}{\omega^2} = \frac{2\Omega_i^2c_s^2 + v_A^2(\Omega_i^2 - \omega^2)}{v_A^2(\Omega_i^2\cos^2\theta - \omega^2)}$$

Rearrange this in the form $\omega = \omega(k)$ and from this show that, in the long wavelength limit ($k^2c_s^2 \ll \Omega_i^2$), $\omega^2 \simeq \Omega_i^2 + k^2c_s^2\sin^2\theta$. This mode is electrostatic as is the mode in the limit $k^2c_s^2 \gg \Omega_i^2$. Write down the dispersion relation for this case.

6.15 Consider how the ion cyclotron resonance changes when a plasma contains two ion species, as for example in the solar wind which consists of protons with helium ions as the principal minority constituent.

Plot dispersion curves $k^2v_A^2/\omega^2$ versus ω/Ω for a plasma with 80% protons and 20% He^{++} ions for $\theta = 0$ and $\theta = \pi/2$.

Show that as propagation switches from $\theta = 0$ towards $\theta = \pi/2$ the resonances move from the cyclotron frequencies to the frequencies determined by $S = 0$, with one of the resonant frequencies shifted to

$$\omega_{H-He}^2 = \frac{\omega_{pH}^2\Omega_{He}^2 + \omega_{pHe}^2\Omega_H^2}{\omega_{pH}^2 + \omega_{pHe}^2}$$

while the other is displaced to lower frequencies.

6.16 Confirm that (6.120) can be written as a three-wave interaction as $kv_{\mathrm{b}}/\omega_{\mathrm{p}} \to 1$. Show that the maximum growth rate, corresponding to $\Delta k = k - \omega_{\mathrm{p}}/v_{\mathrm{b}} = 0$, is given by (6.121).

7

Collisionless kinetic theory

7.1 Introduction

Much of plasma physics can be adequately described by fluid equations, namely, the MHD or wave equations. However, these are derivative descriptions in which some information about the plasma has been suppressed. In situations where that information matters it is necessary to go to a deeper level of physical description.

The information that gets lost in a fluid model is that relating to the distribution of velocities of the particles within a fluid element, since the fluid variables are functions of position and time but not of velocity. Any physical properties of the plasma that depend on this microscopic detail can be discovered only by a description in six-dimensional $(\mathbf{r}, \mathbf{v})$ space. Thus, instead of starting with the density of particles, $n(\mathbf{r}, \mathbf{t})$, at position $\mathbf{r}$ and time t, we begin with the so-called *distribution function*, $f(\mathbf{r}, \mathbf{v}, t)$, which is the density of particles in $(\mathbf{r}, \mathbf{v})$ space at time t. The evolution of the distribution function is described by *kinetic theory*.

With the additional information on particle velocities *within a volume element* introduced by a phase space description we now have microscopic detail that we did not have before. For that reason, kinetic and fluid theories are identified as *microscopic* and *macroscopic*, respectively.

At the most fundamental level we may define the distribution function in terms of the individual particle positions and velocities by

$$f_{\mathrm{K}}(\mathbf{r}, \mathbf{v}, t) = \sum_{i=1}^{N} \delta[\mathbf{r} - \mathbf{r}_i(t)]\delta[\mathbf{v} - \mathbf{v}_i(t)] \tag{7.1}$$

where the sum is over all particles of a given type. This is the *Klimontovich distribution function* which we have denoted by f_{K} to distinguish it from f. It is a very spiky function being zero throughout $(\mathbf{r}, \mathbf{v})$ space except at the N points $[\mathbf{r} = \mathbf{r}_i(t), \mathbf{v} = \mathbf{v}_i(t)]$ where it is doubly infinite. However, we can generate a smoother function by integrating (7.1) over a volume element $\Delta\mathbf{r}\Delta\mathbf{v}$ about the point $(\mathbf{r}, \mathbf{v})$ which is large enough to contain a number of particles $N_p(\mathbf{r}, \mathbf{v}, t) \gg 1$ but small enough that

$$f(\mathbf{r}, \mathbf{v}, t) = \frac{1}{\Delta\mathbf{r}\Delta\mathbf{v}} \int_{\Delta\mathbf{r}} \mathrm{d}\mathbf{r} \int_{\Delta\mathbf{v}} \mathrm{d}\mathbf{v}\, f_{\mathrm{K}} = \frac{N_p(\mathbf{r}, \mathbf{v}, t)}{\Delta\mathbf{r}\Delta\mathbf{v}} \tag{7.2}$$

does not change significantly over the dimensions of the volume element. Thus, $f(\mathbf{r}, \mathbf{v}, t)$ is the number density of particles in a small volume element centred at the point $(\mathbf{r}, \mathbf{v})$ at time t.

Provided no particles are created or destroyed, f obeys a continuity equation in $(\mathbf{r}, \mathbf{v})$ space which is derived by exactly the same arguments used to derive the continuity equation for the mass density $\rho(\mathbf{r}, t)$ in Chapter 3. There are now two divergence terms arising from the flow of particles through the surfaces of the volume element in both $\mathbf{r}$ and $\mathbf{v}$ space. Thus, we have

$$\frac{\partial f}{\partial t} + \frac{\partial}{\partial \mathbf{r}} \cdot (f\mathbf{v}) + \frac{\partial}{\partial \mathbf{v}} \cdot (f\mathbf{a}) = 0 \tag{7.3}$$

where $\mathbf{a}$ is the acceleration of the particles in the volume element. Since $\mathbf{r}$ and $\mathbf{v}$ are independent variables in (7.3) we may bring $\mathbf{v}$ outside the differential operator and, if in addition $\boldsymbol{\nabla}_v \cdot \mathbf{a} = 0$, we get

$$\frac{\partial f}{\partial t} + \mathbf{v} \cdot \frac{\partial f}{\partial \mathbf{r}} + \frac{\mathbf{F}}{m} \cdot \frac{\partial f}{\partial \mathbf{v}} = 0 \tag{7.4}$$

in which we have replaced $\mathbf{a}$ by $\mathbf{F}/m$ where $\mathbf{F}$ is the force acting on the particles of mass m at the point $(\mathbf{r}, \mathbf{v})$ at time t. Such a partial differential equation describing the evolution of the distribution function is known as a *kinetic equation*.

As a matter of fact, (7.4) is necessarily a *collisionless* kinetic equation since we certainly cannot assume $\boldsymbol{\nabla}_v \cdot \mathbf{a} = 0$ if we want to include the collisional interactions taking place inside the volume element. A proper description of collisions is a formidable problem as we shall see in Chapter 12. However, the transition from (7.4) to a collisional kinetic equation can be made by a simple heuristic argument. Assuming that $\mathbf{F}$ represents all the non-collisional (macroscopic) forces we note that (7.4) states that

$$\begin{aligned}\frac{\mathrm{d}f(\mathbf{r}, \mathbf{v}, t)}{\mathrm{d}t} &= \frac{\partial f}{\partial t} + \frac{\mathrm{d}\mathbf{r}}{\mathrm{d}t} \cdot \frac{\partial f}{\partial \mathbf{r}} + \frac{\mathrm{d}\mathbf{v}}{\mathrm{d}t} \cdot \frac{\partial f}{\partial \mathbf{v}} \\ &= \frac{\partial f}{\partial t} + \mathbf{v} \cdot \frac{\partial f}{\partial \mathbf{r}} + \frac{\mathbf{F}}{m} \cdot \frac{\partial f}{\partial \mathbf{v}} = 0\end{aligned}$$

i.e. in the absence of collisions f is constant along any trajectory in $(\mathbf{r}, \mathbf{v})$ space. Collisions, however, change this so we write

$$\frac{\partial f}{\partial t} + \mathbf{v} \cdot \frac{\partial f}{\partial \mathbf{r}} + \frac{\mathbf{F}}{m} \cdot \frac{\partial f}{\partial \mathbf{v}} = \left(\frac{\partial f}{\partial t}\right)_{\mathrm{c}} \tag{7.5}$$

where $(\partial f/\partial t)_c$ represents the change in f with time due to collisions. This is then the *collisional kinetic equation*, though how we represent the collision term in (7.5) is a problem we defer until the next chapter.

In this chapter we are concerned only with collisionless kinetic theory for the very good reason that most plasmas are essentially collisionless. All of the terms in (7.5) have the dimensions of f times a frequency. The frequency appropriate to the right-hand side is, of course, the collision frequency ν_c while that on the left-hand side depends on the dominant macroscopic force. Since this macroscopic force is typically the Lorentz force due to the self-consistent fields the appropriate frequency is likely to be one of the wave frequencies encountered in the previous chapter. In particular, we have noted the dominance of the electrostatic field in maintaining charge neutrality and causing oscillations at the electron plasma frequency in response to any local charge inequality. As the plasma frequency is usually much greater than the collision frequency, unless we are specifically interested in collisional effects, we can ignore the collision term and take (7.4) as the kinetic equation. We shall study it in order to discover some important properties of plasmas which depend on distributions of the plasma particles in velocity space and which, therefore, are not accessible to fluid descriptions.

7.2 Vlasov equation

Vlasov first solved the collisionless kinetic equation (7.4), now known universally as the Vlasov equation, in the case where $\mathbf{F} = e\mathbf{E}(\mathbf{r}, t)$ and $\mathbf{E}$ is the self-consistent electric field. Interestingly he did not solve (7.4) as an initial value problem and consequently missed its most important property! This was subsequently discovered by Landau and is discussed in the next section. The collisionless kinetic equation is sometimes referred to as the collisionless Boltzmann equation but this is something of a contradiction in terms since the representation of collisions is at the very heart of the Boltzmann equation.

The first thing we shall do with the Vlasov equation is to show its formal equivalence to the equation describing individual particle orbits. The latter is

$$m\ddot{\mathbf{r}} = \mathbf{F} \tag{7.6}$$

having the solution

$$\left.\begin{aligned} \mathbf{r} &= \mathbf{r}(c_1, c_2, \ldots, c_6, t) \\ \mathbf{v} &= \mathbf{v}(c_1, c_2, \ldots, c_6, t) \end{aligned}\right\} \tag{7.7}$$

where $c_1, c_2, \ldots, c_6$ are the six constants of integration, which might for example be the initial values of $\mathbf{r}$ and $\mathbf{v}$. Inverting (7.7) gives the formal solution

$$c_i = c_i(\mathbf{r}, \mathbf{v}, t) \quad (i = 1, 2, \ldots, 6) \tag{7.8}$$

Now any arbitrary function of the c_i

$$f = f(c_1, c_2, \ldots, c_6)$$

is a solution of (7.4) as one can see by direct substitution:

$$\sum_{i=1}^{6} \frac{\partial f}{\partial c_i}\left(\frac{\partial c_i}{\partial t} + \mathbf{v}\cdot\frac{\partial c_i}{\partial \mathbf{r}} + \frac{\mathbf{F}}{m}\cdot\frac{\partial c_i}{\partial \mathbf{v}}\right) = \sum_{i=1}^{6} \frac{\partial f}{\partial c_i}\frac{\mathrm{d}c_i}{\mathrm{d}t} = 0$$

since the c_i are constants of the motion. Thus the general solution of the Vlasov equation is an arbitrary function of the integrals of (7.6), the equation describing orbit theory. This was demonstrated by Jeans in work on stellar dynamics and so is generally known as the *Jeans theorem*.

The formal equivalence of the Vlasov equation and particle orbit theory is rigorous and simple but, in one sense, slightly deceptive. It should not be imagined that solving the Vlasov equation is as easy as finding some of the orbit solutions obtained in Chapter 2. The reason is that, in the Vlasov equation, $\mathbf{F}$ contains both external fields and self-consistent fields arising from the plasma motion. In Chapter 2 we assumed that the latter were negligible compared with any applied fields and so the orbit equations solved there are only approximations to the equation of motion (7.6). We shall see later that there is a direct relationship between solving the *linearized* Vlasov equation and calculating the simple (unperturbed) orbits of the particles in the external fields. We note, in this context, that Jeans' theorem provides a method of obtaining zero-order, equilibrium distribution functions, namely, any function of the constants of the motion in the (zero-order) external fields. Illustrations of this appear in later sections of this chapter.

For the moment, we observe that the full set of equations to be solved, in the general case, is the Vlasov equation (7.4) for the distribution function of each species of particle together with the Maxwell equations:

$$\nabla \times \mathbf{E} = -\frac{\partial \mathbf{B}}{\partial t} \tag{7.9}$$

$$\nabla \times \mathbf{B} = \varepsilon_0\mu_0\frac{\partial \mathbf{E}}{\partial t} + \mu_0\mathbf{j} \tag{7.10}$$

$$\nabla \cdot \mathbf{E} = q/\varepsilon_0 \tag{7.11}$$

$$\nabla \cdot \mathbf{B} = 0 \tag{7.12}$$

where

$$q = \sum_\alpha e_\alpha \int f_\alpha \,\mathrm{d}\mathbf{v} \tag{7.13}$$

$$\mathbf{j} = \sum_\alpha e_\alpha \int \mathbf{v} f_\alpha \,\mathrm{d}\mathbf{v} \tag{7.14}$$

and the sums are over the particle species. The fact that there are at least two Vlasov equations to be solved (one each for ions and electrons) is a relatively minor complication compared with the problem of solving the full Vlasov–Maxwell set of equations self-consistently for f, when f itself is a source of the fields through (7.13) and (7.14). (The equivalent of this latter problem in orbit theory would be finding the orbits in the full self-consistent fields.) Nevertheless, the Vlasov–Maxwell equations are the starting point for most calculations in plasma kinetic theory. They include the principal effect of particle interactions, the self-consistent field, and in the approximation that there are a large number of particles within the Debye sphere ($n\lambda_D^3 \gg 1$), known as the weak coupling approximation (see Exercise 7.3), they provide an adequate description unless we are specifically interested in collisional effects. Their complexity means that one generally has to resort to numerical methods even when a linear solution is sought. In the following sections we present some important solutions of the linearized set of equations which are amenable to analytic methods.

7.3 Landau damping

The most important and fundamental property of the Vlasov equation was discovered by Landau (1946) who solved the linearized electron Vlasov equation for $\mathbf{F} = -e\mathbf{E}$ where $\mathbf{E}$ is the electric field created when a homogeneous plasma in equilibrium is slightly perturbed. It is assumed that the perturbation is in the electron distribution only, so that the ions remain as a steady, homogeneous, neutralizing backgound. This simplifying assumption avoids having to solve two Vlasov equations. Following Landau, we solve the linearized equation as an initial value problem, i.e. the perturbation is introduced at $t = 0$. The alternative in which a perturbation is introduced at $\mathbf{r} = 0$ and its spatial evolution examined is considered in Section 7.5.

Any steady, homogeneous distribution function $f_0(\mathbf{v})$ satisfies (7.4) identically, since the electron density is uniform and equal to the ion density and there is no electric field. If a small perturbation $f_1(\mathbf{r}, \mathbf{v}, t)$ is introduced, we may write

$$f(\mathbf{r}, \mathbf{v}, t) = f_0(\mathbf{v}) + f_1(\mathbf{r}, \mathbf{v}, t) \tag{7.15}$$

and, since the contribution of f_0 to $\mathbf{E}$ is zero, $|\mathbf{E}|$ is of order f_1 and the linearized Vlasov equation is

$$\frac{\partial f_1}{\partial t} + \mathbf{v} \cdot \frac{\partial f_1}{\partial \mathbf{r}} - \frac{e\mathbf{E}}{m} \cdot \frac{\partial f_0}{\partial \mathbf{v}} = 0 \tag{7.16}$$

where, from (7.11) and (7.13),

$$\nabla \cdot \mathbf{E} = -\frac{e}{\varepsilon_0} \int f_1 \, d\mathbf{v} \tag{7.17}$$

Solving (7.16) by means of Fourier and Laplace transforms, we write

$$f_1(\mathbf{r}, \mathbf{v}, t) = \frac{1}{(2\pi)^{3/2}} \int f_1(\mathbf{k}, \mathbf{v}, t) \exp(i\mathbf{k} \cdot \mathbf{r}) \mathrm{d}\mathbf{k} \quad (7.18)$$

$$\mathbf{E}(\mathbf{r}, t) = \frac{1}{(2\pi)^{3/2}} \int \mathbf{E}(\mathbf{k}, t) \exp(i\mathbf{k} \cdot \mathbf{r}) \mathrm{d}\mathbf{k} \quad (7.19)$$

so that (7.16) gives for each Fourier component

$$\frac{\partial f_1(\mathbf{k}, \mathbf{v}, t)}{\partial t} + i\mathbf{k} \cdot \mathbf{v} f_1(\mathbf{k}, \mathbf{v}, t) - \frac{e\mathbf{E}(\mathbf{k}, t)}{m} \cdot \frac{\partial f_0(\mathbf{v})}{\partial \mathbf{v}} = 0 \quad (7.20)$$

Before taking the Laplace transform, (7.20) may be simplified by noting that, since **E** is electrostatic,

$$\nabla \times \mathbf{E}(\mathbf{r}, t) = 0$$

and hence

$$\mathbf{k} \times \mathbf{E}(\mathbf{k}, t) = 0$$

Thus $\mathbf{E}(\mathbf{k}, t)$ is parallel to $\mathbf{k}$ so that, if u is the component of $\mathbf{v}$ along $\mathbf{k}$, (7.20) becomes

$$\frac{\partial f_1(\mathbf{k}, \mathbf{v}, t)}{\partial t} + iku f_1(\mathbf{k}, \mathbf{v}, t) - \frac{eE(\mathbf{k}, t)}{m} \frac{\partial f_0(\mathbf{v})}{\partial u} = 0 \quad (7.21)$$

Taking the Laplace transform of (7.21), that is, multiplying by e^{-pt} and integrating over t from 0 to ∞, we get

$$(p + iku) f_1(\mathbf{k}, \mathbf{v}, p) - \frac{eE(\mathbf{k}, p)}{m} \frac{\partial f_0(\mathbf{v})}{\partial u} = f_1(\mathbf{k}, \mathbf{v}, t = 0) \quad (7.22)$$

where

$$f_1(\mathbf{k}, \mathbf{v}, p) = \int_0^\infty f_1(\mathbf{k}, \mathbf{v}, t) e^{-pt} \, \mathrm{d}t \quad (7.23)$$

$$E(\mathbf{k}, p) = \int_0^\infty E(\mathbf{k}, t) e^{-pt} \, \mathrm{d}t \quad (7.24)$$

From (7.22) f_1 is obtained as a function of E which we can now substitute in the Fourier–Laplace transform of (7.17)

$$ikE(\mathbf{k}, p) = -\frac{e}{\varepsilon_0} \int f_1(\mathbf{k}, \mathbf{v}, p) \mathrm{d}\mathbf{v}$$

to obtain an equation for E alone. Substituting for f_1 from (7.22)

$$ikE(\mathbf{k}, p) = -\frac{e}{\varepsilon_0} \int \frac{f_1(\mathbf{k}, \mathbf{v}, t = 0)}{p + iku} \mathrm{d}\mathbf{v} - \frac{e^2}{\varepsilon_0 m} E(\mathbf{k}, p) \int \frac{\partial f_0 / \partial u}{p + iku} \mathrm{d}\mathbf{v}$$

Hence

$$E(\mathbf{k}, p) = \frac{ie}{\varepsilon_0 k D(k, p)} \int \frac{f_1(\mathbf{k}, \mathbf{v}, t = 0)}{p + iku} \, \mathrm{d}\mathbf{v} \tag{7.25}$$

where

$$D(k, p) \equiv 1 - \frac{ie^2}{\varepsilon_0 mk} \int \frac{\partial f_0/\partial u}{p + iku} \, \mathrm{d}\mathbf{v}$$

is the *plasma dielectric function*; note that this is independent of initial conditions. Carrying out the inverse Laplace and Fourier transforms formally solves the problem. Unfortunately, this is in general no simple matter. The time dependence of the kth Fourier component of the electric field is given by

$$E(\mathbf{k}, t) = \frac{1}{2\pi i} \int_{\sigma - i\infty}^{\sigma + i\infty} E(\mathbf{k}, p) e^{pt} \, \mathrm{d}p \tag{7.26}$$

where the integration is along a line parallel to the imaginary p-axis and to the right of all singularities of the integral as indicated in Fig. 7.1(a).

It is a well-known result of complex variable theory that if $E(\mathbf{k}, p)e^{pt}$ is an analytic function of p except for a finite number of poles in the infinite strip between $\Re p = -\alpha$ and $\Re p = \sigma$ then we may deform the contour of integration to that shown in Fig. 7.1(b), i.e. we integrate along $\Re p = -\alpha$ instead of $\Re p = \sigma$ but take a horizontal detour to go around each of the poles lying between the two vertical lines. The advantage of this deformation of the contour is that now, on its vertical section, the integrand decays with time like $e^{-\alpha t}$ and vanishes asymptotically. The integrations along the horizontal lines are taken one in each direction and therefore cancel out so that we are left with the integrations around the poles which give $2\pi i$ times the sum of the residues at the poles.

For suitable choices of f_0 and $f_1(t = 0)$ (the conditions that $\partial f_0/\partial u$ and $f_1(t = 0)$ are analytic functions of u are sufficient) the only singularities of $E(\mathbf{k}, p)e^{pt}$ in the p-plane are simple poles where the dielectric function vanishes. Choosing $v_x = u$ and defining F_0 by $n_0 F_0(u) = \int f_0(\mathbf{v}) \mathrm{d}v_y \, \mathrm{d}v_z$, the zeros of $D(k, p)$ are given by

$$D(k, p) = 1 - \frac{i\omega_{\mathrm{pe}}^2}{k} \int_{-\infty}^{+\infty} \frac{\mathrm{d}F_0/\mathrm{d}u}{p + iku} \, \mathrm{d}u = 0 \tag{7.27}$$

where $\omega_{\mathrm{pe}} = (n_0 e^2/\varepsilon_0 m)^{1/2}$ is the electron plasma frequency. Thus, if the solutions of (7.27) are denoted by p_j then from (7.26) we get, as $t \to \infty$,

$$E(\mathbf{k}, t) = \sum_j R_j e^{p_j t} \tag{7.28}$$

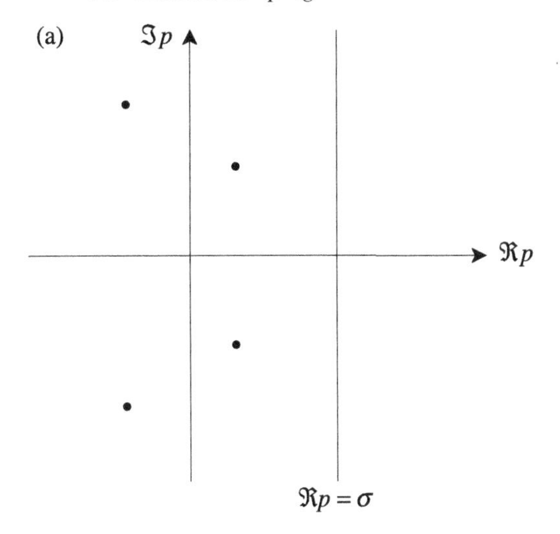

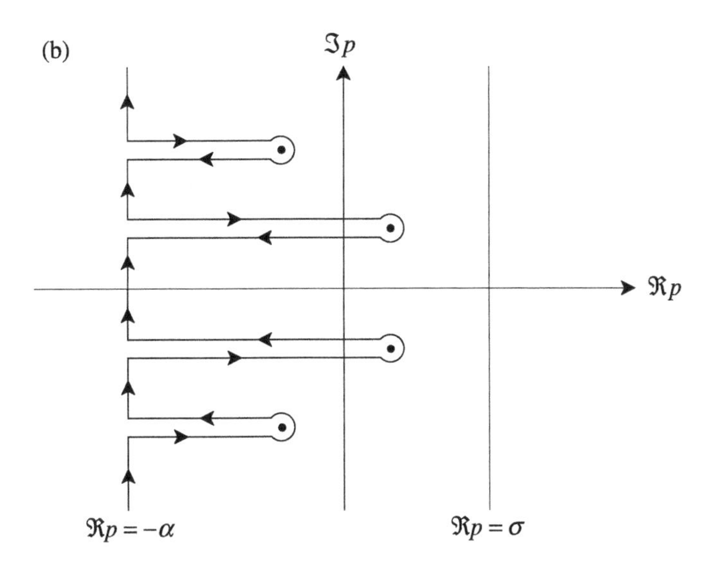

Fig. 7.1. Deformation of contour of integration in complex p-plane.

where

$$R_j = \lim_{p \to p_j} (p - p_j) E(\mathbf{k}, p)$$

is the residue of $E(\mathbf{k}, p)$ at p_j. In general the poles p_j are complex, so writing

$$p_j(\mathbf{k}) = -i\omega_j(\mathbf{k}) + \gamma_j(\mathbf{k}) \tag{7.29}$$

where ω_j and γ_j are real, (7.28) becomes

$$E(\mathbf{k}, t) = \sum_j R_j e^{-i\omega_j(\mathbf{k})t+\gamma_j(\mathbf{k})t} \tag{7.30}$$

If any $\gamma_j > 0$ the field grows exponentially and the linear approximation breaks down, so we shall assume, for the moment, that all poles lie to the left of the imaginary p-axis. Then all terms with $\gamma_j \neq 0$ in (7.30) are exponentially damped oscillations. Note that none are damped as strongly as $e^{-\alpha t}$; in general we are interested in the pole closest to the imaginary p-axis since this corresponds to the smallest damping decrement.

We now investigate the limit of long-wavelength waves, $k \to 0$. To lowest order in this limit (7.27) gives, on integration by parts and using $\int F_0 \, \mathrm{d}u = 1$,

$$p = \pm i\omega_{\mathrm{pe}} \tag{7.31}$$

that is, undamped plasma oscillations. To find the lowest-order k dependence we again integrate by parts and expand $(p + iku)^{-2}$ in powers of (iku/p), giving

$$-\frac{\omega_{\mathrm{pe}}^2}{p^2} \int_{-\infty}^{+\infty} \mathrm{d}u \, F_0(u) \left(1 - \frac{2iku}{p} - \frac{3k^2u^2}{p^2} + \cdots\right) = 1 \tag{7.32}$$

The imaginary term vanishes if $F_0(u)$ is isotropic. This is true for all the imaginary terms in the expansion in (7.32) since they are all odd in u. The first correction to (7.31) arises from the term in k^2. Choosing the *Maxwell distribution*

$$f_0 = n_0(m/2\pi k_{\mathrm{B}} T_{\mathrm{e}})^{3/2} \exp(-mv^2/2k_{\mathrm{B}} T_{\mathrm{e}}) \tag{7.33}$$

it follows that

$$F_0(u) = (m/2\pi k_{\mathrm{B}} T_{\mathrm{e}})^{1/2} \exp(-mu^2/2k_{\mathrm{B}} T_{\mathrm{e}})$$

and from (7.32) we get

$$p = \pm i\omega_{\mathrm{pe}} \left[1 + \frac{3}{2}(k\lambda_{\mathrm{D}})^2\right] \tag{7.34}$$

It is easily verified (see Exercise 7.4) that (7.34) corresponds to the dispersion relation (6.97) for longitudinal electron plasma waves (Langmuir waves) in a warm plasma.

Since all the imaginary terms vanish in the expansion in powers of k, no damping appears in such a solution. To find the damping decrement one must resort to the full expression (7.27). This presents a problem since the integrand contains a pole at $u = ip/k$ which, for pure imaginary p, lies on the path of integration. However $E(\mathbf{k}, p)$ was originally defined on a line in the p-plane to the right of all singularities, that is, for $\Re p > 0$ (see Fig. 7.1). Thus, the integral in the u-plane in (7.27) is also defined for $\Re p > 0$. This means that the pole at $u = ip/k$ lies

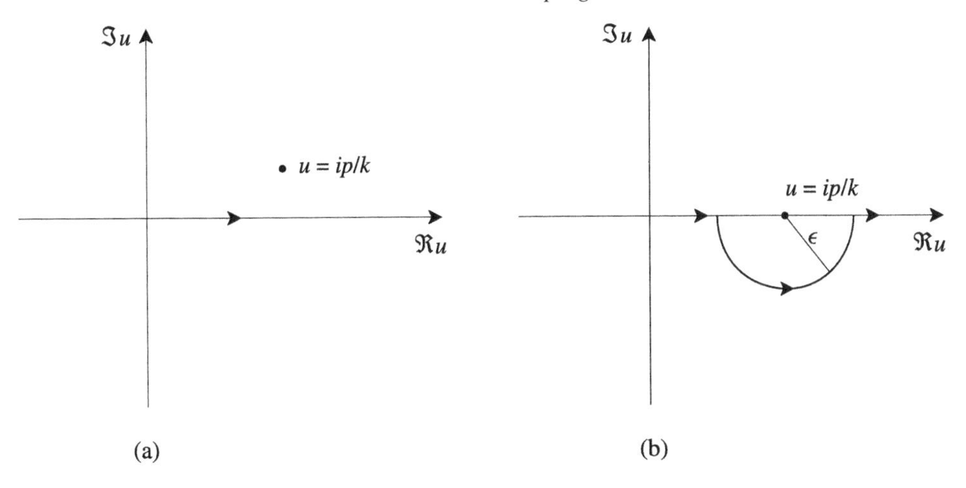

Fig. 7.2. Path of integration along $\Re u$-axis for pole (a) above axis and (b) on axis.

above the real axis, which is the path of integration in Fig. 7.2(a). In the limit $\Re p \to 0$ the pole drops on to the real axis but analytic continuation requires that the path of integration must stay below the pole and so we integrate from $-\infty$ up to $(ip/k) - \epsilon$, then around the semi-circle of radius ϵ *below* the pole, and finally continue along the real axis from $(ip/k) + \epsilon$ to $+\infty$ as shown in Fig. 7.2(b). Thus (7.27) is evaluated by means of the relationship

$$\int_{-\infty}^{+\infty} \frac{\mathrm{d}F_0/\mathrm{d}u}{(u - ip/k)}\,\mathrm{d}u = P \int_{-\infty}^{+\infty} \frac{\mathrm{d}F_0/\mathrm{d}u}{(u - ip/k)}\,\mathrm{d}u + i\pi \left[\frac{\mathrm{d}F_0}{\mathrm{d}u}\right]_{ip/k} \tag{7.35}$$

where the second term on the right-hand side is simply $i\pi$ times the residue of the integrand at the pole. (Had the pole approached the real axis from below and the contour been deformed above it, the semi-circle would then have been described in the negative (clockwise) direction and the sign of this term would be reversed.) The principal part in (7.35) may be approximated by a power series as in (7.32). Thus, (7.27) becomes

$$p^2 = (-i\omega + \gamma)^2 = -\omega_{\mathrm{pe}}^2 \left(1 - \frac{3k^2 k_{\mathrm{B}} T_{\mathrm{e}}}{p^2 m} + \frac{i\pi p^2}{k^2}\left[\frac{\mathrm{d}F_0}{\mathrm{d}u}\right]_{ip/k}\right) \tag{7.36}$$

the imaginary part of which gives for the *Landau damping decrement*

$$\gamma = \frac{\pi \omega_{\mathrm{pe}}^3}{2k^2}\left[\frac{\mathrm{d}F_0}{\mathrm{d}u}\right]_{ip/k} = -\left(\frac{\pi}{8}\right)^{1/2} \frac{\omega_{\mathrm{pe}}}{(k\lambda_{\mathrm{D}})^3} \exp\left[-\frac{1}{2(k\lambda_{\mathrm{D}})^2} - \frac{3}{2}\right] \quad (k\lambda_{\mathrm{D}} \ll 1) \tag{7.37}$$

This result confirms what was said earlier concerning the vanishing of all imaginary terms in a power series expansion in small k; as $k \to 0$, $\gamma \to 0$ faster than any power of k. Numerical solution of (7.27) shows that as $k\lambda_D \to 1$, $|\gamma| \to \omega_{pe}$, that is, the damping time approaches the period of the oscillations. Thus, the Debye shielding distance λ_D is the minimum wavelength at which longitudinal oscillations ($\mathbf{k} \parallel \mathbf{E}$) can occur. This is easily understood when one notes that at $k\lambda_D = 1$ the phase speed of the wave, ω/k, is equal to the mean thermal speed of the electrons. They are easily able to neutralize the space charge, therefore, and so prevent the wave from propagating.

A further observation to be made from (7.37) is the following. The damping decrement arose from the residue at the pole in (7.27). The sign of γ therefore depends critically on the slope of $F_0(u)$ at the pole, as is obvious from the first equality in (7.37). Since we considered a Maxwellian, centred at the origin, the slope was necessarily negative leading to damping. Clearly, the phenomenon of Landau damping has its physical origin in the interaction of those 'resonant' electrons with $u \approx \omega/k$. On reflection this is not surprising. Since u is that component of electron velocity in the direction of propagation of the wave, those electrons with $u \approx \omega/k$ stay roughly in phase with the wave and, therefore, more effectively exchange energy with it. The actual energy exchange between any particular resonant electron and the wave depends on the phase of the wave at the position of the electron. But if a particle with $u < \omega/k$ is accelerated then its interaction with the wave is made more resonant and therefore stronger than if it had been decelerated. Thus, for particles moving slightly slower than the wave, acceleration is a stronger effect than deceleration so that, on average, slower particles gain energy from the wave. Clearly the opposite is true for particles travelling slightly faster than the wave. Figure 7.3 illustrates the cases of the strongly resonant electrons. A negative slope to the distribution function at the resonant speed ($\mathrm{d}F_0(u = \omega/k)/\mathrm{d}u < 0$) means that slower particles outnumber faster ones so that the wave loses more energy than it gains and is therefore damped. It is clear from this argument that kinetic theory is necessary for a description of Landau damping. Integration (or averaging) over velocity space which gives rise to a fluid theory removes the physical mechanism, the microstructure of $F_0(u)$, essential for Landau damping. Dawson (1962) developed the idea of energy exchange between particles and Langmuir waves into a model from which he was able to retrieve Landau's result. Nonetheless, misgivings persisted for a long time as to whether collisionless damping was a real effect.

Had we chosen a distribution function with a range of values of u for which $\mathrm{d}F_0/\mathrm{d}u > 0$, then for waves with phase velocities in that range we should have found $\gamma > 0$ indicating Landau growth rather than damping. Any such unstable waves are also lost in macroscopic theory and are, therefore, known as *micro-instabilities*, some of which we discuss in Section 7.4.

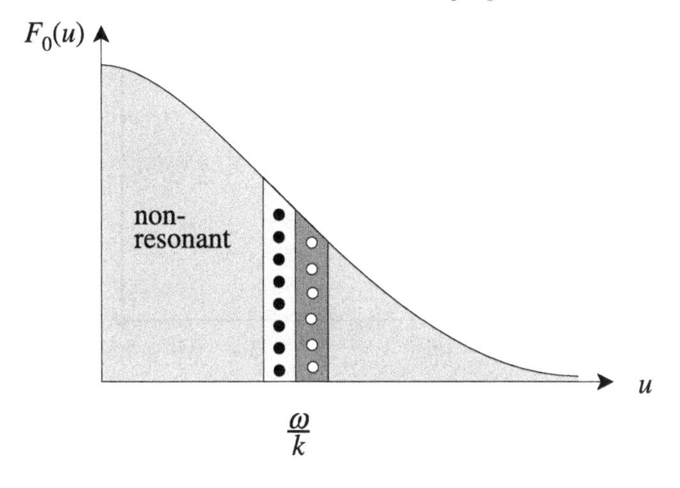

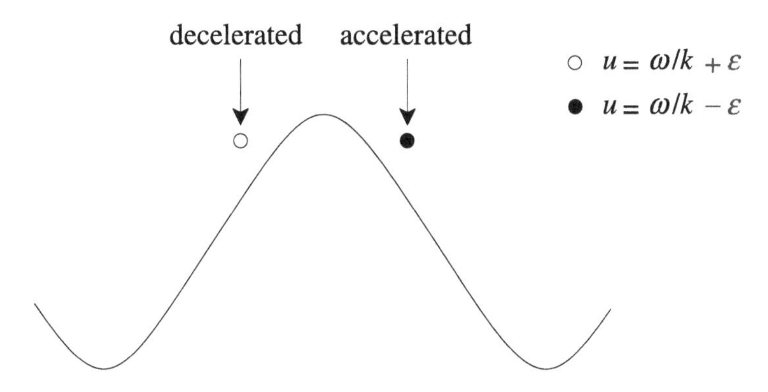

Fig. 7.3. Illustration of interaction of strongly resonant electrons with wave. Filled circles represent electrons with speed $u < \omega/k$ which take energy from the wave. Open circles represent electrons with speed $u > \omega/k$ which give energy to the wave.

7.3.1 Experimental verification of Landau damping

Any lingering doubts about the reality of Landau damping were dispelled by definitive experiments by Malmberg and Wharton (1964, 1966) who showed that the measured spatial attenuation of Langmuir waves agreed remarkably well with Landau's result. The appropriate formulation of the Landau problem for comparison with the measured damping is one in which ω is taken to be real and the dispersion relation is solved for complex k. In this case we have

$$\frac{\Im k}{\Re k} \propto \exp\left(-\frac{1}{2k^2\lambda_{\mathrm{D}}^2}\right)$$

In these experiments the plasma was, to a good approximation, collisionless (see Exercise 7.5). Two probes were used, one of which, the transmitter, was set at a

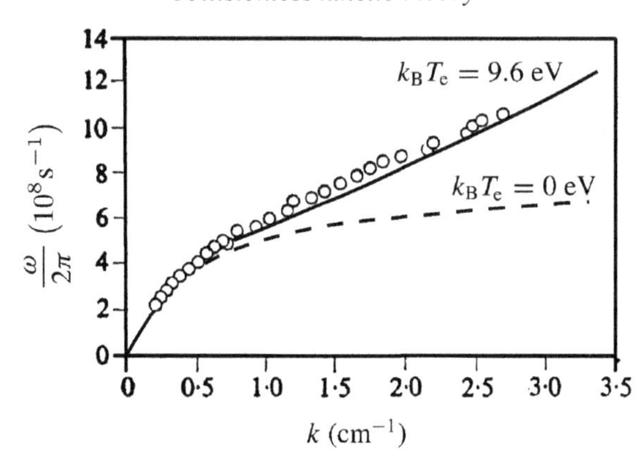

Fig. 7.4. Comparison of experimental results (circles) with theoretical dispersion curve for electron plasma waves. The solid line corresponds to a calculation using the measured temperature while the dashed line is for a cold plasma (after Malmberg and Wharton (1966)).

series of fixed frequencies while the receiving probe, at each setting, was moved longitudinally. From the data, the real and imaginary parts of the wavenumber were obtained as functions of frequency. A dispersion plot is shown in Fig. 7.4. In comparing this result with the theoretical dispersion relation, $k = k(\omega)$, Malmberg and Wharton chose a value of electron density which normalized the theoretical curve to the experimental data at low frequencies; in this region (high phase velocities) temperature corrections to the dispersion relation should be negligible. The measured dispersion plot in Fig. 7.4 shows excellent agreement with theory. Note however that the theoretical curve is not simply a plot of (7.34); we recall that (7.34) shows just the leading terms in an asymptotic series. For the conditions in this experiment, the series is only weakly convergent so that additional terms have to be retained. Note too that as $k \to 0$, $\omega \to 0$ rather than ω_{pe}; this departure from the dispersion relation arises on account of the finite length of the plasma which cuts off long wavelengths.

Figure 7.5 shows the measured damping compared with that predicted by theory. The ordinate is $\Im k/\Re k$ and the abscissa $(\omega/kV_e)^2 \equiv (v_p/V_e)^2$ where V_e is the electron thermal velocity. The ratio $\Im k/\Re k$ and the phase velocity ω/k are found directly from experiment. Since the electron velocity distribution was shown to be Maxwellian, T_e was known experimentally. It is clear that in a collisionless plasma, electron plasma waves suffer exponential damping. The observed damping lengths range from 0.02 to 0.5 m, very much shorter than the electron mean free path. The magnitude of this damping, together with its dependence on phase velocity and on electron temperature, confirms the behaviour predicted for Landau damping.

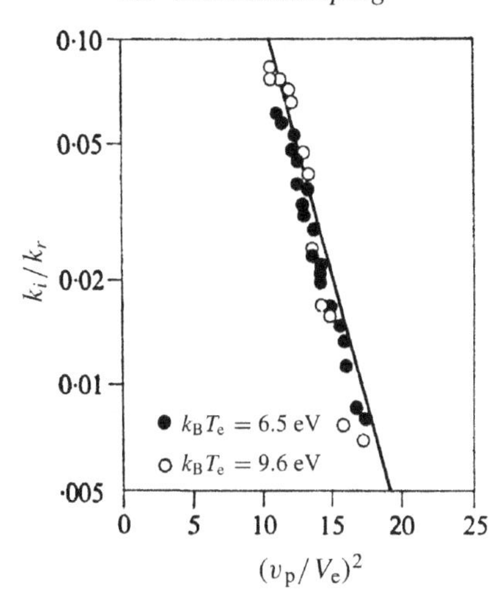

Fig. 7.5. Comparison of experimental results with theoretically predicted Landau damping of electron plasma waves (after Malmberg and Wharton (1966)).

7.3.2 Landau damping of ion acoustic waves

Landau damping is not restricted to Langmuir waves nor is it solely an electron phenomenon. Any wave with phase velocity close to either of the particle thermal velocities will suffer Landau damping. Ion acoustic waves with phase velocity c_s lying between V_i and V_e provide a particularly interesting example.

In general, $V_i \ll V_e$ so $F_{0i}(u)$ is a squeezed version of $F_{0e}(u)$ and, for $T_i \ll T_e$, Fig. 7.6(a) shows that there is weak Landau damping of the waves due to both ions and electrons. The damping is weak on the part of the ions because $c_s \gg V_i$ and so there are very few ions in resonance with the wave. On the other hand, electron Landau damping is weak because $c_s \ll V_e$ and so $\mathrm{d}F_{0e}/\mathrm{d}u \approx 0$ at $u = c_s$. For the case that $T_i \approx T_e$, $c_s \sim V_i$ as shown in Fig. 7.6(b). Now, although electron Landau damping is still weak, ion damping is strong. In fact a numerical solution of the kinetic dispersion relation (see Fig. 7.7(a)) shows that $|\gamma| \sim \omega_r$.

To examine the dispersion and damping characteristics of ion acoustic waves in detail we need a Vlasov equation for each species. The counterpart to the dispersion relation (7.27) now has an ion contribution so that

$$D(k, \omega) = 1 - \frac{\omega_{pe}^2}{k^2} \int_{-\infty}^{\infty} \frac{F'_{0e}(u)}{u - \omega/k} \mathrm{d}u - \frac{\omega_{pi}^2}{k^2} \int_{-\infty}^{\infty} \frac{F'_{0i}(u)}{u - \omega/k} \mathrm{d}u = 0 \qquad (7.38)$$

It is often convenient to write the dispersion relation for electrostatic waves in terms of the *plasma dispersion function* $Z(\zeta)$, defined by

$$Z(\zeta) = \frac{1}{\sqrt{\pi}} \int_{-\infty}^{\infty} \frac{e^{-\xi^2}}{\xi - \zeta} \mathrm{d}\xi \qquad \Im\, \zeta > 0 \qquad (7.39)$$

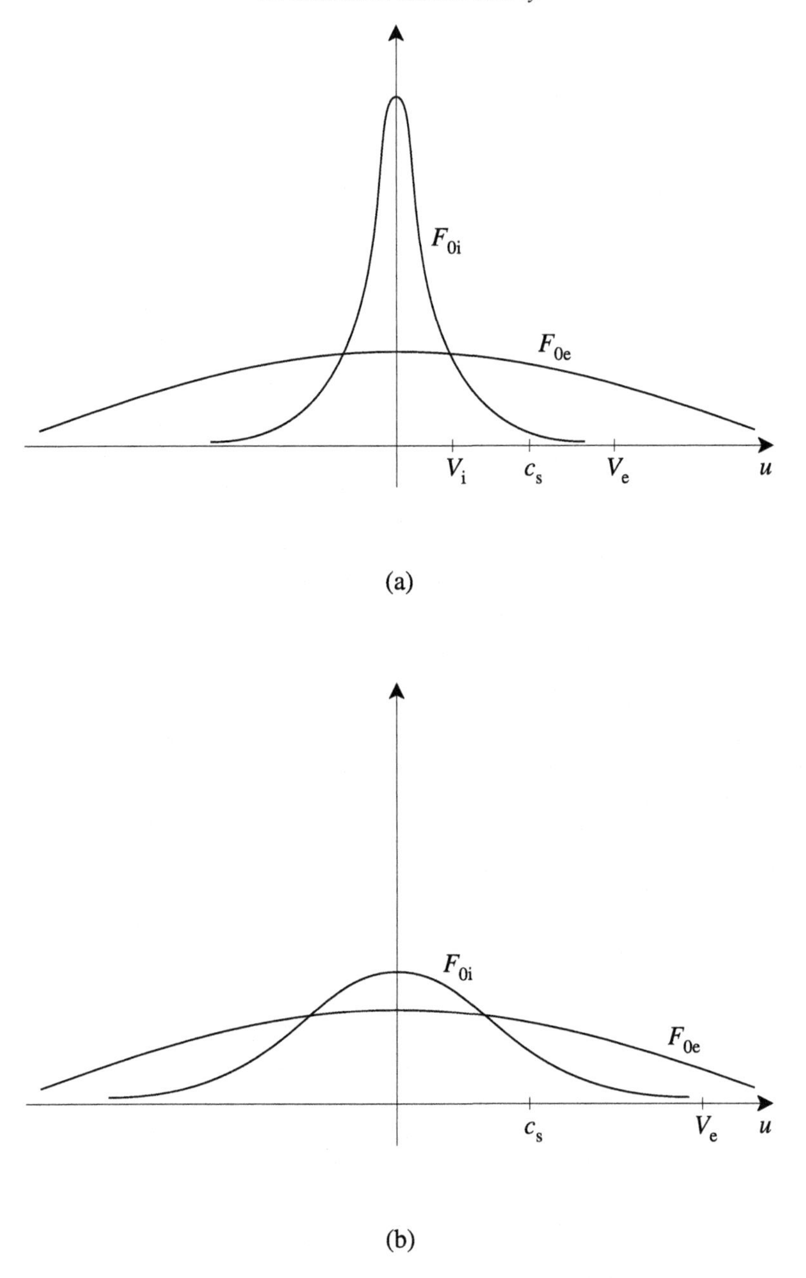

Fig. 7.6. Landau damping of ion acoustic waves for (a) $T_i \ll T_e$ and (b) $T_i \sim T_e$.

This function satisfies the differential equation

$$Z'(\zeta) = -2\,[1 + \zeta Z(\zeta)] \tag{7.40}$$

and has an asymptotic series representation

$$Z(\zeta) = -\frac{1}{\zeta}\left(1 + \frac{1}{2\zeta^2} + \frac{3}{4\zeta^4} + \cdots\right) + is\sqrt{\pi}e^{-\zeta^2} \tag{7.41}$$

with

$$s = \begin{cases} 0 & \Im\zeta > 0 \\ 1 & \Im\zeta = 0 \\ 2 & \Im\zeta < 0 \end{cases}$$

and a power series representation

$$Z(\zeta) = is\sqrt{\pi}e^{-\zeta^2} - 2\zeta\left(1 - \frac{2}{3}\zeta^2 + \frac{4}{15}\zeta^4 - \cdots\right) \tag{7.42}$$

The dispersion relation may be represented in terms of the plasma dispersion function as

$$2k^2\lambda_D^2 = Z'(\omega/\sqrt{2}kV_e) + \left(\frac{Z_a T_e}{T_i}\right) Z'(\omega/\sqrt{2}kV_i) \tag{7.43}$$

where Z_a is the ion atomic number. Assuming $V_i \ll \omega/k \ll V_e$ it follows that $\zeta_e \equiv (\omega/\sqrt{2}\,kV_e) \ll 1$ while $\zeta_i \equiv (\omega/\sqrt{2}\,kV_i) \gg 1$, which allows us to use the power series representation (7.42) for $Z(\omega/\sqrt{2}kV_e)$ and the asymptotic series representation (7.41) for $Z(\omega/\sqrt{2}kV_i)$ to write an approximate dispersion relation

$$2k^2\lambda_D^2 + [2 + 2i\sqrt{\pi}\zeta_e] - Z_a\frac{T_e}{T_i}\left[\frac{1}{\zeta_i^2} + \frac{3}{2\zeta_i^4} - 2i\sqrt{\pi}\zeta_i e^{-\zeta_i^2}\right] = 0 \tag{7.44}$$

The real part of (7.44) reproduces the fluid dispersion relation (6.98), i.e.

$$\frac{\omega_r^2}{k^2} = \frac{Z_a k_B T_e/m_i}{(1 + k^2\lambda_D^2)} + \frac{3k_B T_i}{m_i} \tag{7.45}$$

The imaginary part of (7.44) determines the Landau damping of the ion acoustic mode which is approximately

$$\frac{\gamma}{\omega_{pi}} \simeq -\left(\frac{\pi}{8}\right)^{1/2}\frac{\omega_r}{\omega_{pi}}\frac{1}{(1 + k^2\lambda_D^2)^{3/2}}\left[\left(\frac{m_e}{m_i}\right)^{1/2} + \left(\frac{T_e}{T_i}\right)^{3/2}\exp\left(-\frac{T_e/2T_i}{1 + k^2\lambda_D^2} - \frac{3}{2}\right)\right] \tag{7.46}$$

The terms in square brackets denote electron and ion Landau damping, respectively. Expressed in this form we see that for $T_e/T_i \gg 1$, ion Landau damping can be neglected compared with the electron contribution. However, in practice this condition is rarely satisfied sufficiently strongly so that both contributions are needed. Moreover, though (7.46) is useful in that it shows the parametric dependence of both electron and ion contributions, to get an accurate picture of the damping it is necessary to solve complex dispersion relations numerically. Results of a numerical solution of (7.43) are shown in Fig. 7.7.

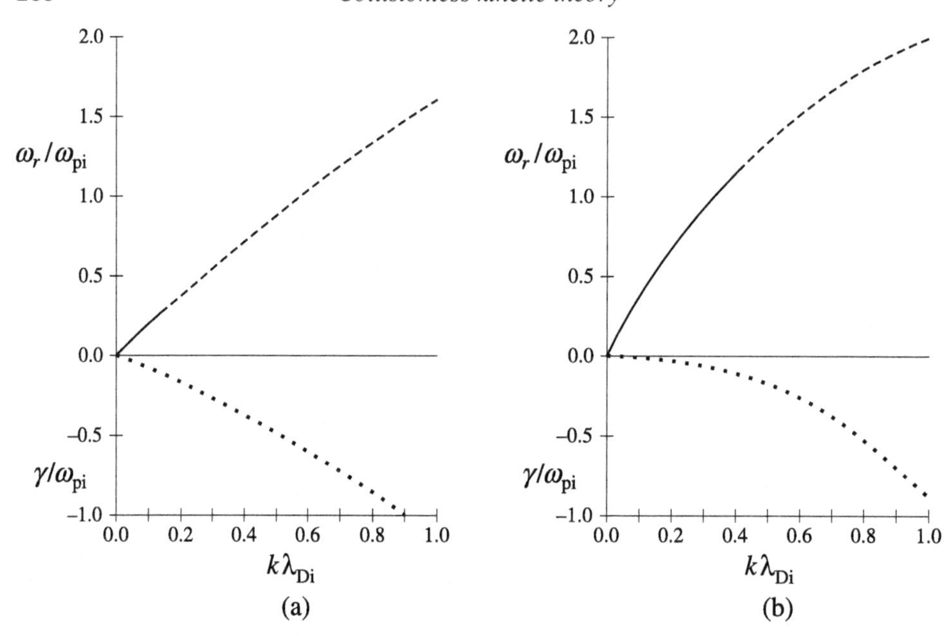

Fig. 7.7. The real part of the frequency and the damping rate (dots) for an ion acoustic wave as functions of wavenumber for a hydrogen plasma with (a) $T_{\mathrm{e}}/T_{\mathrm{i}} = 1$, (b) $T_{\mathrm{e}}/T_{\mathrm{i}} = 10$. Dashed lines indicate that the wave is heavily Landau damped.

7.4 Micro-instabilities

Fluid instabilities are macroscopic in the sense that their growth depends on certain fluid parameters and if growth occurs it involves all of the plasma within some region in which the relevant parameters have the appropriate values. Micro-instabilities, on the other hand, are driven by the interaction of a wave with only a relatively small fraction of the particle population, namely those that are in resonance with the wave. The instability is 'localized' in velocity (**v**) space rather than in coordinate (**r**) space. Even within a fluid element not all of the particles are directly involved in the instability and so there is no bulk motion of the plasma such as one sees with fluid instabilities. Nevertheless, micro-instabilities can have significant effects on the properties of a plasma. For example, the enhanced fluctuation levels of naturally occurring, or externally excited, waves may alter the transport properties of the plasma giving rise to anomalous or turbulent (wave–particle) transport rather than classical (collisional) transport.

The simplest example of a micro-instability is the so-called 'bump-on-tail' instability (BTI). Instead of the single-humped Maxwellian $F_0(u)$ that we considered in the last section we suppose that a few of the electrons have been removed from the main body of the plasma and re-inserted as a small flux of hot particles out in the tail of the distribution as shown in Fig. 7.8. If we carry out the same analysis

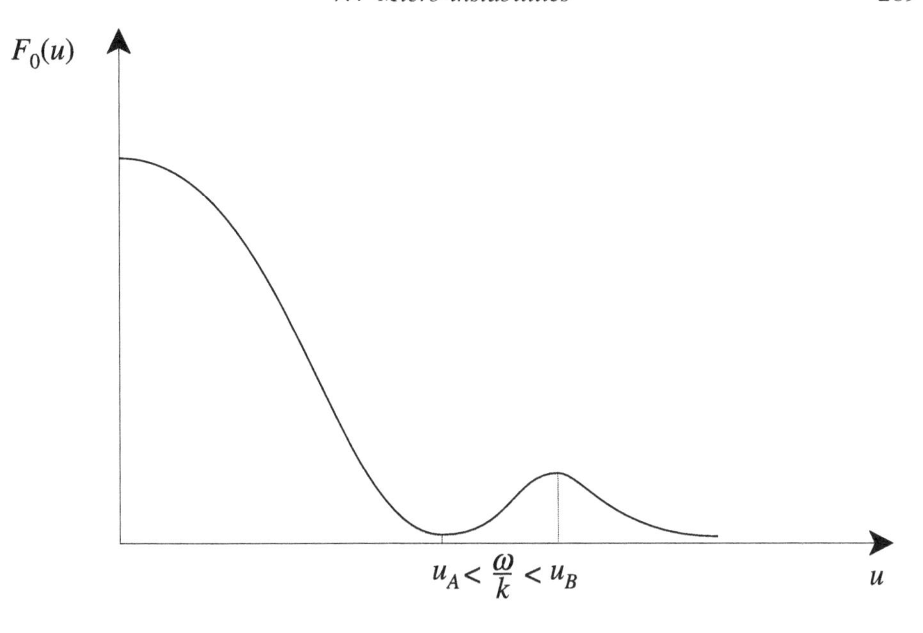

Fig. 7.8. Bump-on-tail distribution function.

that led to Landau damping we would expect that $\Im p$ would still be given by (7.34) since the small bump would have no significant effect on the integral in (7.32) but, of course, $\Re p = \gamma$ would change sign for waves with phase velocities lying between u_A and u_B because $\mathrm{d}F_0/\mathrm{d}u > 0$ between these limits.

It might be thought that whenever the distribution function $F_0(u)$ has a range of values of u for which $\mathrm{d}F_0/\mathrm{d}u > 0$ some waves will grow and the equilibrium will be unstable. However, that would overlook the simplifying assumption made to obtain the results (7.34) and (7.37), namely, that $k \to 0$, in which case $|\gamma| \ll |\Im p|$. In general, the equilibrium is unstable if there is a solution of (7.27) for which $\Re p > 0$ (see Fig. 7.1). Now there is a powerful theorem in complex analysis which provides a method of determining this. Taking k as real and positive and re-writing (7.27) as

$$D(V) = 1 - \frac{\omega_{\mathrm{pe}}^2}{k^2} \int_{-\infty}^{+\infty} \frac{F_0'(u)\mathrm{d}u}{u - V} = 0 \tag{7.47}$$

where $V = (\omega_r + i\gamma)/k$, we wish to know whether there are any values of V which satisfy this equation and for which $\gamma/k > 0$. The *argument principle* tells us that if we draw any closed contour C in the complex V-plane and trace its image C_D in the complex D-plane then the number of zeros minus the number of poles of $D(V)$ inside C is equal to the number of times C_D encircles the origin in the D-plane. The contour C that we wish to investigate is shown in Fig. 7.9. As $R \to \infty$ this

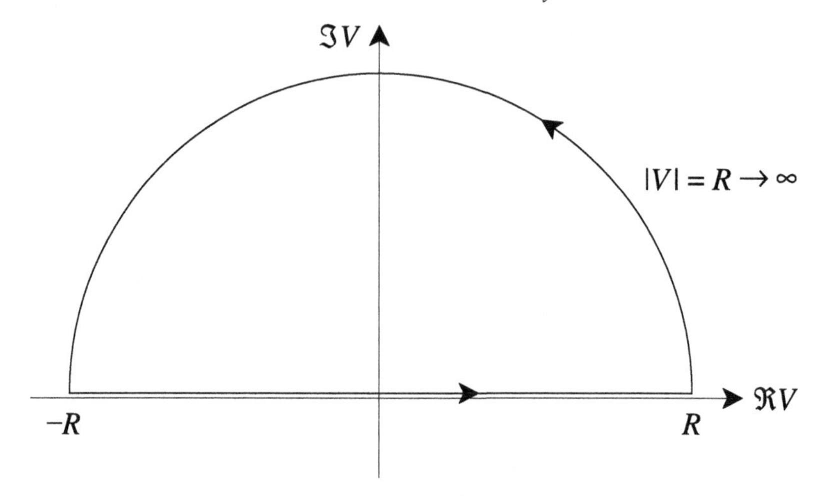

Fig. 7.9. Semi-circular contour of integration in upper half-plane.

encompasses the whole of the upper half-plane, in which $\Im V > 0$, i.e. $\gamma > 0$. Now, by its definition, $D(V)$ is analytic in the upper half-plane so it has no poles and the argument principle simply tells us how many zeros of $D(V)$ there are.

The image contours C_D in the D-plane are called Nyquist diagrams, examples of which are shown in Fig. 7.10. Figure 7.10(a) shows schematically what the C_D contour might look like for the stable case of $F_0(u)$ given by the Maxwell distribution. In the limit of $R \to \infty$ the semi-circle, $|V| \to \infty$, maps on to the point $D = 1$ as we can easily see from (7.47). For the rest of the contour, i.e. $-\infty < \Re V < +\infty$, (7.35) may be used to evaluate the integral in (7.47) from which we see that there is only one other point where $\Im D = 0$ and that is at $V = 0$ where $\mathrm{d}F_0/\mathrm{d}u = 0$. At this point the second term in (7.47) is positive and so $D > 1$. Also, on the contour, $\Im D$ has the same sign as $\Re V$ so that the curve is traced in the manner shown. This contour does not encircle the origin confirming that there are no zeros of (7.47) in the upper half V-plane.

Figures 7.10(b),(c) show possible contours C_D for a double-humped distribution where there are now three values of V for which $\Im D = 0$. For the bump-on-tail distribution (see Fig. 7.8) these are $V = 0, u_A, u_B$. Figure 7.10(b) corresponds to a stable double-humped distribution because C_D still does not encircle the origin. On the other hand, in Fig. 7.10(c) the origin is encircled once so there is an unstable root of the dispersion relation (7.47).

Penrose (1960) showed that there is a simple criterion which can be applied to determine stability without the need to construct the Nyquist diagram. First we note that, if there is to be an unstable root, C_D must cross the $\Re D$-axis in the left-half

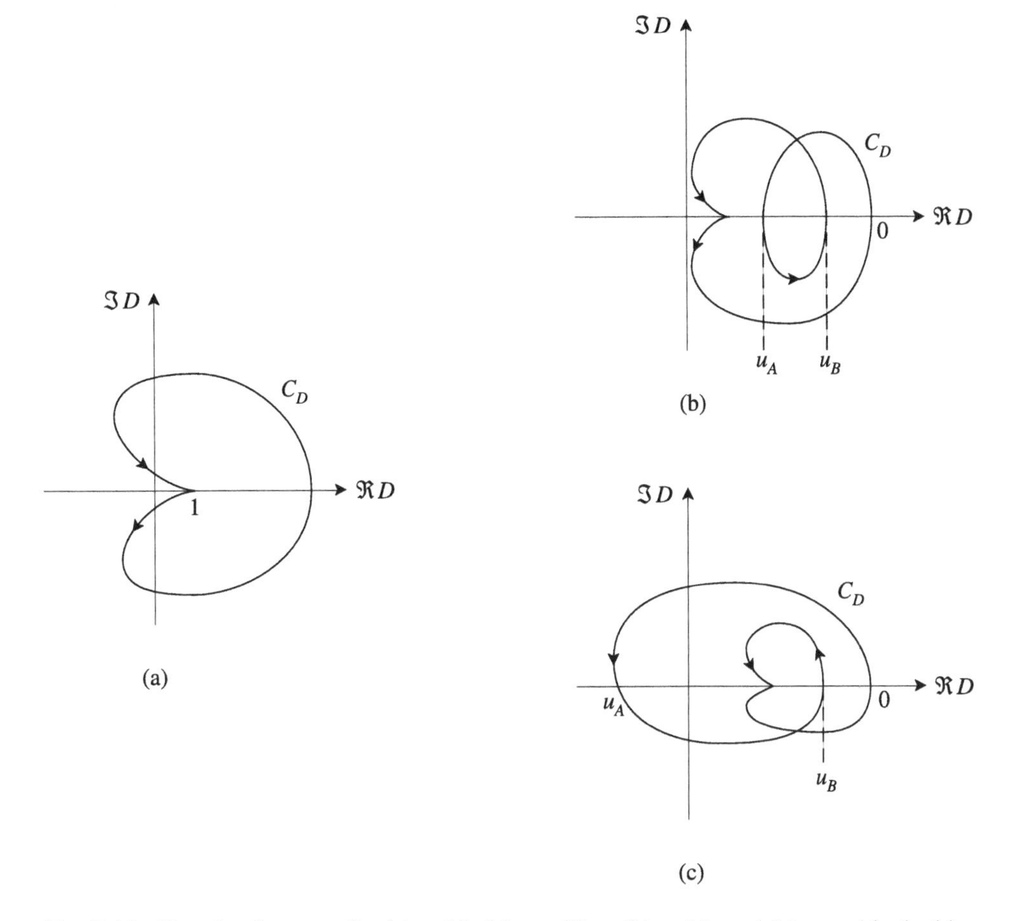

Fig. 7.10. Nyquist diagrams for (a) stable Maxwellian, (b) stable and (c) unstable double-humped distributions.

plane, that is $D(u_A) < 0$. Using (7.35) and noting that $\mathrm{d}F_0(u_A)/\mathrm{d}u = 0$†

$$D(u_A) = 1 - \frac{\omega_{\mathrm{pe}}^2}{k^2} \int_{-\infty}^{+\infty} \frac{F_0'(u) - F_0'(u_A)}{u - u_A} \,\mathrm{d}u < 0$$

On integration by parts this becomes

$$1 - \frac{\omega_{\mathrm{pe}}^2}{k^2} \int_{-\infty}^{+\infty} \frac{F_0(u) - F_0(u_A)}{(u - u_A)^2} \,\mathrm{d}u < 0 \tag{7.48}$$

† Note that since both numerator and denominator vanish at $u = u_A$ we do not need to take the principal part of the integral.

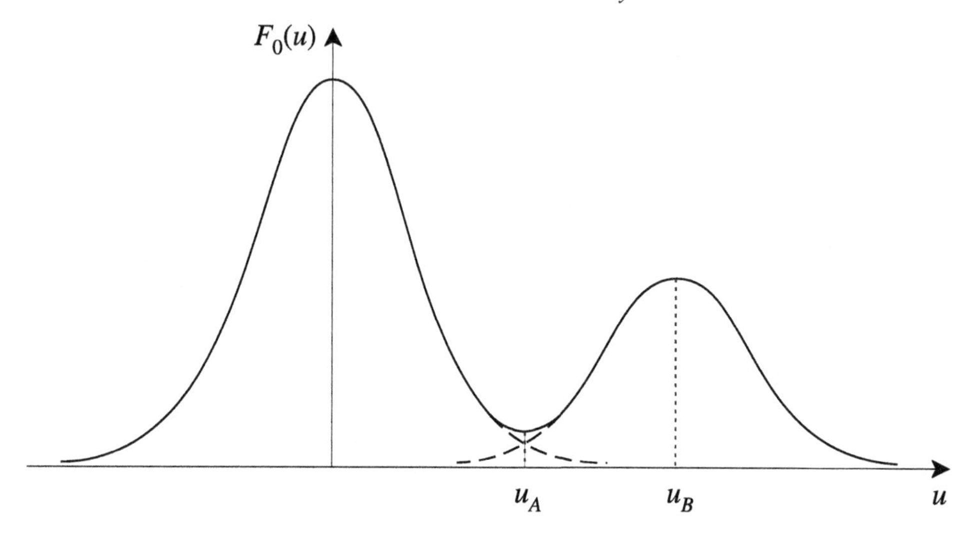

Fig. 7.11. Beam–plasma distribution function.

But since k can vary from 0 to ∞ there will be some value of k for which (7.48) is satisfied provided only that

$$\int_{-\infty}^{+\infty} \frac{F_0(u) - F_0(u_A)}{(u - u_A)^2} \, du > 0 \tag{7.49}$$

This is a necessary and sufficient condition for instability and is known as the *Penrose criterion*. Near $u = u_A$ the integrand in (7.49) is approximately equal to $F''(u_A)$, which is positive since $F(u_A)$ is a minimum. However, the existence of a minimum is not, of itself, a sufficient condition for instability since all particles interact with the wave, giving or taking energy depending upon their relative phase. In effect, the Penrose criterion says that the minimum must be deep enough that the net effect of the givers outweighs that of the takers.

If u_B is sufficiently large then the hot electron beam becomes completely separated in velocity space from the main distribution function and rather than a 'bump-on-tail' instability we have a beam–plasma instability. In this case, illustrated in Fig. 7.11, $F_0(u_A) \to 0$ and so (7.49) is certainly satisfied ($F_0(u) > 0$ for all u) but since *all* the beam electrons now contribute to the instability, it is no longer a resonant, micro-instability but a macroscopic instability. It ought, therefore, to be describable by the fluid equations. The link between the two is explored in the following section.

7.4.1 Kinetic beam–plasma and bump-on-tail instabilities

In Section 6.5 we found the characteristics of the two-stream instability (TSI) and the beam–plasma instability (BPI) from a cold fluid model. In both cases instability was caused by a feedback that produced charge bunching, with one system reacting back on the other. However, there are other cases in which only nearly resonant particles in the distribution are involved and which consequently cannot be described by a fluid model, since one is not then able to identify separate systems. Resonant instabilities have to be described using kinetic theory. Instabilities often exist in both reactive and resonant forms. For example, the bump-on-tail instability (BTI) is the kinetic counterpart of the reactive BPI. To explore the relation between the two we look again at BPI characteristics, this time from a kinetic standpoint. With the beam electrons described by the distribution function

$$f(\mathbf{v}) = \frac{n_{\mathrm{b}}}{(2\pi)^{3/2}V_{\mathrm{b}}^3}\exp\left[-\frac{(\mathbf{v}-\mathbf{v}_{\mathrm{b}})^2}{2V_{\mathrm{b}}^2}\right]$$

where n_{b}, $\mathbf{v}_{\mathrm{b}}$, V_{b} denote density, streaming and thermal velocities of the beam particles, it is straightforward to recover (6.120) in the cold plasma limit. For the reactive instability, $\gamma_{\mathrm{max}} \sim (n_{\mathrm{b}}/n_0)^{1/3}\omega_{\mathrm{p}}$. The physical effect of a finite beam electron temperature is to produce a spread of the beam electron velocities about v_{b} with consequent reduction of the BPI growth rate and ultimate suppression of the instability. Thermal effects may be ignored provided $|\omega - \mathbf{k}\cdot\mathbf{v}_{\mathrm{b}}| \gg \sqrt{2}kV_{\mathrm{b}}$. With $k \simeq \omega_{\mathrm{p}}/v_{\mathrm{b}}$ this reduces to

$$\left(\frac{n_{\mathrm{b}}}{n_0}\right)^{1/3} \gg \frac{V_{\mathrm{b}}}{v_{\mathrm{b}}} \tag{7.50}$$

For the BTI with $v_{\mathrm{b}} \gg V_{\mathrm{b}}$ we see by comparison with Fig. 7.8 that growth occurs in the region over which the slope of the distribution function is positive, when ω/k is within the range $v_{\mathrm{b}} - V_{\mathrm{b}} < \omega/k < v_{\mathrm{b}}$. The maximum growth rate is

$$|\gamma_{\mathrm{max}}| \simeq \left(\frac{\pi}{2e}\right)^{1/2}\frac{n_{\mathrm{b}}}{n_0}\left(\frac{v_{\mathrm{b}}}{V_{\mathrm{b}}}\right)^2\omega_{\mathrm{p}} \tag{7.51}$$

in which $e = \exp(1)$. The bandwidth $\Delta\omega$ across which growth is optimal is such that $\Delta\omega \sim kV_{\mathrm{b}}$ and hence

$$\frac{\Delta\omega}{\omega_{\mathrm{p}}} \simeq \frac{V_{\mathrm{b}}}{v_{\mathrm{b}}} \tag{7.52}$$

Thus the growth rate is less than the bandwidth for optimal BTI growth provided

$$\left(\frac{n_{\mathrm{b}}}{n_0}\right)^{1/3} \leq \frac{V_{\mathrm{b}}}{v_{\mathrm{b}}} \tag{7.53}$$

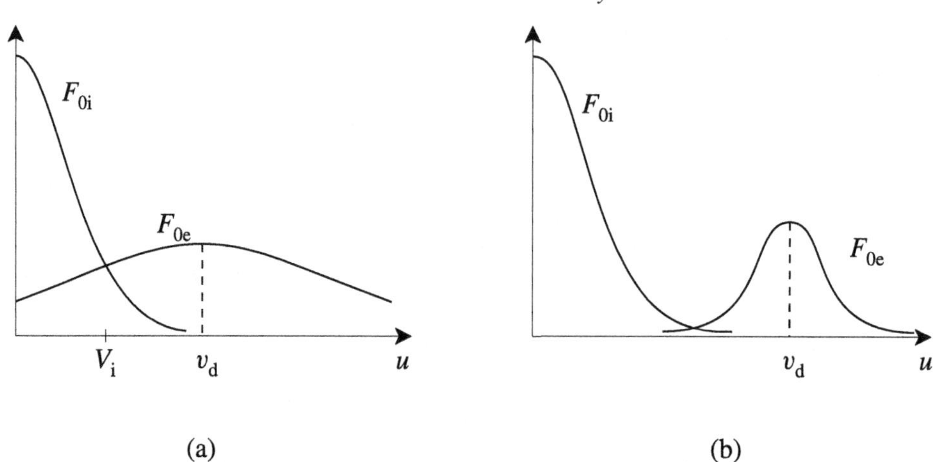

Fig. 7.12. Ion and electron distribution functions in a current-carrying plasma subject to (a) ion acoustic instability and (b) Buneman instability.

The conditions (7.50), (7.53) serve to distinguish the reactive BPI from its kinetic counterpart.

7.4.2 Ion acoustic instability in a current-carrying plasma

By allowing electrons to drift relative to ions so that the plasma is now current-carrying, we introduce a source of free energy which will counter Landau damping and, if strong enough, may drive the mode unstable. Before analysing this ion acoustic drift instability in terms of the dispersion relation, it is easy to see qualitatively how instability arises from a picture of the distribution functions. Representing the current by a net drift velocity v_d between ions and electrons, for the velocity component parallel to the current the distribution functions are as shown in Fig. 7.12(a). Provided $v_d \gg V_i$ there is a range of phase velocities ($V_i \ll c_s < v_d$) for which ion Landau damping is negligible but Landau growth takes place ($dF_{0e}/du > 0$) due to interaction with the resonant electrons.

Here again there is a smooth transition from the resonant micro-instability to the macroscopic, *Buneman instability*. Either by increasing v_d or decreasing T_e we can separate the ion and electron distributions in velocity space as shown in Fig. 7.12(b), thus strengthening the instability and converting it from resonant to reactive.

Returning to the dispersion relation (7.43) it is straightforward to modify it to allow for electrons drifting relative to ions with a drift velocity v_d. We prescribe an ordering $v_d \ll V_e$ in addition to the requirement that $T_e/T_i \gg 1$ so that ion Landau

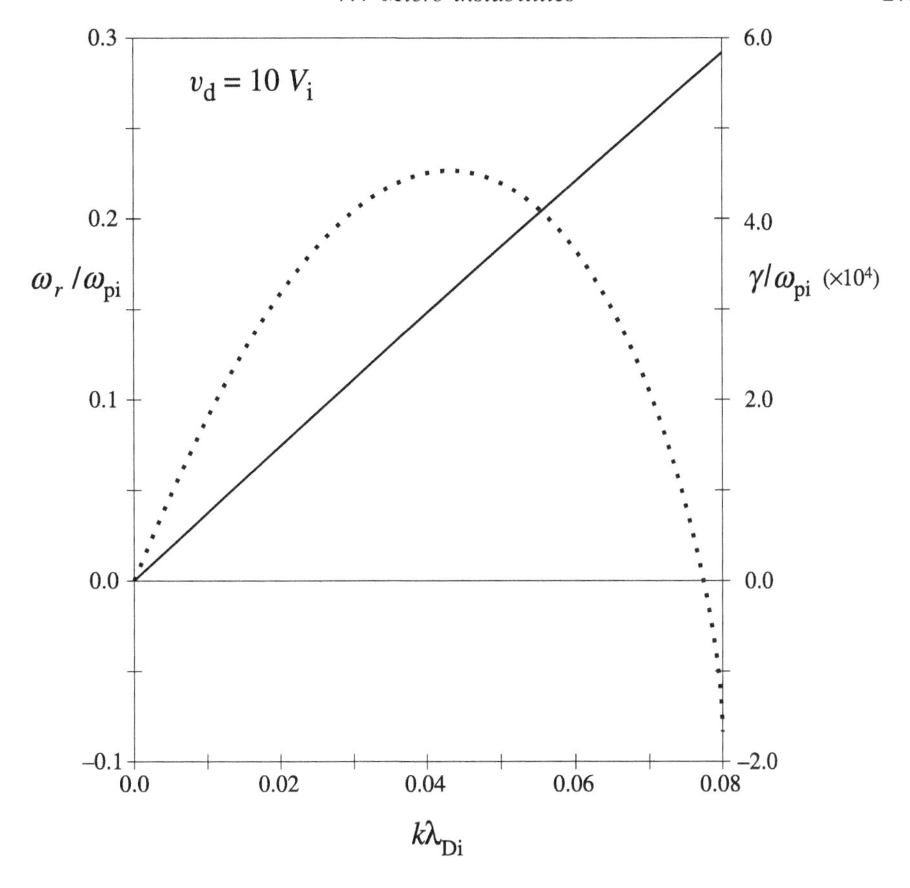

Fig. 7.13. The real part of the frequency and the growth rate (dots) for the current-driven ion acoustic instability as functions of the wavenumber for a hydrogen plasma with $T_e/T_i = 10$.

damping can be ignored. It follows that

$$\frac{\gamma}{\omega_{pi}} \simeq \left(\frac{\pi}{8}\right)^{1/2} \left(\frac{\omega_r}{k}\right)^3 \left(\frac{m_i}{T_e}\right) \frac{1}{\omega_{pi}} \frac{(kv_d - \omega_r)}{V_e} \exp\left[-\frac{(\omega_r/k - v_d)^2}{2V_e^2}\right] \tag{7.54}$$

We see at once that electron drift reduces the electron Landau damping of the mode and for $v_d > c_s$ instability will develop. Instability threshold is determined by (see Exercise 7.12)

$$v_d \simeq \frac{\omega_r}{k}\left[1 + \left(\frac{T_e}{T_i}\right)^{3/2} \left(\frac{m_i}{m_e}\right)^{1/2} \exp\left(-\frac{T_e}{2T_i} - \frac{3}{2}\right)\right]$$

which is a sensitive function of the species temperature ratio. Again in practice it is essential to solve the exact dispersion relation numerically (see Fig. 7.13).

7.5 Amplifying waves

In the light of the Landau analysis developed in this chapter we return to the question of amplifying waves and convective and absolute instabilities, first discussed in Section 6.6. In our derivation of Landau damping (and of Landau growth when a source of free energy is available from a suitable distribution of electrons or ions) the dispersion characteristics were described in terms of their evolution in time, i.e. for real values of k, solutions were found for $\omega \equiv \omega(k)$ with ω complex. The Landau analysis provided the response in time of the plasma to an initial perturbation $f_1(x, v, t = 0)$. Equation (7.25) for the electric field $\mathbf{E}(\mathbf{k}, \omega)$ produced by the perturbation has the form

$$E(\mathbf{k}, \omega) = \frac{g(\mathbf{k}, \omega)}{D(\mathbf{k}, \omega)} \tag{7.55}$$

where $g(\mathbf{k}, \omega)$ is determined by the initial perturbation and $D(\mathbf{k}, \omega)$, the plasma dielectric function, is a characteristic of the unperturbed plasma. In Section 7.3 we supposed that the only singularities of $E(\mathbf{k}, \omega)$ in the complex ω-plane were poles where $D(\mathbf{k}, \omega) = 0$. These complex roots determine the time-asymptotic behaviour of a perturbation with prescribed (real) k. We now want to turn to other considerations.

In discussing weakly coupled waves in Chapter 6 we found that conditions under which amplifying waves were present corresponded to conditions for convective instability. To determine whether or not a plasma is convectively unstable one has to examine the evolution of some initial perturbation in both time and space. The consideration of spatially amplifying waves, on the other hand, is akin to the Landau analysis of Section 7.3. Here we need to determine the spatial response to an initial perturbation at some point in the plasma, namely $f_1(x = 0, v, t)$, rather than the response in time. This means we now have to allow the wavenumber k to be complex. We can see at once that this presents a contrast to the Landau case since clearly the sign of $\Im k$ cannot of itself provide a criterion for distinguishing amplification on the one hand from attenuation on the other, since a change in the direction of propagation results in k changing sign.

To determine whether amplification takes place in a plasma we examine the spatial development of a perturbation at $x = 0$ oscillating in time,

$$g(x, t) = \begin{cases} 0 & t < 0 \\ g_0 \delta(x) e^{-i\omega_0 t} & t > 0 \end{cases} \tag{7.56}$$

where g_0 is a constant. The response of the plasma to this perturbation will be determined by the response function $R(x, t)$ where

$$R(x,t) = \frac{g_0}{2\pi} \int_{-\infty+i\sigma}^{\infty+i\sigma} \frac{1}{2\pi \mathrm{i}} \int_{-\infty}^{\infty} \frac{e^{i(kx-\omega t)}}{(\omega-\omega_0)D(k,\omega)} \,\mathrm{d}k\,\mathrm{d}\omega \tag{7.57}$$

We wish to determine the asymptotic ($|x| \to \infty$) behaviour of $R(x, t)$ as ($t \to \infty$). A perturbation tending to zero as $x \to \pm\infty$ is evanescent; one that increases in either direction corresponds to amplification. To decide the asymptotic behaviour we must first determine the asymptotic response in time before letting $|x| \to \infty$ since $R(|x| \to \infty, t) \to 0$. To find the time-asymptotic behaviour we lower the ω-contour. The plasma being at most only C-unstable, there is no singularity from $D(k, \omega)$ in the upper half ω-plane so that the uppermost singularity in the integrand in (7.57) is the pole on the real axis at $\omega = \omega_0$. Thus

$$R(x, t \to \infty) = \frac{g_0}{2\pi i} \int_C \frac{e^{i(kx-\omega_0 t)}}{D(k,\omega_0)} \,\mathrm{d}k$$

In lowering the ω-contour the singularities of the response function will move in the complex k-plane. Should one or other of these singularities cross the $\Re k$-axis the contour has to be displaced to ensure that the deformed contour passes below singularities originating in the upper half-plane and above any that have crossed the $\Re k$-axis from below. Amplifying waves are described by the poles of the response function that cross the $\Re k$-axis as $\Im\omega \to 0$. This gives

$$\begin{aligned} R(x, t \to \infty) &= \sum_{k_+} \frac{H(x)}{(\partial D/\partial k)_{k_+(\omega_0)}} e^{i(k_+(\omega_0)x-\omega_0 t)} \\ &\quad - \sum_{k_-} \frac{H(-x)}{(\partial D/\partial k)_{k_-(\omega_0)}} e^{i(k_-(\omega_0)x-\omega_0 t)} \end{aligned} \tag{7.58}$$

where $H(x)$ is the Heaviside step function. From this it follows that waves with $\Im k_+(\omega_0) < 0$ are spatially growing for $x > 0$ and those with $\Im k_-(\omega_0) > 0$ are spatially growing for $x < 0$.

A more complete discussion of wave amplification may be found in Briggs (1964).

7.6 The Bernstein modes

Although, in Section 7.2, we wrote down the full set of Vlasov–Maxwell equations, so far we have discussed unmagnetized plasmas only. Consequently, our investigations have been restricted to solutions of the Vlasov–Poisson equations. In a classic paper, Bernstein (1958) solved the Vlasov–Maxwell set of equations, by the Landau procedure used in Section 7.3, but including an equilibrium magnetic field, $\mathbf{B}_0$, and allowing for transverse as well as longitudinal waves. He also included the ion dynamics but this extension is, mathematically, fairly trivial compared with the other two; physically it is not trivial in that it introduces further (low frequency) waves. Bernstein's general dispersion relation reproduced the various waves previously discovered using fluid equations but it also included ion and electron modes propagating without growth or damping across the magnetic field lines ($\mathbf{k} \perp \mathbf{B}_0$), which have become known as the *Bernstein modes*.

It is the inclusion of the equilibrium field $\mathbf{B}_0$ which particularly complicates the calculation but we can omit the ion motion and exclude transverse waves without losing the electron Bernstein modes so, for clarity, we shall adopt both of these simplifications. Furthermore, instead of using a Laplace transform it is now more common to use a Fourier transform in time as well as in space so that all perturbations vary as $\exp i(\mathbf{k} \cdot \mathbf{r} - \omega t)$; see Section 7.5. Assuming that the imaginary part of ω is positive so that all perturbations vanish as $t \to -\infty$, it can be shown that analytic continuation of the dispersion relation into the lower half ω-plane is then equivalent to the Landau procedure. Physically, this can be thought of as switching on the perturbation at an infinitesimally slow rate.

The linearized Vlasov equation is now

$$\frac{\partial f_1}{\partial t} + \mathbf{v} \cdot \frac{\partial f_1}{\partial \mathbf{r}} - \frac{e}{m}(\mathbf{v} \times \mathbf{B}_0) \cdot \frac{\partial f_1}{\partial \mathbf{v}} = \frac{e}{m}\mathbf{E} \cdot \frac{\partial f_0}{\partial \mathbf{v}} \tag{7.59}$$

where

$$f(\mathbf{r}, \mathbf{v}, t) = f_0 + f_1(\mathbf{r}, \mathbf{v}, t)$$

and we shall assume the equilibrium distribution function f_0 to be the Maxwellian (7.33). In (7.59) the equilibrium electric field $\mathbf{E}_0$ is taken to be zero as before and the magnetic field perturbation $\mathbf{B}_1$ is ignored since only longitudinal waves are to be examined. Equation (7.59) is then solved by the method of characteristics or, in plasma terms, by integration over unperturbed orbits. The essence of the method is that

$$\frac{\mathrm{d}f_1(\mathbf{r}, \mathbf{v}, t)}{\mathrm{d}t} = \frac{\partial f_1}{\partial t} + \mathbf{v} \cdot \frac{\partial f_1}{\partial \mathbf{r}} + \frac{\mathrm{d}\mathbf{v}}{\mathrm{d}t} \cdot \frac{\partial f_1}{\partial \mathbf{v}}$$

and if we use

$$\frac{\mathrm{d}\mathbf{v}}{\mathrm{d}t} = -\frac{e}{m}(\mathbf{v} \times \mathbf{B}_0) \tag{7.60}$$

which is the equation of motion of the electron in the equilibrium (or unperturbed) field, then we may write (7.59) as

$$\frac{\mathrm{d}f_1(\mathbf{r}, \mathbf{v}, t)}{\mathrm{d}t} = \frac{e}{m}\mathbf{E} \cdot \frac{\partial f_0}{\partial \mathbf{v}} = g(\mathbf{r}, \mathbf{v}, t) \tag{7.61}$$

say. In (7.61) $\mathbf{r} = \mathbf{r}(t)$ and $\mathbf{v} = \mathbf{v}(t)$ are the solutions of (7.60) given by (see

Section 2.2)

$$\left.\begin{aligned}\mathbf{v}(t) &= \{v_\perp \cos[\Omega_e(t-t_0)+\theta],\, v_\perp \sin[\Omega_e(t-t_0)+\theta],\, v_\parallel\} \\ \mathbf{r}(t)-\mathbf{r}(t_0) &= \{(v_\perp/\Omega_e)(\sin[\Omega_e(t-t_0)+\theta]-\sin\theta), \\ &\quad -(v_\perp/\Omega_e)(\cos[\Omega_e(t-t_0)+\theta]-\cos\theta),\, v_\parallel(t-t_0)\}\end{aligned}\right\} \tag{7.62}$$

Hence, the solution of (7.61) is

$$\begin{aligned} f_1(\mathbf{r}(t),\mathbf{v}(t),t) &= \int_{-\infty}^{t} g(\mathbf{r}(t'),\mathbf{v}(t'),t')\mathrm{d}t' \\ &= \frac{e}{m}\int_{-\infty}^{t} \mathbf{E}(\mathbf{r}(t'),t')\cdot\frac{\partial f_0(v^2(t'))}{\partial \mathbf{v}(t')}\mathrm{d}t' \\ &= \frac{e}{k_B T}\int_{-\infty}^{t} \frac{\partial\phi(\mathbf{r}(t'),t')}{\partial \mathbf{r}(t')}\cdot\mathbf{v}(t') f_0(v^2(t'))\mathrm{d}t' \\ &= \frac{e f_0(v^2(t))}{k_B T}\int_{-\infty}^{t}\left[\frac{\mathrm{d}\phi(\mathbf{r}(t'),t')}{\mathrm{d}t'}-\frac{\partial\phi(\mathbf{r}(t'),t')}{\partial t'}\right]\mathrm{d}t' \end{aligned}$$

where $\mathbf{E}$ has been replaced by $-\boldsymbol{\nabla}\phi$ and in the last step we have used, from (7.62),

$$v^2(t) = v_\perp^2 + v_\parallel^2 = v^2(t')$$

and

$$\frac{\mathrm{d}\phi(\mathbf{r},t)}{\mathrm{d}t} = \frac{\partial\phi}{\partial t} + \mathbf{v}(t)\cdot\frac{\partial\phi}{\partial \mathbf{r}}$$

Now, from Poisson's equation

$$\begin{aligned} \nabla^2\phi &= \frac{e}{\varepsilon_0}\int \mathrm{d}\mathbf{v}\, f_1(\mathbf{r},\mathbf{v},t) \\ &= \frac{n_0 e^2\phi(\mathbf{r},t)}{\varepsilon_0 k_B T} - \frac{e^2}{\varepsilon_0 k_B T}\int \mathrm{d}\mathbf{v}\, f_0(v^2)\int_{-\infty}^{t}\frac{\partial\phi(\mathbf{r}(t'),t')}{\partial t'}\mathrm{d}t' \end{aligned}$$

and with $\phi(\mathbf{r},t) \propto \exp(i(\mathbf{k}\cdot\mathbf{r}-\omega t))$ this yields the dispersion relation

$$\begin{aligned} -k^2 = \frac{1}{\lambda_D^2}\Bigg[1 + i\omega\left(\frac{m}{2\pi k_B T}\right)^{3/2} \\ \int \mathrm{d}\mathbf{v}\, e^{-mv^2/2k_B T}\int_{-\infty}^{t}\mathrm{d}t' \exp\{i[\mathbf{k}\cdot(\mathbf{r}(t')-\mathbf{r}(t))-\omega(t'-t)]\}\Bigg] \end{aligned}$$

Since we want to consider waves propagating perpendicular to $\mathbf{B}_0$ it is convenient to choose $\mathbf{k} = (k, 0, 0)$ and the dispersion relation then reduces to

$$1 + k^2\lambda_D^2 + i\omega\left(\frac{m}{2\pi k_B T}\right)$$
$$\int_0^\infty v_\perp \, dv_\perp e^{-mv_\perp^2/2k_B T} \int_0^{2\pi} d\theta \int_0^\infty d\tau e^{-[ikv_\perp/\Omega_e][\sin(\Omega_e\tau+\theta)-\sin\theta]+i\omega\tau} = 0$$

The next step is to use the relationship

$$e^{ix\sin y} = \sum_{n=-\infty}^{\infty} J_n(x)e^{iny}$$

where the J_n are the Bessel functions of the first kind. Then after the θ and τ integrations the result is

$$1 + k^2\lambda_D^2 - \frac{\omega m}{k_B T}\int_0^\infty v_\perp \, dv_\perp e^{-mv_\perp^2/2k_B T} \sum_{n=-\infty}^{\infty} \frac{J_n^2(kv_\perp/\Omega_e)}{\omega - n\Omega_e} = 0 \qquad (7.63)$$

Finally, using another Bessel function relationship

$$\int_0^\infty e^{-\gamma x^2} J_n(\alpha x) J_n(\beta x) x \, dx = \frac{1}{2\gamma} e^{-(\alpha^2+\beta^2)/4\gamma} I_n\left(\frac{\alpha\beta}{2\gamma}\right)$$

where I_n is the modified Bessel function, this becomes

$$1 + k^2\lambda_D^2 = \omega e^{-\lambda} \sum_{n=-\infty}^{\infty} \frac{I_n(\lambda)}{\omega - n\Omega_e} \qquad (7.64)$$

where $\lambda = (k^2 k_B T)/m\Omega_e^2$. Making use of a Bessel function sum rule

$$\sum_{n=-\infty}^{\infty} I_n(\lambda)e^{-\lambda} = 1$$

allows (7.64) to be written as

$$1 - \frac{2\Omega_e^2 e^{-\lambda}}{k^2\lambda_D^2} \sum_{n=1}^{\infty} \frac{n^2 I_n(\lambda)}{(\omega^2 - n^2\Omega_e^2)} = 0 \qquad (7.65)$$

The dispersion relation (7.65) was found by Bernstein who showed that it has real solutions so that the waves neither grow nor damp. By appealing to the small and large λ approximations for $I_n(\lambda)$, namely

$$\begin{aligned} I_n(\lambda) &\simeq (\lambda/2)^n/n! & \lambda \to 0 \\ I_n(\lambda) &\simeq e^\lambda/\sqrt{2\pi\lambda} & \lambda \to \infty \end{aligned}$$

we can see at once that resonances ($k \to 0$) occur at harmonic frequencies with the exception of the $n = 1$ term for which the cut-off is at the upper hybrid frequency

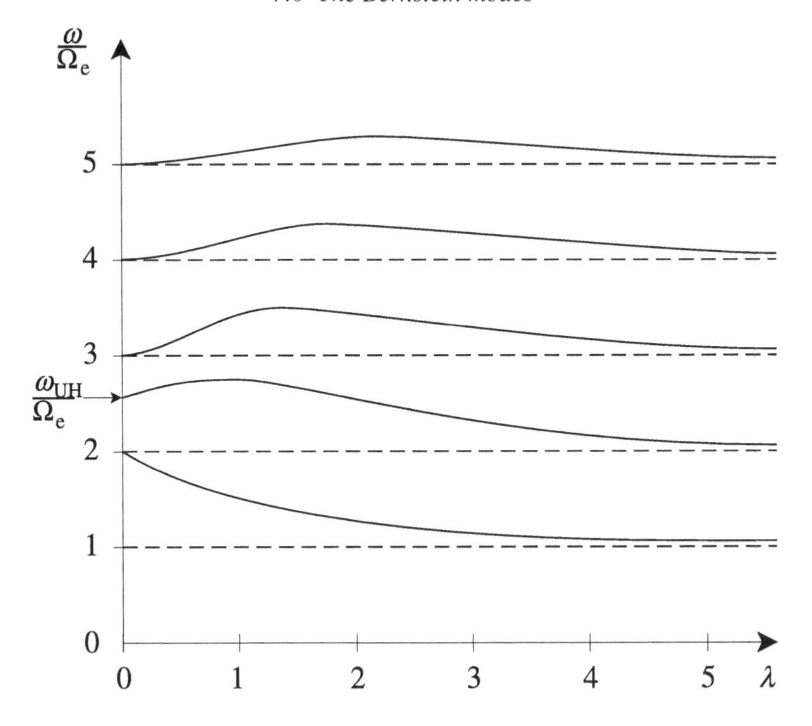

Fig. 7.14. Dispersion curves for electron Bernstein modes.

$\omega_{\rm UH} = (\omega_{\rm pe}^2 + \Omega_{\rm e}^2)^{1/2}$, corresponding to the cold plasma result in Section 6.3.3. Numerical solution of the full dispersion relation (7.65) for $2|\Omega_{\rm e}| < \omega_{\rm UH} < 3|\Omega_{\rm e}|$ produces the dispersion curves in Fig. 7.14. Finite Larmor radius effects included in the kinetic theory model lead to complex dispersion characteristics. In particular the nature of the dispersion curves changes on passing through the upper hybrid frequency. For $\lambda \gg 1$ the dispersion curves approach the various harmonic frequencies from above. For $n^2 < \omega_{\rm UH}^2/\Omega_{\rm e}^2$, the dispersion curves start from $\omega^2 = n^2\Omega_{\rm e}^2$ at $\lambda = 0$ and tend to $\omega^2 = (n-1)^2\Omega_{\rm e}^2$ as $\lambda \to \infty$, whereas for $n^2 > \omega_{\rm UH}^2/\Omega_{\rm e}^2$, the characteristic through $\omega^2 = n^2\Omega_{\rm e}^2$ at $\lambda = 0$ first increases with λ, passing through some maximum, before again tending to $\omega^2 = n^2\Omega_{\rm e}^2$ as $\lambda \to \infty$.

The reason Bernstein modes ($\omega \approx n\Omega_{\rm e}$) are not found by fluid theory is that the propagation of these waves depends on the cyclotron motion of the electrons about the field lines. Fluid theory, which averages over the Larmor orbits, therefore loses these modes. The Larmor orbits also hold the key to understanding why Bernstein modes are not Landau damped. Since all particles must travel in circular orbits about the field lines, they are unable to stay in phase with the wave propagating across the field lines. On the other hand, lifting the restriction of perpendicular propagation, allowing a component, $k_\parallel$, of the propagation vector, $\mathbf{k}$, parallel to

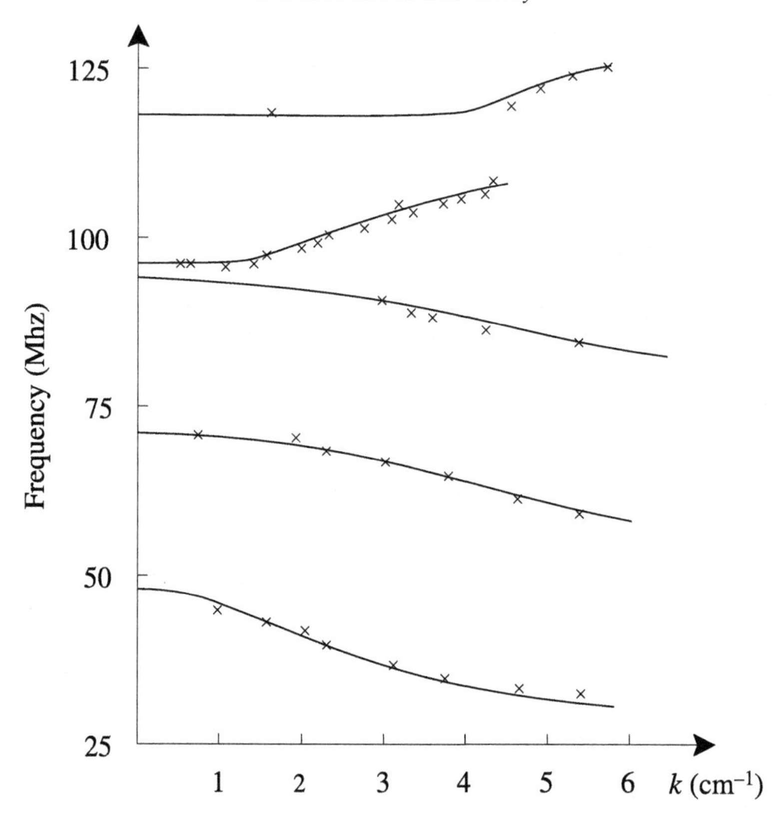

Fig. 7.15. Dispersion curves for electron Bernstein modes compared with measured characteristics (after Armstrong *et al.* (1981)); $\omega_{\mathrm{pe}}/2\pi = 93.7\,\mathrm{MHz}$, $\Omega_{\mathrm{e}}/2\pi = 23.7\,\mathrm{MHz}$, $\omega_{\mathrm{UH}}/2\pi = 96.7\,\mathrm{MHz}$.

the magnetic field, permits Landau damping. This is because particle motion along the field lines is unrestricted so that resonant interaction in this direction can take place. Thus, propagation of Bernstein modes is confined within a fan about the perpendicular direction. A rough measure of the half-angle of this fan is given by the condition that, for negligible Landau damping, the phase speed along the field lines (where the damping takes place) must be much greater than the electron thermal speed,

$$\frac{\omega}{k_{\parallel}} \gg \sqrt{\frac{k_{\mathrm{B}}T}{m}}$$

or, if θ is the angle between $\mathbf{k}$ and $\mathbf{B}_0$,

$$\cos\theta \ll \frac{\omega}{k}\left(\frac{m}{k_{\mathrm{B}}T}\right)^{1/2} \approx \frac{n\Omega_{\mathrm{e}}}{k}\left(\frac{m}{k_{\mathrm{B}}T}\right)^{1/2} = \frac{n}{\lambda^{1/2}} = \frac{n}{kr_{\mathrm{L}}}$$

where n is the harmonic number and r_{L} is the electron Larmor radius.

Electron Bernstein waves have been characterized across a range of plasmas in the laboratory. Figure 7.15 shows one example of the measured dispersion characteristics for propagation orthogonal to the magnetic field compared with theory (Armstrong *et al.* (1981)) over a limited range of wavenumber.

7.7 Inhomogeneous plasma

The calculation in the last section demonstrates once again the close relationship between particle orbit theory and collisionless kinetic theory. The characteristics of the partial differential equation we solved were the electron orbits in the equilibrium fields. In fact, since we assumed no electric field and a constant magnetic field, these were the simplest orbits with each electron free to move along the field lines but restricted to circular motion about the field lines in the plane perpendicular to $\mathbf{B}_0$. Knowing that magnetic fields, especially in confined plasmas, are almost never uniform and that inhomogeneity introduces grad B drifts in opposite directions for ions and electrons we may wonder what might be the consequences of such practical considerations. This opens a huge field of investigations, including a whole 'zoo' of drift instabilities (see Gary (1993)), which we have not space to discuss. Instead, we note a few general principles and illustrate the use of Jeans' theorem in defining equilibrium distribution functions.

The first point to note is that the equilibrium current density given by $\mathbf{j} = \nabla \times \mathbf{B}$ is not due to the electron grad B drift (it is in the wrong direction) and so must be established by some compensating plasma inhomogeneity. This was illustrated in Section 2.5 for the case where the equilibrium is maintained by oppositely directed plasma and magnetic pressures

$$\nabla(P + B^2/2\mu_0) = 0$$

A second general point is that the equilibrium distribution function cannot be a simple Maxwellian but must contain either a density or temperature gradient or, indeed, both. Suppose for simplicity that all variations are in the x direction only; then we need an f_0 which is x dependent. We therefore construct f_0 using a constant of the motion which includes x. It is easily verified that the constants of the motion for electron orbits in the equilibrium fields, $\mathbf{E}_0 = 0$, $\mathbf{B}_0 = (0, 0, B(x))$ are

$$\begin{aligned} W_\perp &= m(v_x^2 + v_y^2)/2 \\ p_z &= m v_z \\ p_x &= m\left[v_x + \frac{e}{m}\int B(x)\,\mathrm{d}y\right] \end{aligned}$$

$$p_y = m\left[v_y - \frac{e}{m}\int B(x)\,\mathrm{d}x\right]$$

Now we have discussed orbit theory only under the assumption that gradient length scales are small compared with the Larmor radius so we may treat $B(x)$ as constant in the integral in p_y and obtain $x - v_y/\Omega$ as an approximate constant of the motion. Thus, for the case of a density gradient

$$n(x) = n_0(1 + \epsilon_n x) \tag{7.66}$$

but no temperature gradient, a suitable choice of equilibrium distribution function is

$$f_0 = n_0\left(\frac{m}{2\pi k_B T}\right)^{3/2}[1 + \epsilon_n(x - v_y/\Omega)]e^{-mv^2/2k_B T}$$

On taking the velocity moment of this equation we find a macroscopic drift velocity

$$v_n = \int v_y f_0\,\mathrm{d}\mathbf{v} = -\epsilon_n k_B T/m\Omega = -\epsilon_n V_e^2/\Omega$$

From the pressure-balance equation we get

$$v_n + 2\bar{v}_B/\beta = 0 \tag{7.67}$$

where $\bar{v}_B = \epsilon_B k_B T/m\Omega$ is some sort of average grad B drift velocity in the field

$$B(x) = B_0(1 + \epsilon_B x)$$

and β is the ratio of plasma and magnetic pressures. Note that $\bar{v}_B$ is *not* the grad B drift velocity which appeared in the particle orbits in Section 2.4.1. That is given by $v_B = \epsilon_B v_\perp^2/2\Omega$ and is different for each particle. We shall return to this point shortly.

Extending the argument to include a temperature gradient given by

$$T(x) = T_0(1 + \epsilon_T x) \tag{7.68}$$

as well as the density gradient we take

$$f_0 = n_0\left(\frac{m}{2\pi k_B T_0}\right)^{3/2}\{1 + (x - v_y/\Omega)[\epsilon_n + \epsilon_T(mv^2/2k_B T_0 - 3/2)]\}e^{-mv^2/2k_B T_0} \tag{7.69}$$

Here the energy moment gives (7.68) and the velocity moment (see Exercise 7.10)

$$\int v_y f_0\,\mathrm{d}\mathbf{v} = v_n + v_T$$

where $v_T = -\epsilon_T k_B T_0/m\Omega$ so that (7.67) becomes

$$v_n + v_T + 2\bar{v}_B/\beta = 0 \tag{7.70}$$

Now let us examine the roles of these macroscopic drift velocities and compare them with the microscopic v_B which is velocity dependent and actually appears in the orbit equations. Since the magnetic field gradient determines the current by Ampère's law, it follows that the main role of $\bar{v}_B$ is to determine the net drift velocity between the ions and electrons. For simplicity let us treat the ions as a static neutralizing background so that the current is carried entirely by the electrons. Then Ampère's law gives the drift velocity v_d as

$$v_d = -2\bar{v}_B/\beta = v_n + v_T$$

This can give rise to drift wave instabilities as we saw earlier when discussing ion acoustic waves.

Whilst the sum of v_n and v_T is fixed by v_d it turns out that v_T is in general more destabilizing than v_n because, for a given v_d, it produces a more distorted f_0. Within the approximation of weak gradients it is easily seen that a density gradient moves the peak of $f_0(v_y)$ only slightly away from $v_y = 0$. On the other hand, the v_y^3 dependence of the ϵ_T term in (7.69) shifts the peak much further from $v_y = 0$ as illustrated in Fig. 7.16.

An interesting example occurs in the physics of shock waves. In laminar, perpendicular shocks, for which the magnetic field is at right angles to the shock normal, all three gradients are in the same direction, along the normal, and the equilibrium is maintained by an electric field opposing the combined magnetic and plasma pressure. The macroscopic drifts now include the $\mathbf{E} \times \mathbf{B}$ drift, $v_E = E_0/B_0$, and obey the equation

$$v_d = v_E - (v_n + v_T) = 2\bar{v}_B/\beta$$

Priest and Sanderson (1972) showed that in this case the density gradient has no significant effect, merely increasing v_E to maintain v_d which is determined by the magnetic field gradient through $\bar{v}_B$. However, the distortion of f_0 introduced by a temperature gradient moves the peak of f_0 from v_d to $v_d + 3v_T/2$, as shown in Fig. 7.17, and can produce a very significant increase in instability. Allan and Sanderson (1974) showed that this effect can drive the ion acoustic instability even in the case of zero net drift velocity ($v_d = 0$) and $T_i \sim T_e$. Note that although v_E is a microscopic drift, since it appears in the orbits, as well as a macroscopic drift, because it is the same for all electrons, it is the net drift v_d which matters. The equilibrium equation must be obeyed and in the absence of a pressure gradient $v_E = v_d$. Introduction of density and temperature gradients then increase v_E but in such a way as to maintain the same net drift velocity.

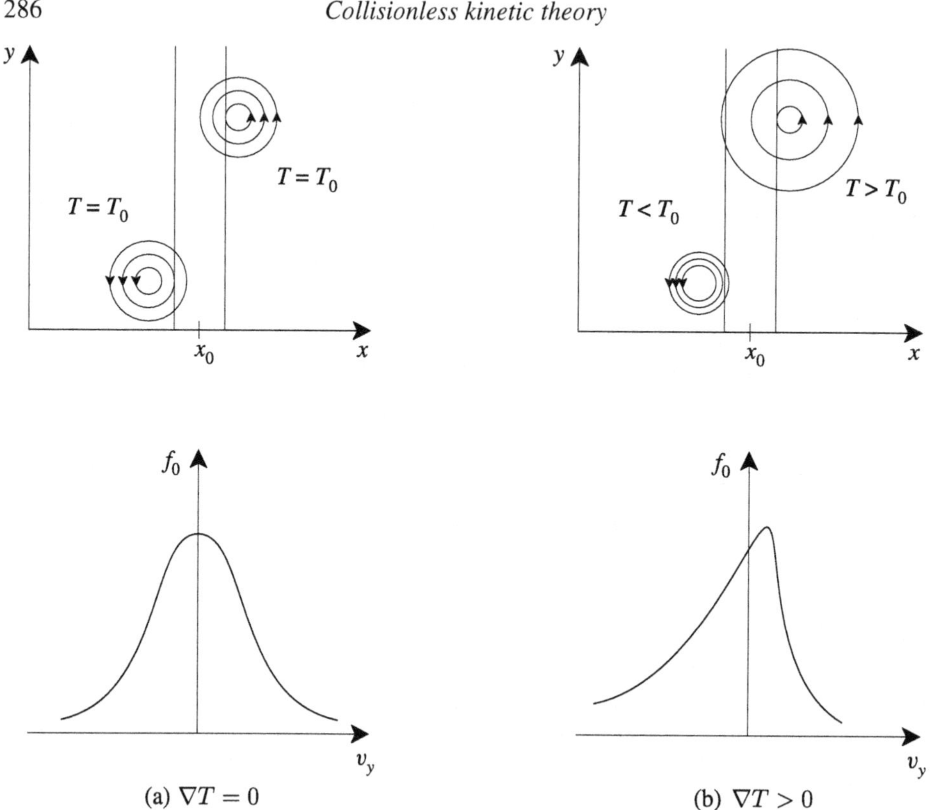

Fig. 7.16. Schematic representation of temperature gradient distortion. Electrons on the left-hand sides ($x < x_0$) of the upper figures fill the right-hand sides ($v_y > 0$) of the lower figures and vice versa. The peak in (b) moves to the right to preserve particle number (after Priest and Sanderson (1972)).

Finally, let us consider the effect of the other microscopic drift velocity $v_\mathrm{B} = \epsilon_\mathrm{B} v_\perp^2/2\Omega$. This complicates calculations because it appears in the orbits and, unlike v_E, is not the same for all electrons. For low β plasmas it is smaller than the other drift velocities and, on these grounds, is usually ignored. Even for low β plasmas, however, it can have a significant effect on resonant wave phenomena because the $v_\perp^2$ dependence spreads the resonance over a range of velocities. In the case of the Bernstein modes, for example, the resonant denominator $\omega - n\Omega_\mathrm{e}$ in (7.63) is replaced by $\omega - n\Omega_\mathrm{e} - kv_\mathrm{B}(v_\perp^2)$ and the integration over velocity no longer produces the sharp resonances at the cyclotron harmonics seen in (7.64). This smearing out of the resonances means that the effect of v_B should be to reduce Bernstein wave instability. For perpendicular shocks this was demonstrated analytically by Sanderson and Priest (1972) confirming earlier numerical calculations by Gary (1970).

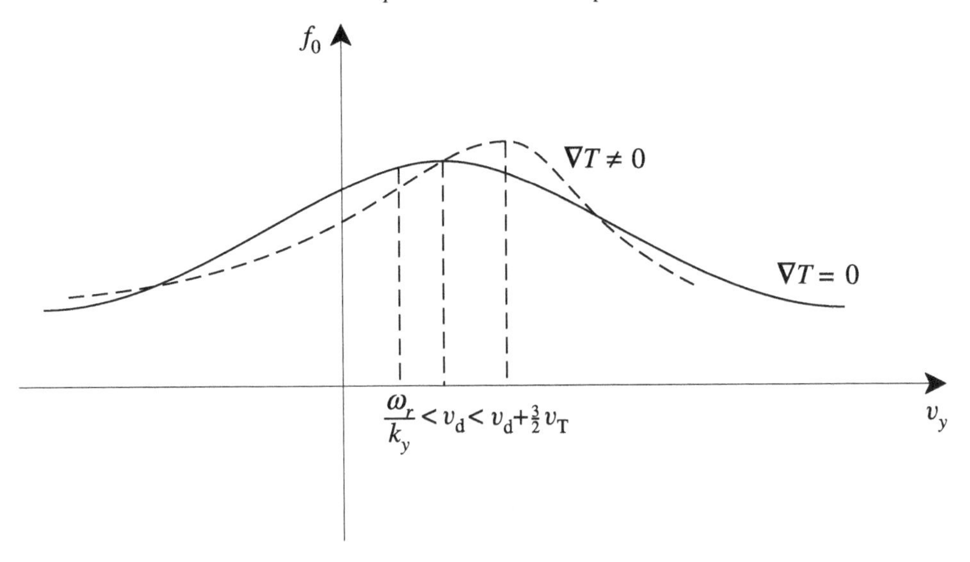

Fig. 7.17. Temperature gradient distortion of the electron distribution function. The slope of f_0 at $v_y = \omega_r/k_y$ is proportional to $(k_y v_d - \omega_r)$ for $v_T = 0$ and to $(k_y(v_d + 3v_T/2) - \omega_r)$ for $v_T \neq 0$ (after Priest and Sanderson (1972)).

7.8 Test particle in a Vlasov plasma

In the next chapter we deal with the kinetic theory of plasmas allowing for collisional effects. We shall see there that the important collisional effects in plasmas are long range many-body interactions rather than short range binary collisions. In this section we outline an approach to particle–plasma interactions that focuses on the interaction between a discrete charged particle, or *test particle* and the other charges in the plasma using the Vlasov–Poisson equations. We begin by isolating a single particle of charge q_T which is injected at $t = 0$ (at $\mathbf{r}_0 = 0$) and moves through the plasma with velocity $\mathbf{u}_0$, assumed constant. We suppose that our test charge causes only a small perturbation in the plasma electron density so that we may reasonably describe the effect of the test charge on the rest of the plasma electrons using the linearized Vlasov equation (7.22)

$$(p + i\mathbf{k} \cdot \mathbf{v}) f_1(\mathbf{k}, \mathbf{v}, p) = -\frac{ie}{m} \left[\mathbf{k} \cdot \frac{\partial f_0(\mathbf{v})}{\partial \mathbf{v}} \right] \phi(\mathbf{k}, p) \tag{7.71}$$

where we now take $f_1(\mathbf{k}, \mathbf{v}, t = 0) = 0$. From Poisson's equation

$$k^2 \phi(\mathbf{k}, p) = -\frac{e}{\varepsilon_0} \int f_1(\mathbf{k}, \mathbf{v}, p) \mathrm{d}\mathbf{v} + \frac{1}{\varepsilon_0} \frac{q_T}{(p + \mathbf{k} \cdot \mathbf{u}_0)} \tag{7.72}$$

From (7.71) and (7.72) proceeding in parallel with Section 7.3 we find

$$\phi(\mathbf{k}, p) = \frac{q_{\mathrm{T}}}{\varepsilon_0} \frac{1}{k^2 D(k, p)(p + i\mathbf{k} \cdot \mathbf{u}_0)} \tag{7.73}$$

When it comes to inverting the Laplace time transform we have now an additional pole at $p = -i\mathbf{k} \cdot \mathbf{u}_0$. As before we consider the long time behaviour of $\phi(\mathbf{k}, t)$ and, since we are principally concerned with the effect of the test electron on the plasma, only the contribution from the pole at $p = -i\mathbf{k} \cdot \mathbf{u}_0$ is taken into account so that

$$\phi(\mathbf{k}, t) = \frac{q_{\mathrm{T}}}{\varepsilon_0} \frac{e^{-i\mathbf{k}\cdot\mathbf{u}_0 t}}{k^2 D(k, \omega = \mathbf{k} \cdot \mathbf{u}_0)} \tag{7.74}$$

In the special case $u_0 \ll V_{\mathrm{e}}$ we find $\phi(\mathbf{k}, t) = (q_{\mathrm{T}}/\varepsilon_0)(k^2 + k_{\mathrm{D}}^2)^{-1}$ where k_{D} is the reciprocal of the Debye length. From this it is straightforward to retrieve the Debye shielding potential (see Exercise 7.14)

$$\phi(\mathbf{r}, t \to \infty) = \frac{q_{\mathrm{T}}}{4\pi\varepsilon_0} \frac{\exp(-r/\lambda_{\mathrm{D}})}{r} \tag{7.75}$$

7.8.1 Fluctuations in thermal equilibrium

When we discuss collective effects in radiation from plasmas in Chapter 9 we shall need a representation for electric field fluctuations in a plasma. Since the electric field itself is a random variable in both space and time what we want is the ensemble average of the energy density of the electric field. Using the concept of a test charge we allow each plasma particle in turn to take the role of the test particle and sum the contributions of each. From (7.73),

$$\mathbf{E}(\mathbf{r}, t) = -\frac{iq_{\mathrm{T}}}{\varepsilon_0} \int \frac{\mathbf{k} e^{i\mathbf{k}\cdot[\mathbf{r}-\mathbf{r}_0(t)]}}{k^2 D(k, \omega = \mathbf{k} \cdot \mathbf{u}_0)} \, \mathrm{d}\mathbf{k} \tag{7.76}$$

where $\mathbf{r}_0(t) = \mathbf{u}_0 t$. Allowing each particle in turn to be the test particle we then find the ensemble average of the electric field by introducing the distribution function $f_0(\mathbf{r}_0, \mathbf{v}_0)$ which is the probability density for particles to have velocity $\mathbf{v}_0$ at position $\mathbf{r}_0$, i.e.

$$\langle \mathbf{E}(\mathbf{r}, t) \rangle = \iint \mathbf{E}(\mathbf{r}, t) f_0(\mathbf{r}_0, \mathbf{v}_0) \, \mathrm{d}\mathbf{r}_0 \, \mathrm{d}\mathbf{v}_0$$

For a uniform isotropic plasma clearly $\langle \mathbf{E}(\mathbf{r}, t) \rangle = 0$.

On the other hand for the ensemble average of the energy density of the electric field we have

$$W = \frac{\varepsilon_0}{2} \iint [\mathbf{E}(\mathbf{r}, t) \cdot \mathbf{E}^*(\mathbf{r}, t)] f_0(\mathbf{r}_0, \mathbf{v}_0) \, \mathrm{d}\mathbf{r}_0 \, \mathrm{d}\mathbf{v}_0$$

Introducing the Fourier–Laplace transform of the ensemble average energy density $W(\mathbf{k}, \omega)$ it is straightforward to show (see Exercise 7.14)

$$W(\mathbf{k}, \omega) = \frac{2n_0 e^2 F_0(\omega/k)}{\varepsilon_0 k^3 |D(k, \omega)|^2} \tag{7.77}$$

Exercises

7.1 Show that $\nabla_v \cdot \mathbf{a} = 0$ when the acceleration $\mathbf{a}$ is due to the self-consistent electromagnetic field, $\mathbf{E} + \mathbf{v} \times \mathbf{B}$. Why can we not assume that $\nabla_v \cdot \mathbf{a} = 0$ when the acceleration is caused by collisional interactions with neighbouring particles?

7.2 Show that the Maxwell distribution function satisfies the Vlasov equation identically. Explain this in terms of (i) constants of the motion and (ii) the Maxwell distribution being the asymptotic solution of the collisional kinetic equation.

7.3 In the weak coupling approximation the potential energy of particle interactions is very much smaller than particle kinetic energy. Show that this approximation is equivalent to the condition for the number of particles in the Debye sphere being very large.

7.4 Show that the dispersion relation (6.97) for electron plasma waves, derived from the warm plasma wave equations, is equivalent to the result (7.34) obtained from kinetic theory. What assumptions and approximations have to be made to obtain this equivalence?

Obtain (7.37) from (7.36). Explain mathematically and physically why this result cannot be obtained from the warm plasma wave equations.

7.5 The plasma used in the measurements of the dispersion characteristics and Landau damping of Langmuir waves in Figs. 7.4 and 7.5 formed a column 2.3 m long with an axial electron density typically 10^{14}–$10^{15}\,\text{m}^{-3}$. Electron temperature ranged between 5 and 20 eV and the pressure of the background gas (mostly hydrogen) was $\sim 10^{-3}$ pascals.

Estimate λ_D and $n\lambda_D^3$. Determine the mean free path for both electron–ion and electron–neutral collisions.

Note that in this experiment Malmberg and Wharton did not measure electron density directly but chose a value which normalized the theoretical dispersion curve to the data points at low frequencies. Why is this justified? Why does it appear that in the limit of small k, $\omega \to 0$ rather than ω_{pe}?

Plot ω versus k using (7.36) with $T_e = 9.6\,\text{eV}$ and compare your results with Fig. 7.4. Interpret the discrepancy between this result and the corresponding line in Fig. 7.4.

In the experiment two probes were used, one serving as transmitter while the detector was moved along the axis. The data recorded consisted of the real and imaginary parts of k. Solve the dispersion relation in the form $k = k(\omega)$ for real ω and obtain an expression for the attenuation of Langmuir waves as a result of Landau damping. Plot $\Im k/\Re k$ as a function of $(v_p/V_e)^2$, where v_p denotes the phase velocity, and compare your results with Fig. 7.5.

7.6 Dawson (1962) devised a physical model of Landau damping based on considerations of wave–particle interactions. The rate of change of the kinetic energy T_k of particles resonant with the wave, computed in the wave frame, in which the wave electric field is represented as $E = E_0 \sin kx$, is

$$\frac{dT_k}{dt} = \frac{n_0 m}{2} \int_{-\infty}^{\infty} \frac{\partial \langle f \rangle}{\partial t} \left(v + \frac{\omega}{k} \right)^2 dv \tag{E7.1}$$

where $\langle f \rangle$ denotes the spatially averaged distribution function. It follows from the Vlasov equation that

$$\frac{\partial \langle f \rangle}{\partial t} = \frac{eE_0}{m} \left\langle \frac{\partial f}{\partial v} \sin kx \right\rangle \tag{E7.2}$$

The solution to the *linearized* Vlasov equation with the initial condition $f_1(x, v, 0) = f_1(v, 0) \cos kx$ is

$$f_1(x, v, t) = f_1(v, 0) \cos(kx - vt) + \frac{eE_0}{mkv} \frac{\partial_0}{\partial v} [\cos k(x - vt) - \cos kx] \tag{E7.3}$$

Use (E7.3) in (E7.2) to determine $\partial \langle f \rangle / \partial t$ and substitute this in (E7.1) to show that

$$\frac{dT_k}{dt} = -\frac{1}{2} neE_0 \int_{-\infty}^{\infty} f_1(v, 0) \left(v + \frac{\omega}{k} \right) \sin kvt \, dv$$
$$- \frac{1}{2m} e^2 E_0^2 \int_{-\infty}^{\infty} \frac{\partial f_0}{\partial v} \left(v + \frac{\omega}{k} \right) \frac{\sin kvt}{kv} dv \tag{E7.4}$$

The first term in (E7.4) decays through phase mixing. In the second term the only particles that make a contribution to dT_k/dt are those moving slowly in the wave frame and $\lim_{v \to 0} \sin kvt/kv \to \pi \delta(kv)$. Show that

$$\frac{dT_k}{dt} = -\pi \omega \frac{\omega_p^2}{k^2} \left. \frac{\partial f_0}{\partial v} \right|_{v=0} W \tag{E7.5}$$

where $W = \frac{1}{2} \varepsilon_0 E^2$ is the wave energy density. Now

$$\frac{dT_k}{dt} = -\frac{dW}{dt} = -2\gamma_L W \tag{E7.6}$$

so that the linear damping rate for the electrostatic field set up by the initial perturbation is, after transforming back to the laboratory frame,

$$\gamma_{\mathrm{L}} = \frac{\pi}{2}\omega\frac{\omega_{\mathrm{p}}^2}{k^2}\left.\frac{\partial f_0}{\partial v}\right|_{v=\omega/k} \tag{E7.7}$$

This is the Landau damping rate (7.37).

7.7 Obtain the dispersion relation (7.45) for ion acoustic waves and the approximation to the Landau damping decrement (7.46). Compare results from (7.45), (7.46) with the numerical solution to the dispersion relation in Fig. 7.7.

Consider a plasma consisting of two ion species and electrons. Suppose that the temperature of the two ion populations is the same and that one of the ions is much heavier than the other. Obtain a dispersion relation for ion acoustic waves in this plasma.

7.8 Definitive measurements of the dispersion characteristics of ion acoustic waves were made in stable plasmas in Q machines (Wong, Motley and D'Angelo (1964)). In this work an alkali metal plasma was created by contact ionization. The plasma was almost completely ionized and close to thermal equilibrium. Magnetic fields of the order of 1 T meant that $\Omega_{\mathrm{i}} \sim$ 100 MHz which is very much higher than the mode frequencies studied. Measurements were made using potassium and caesium plasmas with $T_{\mathrm{e}} = T_{\mathrm{i}} \simeq 2300\,\mathrm{K}$.

Confirm that to a good approximation the ions are collisionless. Since the plasma is produced at one end of the column, recombination losses induce a drift of plasma away from the producing plate. Accordingly, phase velocities of waves moving both upstream and downstream were measured:

$$\left(\frac{\omega}{k}\right)_{\mathrm{K}}^{\mathrm{up}} = 1.3 \times 10^3\,\mathrm{m\,s^{-1}} \qquad \left(\frac{\omega}{k}\right)_{\mathrm{Cs}}^{\mathrm{up}} = 0.9 \times 10^3\,\mathrm{m\,s^{-1}}$$

$$\left(\frac{\omega}{k}\right)_{\mathrm{K}}^{\mathrm{down}} = 2.5 \times 10^3\,\mathrm{m\,s^{-1}} \qquad \left(\frac{\omega}{k}\right)_{\mathrm{Cs}}^{\mathrm{down}} = 1.3 \times 10^3\,\mathrm{m\,s^{-1}}$$

Allowing for a drift velocity V_0 (negligible compared with *electron* thermal velocity), show that the upstream ($-$) and downstream ($+$) phase velocities

$$\frac{\omega}{k_r} = 2.05\left(\frac{k_{\mathrm{B}}T_{\mathrm{i}}}{m_{\mathrm{i}}}\right)^{1/2} \pm V_0 + \frac{0.72(k_{\mathrm{B}}T_{\mathrm{i}}/m_{\mathrm{i}})^{1/2}}{2.05 \pm V_0/(k_{\mathrm{B}}T_{\mathrm{i}}/m_{\mathrm{i}})^{1/2}}$$

($k = k_r + ik_i$). Use this expression to compare predicted and measured phase velocities.

Table 7.1. *Ion wave damping*

	$\left(\frac{\delta}{\lambda}\right)_{\mathrm{K}}^{\mathrm{theory}}$	$\left(\frac{\delta}{\lambda}\right)_{\mathrm{K}}^{\mathrm{expt}}$	$\left(\frac{\delta}{\lambda}\right)_{\mathrm{Cs}}^{\mathrm{theory}}$	$\left(\frac{\delta}{\lambda}\right)_{\mathrm{Cs}}^{\mathrm{expt}}$
Downstream wave		0.65		0.55
Upstream wave		0.14		0.25

Damping of the ion wave was also measured. The damping distance δ is that distance over which the wave amplitude is attenuated by a factor $e = \exp(1)$. The damping constant was calculated and found to be

$$\frac{\delta}{\lambda} = \frac{1}{2\pi}\frac{k_r}{k_i} = 0.39 \pm 0.19 \frac{V_0}{(k_{\mathrm{B}}T_{\mathrm{i}}/m_{\mathrm{i}})^{1/2}}$$

Verify this result and use it to complete Table 7.1.

7.9 Many plasmas both in the laboratory and in space are characterized by non-Maxwellian distributions. As an example consider a two-component electron distribution function, $f = f_{\mathrm{h}} + f_{\mathrm{c}}$ where both hot (h) and cold (c) components are Maxwellian.

Show that for $|\zeta_{\mathrm{c}}| \ll 1$, $|\zeta_{\mathrm{h}}| \ll 1$ (7.45) remains valid provided T_{e} is replaced by an effective temperature defined by

$$T_{\mathrm{eff}} = \frac{n_{\mathrm{e}}T_{\mathrm{c}}T_{\mathrm{h}}}{n_{\mathrm{h}}T_{\mathrm{c}} + n_{\mathrm{c}}T_{\mathrm{h}}}$$

If $n_{\mathrm{h}} \simeq n_{\mathrm{c}}$ and $T_{\mathrm{h}}/T_{\mathrm{c}} \gg 1$ another acoustic-like mode, the *electron acoustic mode*, appears. Refer to Gary (1993) (Section 2.2.3) for a summary of the characteristics of this mode.

7.10 By taking zero-, first-, and second-order moments of (7.69), obtain (7.66), (7.68) and (7.70), respectively. [Hint: Keep only linear terms in small quantities.]

7.11 How is the growth rate for the BTI instability changed when $\mathbf{k}$ and $\mathbf{v}_{\mathrm{b}}$ are not parallel?

We shall find in Chapter 10 that the one-dimensional BTI evolves to produce a plateau distribution across a range of velocities

$$\begin{aligned} f(v_0 < v < v_{\mathrm{b}}) &= \text{const.} \\ f(v > v_{\mathrm{b}}) &= 0 \end{aligned}$$

Show that this distribution is unstable in the case of Langmuir waves propagating at an angle θ ($\neq 0$) to the direction of the beam.

7.12 In a plasma in which electrons drift relative to ions with a drift velocity $v_{\mathrm{d}} \ll V_{\mathrm{e}}$, the electron thermal velocity, show that the growth rate of the ion acoustic instability is given by (7.54). Determine the instability threshold.

7.13 In magnetized plasmas there is an ion counterpart to electron Bernstein modes, the *ion Bernstein modes*. In this case we have to distinguish between two possible outcomes. One corresponds to the electron mode in that $k_z \simeq 0$ and these so-called *pure ion Bernstein modes* mirror the electron modes having no damping for exact orthogonal propagation. Moreover, in the fluid limit they collapse to the lower hybrid mode. However, unlike the electron case with finite k_z such that $\omega/k_z V_{\mathrm{e}} \ll 1$, electrons can now maintain a Boltzmann distribution by flowing along the magnetic field lines to cancel charge separation.

Show that the dispersion relation for these neutralized ion Bernstein waves may be written

$$1 + k^2\lambda_{\mathrm{De}}^2 = \frac{T_{\mathrm{e}}}{T_{\mathrm{i}}} \sum_{n=1}^{\infty} \frac{2n^2\Omega_{\mathrm{i}}^2}{(\omega^2 - n^2\Omega_{\mathrm{i}}^2)} e^{-\lambda_{\mathrm{i}}} I_n(\lambda_{\mathrm{i}}) \tag{E7.8}$$

Take the fluid limit of (E7.8), i.e. $\lambda_{\mathrm{i}} \to 0$, and assuming quasi- neutrality, $k\lambda_{\mathrm{De}} \ll 1$, retrieve the dispersion relation for electrostatic ion cyclotron waves

$$\omega^2 = \Omega_{\mathrm{i}}^2 + 2k^2 \left(\frac{k_{\mathrm{B}} T_{\mathrm{e}}}{m_{\mathrm{i}}} \right) \tag{E7.9}$$

7.14 Obtain (7.75).

In Chapter 9 it will prove helpful to decompose $\phi(\mathbf{k}, p)$ in (7.73) into two parts, the 'self-field' of the test charge and the field due to polarization induced in the plasma by the test charge. Show that the induced field is determined by

$$\phi_{\mathrm{ind}} = \frac{q_{\mathrm{T}}}{\varepsilon_0} \frac{1}{k^2(p + i\mathbf{k}\cdot\mathbf{u}_0)} \left[\frac{1}{D(k, p)} - 1 \right]$$

Establish the expression for the ensemble-averaged energy density $W(\mathbf{k}, \omega)$ in (7.77).

7.15 Plasma kinetic theory developed in this chapter on the basis of the Vlasov–Maxwell equations relies on linearization, and even then one generally has to solve dispersion relations numerically. By way of illustration we

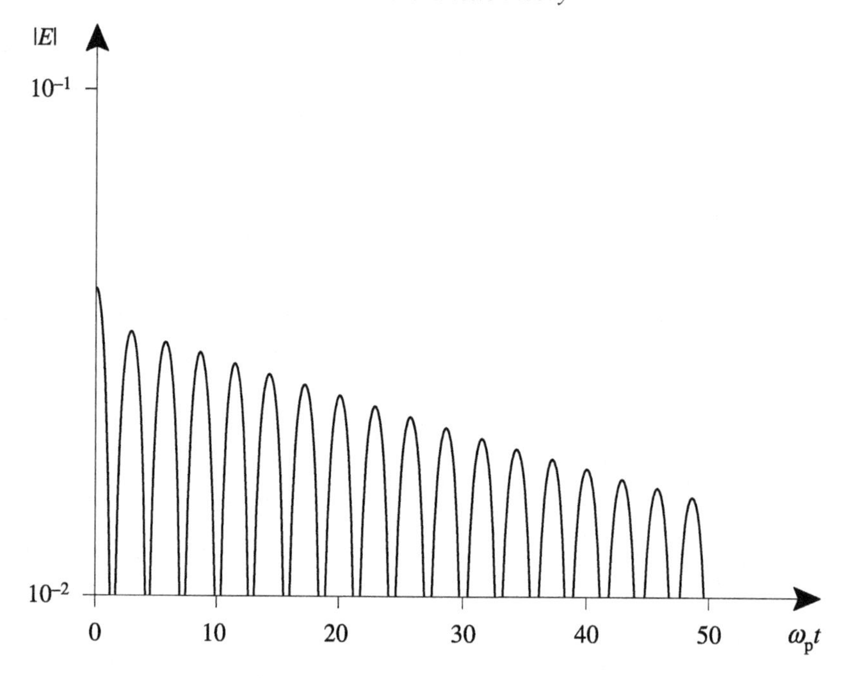

Fig. 7.18. Time evolution of a Langmuir wave from a 1D Vlasov code.

consider a direct numerical integration of the 1D normalized Vlasov–Poisson equations

$$\frac{\partial f}{\partial t} + v\frac{\partial f}{\partial x} - E\frac{\partial f}{\partial v} = 0 \tag{E7.10}$$

$$\frac{\partial E}{\partial x} = 1 - \int_{-\infty}^{\infty} f \, dv \tag{E7.11}$$

The Vlasov equation (E7.10) is an advective equation in the 2D phase space (x, v). If we set about differencing (E7.10) directly we find that the numerical diffusion that is characteristic of difference schemes in configuration space now gives rise to diffusion in velocity space as well. We shall see in the following chapter that velocity space diffusion is a property of the Fokker–Planck collision operator. Clearly a numerical solution to the Vlasov equation that mimics collisional effects is to be avoided.

An alternative approximation first used by Cheng and Knorr (1976) introduces a splitting technique in which the Vlasov equation is replaced by the pair of equations

$$\frac{\partial f}{\partial t} + v\frac{\partial f}{\partial x} = 0 \qquad \frac{\partial f}{\partial t} + E(x,t)\frac{\partial f}{\partial v} = 0 \tag{E7.12}$$

The integration of (E7.12) was reduced to a shifting of the distribution function:

$$f^*(x, v) = f^n(x - \frac{v\Delta t}{2}, v) \tag{E7.13}$$

$$f^{**}(x, v) = f^n(x, v - E(x)\Delta t) \tag{E7.14}$$

$$f^{n-1}(x, v) = f^{**}(x - \frac{v\Delta t}{2}, v) \tag{E7.15}$$

in which f^n denotes the distribution function evaluated at $t = n\Delta t$. The shifts are generated by interpolating f in both x and v. Periodic boundary conditions are applied at boundaries in the x direction while in v the distribution function is assumed to vanish at the boundaries. The procedure is straightforward for a 2D phase space.

Construct a 1D Vlasov code and use it to study the evolution of a Langmuir wave in time. In particular estimate the Landau damping and compare this estimate with the predicted damping. Figure 7.18 is output from a 1D code showing the time evolution of a Langmuir wave.

8

Collisional kinetic theory

8.1 Introduction

In Section 7.1 we used a simple heuristic argument to obtain the collisional kinetic equation

$$\frac{\partial f}{\partial t} + \mathbf{v} \cdot \frac{\partial f}{\partial \mathbf{r}} + \frac{\mathbf{F}}{m} \cdot \frac{\partial f}{\partial \mathbf{v}} = \left(\frac{\partial f}{\partial t} \right)_{\mathrm{c}} \tag{8.1}$$

in which the left-hand side is the same as in the Vlasov equation and the right-hand side, in some way yet to be determined, represents the rate of change of the distribution function, f, due to collisions. We then argued that, since in plasmas the force, $\mathbf{F}$, includes the self-consistent electric field which gives rise to plasma oscillations, the frequency typical of non-collisional changes in f is ω_{p}. Thus, a dimensional comparison of the left- and right-hand sides of (8.1) suggests that collisional effects may be ignored provided that the collision frequency $\nu_{\mathrm{c}} \ll \omega_{\mathrm{p}}$, which is almost always the case. This is, of course, no more than a hand-waving argument and, in this chapter, we shall examine this matter more carefully.

To begin with we may ask what is meant by the collision frequency. We shall discover that there are, in fact, several collision frequencies, differing by many orders of magnitude, and that the choice of an appropriate one depends on what kind of collisions have the greatest influence on the physical effects under investigation.

One effect, for which collisions are crucial, is the establishment of thermodynamic equilibrium in a plasma. This was touched on in Chapter 3 where ν_{c}^{-1} was the time scale for the distribution function to relax to its minimum energy state, the Maxwell distribution. Aside from this, the main area of plasma physics in which collisions are important is transport theory. Matter, momentum and energy may be transferred by the action of collisions. Since plasma heating and containment are crucial goals of controlled thermonuclear reactor physics, a proper understanding of plasma transport is fundamental to the success of this programme.

In fact, the development of plasma transport theory has proved to be one of the most challenging problems in plasma physics. In this chapter our aim is to do no more than give a short introduction to the topic by means of basic physical arguments and approximate mathematical models. It is preferable to begin with this admittedly oversimplified transport theory and so, for the moment, we continue to beg the question as to what is the collision frequency. Later on in the chapter, using a more sophisticated model for the collision term in (8.1), we derive parametric expressions for various collision frequencies and thereby identify which frequency is appropriate for each transport process.

8.2 Simple transport coefficients

In the equation relating a flux to the thermodynamic force driving the flux, the constant of proportionality is called the *transport coefficient*. For example, in *Fourier's law*

$$\mathbf{q} = -\kappa \boldsymbol{\nabla} T$$

relating the heat flux $\mathbf{q}$ to the temperature gradient $\boldsymbol{\nabla} T$, the constant of proportionality κ is called the coefficient of thermal conductivity. Our aim is to derive from (8.1) expressions for the most important transport coefficients. We do this by adopting a very simple model for the collision term, $(\partial f/\partial t)_{\mathrm{c}}$, and then taking velocity moments of the equation to obtain the relationships between various fluxes and their thermodynamic forces. The model simulates the effect of close binary collisions where particles experience sudden, local velocity changes and so its application to plasmas, strictly speaking, ought to be limited to the *Lorentz gas model*, which assumes that the electrons scatter off infinitely heavy, stationary ions.

Suppose that a plasma, initially in thermal equilibrium, is disturbed by various external perturbations giving rise to small, steady state fluxes of matter, momentum or energy. Since collisions drive the distribution function towards a Maxwellian on a time scale of order ν_{c}^{-1}, the simplest representation of the collision term in the kinetic equation is

$$\left(\frac{\partial f}{\partial t}\right)_{\mathrm{c}} = -\nu_{\mathrm{c}}(f - f_0) \tag{8.2}$$

where f_0 is a local Maxwellian given in general by

$$f_0(\mathbf{r}, \mathbf{v}) = n(\mathbf{r}) \left(\frac{m}{2\pi k_{\mathrm{B}} T(\mathbf{r})}\right)^{3/2} \exp\left\{-m[\mathbf{v} - \mathbf{u}(\mathbf{r})]^2/2k_{\mathrm{B}} T(\mathbf{r})\right\} \tag{8.3}$$

In (8.2), known after Bhatnagar, Gross and Krook (1954) as the BGK model, ν_{c} is a constant, which we shall subsequently identify with an appropriate collision frequency, and $(f - f_0)$ is, of course, the difference between the actual value of the distribution function and the local Maxwellian. The BGK model supposes that collisions act in such a way as to decrease f where it is greater than f_0 and increase f where it is less than f_0 at a rate which is proportional to $|f - f_0|$. The method assumes all perturbations and perturbing forces are small and solves the linearized kinetic equation. Also, as simple a version of (8.3) is used as is consistent with the description of the transport process under consideration. We wish to derive expressions valid for both ions and electrons but to avoid cluttered notation we suppress the species label and write the particle charge as Ze, where $Z = -1$ for electrons.

As a first example let us find an expression for the *electrical conductivity* σ. Here a uniform plasma may be assumed and we may take the unperturbed plasma to be at rest so that f_0 is the Maxwell distribution

$$f_0 = f_\mathrm{M} \equiv n\left(\frac{m}{2\pi k_\mathrm{B}T}\right)^{3/2} e^{-mv^2/2k_\mathrm{B}T} \tag{8.4}$$

The plasma is then subjected to a weak electric field $\mathbf{E}$ which is both constant and uniform. This perturbs the plasma so that $f = f_0 + f_1$ and the linearized kinetic equation is

$$\frac{Ze\mathbf{E}}{m}\cdot\frac{\partial f_0}{\partial \mathbf{v}} = -\nu_\mathrm{c} f_1 \tag{8.5}$$

Since

$$\sigma\mathbf{E} = \mathbf{j} = Ze\int \mathbf{v} f \,\mathrm{d}\mathbf{v} = Ze\int \mathbf{v} f_1 \,\mathrm{d}\mathbf{v} \tag{8.6}$$

it follows from (8.5), on multiplying by $Ze\mathbf{v}$ and integrating over velocity space, that

$$\nu_\mathrm{c}\mathbf{j} = \frac{Z^2e^2}{k_\mathrm{B}T}\mathbf{E}\cdot\int \mathbf{v}\mathbf{v} f_\mathrm{M}\,\mathrm{d}\mathbf{v} \tag{8.7}$$

Taking $\mathbf{E}$ to define the x direction, j_y and j_z are zero and we find from (8.6) and (8.7)

$$\sigma = \frac{nZ^2e^2}{m\nu_\mathrm{c}} \tag{8.8}$$

The electrical conductivity is the current per unit electric field. The *mobility* of a particle, μ_m, is defined as its velocity per unit field and hence

$$\mu_\mathrm{m} = \frac{\sigma}{nZe} = \frac{Ze}{m\nu_\mathrm{c}} \tag{8.9}$$

Next consider the *diffusion coefficient* D. This is defined as the particle flux caused by unit density gradient. Here, assuming uniform temperature, we may take

$$f_0 = n(\mathbf{r})\left(\frac{m}{2\pi k_B T}\right)^{3/2} e^{-mv^2/2k_B T}$$

so that with no external forces the linearized kinetic equation is

$$\mathbf{v}\cdot\frac{\partial f_0}{\partial \mathbf{r}} = \frac{f_0}{n}\mathbf{v}\cdot\frac{\partial n}{\partial \mathbf{r}} = -\nu_c f_1 \tag{8.10}$$

Since the particle flux $\mathbf{\Gamma} = \int \mathbf{v} f \,\mathrm{d}\mathbf{v}$, we find on multipling (8.10) by $\mathbf{v}$ and integrating over velocity space

$$\mathbf{\Gamma} = -\frac{1}{n\nu_c}\int \mathbf{v} f_0 \mathbf{v}\cdot\nabla n \,\mathrm{d}\mathbf{v} = -\frac{k_B T}{m\nu_c}\nabla n$$

and hence

$$D = \frac{k_B T}{m\nu_c} \tag{8.11}$$

From (8.9) and (8.11)

$$\frac{D}{\mu_m} = \frac{k_B T}{Ze} \tag{8.12}$$

which is known as the *Einstein relation.*

The *thermal conductivity* κ is usually defined for constant pressure so that, since

$$p = nk_B T \tag{8.13}$$

one must allow both n and T to be inhomogeneous and take

$$f_0 = n(\mathbf{r})\left(\frac{m}{2\pi k_B T(\mathbf{r})}\right)^{3/2} \exp[-mv^2/2k_B T(\mathbf{r})]$$

Then the steady state, force-free kinetic equation gives, using (8.13),

$$f_0\mathbf{v}\cdot\frac{\partial T}{\partial \mathbf{r}}\left(\frac{mv^2}{2k_B T^2} - \frac{5}{2T}\right) = -\nu_c f_1 \tag{8.14}$$

The heat flux $\mathbf{q}$ is given by

$$\mathbf{q} = \int \frac{1}{2}mv^2\mathbf{v} f \,\mathrm{d}\mathbf{v} = \frac{1}{2}m\int v^2\mathbf{v} f_1 \,\mathrm{d}\mathbf{v}$$

so that multiplication of (8.14) by $\frac{1}{2}mv^2\mathbf{v}$ and integration over velocity space gives

$$\begin{aligned}\nu_c\mathbf{q} &= \frac{5m}{4T}\int v^2\mathbf{v}\mathbf{v}\cdot\nabla T f_0 \,\mathrm{d}\mathbf{v} - \frac{m^2}{4k_B T^2}\int v^4\mathbf{v}\mathbf{v}\cdot\nabla T f_0 \,\mathrm{d}\mathbf{v} \\ &= -\frac{5nk_B^2 T}{2m}\nabla T\end{aligned}$$

Hence the thermal conductivity

$$\kappa = \frac{5nk_{\mathrm{B}}^2T}{2m\nu_{\mathrm{c}}}$$

The *coefficient of viscosity* μ is defined as the shear stress produced by unit velocity gradient. Taking the local flow velocity $\mathbf{u}$ in the x direction and its gradient in the z direction we may write

$$f_0 = n\left(\frac{m}{2\pi k_{\mathrm{B}}T}\right)^{3/2} \exp\left\{-\frac{m}{2k_{\mathrm{B}}T}[(v_x - u(z))^2 + v_y^2 + v_z^2]\right\}$$

and substitution in the kinetic equation gives

$$\frac{mv_z}{k_{\mathrm{B}}T}\frac{\mathrm{d}u}{\mathrm{d}z}(v_x - u)f_0 = -\nu_{\mathrm{c}}f_1 \tag{8.15}$$

From the definition of μ

$$\mu\frac{\mathrm{d}u}{\mathrm{d}z} = -\int m(v_x - u)v_z f\,\mathrm{d}\mathbf{v} = -m\int (v_x - u)v_z f_1\,\mathrm{d}\mathbf{v} \tag{8.16}$$

Thus, multiplying (8.15) by $m(v_x - u)v_z$ and integrating gives, using (8.16),

$$\mu = nk_{\mathrm{B}}T/\nu_{\mathrm{c}}$$

Note that there is no explicit mass dependence in this result. A consequence of this is that the viscosity of a plasma is determined by the ions since, as we shall see in Section 8.5, the ion collision frequency is smaller than that of the electrons by the square-root of the mass ratio. For all the other transport coefficients calculated so far, this consideration is outweighed by the explicit appearance of the particle mass in the denominator so that the electrons dominate these transport processes.

8.2.1 Ambipolar diffusion

So far we have investigated transport under the simplest possible conditions. To complete this discussion we examine some important practical considerations. For example, in the presence of a density gradient the diffusion of electrons and ions will occur in general at different rates. When particle temperatures are approximately equal, the electrons diffuse more rapidly than the ions and if the containing walls are insulated, a space charge is set up due to the accumulation of excess electrons near the wall. This has the effect of simultaneously decreasing electron mobility μ_{e} and increasing ion mobility μ_{i}. The ion and electron fluxes are determined by

$$\left.\begin{aligned} \boldsymbol{\Gamma}_{\mathrm{i}} &= -D_{\mathrm{i}}\nabla n_{\mathrm{i}} + n_{\mathrm{i}}\mu_{\mathrm{i}}\mathbf{E} \\ \boldsymbol{\Gamma}_{\mathrm{e}} &= -D_{\mathrm{e}}\nabla n_{\mathrm{e}} + n_{\mathrm{e}}\mu_{\mathrm{e}}\mathbf{E} \end{aligned}\right\} \tag{8.17}$$

where $\mathbf{E}$ is the field due to the space charge. A steady state is reached when $\mathbf{j} = 0$ so that there is no further build-up of space charge. Then $Z\mathbf{\Gamma}_\mathrm{i} = \mathbf{\Gamma}_\mathrm{e} = \mathbf{\Gamma}$, say, and eliminating $\mathbf{E}$ from (8.17), assuming $Zn_\mathrm{i} = n_\mathrm{e} = n$, gives

$$\mathbf{\Gamma} = -D_\mathrm{a}\nabla n$$

where the *ambipolar diffusion coefficient*

$$D_\mathrm{a} = \frac{\mu_\mathrm{i} D_\mathrm{e} - \mu_\mathrm{e} D_\mathrm{i}}{\mu_\mathrm{i} - \mu_\mathrm{e}}$$

Using the Einstein relation (8.12) and assuming $T_\mathrm{i} = T_\mathrm{e}$ this becomes

$$D_\mathrm{a} = \frac{(Z+1)D_\mathrm{i} D_\mathrm{e}}{ZD_\mathrm{i} + D_\mathrm{e}} \approx (Z+1)D_\mathrm{i}$$

since $D_\mathrm{e} \gg ZD_\mathrm{i}$. Thus, the resultant ambipolar diffusion is determined by the slower ion rate.

In the steady state the field set up by the space charge is, from (8.17),

$$\mathbf{E} = \frac{(D_\mathrm{i} - D_\mathrm{e})}{(\mu_\mathrm{i} - \mu_\mathrm{e})} \frac{\nabla n}{n} \approx -\frac{k_\mathrm{B} T}{ne} \nabla n \tag{8.18}$$

Since

$$\nabla \cdot \mathbf{E} = \frac{e}{\varepsilon_0}(Zn_\mathrm{i} - n_\mathrm{e}) \tag{8.19}$$

the quasi-neutrality condition $Zn_\mathrm{i} \approx n_\mathrm{e}$ implies from (8.18) and (8.19) that

$$\frac{Zn_\mathrm{i} - n_\mathrm{e}}{n} \sim \frac{\varepsilon_0 k_\mathrm{B} T}{ne^2 L^2} = \left(\frac{\lambda_\mathrm{D}}{L}\right)^2 \ll 1$$

where L is the length scale of the boundary layer over which the field and density gradients exist. Thus, quasi-neutrality is established within a few Debye lengths of the insulating wall; this defines the sheath thickness.

8.2.2 Diffusion in a magnetic field

Finally, let us consider diffusion in a magnetized plasma. For a plasma in a steady state with no electric field we have from the kinetic equation

$$\mathbf{v} \cdot \frac{\partial f}{\partial \mathbf{r}} + \frac{Ze}{m}(\mathbf{v} \times \mathbf{B}) \cdot \frac{\partial f}{\partial \mathbf{v}} = -\nu_\mathrm{c} f_1$$

which becomes on linearization

$$\mathbf{v} \cdot \frac{\partial f_0}{\partial \mathbf{r}} + \frac{Ze}{m}(\mathbf{v} \times \mathbf{B}_0) \cdot \frac{\partial f_1}{\partial \mathbf{v}} = -\nu_\mathrm{c} f_1 \tag{8.20}$$

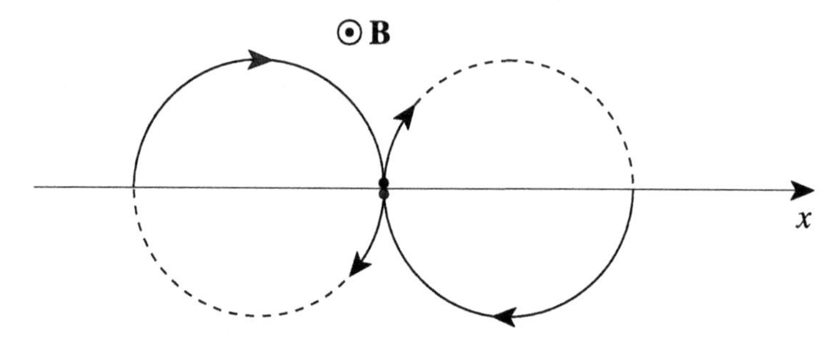

Fig. 8.1. Collisional diffusion in a strong magnetic field.

If we suppose that $\mathbf{B}_0$ is in the z direction and the density gradient is in the z and x directions, then we choose f_0 to be the *bi-Maxwellian distribution*

$$f_0 = n(x, z)\left(\frac{m}{2\pi k_B T_\perp}\right)\left(\frac{m}{2\pi k_B T_\parallel}\right)^{1/2} \exp\left\{-\frac{m(v_x^2 + v_y^2)}{2k_B T_\perp} - \frac{mv_z^2}{2k_B T_\parallel}\right\}$$

since a plasma may have, in general, different parallel and perpendicular temperatures. The v_x, v_y, and v_z moments of (8.20) give

$$\left.\begin{aligned} (k_B T_\perp/m)(\partial n/\partial x) - \Omega\Gamma_y &= -\nu_c\Gamma_x \\ \Omega\Gamma_x &= -\nu_c\Gamma_y \\ (k_B T_\parallel/m)(\partial n/\partial z) &= -\nu_c\Gamma_z \end{aligned}\right\} \tag{8.21}$$

where $\Omega = ZeB/m$ is the cyclotron frequency.

Defining $D_\perp$ and $D_\parallel$ by

$$\Gamma_x = -D_\perp\frac{\partial n}{\partial x} \qquad \Gamma_z = -D_\parallel\frac{\partial n}{\partial z}$$

we find from (8.21)

$$D_\perp = \frac{k_B T_\perp}{m\nu_c[1 + (\Omega/\nu_c)^2]} \qquad D_\parallel = \frac{k_B T_\parallel}{m\nu_c}$$

showing that diffusion along the magnetic field is unaffected by the field, while diffusion across the field is reduced by the factor $(1 + (\Omega/\nu_c)^2)^{-1}$.

In the limit $(\Omega/\nu_c)^2 \to 0$, we recover, as expected, the unmagnetized result for $D_\perp = k_B T_\perp/m\nu_c = \lambda_c^2\nu_c$, where λ_c is the collisional mean free path for motion across the field. In the opposite limit $D_\perp \approx k_B T_\perp\nu_c/m\Omega^2 = r_L^2\nu_c$, where r_L is the Larmor radius. Thus, the Larmor radius replaces the mean free path as the length scale for diffusion but the time scale is still the collision time.

This is explained in Fig. 8.1 which shows the effect of collisions on gyrating particles. For simplicity, a head-on collision is considered between particles with equal speeds as a result of which the particles exchange orbits allowing each to

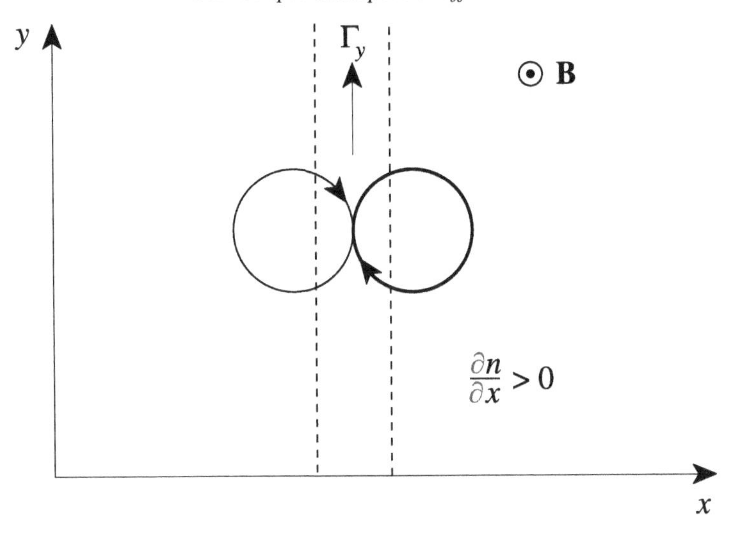

Fig. 8.2. Gyro-magnetic particle flux.

progress in opposite directions along the x-axis. In the absence of a density gradient there would, of course, be no net flux but, for $\partial n/\partial x > 0$, there are more particles travelling to the left than to the right and hence there is a net flux down the density gradient. Since the flux in the x direction is entirely dependent upon collisions and particles describe their Larmor orbits in search of a collision partner, it follows that the diffusion length scale is now determined by the Larmor radius rather than the collisional mean free path.

Note that there is a flux in the y direction as well,

$$\Gamma_y = -\frac{\Omega}{\nu_c}\Gamma_x = \frac{\Omega D_\perp}{\nu_c}\frac{\partial n}{\partial x} \to r_L^2 \Omega \frac{\partial n}{\partial x}$$

in the limit $(\Omega/\nu_c)^2 \to \infty$. This gyro-magnetic flux is the dominant flux in this limit and is independent of the collision frequency! Applying the argument used in Fig. 8.1 to collisions occurring along the y-axis yields no net flux because there is no density gradient in this direction. On the other hand, in a thin sheet in the yz-plane there are more particles to the right than to the left so that contributions to Γ_y from particles whose Larmor orbits intersect the sheet do not cancel out, as demonstrated in Fig. 8.2. Collisions play no part in this flux which is entirely gyro-magnetic.

8.3 Neoclassical transport

The model used in the preceding section to calculate transport coefficients avoids various complications and interdependences. Nevertheless, for collisional plasma transport in a *uniform* magnetic field, generally referred to as *classical transport*, it gives the correct parametric dependence. However, the magnetic fields needed for toroidal confinement are both curved and inhomogeneous so it is essential to see what modifications to classical transport theory are needed as a result. This development of the theory is known as *neoclassical transport*.

As for classical transport, a rigorous treatment requires solution of the kinetic equations and is very complicated. However, order of magnitude expressions can be obtained by simple heuristic arguments beginning with the expressions obtained for diffusion in a uniform magnetic field. Diffusion coefficients have dimensions of $(\text{length})^2/\text{time}$ and for $D_\parallel$, where the magnetic field has no effect, we have shown that

$$D_\parallel = \lambda_c^2/\tau_c$$

since the collisional mean free path $\lambda_c = V_{th}\tau_c$ and τ_c is the interval between collisions. On the other hand $D_\perp$, for the strong field case $|\Omega| \gg \nu_c$, is expressed as

$$D_\perp = r_L^2/\tau_c$$

Each of these results may be interpreted in terms of a *random walk* model of diffusion. For parallel diffusion the particle travels, on average, a distance λ_c before a collision randomly alters its direction, the average interval for such random changes being τ_c. The time interval is the same for perpendicular diffusion since it is still collisions which cause the random realignments but particles restricted to Larmor orbits cannot travel a mean free path in the perpendicular direction and so λ_c must be replaced by r_L.

This simple picture changes fundamentally once the field becomes inhomogeneous because particle guiding centres are no longer attached to field lines but drift across them. The Larmor orbits are of no significance in this case since the perpendicular migration is determined by the guiding centre motion. To find the appropriate length scale for perpendicular diffusion we need to consider the global geometry of the field. In a toroidal plasma, as we know from the discussion in Section 4.3.2, the field lines turn in the poloidal direction as they wind around the torus so that one 'cycle', defined by the line returning to its starting point, is completed after travelling a distance qR, where q is the safety factor and R is the (major) radial coordinate of the guiding centre. According to the random walk model the time for the particle guiding centre to travel this distance is given by

$$t = (qR)^2/D_\parallel = (qR)^2/V_{\text{th}}^2\tau_c$$

assuming parallel diffusion is collisional. During this time the guiding centre migrates in the perpendicular direction a distance of order $v_d t$, where v_d is the drift speed, and

$$D_\perp^{\text{col}} = \frac{(v_d t)^2}{t} = \left(\frac{v_d q R}{V_{\text{th}}}\right)^2 / \tau_c$$

In general, the actual drift velocity, being the resultant of both grad B and curvature drifts, is given, in order of magnitude, by

$$v_d \sim \frac{V_{\text{th}}^2}{|\Omega| R}$$

and hence

$$D_\perp^{\text{col}} \approx q^2 r_L^2/\tau_c$$

Typically, the safety factor $q \sim 3$ so this is an order of magnitude greater than for a uniform field.

This result is valid provided $qR > \lambda_c$ for we have assumed that diffusion in the parallel direction is collisional. In fact, it is frequently the case that $qR < \lambda_c$ and then a further modification is needed. Because the particles migrate across the poloidal cross-section they are subject to a varying toroidal field given by

$$B = \frac{B_0 R_0}{R} = \frac{B_0 R_0}{R_0 + r\cos\theta} \approx B_0(1 - \epsilon\cos\theta)$$

where R_0 is the major radius, B_0 the field strength on the minor axis and $\epsilon = r/R_0$ the inverse aspect ratio, assumed to be small. It follows that those particles with small enough $v_\parallel$ may be reflected by the magnetic mirror effect. Such particles are trapped in the banana-shaped orbits discussed in Section 2.10 where it was shown that these particles have velocities satisfying the condition

$$|v_\parallel(0)|/v_\perp(0) \leq (2\epsilon)^{1/2}$$

If n_b is the number density of such particles, for an isotropic velocity distribution,

$$\frac{n_b}{n} \propto \frac{v_\parallel(0)}{v_\perp(0)} \propto \epsilon^{1/2}$$

is small. However, the effect on diffusion is significant because $v_\parallel$ is also small and so the time for the trapped particles to traverse the orbit is correspondingly long, as is the resultant perpendicular migration.

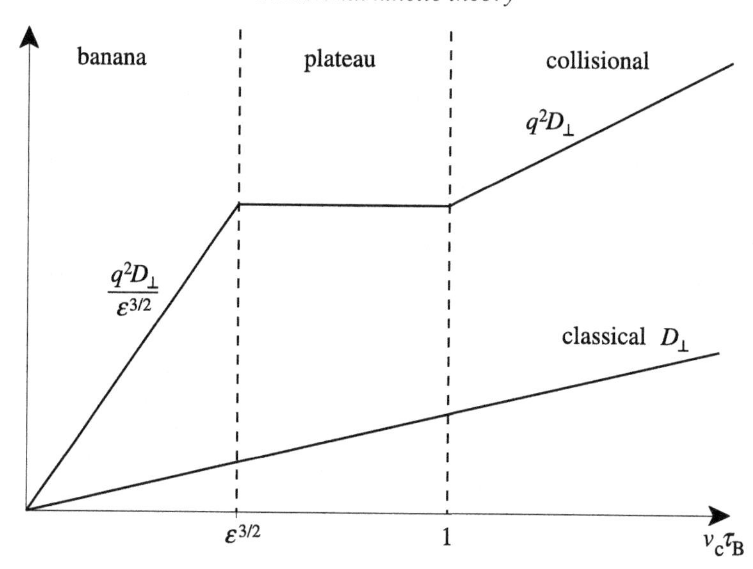

Fig. 8.3. Variation of perpendicular diffusion coefficient with collision frequency.

Repeating the random walk calculation for the perpendicular diffusion of the trapped particles we now have a length scale given by

$$v_d \left(\frac{\tau_B V_{th}}{v_\parallel} \right) = \left(\frac{V_{th}^2}{|\Omega| R} \right) \left(\frac{q R}{\epsilon^{1/2} V_{th}} \right) = \frac{V_{th} q}{\epsilon^{1/2} |\Omega|}$$

where $\tau_B = (qR/V_{th})$ is the *bounce time*. The particle will remain trapped in the banana orbit until it is scattered out of the trapped velocity band by collisions. We shall show in Section 8.5 that the effective collision frequency for vanishingly small velocity V varies as V^{-2} and hence the effective collision time in this case is

$$\tau_{eff} = \epsilon \tau_c$$

Bearing in mind that only a fraction $\propto \epsilon^{1/2}$ of the particles are trapped in banana orbits, it follows that the effective diffusion coefficient across the field lines is given by

$$D_\perp^{ban} = \epsilon^{1/2} \left(\frac{q V_{th}}{\epsilon^{1/2} \Omega} \right)^2 / \epsilon \tau_c = \frac{q^2 r_L^2}{\epsilon^{3/2} \tau_c}$$

which is a factor $\epsilon^{-3/2}$ greater than that obtained for untrapped (or passing) particles. This represents a further order of magnitude increase over the uniform field result for typical aspect ratios.

These results are represented in Fig. 8.3 which shows schematically the variation of the perpendicular diffusion coefficient with collision frequency $\nu_c = \tau_c^{-1}$. At lowest collision frequencies we are in the banana regime for which $\tau_{eff} > \tau_B/\epsilon^{1/2}$, that is $\tau_B < \epsilon^{3/2}\tau_c$. In the collisional (or Pfirsch–Schlüter) regime we have $|\Omega|^{-1} \ll \tau_c < \tau_B$ and it follows that there is an intermediate regime in which $\tau_B < \tau_c < \tau_B/\epsilon^{3/2}$. This is the most difficult regime to analyse but, as one can see from the figure, the perpendicular diffusion coefficient has the same order of magnitude at either end of the interval $\epsilon^{3/2} < \nu_c\tau_B < 1$ and hence this is known as the *plateau regime*.

8.4 Fokker–Planck equation

The BGK model for the collision term in (8.1), while useful for the sort of calculation carried out in Section 8.2, is too simple to give a realistic representation of collisional effects in quantitative calculations. It does not, for example, conserve particle number, momentum or energy. Also, as already noted, it assumes that there is a given collision frequency so we cannot use the model to discover the properties (e.g. parametric dependence) of collision frequencies. We must, therefore, turn to more sophisticated models.

At first, plasma physicists used the well-known Boltzmann collision integral for $(\partial f/\partial t)_c$ even though it was recognized that logically this was an unsatisfactory way to proceed. The Boltzmann derivation assumes short range, binary collisions whereas in a plasma there may be typically a thousand particles in the Debye sphere, all of which are interacting with each other simultaneously so that collisions are characteristically long range (compared with the mean interparticle separation) and many-body. Most of these collisions are 'weak' in the sense that the potential energy of the interaction ($\sim e^2/\lambda_D$) is very much less than the mean thermal energy ($\sim k_BT$) and it can easily be shown (see Exercise 8.4) that the cumulative effect of the many weak collisions far outweighs the effect of the rare strong interactions for which $e^2/r \sim k_BT$, that is $r/\lambda_D \sim (n\lambda_D^3)^{-1} \ll 1$.

In these circumstances, akin to those met in Brownian motion, the Fokker–Planck approach is more appropriate. Here, one supposes that a function $\psi(\mathbf{v}, \Delta\mathbf{v})$ may be defined such that ψ is the probability that a particle with velocity $\mathbf{v}$ acquires a small increment $\Delta\mathbf{v}$ in a time Δt. It then follows that

$$f(\mathbf{r}, \mathbf{v}, t) = \int f(\mathbf{r}, \mathbf{v} - \Delta\mathbf{v}, t - \Delta t)\psi(\mathbf{v} - \Delta\mathbf{v}, \Delta\mathbf{v})\mathrm{d}(\Delta\mathbf{v}) \tag{8.22}$$

since this equation simply states that we arive at $f(\mathbf{r}, \mathbf{v}, t)$ by 'summing over' all possible increments $\Delta\mathbf{v}$ which were likely to occur Δt seconds earlier. Note that $\psi(\mathbf{v}, \Delta\mathbf{v})$ is assumed independent of t, i.e. the collisional process has no 'memory' of earlier collisions; a process having this property is said to be *Markovian*. This is discussed further in Section 12.6.2.

Since the 'increments' $\Delta\mathbf{v}$ are small the integral in (8.22) may be expanded to give

$$f(\mathbf{r},\mathbf{v},t) = \int \mathrm{d}(\Delta\mathbf{v})\left\{ f(\mathbf{r},\mathbf{v},t-\Delta t)\psi(\mathbf{v},\Delta\mathbf{v}) - \Delta\mathbf{v}\cdot\frac{\partial}{\partial\mathbf{v}}(f\psi) + \frac{1}{2}\Delta\mathbf{v}\Delta\mathbf{v} : \frac{\partial^2(f\psi)}{\partial\mathbf{v}\partial\mathbf{v}} + \cdots \right\} \tag{8.23}$$

Clearly, the total probability of all possible deflections must be unity:

$$\int \psi\, \mathrm{d}(\Delta\mathbf{v}) = 1$$

Then, defining the rate of change of f due to collisions by

$$\left(\frac{\partial f}{\partial t}\right)_{\mathrm{c}} = \frac{f(\mathbf{r},\mathbf{v},t) - f(\mathbf{r},\mathbf{v},t-\Delta t)}{\Delta t}$$

we find from (8.23)

$$\left(\frac{\partial f}{\partial t}\right)_{\mathrm{c}} = -\frac{\partial}{\partial\mathbf{v}}\cdot\left(\frac{f\langle\Delta\mathbf{v}\rangle}{\Delta t}\right) + \frac{1}{2}\frac{\partial^2}{\partial\mathbf{v}\partial\mathbf{v}} : \left(\frac{f\langle\Delta\mathbf{v}\Delta\mathbf{v}\rangle}{\Delta t}\right) \tag{8.24}$$

where

$$\left\{ \begin{array}{c} \langle\Delta\mathbf{v}\rangle \\ \langle\Delta\mathbf{v}\Delta\mathbf{v}\rangle \end{array} \right\} = \int \psi(\mathbf{v},\Delta\mathbf{v}) \left\{ \begin{array}{c} \Delta\mathbf{v} \\ \Delta\mathbf{v}\Delta\mathbf{v} \end{array} \right\} \mathrm{d}(\Delta\mathbf{v})$$

are the average changes in $\Delta\mathbf{v}$ and $\Delta\mathbf{v}\Delta\mathbf{v}$ in time Δt. It is important to note here that both of these average changes are proportional to Δt whereas third and higher order terms in the Taylor expansion in (8.23) are of higher order in Δt and have, therefore, been dropped. The reason why $\langle\Delta\mathbf{v}\Delta\mathbf{v}\rangle$ is of the same order in Δt as $\langle\Delta\mathbf{v}\rangle$ is that collisions are treated as a random walk process in which mean square displacements increase linearly with time.

Substitution of (8.24) in (8.1) gives the *Fokker–Planck equation*. Until we define the probability function $\psi(\mathbf{v},\Delta\mathbf{v})$, however, it remains a formal statement. Various forms of the Fokker–Planck equation have been derived for a plasma including attempts to describe many-particle collisions in terms of rapidly oscillating electric fields (Gasiorowicz, Neuman and Riddell (1956)) and charge density fluctuations (Kaufman (1960)). We shall follow the derivation of Rosenbluth, MacDonald and Judd (1957) who, using heuristic arguments like those of Landau (1946), assumed that multiple collisions could be treated as sequences of binary collisions and calculated $\langle\Delta\mathbf{v}\rangle/\Delta t$ and $\langle\Delta\mathbf{v}\Delta\mathbf{v}\rangle/\Delta t$ on the basis of the dynamics of Coulomb collisions.

We consider collisions between a particle of mass m with initial velocity $\mathbf{v}$ and a 'scattering' particle of mass m_s and initial velocity $\mathbf{v}_s$. It is convenient to work in the centre of mass frame of reference and define the initial relative velocity

$$\mathbf{g} = \mathbf{v} - \mathbf{v}_s$$

and the centre of mass velocity

$$\mathbf{V} = \frac{m\mathbf{v} + m_s\mathbf{v}_s}{m + m_s}$$

We ignore the effect of any macroscopic forces over the duration of a collision on the assumption that they act over length scales much greater than the Debye length. Then, denoting final velocities by primed variables, we may write

$$\mathbf{v} = \mathbf{V} + \frac{m_s}{m + m_s}\mathbf{g} \qquad \mathbf{v}' = \mathbf{V}' + \frac{m_s}{m + m_s}\mathbf{g}'$$

and conservation of momentum and energy gives

$$\mathbf{V} = \mathbf{V}' \qquad |\mathbf{g}| = |\mathbf{g}'|$$

Thus,

$$\Delta\mathbf{v} = \mathbf{v}' - \mathbf{v} = \frac{m_s}{m + m_s}(\mathbf{g}' - \mathbf{g}) = \frac{m_s}{m + m_s}\Delta\mathbf{g} \tag{8.25}$$

Now if the differential scattering cross-section is $\sigma(|\mathbf{g}|, \theta)$ then the probability in time Δt of collisions with scattering angle θ is proportional to $\Delta t f_s(\mathbf{v}_s)|\mathbf{g}|\sigma(|\mathbf{g}|, \theta)$, where f_s is the distribution function of scattering particles, and so the average value of $\Delta\mathbf{v}$ is given by

$$\begin{aligned}\langle\Delta\mathbf{v}\rangle &= \Delta t \int \mathrm{d}\mathbf{v}_s\, \mathrm{d}\Omega\, f_s(\mathbf{v}_s)|\mathbf{g}|\sigma(|\mathbf{g}|, \theta)\Delta\mathbf{v} \\ &= \frac{m_s}{m + m_s}\Delta t \int \mathrm{d}\mathbf{v}_s\, \mathrm{d}\Omega\, f_s(\mathbf{v}_s)|\mathbf{g}|\sigma(|\mathbf{g}|, \theta)\Delta\mathbf{g}\end{aligned} \tag{8.26}$$

where the integration is over the solid angle Ω and all scattering velocities $\mathbf{v}_s$.

With reference to Fig. 8.4 we see that all particles passing through the element of area $2\pi b\, \mathrm{d}b$ are scattered into the element of solid angle $\mathrm{d}\Omega = 2\pi \sin\theta\, \mathrm{d}\theta$ and so the differential scattering cross-section

$$\sigma(|\mathbf{g}|, \theta) = -\frac{2\pi b\, \mathrm{d}b}{\mathrm{d}\Omega} = -\frac{b\, \mathrm{d}b}{\sin\theta\, \mathrm{d}\theta}$$

where the minus sign is introduced to make σ a positive quantity since $\mathrm{d}b/\mathrm{d}\theta$ is negative. The fundamental relationship between the impact parameter b and the scattering angle θ for Coulomb interactions is (see Goldstein (1959))

$$b = b_0 \cot\theta/2 \tag{8.27}$$

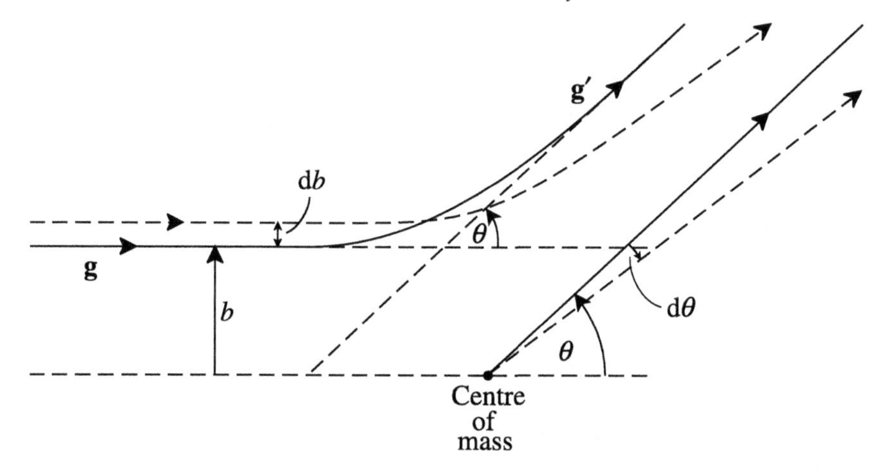

Fig. 8.4. Scattering in centre of mass frame.

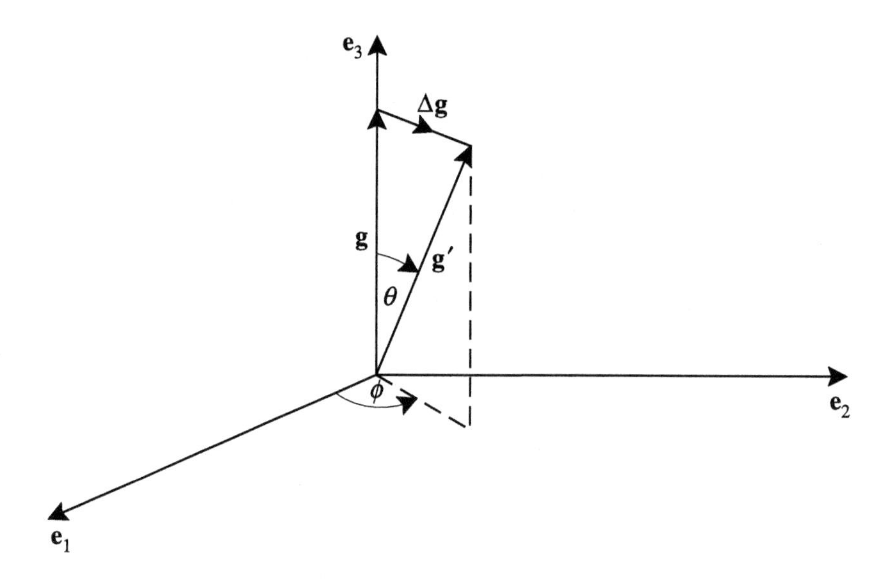

Fig. 8.5. Resolution of $\mathbf{g}'$.

where

$$b_0 = \frac{z z_s e^2 (m + m_s)}{4\pi \varepsilon_0 m m_s |\mathbf{g}|^2}$$

is the impact parameter for right-angle scattering and z, z_s are the atomic numbers† of the scattered and scattering particles, respectively.

† We use lower case z here since we want to allow for $z = Z$ (ions) and $z = -1$ (electrons).

To carry out the integration over solid angle we first resolve $\mathbf{g}'$ into its components in a rectangular coordinate system with polar axis parallel to $\mathbf{g}$. This is illustrated in Fig. 8.5 and leads to the result

$$\Delta\mathbf{g} = \mathbf{g}' - \mathbf{g} = g[\sin\theta\cos\phi\,\mathbf{e}_1 + \sin\theta\sin\phi\,\mathbf{e}_2 + (\cos\theta - 1)\,\mathbf{e}_3]$$

On integration over the azimuthal angle ϕ, components perpendicular to $\mathbf{g}$ vanish (scattering is equally likely in all perpendicular directions) and one finds

$$\frac{\langle\Delta\mathbf{v}\rangle}{\Delta t} = -\Gamma\sum_s z_s^2\left(\frac{m+m_s}{m_s}\right)\int\frac{\mathbf{g}}{g^3}f_s(\mathbf{v}_s)\,\mathrm{d}\mathbf{v}_s \tag{8.28}$$

where the sum is over all types of scattering particles and we have assumed

$$\Gamma = \frac{z^2e^4}{4\pi\varepsilon_0^2m^2}\ln(\lambda_\mathrm{D}/b_0) \approx \frac{z^2e^4}{4\pi\varepsilon_0^2m^2}\ln\Lambda$$

where $\ln\Lambda$ is the Coulomb logarithm. In deriving this result we have applied a cut-off at small scattering angles corresponding to $\theta_{\min} = 2b_0/\lambda_\mathrm{D}$ which is equivalent to a maximum impact parameter $b_{\max} = \lambda_\mathrm{D}$, as can be seen from (8.27) in the limit of small θ. Strictly speaking, this means that Γ should have a subscript s, indeed is a function of $\mathbf{v}_s$, but since this dependence lies entirely inside the argument of the logarithm it is customary to ignore it. Thus, we substitute the plasma Λ for λ_D/b_0, and treat Γ as a constant. In evaluating $\langle\Delta\mathbf{v}\Delta\mathbf{v}\rangle$ only the cross terms in $\langle\Delta\mathbf{g}\Delta\mathbf{g}\rangle$ now vanish on integration over ϕ. However, since weak collisions (small θ) dominate, the $\mathbf{e}_3\mathbf{e}_3$ term, which is of order θ^4 (compared with the $\mathbf{e}_1\mathbf{e}_1$ and $\mathbf{e}_2\mathbf{e}_2$ terms which are of order θ^2) turns out to be smaller by a factor $\ln(\lambda_\mathrm{D}/b_0)$ and we neglect it with the result

$$\frac{\langle\Delta v_i\Delta v_j\rangle}{\Delta t} = \Gamma\sum_s z_s^2\int\frac{g^2\delta_{ij} - g_ig_j}{g^3}f_s(\mathbf{v}_s)\,\mathrm{d}\mathbf{v}_s \tag{8.29}$$

Noting that

$$\frac{\mathbf{g}}{g^3} = -\frac{\partial}{\partial\mathbf{v}}\left(\frac{1}{g}\right) \tag{8.30}$$

and

$$\frac{g^2\delta_{ij} - g_ig_j}{g^3} = \frac{\partial^2 g}{\partial v_i\partial v_j} \tag{8.31}$$

we may express the Fokker–Planck coefficients in terms of the *Rosenbluth potentials* defined by

$$G(\mathbf{v}) = \sum_s z_s^2\int|\mathbf{v} - \mathbf{v}_s|\,f_s(\mathbf{v}_s)\,\mathrm{d}\mathbf{v}_s \tag{8.32}$$

and

$$H(\mathbf{v}) = \sum_s z_s^2 \left(\frac{m + m_s}{m_s}\right) \int \frac{f_s(\mathbf{v}_s)}{|\mathbf{v} - \mathbf{v}_s|} \, d\mathbf{v}_s \tag{8.33}$$

From (8.28) and (8.30) we have

$$\frac{\langle \Delta \mathbf{v} \rangle}{\Delta t} = \Gamma \frac{\partial H}{\partial \mathbf{v}}$$

and from (8.29) and (8.31)

$$\frac{\langle \Delta \mathbf{v}_i \Delta \mathbf{v}_j \rangle}{\Delta t} = \Gamma \frac{\partial^2 G}{\partial v_i \partial v_j}$$

which, on substitution in (8.24), gives

$$\left(\frac{\partial f}{\partial t}\right)_c = -\Gamma \frac{\partial}{\partial \mathbf{v}} \cdot \left(f \frac{\partial H}{\partial \mathbf{v}}\right) + \frac{1}{2} \Gamma \frac{\partial^2}{\partial \mathbf{v} \partial \mathbf{v}} : \left(f \frac{\partial^2 G}{\partial \mathbf{v} \partial \mathbf{v}}\right) \tag{8.34}$$

This is the usual form of the Fokker–Planck collision term. As we shall see in the next section, the first term in (8.34) produces a deceleration of a test particle and is known as the *coefficient of dynamical friction*. The second term accounts for the spreading of a unidirectional beam throughout velocity space and is called the *coefficient of diffusion*. A complication of the Fokker–Planck coefficients is their non-linear dependence on the distribution function f which appears explicitly and implicitly in the Rosenbluth potentials. An approximation frequently made is to use Maxwell distributions in the Rosenbluth potentials though this is really justified only for near-equilibrium plasmas.

Another interesting and useful form of the Fokker–Planck collision term is obtained (see Exercise 8.5) by noting that

$$\frac{\partial}{\partial \mathbf{v}} \cdot \left(\frac{\partial^2 g}{\partial \mathbf{v} \partial \mathbf{v}}\right) = -\frac{2\mathbf{g}}{g^3} = -\frac{\partial}{\partial \mathbf{v}_s} \cdot \left(\frac{\partial^2 g}{\partial \mathbf{v} \partial \mathbf{v}}\right) \tag{8.35}$$

Substituting this expression for $\mathbf{g}/g^3$ in (8.28) and integrating by parts with respect to $\mathbf{v}_s$ we obtain

$$\left(\frac{\partial f}{\partial t}\right)_c = \frac{\Gamma m}{2} \frac{\partial}{\partial \mathbf{v}} \cdot \sum_s z_s^2 \int d\mathbf{v}_s \frac{\partial^2 |\mathbf{v} - \mathbf{v}_s|}{\partial \mathbf{v} \partial \mathbf{v}} \cdot \left(\frac{f_s(\mathbf{v}_s)}{m} \frac{\partial f(\mathbf{v})}{\partial \mathbf{v}} - \frac{f(\mathbf{v})}{m_s} \frac{\partial f_s(\mathbf{v}_s)}{\partial \mathbf{v}_s}\right) \tag{8.36}$$

It can be shown (see Hinton (1983)) that the Fokker–Planck equation has the following desirable properties:

- the distribution function cannot become negative – collisions act to fill holes in velocity space;
- particle number, momentum and energy are conserved;

- it satisfies Boltzmann's H-theorem, i.e. the only time-independent distribution functions satisfying $(\partial f/\partial t)_c = 0$ are Maxwellians.

8.5 Collisional parameters

By taking velocity moments of the Fokker–Planck equation we may define and find estimates for various collisional parameters. The sort of parameters of interest are the time scales for an arbitrary distribution of velocities to become Maxwellian and for the equalization of ion and electron temperatures. A very simple model that permits the calculation of rough estimates of such time scales is the *test particle model* in which a single test particle (either an electron or an ion) travels through a uniform, field-free plasma in thermal equilibrium. Then the Fokker–Planck equation is simply

$$\frac{\partial f}{\partial t} = -\Gamma \frac{\partial}{\partial \mathbf{v}} \cdot \left(f \frac{\partial H}{\partial \mathbf{v}} \right) + \frac{1}{2} \Gamma \frac{\partial^2}{\partial \mathbf{v} \partial \mathbf{v}} : \left(f \frac{\partial^2 G}{\partial \mathbf{v} \partial \mathbf{v}} \right) \tag{8.37}$$

Its moments give expressions for the rates of change of velocity, energy, etc. and the distribution functions are simple enough for the collision moments to be evaluated.

The distribution function of the test particle is

$$f(\mathbf{v}, t) = \delta[\mathbf{v} - \mathbf{V}(t)] \tag{8.38}$$

where $\mathbf{V}(t)$ is the particle velocity at time t and we find estimates for various collisional parameters by evaluating the velocity moments at $t = 0$. A more rigorous approach is presented by Hinton (1983) but the results are the same.

Multiplying (8.37) by $\mathbf{v}$ and integrating gives

$$\frac{\partial \mathbf{V}}{\partial t} = \Gamma \frac{\partial H(\mathbf{V})}{\partial \mathbf{V}} \tag{8.39}$$

where the term in G has vanished on integration by parts twice. This confirms the statement in the previous section that the first Fokker–Planck coefficient represents the dynamical friction decelerating the test particle.

Since the plasma particles, the scatterers, are assumed to be in thermal equilibrium

$$f_s(\mathbf{v}_s) = f_{\mathrm{M}}(v_s) = \frac{n_s a_s^3}{\pi^{3/2}} e^{-a_s^2 v_s^2} \tag{8.40}$$

where

$$a_s^2 = \frac{m_s}{2 k_{\mathrm{B}} T_s}$$

and it follows that

$$H(\mathbf{v}) = \sum_s z_s^2 \left(\frac{m + m_s}{m_s} \right) \frac{n_s}{v} \phi(a_s v) \tag{8.41}$$

where $\phi(x)$ is the error function

$$\phi(x) = \frac{2}{\sqrt{\pi}} \int_0^x e^{-y^2}\, dy$$

Noting that $H(\mathbf{v})$ is an isotropic function (a direct consequence of assuming f_s to be isotropic) it follows that (8.39) may be written

$$\frac{\partial V}{\partial t} = \Gamma \frac{\partial H(V)}{\partial V} = -\nu_{\mathrm{f}}(V) V$$

say, defining the *frictional coefficient* ν_{f}. Hence,

$$\nu_{\mathrm{f}} = \frac{2\Gamma}{V} \sum_s z_s^2 \left(\frac{m + m_s}{m_s} \right) n_s a_s^2 \Psi(a_s V) \tag{8.42}$$

where

$$\Psi(x) = \frac{\phi(x) - x\phi'(x)}{2x^2}$$

The first observation from (8.42) is that, for given V, ν_{f} decreases as the plasma density decreases or its temperature increases; in other words, collisions are less effective under low density, high temperature conditions. Also, using the limiting values

$$\left. \begin{array}{lll} \phi(x) \to 2x/\sqrt{\pi} & \Psi(x) \to 2x/3\sqrt{\pi} & \text{as } x \to 0 \\ \phi \to 1 & \Psi(x) \to 1/2x^2 & \text{as } x \to \infty \end{array} \right\} \tag{8.43}$$

we see that for very fast test particles ($a_s V \to \infty$) $\nu_{\mathrm{f}} \propto V^{-3}$, while in the opposite limit ($a_s V \to 0$), where V is much less than the thermal speeds, ν_{f} is independent of V. Thus, the frictional deceleration of a test particle increases with V at low speeds but decreases with V at sufficiently high speeds. A consequence of this is that the current in a plasma tends to be carried predominantly by the electrons in the tail of the velocity distribution.

Another possible consequence is a phenomenon known as *electron runaway*. If one imagines an electron subjected to a constant accelerating force then a balance will be achieved at low velocities but not at those velocities for which the frictional force is less than the accelerating force and decreasing. The situation is represented schematically in Fig. 8.6 where A is the acceleration and $F(V) = \nu_{\mathrm{f}}(V)V$ is the frictional deceleration. The equilibrium point at V_1 is stable but that at V_2 is unstable and an electron with $V > V_2$ is continually accelerated. In practice this is likely to be limited by instabilities, such as the two-stream instability, driven by the free energy in the runaway electrons.

Although the frictional coefficient is an important parameter for such matters as the slowing down of particle beams, collision frequencies, representing all manner

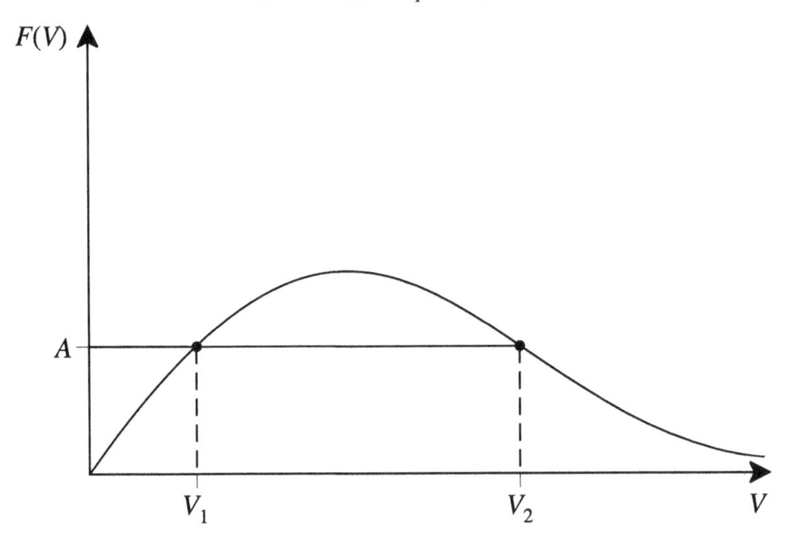

Fig. 8.6. Schematic illustration of electron runaway.

of collisional effects, must be defined more generally. By convention the *collision frequency* is taken to be the inverse of the mean time for a particle to be deflected through a right-angle. To find a measure of this using the test particle model we take the $v_\perp^2$ moment of (8.37) where $v_\perp^2$ is the sum of the squares of the components of $\mathbf{v}$ perpendicular to $\mathbf{V}(0)$; it is necessary to use a mean square deviation since, in an isotropic plasma, the mean deviation is zero. Here we find (see Exercise 8.6) that only the second term in (8.37) gives a non-zero contribution (as predicted in the previous section) which may be written as

$$\frac{\partial V_\perp^2}{\partial t} = \frac{2\Gamma}{V}\frac{\partial G}{\partial V} \tag{8.44}$$

Substituting (8.40) in (8.32) gives

$$G(v) = \sum_s z_s^2 n_s \left[\left(v + \frac{1}{2a_s^2 v}\right)\phi(a_s v) + \frac{e^{-a_s^2 v^2}}{a_s\sqrt{\pi}}\right] \tag{8.45}$$

and

$$\frac{\partial G}{\partial v} = \sum_s z_s^2 n_s [\phi(a_s v) - \Psi(a_s v)] \tag{8.46}$$

It is convenient to define separate collision frequencies for the scatterers so we write (8.44) as

$$\frac{\partial V_\perp^2}{\partial t} = \frac{2\Gamma}{V}\frac{\partial G}{\partial V} = \sum_s \nu_s V^2$$

and using (8.46) we have

$$\nu_s(V) = \frac{2\Gamma}{V^3} z_s^2 n_s [\phi(a_s V) - \Psi(a_s V)] \tag{8.47}$$

This parameter gives a measure of the time scale for relaxation to isotropy of an initially anisotropic distribution which is perhaps the chief reason for its adoption as the collision frequency. Using the asymptotic expansions (8.43) we see that $\nu_s \propto V^{-2}$ for $a_s V \to 0$ and $\nu_s \propto V^{-3}$ for $a_s V \to \infty$.

It is of particular interest to compare electron–electron, ion–ion, electron–ion and ion–electron collision frequencies which we label ν_{ab}, where a indicates the scattered (test) particle and b the scattering particles. From (8.47) we find

$$\begin{aligned}
\nu_{\text{ee}} &= \frac{n_{\text{e}} e^4 \ln \Lambda}{2\pi \varepsilon_0^2 m_{\text{e}}^2 V^3} [\phi(a_{\text{e}} V) - \Psi(a_{\text{e}} V)] \\
\nu_{\text{ii}} &= \frac{n_{\text{i}} Z^4 e^4 \ln \Lambda}{2\pi \varepsilon_0^2 m_{\text{i}}^2 V^3} [\phi(a_{\text{i}} V) - \Psi(a_{\text{i}} V)] \\
\nu_{\text{ei}} &= \frac{n_{\text{i}} Z^2 e^4 \ln \Lambda}{2\pi \varepsilon_0^2 m_{\text{e}}^2 V^3} [\phi(a_{\text{i}} V) - \Psi(a_{\text{i}} V)] \\
\nu_{\text{ie}} &= \frac{n_{\text{e}} Z^2 e^4 \ln \Lambda}{2\pi \varepsilon_0^2 m_{\text{i}}^2 V^3} [\phi(a_{\text{e}} V) - \Psi(a_{\text{e}} V)]
\end{aligned}$$

where Ze is the ionic charge so that $n_{\text{e}} = Zn_{\text{i}}$. To estimate relaxation times we take the test particle speed V to be thermal in which case the terms in square brackets in ν_{ee} and ν_{ii} become constants of order one but for ν_{ei} and ν_{ie} we must take respectively the large and small argument limits in (8.43) with the result

$$\nu_{\text{ee}} : \nu_{\text{ii}} : \nu_{\text{ei}} : \nu_{\text{ie}} \sim 1 : Z^3 \left(\frac{m_{\text{e}}}{m_{\text{i}}}\right)^{1/2} \left(\frac{T_{\text{e}}}{T_{\text{i}}}\right)^{3/2} : Z : Z^2 \left(\frac{m_{\text{e}}}{m_{\text{i}}}\right) \left(\frac{T_{\text{e}}}{T_{\text{i}}}\right)$$

Except when Z or $T_{\text{e}}/T_{\text{i}}$ is large, it is the mass ratio which dominates this comparison of collision frequencies and we see that $\nu_{\text{ei}} \sim Z\nu_{\text{ee}} \gg \nu_{\text{ii}} \gg \nu_{\text{ie}}$. The first of these gross inequalities arises because thermal ion speeds are less than thermal electron speeds, by $(m_{\text{e}}/m_{\text{i}})^{1/2}$ if $T_{\text{e}} \approx T_{\text{i}}$, and so ions take longer to meet each other. The second reflects the fact that the electrons are not very effective in deflecting the much heavier ions.

Another set of useful collision parameters is obtained by considering the energy exchange between the test particle and the scatterers. Here again we use a mean square deviation to define the parameters because the test particle is both losing energy due to its deceleration in the forward direction and gaining energy in the perpendicular direction by its deflection. If W is the energy of the test particle, $\Delta W = m[v^2 - V^2(0)]/2$ and we take the $(\Delta W)^2$ moment of (8.37). Denoting this by $\overline{(\Delta W)^2}$, we find

$$\begin{aligned}\frac{\partial \overline{(\Delta W)^2}}{\partial t} &= \Gamma m^2 V^2 \frac{\partial^2 G}{\partial V^2} \\ &= 2\Gamma m^2 V \sum_s n_s z_s^2 \Psi(a_s V) \\ &= \sum_s \nu_s^E \left(\frac{mV^2}{2}\right)^2 \end{aligned} \tag{8.48}$$

say, defining collision coefficients ν_s^E for energy exchange. Thus,

$$\nu_s^E(V) = \frac{8\Gamma n_s z_s^2 \Psi(a_s V)}{V^3} \tag{8.49}$$

Since $[\phi(1) - \Psi(1)] \approx 0.6$ and $\Psi(1) \approx 0.2$, comparing (8.47) and (8.49) we see that for thermal speeds $\nu_{\mathrm{ee}}^E \sim \nu_{\mathrm{ee}}$ and $\nu_{\mathrm{ii}}^E \sim \nu_{\mathrm{ii}}$. On the other hand, energy exchange between ions and electrons is the least efficient process since at most a fraction $(m_{\mathrm{e}}/m_{\mathrm{i}})$ of the kinetic energy involved in a collision can be transferred from one particle to the other. For thermal speeds, low Z and $T_{\mathrm{e}} \sim T_{\mathrm{i}}$ we find

$$\nu_{\mathrm{ee}}^E : \nu_{\mathrm{ii}}^E : \nu_{\mathrm{ei}}^E (\sim \nu_{\mathrm{ie}}^E) \sim 1 : \left(\frac{m_{\mathrm{e}}}{m_{\mathrm{i}}}\right)^{1/2} : \left(\frac{m_{\mathrm{e}}}{m_{\mathrm{i}}}\right)$$

Thus, in an anisotropic plasma with unequal electron and ion temperatures, the electrons will relax to a Maxwellian distribution within a few electron–electron collision times followed by the ions in a few ion–ion collision times and finally, after a time of order $(m_{\mathrm{i}}/m_{\mathrm{e}})\nu_{\mathrm{ee}}^{-1}$, equilibration of the electron and ion temperatures takes place.

8.6 Collisional relaxation

The collision frequencies derived in the previous section are useful for making order of magnitude calculations but their velocity dependence means that the actual time taken for the relaxation of the high velocity part of a distribution function to a Maxwellian can be very much greater than the 'thermal speed' estimate. Numerical studies have shown that typically these estimates are good for the bulk of the distribution out to approximately twice the thermal speed but beyond that relaxation is progressively much slower. The assumption, usually made even for collisionless plasmas, that the distribution function is approximately Maxwellian is, therefore, often not sustainable. Consequently, there has long been an interest in feasible alternative distribution functions.

One important example is the *self-similar* distribution function of order s defined as

$$f_s(v) = \frac{c_s n}{\pi^{3/2} v_s^3} \exp(-v^s / v_s^s)$$

The constants c_s and v_s are given by

$$\begin{aligned} c_s &= \frac{3\pi^{1/2}}{4\Gamma(1+3/s)} \\ v_s^2 &= \frac{k_B T s \Gamma(1+3/s)}{m\Gamma(5/s)} \end{aligned}$$

where n is the particle number density and $\Gamma(x)$ is the gamma function. It is easily seen that $s = 2$ corresponds to the Maxwellian distribution.

Jones (1980), Langdon (1980) and Balescu (1982) have shown that laser-irradiated high Z plasmas reach a self-similar state with $s = 5$ for the electrons and $s = 2$ for the ions. Compared with the Maxwellian, self-similar distributions with $s > 0$ (also known as super-Gaussian distributions) are flat-topped and for this reason they have been of interest to space-plasma physicists in explaining the frequently observed electron distributions at the Earth's bow shock (Feldman *et al.* (1983)). In weak turbulence theory, anomalous transport coefficients have been derived from self-similar distributions (see Dum (1978)).

As we have seen in this chapter and will discuss more extensively in Chapter 12, the derivation of transport coefficients is directly related to the relaxation of the distribution function and involves the calculation of its velocity moments. For this reason in particular there has been considerable analytical and numerical investigation of the collisional relaxation of non-Maxwellian distribution functions. Analytical progress depends upon some simplifying assumption, usually that the distribution function remains in the same class of function (e.g. remains self-similar but with varying s) throughout the relaxation. Given our remarks in the opening paragraph of this section, this is hardly likely to be the case and numerical studies based on the Fokker–Planck equation have confirmed this suspicion. As an illustration we shall consider two examples of temperature equalization in a single species plasma. Plasmas with two-temperature velocity distributions are created in the laboratory in heating processes and occur naturally in space.

The *isotropic two-temperature distribution function*

$$f(v^2) = f_c(v^2; n_c, T_c) + f_h(v^2; n_h, T_h)$$

where both f_c and f_h are Maxwell distributions (8.4) and the subscripts denote cold and hot components, is used for plasmas created, for example, by the injection of a hot plasma into a cooler, background plasma. It has been noted in numerical simulations of such plasmas that the temperatures of the separate components

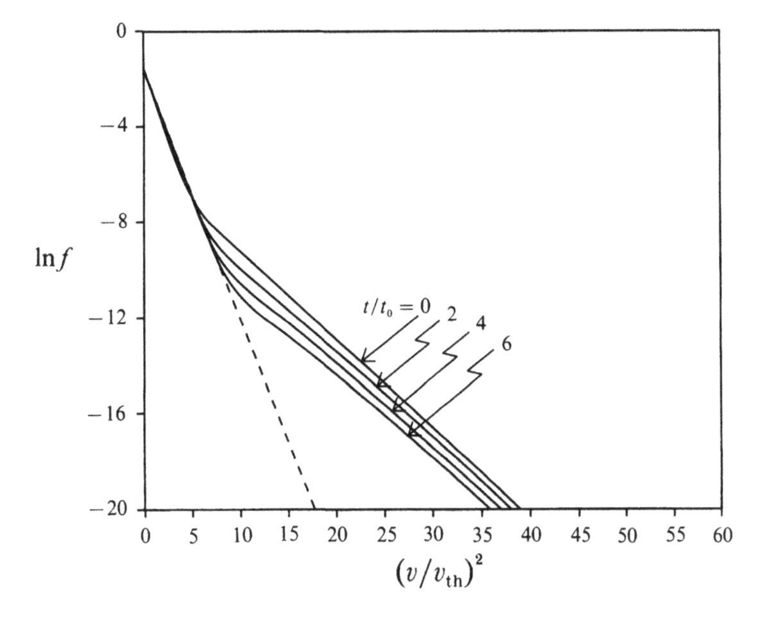

Fig. 8.7. Relaxation of isotropic two-temperature plasma (after McGowan and Sanderson (1992)).

change much more slowly than their densities. McGowan and Sanderson (1992) modelled their evolution by

$$f(v^2, t) = f(v^2; n(t), T) + f_c(v^2; n_c(t), T_c) + f_h(v^2; n_h(t), T_h)$$

where all three components are Maxwellians with constant temperatures but variable densities. The total number density n_0 and the average (or final) temperature T are then given by

$$\begin{aligned} n_0 &= n(t) + n_c(t) + n_h(t) \\ n_0 T &= n(t)T + n_c(t)T_c + n_h(t)T_h \end{aligned}$$

representing the conservation of particles and energy. Initially $n(0) = 0$ and finally $n(\infty) = n_0$, this component having grown at the expense of the other two for which $n_c(\infty) = n_h(\infty) = 0$. The collisional relaxation of such a plasma was shown to be well-represented by this model and is illustrated in Fig. 8.7, where the hot component appears as a succession of straight lines of constant slope (constant T_h).

The anisotropic, two-temperature distribution most widely used is the *bi-Maxwellian distribution*, introduced in Section 8.2.2,

$$f(\mathbf{v}) = n_0 \left(\frac{m}{2\pi k_B T_\perp}\right) \left(\frac{m}{2\pi k_B T_\parallel}\right)^{1/2} \exp\left(-\frac{mv_\perp^2}{2k_B T_\perp} - \frac{mv_\parallel^2}{2k_B T_\parallel}\right)$$

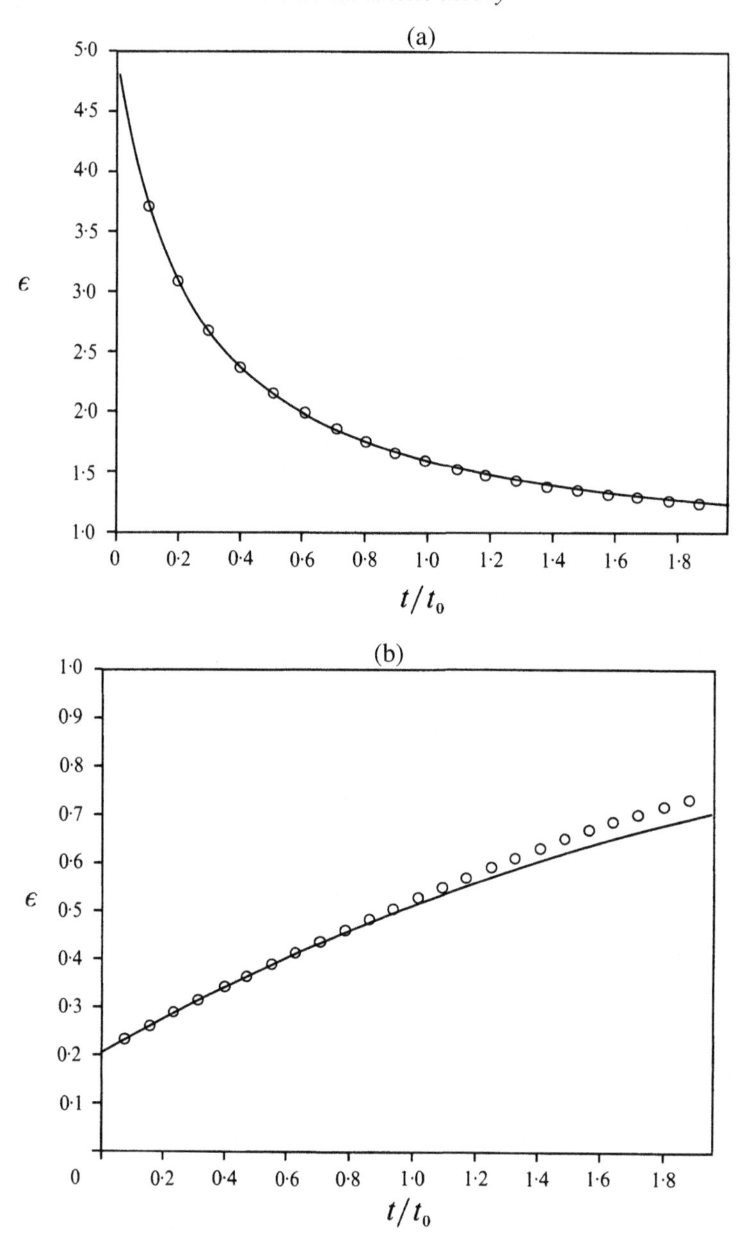

Fig. 8.8. Relaxation of a bi-Maxwellian plasma for (a) $\epsilon = 5$ (oblate distribution) and (b) $\epsilon = 0.2$ (prolate distribution) (after McGowan and Sanderson (1992)).

Here the anisotropy introduced by a strong magnetic field enables the velocity distributions for motion along the field lines ($v_\parallel$) and across the field lines ($\mathbf{v}_\perp$) to differ significantly.

Schamel *et al.* (1989), following earlier work by Kogan (1961) derived an evolution equation for the relaxation of a bi-Maxwellian distribution in terms of the *anisotropy parameter* $\epsilon = T_\perp / T_\parallel$. The equation is

$$\frac{d\epsilon}{d\tau} = \frac{(1+2\epsilon)^{5/2}}{4(1-\epsilon)}[(2+\epsilon)g(\epsilon) - 3] \tag{8.50}$$

where

$$\begin{aligned} g(\epsilon) &= \frac{\tanh^{-1}|\epsilon - 1|^{1/2}}{|\epsilon - 1|^{1/2}} \\ \tau &= \nu_E t \\ \nu_E &= \frac{8\pi^{1/2} n (Ze)^4 \ln \Lambda}{(3mk_B^3 T^3)^{1/2} \varepsilon_0^2} \end{aligned}$$

However, this assumes that the distribution function remains bi-Maxwellian. Numerical studies by Jorna and Wood (1987) and McGowan and Sanderson (1992) have shown that this is not a valid assumption, especially for the prolate case $\epsilon < 1$. Despite this, the evolution equation, which involves moments of the distribution function, does give reliable results for the oblate case ($\epsilon > 1$) where the moments are less sensitive to the departures from bi-Maxwellian form. In Fig. 8.8 solid lines denote the numerical results while the circles are points determined analytically from (8.50).

Exercises

8.1 Repeat the calculation of electrical conductivity σ carried out in Section 8.2 but allowing for a velocity dependent collision frequency. Show that the result is

$$\sigma = \frac{Z^2 e^2}{12\pi \varepsilon_0 k_B T} \int \frac{v^2 f_M \, d\mathbf{v}}{\nu_c(v)}$$

8.2 Explain physically why you would expect the ions to determine the coefficient of viscosity in a plasma but the electrons to determine the electrical and thermal transport coefficients.

What property of a plasma leads to ion determination of the ambipolar diffusion coefficient?

8.3 In neoclassical transport show that the fraction of particles performing banana orbits is proportional to the square root of the inverse aspect ratio.

8.4 Show that the cumulative effect of weak collisions in a plasma outweighs the effect of rare strong interactions by comparing the mean time for a single $\pi/2$ deflection with the plasma collision time ν_c^{-1}.

8.5 Verify (8.35) and hence obtain (8.36).

8.6 By carrying out the integrations by parts, show that in (8.37) the term in G gives no contribution to dynamical friction and that in H gives none to diffusion.

8.7 By adding a driving term $-(e\mathbf{E}/m)\cdot\partial f/\partial\mathbf{v}$ to the left-hand side of (8.37) and taking the first-order moment, as in the derivation of (8.39), find the relationship between the velocity $\mathbf{V}$ and field strength $\mathbf{E}$ for electron runaway (i.e. such that $\partial\mathbf{V}/\partial t > 0$).

8.8 Show that the Fokker–Planck equation that describes the evolution of the electron velocity distribution function $f_e(\mathbf{v}, t)$ for collisions with stationary massive ions may be written

$$\left(\frac{\partial f_e}{\partial t}\right)_c = \frac{n_e Z e^4 \ln\Lambda}{8\pi\varepsilon_0^2 m^2}\frac{\partial}{\partial\mathbf{v}}\cdot\left(\frac{v^2\mathbf{I}-\mathbf{vv}}{v^3}\cdot\frac{\partial f_e}{\partial\mathbf{v}}\right)$$

Using this obtain an expression for the resistivity when the plasma is perturbed by a time-dependent electric field $\mathbf{E} = \mathbf{E}_0\cos\omega t$. To do this write $f_e(\mathbf{v}) = f_0(v) + f_1(v)\cos\theta$, where θ is the angle between $\mathbf{E}_0$ and $\mathbf{v}$, and show that

$$f_1(v) = \frac{ieE_0}{m}\frac{\partial f_0}{\partial v}\left[\omega + \frac{in_e Z e^4\ln\Lambda}{4\pi\varepsilon_0^2 m^2 v^3}\right]^{-1}$$

From $f_1(v)$ write down the perturbed current density $\mathbf{j}_1$ and form $\langle\mathbf{j}_1\cdot\mathbf{E}\rangle$ to determine the average rate at which energy is absorbed by the plasma. Balancing this against the rate of loss of field energy, $\nu\varepsilon_0 E^2/2$ where ν denotes the energy damping rate and assuming that in absorbing energy the electron distribution remains Maxwellian, show that $\nu = (\omega_{pe}^2/\omega^2)\nu_{ei}$ where

$$\nu_{ei} = \frac{Zn_e e^4\ln\Lambda}{4\sqrt{2\pi}\,\varepsilon_0^2 m^{1/2} T_e^{3/2}}$$

8.9 Plasmas are often heated by ion beams (injected as neutral atoms) with energies in the 100 keV range, i.e. an order of magnitude greater than the plasma thermal energy. Within this range the beam velocity V is intermediate between ion and electron thermal velocities, i.e. $V_i < V < V_e$. Collisions between beam ions and plasma ions and electrons slow the beam through frictional drag and scattering off plasma ions deflects the beam.

Assuming that frictional drag dominates over velocity diffusion show that the Fokker–Planck equation for the distribution of beam ions, $f_b(\mathbf{v}, t)$, takes the form

$$\frac{\partial f_b}{\partial t} = A\frac{\partial}{\partial\mathbf{V}}\cdot\left[\frac{\mathbf{V}}{V^3}\left(1+\frac{V^3}{V_c^3}\right)f_b\right]$$

where $A = ZZ_b^2 n_e e^4 \ln\Lambda / 4\pi\varepsilon_0^2 m_i m_b$ and $V_c = (3\sqrt{\pi}Z/2)^{1/3} \times (T_e^3/m_e m_i^2)^{1/6}$.

8.10 Can you suggest why, in the relaxation of a bi-Maxwellian distribution function, the oblate case ($\epsilon > 1$) is less sensitive than the prolate case ($\epsilon < 1$) to the fact that the distribution does not remain bi-Maxwellian?

9
Plasma radiation

9.1 Introduction

We know from classical electrodynamics that accelerated charged particles are sources of electromagnetic radiation. Particles accelerated in electric or magnetic fields radiate with distinct characteristics. Electric micro-fields present in the plasma result in *bremsstrahlung* emission by plasma electrons. External radiation fields interacting with the plasma give rise to *scattered radiation*. Charged particles moving in magnetic fields emit *cyclotron* or *synchrotron radiation*, depending on the energy range of the particles.

The interaction of radiation with plasmas in all its aspects – emission, absorption, scattering and transport – is a key to understanding many effects in both laboratory and natural plasmas. Laboratory plasmas in particular do not radiate as black bodies so that an integrated treatment of emission, absorption and transport of radiation is usually needed. Core plasma parameters such as electron and ion temperatures and densities as well as plasma electric and magnetic fields may all be determined spectroscopically, in the most general sense of the term. Rather arbitrarily we shall confine our discussion to radiation from fully ionized plasmas thus excluding line radiation on which many diagnostic procedures are based. To some extent alternative spectroscopic techniques, in particular light scattering, have replaced if not entirely supplanted measurements of line radiation as preferred diagnostics of some key parameters in fusion plasmas (see Hutchinson (1988)). In the course of this chapter we shall outline the basis of some of these diagnostics, notably those that rely on bremsstrahlung and cyclotron radiation as well as those involving light scattering. We shall limit our discussion of radiation to plasmas in thermal equilibrium, with few exceptions. Non-thermal emission, while an important issue in practice, is in many instances still relatively poorly understood.

9.2 Electrodynamics of radiation fields

We begin with a statement of the results of Maxwellian electrodynamics essential to an understanding of radiation fields. Details of the derivations leading to these results are included in Exercise 9.1. The potentials $\mathbf{A}$ and ϕ are determined by

$$\left.\begin{aligned} \left[\nabla^2 - \frac{1}{c^2}\frac{\partial^2}{\partial t^2}\right]\mathbf{A} &= -\mu_0\mathbf{j} \\ \left[\nabla^2 - \frac{1}{c^2}\frac{\partial^2}{\partial t^2}\right]\phi &= -\frac{q}{\varepsilon_0} \end{aligned}\right\} \tag{9.1}$$

with specified current and charge sources together with the Lorentz gauge condition

$$\boldsymbol{\nabla}\cdot\mathbf{A} + \frac{1}{c^2}\frac{\partial\phi}{\partial t} = 0 \tag{9.2}$$

The solutions to (9.1) are expressed in terms of the *retarded potentials*:

$$\left.\begin{aligned} \mathbf{A}(\mathbf{r}, t) &= \frac{\mu_0}{4\pi}\int\frac{[\mathbf{j}(\mathbf{r}', t')]_{t'=t-|\mathbf{r}-\mathbf{r}'|/c}}{|\mathbf{r}-\mathbf{r}'|}\,\mathrm{d}\mathbf{r}' \\ \phi(\mathbf{r}, t) &= \frac{1}{4\pi\varepsilon_0}\int\frac{[q(\mathbf{r}', t')]_{t'=t-|\mathbf{r}-\mathbf{r}'|/c}}{|\mathbf{r}-\mathbf{r}'|}\,\mathrm{d}\mathbf{r}' \end{aligned}\right\} \tag{9.3}$$

Consider now a source consisting of a single particle of charge e moving arbitrarily with velocity $\dot{\mathbf{r}}_0(t)$ at a point $\mathbf{r}_0(t)$. Then

$$\mathbf{j}(\mathbf{r}, t) = e\dot{\mathbf{r}}_0(t)\delta(\mathbf{r}-\mathbf{r}_0(t)) \qquad q(\mathbf{r}, t) = e\delta(\mathbf{r}-\mathbf{r}_0(t)) \tag{9.4}$$

Substituting in (9.3) we find

$$\mathbf{A}(\mathbf{r}, t) = \frac{\mu_0}{4\pi}\left[\frac{e\mathbf{v}c}{cR - \mathbf{v}\cdot\mathbf{R}}\right]_{t'=t-R(t')/c} \tag{9.5}$$

$$\phi(\mathbf{r}, t) = \frac{1}{4\pi\varepsilon_0}\left[\frac{ec}{cR - \mathbf{v}\cdot\mathbf{R}}\right]_{t'=t-R(t')/c} \tag{9.6}$$

where $R(t') = |\mathbf{r} - \mathbf{r}_0(t')|$ and $\mathbf{v}(t') = \dot{\mathbf{r}}_0(t')$. These expressions are the Liénard–Wiechert potentials. Using the retarded potentials the electric field $\mathbf{E}(\mathbf{r}, t)$ may be expressed in a form due to Feynman:

$$\mathbf{E}(\mathbf{r}, t) = \frac{e}{4\pi\varepsilon_0}\left[\frac{\mathbf{n}}{R^2} + \frac{R}{c}\frac{\mathrm{d}}{\mathrm{d}t}\left(\frac{\mathbf{n}}{R^2}\right) + \frac{1}{c^2}\frac{\mathrm{d}^2\mathbf{n}}{\mathrm{d}t^2}\right]_{\text{ret}} \tag{9.7}$$

where $\mathbf{n}(t')$ is the unit vector from the source to the field point in Fig. 9.1 and ret denotes that the expression within the square brackets must be evaluated at the retarded time $t' = t - R(t')/c$. The first term in (9.7) represents the Coulomb field of the charge e at its retarded position. The second is a correction to the

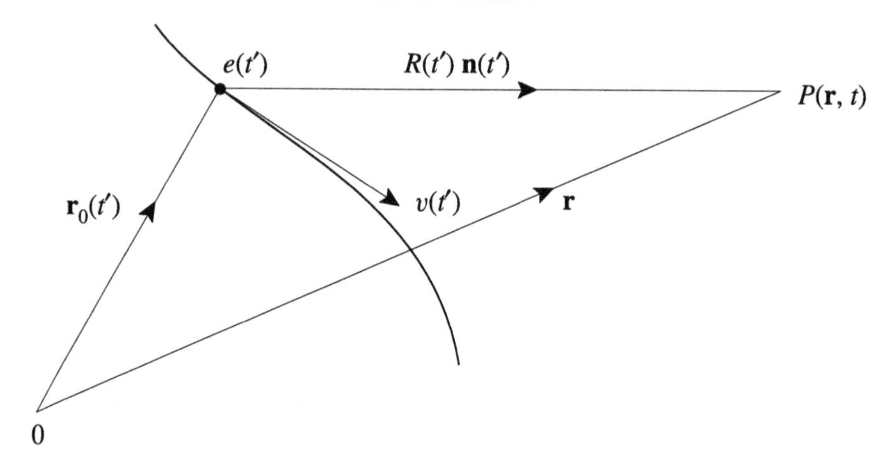

Fig. 9.1. Source–observer geometry for a radiating charged particle.

retarded Coulomb field, being the product of the rate of change of this field and the retardation delay time R/c. Thus the first and the second terms together correspond to the retarded Coulomb field advanced in time by R/c, namely to the observer's time t. In other words, for fields varying slowly enough these two terms represent the *instantaneous* Coulomb field of the charge. The final term, the second time derivative of the unit vector from the retarded position of the charge to the observer, contains the radiation electric field $\mathbf{E}^{\rm rad} \propto R^{-1}$, i.e.

$$\mathbf{E}^{\rm rad}(\mathbf{r}, t) = \frac{e}{4\pi\varepsilon_0}\left[\frac{\mathbf{n} \times \{(\mathbf{n} - \boldsymbol{\beta}) \times \dot{\boldsymbol{\beta}}\}}{cg^3 R}\right]_{\rm ret} \tag{9.8}$$

where $g = (1 - \mathbf{n} \cdot \boldsymbol{\beta})$. The total electric field is

$$\mathbf{E}(\mathbf{r}, t) = \frac{e}{4\pi\varepsilon_0}\left[\frac{(1 - \beta^2)(\mathbf{n} - \boldsymbol{\beta})}{g^3 R^2}\right]_{\rm ret} + \mathbf{E}^{\rm rad}(\mathbf{r}, t) \tag{9.9}$$

9.2.1 Power radiated by an accelerated charge

Once the radiation field is known, we can construct the Poynting vector $\mathbf{S}$ and so determine the instantaneous flux of energy (see Section 6.2)

$$\mathbf{S} = \mathbf{E} \times \mathbf{H} = c\varepsilon_0 |\mathbf{E}^{\rm rad}|^2 \mathbf{n} \tag{9.10}$$

with $\mathbf{E}^{\rm rad}$ given by (9.8). Thus the power P, radiated per unit solid angle Ω, is

$$\frac{\mathrm{d}P(t)}{\mathrm{d}\Omega} = (\mathbf{S} \cdot \mathbf{n}) R^2 \tag{9.11}$$

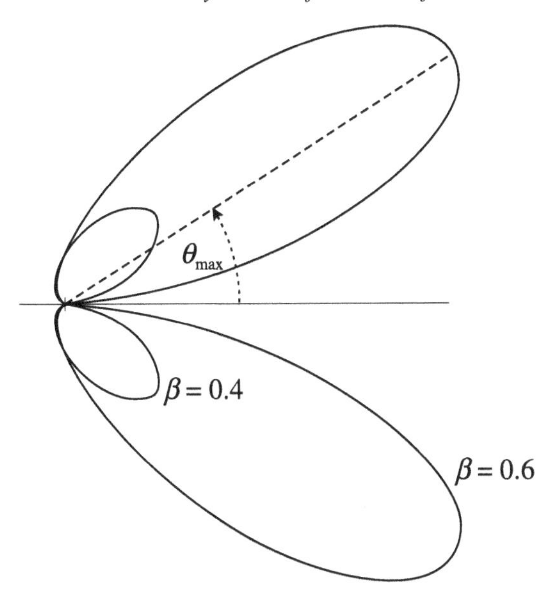

Fig. 9.2. Polar diagram of the instantaneous power radiated by an accelerated charged particle; $\theta_{\rm max}$ denotes the angle at which peak power is radiated.

where

$$\mathbf{S}\cdot\mathbf{n} = \frac{e^2}{16\pi^2\varepsilon_0 c}\left[\frac{|\mathbf{n}\times\{(\mathbf{n}-\boldsymbol{\beta})\times\dot{\boldsymbol{\beta}}\}|^2}{g^6 R^2}\right]_{\rm ret} \tag{9.12}$$

Note that $(\mathrm{d}P(t)/\mathrm{d}\Omega)\,\mathrm{d}\Omega$ denotes the radiated power measured by the observer at time t due to emission by the charge at time t'. It is often useful to consider power as a function of the retarded time t'. Then

$$\frac{\mathrm{d}P(t')}{\mathrm{d}\Omega} = (\mathbf{S}\cdot\mathbf{n})R^2\frac{\mathrm{d}t}{\mathrm{d}t'} = g(\mathbf{S}\cdot\mathbf{n})R^2 = \frac{e^2}{16\pi^2\varepsilon_0 c}\frac{|\mathbf{n}\times\{(\mathbf{n}-\boldsymbol{\beta})\times\dot{\boldsymbol{\beta}}\}|^2}{g^5} \tag{9.13}$$

Relativistic effects appear both in the numerator and, through g, in the denominator. In the ultra-relativistic limit ($\beta \to 1$) the effect of the denominator is dominant in determining the radiation pattern; the dipole distribution familiar from the non-relativistic limit deforms with the lobes inclined increasingly forward as in Fig. 9.2. In the non-relativistic limit $g \to 1$ and we recover the dipole distribution

$$\frac{\mathrm{d}P}{\mathrm{d}\Omega} = \frac{e^2\dot{\beta}^2}{16\pi^2\varepsilon_0 c}\sin^2\theta \tag{9.14}$$

where θ is the angle between $\dot{\boldsymbol{\beta}}$ and $\mathbf{n}$.

Larmor's formula for the power radiated in *all* directions follows on integrating over solid angle, i.e.

$$P = \int \frac{\mathrm{d}P}{\mathrm{d}\Omega}\,\mathrm{d}\Omega = \frac{e^2 \dot{v}^2}{6\pi \varepsilon_0 c^3} \tag{9.15}$$

The corresponding relativistic expression for the radiated power may be found by replacing (9.15) by its covariant form, giving

$$P = \frac{e^2}{6\pi \varepsilon_0 m^2 c^3}\left[\frac{\mathrm{d}p_\mu}{\mathrm{d}\tau}\frac{\mathrm{d}p^\mu}{\mathrm{d}\tau}\right] = \frac{e^2}{6\pi \varepsilon_0 c}\frac{[(\dot{\boldsymbol{\beta}})^2 - (\boldsymbol{\beta} \times \dot{\boldsymbol{\beta}})^2]}{(1-\beta^2)^3} \tag{9.16}$$

Much of our discussion will focus on the distinct characteristics of radiation from particles accelerated in the plasma micro-electric fields and in any magnetic fields present. Where plasmas are subject to external electromagnetic fields the incident radiation is scattered, with the scattering governed by the *Thomson cross-section*

$$\sigma_{\mathrm{T}} = \frac{8\pi}{3} r_{\mathrm{e}}^2 \tag{9.17}$$

where the classical electron radius $r_{\mathrm{e}} = (e^2/4\pi \varepsilon_0 m c^2)$.

9.2.2 Frequency spectrum of radiation from an accelerated charge

We consider next how the radiated energy is distributed in frequency. Since the spectrum is represented in terms of frequencies at a detector it is natural to revert to using time t (the observer's time). Then

$$\frac{\mathrm{d}P(t)}{\mathrm{d}\Omega} = \frac{e^2}{16\pi^2 \varepsilon_0 c}\left[\frac{|\mathbf{n} \times \{(\mathbf{n} - \boldsymbol{\beta}) \times \dot{\boldsymbol{\beta}}\}|^2}{g^6}\right]_{\mathrm{ret}} \equiv |\mathbf{a}(t)|^2 \tag{9.18}$$

The energy radiated per unit solid angle $\mathrm{d}W/\mathrm{d}\Omega$ is found by integrating (9.18) over time, giving

$$\frac{\mathrm{d}W}{\mathrm{d}\Omega} = \int_{-\infty}^{\infty} |\mathbf{a}(t)|^2\,\mathrm{d}t \tag{9.19}$$

Introducing the Fourier transform of $\mathbf{a}(t)$

$$\mathbf{a}(\omega) = \frac{1}{\sqrt{(2\pi)}}\int_{-\infty}^{\infty} \mathbf{a}(t) e^{i\omega t}\,\mathrm{d}t \tag{9.20}$$

and using Parseval's theorem allows us to represent the energy radiated per unit solid angle as

$$\frac{\mathrm{d}W}{\mathrm{d}\Omega} = \int_{-\infty}^{\infty} |\mathbf{a}(\omega)|^2\,\mathrm{d}\omega \tag{9.21}$$

Thus

$$\frac{\mathrm{d}W}{\mathrm{d}\Omega} = \int_0^\infty [|\mathbf{a}(\omega)|^2 + |\mathbf{a}(-\omega)|^2]\,\mathrm{d}\omega = 2\int_0^\infty |\mathbf{a}(\omega)|^2\,\mathrm{d}\omega \tag{9.22}$$

The energy radiated per unit solid angle per unit angular frequency interval is then

$$\frac{\mathrm{d}^2 W}{\mathrm{d}\Omega\,\mathrm{d}\omega} = 2|\mathbf{a}(\omega)^2| \tag{9.23}$$

Finding $\mathbf{a}(\omega)$ from (9.20) is simplified if we change the variable of integration from t to t', so removing the evaluation of the term in square brackets at the retarded time; then

$$\mathbf{a}(\omega) = \left(\frac{e^2}{32\pi^3\varepsilon_0 c}\right)^{1/2} \int_{-\infty}^{\infty} \exp\left\{i\omega\left(t' + \frac{R(t')}{c}\right)\right\} \left[\frac{\mathbf{n}\times\{(\mathbf{n}-\boldsymbol{\beta})\times\dot{\boldsymbol{\beta}}\}}{g^2}\right]\mathrm{d}t' \tag{9.24}$$

Since we wish to determine the spectrum in the radiation zone ($r \gg r_0$ in Fig. 9.1), $\mathbf{n}$ is effectively time-independent and $R(t') \simeq r - \mathbf{n}\cdot\mathbf{r}_0(t')$ so that

$$\frac{\mathrm{d}^2 W(\omega,\mathbf{n})}{\mathrm{d}\Omega\,\mathrm{d}\omega} = \frac{e^2}{16\pi^3\varepsilon_0 c}\left|\int_{-\infty}^{\infty} \exp\left\{i\omega\left(t' - \frac{\mathbf{n}\cdot\mathbf{r}_0(t')}{c}\right)\right\}\left[\frac{\mathbf{n}\times\{(\mathbf{n}-\boldsymbol{\beta})\times\dot{\boldsymbol{\beta}}\}}{g^2}\right]\mathrm{d}t'\right|^2 \tag{9.25}$$

Thus the energy radiated per unit solid angle per unit frequency interval is determined as a function of ω and $\mathbf{n}$ once $\mathbf{r}_0(t')$ is prescribed.

For purposes of calculation, we cast (9.25) in a slightly different form. Using the representation (see Exercise 9.1)

$$\frac{\mathbf{n}\times\{(\mathbf{n}-\boldsymbol{\beta})\times\dot{\boldsymbol{\beta}}\}}{g^2} = \frac{\mathrm{d}}{\mathrm{d}t'}\left[\frac{\mathbf{n}\times(\mathbf{n}\times\boldsymbol{\beta})}{g}\right]$$

we integrate (9.25) by parts to find, in the radiation zone,

$$\frac{\mathrm{d}^2 W(\omega,\mathbf{n})}{\mathrm{d}\Omega\,\mathrm{d}\omega} = \frac{e^2\omega^2}{16\pi^3\varepsilon_0 c}\left|\int_{-\infty}^{\infty} \exp\left\{i\omega\left(t' - \frac{\mathbf{n}\cdot\mathbf{r}_0(t')}{c}\right)\right\}[\mathbf{n}\times(\mathbf{n}\times\boldsymbol{\beta})]\mathrm{d}t'\right|^2 \tag{9.26}$$

The results summarized in this section provide a basis for the formalism needed to describe radiation emitted by charged particles. Much of the rest of the chapter is taken up with the characteristics of emission from particles moving in particular fields. Emission is of course only part of the story. Plasmas in thermal equilibrium that emit radiation absorb it as well and details of absorption mechanisms are crucial for the radiative heating of plasmas as we shall see in Chapter 11. Before discussing the emission of radiation we first summarize some ideas central to radiation transport in plasmas.

9.3 Radiation transport in a plasma

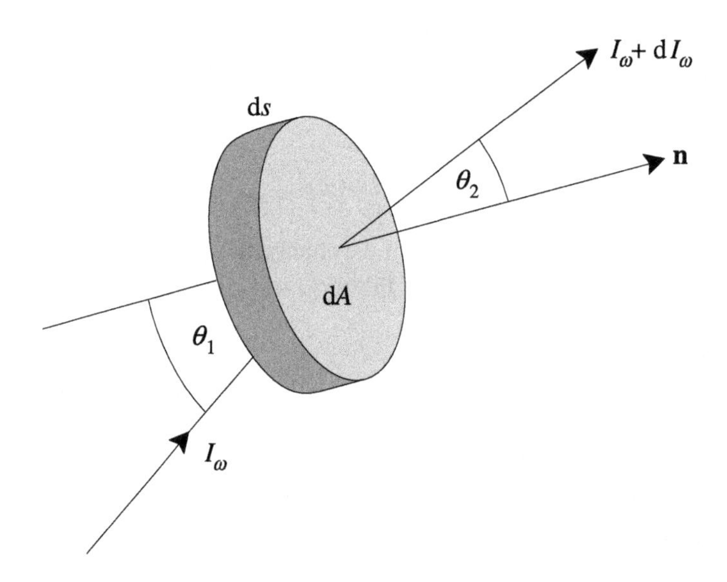

Fig. 9.3. Pencil of radiation refracted across an element of plasma.

The general problem of radiation transport in plasmas is complicated but fortunately for our purposes does not need to be discussed in detail. For simplicity we ignore scattering and take account of emission and absorption in the transport equation. This is strictly valid only under conditions of local thermodynamic equilibrium (LTE), a concept to be introduced later in this section.

The equation of radiative transfer can be thought of as an expression of energy conservation in terms of geometric optics. If $\mathbf{F}_\omega$ denotes the *spectral density of the radiation flux* then, by energy conservation in steady state,

$$\nabla \cdot \mathbf{F}_\omega = 0 \tag{9.27}$$

Note that $\mathbf{F}_\omega$ has dimensions of power per unit area per angular frequency interval. The principal assumption in geometric optics is that the properties of the medium vary *slowly* with position; that is, the scale length of variations is very much greater than the wavelength of radiation in the medium. In this approximation, one may picture the radiation being transported along *rays*. Figure 9.3 shows an element of plasma of cross-section dA, thickness ds and $\mathbf{n}$ is the unit normal outwards from the surface. The net radiation flux across this element is

$$\mathrm{d}\mathbf{F}_\omega \cdot \mathbf{n} = \mathrm{d}F_\omega \cos\theta = I_\omega(\mathbf{s}) \cos\theta \, \mathrm{d}\Omega \tag{9.28}$$

$I_\omega(\mathbf{s})$ is the *intensity* of the radiation and $\mathbf{s}$ denotes displacement along the ray. Its importance in radiation transport is due to the fact that it can be measured more or less directly. $I_\omega(\mathbf{s})$ is defined by

$$\mathrm{d}P_\omega(\mathbf{s}) = I_\omega(\mathbf{s})\cos\theta\,\mathrm{d}\Omega\,\mathrm{d}\omega\,\mathrm{d}A$$

where $\mathrm{d}P_\omega$ is the time-averaged power in the spectral range $\mathrm{d}\omega$ crossing an area $\mathrm{d}A$ within a cone of solid angle $\mathrm{d}\Omega$. I_ω is expressed in units of watts per square metre per steradian per unit angular frequency. In general, the intensity is a function both of direction and position in the medium. When it is a function of position alone, the radiation is said to be isotropic.

Suppose the plasma through which the radiation is passing is loss-free and isotropic but slightly inhomogeneous so that a ray, on passing through the element of plasma shown in Fig. 9.3, suffers bending. Then, by energy conservation,

$$(I_\omega + \mathrm{d}I_\omega)\cos\theta_2\,\mathrm{d}\Omega_2\,\mathrm{d}\omega\,\mathrm{d}A - I_\omega\cos\theta_1\,\mathrm{d}\Omega_1\,\mathrm{d}\omega\,\mathrm{d}A = 0 \tag{9.29}$$

supposing no reflection of energy at the interface takes place. Now from Snell's law, $n\sin\theta$ is constant (where n is the refractive index) along the ray. Then, since

$$\frac{\mathrm{d}\Omega_2}{\mathrm{d}\Omega_1} = \frac{\sin\theta_2\,\mathrm{d}\theta_2}{\sin\theta_1\,\mathrm{d}\theta_1} = \frac{\sin\theta_2\,n_1\cos\theta_1}{\sin\theta_1\,n_2\cos\theta_2} = \left(\frac{n_1}{n_2}\right)^2\frac{\cos\theta_1}{\cos\theta_2}$$

(9.29) leads to

$$I_{\omega 2}\left(\frac{n_1}{n_2}\right)^2\cos\theta_1\,\mathrm{d}\Omega_1 = I_{\omega 1}\cos\theta_1\,\mathrm{d}\Omega_1$$

so that

$$\frac{I_\omega}{n^2} = \text{const.} \tag{9.30}$$

along a ray path in a slowly varying inhomogeneous, isotropic, transparent medium. At frequencies much greater than the plasma frequency, $n^2 \simeq 1$ and (9.30) simplifies to $I_\omega = \text{const.}$ along a ray path. The result for an anisotropic plasma is more complicated since Snell's law is no longer obeyed in general but holds only for waves propagating in certain directions relative to the magnetic field.

Next we relax the requirement that the plasma be both source-free and loss-free. Within the geometrical optics representation we introduce absorption and emission terms on the right-hand side of (9.27). Let α_ω be the absorption coefficient per unit path length in the plasma, so that a radiative flux $I_\omega\,\mathrm{d}A\,\mathrm{d}\Omega$ suffers a loss $\alpha_\omega I_\omega\,\mathrm{d}A\,\mathrm{d}\Omega\,\mathrm{d}s$ in travelling a distance $\mathrm{d}s$. Similarly we introduce an emission coefficient ϵ_ω defined so that $\epsilon_\omega\,\mathrm{d}A\,\mathrm{d}\Omega\,\mathrm{d}s$ is the emission from the elemental volume into solid angle $\mathrm{d}\Omega$ in the direction of the ray. Then,

$$\frac{\mathrm{d}I_\omega}{\mathrm{d}s} = \frac{\partial I_\omega}{\partial s} - \alpha_\omega I_\omega + \epsilon_\omega \tag{9.31}$$

Now, $\partial I_\omega/\partial s$ is the rate of change of I_ω due to the change in refractive index along a ray path; from (9.30)

$$\frac{\partial I_\omega}{\partial s} = \frac{2I_\omega}{n}\frac{\mathrm{d}n}{\mathrm{d}s}$$

so that

$$n^2\frac{\mathrm{d}}{\mathrm{d}s}\left(\frac{I_\omega}{n^2}\right) = \epsilon_\omega - \alpha_\omega I_\omega \tag{9.32}$$

This is the *radiative transport equation.*

To solve the transfer equation, we introduce a *source function*

$$S_\omega = \frac{1}{n^2}\frac{\epsilon_\omega}{\alpha_\omega} \tag{9.33}$$

and define an *optical depth* τ

$$\tau = \int^\tau \mathrm{d}\tau = -\int^s \alpha_\omega \,\mathrm{d}s \tag{9.34}$$

in which the minus sign denotes that the optical depth is measured back into the plasma along the ray path. Then (9.32) reads

$$\frac{\mathrm{d}}{\mathrm{d}\tau}\left(\frac{I_\omega}{n^2}\right) = \frac{I_\omega}{n^2} - S_\omega \tag{9.35}$$

Integrating along the ray path in the plasma between points A and B one has,

$$\frac{I_\omega(A)}{n^2(A)}e^{-\tau(A)} = \frac{I_\omega(B)}{n^2(B)}e^{-\tau(B)} + \int_{\tau(A)}^{\tau(B)} S_\omega(\tau)e^{-\tau}\,\mathrm{d}\tau \tag{9.36}$$

In practice it may be permissible to ignore curvature of the ray path where changes in refractivity are negligible. Where A, B are points on a plasma–vacuum boundary as in Fig. 9.4 then $\tau(A) = 0$, $\tau(B) = \tau_0$, the total optical depth of the plasma, and $n(A) = 1 = n(B)$. Thus, neglecting reflection at the boundaries, the emergent intensity is

$$I_\omega^{\mathrm{em}} = I_\omega^{\mathrm{inc}}e^{-\tau_0} + \int_0^{\tau_0} S_\omega(\tau)e^{-\tau}\,d\tau \tag{9.37}$$

The first term on the right-hand side takes account of absorption of the incident radiation while the second represents contributions from sources within the plasma, again allowing for absorption of radiation in transit from its origin to the point A. When $I_\omega^{\mathrm{inc}} = 0$,

$$I_\omega^{\mathrm{em}} = \int_0^{\tau_0} S_\omega(\tau)e^{-\tau}\,\mathrm{d}\tau \tag{9.38}$$

Two important limiting cases of this result correspond to $\tau_0 \ll 1$, when the plasma is said to be *optically thin* and the opposite limit $\tau_0 \gg 1$, when it is *optically*

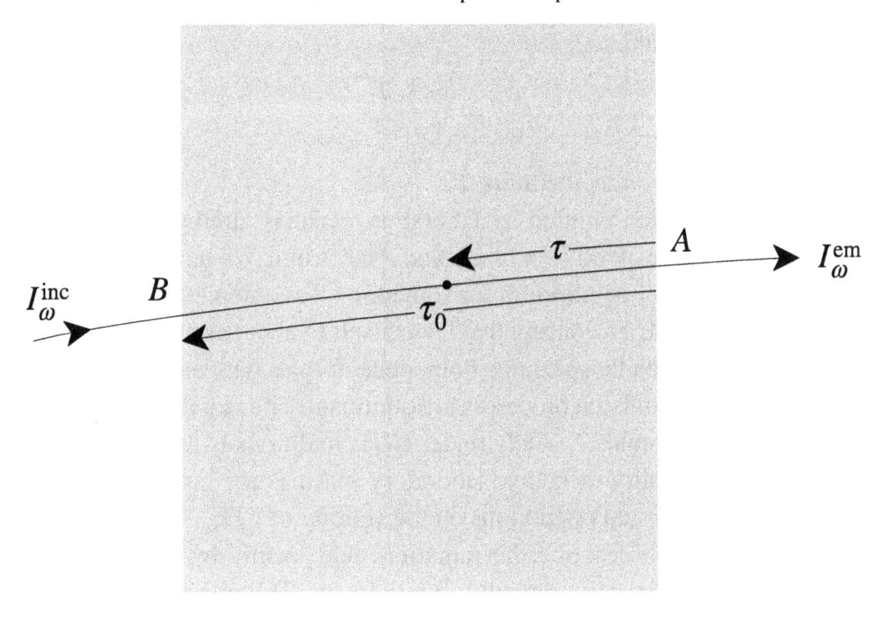

Fig. 9.4. Ray path through the plasma.

thick. In the optically thin limit, absorption along a ray path is negligible so that the emergent intensity is simply the sum of contributions along the ray, i.e.,

$$I_\omega^{\rm em} \simeq \int_{s_B}^{s_A} \epsilon_\omega(s)\mathrm{d}s \tag{9.39}$$

In the optically thick limit $\tau_0 \to \infty$ so that after integrating (9.38) by parts

$$I_\omega^{\rm em} = \frac{\epsilon_\omega(s_A)}{\alpha_\omega(s_A)} \tag{9.40}$$

In other words the intensity affords a direct measure of the source function. Between these two limits one has to solve the radiative transfer equation to determine the intensity of the radiation observed at a detector.

A medium in thermal equilibrium that is perfectly absorbing emits radiation that is Planckian, i.e. the intensity $I(\omega) = B(\omega)$ where

$$B(\omega) = \frac{\hbar\omega^3}{8\pi^3c^2}[\exp(\hbar\omega/k_{\rm B}T) - 1]^{-1} \tag{9.41}$$

is the black body intensity (for a single polarization) and $\hbar = h/2\pi$ where h is Planck's constant. In the classical limit (9.41) reduces to the Rayleigh–Jeans form

$$B(\omega) = \frac{\omega^2 k_{\rm B}T}{8\pi^3c^2} \tag{9.42}$$

Then from (9.40) we find

$$\frac{\epsilon_\omega}{\alpha_\omega} = \frac{\omega^2 k_B T_r}{8\pi^3 c^2} \tag{9.43}$$

which defines a *radiation temperature* T_r.

By and large radiation emitted by laboratory plasmas, unlike that from stellar sources, does not correspond to a black body spectrum. We have to abandon the notion of global thermal equilibrium for something less complete, *local thermodynamic equilibrium* (LTE), a concept that lends itself to about as many definitions as it finds application! Broadly speaking, homogeneous plasmas can be assumed to be in an LTE state when collision processes are dominant. The radiation field is locally Planckian with temperature T_e. Only under LTE conditions is the source function $\epsilon_\omega/\alpha_\omega = B(\omega)$. In reality of course laboratory plasmas are rarely homogeneous and this imposes additional restrictions on the validity of LTE.

It is still possible to describe the radiation field locally by a temperature T_e even when the temperature is globally non-uniform. This means that the region concerned has to be sufficiently local for the temperature to be considered uniform while at the same time extensive enough for thermodynamics to be valid. The LTE approximation breaks down when the source function is no longer a local function of electron temperature but depends on the radiative flux from other regions of the plasma.

9.4 Plasma bremsstrahlung

We turn next to the principal sources of radiation from fully ionized plasmas, bremsstrahlung and, with magnetic fields present, cyclotron or synchrotron radiation. We shall deal with these separately, since the spectral characteristics in each case are quite distinct. The spectral range of bremsstrahlung is very wide, extending from just above the plasma frequency into the X-ray continuum for typical plasma temperatures. By contrast the cyclotron spectrum is characterized by line emission at low harmonics of the Larmor frequency. Synchrotron spectra from relativistic electrons display distinctive characteristics as we shall see later on. Moreover, whereas cyclotron and synchrotron radiation can be dealt with classically, the dynamics being treated relativistically in the case of synchrotron radiation, bremsstrahlung from plasmas has to be interpreted quantum mechanically, though not usually relativistically. Bremsstrahlung results from electrons undergoing transitions between two states of the continuum in the field of an ion (or atom). Oppenheimer (1970) has described bremsstrahlung graphically as the shaking off of quanta from the field of an electron that suffers a sudden jerk.

In place of a full quantum mechanical treatment we opt instead for a semi-classical model of bremsstrahlung which turns out to be adequate for most plasmas. Classically we can think of bremsstrahlung in terms of the emisssion of radiation by an electron undergoing acceleration in the field of a positive ion. The classical emission spectrum can then be massaged to agree with the quantum mechanical spectrum by multiplying by a correction factor, the *Gaunt factor*.

To see what is involved let us make an estimate of plasma bremsstrahlung from a simple model in which an electron moves in the Coulomb field of a single stationary ion of charge Ze. Then

$$|\dot{\mathbf{v}}| = \frac{Ze^2}{4\pi\varepsilon_0 m r^2}$$

and substituting in Larmor's formula (9.15), the power radiated by the electron is given by

$$P_{\mathrm{e}} = \frac{e^2}{6\pi\varepsilon_0 c^3}\left(\frac{Ze^2}{4\pi\varepsilon_0 m r^2}\right)^2 \tag{9.44}$$

If we take the spatial distribution of the plasma electrons about the ion to be uniform, then the contribution to the bremsstrahlung from all electrons in encounters with this test ion is found by summing the individual contributions to give,

$$P = \frac{8\pi Z^2 e^6 n_{\mathrm{e}}}{3(4\pi\varepsilon_0)^3 m^2 c^3}\int_{r_{\min}}^{\infty}\frac{\mathrm{d}r}{r^2} = \frac{8\pi Z^2 e^6 n_{\mathrm{e}}}{3(4\pi\varepsilon_0)^3 m^2 c^3 r_{\min}}$$

The cut-off at $r = r_{\min}$ is needed to avoid divergence. A value for $r_{\min}$ may be chosen in a number of ways and we shall see later that plasma bremsstrahlung is not sensitive to this choice. For present purposes we take $r_{\min} \simeq \lambda_{\mathrm{deB}}$, the de Broglie wavelength, the distance over which an electron may no longer be regarded as a classical particle. For a thermal electron

$$\lambda_{\mathrm{deB}} \sim \hbar/(m k_{\mathrm{B}} T_{\mathrm{e}})^{1/2}$$

where T_{e} is the electron temperature. Thus

$$P \simeq \frac{8\pi Z^2 e^6 n_{\mathrm{e}}}{3(4\pi\varepsilon_0)^3 m c^3 \hbar}\left(\frac{k_{\mathrm{B}} T_{\mathrm{e}}}{m}\right)^{\frac{1}{2}}$$

If n_{i} denotes the ion density, the total bremsstrahlung power radiated per unit volume of plasma, P_{ff}, is then

$$P_{\mathrm{ff}} = \frac{8\pi}{3}\frac{Z^2 n_{\mathrm{e}} n_{\mathrm{i}}}{m c^3 \hbar}\left(\frac{e^2}{4\pi\varepsilon_0}\right)^3\left(\frac{k_{\mathrm{B}} T_{\mathrm{e}}}{m}\right)^{\frac{1}{2}} = 5.34\times 10^{-37} Z^2 n_{\mathrm{e}} n_{\mathrm{i}} T_{\mathrm{e}}^{1/2}(\mathrm{keV})\ \mathrm{W\,m^{-3}} \tag{9.45}$$

We see that the power radiated as bremsstrahlung is proportional to the product of electron and ion densities and to Z^2. Thus any high Z impurities present will contribute bremsstrahlung losses disproportionate to their concentrations. Note that since electron–electron collisions do not alter the total electron momentum they make no contribution to bremsstrahlung in the dipole approximation.

9.4.1 Plasma bremsstrahlung spectrum: classical picture

The exact classical treatment of an electron moving in the Coulomb field of an ion is a standard problem in electrodynamics. Provided the energy radiated as bremsstrahlung is a negligibly small fraction of the electron energy (we treat the ion as stationary) the electron orbit is hyperbolic and the power spectrum $\mathrm{d}P(\omega)/\mathrm{d}\omega$ from a test electron colliding with plasma ions of density n_i may be shown to be

$$\frac{\mathrm{d}P(\omega)}{\mathrm{d}\omega} = \frac{16\pi}{3\sqrt{3}} \frac{Z^2 n_\mathrm{i}}{m^2 c^3} \left(\frac{e^2}{4\pi\varepsilon_0}\right)^3 \frac{1}{v} G(\omega b_0/v) \tag{9.46}$$

where $b_0 = Ze^2/4\pi\varepsilon_0 m v^2$ is the impact parameter for 90° scattering, v the incident velocity of the electron and $G(\omega b_0/v)$ a dimensionless factor, known as the Gaunt factor, which varies only weakly with ω.

Most of the bremsstrahlung is emitted at peak electron acceleration, i.e. at the distance of closest approach to the ion. Collisions described by a small impact parameter produce hard photons; less energetic photons come from distant encounters, with correspondingly large impact parameters. Denoting the impact parameter by b, collisions producing hard photons correspond to $b \ll b_0$ and the electron orbit is approximately parabolic. In the opposite limit $b \gg b_0$, the electron trajectory is more or less linear and in reality it is only in this limit that a classical picture of bremsstrahlung is justified. For an electron following the linear trajectory shown in Fig. 9.5, $r^2(t) = (vt)^2 + b^2$. The components of the acceleration normal and parallel to the trajectory are,

$$\dot{v}_\perp(t) = \frac{Ze^2}{4\pi\varepsilon_0 m} \frac{b}{[(vt)^2 + b^2]^{3/2}} \qquad \dot{v}_\parallel(t) = \frac{Ze^2}{4\pi\varepsilon_0 m} \frac{vt}{[(vt)^2 + b^2]^{3/2}}$$

Integrating (9.26) over the solid angle allows us to express the energy radiated per unit frequency interval in the non-relativistic limit as

$$\frac{\mathrm{d}W(\omega)}{\mathrm{d}\omega} = \frac{e^2}{6\pi^2\varepsilon_0 c^3} \left| \int_{-\infty}^{\infty} \dot{\mathbf{v}}(t) e^{i\omega t}\, \mathrm{d}t \right|^2$$

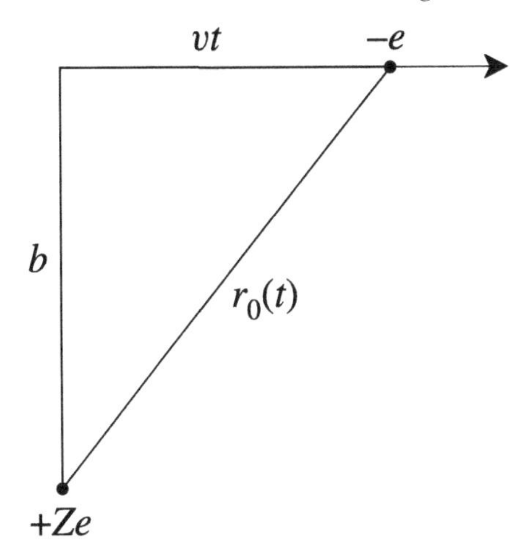

Fig. 9.5. Linear electron trajectory.

$$
= \frac{e^2}{6\pi^2\varepsilon_0 c^3}\left(\frac{Ze^2}{4\pi\varepsilon_0 m}\right)^2 \left|\int_{-\infty}^{\infty} \frac{b\mathbf{e}_\perp + vt\mathbf{e}_\parallel}{[(vt)^2+b^2]^{3/2}} e^{i\omega t}\,\mathrm{d}t\right|^2
$$

$$
= \frac{2e^2\omega^2}{3\pi^2\varepsilon_0 c^3}\left(\frac{Ze^2}{4\pi\varepsilon_0 m v^2}\right)^2 \left[K_1^2\left(\frac{\omega b}{v}\right) + K_0^2\left(\frac{\omega b}{v}\right)\right] \qquad (9.47)
$$

where K_0 and K_1 are modified Bessel functions of the second kind (see Exercise 9.5).

The bremsstrahlung emitted by the test electron from distant encounters with all the plasma ions is then found by integrating over a suitably defined range of the impact parameter. The number of encounters with ions per second having impact parameters between b and $b+\mathrm{d}b$ is $2\pi n_i v b\,\mathrm{d}b$, so that the power radiated (per unit frequency interval) by a single electron is

$$
\frac{\mathrm{d}P(\omega)}{\mathrm{d}\omega} = 2\pi \int_{b_{\min}}^{b_{\max}} \frac{\mathrm{d}W(\omega)}{\mathrm{d}\omega} n_i v b\,\mathrm{d}b
$$

Choices have to be made for the limits to the impact parameter. For consistency with the approximation by a linear trajectory we identify $b_{\min}$ with b_0 and take $b_{\max} = v/\omega$ corresponding to the width of the bremsstrahlung spectrum. The bremsstrahlung from all plasma electrons is found by integrating over the electron distribution function. Assuming Maxwellian electrons we can carry out the integrations over impact parameter and electron velocity to determine the bremsstrahlung emission coefficient. The emission coefficient ϵ_ω is the power radiated per unit volume per unit solid angle per unit (angular) frequency and, in the low frequency limit, is given by

$$\epsilon_\omega = \frac{8}{3\sqrt{3}}\frac{Z^2 n_e n_i}{m^2 c^3}\left(\frac{e^2}{4\pi\varepsilon_0}\right)^3\left(\frac{m}{2\pi k_B T_e}\right)^{\frac{1}{2}}\bar{g} \tag{9.48}$$

and $\bar{g}$ is the Maxwellian-averaged Gaunt factor. In our case

$$\bar{g}(\omega, T_e) = \frac{\sqrt{3}}{\pi}\ln\left|\frac{2m}{\zeta\omega}\frac{4\pi\varepsilon_0}{Ze^2}\left(\frac{2k_B T_e}{\zeta m}\right)^{\frac{3}{2}}\right| \tag{9.49}$$

where $\ln\zeta = 0.577$ is Euler's constant and the factor $(2/\zeta) \simeq 1.12$ in the argument of the logarithm has been included to make $\bar{g}$ in (9.49) agree with the exact low frequency limit determined from (9.46). The factor $\sqrt{3}/\pi$ is introduced into (9.49) to conform with the conventional definition of the Gaunt factor in the quantum mechanical treatment.

9.4.2 Plasma bremsstrahlung spectrum: quantum mechanical picture

While the classical description of bremsstrahlung is useful in the low frequency range, at high frequencies a quantum mechanical formulation is needed. For present purposes it is enough to treat the electron as a wave packet. In the same spirit, the quantum nature of radiation is allowed for through the photon limit. In the simple model used at the start of Section 9.4 to illustrate the dependence of bremsstrahlung on plasma parameters we chose the de Broglie wavelength as a cut-off impact parameter. This choice is dictated by the Uncertainty Principle since an electron with momentum p can be determined only to within an uncertainty $\Delta x \sim \hbar/p \sim \lambda_{deB}$, the de Broglie wavelength. So for impact parameters $b \leq \Delta x$ we need a quantal picture of bremsstrahlung.

In Section 9.4.1 we found the plasma bremsstrahlung emissivity by averaging over a Maxwellian distribution function. Here we have to allow for the fact that there can be no bremsstrahlung emission at frequencies above the photon limit, $\omega = mv^2/2\hbar$; in other words a photon of energy $\hbar\omega$ can only be emitted by an electron with energy at least $\hbar\omega$. Consequently averaging the Gaunt factor $G(\omega, v)$ over a Maxwellian distribution gives

$$\bar{g}(\omega, T_e) = \int_0^\infty g(\omega, v) f(v) v \, dv = \int_0^\infty g(\omega, E) e^{-E/k_B T_e} \, d(E/k_B T_e) \tag{9.50}$$

and with proper allowance made for the photon limit, we require $g(\omega, E) = 0$ for $E < \hbar\omega$. Writing $E = E' + \hbar\omega$, and normalizing so that $\epsilon = E/k_B T_e$ we find from (9.50)

$$\bar{g}(\omega, T_e) = \left[\int_0^\infty g\left(\omega, \epsilon' + \frac{\hbar\omega}{k_B T_e}\right) e^{-\epsilon'} \, d\epsilon'\right] e^{-\hbar\omega/k_B T_e}$$

It is customary to show the exponential dependence on electron temperature explicitly in the expression for the emission coefficient so that the Maxwellian-averaged Gaunt factor is then defined as

$$\bar{g}(\omega, T_e) = \int_0^\infty g\left(\omega, \epsilon' + \frac{\hbar\omega}{k_B T_e}\right) e^{-\epsilon'}\, d\epsilon' \tag{9.51}$$

Simple analytic representations can be found in limiting cases. At low frequencies and high electron temperatures (but not so high that electron thermal velocities become relativistic)

$$\bar{g}(\omega, T_e) = \frac{\sqrt{3}}{\pi} \ln\left|\frac{4}{\zeta}\frac{k_B T_e}{\hbar\omega}\right| \tag{9.52}$$

At high frequencies a Born (plane wave) approximation (equivalent to representing the electron trajectory as a straight line) is commonly used. A more general expression has been given by Elwert (see Griem (1964)). The Born approximation results in a Gaunt factor

$$\bar{g}_B(\omega, T_e) = \frac{\sqrt{3}}{\pi} K_0\left(\frac{\hbar\omega}{2k_B T_e}\right) e^{\hbar\omega/2k_B T_e} \tag{9.53}$$

This reduces to (9.52) in the low frequency limit.

The bremsstrahlung emission coefficient is represented in terms of the Gaunt factor (in whatever approximation) as

$$\epsilon_\omega(T_e) = \frac{8}{3\sqrt{3}} \frac{Z^2 n_e n_i}{m^2 c^3} \left(\frac{e^2}{4\pi\varepsilon_0}\right)^3 \left(\frac{m}{2\pi k_B T_e}\right)^{1/2} \bar{g}(\omega, T_e) e^{-\hbar\omega/k_B T_e} \tag{9.54}$$

The Gaunt factor is a relatively slowly varying function of $\hbar\omega/k_B T_e$ over a wide range of parameters which means that the dependence of bremsstrahlung emission on frequency and temperature is largely governed by the factor $(m/2\pi k_B T_e)^{1/2} \exp(-\hbar\omega/k_B T_e)$ in (9.54). For laboratory plasmas with electron temperatures in the keV range, the bremsstrahlung spectrum extends into the X-ray region of the spectrum.

9.4.3 Recombination radiation

Although we have excluded line radiation from our discussion of plasma radiation we need to consider briefly *free–bound* transitions leading to recombination radiation. The final state of the electron is now a bound state of the atom (or ion, if the ion was initially multiply ionized). The kinetic energy of the electron together with the difference in energy between the final quantum state n and the ionization energy of the atom or ion now appears as photon energy. This event involving electron capture is known as *radiative recombination* and the emission as *recombination radiation*. In certain circumstances, recombination radiation may dominate over bremsstrahlung.

It is again useful to follow a semi-classical argument to arrive at an emission coefficient for recombination radiation. Essentially one takes the corresponding bremsstrahlung coefficient and applies the *correspondence principle* to introduce the bound final state in place of a continuum level. The correspondence principle is essentially a statement that quantum mechanical results must reduce to their classical limits when the density of quantum states is high. In that event we can think of a free–bound transition in terms of bremsstrahlung formalism adjusted to allow for the contribution to the photon energy of the additional energy released through recombination. The free–bound spectrum consists of lines corresponding to $\hbar\omega_n = mv^2/2 + Z^2 R_y/n^2$ where R_y is the Rydberg constant. The correspondence principle attributes the power radiated classically to the line spectrum as opposed to the continuum in the case of bremsstrahlung. The energy corresponding to a transition from the continuum to a quantum state n may be shown to be

$$\hbar\Delta\omega_n \simeq \frac{2Z^2 R_y}{n^3} \tag{9.55}$$

for large n (see Exercise 9.5).

The energy emitted in a transition to a quantum level n is then written $(\mathrm{d}W/\mathrm{d}\omega)_{\mathrm{class}} \times \Delta\omega_n$ where $(\mathrm{d}W/\mathrm{d}\omega)_{\mathrm{class}}$ is the same as the expression used to calculate the bremsstrahlung emission. As in that case we can then proceed to integrate over the impact parameter. The power emitted as recombination radiation to level n by the plasma electrons is found by integrating over the distribution function. For a Maxwellian distribution this amounts to evaluating

$$\int_0^\infty \exp\left(-\frac{E}{k_B T_e}\right)\delta\left[E - \left(\hbar\omega - \frac{Z^2 R_y}{n^2}\right)\right]\mathrm{d}E$$

The emission coefficient for recombination radiation to level n for a thermal plasma is then

$$\begin{aligned}\epsilon_n(\omega, T_e) = {} & \frac{8}{3\sqrt{3}}\frac{Z^2 n_e n_i}{m^2 c^3}\left(\frac{e^2}{4\pi\varepsilon_0}\right)^3\left(\frac{m}{2\pi k_B T_e}\right)^{1/2}\exp(-\hbar\omega/k_B T_e) \\ & \times\left[\frac{Z^2 R_y}{k_B T_e}\frac{2g_n}{n^3}\exp(Z^2 R_y/n^2 k_B T_e)\right]\end{aligned} \tag{9.56}$$

This is identical to the expression found for bremsstrahlung emissivity (9.54) with $\bar{g}(\omega, T_e)$ replaced by the factor in the square bracket provided $\hbar\omega > Z^2 R_y/n^2$; if this is not satisfied the Gaunt factor $g_n \equiv 0$. In other words recombination radiation only contributes to the plasma emissivity for photon energies greater than the

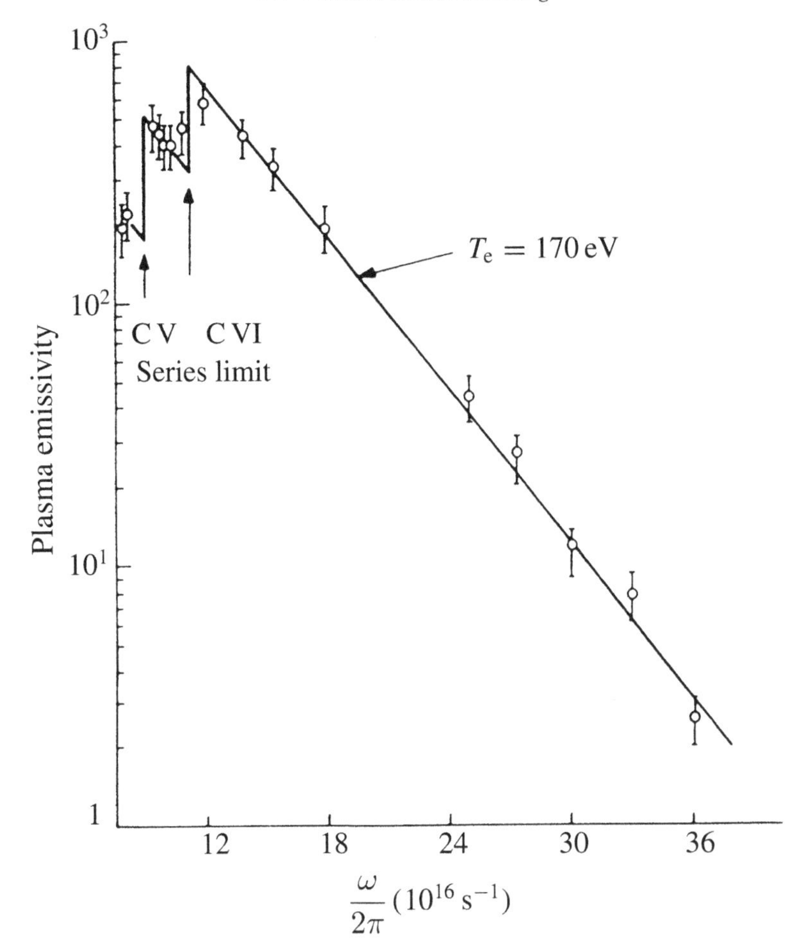

Fig. 9.6. Plasma emissivity showing contributions from recombination radiation superposed on the bremsstrahlung spectrum as a function of photon frequency (after Galanti and Peacock (1975)).

ionization energy of the quantum state involved. This is seen in the characteristic step at the recombination edge, $\hbar\omega = Z^2 R_y/n^2$ (see Fig. 9.6).

9.4.4 Inverse bremsstrahlung: free–free absorption

The process inverse to bremsstrahlung, free–free absorption, occurs when a photon is absorbed by an electron in the continuum. Its macroscopic equivalent is the collisional damping of electromagnetic waves (see Exercise 6.4). For a plasma in local thermal equlibrium, having found the bremsstrahlung emission (9.54), we may then appeal to Kirchhoff's law to find the free–free absorption coefficient α_ω.

In the Rayleigh–Jeans limit this gives

$$\alpha_\omega(T_e) = \frac{64\pi^4}{3\sqrt{3}} \frac{Z^2 n_e n_i}{m^3 c \omega^2} \left(\frac{e^2}{4\pi\varepsilon_0}\right)^3 \left(\frac{m}{2\pi k_B T_e}\right)^{3/2} \bar{g}(\omega, T_e) \tag{9.57}$$

where $\bar{g}(\omega, T_e)$ is defined by (9.49). It is instructive to note that the parametric dependences in this expression for inverse bremsstrahlung may be retrieved from a quite different approach. If we return to the result of Exercise 6.4 expressing the collisional damping of electromagnetic waves and use this to obtain the absorption coefficient we recover (9.57) with the Coulomb logarithm in place of the Maxwell-averaged Gaunt factor, a difference that reflects the distinction between these separate approaches. Whereas inverse bremsstrahlung is identified with incoherent absorption of photons by thermal electrons, the result in Exercise 6.4 is macroscopic in that it derives from a transport coefficient, namely the plasma conductivity. At the macroscopic level, electron momentum is driven by an electromagnetic field before being dissipated by means of collisions with ions.

Absorption of radiation by inverse bremsstrahlung as expressed by (9.57) is most effective at high densities, low electron temperature and for low frequencies. The mechanism is important for the efficient absorption of laser light by plasmas. We expect absorption to be strongest in the region of the critical density n_c, since this is the highest density to which incident light can penetrate. In the neighbourhood of the critical density $Zn_e n_i \sim n_c^2 = (m\varepsilon_0/e^2)^2\omega_L^4$, where ω_L denotes the frequency of the laser light, so that free–free absorption is sensitive to the wavelength of the incident laser light.

9.4.5 Plasma corrections to bremsstrahlung

Up to now we have ignored plasma effects in discussing bremsstrahlung emission and its transport through the plasma. To deal with the second issue we need to refer to our discussion of radiative transport in Section 9.3 where plasma dielectric effects were allowed for. For an isotropic plasma the emission coefficient (9.48) is valid only for frequencies $\omega \gg \omega_p$ and otherwise needs to be corrected when the refractive index is no longer approximately unity. This in turn amounts to abandoning the particle model for a full kinetic theory formulation of wave propagation which is beyond the scope of this discussion.

The bremsstrahlung emission described by (9.48) was determined on the basis of binary encounters between electrons and ions. However as we saw in Section 8.4, collisions in plasmas are predominantly many-body rather than binary. For frequencies around ω_p there is time for an electron cloud to screen the positive ion so that an electron no longer feels a simple Coulomb field. This suggests that the cut-off in the Gaunt factor introduced in Section 9.4.1 should be taken as

$b_{max} \sim \lambda_D$, the Debye length. However as we have seen already, bremsstrahlung emissivity is not especially sensitive to the choices made for the impact parameter cut-off.

For frequencies close to ω_p it is no longer correct to neglect correlations between electrons. Dawson and Oberman (1962) showed that the correction to the Gaunt factor in the region $\omega \simeq \omega_p$ due to Langmuir wave generation was insignificant for a plasma in thermal equilibrium. However for non-thermal plasmas, emission in the neighbourhood of the plasma frequency may be many orders greater than thermal levels. A brief account of one aspect of radiation by Langmuir waves is given in Section 11.6.

9.4.6 Bremsstrahlung as plasma diagnostic

Bremsstrahlung emissivity through its dependence on electron temperature, plasma density and atomic number clearly has potential as a plasma diagnostic. In the first place the exponential dependence in (9.54) means that for $\hbar\omega \geq k_B T_e$ the slope of a log-linear plot of the bremsstrahlung emissivity provides a direct measure of T_e. Next, the strong dependence of the emission on the atomic number of the plasma ions in principle allows the impurity content in a hydrogen plasma to be determined. Moreover, if the plasma electron temperature is known independently, the level of bremsstrahlung could be used to estimate the plasma density.

In practice the picture is less clear. Even for thermal plasmas, for which bremsstrahlung losses do not result in significant modification of the distribution function, unfolding the electron temperature from the bremsstrahlung spectrum is not as straightforward as might first appear. Limited spectral resolution may result in the true slope being masked by recombination edge effects or suffering distortion from discrete lines in the spectrum. Bremsstrahlung from a tokamak plasma with a modest content of high Z impurities such as nickel and molybdenum, will be affected by contributions from these impurities.

Moreover, the assumption of a Maxwellian or near-Maxwellian electron distribution may not be justified. Since the temperature is deduced from the X-ray spectrum in the region $\hbar\omega/k_B T_e > 1$, any non-Maxwellian component will lead to errors in the measurements. Non-Maxwellian electron distributions in both space and laboratory plasmas are commonplace. It may happen for example that a Maxwellian distribution that describes the bulk electrons is modified by a high-energy tail of suprathermal electrons. Even though the population of suprathermals is only a fraction of that of the bulk electrons, they may nevertheless exercise an influence on the overall electron dynamics disproportionate to their numbers.

9.5 Electron cyclotron radiation

We consider next radiation by an electron moving in a uniform, static magnetic field. For electron energies no more than moderately relativistic, radiation is emitted principally at the cyclotron frequency with contributions at low harmonics of this frequency and is generally referred to as *cyclotron radiation*. Emission from highly relativistic electrons differs in that higher harmonics now contribute significantly to the spectrum and the harmonic structure is smoothed on account of harmonic overlap. In this limit the emission is referred to as *synchrotron radiation*. We discuss the two limits separately.

In solving for the motion of a charged particle in a static uniform magnetic field $\mathbf{B}_0$ in Chapter 2 we neglected radiation, so that the total energy of the particle was a constant of the motion. We shall assume – and later justify – that the energy is effectively constant, that is, the energy radiated per complete orbit is negligible compared with $E = [m^2c^4 + p^2c^2]^{1/2}$, where m is the rest mass and $\mathbf{p}$ the electron momentum. We saw from Section 2.2 that the solution to the Lorentz equation corresponds to a helical trajectory, with the axis of the helix parallel to $\mathbf{B}_0$ as in Fig. 9.7. If we now take the z-axis as the path of the guiding centre and set $\alpha = 0$, the electron velocity $\mathbf{v}$ and trajectory $\mathbf{r}_0$ are

$$\left.\begin{aligned} \mathbf{v} &= \hat{\mathbf{x}}v_\perp \cos\Omega t - \hat{\mathbf{y}}v_\perp \sin\Omega t + \hat{\mathbf{z}}v_\parallel \\ \mathbf{r}_0 &= \hat{\mathbf{x}}(v_\perp/\Omega)\sin\Omega t + \hat{\mathbf{y}}(v_\perp/\Omega)\cos\Omega t + \hat{\mathbf{z}}v_\parallel t \end{aligned}\right\} \tag{9.58}$$

where

$$\Omega = eB_0/\gamma m = \Omega_0/\gamma \tag{9.59}$$

is the relativistic Larmor frequency. Our task is then to use (9.58) in (9.26) to calculate the power radiated by an electron per unit solid angle per unit frequency interval. If we suppose the axes are oriented so that radiation is detected by an observer in the Oxz plane then $\mathbf{n} = (\sin\theta, 0, \cos\theta)$. The steps involved are outlined in Exercise 9.6.

The expression found for the cyclotron power radiated by an electron is then

$$\frac{\mathrm{d}^2P}{\mathrm{d}\Omega\,\mathrm{d}\omega} = \frac{e^2\omega^2}{8\pi^2\varepsilon_0 c}\sum_{l=1}^{\infty}\left[\left(\frac{\cos\theta - \beta_\parallel}{\sin\theta}\right)^2 J_l^2(x) + \beta_\perp^2 J_l'^2(x)\right] \times \delta(l\Omega - \omega[1 - \beta_\parallel\cos\theta]) \tag{9.60}$$

where $x = (\omega/\Omega)\beta_\perp \sin\theta$ and the J_l denote Bessel functions of the first kind. The spectrum of the emitted radiation consists of lines at frequencies

$$\omega_l = \frac{l\Omega}{1 - \beta_\parallel \cos\theta} = \frac{l\Omega_0(1 - \beta_\perp^2 - \beta_\parallel^2)^{1/2}}{1 - \beta_\parallel\cos\theta} \tag{9.61}$$

The emission lines are shifted from the cyclotron resonances on two accounts, a

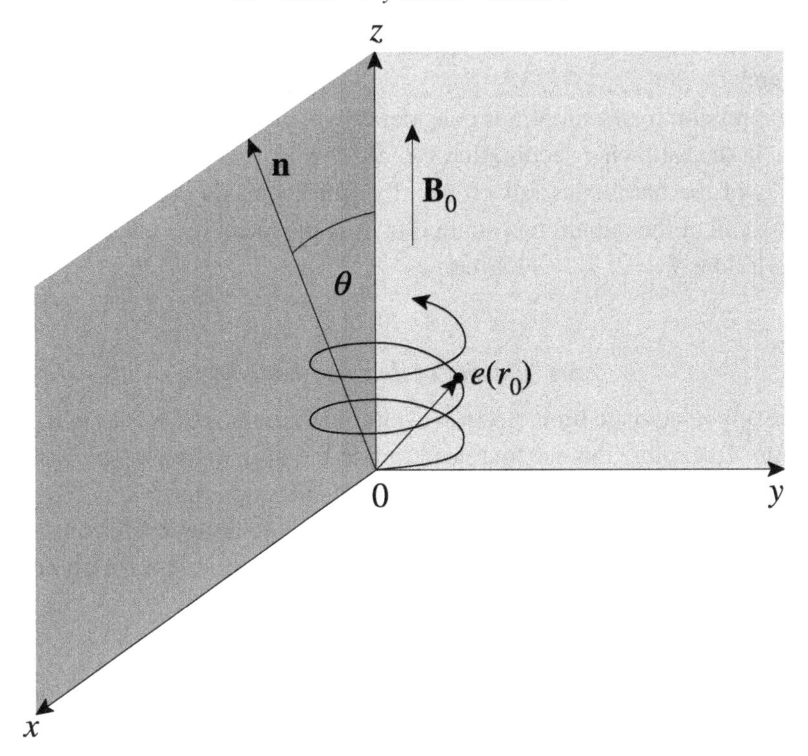

Fig. 9.7. Source–observer geometry for electron radiating in a uniform magnetic field.

relativistic shift from the numerator and a Doppler shift from the denominator. For weakly relativistic electron energies, the Doppler shift is dominant other than for angles $\theta \to \pi/2$. A point to bear in mind is that (9.60) expresses the rate of emission *at the source*; to obtain the power seen by an observer we have to multiply (9.60) by $(1 - \beta_\parallel \cos\theta)^{-1}$.

The total power P_l in a given harmonic line l follows on integrating over all directions. This is best done by transforming to the guiding centre frame and then using a Lorentz transformation to find the radiated power in the laboratory frame. The procedure is outlined in Exercise 9.7, the result being

$$P_l = \frac{e^2\Omega_0^2(1-\beta^2)}{2\pi\varepsilon_0 c\beta_\perp(1-\beta_\parallel^2)^{3/2}} \times \left[l\beta_\perp^2 J_{2l}'\left(\frac{2l\beta_\perp}{(1-\beta_\parallel^2)^{1/2}}\right) - l^2(1-\beta^2)\int_0^{\beta_\perp/(1-\beta_\parallel^2)^{1/2}} J_{2l}(2lt)\,\mathrm{d}t \right] \quad (9.62)$$

This result has an interesting history and was first found by Schott (1912) determining the power radiated by a ring of n electrons. Schott's results were subsequently rediscovered decades later in descriptions of synchrotron radiation from particle accelerators.

The expression for P_l simplifies considerably in the weakly relativistic limit. We shall see in the following section that provided $l\beta \ll 1$, $P_{l+1}/P_l \sim \beta_\perp^2$ so that the intensities of the harmonics fall off rapidly with increasing harmonic number so that almost all of the radiation is emitted in the fundamental, or *cyclotron emission line*, and in low ($l = 2, 3, \ldots$) harmonics.

9.5.1 Plasma cyclotron emissivity

In the weakly relativistic limit electron cyclotron emission (ECE) has potential as a diagnostic. To explore this we first need to find the *plasma cyclotron emissivity* by averaging the power radiated over the electron distribution function, $f(\beta_\perp, \beta_\parallel)$. On the assumption that the electrons are uncorrelated, the plasma cyclotron emissivity $\epsilon_l(\omega)$, defined as the rate of emission of cyclotron radiation in the harmonic line l per unit volume of plasma per unit solid angle, is given by

$$\epsilon_l(\omega) = 2\pi c^3 \int \frac{\mathrm{d}^2 P}{\mathrm{d}\Omega\,\mathrm{d}\omega} f(\beta_\perp, \beta_\parallel)\beta_\perp\,\mathrm{d}\beta_\perp\,\mathrm{d}\beta_\parallel \tag{9.63}$$

with $\mathrm{d}^2 P/\mathrm{d}\Omega\,\mathrm{d}\omega$ defined by (9.60). In the weakly relativistic limit, with $l\beta \ll 1$ and using the small argument approximation for the Bessel function, $J_l(x) \sim (x/2)^l/l!$ in (9.60), (9.63) reduces to

$$\begin{aligned}\epsilon_l(\omega, \theta) &= \frac{e^2c^2\omega^2}{2\pi\varepsilon_0}\frac{l^{2l}}{[l!]^2}(\cos^2\theta + 1)\sin^{2(l-1)}\theta \\ &\times \int_0^\infty \int_{-\infty}^\infty \left(\frac{\beta_\perp}{2}\right)^{2l+1} f(\beta_\perp, \beta_\parallel)\delta(l\Omega - \omega[1 - \beta_\parallel\cos\theta])\mathrm{d}\beta_\perp\,\mathrm{d}\beta_\parallel\end{aligned} \tag{9.64}$$

For a Maxwellian distribution function integration over velocity space gives

$$\begin{aligned}\epsilon_l(\omega, \theta) &= \frac{ne^2l^2\Omega^2}{16\pi^2\varepsilon_0 c}\frac{l^{2l}}{[l!]^2}(\cos^2\theta + 1)\sin^{2(l-1)}\theta \\ &\times \left\{\left(\frac{k_\mathrm{B}T_\mathrm{e}}{2mc^2}\right)^l \left(\frac{mc^2}{2\pi k_\mathrm{B}T_\mathrm{e}}\right)^{1/2}(\omega_l\cos\theta)^{-1}\exp\left[-\frac{mc^2}{2k_\mathrm{B}T_\mathrm{e}}\frac{(\omega-\omega_l)^2}{\omega_l^2\cos^2\theta}\right]\right\}\end{aligned} \tag{9.65}$$

where ω_l is defined by (9.61). In contrast to bremsstrahlung, cyclotron radiation appears as a line spectrum so that the question of line broadening is critical to determining line profiles.

We see from this expression for the plasma cyclotron emissivity that the emission is anisotropic, in that the observed intensity depends on the direction of the detector relative to the source, and consists of a fundamental at the electron cyclotron frequency together with harmonics. The relative intensities of the harmonics depend on electron temperature. The term in braces, the *shape function*, governs line shapes with the line width $\Delta\omega$ determined by Doppler broadening

$$\Delta\omega \sim l\Omega(2k_B T_e/mc^2)^{1/2}\cos\theta$$

Thus by measuring the width of the cyclotron line or its harmonics we may determine the electron temperature. Note that when $\cos\theta < \beta$, line widths will be determined by the relativistic mass increase. The shape function for a relativistically broadened line is distinct from the Doppler line shape, being both narrower and asymmetric. Other mechanisms can and do contribute to line broadening across the range of plasma parameters. However, if we confine our interest to ECE from fusion plasmas we may disregard *radiation broadening* due to loss of energy by an electron as it radiates and *collision broadening*, since collisions contribute in only a small way to line widths in hot plasmas.

9.5.2 ECE as tokamak diagnostic

When it comes to using ECE as a diagnostic for electron temperature in tokamaks, other considerations come into play. Two in particular need to be taken into account. The first concerns line broadening due to the spatially inhomogeneous magnetic field of a tokamak which may dominate Doppler and relativistic broadening in determining the line width. Tokamak magnetic fields are determined largely by the toroidal component $B_t(R) \propto R^{-1}$. The fact that this field is known accurately means that emission at $\omega = l\Omega(R)$ is characterized by good spatial resolution. Thus the electron temperature profile can be determined. Normally only the first few harmonic lines are used in contemporary tokamaks and these may be optically thick or optically thin or somewhere in between. For optically thick lines the temperature profile is determined directly (from (9.42)), i.e.

$$k_B T_e(R) = \frac{8\pi^3 c^2}{l^2\Omega^2(R_0)}\left(\frac{R}{R_0}\right)^2 I\left(\frac{lR_0\Omega(R_0)}{R}\right) \tag{9.66}$$

A second consideration comes from the need to allow for aspects of wave propagation in plasmas introduced in Chapters 6 and 7. Effects in inhomogeneous and bounded plasmas will be discussed in Chapter 11. More particularly, a kinetic theory formulation is needed to deal with absorption in hot plasmas. Moreover, account needs to be taken of reflection at the walls and at the divertor so that in practice a detailed picture of propagation has to be found using a toroidal ray-tracing code.

In principle, ECE measurements using optically thin harmonics allow the plasma density to be inferred once the temperature has been found from an optically thick harmonic measurement. This follows since the optical depth is a function both of the temperature and density. However multiple reflection of the optically thin radiation at the walls makes this less straightforward in practice than might at first appear.

ECE offers further diagnostic potential through polarization measurements that allow determination of the direction of the magnetic field inside the plasma at the position at which the radiation is emitted. However, should the plasma density be large enough to cause strong birefringence, the radiation, instead of retaining its source polarization, reflects the polarization that characterizes the field point. If that happens what we end up with is the direction of the magnetic field at the edge of the plasma, rather than at the point of emission.

For tokamaks operating at higher temperatures, the use of second harmonic ECE to measure the temperature profile suffers from harmonic overlap. As the temperature increases, emission at higher harmonics contributes increasingly to the spectrum so that the weakly relativistic condition $l\beta \ll 1$ may no longer be satisfied. Higher harmonic contributions then change the characteristics of the spectrum, as we shall find in the following section. Moreover, the presence of even a small population of suprathermal electrons leads to changes in the cyclotron emission, disproportionate to the numbers involved. For non-thermal plasmas, emission and absorption are no longer related by Kirchhoff's law and have to be determined independently. In cases where the electron distribution is characterized by a hot electron tail on a bulk Maxwellian distribution, it is in principle possible to discriminate between thermal and suprathermal contributions to ECE by measurements at right angles to the magnetic field using an optically thick harmonic.

9.6 Synchrotron radiation

We shall separate our discussion of synchrotron radiation into two ranges, one characterized by electron energies ranging from some tens to a few hundred keV and the other in which electron energies are ultra-relativistic. The moderately relativistic range is of interest in that it includes, at the lower end, electron energies expected in the next generation of tokamaks. The ultra-relativistic range is largely of astrophysical interest. Analyzing spectra in both relativistic regimes is possible only by making various approximations and in general the synchrotron radiation spectrum has to be found numerically.

9.6.1 Synchrotron radiation from hot plasmas

We return to the general expression for the spectral power density given in (9.60) and for simplicity set $\theta = \pi/2$, since this choice corresponds to peak synchrotron emission. We identify the contributions from the $O(\mathbf{E} \parallel \mathbf{B}_0)$ and $X(\mathbf{E} \perp \mathbf{B}_0)$ modes, namely

$$\frac{\mathrm{d}P^O(\omega, \pi/2)}{\mathrm{d}\omega} = \frac{e^2\omega^2}{8\pi^2\varepsilon_0 c} \sum_{l=1}^{\infty} \beta_\parallel^2 J_l^2\left(\frac{\omega\beta_\perp}{\Omega}\right) \delta(\omega - l\Omega) \tag{9.67}$$

$$\frac{\mathrm{d}P^X(\omega, \pi/2)}{\mathrm{d}\omega} = \frac{e^2\omega^2}{8\pi^2\varepsilon_0 c} \sum_{l=1}^{\infty} \beta_\perp^2 J_l'^2\left(\frac{\omega\beta_\perp}{\Omega}\right) \delta(\omega - l\Omega) \tag{9.68}$$

The synchrotron emission coefficient $\epsilon_S(\omega)$ may be found from (9.67), (9.68) by averaging over the distribution function $f(\mathbf{p})$. If we assume an isotropic distribution then, dropping the $\pi/2$ signature

$$\epsilon_S^{(O,X)}(\omega) = \int_0^\infty \left\langle \frac{\mathrm{d}P^{(O,X)}(\omega)}{\mathrm{d}\omega} \right\rangle f(p) p^2\, \mathrm{d}p \tag{9.69}$$

with

$$\begin{aligned} \left\langle \frac{\mathrm{d}P^{(O,X)}(\omega)}{\mathrm{d}\omega} \right\rangle &= 2\pi \int_0^\pi \frac{\mathrm{d}P^{(O,X)}(\omega, \vartheta)}{\mathrm{d}\omega} \sin\vartheta\, \mathrm{d}\vartheta \\ &= \frac{e^2\omega^2}{8\pi^2\varepsilon_0 c} \sum_{l=1}^{\infty} A_l^{(O,X)}(\gamma)\delta(\omega - l\Omega) \end{aligned} \tag{9.70}$$

where

$$A_l^{(O,X)}(\gamma) = 2\pi\beta^2 \int_0^\pi \begin{bmatrix} J_l^2\left(\frac{\omega}{\Omega}\beta \sin\vartheta\right) \cos^2\vartheta \\ J_l'^2\left(\frac{\omega}{\Omega}\beta \sin\vartheta\right) \sin^2\vartheta \end{bmatrix} \sin\vartheta\, \mathrm{d}\vartheta \tag{9.71}$$

and ϑ denotes the angle between $\mathbf{B}_0$ and $\mathbf{p}$. Explicit forms for $A_l^{(O,X)}(\gamma)$ were found by Trubnikov (1958) for three ranges of electron energy: non-relativistic ($l\beta \ll 1$), moderately relativistic ($\gamma^3 \ll l$) and ultra-relativistic ($\gamma \gg 1, l \gg 1$).

To determine the plasma emission coefficient for moderately relativistic electrons we use (9.69) with the relativistic Maxwellian distribution function

$$f(p) = \frac{N \exp[-(p^2c^2 + m^2c^4)^{1/2}/k_\mathrm{B}T_\mathrm{e}]}{4\pi(k_\mathrm{B}T_\mathrm{e})^2(m/c)yK_2(y)} \tag{9.72}$$

where $y = mc^2/k_\mathrm{B}T_\mathrm{e}$ and $K_2(y)$ is a modified Bessel function. The plasma emission coefficient is then

$$\epsilon_S^{(O,X)}(\omega) = \frac{e^2\omega^2}{8\pi^2\varepsilon_0 c} \int_0^\infty \sum_{l=1}^{\infty} A_l^{(O,X)}(\gamma)\, \delta[\omega - l\Omega]\, f(p)\, p^2\, \mathrm{d}p \tag{9.73}$$

with the representation for $A^{(O,X)}$ appropriate to this energy range ($\gamma^3 \ll l$) given by Trubnikov (1958). Evaluating (9.73) is straightforward and with an upper

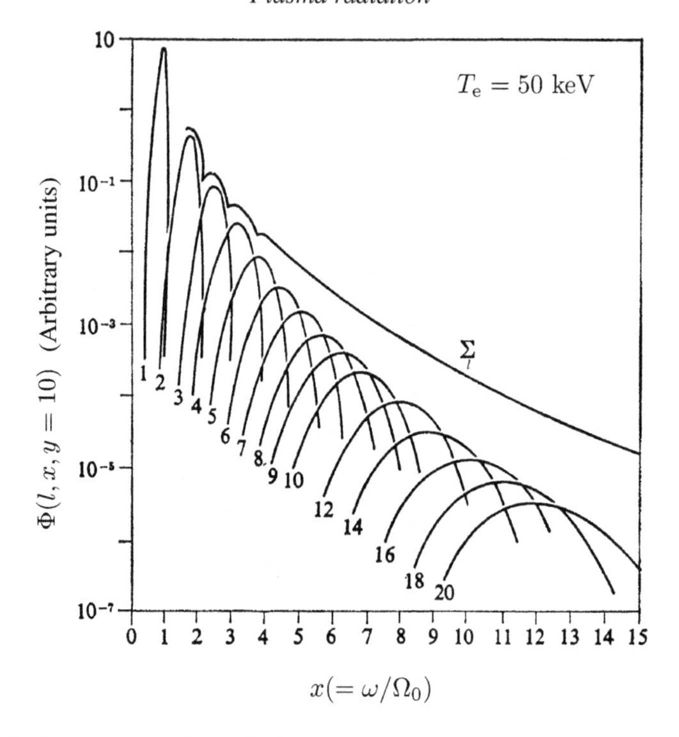

Fig. 9.8. Emission spectrum for radiation by moderately relativistic electrons before and after summation over the harmonics (after Hirshfield, Baldwin and Brown (1961)).

limit for electron energies consistent with the assumption $y \gg 1$, the synchrotron emission at $\theta = \pi/2$ is given by

$$\epsilon_S^{(O,X)}(\omega, \frac{\pi}{2}) = \left(\frac{\omega_p^2}{\Omega_0 c}\right)\left(\frac{\omega^2 k_B T}{8\pi^3 c^2}\right)\sum_{l=1}^{\infty}\Phi(l, x, y) \tag{9.74}$$

where

$$\Phi(l, x, y) = \sqrt{2\pi}\, y^{5/2}\frac{l^2}{x^4}(l^2 - x^2)^{1/2} A_l^{(O,X)}\left(\frac{l}{x}\right)\exp\left[-y\left(\frac{l}{x} - 1\right)\right] \tag{9.75}$$

with $x = \omega/\Omega_0$, denotes the Trubnikov function.

The emission spectrum is governed by the Trubnikov function. Figure 9.8 plots $\Phi(l, x, 10)$ for the first twenty harmonics of the X-mode as a function of x; $y = 10$ corresponds to an electron temperature of about 50 keV. The emission spectrum shows discrete harmonic structure below some critical value l_c beyond which structure is smoothed on account of harmonic overlap. Individual harmonics are now only approximately Gaussian near maximum intensities with a half-width $\Delta\omega_l \simeq l^{3/2}(k_B T/mc^2)\Omega_0$ for small l. The lines are subject to a *relativistic broaden-*

ing (due to the relativistic change of mass) which increases with harmonic number and which has the effect of making the shape function asymmetric.

For electron temperatures above about 10 keV ECE becomes less useful as a diagnostic on account of harmonic overlap. Fidone and Granata (1994) have proposed using synchrotron radiation (ESE) as an alternative diagnostic for next-generation tokamaks. ESE offers several advantages in that high harmonics are unaffected by cut-offs and the ray paths are effectively straight lines.

9.6.2 Synchrotron emission by ultra-relativistic electrons

Synchrotron radiation by electrons of ultra-relativistic energies, $\gamma \gg 1$, was first suggested as a source of cosmic radio waves by Alfvén and Herlofson (1950) and by Kiepenheuer (1950) and later used by Shklovsky (1953) to interpret the radio spectrum from the Crab nebula. It is helpful to see first of all how the principal characteristics of the emission in this regime can be established qualitatively from a simple model. In particular it is easy to show that the radiation is focused within a cone of aperture angle $\sim \gamma^{-1}$ about the direction of the instantaneous velocity of the electron. From the Liénard–Wiechert potential (9.5) it is clear that the denominator becomes small as $\theta = \cos^{-1}(\hat{\mathbf{v}} \cdot \hat{\mathbf{R}})$ tends to zero. This property determines the character of the radiation field which has the form of a sequence of pulses emitted at that point on the orbit at which the radiation is beamed towards the observer. Figure 9.9 illustrates the essential feature, namely that an observer sees the electron only through flashes of radiation emitted as it transits a small segment of its orbit. The radiation field is governed by the maximum value of $\mathbf{A}_{\perp}$, the component of $\mathbf{A}$ perpendicular to $\mathbf{n}$:

$$|\mathbf{A}_{\perp}| = \frac{\mu_0}{4\pi R} \left[\frac{evc \sin\theta}{c - v\cos\theta} \right]$$

These maxima occur at $\theta_{\max} \simeq \pm\gamma^{-1}$ and since $\gamma \gg 1$ for highly relativistic electrons, it follows that the synchrotron emission is beamed strongly into a narrow cone about the forward direction in what is sometimes referred to as the 'lighthouse effect'.

This result enables us to determine the pulse width $\Delta t'$. As seen by the observer, the pulse switches on at time $t_1 = (R + v\Delta t')/c$, and ends at the later time $t_2 = 2\Delta t' + (R - v\Delta t')/c$. The *observed* pulse width Δt is therefore

$$\Delta t = 2\Delta t' \left(1 - \frac{v}{c}\right) = 2\Delta t'/\gamma^2 \left(1 + \frac{v}{c}\right)$$

Now $\Delta t' = \theta_m/\Omega = 1/\gamma\Omega$ and since $v/c \sim 1$

$$\Delta t \simeq (\gamma^3 \Omega)^{-1} = (\gamma^2 \Omega_0)^{-1}$$

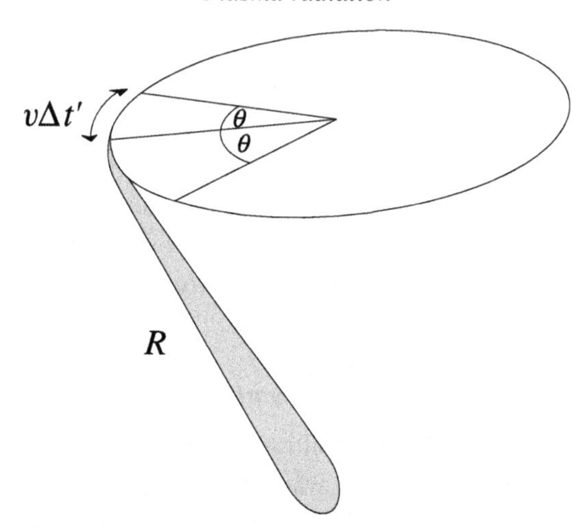

Fig. 9.9. Instantaneous synchrotron radiation beamed from an ultra-relativistic electron (the 'lighthouse effect').

The radiation spectrum is made up from a sequence of pulses at intervals $\tau = 2\pi/\Omega \gg \Delta t$. The emission is linearly polarized and, on the basis of the argument presented, we expect the emission to peak at a frequency

$$\nu_c = \frac{\omega_c}{2\pi} \simeq \frac{1}{2\pi}\gamma^3\Omega \tag{9.76}$$

For frequencies above this the spectrum should show a cut-off so that (9.76) provides a measure of the width of the synchrotron spectrum.

Returning to the general result for the synchrotron power radiated by an electron, it is possible to apply the ultra-relativistic criterion together with the requirement $l \gg 1$ to show that in this limit with $\beta_\parallel = 0$ one can reduce (9.62) to

$$P_l^{\mathrm{UR}} = \frac{e^2\Omega_0^2}{4\pi^2\sqrt{3}\varepsilon_0 c}\frac{l}{\gamma^4}\int_{2l/3\gamma^3}^{\infty} K_{5/3}(y)\,\mathrm{d}y \tag{9.77}$$

Here $K_{5/3}(y)$ is a modified Bessel function. Since harmonics are now very closely spaced it makes more sense to recast (9.77) to show the power radiated per unit frequency rather than per harmonic. This gives

$$\frac{\mathrm{d}P^{\mathrm{UR}}(\omega)}{\mathrm{d}\omega} = \frac{\sqrt{3}e^2\Omega_0}{8\pi^2\varepsilon_0 c}\left(\frac{\omega}{\omega_c}\right)\int_{\omega/\omega_c}^{\infty} K_{5/3}(y)\,\mathrm{d}y \tag{9.78}$$

For $\omega/\omega_c \ll 1$, $\mathrm{d}P^{\mathrm{UR}}(\omega)/\mathrm{d}\omega \propto (\omega/\omega_c)^{1/3}$ while in the opposite limit $\omega/\omega_c \gg 1$, $\mathrm{d}P^{\mathrm{UR}}(\omega)/\mathrm{d}\omega \propto (\omega/\omega_c)^{1/2}\exp(-\omega/\omega_c)$. The shape function determining the spectrum across the range is shown in Fig. 9.10.

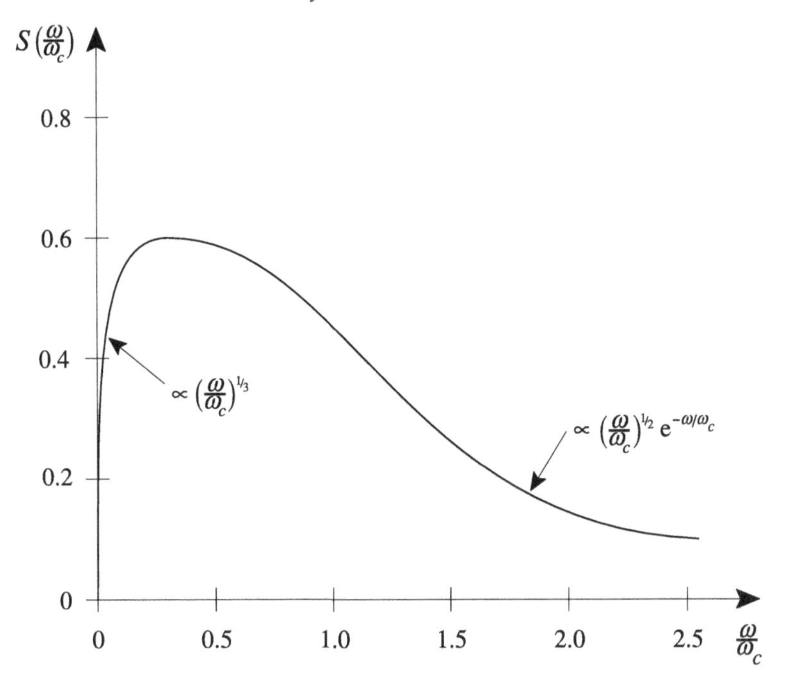

Fig. 9.10. Shape function $S(\omega/\omega_c)$ for synchrotron radiation spectrum in the ultra-relativistic limit.

To determine the spectrum in this limit we need to know the electron energy distribution. A clue to the distribution follows from the observation that energetic electrons in supernovae remnants are sources of cosmic rays and the observed cosmic ray energy spectrum is well represented by a power law. If we suppose that this reflects the electron energy distribution in the source, then

$$N(E)\mathrm{d}E = CE^{-\delta}\,\mathrm{d}E \tag{9.79}$$

where $N(E)\mathrm{d}E$ denotes the number of electrons with energy in the range $(E,\ E + \mathrm{d}E)$ and δ is a constant. Balancing the radiation emitted against electron energy loss, it is straightforward to show that the intensity of synchrotron emission is given by

$$I(\nu)\mathrm{d}\nu = K_1 B_\perp^{(\delta+1)/2} \nu^{-(\delta-1)/2}\,\mathrm{d}\nu \tag{9.80}$$

We conclude from this analysis that an electron energy spectrum with a power law dependence $E^{-\delta}$ generates a synchrotron spectrum $I(\nu) \propto \nu^{-\alpha}$ with a spectral index

$$\alpha = \frac{1}{2}(\delta - 1)$$

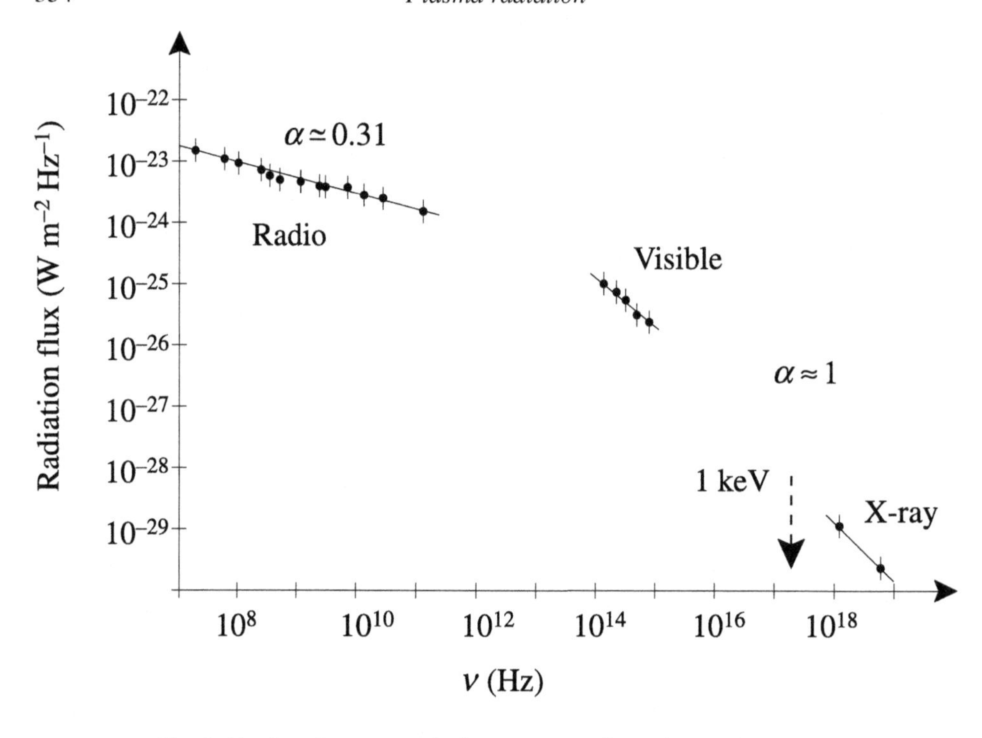

Fig. 9.11. Synchrotron emission spectrum from the Crab nebula.

The observed power law dependence of radio spectra together with the strong linear polarization of the radiation make a compelling case for interpreting the emission in terms of a synchrotron mechanism.

A typical emission spectrum from the Crab nebula is shown in Fig. 9.11. Comparison with the bremsstrahlung spectrum shows that the source is non-thermal; in general the contribution from thermal sources to radiation emitted by supernova remnants is negligible. The observed radio emission is represented by a power law dependence, $I_\nu \propto \nu^{-\alpha}$, over a wide range of frequencies. The spectral index generally lies in the range 0.3–1.5 depending on the source. The spectral range is very wide, extending up to frequencies in the hard gamma-ray region. The gaps are due to absorption in the atmosphere. The marked change in spectral index at $\nu \sim 10^{13}$ Hz is indication of a difference in the energy distribution of electrons from those responsible for the radio spectrum. This in turn reflects the different populations, one associated with the supernova explosion itself, the source of the other being the rotating magnetic neutron star. The pulsar injects highly relativistic electrons into the nebula. Synchrotron self-absorption, unimportant for the X-ray region, becomes significant at radio frequencies and distorts the spectrum from a simple power law representation.

9.7 Scattering of radiation by plasmas

We next consider ways in which radiation is scattered by plasmas. A plane monochromatic electromagnetic wave incident on a free electron at rest is scattered, the scattered wave having the same frequency as the incident radiation; the scattering cross-section is defined by the Thomson cross-section, (9.17). For scattering by electrons the Thomson cross-section has the value $6.65 \times 10^{-29}\,\mathrm{m}^2$; scattering by ions, being at least six orders of magnitude smaller, rarely matters in practice.

Thomson scattering results in a force driving the electron in the direction of the incident wave (see Landau and Lifschitz (1962)). The electron 'absorbs' energy from the wave at the average rate $\langle S\rangle\sigma_{\mathrm{T}}$. Thus, the electron gains momentum from the wave at a rate $\langle \mathbf{S}\rangle\sigma_{\mathrm{T}}/c$. On the other hand, the rate at which momentum is radiated by the electron is $c^{-1}\int\langle \mathrm{d}P/\mathrm{d}\Omega\rangle\mathbf{n}\,\mathrm{d}\Omega$, so that from (9.14) the total momentum of the scattered wave is zero. The final result, therefore, is equivalent to a force on the electron of magnitude

$$\frac{\langle S\rangle\sigma_{\mathrm{T}}}{c} = \frac{4\pi}{3}\varepsilon_0 E_0^2 r_{\mathrm{e}}^2$$

in the direction of the incident wave. This phenomenon is known as *radiation pressure* and provides a small correction to the Lorentz equation.

9.7.1 Incoherent Thomson scattering

In practice we have to find an expression for the Thomson scattering cross-section from all the electrons contained within a finite volume of plasma. Since Thomson scattering is a key diagnostic for high temperature plasmas this needs in general to be treated relativistically. Nevertheless, the non-relativistic limit is widely used for electron temperatures up to several keV and is simpler to deal with analytically. In this limit of (9.8) the scattered electric field $\mathbf{E}_s$ is given in terms of an incident electromagnetic field $\mathbf{E}_i(\mathbf{r}, t) = \mathbf{E}_0 \exp i(\mathbf{k}_0\cdot\mathbf{r} - \omega_0 t)$ by the dipole approximation

$$\mathbf{E}_s(\mathbf{r}, t) = \left[\frac{r_{\mathrm{e}}}{R}\mathbf{n}\times(\mathbf{n}\times\mathbf{E}_i)\right]_{\mathrm{ret}} = [(r_e/R)(\mathbf{nn} - \mathbf{I})\cdot\mathbf{E}_i]_{\mathrm{ret}} \tag{9.81}$$

where $\mathbf{I}$ denotes the unit dyadic. In a relativistic formulation we may retain the format of (9.81), substituting a generalized polarization dyadic $\mathcal{P}$ for $(\mathbf{nn} - \mathbf{I})$ so that

$$\mathbf{E}_s(\mathbf{r}, t) = \left[\frac{r_{\mathrm{e}}}{R}\mathcal{P}\cdot\mathbf{E}_i\right]_{\mathrm{ret}} \tag{9.82}$$

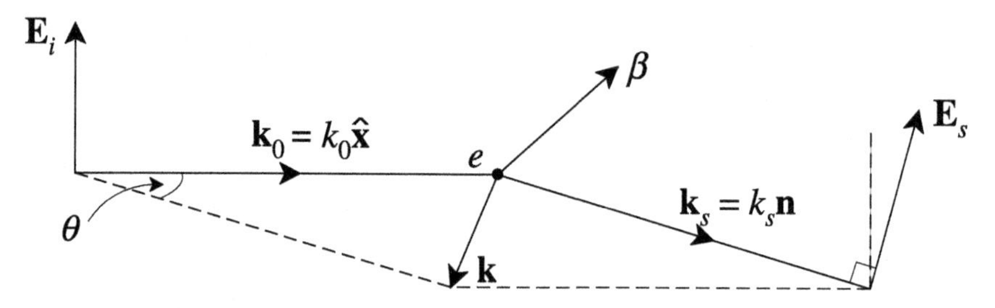

Fig. 9.12. Scattering geometry.

To find a representation for $\mathcal{P}$ we have to return to the relativistic expression (9.8) with $\dot{\boldsymbol{\beta}}$ determined by the Lorentz equation as in Exercise 2.9. The Fourier time transform of (9.82) determines the frequency spectrum of the scattered field over an interval T:

$$\mathbf{E}_s(\mathbf{r}, \omega_s) = \frac{1}{\sqrt{2\pi}} \int_{-T/2}^{T/2} \left[\frac{r_e}{R} \mathcal{P} \cdot \mathbf{E}_i \right]_{\text{ret}} e^{i\omega_s t} \, dt \tag{9.83}$$

To proceed it is simpler to re-cast (9.83) in terms of the retarded time, t', which in the usual radiation (far-field) approximation becomes $t' \simeq t - (r - \mathbf{n} \cdot \mathbf{r}_0(t'))/c$ (see Fig. 9.1). In this approximation, $R \simeq r$ and the scattered field may then be represented as

$$\mathbf{E}_s(\mathbf{r}, \omega_s) = \frac{r_e e^{i\mathbf{k}_s \cdot \mathbf{r}}}{\sqrt{2\pi} r} \int_{-T'/2}^{T'/2} g\mathcal{P}(t') \cdot \mathbf{E}_0 e^{i(\omega t' - \mathbf{k} \cdot \mathbf{r}_0(t'))} \, dt' \tag{9.84}$$

where $\mathbf{k}_s = \omega_s \mathbf{n}/c$, $\omega = \omega_s - \omega_0$ and $\mathbf{k} = \mathbf{k}_s - \mathbf{k}_0$. We now assume that over the interval T' the particle velocity $\mathbf{v}$ is effectively constant so that $\mathbf{r}_0(t') = \mathbf{r}_0(0) + \mathbf{v}t'$. If we disregard the initial displacement, (9.84) becomes

$$\mathbf{E}_s(\mathbf{r}, \omega_s) = \sqrt{2\pi} \frac{r_e}{r} e^{i\mathbf{k}_s \cdot \mathbf{r}} \, g(\mathcal{P} \cdot \mathbf{E}_0) \, \delta(\omega - \mathbf{k} \cdot \mathbf{v}) \tag{9.85}$$

Explicitly, the scattered frequency $\omega_s = \omega_0 + (\mathbf{k}_s - \mathbf{k}_0) \cdot \mathbf{v}$ so that for an incident wave propagating in the x-direction (see Fig. 9.12),

$$\omega_s = \frac{(1 - \hat{\mathbf{x}} \cdot \boldsymbol{\beta})}{(1 - \mathbf{n} \cdot \boldsymbol{\beta})} \omega_0$$

or, to first order in β,

$$\omega_s = \omega_0 (1 - (\hat{\mathbf{x}} - \mathbf{n}) \cdot \boldsymbol{\beta}) \equiv \omega_D \tag{9.86}$$

The Doppler-shifted frequency ω_D of the incident wave combines the effect of electron motion both along the propagation vector of the incident wave (down-shifting the incident frequency) and along the direction of propagation of the scattered wave (up-shifting the frequency).

The next step is determining the scattered power. Making use of the property of the delta function expressed in Exercise 9.1 we may invert (9.85) to get

$$\mathbf{E}_s(\mathbf{r}, t) = \frac{r_\mathrm{e}}{r}\mathcal{P} \cdot \mathbf{E}_0 e^{i(\mathbf{k}_s \cdot \mathbf{r} - \omega_\mathrm{D} t)} \tag{9.87}$$

The corresponding Poynting flux is

$$\mathbf{S}_s = \frac{1}{2} c\varepsilon_0 \left(\frac{r_\mathrm{e}}{r}\right)^2 |\mathcal{P} \cdot \mathbf{E}_0|^2 \mathbf{n}$$

The power per unit solid angle per unit frequency in the scattered radiation is then

$$\frac{\mathrm{d}^2 P}{\mathrm{d}\Omega\, \mathrm{d}\omega_s} = r_\mathrm{e}^2 \langle S_0 \rangle |\mathcal{P} \cdot \mathbf{e}|^2 \delta(\omega_s - \omega_\mathrm{D}) \tag{9.88}$$

where $\langle S_0 \rangle$ is the mean incident Poynting flux and $\mathbf{e} = \mathbf{E}_0/|\mathbf{E}_0|$.

Provided correlations between plasma electrons and ions are unimportant the total scattered power is found by adding the contributions from individual electrons. We shall see in Section 9.8 that this is permissible provided $k\lambda_\mathrm{D} \gg 1$. For present purposes it is enough to note that for $k\lambda_\mathrm{D} \gg 1$, the phase difference between the radiation fields from a test electron and from electrons in its Debye cloud is large and in consequence the radiation fields of the test electron and its Debye cloud are incoherent. Under these conditions the total scattered power is found by averaging (9.88) over the electron distribution.

At this point we need to exercise some care. In computing the total power in the radiation scattered by electrons in some volume element $\mathbf{dr}$ we add the contributions from individual electrons which means that we must measure time at the *source*, not at the detector. As we have already seen in Section 9.2.2 this has the effect of introducing a factor $\mathrm{d}t/\mathrm{d}t' \equiv g$. Sheffield (1975) has given an instructive physical argument for this *finite transit time effect*, the need for which was identified by Pechacek and Trivelpiece (1967). The scattered power from a distribution of electrons $f(\mathbf{v})$ in a scattering volume V is then

$$\frac{\mathrm{d}^2 P}{\mathrm{d}\Omega\, \mathrm{d}\omega_s} = r_\mathrm{e}^2 \int\!\!\int_\mathrm{V} \langle S_0 \rangle |\mathcal{P} \cdot \mathbf{e}|^2 f(\mathbf{v}) g^2 \delta(\omega - \mathbf{k} \cdot \mathbf{v}) \mathbf{dr}\, \mathbf{dv} \tag{9.89}$$

showing both g factors, one from (9.85), the other from the transit time effect.

In the non-relativistic limit these distinctions disappear and (9.89) is greatly simplified by discarding g^2 and taking the polarization dyadic out of the integrand since it is velocity-independent in this limit. Then (9.89) reduces to

$$\frac{\mathrm{d}^2 P}{\mathrm{d}\Omega\, \mathrm{d}\omega_s} = \frac{r_\mathrm{e}^2}{k} \left\{ |(\mathbf{nn} - \mathbf{I}) \cdot \mathbf{e}|^2 \int_V \langle S_0 \rangle \mathbf{dr} \right\} f_k \left(\frac{\omega}{k}\right) \tag{9.90}$$

where $f_k(v_k)$ is the projection of the velocity distribution along $\mathbf{k}$, i.e.

$$f_k(v_k) = \int f(\mathbf{v}_\perp, v_k)\mathrm{d}\mathbf{v}_\perp$$

For a Maxwellian distribution the differential scattering cross-section per unit volume per unit frequency interval is then

$$\frac{\mathrm{d}^2\sigma}{\mathrm{d}\Omega\,\mathrm{d}\omega_s} = r_\mathrm{e}^2 S(\mathbf{k}, \omega)\sin^2\theta \tag{9.91}$$

where $\theta = \cos^{-1}(\mathbf{n}\cdot\mathbf{e})$ and the scattering 'form factor' $S(\mathbf{k}, \omega)$ is given by the Gaussian

$$S(\mathbf{k}, \omega) = \frac{n_\mathrm{e}}{k}\left(\frac{m}{2\pi k_\mathrm{B}T_\mathrm{e}}\right)^{1/2}\exp\left[-\frac{m\,\omega^2}{2k^2k_\mathrm{B}T_\mathrm{e}}\right] \tag{9.92}$$

In principle the form factor affords a direct measure of the electron distribution function projected on to $\mathbf{k}$. In practice mapping the distribution function from measurements of scattered radiation is not usually sufficiently accurate to be useful. It does however serve to discriminate non-Maxwellian distributions from Maxwellian. Thomson scattering measurements do allow electron temperature and electron density to be determined in equilibrium plasmas. The height of the spectrum determines the electron density while the half width, or more exactly the full width at half maximum (FWHM), $\Delta\omega$, given by

$$\Delta\omega = 4\omega_0\left(\frac{2k_\mathrm{B}T_\mathrm{e}}{mc^2}\ln 2\right)^{1/2}\sin\frac{\theta}{2} \tag{9.93}$$

provides a direct measure of electron temperature.

9.7.2 Electron temperature measurements from Thomson scattering

Thomson scattering of laser light is a widely used diagnostic in fusion plasmas, providing good spatial resolution. The scattered power is proportional to $\sigma_\mathrm{T} n_\mathrm{e} L$ where L denotes the length of plasma traversed by the light beam. Given the smallness of σ_T this means that only a minute fraction of incident photons will be scattered. For a density $n_\mathrm{e} \sim 10^{20}\,\mathrm{m}^{-3}$ and taking $L \sim 0.01$ m, the fraction of photons scattered is only about 10^{-10}. Various factors reduce this yet further.

Aside from the requirement that the source of the incident radiation be both monochromatic and of high brightness, a light scattering diagnostic needs optics to carry the laser light into the plasma and dispose of it in a beam dump, optics to collect the scattered light and separate out the different wavelengths, and a detector. Allowing for the geometry of the collection optics, the efficiency of the optical system and the quantum efficiency of the detector means that in practice

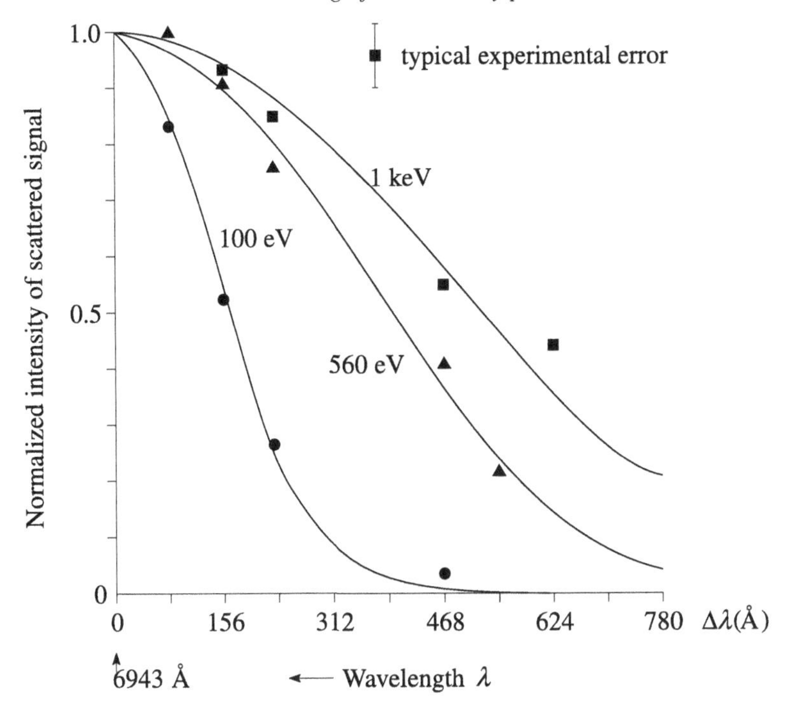

Fig. 9.13. Thomson scattering measurements of electron temperature in the T3-A tokamak (after Peacock *et al.* (1969)).

the fraction of incident photons collected may be no more than 10^{-15}. A key issue in light scattering experiments is the need to prevent stray light from reaching the detector, given that the ratio of the scattered power that is detected to the incident power is so small. One way round this problem is by wavelength discrimination since stray light, as distinct from scattered light, is not shifted in wavelength. Other precautions include an arrangement of baffles to reduce light scattered from the walls and entry port of the input beam and the beam dump. Apart from quantum noise from both the detector and amplifying electronics, plasma bremsstrahlung and recombination radiation will be present as background noise at the detector. Moreover unlike the scattered light, this emission is generated over the plasma as a whole, not simply from the scattering volume.

A landmark in Thomson scattering experiments was the measurement of electron temperature in the tokamak T3-A by Peacock *et al.* (1969). Prior to this experiment, measurements of the plasma energy had been made using a diagmagnetic diagnostic which is subject to a number of limitations. Thomson scattering established beyond doubt that temperatures of the order of 1 keV were reached in early tokamak experiments. Results from these measurements are shown in Fig. 9.13.

9.7.3 *Effect of a magnetic field on the spectrum of scattered light*

The effect of a magnetic field on the scattering form factor (Salpeter (1961), Hagfors (1961) and Dougherty, Barron and Farley (1961)) appears only when the wave vector **k** is almost orthogonal to the magnetic field **B**. With this alignment the spectrum is modulated at integral multiples of the electron cyclotron frequency. As the orthogonality of **k** with **B** weakens, the component of electron motion along the magnetic field gives rise to a Doppler broadening of the gyro-resonances. Once the Doppler line width $2(\mathbf{k}\cdot\mathbf{b})V_e$, where V_e is the electron thermal velocity, exceeds the spacing between the resonances, Ω_e, the resonances will be smeared out. So a necessary condition for observing magnetic fine structure is $2kV_e\cos(\hat{\mathbf{k}}\cdot\mathbf{b}) \lesssim \Omega_e$. In practice spatial variation in the magnetic field over the scattering volume may also result in demodulation.

Magnetic modulation of the spectrum was first detected by Evans and Carolan (1970) in an experiment using a relatively dense ($n_e \sim 10^{21}\,\mathrm{m}^{-3}$), cool ($T_e \approx 20\,\mathrm{eV}$) theta-pinch plasma as a source. With the scattering vector **k** almost perpendicular to **B** the fine structure shown in Fig. 9.14 was resolved. The regularly spaced peaks show a separation approximately equal to the electron cyclotron frequency, measured independently by Faraday rotation and corresponding to a magnetic field of about 1.5 T. Forrest, Carolan and Peacock (1978) subsequently made use of the sensitivity of the depth of modulation at the cyclotron frequency to the orthogonality of **k** and **B** to measure the direction of the poloidal magnetic

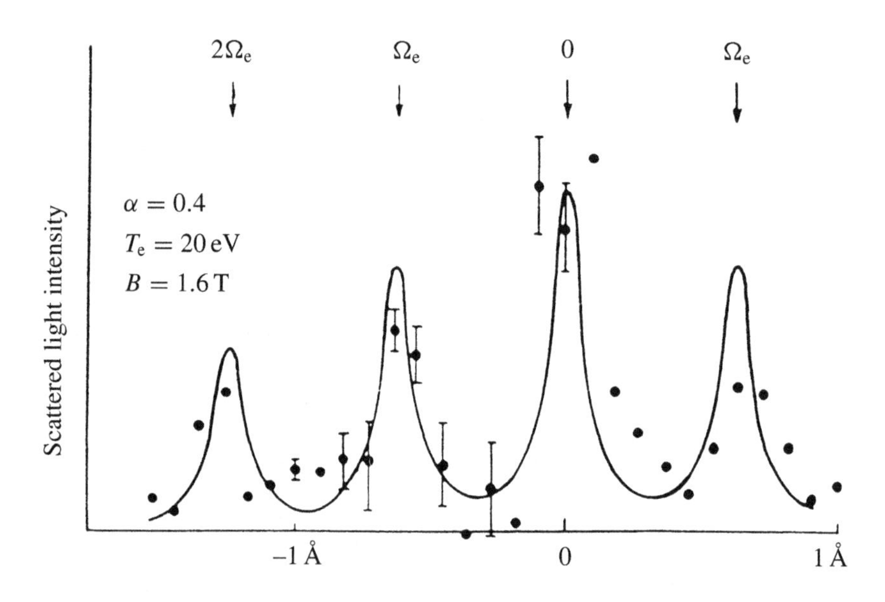

Fig. 9.14. Magnetic modulation of scattered light (after Evans and Carolan (1970)).

field in a tokamak. For the relatively lower densities and much higher temperatures of tokamak plasmas the flux of scattered light in individual harmonics is too weak to allow harmonic structure in the scattered light spectrum to be resolved. This difficulty was overcome using a multiplexing technique due to Sheffield (1975), in which the free spectral range of a Fabry–Perot interferometer is set equal to the cyclotron harmonic frequency. The sensitivity of the modulation to the orthogonality condition allowed the pitch of the field line to be determined to within $0.15°$.

9.8 Coherent Thomson scattering

In treating incoherent Thomson scattering the expression for the total power was found by summing the contributions from individual electrons, a procedure permissible as long as $k\lambda_D \gg 1$. This condition ensures that correlations between plasma particles are unimportant. The next task is to determine the scattering form factor when collective effects have to be taken into account; this happens when $k\lambda_D \leq 1$. The summation over the plasma particles has now to be carried out making proper allowance for correlations. The need to allow for correlations became apparent from observations of ionospheric backscatter. Gordon (1958) suggested that it should be possible to detect Thomson scattering of radar pulses from the ionosphere. Bowles (1958) found that while the scattered power was in broad agreement with that predicted, the bandwidth of the scattered signal proved to be much less than the Doppler width determined by thermal electrons. It appeared that although the radar pulse was scattered by electrons, their behaviour in turn was governed by ion dynamics, a result borne out by subsequent studies of the scattering of radiation by plasma (see Dougherty and Farley (1960), Fejer (1960) and Salpeter (1960)).

9.8.1 Dressed test particle approach to collective scattering

A helpful way of dealing with coherent Thomson scattering makes use of the concept of test particles. Before dealing with this let us first consider what needs to be done. When a test charge is introduced into a plasma we know that plasma electrons and ions react to its presence by forming a shielding cloud around it, resulting in the test particle being 'dressed' by this cloud of particles. We saw in Section 7.8 how to describe this response in terms of the plasma dielectric function. Our objective is to determine the spectrum of density fluctuations which governs the form factor by superposing the fields of all dressed test particles in the plasma. In other words each electron and ion in turn takes on the role of test particle and the individual contributions are added up. Since particle correlations have been allowed for by means of dressed test particles, these particles themselves are then *uncorrelated* so that no further consideration of correlations is needed. A distinction is drawn between test electrons and test ions. For electrons we have to take account of the electron itself as well as the electron cloud dressing it. Test ions on the other hand need not be counted as such since their contribution to scattering is insignificant, so that only the electron cloud contributes in the case of ions.

Using the results from Section 7.8 a Klimontovich density defined as $N_e \equiv \int f_{Ke}(\mathbf{r}, \mathbf{v})\, d\mathbf{v}$, given by

$$\begin{aligned} N_e(\mathbf{k}, \omega) &= \frac{1}{2\pi} \sum_j^{electrons} \left[1 - \frac{\chi_e}{1 + \chi_e + \chi_i} \right] \delta(\omega - \mathbf{k} \cdot \mathbf{v}_j) \\ &+ \frac{Z}{2\pi} \sum_l^{ions} \left[\frac{\chi_e}{1 + \chi_e + \chi_i} \right] \delta(\omega - \mathbf{k} \cdot \mathbf{v}_l) \end{aligned} \tag{9.94}$$

in which χ_α denotes the susceptibility of species α. Since we equate *coherent* scattering from the plasma with *incoherent* scattering from a gas of non-interacting dressed ions and electrons, the spectral power density is now determined by forming the quantity $\langle |N_e(\mathbf{k}, \omega)|^2 \rangle$. Introducing the scattering form factor

$$S(\mathbf{k}, \omega) = \lim_{\substack{T \to \infty \\ V \to \infty}} \frac{(2\pi)^4}{TV} \frac{\langle |N_e(\mathbf{k}, \omega)|^2 \rangle}{n_e} \tag{9.95}$$

and evaluating (9.95) using (9.94) gives

$$\begin{aligned} S(\mathbf{k}, \omega) &= \left\langle \frac{2\pi}{n_e V} \left[\sum_j^{electrons} \left| 1 - \frac{\chi_e}{1 + \chi_e + \chi_i} \right|^2 \delta(\omega - \mathbf{k} \cdot \mathbf{v}_j) \right. \right. \\ &\left. \left. + \sum_l^{ions} \left| \frac{\chi_e}{1 + \chi_e + \chi_i} \right|^2 Z^2 \delta(\omega - \mathbf{k} \cdot \mathbf{v}_l) \right] \right\rangle \end{aligned} \tag{9.96}$$

where the square of the delta functions has been represented in the same way as in Exercise 9.6. Evaluating the summations of the delta functions in (9.96) using the representation of the Klimontovich distribution in (7.1) results in contributions $(V/k) \langle F_{K\alpha k}(\omega/k) \rangle = (V/k) f_{\alpha k}(\omega/k)$, where $\alpha = e, i$ and the subscript k denotes the projection of the velocity distribution along $\mathbf{k}$ defined in Section 9.7.1. Then the scattering form factor reduces to

$$S(\mathbf{k}, \omega) = \frac{2\pi}{n_e k} \left\{ \left| 1 - \frac{\chi_e}{1 + \chi_e + \chi_i} \right|^2 f_{ek}\left(\frac{\omega}{k}\right) + Z^2 \left| \frac{\chi_e}{1 + \chi_e + \chi_i} \right|^2 f_{ik}\left(\frac{\omega}{k}\right) \right\} \tag{9.97}$$

It remains to evaluate the susceptibilities and hence determine the form factor explicitly. In the case of a Maxwellian plasma it is straightforward to show that

$$\begin{aligned}\chi_\alpha &= \frac{\omega_{\mathrm{p}\alpha}^2}{kV_\alpha^2}\frac{1}{\sqrt{2\pi}}\int\left(\frac{v}{V_\alpha}\right)\frac{\exp(-v^2/2V_\alpha^2)}{\omega - kv}\,\mathrm{d}v \\ &= \frac{1}{k^2\lambda_\mathrm{D}^2}\left(\frac{Z_\alpha n_\alpha T_\mathrm{e}}{n_\mathrm{e}T_\alpha}\right)w(\xi_\alpha)\end{aligned} \tag{9.98}$$

where $V_\alpha = (k_\mathrm{B}T_\alpha/m_\alpha)^{1/2}$ is the species thermal speed, $\xi_\alpha = \omega/(\sqrt{2}kV_\alpha)$ and

$$w(\xi) = \frac{1}{\sqrt{\pi}}\int_C \frac{\zeta e^{-\zeta^2}\,\mathrm{d}\zeta}{\xi - \zeta}$$

is related to the plasma dispersion function defined in Section 7.3.2. For an appropriate contour C, $w(\xi)$ may be evaluated to give

$$w(\xi) = 1 - 2\xi e^{-\xi^2}\int_0^\xi e^{\zeta^2}\,\mathrm{d}\zeta + i\sqrt{\pi}\xi e^{-\xi^2} \tag{9.99}$$

From (9.98) it is evident that for $k\lambda_\mathrm{D} \gg 1$, $\chi_\alpha \ll 1$ and consequently the form factor is determined by the $O(1)$ contribution from the first term in (9.97). This is just the incoherent Thomson scattering discussed in Section 9.7.1. However for $k\lambda_\mathrm{D} \lesssim 1$, $\chi_\mathrm{e} \gtrsim 1$ and contributions to $S(k, \omega)$ from coherent scattering become important. For $\chi_\mathrm{e} \gg 1$ the entire electron term may be ignored and the sole contribution to the form factor comes from the ion component.

The spectrum of scattered radiation reflects the collective effects exhibited by the form factor. For an unmagnetized plasma the resonant denominator corresponds to electron plasma waves in the region of $\omega = \omega_\mathrm{p}$ and to ion acoustic waves in the low frequency range $\omega \sim kc_\mathrm{s}$, where c_s is the ion acoustic speed. For both features the shape of the resonance is determined by Landau and collisional damping. Electron plasma waves are strongly Landau damped unless $k\lambda_\mathrm{D} \ll 1$; in this case one expects a resonant feature at the Langmuir frequency. Under conditions where Landau damping becomes significant the electron resonance will be broadened correspondingly. The ion feature in turn reflects the characteristics of ion acoustic waves which are governed by both ions and electrons. In particular the ion feature will be weak unless $T_\mathrm{e} \gg T_\mathrm{i}$ to ensure that ion Landau damping is not severe.

The fact that the two contributions to the spectrum are well separated allows a simplification to be made in the general expression for $S(\mathbf{k}, \omega)$ (see Salpeter

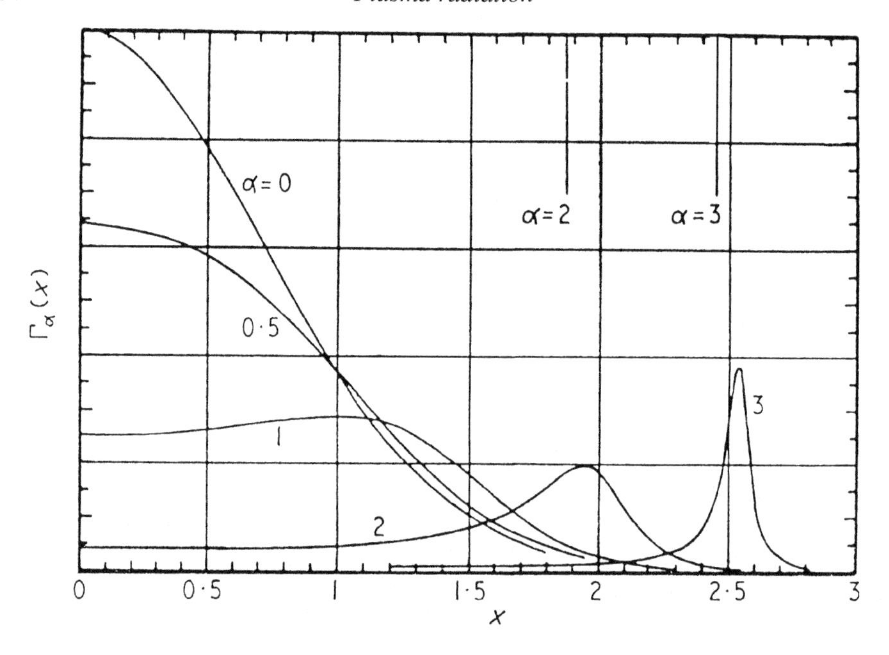

Fig. 9.15. Salpeter shape function.

(1960)). In the ion term Salpeter set $\chi_e \sim 1/(k\lambda_D)^2$ and in the electron term $\chi_i \sim 0$. Then (see Exercise 9.14)

$$S(\mathbf{k}, \omega) = \frac{(2\pi)^{1/2}}{V_e}\Gamma_{\alpha e}(\xi_e) + \frac{(2\pi)^{1/2}}{V_i} Z\left[\frac{1}{1+k^2\lambda_D^2}\right]^2 \Gamma_{\alpha i}(\xi_i) \tag{9.100}$$

in which the *shape function*

$$\Gamma_\alpha(\xi) = \frac{\exp(-\xi^2)}{|1+\alpha^2 w(\xi)|^2} \tag{9.101}$$

has the same functional form for both the electron and ion features and

$$\alpha_e^2 = \frac{1}{k^2\lambda_D^2} \qquad \alpha_i^2 = Z\frac{T_e}{T_i}\left[\frac{1}{1+k^2\lambda_D^2}\right]$$

The shape function is represented in Fig. 9.15. The *scattering parameter* α_e is an important index in scattering from plasmas; $\alpha_e \ll 1$ corresponds to incoherent Thomson scattering and $\alpha_e \gtrsim 1$ to cooperative scattering. In terms of the scattering angle $\theta = \cos^{-1}(\hat{\mathbf{k}}_0 \cdot \hat{\mathbf{k}}_s)$,

$$\alpha_e = \frac{1}{(k_0^2 - 2k_0k_s\cos\theta + k_s^2)^{1/2}\lambda_D} \simeq \frac{\lambda_0}{4\pi\lambda_D \sin\theta/2} \tag{9.102}$$

where λ_0 is the wavelength of the incident radiation.

9.9 Coherent Thomson scattering: experimental verification

It is clear from (9.102) that by varying the scattering angle θ one can pass from a regime of incoherent Thomson scattering ($\alpha_e \ll 1$) to one of coherent (or collective) Thomson scattering for which $\alpha_e > 1$. Alternatively, for a fixed scattering geometry one could sweep through $\alpha_e = 1$ by switching to longer wavelength light. In practice this is not usually an option. Substituting values of λ_D typical of a moderately dense laboratory plasma with electron temperature of 1 keV, and choosing $\lambda_0 = 1.06\,\mu\text{m}$ corresponding to neodymium laser light, it follows from (9.102) that to observe coherent Thomson scattering one has to look close to the forward direction. In this case the shrinking solid angle sets a limit in practice. Moreover, stray light problems are exacerbated for small θ. For realistic choices of scattering angle and laser wavelength, the condition for coherent Thomson scattering translates into a condition that the electron density $n_e \gtrsim 10^{22}\,\text{m}^{-3}$.

Both low and high frequency features in the scattered light spectrum have diagnostic potential. The height of electron line provides a measure of electron density while from the ion resonance we may in principle deduce the ratio of electron to ion temperatures and hence determine T_i if T_e is known from other measurements. However, as we shall see, the presence of impurity ions may introduce ambiguities into this measurement.

The first identification of the ion feature in a laboratory plasma was made by DeSilva, Evans and Forrest (1964) from studies of ruby laser light scattered by a hydrogen arc plasma. The electron density, needed to characterize α_e, was measured independently from Stark broadening of the H_β line. By detecting scattered light at two angles it was possible to isolate both incoherent ($\alpha_e \ll 1$) and coherent ($\alpha_e > 1$) Thomson regimes.

Coherent Thomson scattering diagnostics have been used to advantage in laboratory plasmas, notably in laser-produced plasmas where the high electron density eases the other constraints, despite difficulties over and above those already outlined in Section 9.7.2. Problems may arise on account of the sensitivity of the ion feature to a number of effects, for example becoming asymmetric due to electron drift velocities and the presence of impurities in the plasma. In thermal plasmas, the plasma lines are usually weak features in the spectrum and hence difficult to resolve. These difficulties notwithstanding, various groups, for example Baldis, Villeneuve and Walsh (1986), Baldis *et al.* (1996), Labaune *et al.* (1995, 1996)), have used coherent Thomson scattering to characterize ion acoustic waves and Langmuir waves in laser-produced plasmas.

The distribution worldwide of a number of powerful radar backscatter facilities has allowed a range of parameters characterizing the ionospheric plasma to be determined from measurements of Thomson scattering. These include not only

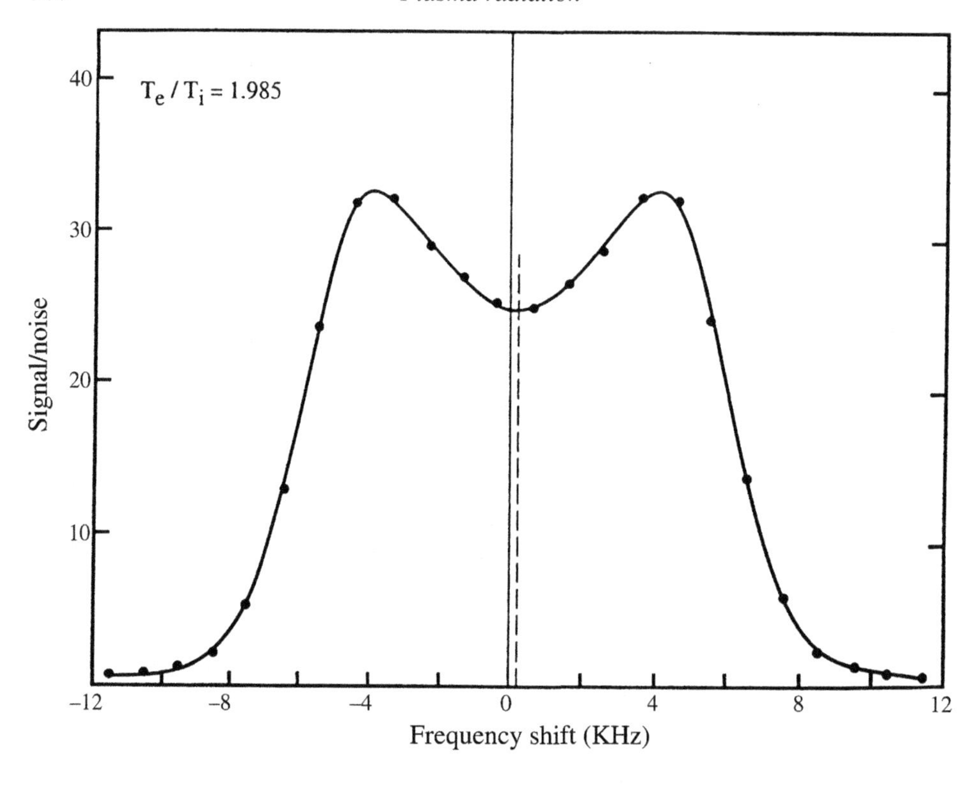

Fig. 9.16. Radar backscatter spectrum from the ionosphere, taken from an altitude of 300 km on 31 March 1971. The solid line represents the Salpeter spectrum fitted to the data points (courtesy of J.M. Holt).

electron density and temperature but ion density, ion mass, composition of the plasma, mean drift velocity and the ion–neutral collision frequency. Figure 9.16 plots radar backscatter data as a function of the frequency shift with the line showing the predicted spectrum. The dashed line denotes the spectral shift resulting from the mean motion of the ionospheric plasma. This spectrum was recorded from a height of 300 km, where O^+ is the only ion of significance and this helped minimize deviations from the Salpeter spectrum due to other ions being present. Above this height, protons and He^+ ions from the solar wind increasingly affect the composition of plasma in the high ionosphere. In lower regions plasma composition is governed by the photochemistry of the ionosphere. The effects of plasma composition on the scattering form factor were first discussed by Moorcroft (1963).

9.9.1 Deviations from the Salpeter form factor for the ion feature: impurity ions

Impurity ions are present to some extent in all laboratory plasmas. Given their greater mass and consequently lower thermal velocities, impurities serve to en-

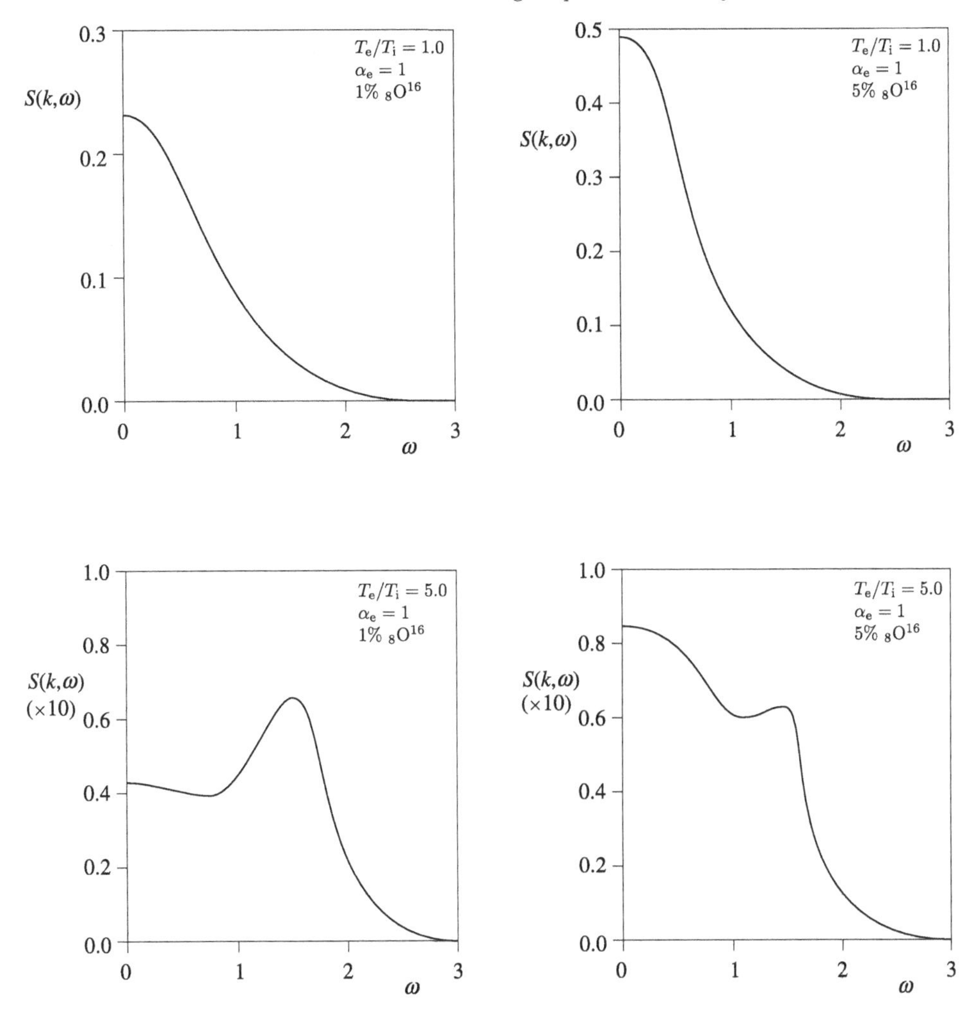

Fig. 9.17. Effect of impurities on the scattering form factor.

hance the central part of the ion feature, so distorting the spectrum. In practice this makes for difficulties in discriminating between a change in ionic composition and one in which the temperature ratio T_e/T_i changes.

It is straightforward to generalize the scattering form factor (9.97) to include different species of ions. Evans (1970) considered the effects of increasing the concentration of oxygen impurity ions ($_8O^{16}$) in a hydrogen plasma, finding a central feature altering the spectrum from that for a pure hydrogen plasma. The height of this feature increased with increasing relative abundance of oxygen ions. For $T_e/T_i = 5$ with 5% oxygen present, Fig. 9.17 shows that scattering from the impurity ions dominates the ion feature, which is well-defined for an impurity

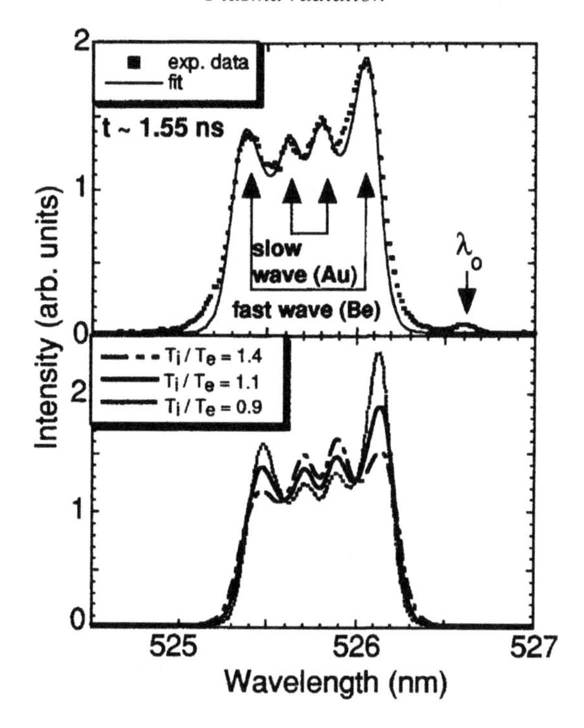

Fig. 9.18. Ion feature in the coherent Thomson spectrum for a two-species plasma showing contributions from Au and Be ions (after Glenzer *et al.* (1999)).

content of 1%. The effect of the impurity becomes even more marked at lower T_e/T_i ratios. Evans also examined the effect of increasing the charge Z of the impurity ion in a hydrogen plasma contaminated by 1% of $_Z\mathrm{Fe}^{56}$, where $0 < Z \leq 15$. Increasing the effective charge of the impurity ion produced a scattered spectrum broadly similar to that in which the impurity abundance is increased. In principle it is feasible to deduce impurity concentrations from the scattered spectra, though for most laboratory plasmas it would prove an unwieldy diagnostic.

Glenzer *et al.* (1999) have reported coherent Thomson scattering from dense ICF plasmas with more than one ion species present. In particular, they formed a plasma by irradiating targets in which the composition was controlled by coating discs with multilayers of Au and Be of varying thickness. Figure 9.18 shows the ion feature in the Thomson spectrum from a plasma containing 4% Au and 96% Be with a pair of ion acoustic waves from each species clearly resolved.

The relative intensities of the ion components are determined by the damping of these waves. For the spectrum in Fig. 9.18 gradients in plasma parameters are not important since it corresponds to a time after the heating beam has been switched off and in addition the low-Z blow-off plasma tends to be isothermal.

9.9.2 Deviations from the Salpeter form factor for the ion feature: collisions

For dense, relatively cold plasmas Coulomb collisions become more important than Landau damping in determining the line shapes. For the ion feature this change appears whenever the mean free path for ion–ion collisions becomes comparable to the wavelength of ion acoustic fluctuations, i.e. for $\nu_{ii}/kc_s \approx 1$ where ν_{ii} is the ion–ion collision frequency.

The effect of Coulomb collisions on the low frequency region of the spectrum was determined by Kivelson and DuBois (1964) from the Balescu–Lenard kinetic equation (see Section 12.2.1) and by Boyd (1966) using a fluid model. The fluid model leads to a low frequency spectrum with two features. In addition to the ion acoustic resonance there is another at zero frequency due to non-propagating entropy fluctuations. The width of the ion resonance is determined by the thermal conductivity and viscosity coefficients; that of the entropy fluctuation resonance depends only on thermal conductivity.

Mostovych and DeSilva (1984) measured the scattered light spectrum from a dense low temperature source for which both density and temperature were well characterized. Their scattered spectrum for an argon plasma is shown in Fig. 9.19. The finite pulse length meant that the entropy fluctuation contribution to the spectrum was not observed. Figure 9.19 plots the Lorentzian line from the fluid model using parameters from the experiment, showing generally satisfactory agreement.

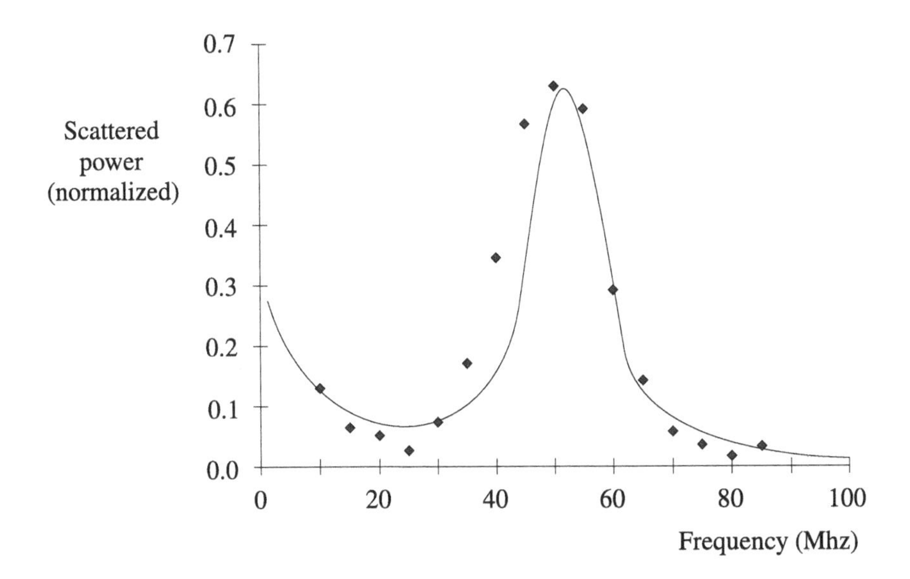

Fig. 9.19. Effect of collisions on the scattering form factor. The data correspond to the spectrum obtained by Mostovych and DeSilva (1984); the curve is the line shape from the fluid model (Boyd (1966)).

Subsequent work by Zhang, DeSilva and Mostovych (1989) succeeded in resolving the contribution to the spectrum due to entropy fluctuations.

Exercises

9.1 Section 9.2 summarizes some key results in the electrodynamics of radiation from charged particles. Solutions to (9.1) are found in terms of the retarded Green function (see Jackson (1975) and Boyd and Sanderson (1969) for details).

In establishing Feynman's result (9.7) one needs the delta function property

$$\int_{-\infty}^{\infty} \delta[h(x)]\gamma(x)\mathrm{d}x = \sum_i \gamma(x_i) \Big/ \left|\frac{\mathrm{d}h(x_i)}{\mathrm{d}x}\right|$$

in which the x_i are the roots of $h(x) = 0$. Starting from (9.3) show that

$$\mathbf{E}(\mathbf{r}, t) = \frac{e}{4\pi\varepsilon_0}\left[\frac{\mathbf{n}}{gR^2} + \frac{1}{cg}\frac{\mathrm{d}}{\mathrm{d}t'}\left(\frac{\mathbf{n}-\boldsymbol{\beta}}{gR}\right)\right]_{\mathrm{ret}}$$

where $g = 1 - \mathbf{n}\cdot\boldsymbol{\beta}$. To cast this in Feynman form first show that

$$\frac{1}{c}\frac{\mathrm{d}\mathbf{n}}{\mathrm{d}t'} = \frac{\mathbf{n}\times(\mathbf{n}\times\boldsymbol{\beta})}{R} \tag{E9.1}$$

Show that the expression (9.8) for $\mathbf{E}^{\mathrm{rad}}(\mathbf{r}, t)$ follows from the $\left[\frac{\mathrm{d}^2\mathbf{n}}{\mathrm{d}t^2}\right]_{\mathrm{ret}}$ term in (9.7). In the non-relativistic limit show that, at large distances from the source, the radiation field from a particle of charge e moving with velocity $\dot{\mathbf{r}}_0(t)$ is given by $(e/4\pi\varepsilon_0 c^2 R)\ddot{\mathbf{r}}_{0\perp}(t - R/c)$ where $\ddot{\mathbf{r}}_{0\perp}$ is the acceleration transverse to the line of sight.

Show that

$$\frac{\mathbf{n}\times\{(\mathbf{n}-\boldsymbol{\beta})\times\dot{\boldsymbol{\beta}}\}}{g^2} \equiv \frac{\mathrm{d}}{\mathrm{d}t'}\left[\frac{\mathbf{n}\times(\mathbf{n}\times\boldsymbol{\beta})}{g}\right]$$

assuming that $\mathbf{n}$ is only weakly time-dependent.

9.2 Using (9.13), consider a charged particle executing instantaneously circular motion and show that

$$\frac{\mathrm{d}P(t')}{\mathrm{d}\Omega} = \frac{e^2\dot{\beta}^2}{16\pi^2\varepsilon_0 c}\frac{1}{(1-\beta\cos\theta)^3}\left[1 - \frac{(1-\beta^2)\sin^2\theta\cos^2\phi}{(1-\beta\cos\theta)^2}\right]$$

Compare this angular distribution with that found in the case in which particle velocity and acceleration are collinear; in particular note that here too the peak power is radiated in the forward direction.

[Hint: Choose a coordinate system with $\boldsymbol{\beta}$ instantaneously along Oz and $\dot{\boldsymbol{\beta}}$ along Ox; θ, ϕ are the usual polar angles.]

9.3 Establish the relativistic generalization of Larmor's formula, (9.16).

[Hint: Rather than the lengthy exercise that a direct integration of (9.13) involves, use the covariant form

$$P = \frac{e^2}{6\pi\varepsilon_0 m_0^2 c^3}\left[\frac{\mathrm{d}p_\mu}{\mathrm{d}\tau}\frac{\mathrm{d}p^\mu}{\mathrm{d}\tau}\right]$$

where p_μ is the momentum–energy four-vector and τ is the proper time. On evaluating the scalar product of the four-vectors, the result follows directly.]

For a particle with energy E moving in a uniform magnetic field $\mathbf{B}$ show that

$$\frac{\mathrm{d}E}{\mathrm{d}t} = -KE^2B_\perp^2$$

where K is a constant.

9.4 By integrating (9.38) successively in the optically thick limit ($\tau_0 \to \infty$) establish that the spectrum of the outgoing radiation is that of a black body provided the temperature does not vary along the line of sight.

9.5 Establish (9.47) for the energy radiated by a classical electron and show that the plasma bremsstrahlung emission coefficient is described by (9.48). For this you will need the Bessel function relationships

$$\begin{aligned} x[K_1^2(x) + K_0^2(x)] &= -(\mathrm{d}/\mathrm{d}x)[xK_1(x)K_0(x)] \\ \lim_{x\to 0} K_0(x) &= -\ln x \qquad \lim_{x\to 0} K_1(x) = 1/x \end{aligned}$$

Confirm the results (9.55) and (9.56).

By comparing the free–bound emissivity ϵ_{fb} to that from bremsstrahlung ϵ_{ff} show that

$$\frac{\epsilon_{\mathrm{fb}}}{\epsilon_{\mathrm{ff}}} \sim \left(\frac{Z^2R_{\mathrm{y}}}{k_{\mathrm{B}}T_{\mathrm{e}}}\right)\frac{2}{n^3}\frac{G_n}{g}\exp\left(\frac{Z^2R_{\mathrm{y}}}{n^2k_{\mathrm{B}}T_{\mathrm{e}}}\right)$$

Under what conditions is ϵ_{fb} negligible compared to ϵ_{ff}?

Verify the expression (9.57) for the free–free absorption coefficient.

9.6 To deduce (9.60) starting from (9.26) with the geometry of Fig. 9.7 use the representation $\exp(ix\sin y) = \sum_{l=-\infty}^{\infty} J_l(x)\exp(ily)$ so that

$$\exp\left\{i\omega\left[t - \frac{\mathbf{n}\cdot\mathbf{r}_0(t)}{c}\right]\right\} = \sum_{l=-\infty}^{\infty} J_l\left(\frac{\omega}{\Omega}\beta_\perp \sin\theta\right) \times \exp[i(\omega - l\Omega - \omega\beta_\parallel\cos\theta)t]$$

From the geometry of Fig. 9.7 show that

$$\begin{aligned}\mathbf{n}\times(\mathbf{n}\times\boldsymbol{\beta}) &= \hat{\mathbf{x}}[-\beta_\perp\cos\Omega t\cos^2\theta + \beta_\parallel\sin\theta\cos\theta] + \hat{\mathbf{y}}[\beta_\perp\sin\Omega t] \\ &\quad + \hat{\mathbf{z}}[\beta_\perp\cos\Omega t\sin\theta\cos\theta - \beta_\parallel\sin^2\theta] \\ &\equiv X\hat{\mathbf{x}} + Y\hat{\mathbf{y}} + Z\hat{\mathbf{z}}\end{aligned} \tag{E9.2}$$

Use these results in (9.26) to show that

$$\frac{\mathrm{d}^2W(\omega)}{\mathrm{d}\Omega\,\mathrm{d}\omega} = \frac{e^2\omega^2}{16\pi^3\varepsilon_0 c}\left|\int_{-\infty}^{\infty}\sum_{l=-\infty}^{\infty} J_l\left(\frac{\omega}{\Omega}\beta_\perp\sin\theta\right) \times \exp[i(\omega - l\Omega - \omega\beta_\parallel\cos\theta)t][X^2+Y^2+Z^2]\mathrm{d}t\right|^2$$

Integrating over time and using Bessel recurrence relations, show that

$$\frac{\mathrm{d}^2W}{\mathrm{d}\Omega\,\mathrm{d}\omega} = \frac{e^2\omega^2}{8\pi^2\varepsilon_0 c}\left(\frac{T}{2\pi}\right)\sum_{l=1}^{\infty}\left|\begin{array}{c}\hat{\mathbf{x}}(\beta_\parallel - \cos\theta)\cot\theta\, J_l(x) \\ +\hat{\mathbf{y}}i\beta_\perp J_l'(x) \\ +\hat{\mathbf{z}}(\cos\theta - \beta_\parallel)J_l(x)\end{array}\right|^2 \times \delta(l\Omega - \omega[1 - \beta_\parallel\cos\theta])$$

where T is the radiation emission time. To arrive at this expression the term involving $\delta^2(\omega)$ has to be interpreted as $\lim_{T\to\infty}(T/2\pi)\delta(\omega)$.

9.7 Integrate (9.60) over all directions to obtain (9.62). The algebra involved is daunting if one goes about this directly. It is best to determine P_l in the guiding centre frame and then apply a Lorentz transformation to the observer's frame.

[Hint: The integration over θ uses properties of Bessel functions. The argument is due to Schott (1912). Making use of the representation $J_l^2(x) = \frac{1}{\pi}\int_0^\pi J_0(2x\sin\alpha)\cos 2l\alpha\,\mathrm{d}\alpha$ leads to

$$\frac{\mathrm{d}P_l}{\mathrm{d}\Omega} = \frac{e^2l^2\Omega^2}{8\pi^2\varepsilon_0 c}\int_0^\pi J_0(2l\beta_\perp\sin\theta\sin\alpha)\left[\beta_\perp^2\cos 2\alpha - 1\right]\cos 2l\alpha\,\mathrm{d}\alpha$$

Then apply the result

$$\int_0^{\pi/2} J_0(y \sin\theta) \sin\theta \, d\theta = \sqrt{\frac{\pi}{2y}} J_{1/2}(y) = \frac{\sin y}{y}$$

to obtain an expression for P_l in the guiding centre frame. Finally Lorentz transform to the observer's frame to get (9.62).]

Use (9.16) to show that the total power radiated is

$$P^{\text{tot}} = \frac{e^2 \Omega_0^2}{6\pi \varepsilon_0 c} \left(\frac{\beta_\perp^2}{1 - \beta^2} \right)$$

Check the assumption made in the development of the theory of cyclotron radiation that the energy radiated is negligible compared with the total energy of the radiating electron.

Establish (9.65) for the plasma cyclotron emissivity.

9.8 Show that the power radiated as cyclotron radiation compared with that radiated as bremsstrahlung is

$$\frac{P_{\text{cyc}}}{P_{\text{b}}} \sim \frac{2 \times 10^{16} T_{\text{e}}^{1/2}(\text{eV}) B^2}{Z^2 n_{\text{e}} n_{\text{i}}}$$

9.9 Establish (9.74) for the synchrotron emissivity from moderately relativistic electrons.

Using the physical model for synchrotron radiation outlined in Section 9.6.2 show that the peak emission occurs at angles of $\theta_m \sim \pm\gamma^{-1}$ for ultra-relativistic electrons ($\gamma \gg 1$).

Establish (9.77), (9.78) for the power radiated by an electron in the ultra-relativistic limit $\beta_\perp \equiv \beta \to 1$ and $l \gg 1$.

[Hint: The following results from the theory of Bessel functions are needed:

$$\begin{aligned}
J_{2l}(2l\beta) &= \frac{1}{\pi\sqrt{3}\gamma} K_{1/3}(R) \\
J'_{2l}(2l\beta) &= \frac{1}{\pi\sqrt{3}\gamma^2} K_{2/3}(R) \\
\int_0^{2l\beta} J_{2l}(x)\mathrm{d}x &= \frac{1}{\pi\sqrt{3}} \int_{2l/3\gamma^3}^{\infty} K_{1/3}(t)\mathrm{d}t \\
2K_{2/3}\left(\frac{2l}{3\gamma^3}\right) - \int_{2l/3\gamma^3}^{\infty} K_{1/3}(t)\mathrm{d}t &= \int_{2l/3\gamma^3}^{\infty} K_{5/3}(t)\mathrm{d}t]
\end{aligned}$$

Find approximate expressions for the radiated power in the limits $\omega/\omega_c \ll 1$, $\omega/\omega_c \gg 1$ respectively. From (9.69) find an expression for the synchrotron emissivity in the ultra-relativistic limit.

Establish (9.80)

9.10 Estimate the electron energy needed to produce 4 keV photons assuming a synchrotron source for X-ray emission from the Crab nebula.

9.11 Consider a cosmic radio source in isolation so that its population of high-energy electrons is not renewed. Suppose the magnetic field present is estimated at 2×10^{-9} T and use this to show that the energy ϵ of the electrons contributing to radio emission at 1 m wavelength is of the order of 1 GeV.

Show that the lifetime of the radiating electrons is proportional to ϵ^{-1} and estimate this lifetime for 1 GeV electrons. How would you expect the radio spectrum to change with time for this source?

9.12 The non-relativistic equation of motion of a charge in an electromagnetic wave is

$$m\dot{\mathbf{v}} = e\mathbf{E} + e\mathbf{v} \times \mathbf{B} + \frac{e^2}{6\pi\varepsilon_0 c^3}\ddot{\mathbf{v}}$$

where the last term is a small correction to the Lorentz equation due to radiation. Show that the time average of the damping force is $(6\pi\varepsilon_0)^{-1}(e^2 E_0/mc^2)^2$ where E_0 is the wave amplitude.

9.13 Show that in an equilibrium plasma for which the electron distribution function is Maxwellian, the line profile of Thomson scattered radiation is determined by the form factor

$$S(\mathbf{k}, \omega) = \frac{n}{k}\left(\frac{m}{2\pi k_B T_e}\right)^{1/2} \exp\left[-\frac{m\omega^2}{2k^2 k_B T_e}\right]$$

Use the experimentally determined half-width from Fig. 9.13 to compute the plasma electron temperature.

9.14 Establish the Salpeter expression for the form factor in (9.100).

9.15 From a consideration of the form factor in a plasma in which an electron drift relative to the ions is present, show that the effect of the drift is to introduce an asymmetry into the ion feature. Interpret this result.

9.16 In Section 2.13.1 considerations of the relativistic dynamics of an electron in an electromagnetic field showed that the electron described a figure-of-eight trajectory. Confirmation of this has been provided in an elegant experiment in which non-linear Thomson scattering was observed by Chen, Maksimchuk and Umstadter (1998). Non-linear Thomson scattering generates harmonic spectra with characteristics that distinguish them from a

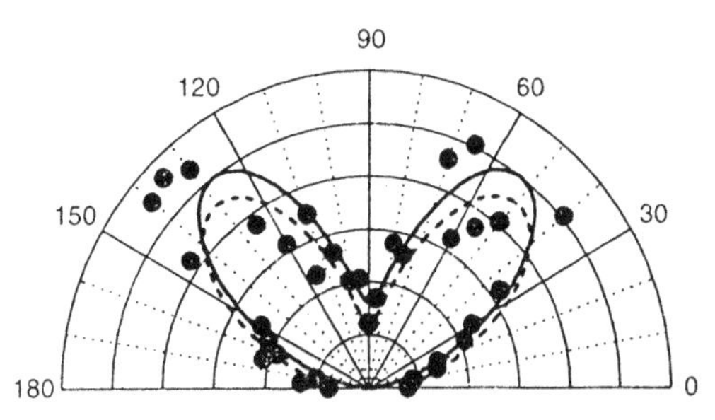

Fig. 9.20. Angular pattern of second-harmonic Thomson scattered light (after Chen, Maksimchuk and Umstadter (1998)).

number of other possible harmonic sources. The polar diagram in Fig. 9.20 shows the dependence of the intensity of second harmonic light (in arbitrary units) on azimuthal angle ϕ in degrees. Circles denote experimental data points and the solid (dashed) lines show predicted dependences for zero drift and a drift velocity $v_{\mathrm{d}} = 0.2c$ respectively. Read the paper by Chen, Maksimchuk and Umstadter for a full discussion of their results.

10

Non-linear plasma physics

10.1 Introduction

Linearization gives rise to such simplification that in many cases it is pushed to its limits and sometimes beyond in the hope that by understanding the linear problem we may gain some insight into the non-linear physics. Perhaps the clearest example of the progress that can be made by analysing linearized equations is in cold plasma wave theory, but linearization, in one form or another, is almost universally applied. For instance, the drift velocities of particle orbit theory are of first order in the ratio of Larmor radius to inhomogeneity scale length. In kinetic theory it is invariably assumed that the distribution function is close to a local equilibrium distribution.

A question of fundamental importance is then, 'How realistic and relevant are linear theories?' Some problems are essentially non-linear in that there is no useful small parameter to allow linearization. Examples of these are sheaths, discussed in Chapter 11, and shock waves. Primarily, our intention is to address the subsidiary question: 'Given that there is a valid linear regime, to what extent need we concern ourselves with non-linear effects?'

Of course, if the linear solution predicts instability then we know that, in time, it will become invalid because the approximation on which the linearization is based no longer holds good. In such cases the aim might be to identify and investigate non-linear processes that come into play and quench the instability. However, an unstable linear regime is emphatically not a pre-requisite for an interest in non-linear phenomena. There are many situations in which the linear equations give only stable solutions but the non-linear equations are secular, i.e. under certain conditions some solutions grow with time. Physically, this comes about because the non-linear coupling of stable, linear modes generates new modes and, if these are natural modes of the system, resonant growth of their amplitudes may occur. Such parametric amplification is widespread throughout physics and engineering and is of particular interest in plasmas, which are well endowed with natural modes.

In each case, whether the non-linear saturation of instabilities or the growth of parametric waves, there are two distinct time scales, that of the rapid oscillation of the initial, linear waves and that of the development of non-linear effects. We shall see that this is crucial to the construction of an analytic non-linear theory. In those cases where linear instability, with growth rate γ_{L}, leads on to non-linear effects, these develop at a rate $\gamma_{\mathrm{NL}} \ll \gamma_{\mathrm{L}}$. Typically, $\gamma_{\mathrm{NL}} \sim (W/nk_{\mathrm{B}}T)\gamma_{\mathrm{L}}$, where W is the energy density associated with the unstable mode.

Computer simulations play an indispensible role in the study of non-linear plasma physics. Since the complexities of non-linear equations severely limit the scope for analytical progress, the usual procedure is to isolate as far as possible the particular non-linear phenomenon one wishes to investigate by suppressing effects which complicate the analysis but do not contribute significantly to the dominant non-linear behaviour. In the main this can be done by averaging over the fast time scale but occasionally it also involves identifying the dominant non-linear term and dropping all others. Progress can sometimes be made by resorting to model equations. Parametric amplification is one example where this has been done to good effect, the model equations serving for an entire class of problems in different branches of physical science.

Our approach in this chapter, therefore, is mainly illustrative. Various non-linear processes are discussed on the basis of the simplest credible mathematical model capable of representing the essential physics of the process.

10.2 Non-linear Landau theory

Linear theories are based on the assumption that perturbations of a steady state or equilibrium are *infinitesimally* small so that all but the linear terms may be ignored. In practice, of course, all perturbations have finite amplitude, however small, and one may begin a non-linear investigation by asking what would be the consequences of recognising this. Assuming small, but *finite*, perturbations and keeping quadratic terms in a perturbation expansion is the basis of weakly non-linear analysis and, for the most part, this will be our approach to the discussion of non-linear plasma phenomena. Various linear theories will be extended in this way into the non-linear regime and we begin with Landau's solution of the Vlasov equation.

10.2.1 Quasi-linear theory

As its name suggests, quasi-linear theory is a kind of halfway stage between linear and non-linear theory and was first developed to deal with the problem which we met when discussing the Landau solution of the Vlasov equation. What happens if some waves experience Landau growth rather than damping? Obviously, wave amplitudes cannot grow indefinitely since the total energy is limited. How then is the growth curtailed? As energy is transferred from particles to waves the distribution of particle velocities must be modified in some way by this growth in wave amplitude. It is just this modification of the distribution function that we seek to describe by quasi-linear theory.

We illustrate this approach by means of the simplest possible problem of unstable, electrostatic waves in an unmagnetized plasma in which we treat the ions as a uniform neutralizing background. We assume that the velocity space modification of the electron distribution function $f(\mathbf{r}, \mathbf{v}, t)$ takes place on a much slower time scale than the fluctuations of the growing waves so that we may separate f into two parts, a slowly varying f_0 which is the value of f when averaged over the fluctuations, and a rapidly varying f_1. For simplicity, we assume also that f_0 is spatially uniform so that

$$f(\mathbf{r}, \mathbf{v}, t) = f_0(\mathbf{v}, t) + f_1(\mathbf{r}, \mathbf{v}, t)$$

The Vlasov equation then reads

$$\frac{\partial f_0}{\partial t} + \frac{\partial f_1}{\partial t} + \mathbf{v} \cdot \frac{\partial f_1}{\partial \mathbf{r}} - \frac{e}{m}\mathbf{E} \cdot \frac{\partial f_0}{\partial \mathbf{v}} - \frac{e}{m}\mathbf{E} \cdot \frac{\partial f_1}{\partial \mathbf{v}} = 0 \tag{10.1}$$

and Poisson's equation becomes

$$\nabla \cdot \mathbf{E} = -\frac{e}{\varepsilon_0} \int f_1 \, \mathrm{d}\mathbf{v} \tag{10.2}$$

the electron charge arising from the slowly varying f_0 being neutralized by the ion charge.

Averaging (10.1) over the rapid fluctuations gives

$$\frac{\partial f_0}{\partial t} = \frac{e}{m}\left\langle \mathbf{E} \cdot \frac{\partial f_1}{\partial \mathbf{v}} \right\rangle \tag{10.3}$$

where $\langle X \rangle$ denotes the average value of X; all other terms are linear in f_1 and therefore have zero averages. This is the equation describing the slow evolution of f_0. Now subtracting (10.3) from (10.1) we find

$$\frac{\partial f_1}{\partial t} + \mathbf{v} \cdot \frac{\partial f_1}{\partial \mathbf{r}} - \frac{e}{m}\mathbf{E} \cdot \frac{\partial f_0}{\partial \mathbf{v}} = \frac{e}{m}\left(\mathbf{E} \cdot \frac{\partial f_1}{\partial \mathbf{v}} - \left\langle \mathbf{E} \cdot \frac{\partial f_1}{\partial \mathbf{v}} \right\rangle\right) \tag{10.4}$$

which describes the rapid variation of f_1. If saturation of the instability takes place in such a way that f_1 remains small compared with f_0 we might plausibly argue that in these equations only the non-linear term on the right-hand side of (10.3) need be kept since this determines the rate of change of f_0 whilst those on the right-hand side of (10.4) may be neglected compared with the linear terms on the left-hand side. It is in this sense that the theory is quasi-linear; (10.4) is linearized but not (10.3). Thus, we replace (10.4) by

$$\frac{\partial f_1}{\partial t} + \mathbf{v} \cdot \frac{\partial f_1}{\partial \mathbf{r}} - \frac{e}{m}\mathbf{E} \cdot \frac{\partial f_0}{\partial \mathbf{v}} = 0 \tag{10.5}$$

This is very like the linearized Vlasov equation (7.16), but not quite the same because here f_0 is time-dependent. However, since the rate of change of f_0 is slow compared with f_1 we may treat f_0 as constant in solving the coupled equations (10.2) and (10.5) and apply the results of Section 7.3.

The Fourier transform of the electric field $\mathbf{E}(\mathbf{k}, t)$ is given by (7.28) where R_j is the residue of the pole at $p = p_j$ in the integral (7.26). For simplicity, we assume that there is only one solution of the dispersion relation (7.27) which gives rise to a pole with $\Re p > 0$ and that this occurs at

$$p = -i\omega = -i\omega_0 = -i\omega_r + \gamma$$

Then all other terms in (7.28) are transient and may be dropped. Evaluating the residue using L'Hôpital's rule we find

$$\mathbf{E}(\mathbf{k}, t) = \frac{iee^{-i\omega_0 t}\mathbf{k}}{\varepsilon_0 k^2 (\partial\epsilon(\mathbf{k}, \omega)/\partial\omega)_{\omega_0}} \int \frac{f_1(\mathbf{k}, \mathbf{v}, 0)}{(\omega_0 - \mathbf{k} \cdot \mathbf{v})} \mathrm{d}\mathbf{v} \tag{10.6}$$

where, since it is more convenient to work in terms of vector variables, we have replaced E and u using $\mathbf{E} = E\mathbf{k}/k$ and $ku = \mathbf{k} \cdot \mathbf{v}$. Also, using the transformation $p \to -i\omega = -i\omega_0$ we have expressed the Laplace transform of the plasma dielectric function $D(\mathbf{k}, p)$ in terms of its Fourier transform

$$\epsilon(\mathbf{k}, \omega_0) = 1 + \frac{e^2}{\varepsilon_0 m k^2} \int \frac{\mathbf{k} \cdot \partial f_0/\partial \mathbf{v}}{(\omega_0 - \mathbf{k} \cdot \mathbf{v})} \mathrm{d}\mathbf{v} = 0 \tag{10.7}$$

which determines $\omega_0(\mathbf{k}, t)$.

We did not display f_1 explicitly in Section 7.3 but it is obtained from (7.22) as

$$f_1(\mathbf{k}, \mathbf{v}, p) = \frac{1}{p + i\mathbf{k} \cdot \mathbf{v}} \left[\frac{e\mathbf{E}(\mathbf{k}, p)}{m} \cdot \frac{\partial f_0}{\partial \mathbf{v}} + f_1(\mathbf{k}, \mathbf{v}, 0) \right] \tag{10.8}$$

Inverting the Laplace transform then gives contributions from the pole at $p = -i\mathbf{k} \cdot \mathbf{v}$ as well as that at $p = -i\omega_0$. However, the former varies like $\exp(-i\mathbf{k} \cdot \mathbf{v}t)$ and as $t \to \infty$ it becomes highly oscillatory in $\mathbf{k}$ and $\mathbf{v}$ space. This term is called the *ballistic* term and we shall return to it later. For the moment we note that since the inverse Fourier transform involves an integral over $\mathbf{k}$ its contribution vanishes as $t \to \infty$ and so we drop it. Here again, therefore, we keep only the term arising from the pole at $p = -i\omega_0$ with the result

$$f_1(\mathbf{k}, \mathbf{v}, t) = \frac{ie\mathbf{E}(\mathbf{k}, t)}{m(\omega_0 - \mathbf{k} \cdot \mathbf{v})} \cdot \frac{\partial f_0}{\partial \mathbf{v}} \tag{10.9}$$

Substitution of (10.6) and (10.9) in (10.3) then gives the evolution equation for f_0 in the form of a diffusion equation

$$\frac{\partial f_0}{\partial t} = \frac{\partial}{\partial V_{\mathrm{i}}} D_{ij} \frac{\partial f_0}{\partial v_j} \tag{10.10}$$

where

$$D_{ij} = \frac{ie^2}{m^2 V} \int \frac{\mathrm{d}\mathbf{k} \langle E_i(-\mathbf{k}, t) E_j(\mathbf{k}, t) \rangle}{(\omega_0 - \mathbf{k} \cdot \mathbf{v})} \tag{10.11}$$

is the diffusion coefficient and V is the volume of the plasma. In deriving (10.10) we have taken the spatial average of the right-hand side since the left-hand side is assumed independent of $\mathbf{r}$. Also, although ω_0 is time-dependent it is only slowly varying as a function of f_0 and so has been taken outside the time average over the rapid fluctuations.

Defining the spectral energy density of the electrostatic field by

$$\mathcal{E}(\mathbf{k}, t) = \frac{1}{V} \langle \mathbf{E}(-\mathbf{k}, t) \cdot \mathbf{E}(\mathbf{k}, t) \rangle = \frac{1}{V} \langle \mathbf{E}^*(\mathbf{k}, t) \cdot \mathbf{E}(\mathbf{k}, t) \rangle \tag{10.12}$$

and noting that

$$\frac{\partial \mathbf{E}(\mathbf{k}, t)}{\partial t} = -i\omega_0 \mathbf{E}(\mathbf{k}, t)$$

it follows that

$$\frac{\partial \mathcal{E}(\mathbf{k}, t)}{\partial t} = 2\gamma \mathcal{E}(\mathbf{k}, t) \tag{10.13}$$

Thus, the coupled equations of quasi-linear theory are (10.10), (10.13) and (10.6). From (10.13) we see that for $\gamma(\mathbf{k}, t) > 0$ the wave amplitude grows thereby increasing the diffusion coefficient which in turn decreases the slope of f_0 and thus reduces γ.

This may be illustrated for the one-dimensional 'bump-on-tail' plasma distribution. In this case (10.10) is

$$\frac{\partial f_0}{\partial t} = \frac{\partial}{\partial v} \frac{ie^2}{m^2} \int \frac{\mathrm{d}k \mathcal{E}(\mathbf{k}, t)}{(\omega_0 - kv)} \frac{\partial f_0}{\partial v}$$

In the limit $\gamma \to 0$ we may evaluate the integral over k in the same way as the integral over u was evaluated in Section 7.3. The principal part vanishes since it is odd in k and $i\pi$ times the residue at the pole gives

$$\frac{\partial f_0}{\partial t} = \frac{\pi e^2}{m^2} \frac{\partial}{\partial v} \frac{\mathcal{E}(\omega_0/v, t)}{v} \frac{\partial f_0}{\partial v} = \frac{\partial}{\partial v} A(v) \mathcal{E}(\omega_0/v, t) \frac{\partial f_0}{\partial v} \tag{10.14}$$

where $A(v) = (\pi e^2/m^2 v)$. Also since $\gamma \propto \partial f_0/\partial v$ we may write (10.13) as

$$\frac{\partial \mathcal{E}(\omega_0/v, t)}{\partial t} = B(v) \mathcal{E}(\omega_0/v, t) \frac{\partial f_0}{\partial v} \tag{10.15}$$

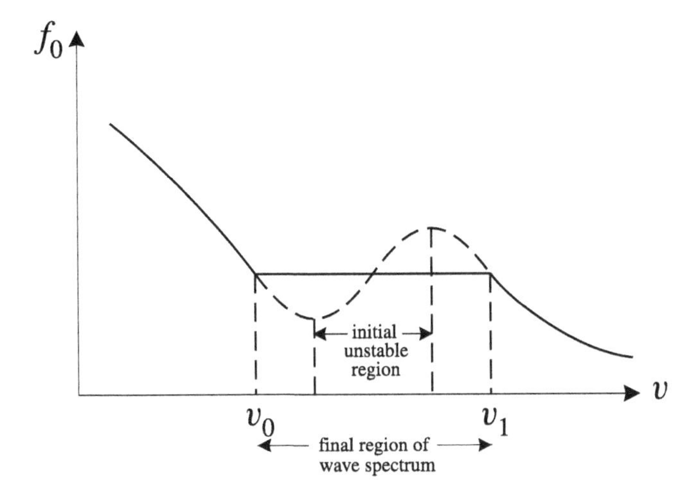

Fig. 10.1. Illustration of quasi-linear evolution of f_0 for the bump-on-tail instability.

where A and B are both positive. Combining (10.14) and (10.15) we have

$$\frac{\partial f_0}{\partial t} = \frac{\partial}{\partial v}\left(\frac{A}{B}\frac{\partial \mathcal{E}}{\partial t}\right)$$

i.e.

$$\frac{\partial}{\partial t}\left[f_0 - \frac{\partial}{\partial v}\left(\frac{A\mathcal{E}}{B}\right)\right] = 0$$

Then if $\mathcal{E}$ is negligible at $t = 0$

$$\frac{\partial}{\partial v}\left(\frac{A\mathcal{E}}{B}\right) = f_0(v, 0) - f_0(v, t)$$

We seek asymptotically steady state solutions, that is $\partial\mathcal{E}/\partial t$, $\partial f_0/\partial t \to 0$ as $t \to \infty$. If $\partial\mathcal{E}/\partial t = 0$, from (10.15) either $\mathcal{E} = 0$ or $\partial f_0/\partial v = 0$. Suppose $\partial f_0/\partial v = 0$ for $v_0 < v < v_1$ and $\mathcal{E} = 0$ for all other v. Then for $v_0 < v < v_1$, $f_0(v, \infty)$ is constant and for all other v, $f_0(v, \infty) = f_0(v, 0)$. These results are shown schematically in Fig. 10.1. The unstable region is initially defined by the range of v for which $f_0' > 0$ but as the diffusion progresses this region expands since f_0' decreases within the unstable range but becomes positive just outside it. The instability is quenched when f_0 is constant across the final range of the wave spectrum between v_0 and v_1. The increase in energy in the waves is compensated for by the net loss of energy of the particles; faster particles in the initial bump have been replaced by slower particles filling the initial trough.

Quasi-linear theory has been widely used with some success despite the arbitrariness of its assumptions and some lack of consensus on the conditions for its validity.

10.2.2 Particle trapping

Another important wave–particle effect that leads to the quenching of instabilities is particle trapping. It comes about because waves have finite amplitudes and particles with insufficient energy to surmount the wave peaks oscillate back and forth in the wave troughs.

To investigate this we consider a single wave and assume that its amplitude grows or decays very slowly compared with the rate at which the wave oscillates. Then, in a frame of reference moving with the wave speed ω/k, the particles see a constant wave profile which is a function of x only. The equation of motion of an electron is

$$m\ddot{x} = -eE(x) = e\frac{\mathrm{d}\phi(x)}{\mathrm{d}x} \tag{10.16}$$

where $\phi(x)$ is the electrostatic potential. A first integral is the energy equation

$$\frac{1}{2}m\dot{x}^2 - e\phi(x) = E_0$$

where E_0 is a constant equal to the total energy of the electron. Clearly, if $E_0 > -e\phi(x)$ for all x the kinetic energy is positive for all x and the electron is untrapped. On the other hand, all electrons with values of E_0 below the wave peaks are trapped and oscillate in the wave troughs between the points at which $E_0 = -e\phi(x)$. The energy diagram and the phase space trajectories are shown in Fig. 10.2; in the laboratory frame the trajectories move to the right with the wave speed ω/k. Note that it is the *resonant* electrons, those with $v \approx \omega/k$, that are trapped.

Electron trapping imposes a severe restriction on the validity of linear Landau damping theory which, as we saw in Chapter 7, is equivalent to integration over unperturbed orbits, namely the straight line trajectories: $x(t) = x(0) + v(0)t$. The trajectories of the trapped (resonant) electrons have a superimposed oscillatory motion governed by (10.16) and if $E(x) = \tilde{E}\sin kx \simeq \tilde{E}kx$ for the most strongly trapped electrons (near the bottom of the potential well) we have

$$m\ddot{x} = -m\omega^2 x = -e\tilde{E}kx$$

giving $\omega^2 = e\tilde{E}k/m \equiv \omega_{\mathrm{B}}^2$, where ω_{B} is called the *bounce frequency*. Usually $\omega_{\mathrm{pe}} \gg \omega_{\mathrm{B}}$ but the Landau damping decrement, given by (7.37) and denoted here

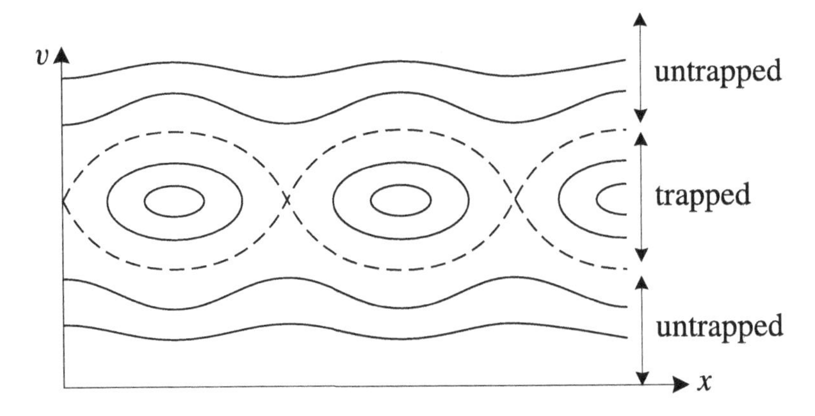

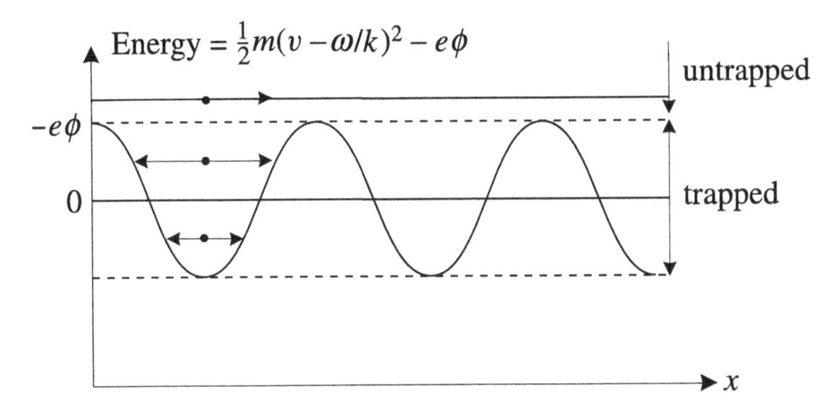

Fig. 10.2. Illustration of particle trapping by finite amplitude waves. The upper figure shows the phase space trajectories and the lower figure the energy diagram.

by γ_{L}, decreases exponentially for long wavelength plasma oscillations. Clearly, if $\omega_{\mathrm{B}} > \gamma_{\mathrm{L}}$ the effect of electron trapping will come into play before appreciable linear Landau damping takes place. This produces, as discussed by Davidson (1972), a non-monotonic decay which is shown schematically in Fig. 10.3. The wave energy decays according to linear Landau theory for $t < 2\pi/\omega_{\mathrm{B}}$, releasing some of the trapped particles, and then oscillates with a frequency of the order of ω_{B}. This oscillation frequency increases with t and for $t \gg 2\pi/\omega_{\mathrm{B}}$ the wave energy tends to a constant value which is lower by a fraction of order $\gamma_{\mathrm{L}}/\omega_{\mathrm{B}}$ times its initial value. The trapped electrons, by continually exchanging energy with the wave, keep it from collapsing and produce the oscillations in the wave energy. This damping is clearly a non-linear process and, although physically quite distinct from linear Landau damping, is often referred to as *non-linear Landau damping*.

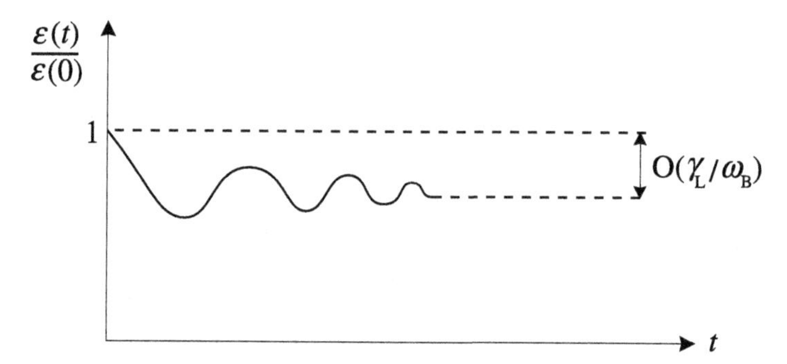

Fig. 10.3. Illustration of non-linear Landau damping. If $\omega_B > \gamma_L$ electron trapping occurs before the wave has decayed significantly and the fractional loss of wave energy is limited to $O(\gamma_L/\omega_B)$.

The theory is equally applicable to the case of growing waves. In this case there is linear Landau growth followed by oscillation and asymptotic approach to a wave energy which is a fraction of order γ_L/ω_B higher than its initial value.

10.2.3 Particle trapping in the beam–plasma instability

We turn next to a consideration of the non-linear phase of the beam–plasma instability (BPI) analysed in its linear phase for the weak-beam case in Section 6.5.2. This showed that three of the four solutions to the linear dispersion relation are of the same magnitude for $kv_b/\omega_p \simeq 1$. The maximum linear growth rate (6.121) is

$$\gamma_{\max} = \frac{\sqrt{3}}{2}\left(\frac{\omega_{pb}^2}{2\omega_p^2}\right)^{1/3}\omega_p \tag{10.17}$$

and the frequency of the growing perturbations is

$$\omega = \omega_p\left[1 - \frac{1}{2}\left(\frac{\omega_{pb}^2}{2\omega_p^2}\right)^{1/3}\right] \equiv \omega_p(1-\delta) \tag{10.18}$$

Restricting ourselves to the weak-beam limit makes possible a quasi-linear extension to the linear result (Drummond *et al.* (1970), O'Neil *et al.* (1971), Gentle and Lohr (1973)). As in the linear model we consider only electron dynamics but now we need to bear in mind that doing so will in general impose restrictions on the validity of the non-linear result.

In the linear regime it is straightforward to show that zero- and first-order variables are related by

$$\frac{n_{b1}}{n_{b0}} = -\frac{1}{\delta}\frac{v_{b1}}{v_{b0}} \qquad \frac{v_{b1}}{v_{b0}} = -\frac{1}{\delta}\frac{v_{p1}}{v_{b0}}$$

which establishes the ordering

$$\left|\frac{v_{p1}}{v_{b0}}\right| \ll \left|\frac{v_{b1}}{v_{b0}}\right| \ll \left|\frac{n_{b1}}{n_{b0}}\right| \tag{10.19}$$

Thus, even if the beam density perturbation is not itself small, the changes in both beam and plasma electron velocities are smaller in order than the beam density perturbation. We saw in Section 6.5.2 that the BPI spectrum was relatively narrow so that we may approximate the wave potential ϕ_1 by a monochromatic sinusoidal wave form,

$$\phi_1(x,t) = \phi_1 \cos\omega_p\left[\frac{x}{v_{b0}} - (1-\delta)t\right] \tag{10.20}$$

even in the non-linear phase, at least until the stage at which significant numbers of beam electrons are trapped by the wave. In the wave frame $\omega/k = v_{b0}(1-\delta)$, the beam electrons experience the potential $\phi_1 = \bar{\phi}\cos kx$, where $\bar{\phi}$ is the time-averaged amplitude. The trajectory of a test electron (labelled j) in this potential is determined by energy conservation from

$$\frac{1}{2}mv_j^2 - e\bar{\phi}\cos kx_j = c_j \tag{10.21}$$

An electron for which $c_j = -e\bar{\phi}$ is trapped at the bottom of the potential well whereas one for which $c_j = e\bar{\phi}$ is on the border between trapped and free electrons. The critical escape velocity is $v_c = (4e\bar{\phi}/m)^{1/2}$. Thus in the quasi-linear model some beam electrons will become trapped once the potential has grown to a level such that $\Delta v = v_{b0} - \omega/k = \delta v_{b0} = v_c$. Then

$$\bar{\phi}_{tr} = \frac{m}{4e}(\Delta v)^2 \tag{10.22}$$

with corresponding energy density at time $t = t_1$, say

$$W(t_1) = \frac{\varepsilon_0}{4}k^2\bar{\phi}_{tr}^2 \simeq 2^{-31/3}\left(\frac{\omega_{pb}^2}{\omega_p^2}\right)^{1/3}\left[\frac{1}{2}n_{b0}mv_{b0}^2\right] \tag{10.23}$$

Note that this is a small fraction of the beam kinetic energy density. Growth of the wave continues until most of the beam electrons have been trapped. By this stage, at $t = t_2$, the beam electrons have lost kinetic energy $\frac{1}{2}n_{b0}m[(v_{b0}+\Delta v)^2 - (v_{b0}-\Delta v)^2] \sim 2n_{b0}mv_{b0}\Delta v$. Since (10.19) guarantees that the plasma electron dynamics is still essentially linear, to a good approximation we may assign one

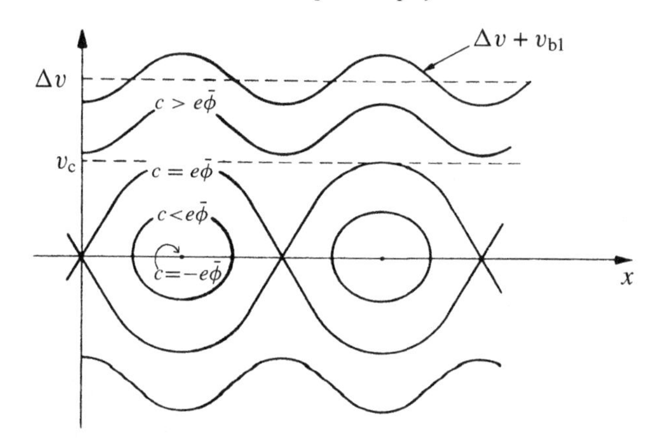

Fig. 10.4. Electron phase space showing trapped and free electron trajectories.

half of the kinetic energy lost to the oscillations and the other half to electrostatic field energy. Then at time $t = t_2$ the average field energy density is

$$W(t_2) \simeq n_{b0} m v_{b0} \Delta v = 2^{-1/3} \left(\frac{\omega_{pb}^2}{\omega_p^2} \right)^{1/3} \left[\frac{1}{2} n_{b0} m v_{b0}^2 \right] \tag{10.24}$$

Comparing (10.23) with (10.24) we see that $W(t_2) = 2^{10} W(t_1)$. Note that for $n_{b0}/n_0 = 0.015$, only 20% of beam kinetic energy is converted to field energy. As particle velocities bounce back up, field energy is reconverted into kinetic energy but since the potential well is not parabolic, particles oscillate at different frequencies. In the phase space representation in Fig. 10.4 electrons rotate round equipotential contours and undergo phase mixing with the result that the position of the beam electrons in phase space is smeared out after some rotations. The field oscillations die out and the field energy settles at a value that is half the difference between the initial beam energy and that of the smeared-out distribution, i.e.

$$W_f(t \gg t_2) = 2^{-4/3} \left(\frac{\omega_{pb}^2}{\omega_p^2} \right)^{1/3} \left[\frac{1}{2} n_{b0} m_e v_{b0}^2 \right]$$

as illustrated in Fig. 10.5.

This simple picture does not provide an accurate representation of beam dynamics. Figure 10.4 illustrates the velocity modulation of the beam so that even when $v_c = \Delta v$ trapping is not complete. One can show that when trapping occurs the beam velocity modulation $v_{b1} = \Delta v$ which in turn implies that $n_{b1}/n_{b0} = 1$; in other words the beam electron dynamics is seriously non-linear, with bunching well developed. The bunches are trapped in the potential well and rotate in

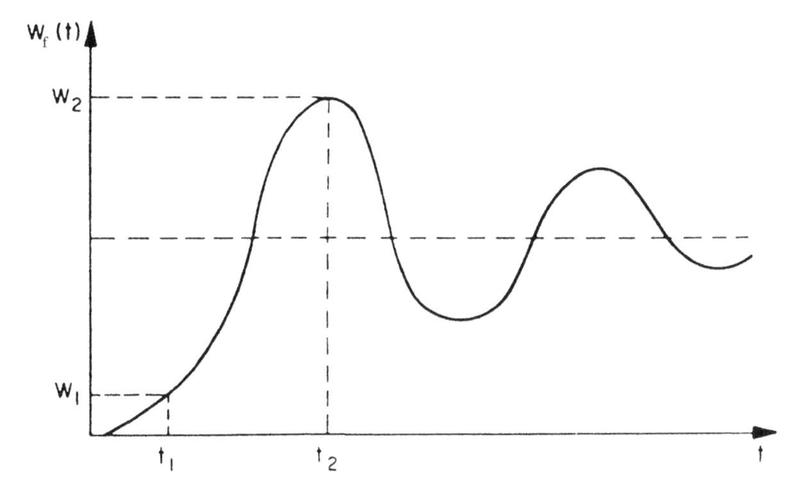

Fig. 10.5. Electrostatic field energy as a function of time. The damping time corresponds to the time characterizing the smearing out of particles in phase space.

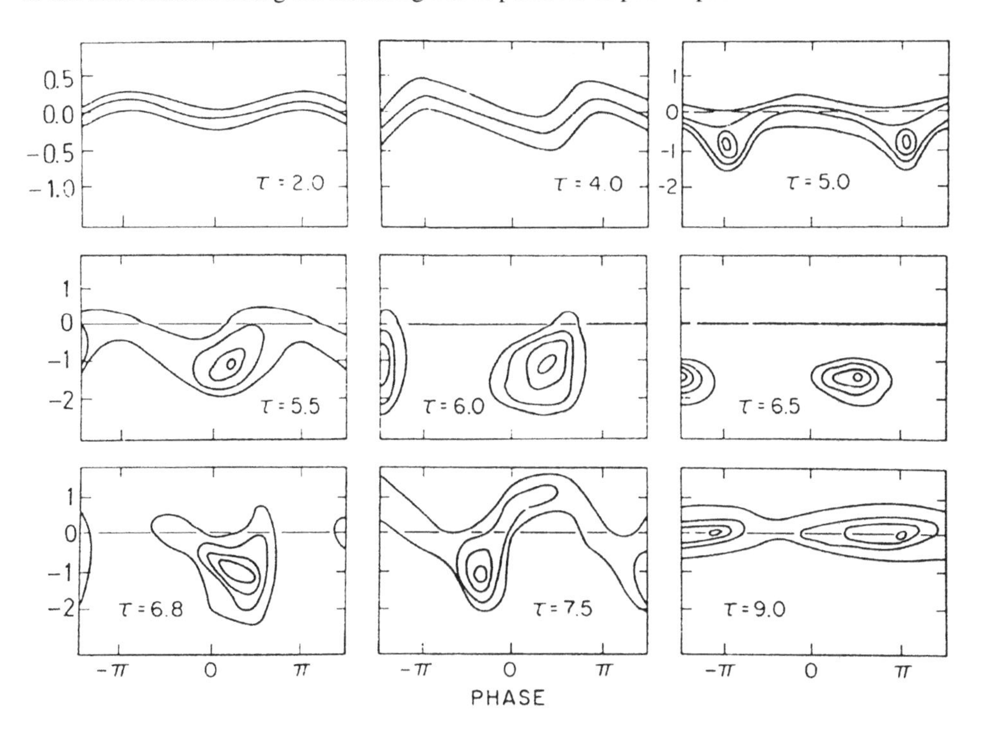

Fig. 10.6. Experimental contour plots of the electron distribution in phase space. At $\tau = 2, 4$ the central line represents the maximum electron density contour, the lines on either side corresponding to the half-maximum value. At later times, the inner contour represents maximum density. The phase reference is arbitrary and differs for each frame (after Gentle and Lohr (1973)).

phase space. Overall the wave energy displays oscillatory behaviour as shown in Fig. 10.5.

The main predictions of the non-linear phase of BPI were confirmed in experiments by Gentle and Lohr (1973) measuring maximum wave amplitude, monochromaticity of the unstable mode and its harmonic content. Measured contour plots of the electron distribution in phase space are shown in Fig. 10.6.

10.2.4 Plasma echoes

A quite remarkable prediction of non-linear Landau theory is the existence of what are called *plasma echoes*. There is no dissipation in a collisionless plasma and therefore no increase in entropy. Consequently, although a wave may effectively disappear as its amplitude decreases through Landau damping, some trace of it remains in the perturbed distribution function f_1. This trace lies in the terms that we discarded in quasi-linear theory on the grounds that they were transient. Specifically, we noted that, on taking the Laplace transform of (10.8), the ballistic term varies like $\exp(-i\mathbf{k}\cdot\mathbf{v}t)$ but we dropped it because it becomes highly oscillatory in $\mathbf{k}\cdot\mathbf{v}$ as $t \to \infty$ and therefore makes a negligible contribution to an integral over $\mathbf{k}$ or $\mathbf{v}$. This is called *phase mixing*. Note, however, that the ballistic term itself does not decay; it is this term that carries the information about the initial perturbation that produced the Landau damped wave.

Now the idea behind the plasma echo is to create two Landau damped waves at different times t_1 and t_2 such that at a later time t_3 a third wave (the echo) may arise from their non-linear interaction. At time t the ballistic term from wave 1 with wavenumber $\mathbf{k}_1$ varies like $\exp(-i\mathbf{k}_1\cdot\mathbf{v}(t-t_1))$ and similarly for wave 2 like $\exp(-i\mathbf{k}_2\cdot\mathbf{v}(t-t_2))$. If we now resurrect the second-order terms on the right-hand side of (10.4), which were neglected in quasi-linear theory, we get contributions varying like $\exp(-i[\mathbf{k}_1\cdot\mathbf{v}(t-t_1)-\mathbf{k}_2\cdot\mathbf{v}(t-t_2)])$. Choosing $\mathbf{k}_1$ and $\mathbf{k}_2$ to be in the same direction we see that the exponent vanishes at

$$t = t_3 = \frac{k_1 t_1 - k_2 t_2}{k_1 - k_2}$$

Consequently, when the velocity integral in (10.2) is performed at $t = t_3$ there is no phase mixing and the echo wave appears.

This effect, which can be discussed in spatial terms also, generating the waves at separate points in a plasma column and observing the echo at a third point further down the column, has been demonstrated experimentally by Wong and Baker (1969).

10.3 Wave–wave interactions

So far our discussion of non-linear effects has been largely by extension of the Landau theory into the non-linear regime. We began with wave–particle interactions and with the discussion of plasma echoes we have moved on to wave–wave interactions though, in this case, one which is realized via the resonant particles. Another example of this kind is *induced scattering* in which the resonant particles interact with the beat wave of two plasma waves. It is interesting to note the sequence of resonance conditions. For linear Landau damping (or growth) we have

$$\omega = \mathbf{k} \cdot \mathbf{v}$$

and this was also the resonance condition for particle trapping where only one wave is involved. The resonance condition for induced scattering is obtained by substituting the relevant ω and $\mathbf{k}$ for the beat wave. When this is between two plasma waves, denoted by frequencies and wavenumbers, $(\omega_1, \mathbf{k}_1)$ and $(\omega_2, \mathbf{k}_2)$, we have

$$\omega_1 - \omega_2 = (\mathbf{k}_1 - \mathbf{k}_2) \cdot \mathbf{v}$$

If the first wave is driven with a finite amplitude the second can arise through the non-linear interaction with the resonant particles and hence the description of this process as induced scattering. Since the beat wave frequency for Langmuir wave scattering is low ($\omega_1, \omega_2 \approx \omega_{\rm pe}, \omega_1 - \omega_2 \ll \omega_{\rm pe}$) the scattering is off the ions and the process is adequately described by ion Landau theory combined with a fluid description of the electron waves as discussed, for example, in Nicholson (1983).

The next stage in this progression is to consider direct wave–wave interactions. It is often the case that wave–particle coupling is sufficiently weak that it is insignificant and yet plasmas can support so many waves that if one wave $(\omega_0, \mathbf{k}_0)$ is propagating and another natural mode $(\omega_1, \mathbf{k}_1)$ spontaneously arises, resonant non-linear coupling may give rise to the beat wave $(\omega_2, \mathbf{k}_2)$. The resonance conditions for this are

$$\omega_0 = \omega_1 + \omega_2 \tag{10.25}$$

$$\mathbf{k}_0 = \mathbf{k}_1 + \mathbf{k}_2 \tag{10.26}$$

and the waves are said to form a *resonant triad*. Of course, we could go on in this way and have four or more waves in resonance but since this involves cubic or higher order terms these are less likely to be of significance than resonant triads.

Derivation of the equations describing non-linear wave coupling may be via the Vlasov–Maxwell equations or the two-fluid wave equations and may treat the waves as coherent or take ensemble averages of systems with many waves having random phases (weak turbulence analysis); for a thorough discussion of this topic see Davidson (1972). Whatever the analysis, the end product is a set of equations of the general form

$$\frac{\partial A_\alpha(\mathbf{k},t)}{\partial t} = \int d\mathbf{k}\, d\mathbf{k}'\, d\mathbf{k}''\delta(\mathbf{k}-\mathbf{k}'-\mathbf{k}'')K_{\alpha\beta\gamma}(\mathbf{k},\mathbf{k}',\mathbf{k}'')$$
$$\times A_\beta(\mathbf{k}',t)A_\gamma(\mathbf{k}'',t)e^{i[\omega_\alpha(\mathbf{k})-\omega_\beta(\mathbf{k}')-\omega_\gamma(\mathbf{k}'')]t} \tag{10.27}$$

where $A_{\alpha,\beta,\gamma}$ are the wave amplitudes and $K_{\alpha\beta\gamma}$ is the interaction kernel for the triplet (α, β, γ). In fact, equations with this structure arise in many branches of physics and engineering and it is only the kernel $K_{\alpha\beta\gamma}$ which varies according to the specific non-linear wave coupling under consideration. To avoid a lot of heavy algebra therefore, we shall derive the equations for a simple system of three coupled harmonic oscillators. In other words, we model the waves by harmonic oscillators and the plasma as the medium which supports them and allows their interaction.

If C is the (constant) coupling coefficient for the three oscillators, the equations of motion are

$$\left.\begin{aligned} \ddot{x}_0 + \omega_0^2 x_0 &= -Cx_1x_2 \\ \ddot{x}_1 + \omega_1^2 x_1 &= -Cx_0x_2 \\ \ddot{x}_2 + \omega_2^2 x_2 &= -Cx_0x_1 \end{aligned}\right\} \tag{10.28}$$

In the linear approximation the solutions are

$$x_j = \frac{1}{2}\left(A_j e^{i\omega_j t} + A_j^* e^{-i\omega_j t}\right) \qquad (j = 0, 1, 2)$$

and if the coupling is weak we may expect the non-linear solutions to be of the form

$$x_j = \frac{1}{2}\left(A_j(t)e^{i\omega_j t} + A_j^*(t)e^{-i\omega_j t}\right) \tag{10.29}$$

where the amplitudes A_j are now slowly varying functions of t such that

$$\left|\frac{\dot{A}_j}{A_j}\right| \ll \omega_j \tag{10.30}$$

Substituting (10.29) in (10.28) we get for the first member

$$(A_0 + 2i\omega_0\dot{A}_0) + (\ddot{A}_0^* - 2i\omega_0\dot{A}_0^*)e^{-2i\omega_0 t}$$
$$= -\frac{C}{2}(A_1e^{i\omega_1 t} + A_1^*e^{-i\omega_1 t})(A_2e^{i\omega_2 t} + A_2^*e^{-i\omega_2 t})e^{-i\omega_0 t}$$

and, when we average over the fast time scale, phase mixing gets rid of the second term on the left-hand side and all terms on the right-hand side except the one whose exponent vanishes because of the resonance condition (10.25). In view of (10.30) we may drop the $\ddot{A}_0$ term as well, leaving

$$\dot{A}_0 = \frac{iC}{4\omega_0} A_1 A_2$$

The corresponding equations for the second and third members of (10.28) are

$$\dot{A}_1 = \frac{iC}{4\omega_1} A_0 A_2^*$$

and

$$\dot{A}_2 = \frac{iC}{4\omega_2} A_0 A_1^*$$

It is convenient to re-define the amplitudes by

$$a_j = A_j \omega_j^{1/2} \tag{10.31}$$

to get the more symmetric set of equations

$$\left.\begin{aligned} \dot{a}_0 &= iKa_1a_2 \\ \dot{a}_1 &= iKa_0a_2^* \\ \dot{a}_2 &= iKa_0a_1^* \end{aligned}\right\} \tag{10.32}$$

where

$$K = C/4(\omega_0\omega_1\omega_2)^{1/2} \tag{10.33}$$

is the common coupling coefficient.

Although the details of the derivation of the equation set (10.32) for the nonlinear wave coupling in a plasma are more complicated, the method is the same. The condition (10.26) on the wavenumbers, expressed by the delta function in (10.27), arises because the waves have spatial as well as temporal harmonic variation, $\exp i(\mathbf{k} \cdot \mathbf{r} - \omega t)$, and is necessary to avoid phase mixing in space.

From (10.25) and (10.32) it is easily shown that

$$\frac{\mathrm{d}}{\mathrm{d}t}\left(\omega_0|a_0|^2 + \omega_1|a_1|^2 + \omega_2|a_2|^2\right) = 0 \tag{10.34}$$

or

$$\frac{\mathrm{d}}{\mathrm{d}t}\left(|A_0|^2 + |A_1|^2 + |A_2|^2\right) = 0$$

which expresses the conservation of energy since wave energy density is proportional to the square of the amplitude.

Directly from (10.32) we get

$$-\frac{\mathrm{d}|a_0|^2}{\mathrm{d}t} = \frac{\mathrm{d}|a_1|^2}{\mathrm{d}t} = \frac{\mathrm{d}|a_2|^2}{\mathrm{d}t}$$

or

$$-\frac{1}{\omega_0}\frac{\mathrm{d}|A_0|^2}{\mathrm{d}t} = \frac{1}{\omega_1}\frac{\mathrm{d}|A_1|^2}{\mathrm{d}t} = \frac{1}{\omega_2}\frac{\mathrm{d}|A_2|^2}{\mathrm{d}t} \tag{10.35}$$

which are the *Manley–Rowe relations*, first discussed in the context of parametric amplification in electronics. They show the rates at which energy is transferred between the waves. An exact solution of (10.32) is obtainable in terms of elliptic functions showing the periodic nature of the interaction.

Generalizations of the theory may be introduced, the most important of which is wave damping. This is done by adding a term $\nu_j a_j$ to the left-hand side of the a_j equation in (10.32), where ν_j is the linear damping rate of the jth wave. As discussed further below, this introduces a threshold for the spontaneous excitation of a natural mode since there are now competing effects and the energy in the excitation must exceed that lost by damping.

Another important generalization is to allow for spatial variation of the wave amplitudes by replacing the time derivative $\mathrm{d}/\mathrm{d}t$ by the convective derivative $(\partial/\partial t + \mathbf{v}_j \cdot \boldsymbol{\nabla})$, where $\mathbf{v}_j$ is the group velocity of the jth wave. This means that the interaction is now between wave packets rather than monochromatic waves, adding a touch of reality. Other extensions of the theory, which we investigate below, allow for frequency and wavenumber mismatch.

10.3.1 Parametric instabilities

The interest of plasma physicists in wave–wave interactions has arisen in the context of plasma heating and particularly in the field of laser–plasma interactions. The laser beam is a large amplitude, transverse, electromagnetic wave being driven through the plasma and is capable, by means of resonant three-wave coupling, of transferring its energy to two other waves. Such a process in which natural modes grow at the expense of the large amplitude wave, usually referred to as the pump wave, is known as a *parametric instability*.

In this class of three-wave interactions we distinguish between the pump wave $(\omega_0, \mathbf{k}_0)$ and the so-called decay waves which are both small amplitude. Thus, from (10.32) to first order, it follows that a_0 is constant and we investigate the growth of a_1 and a_2. To find the threshold condition, damping, which may be Landau or collisional, is included, so the equations are

$$\left.\begin{aligned} \dot{a}_1 + \nu_1 a_1 &= iKa_0 a_2^* \\ \dot{a}_2 + \nu_2 a_2 &= iKa_0 a_1^* \end{aligned}\right\} \tag{10.36}$$

where ν_1 and ν_2 are both positive. Taking the complex conjugate of the second equation in (10.36) and trying a solution $\propto e^{\alpha t}$ we get

$$\begin{aligned}(\alpha+\nu_1)a_1 - iKa_0a_2^* &= 0\\ iKa_0^*a_1 + (\alpha+\nu_2)a_2 &= 0\end{aligned}$$

for which there is a non-trivial solution if

$$(\alpha+\nu_1)(\alpha+\nu_2) - K^2|a_0|^2 = 0 \tag{10.37}$$

Separating real and imaginary parts of this equation we see that α must be real and has a positive root if

$$K^2|a_0|^2 > \nu_1\nu_2 \tag{10.38}$$

This is the *threshold condition* for the instability stating that the combination of the energy in the pump wave and the strength of the non-linear coupling must be sufficient to overcome the damping of the decay waves.

Conditions (10.25) and (10.26) represent perfect matching. In practice we need to explore the consequences of allowing a small frequency mismatch such that

$$\omega_0 - \omega_1 - \omega_2 = \Delta\omega$$

where $|\Delta\omega|$ is very much smaller than any of the wave frequencies. Then instead of (10.36) we have

$$\left.\begin{aligned}\dot{a}_1 + \nu_1 a_1 &= iKa_0a_2^*e^{i\Delta\omega t}\\ \dot{a}_2 + \nu_2 a_2 &= iKa_0a_1^*e^{i\Delta\omega t}\end{aligned}\right\} \tag{10.39}$$

where the residual factor $e^{i\Delta\omega t}$ is a slow variation like $a_1(t)$ and $a_2(t)$ and is best dealt with by absorbing it into the amplitudes by defining $\tilde{a}_j = a_j e^{-i\Delta\omega t/2}$. Now (10.39) becomes

$$\left.\begin{aligned}\mathrm{d}\tilde{a}_1/\mathrm{d}t + (i\Delta\omega/2 + \nu_1)\,\tilde{a}_1 &= iKa_0\tilde{a}_2^*\\ \mathrm{d}\tilde{a}_2/\mathrm{d}t + (i\Delta\omega/2 + \nu_2)\,\tilde{a}_2 &= iKa_0\tilde{a}_1^*\end{aligned}\right\} \tag{10.40}$$

and proceeding as for (10.36) it is easily verified that the resulting auxiliary equation replacing (10.37) is

$$(\alpha+\nu_1+i\Delta\omega/2)(\alpha+\nu_2-i\Delta\omega/2) - K^2|a_0|^2 = 0$$

This equation now has complex roots but at threshold ($\Re\alpha = 0$), on separating real and imaginary parts, we find

$$K^2|a_0|^2 = \nu_1\nu_2 + \frac{\nu_1\nu_2(\Delta\omega)^2}{(\nu_1+\nu_2)^2} \tag{10.41}$$

showing by comparison with (10.38) the increase in pump wave energy required to overcome frequency mismatch.

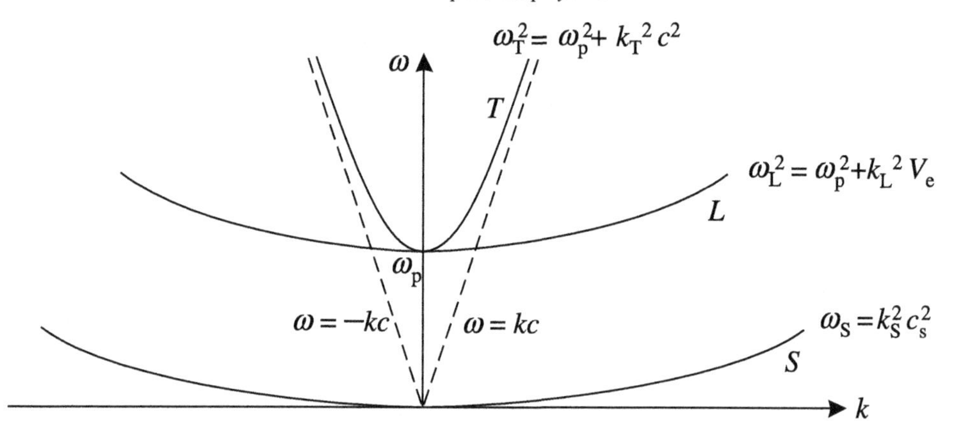

Fig. 10.7. Dispersion relations for transverse electromagnetic, Langmuir and ion acoustic waves.

Wavenumber mismatch has a similar effect reducing growth rate and increasing the threshold for instability. These are important considerations since plasmas, especially laser-produced plasmas, are highly inhomogeneous so the relationship between ω and $\mathbf{k}$ changes as a dispersive wave travels through the plasma. Consequently, the resonant triad conditions between the pump and decay waves will be satisfied only in some restricted region and the parametric instability is likewise restricted. These effects are examined in the following chapter. Here we present a brief qualitative discussion of four parametric instabilities important in laser–plasma interactions.

We consider an unmagnetized plasma, with $T_e \gg T_i$, irradiated by an intense laser beam. In this case the three-wave coupling takes place between various combinations of the transverse, electromagnetic pump wave for which

$$\omega_0^2 = \omega_T^2 = \omega_p^2 + k_T^2 c^2 \tag{10.42}$$

the longitudinal, electrostatic electron plasma (or Langmuir) wave

$$\omega_L^2 = \omega_p^2 + k_L^2 V_e^2 \tag{10.43}$$

and the longitudinal, ion acoustic (or sound) wave

$$\omega_S^2 = k_S^2 c_s^2 \tag{10.44}$$

These dispersion relations were derived in Chapter 6 and are sketched in Fig. 10.7. We consider only positive frequencies but $\mathbf{k}$ values may be of either sign. The decay waves, labelled 1 and 2, may be any of the pairs

$$(1, 2) = (L, S), (L_1, L_2), (T', S), (T', L)$$

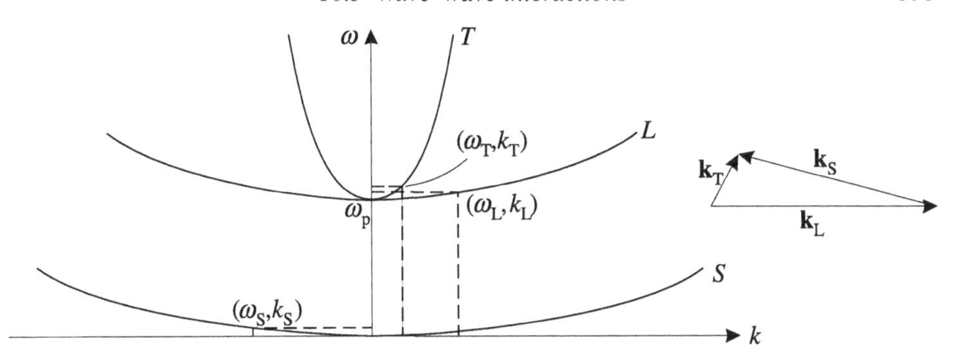

Fig. 10.8. Frequencies and wavenumbers for parametric decay instability.

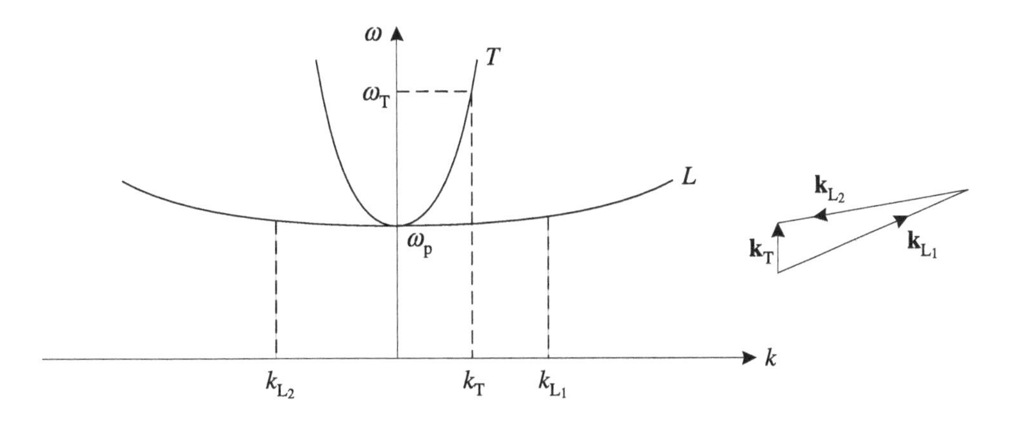

Fig. 10.9. Frequencies and wavenumbers for two plasmon decay instability.

where the second possibility has two Langmuir waves and the third and fourth involve a second (scattered) transverse wave. These are the only three-wave decays allowed in an unmagnetized plasma. Figures 10.8–10.11 illustrate the four cases.

The *parametric decay instability* ($T \rightarrow L + S$) has $\omega_T \approx \omega_L$ since $\omega_S \ll \omega_L$. Also, to avoid strong Landau damping of the Langmuir wave (low threshold) we need $\omega_L \approx \omega_p$. Thus, the instability occurs near the critical surface ($\omega_T \approx \omega_p$). Another consequence of the approximation ($\omega_T \approx \omega_L$) in a non-relativistic plasma is that $|\mathbf{k}_T| \ll |\mathbf{k}_L|$ and hence that $\mathbf{k}_L \approx -\mathbf{k}_S$. Note that although $\mathbf{k}_S$ and $\mathbf{k}_L$ are almost antiparallel they cannot be exactly so because strong coupling requires the electric field of the transverse wave $\mathbf{E}_T$ to be closely aligned to $\mathbf{k}_S$ and $\mathbf{k}_L$ and so $\mathbf{k}_T$ must be approximately perpendicular to them. Enhanced energy absorption results from this instability because the laser energy goes into two plasma waves which propagate only within the plasma and therefore cannot leave it.

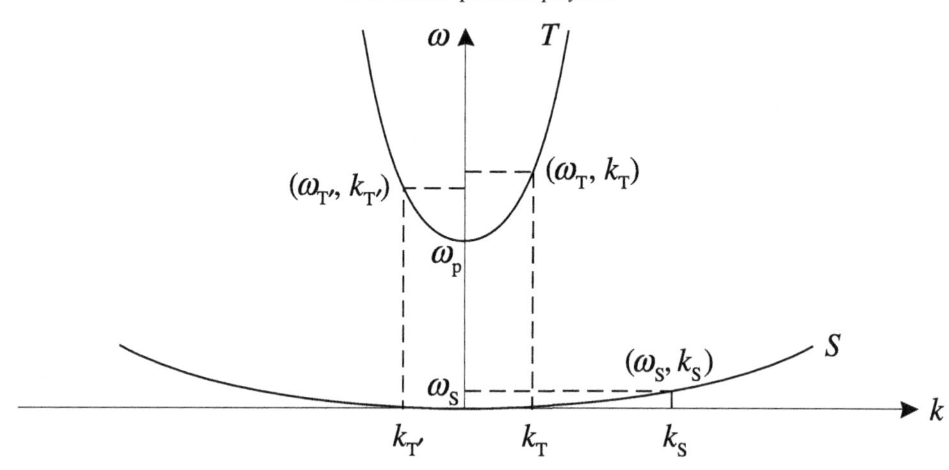

Fig. 10.10. Frequencies and wavenumbers for stimulated Brillouin scattering.

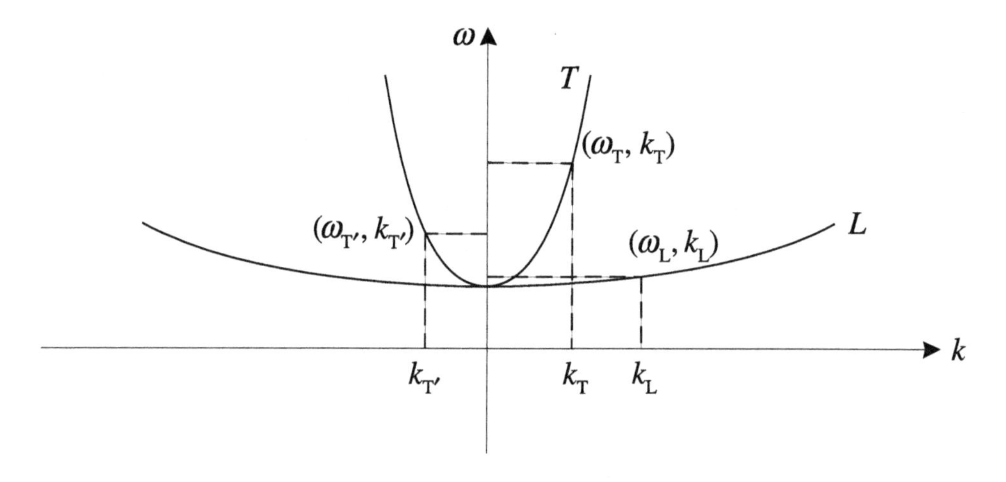

Fig. 10.11. Frequencies and wavenumbers for stimulated Raman scattering.

For the *two plasmon decay instability* ($T \rightarrow L_1 + L_2$) the same argument about avoiding Landau damping for a low threshold means that $\omega_{L_1} \approx \omega_{L_2} \approx \omega_p$ and hence $\omega_T \approx 2\omega_p$. It follows that this instability occurs around quarter-critical density. Here again strong coupling requires $\mathbf{k}_{L_1} \approx -\mathbf{k}_{L_2}$ and is maximized when $\mathbf{k}_T$ makes angles of approximately $\pi/4$ and $5\pi/4$ with $\mathbf{k}_{L_1}$ and $\mathbf{k}_{L_2}$. Enhanced energy absorption results for the same reasons as for the parametric decay instability.

In contrast with the previous cases both of the scattering instabilities may be one dimensional in the sense that the $\mathbf{k}$ vectors can all be collinear. From Fig. 10.10

we see that for *stimulated Brillouin scattering* ($T \rightarrow T' + S$), $\mathbf{k}_{T'}$ is necessarily in the opposite direction to $\mathbf{k}_T$ so that the scattered transverse wave takes energy back out of the plasma. Furthermore, since $\omega_S \ll \omega_T$, it follows from (10.35) that energy from the pump wave goes overwhelmingly into the transverse wave and not the plasma wave. This instability is therefore very detrimental to laser energy absorption by the plasma. The wave matching can occur for any frequency $\omega_T > \omega_p$ and thus the instability may arise anywhere up to the critical surface.

Stimulated Raman scattering ($T \rightarrow T' + L$) is discussed in more detail in the next chapter. Since, as before, we require $\omega_L \approx \omega_p$ for low threshold and $\omega_{T'} \geq \omega_p$, the instability can only occur for $\omega_T \geq 2\omega_p$, i.e. at and below quarter-critical density. As for Brillouin scattering, the scattered wave travels in the opposite direction to the incoming wave and therefore takes energy back out of the plasma. In this case, however, more of the energy gets into the plasma wave.

Quenching of parametric instabilities may come about as a result of:

(i) depletion of the pump wave to below threshold intensity;
(ii) decay of the daughter waves leading to a cascade of modes;
(iii) particle trapping as the electrostatic decay waves achieve large amplitude. Trapped particles may damp the wave at a rate greater than the linear damping; this affects the threshold and may switch off the instability;
(iv) plasma inhomogeneity leading to wavenumber mismatch.

10.4 Zakharov equations

In this section we investigate an important example of the modification of linear wave propagation by the retention of non-linear terms in the wave equations. The coupled equations we shall derive were first obtained by Zakharov (1972) using heuristic arguments to express analytically the physical effects involved in the coupling.

The problem we wish to study is the interaction of electron plasma and ion acoustic waves. The first is a high frequency wave dominated by the electron dynamics and the second a low frequency wave dominated by the ion dynamics. The role of the non-dominant species is to maintain approximate charge neutrality. The separation of these waves in linear theory is a direct result of the high ion to electron mass ratio. On the fast time scale of the electron wave the massive ions are essentially in static equilibrium. For the ion wave, on the other hand, the electrons are in dynamic equilibrium in the sense that their inertia is so small that they respond quickly enough to maintain force balance on the slow time scale. The coupling of ions and electrons via charge neutrality, however, means that the ion waves produce, through ion density fluctuations, a small perturbation of the electron wave dispersion relation. Likewise, the electron waves influence the ion waves by the appearance of the ponderomotive force in the force balance equation.

To find the non-linear interaction of these waves we carry out a two-time scale, perturbation analysis of the warm plasma wave equations given in Table 3.5. The procedure is similar to that applied to the Vlasov equation to obtain the quasi-linear equations in Section 10.2.1. There are two time scales because the electrons can react to the fields much more rapidly than the massive ions. Electron and field perturbations, therefore, have fast and slow components which we denote by subscripts f and s, respectively; ion perturbations, on the other hand, have only slow components denoted by subscript 1.

In equilibrium there are no fields and $Zn_{\mathrm{i}} = n_{\mathrm{e}} = n_0$, say. Even in the perturbed state, the rapid response of the electrons to the strong Coulomb force maintains approximate charge neutrality so that $Zn_1 \approx n_s$ and $\mathbf{u}_1 \approx \mathbf{u}_s$. Thus, we have

$$Zn_{\mathrm{i}} = n_0 + Zn_1 \approx n_0 + n_s \tag{10.45}$$

$$n_{\mathrm{e}} = n_0 + n_s + n_f \tag{10.46}$$

$$\mathbf{u}_{\mathrm{i}} = \mathbf{u}_1 \approx \mathbf{u}_s \tag{10.47}$$

$$\mathbf{u}_{\mathrm{e}} = \mathbf{u}_s + \mathbf{u}_f \tag{10.48}$$

Note that the fast time scale perturbations have time dependence of the form $a(t)e^{-i\omega t}$ where the amplitude $a(t)$ is slowly varying compared with the fluctuations at frequency ω, that is

$$\left|\frac{1}{a}\frac{\mathrm{d}a}{\mathrm{d}t}\right| \ll \omega \tag{10.49}$$

so that when averaging over the fast time scale the amplitude may be treated as constant and we have

$$\left.\begin{aligned} \langle n_{\mathrm{e}} \rangle &= n_0 + n_s \\ \langle \mathbf{u}_{\mathrm{e}} \rangle &= \mathbf{u}_s \end{aligned}\right\} \tag{10.50}$$

These equations define the slow perturbations and then (10.46) and (10.48) define the fast perturbations as

$$\left.\begin{aligned} n_f &= n_{\mathrm{e}} - \langle n_{\mathrm{e}} \rangle \\ \mathbf{u}_f &= \mathbf{u}_{\mathrm{e}} - \langle \mathbf{u}_{\mathrm{e}} \rangle \end{aligned}\right\} \tag{10.51}$$

From

$$q = e(Zn_{\mathrm{i}} - n_{\mathrm{e}})$$

it follows that

$$\begin{aligned} q_s &= e(Zn_1 - n_s) = \epsilon_0 \nabla \cdot \mathbf{E}_s & (10.52) \\ q_f &= -en_f = \varepsilon_0 \nabla \cdot \mathbf{E}_f & (10.53) \end{aligned}$$

where the total field

$$\mathbf{E} = \mathbf{E}_s + \mathbf{E}_f \tag{10.54}$$

Assuming ions and electrons behave like perfect gases and eliminating the partial pressures using the adiabatic gas equation, the warm plasma wave equations are

$$\begin{aligned} \frac{\partial n_\alpha}{\partial t} + \nabla(n_\alpha \mathbf{u}_\alpha) &= 0 & (10.55) \\ \left(\frac{\partial}{\partial t} + \mathbf{u}_\alpha \cdot \nabla\right) \mathbf{u}_\alpha &= \frac{e_\alpha \mathbf{E}}{m_\alpha} - \frac{\gamma_\alpha k_B T_\alpha}{m_\alpha n_\alpha} \nabla n_\alpha & (10.56) \end{aligned}$$

For the electrons these equations contain non-linear terms $n_e \mathbf{u}_e$ and $\mathbf{u}_e \cdot \nabla \mathbf{u}_e$ for which we need to calculate fast and slow components and we do this using the same recipe used for the linear terms (10.50) and (10.51). Thus,

$$\begin{aligned} (n_e \mathbf{u}_e)_s &= \langle n_e \mathbf{u}_e \rangle = \langle (n_0 + n_s + n_f)(\mathbf{u}_s + \mathbf{u}_f) \rangle \\ &= (n_0 + n_s)\mathbf{u}_s + \langle n_f \mathbf{u}_f \rangle & (10.57) \\ (n_e \mathbf{u}_e)_f &= n_e \mathbf{u}_e - \langle n_e \mathbf{u}_e \rangle \\ &= (n_0 + n_s)\mathbf{u}_f + n_f \mathbf{u}_s + (n_f \mathbf{u}_f - \langle n_f \mathbf{u}_f \rangle) & (10.58) \end{aligned}$$

and

$$\begin{aligned} (\mathbf{u}_e \cdot \nabla \mathbf{u}_e)_s &= \langle (\mathbf{u}_s + \mathbf{u}_f) \cdot \nabla (\mathbf{u}_s + \mathbf{u}_f) \rangle \\ &= \mathbf{u}_s \cdot \nabla \mathbf{u}_s + \langle \mathbf{u}_f \cdot \nabla \mathbf{u}_f \rangle & (10.59) \\ (\mathbf{u}_e \cdot \nabla \mathbf{u}_e)_f &= \mathbf{u}_s \cdot \nabla \mathbf{u}_f + \mathbf{u}_f \cdot \nabla \mathbf{u}_s \\ &\quad + (\mathbf{u}_f \cdot \nabla \mathbf{u}_f - \langle \mathbf{u}_f \cdot \nabla \mathbf{u}_f \rangle) & (10.60) \end{aligned}$$

We shall not keep all of the non-linear terms in (10.57)–(10.60) but only those of 'leading order'. To determine which are leading order terms we assume that $|n_f| \ll |n_s|$ and $|\mathbf{u}_s| \ll |\mathbf{u}_f|$. The first of these assumptions is justified on the grounds that n_f is limited by charge neutrality whereas n_s, since it is matched by Zn_1, is not. The second assumption is obvious since $\mathbf{u}_s \approx \mathbf{u}_1$, the ion flow velocity. Note also that the pressure term in (10.56) is treated as a linear term since we shall replace n_α in the denominator by its equilibrium value.

With these preliminaries we now proceed to the fast wave analysis for which the relevant equations are (10.53) and, from (10.55) and (10.56),

$$\frac{\partial n_f}{\partial t} + \nabla \cdot \left[(n_0 + n_s)\mathbf{u}_f\right] = 0 \tag{10.61}$$

$$\frac{\partial \mathbf{u}_f}{\partial t} = -\frac{e\mathbf{E}_f}{m_e} - \frac{\gamma_e k_B T_e}{m_e n_0}\nabla n_f \tag{10.62}$$

In fact, the only non-linear term retained is the leading order term in (10.58), all the non-linear terms in (10.60) being negligible compared with $\partial \mathbf{u}_f/\partial t$ in (10.62). Next, we take the partial time derivative of (10.61), neglect the slow $\partial n_s/\partial t$ term and substitute for n_f from (10.53) and for $\partial \mathbf{u}_f/\partial t$ from (10.62) to get

$$\nabla \cdot \left\{\frac{\partial^2 \mathbf{E}_f}{\partial t^2} + (n_0 + n_s)\left[\frac{e^2}{\varepsilon_0 m_e}\mathbf{E}_f - \frac{\gamma_e k_B T_e}{m_e n_0}\nabla(\nabla \cdot \mathbf{E}_f)\right]\right\} = 0$$

Hence, assuming all perturbations are vanishingly small initially,

$$\frac{\partial^2 \mathbf{E}_f}{\partial t^2} + \omega_{pe}^2 \mathbf{E}_f - \frac{\gamma_e k_B T_e}{m_e}\nabla(\nabla \cdot \mathbf{E}_f) = -\omega_{pe}^2\left(\frac{n_s}{n_0}\right)\mathbf{E}_f \tag{10.63}$$

where the (n_s/n_0) contribution to the pressure term has been dropped.

If, for the moment, we neglect the non-linear term on the right-hand side of (10.63) and assume $\mathbf{E}_f \sim \exp i(\mathbf{k} \cdot \mathbf{r} - \omega t)$, we recover the dispersion relation (6.95) for electron plasma waves

$$\omega^2 = \omega_{pe}^2 + k^2 \gamma_e k_B T_e/m_e = \omega_{pe}^2(1 + \gamma_e k^2/k_D^2)$$

Thus, (10.63) is the equation for the non-linear development of these waves when they interact with the slow waves through the slow time scale perturbation in the electron density. Since electron plasma waves are strongly Landau damped unless $k \ll k_D$ we may take the fast frequency to be approximately ω_{pe} and write

$$\mathbf{E}_f(\mathbf{r}, t) = \mathbf{E}_0(\mathbf{r}, t)e^{-i\omega_{pe}t} \tag{10.64}$$

Substituting (10.64) in (10.63) and neglecting the term in $\partial^2 \mathbf{E}_0/\partial t^2$ gives

$$2i\omega_{pe}\frac{\partial \mathbf{E}_0}{\partial t} + \frac{\gamma_e k_B T_e}{m_e}\nabla(\nabla \cdot \mathbf{E}_0) = \omega_{pe}^2\left(\frac{n_s}{n_0}\right)\mathbf{E}_0 \tag{10.65}$$

This equation for the evolution of the amplitude of the fast wave is the *first Zakharov equation.* To it we must add an equation for the evolution of n_s.

This comes from (10.55) and (10.56) for the ions which are to leading order

$$Z\frac{\partial n_1}{\partial t} + n_0 \nabla \cdot \mathbf{u}_1 = 0 \tag{10.66}$$

$$\frac{\partial \mathbf{u}_1}{\partial t} = \frac{e\mathbf{E}_s}{m_i} - \frac{Z\gamma_i k_B T_i}{m_i n_0}\nabla n_1 \tag{10.67}$$

and the slow time scale equation of force balance for the electrons. In this we must include the ponderomotive force which was derived in Section 2.14 and takes account of electron acceleration due to the slow variation in amplitude of the electric field. Thus, from (2.68) and (10.56) we have

$$\frac{e\mathbf{E}_s}{m_e} + \frac{\gamma_e k_B T_e}{m_e n_0}\nabla n_s + \left(\frac{e}{2m_e\omega_{pe}}\right)^2 \nabla|\mathbf{E}_0|^2 = 0 \tag{10.68}$$

In the ion equations we replace $\mathbf{u}_1$ by $\mathbf{u}_s$, Zn_1 by n_s, and substitute for $\mathbf{E}_s$ from (10.68) to get

$$\frac{\partial \mathbf{u}_s}{\partial t} = -\frac{m_e}{m_i}\left(\frac{e}{2m_e\omega_{pe}}\right)^2 \nabla|\mathbf{E}_0|^2 - \frac{c_s^2}{n_0}\nabla n_s \tag{10.69}$$

where $c_s = [(\gamma_e k_B T_e + \gamma_i k_B T_i)/m_i]^{1/2}$ is the ion acoustic speed. Now taking the partial time derivative of (10.66) and substituting for $\partial\mathbf{u}_s/\partial t$ from (10.69) gives

$$\left(\frac{\partial}{\partial t^2} - c_s^2\nabla^2\right) n_s = \frac{\varepsilon_0}{4m_i}\nabla^2|\mathbf{E}_0|^2 \tag{10.70}$$

which is the *second Zakharov equation* and, together with (10.65), gives a closed, coupled pair of equations for $\mathbf{E}_0$ and n_s.

To understand the physics of this non-linear analysis let us briefly review the essential steps. We have reduced two sets of equations (10.53), (10.61), (10.62) and (10.66)–(10.68) to a pair of coupled equations (10.65) and (10.70). We have retained only one non-linear term in each of these sets, $\nabla \cdot (n_s\mathbf{u}_f)$ in (10.61) and the ponderomotive term in (10.68). Without these non-linear terms we recover the uncoupled linear wave equations for electron plasma waves and ion acoustic waves. The non-linear coupling takes account of the maintenance of approximate charge neutrality so that the slow perturbation in ion density must have a matching slow perturbation in electron density $n_s \approx Zn_1$. This appears as a 'correction' to the electron plasma frequency in (10.63) and subsequently as a moderating term in the evolution equation (10.65) for the wave amplitude.

Similarly, in the slow wave equations we have allowed for the moderation of $\mathbf{E}_s$ caused by the displacement of electrons due to the ponderomotive force. Linear theory says that the dynamic equilibrium of the electrons is maintained by the balance of the electrostatic field and electron pressure gradient; non-linear theory recognizes that there are three forces in balance. Consequently, the ponderomotive term appears in the evolution equation (10.70) for the slow density perturbation.

As noted in Section 2.14, the effect of the ponderomotive force is to drive electrons away from regions of high wave intensity. For the case where the gradient in field amplitude is parallel to the field this is easily explained with the help of

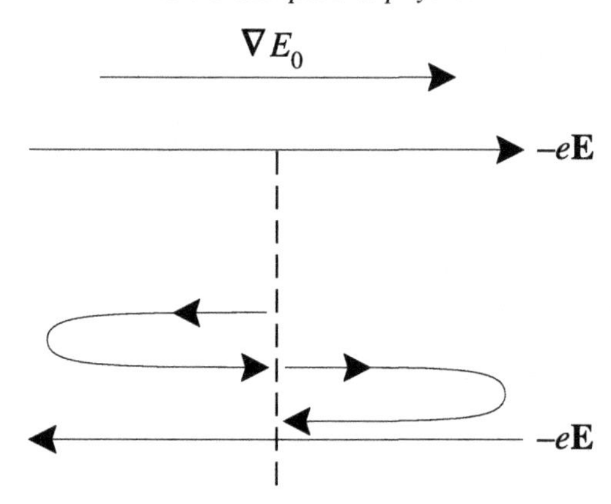

Fig. 10.12. Electron motion in inhomogeneous, oscillating electric field.

Fig. 10.12 which shows successive half-cycles of the electron motion in a field with amplitude increasing to the right; the dashed line represents the mid-point of the electron oscillations. In the half-cycle when the force $-e\mathbf{E}$ is to the right the electron experiences a weaker force than in the next half-cycle when the force is to the left. Consequently, the net effect is to cause the electron to migrate to the left, i.e. in the direction of the weaker field. The result is the same when the gradient in amplitude is perpendicular to the field but in this case the ponderomotive acceleration arises from the $\mathbf{v} \times \mathbf{B} \sim \mathbf{v} \times \nabla \times \mathbf{E}_0$ term. In both cases the acceleration is given by (2.68). The migrating electrons drag the ions with them creating plasma cavities in regions of high field intensity and leading to a new kind of instability, known as the modulational instability.

10.4.1 Modulational instability

Following Nicholson (1983), we may write the one-dimensional Zakharov equations in terms of dimensionless variables (see Exercise 10.7) as

$$i\frac{\partial E}{\partial \tau} + \frac{\partial^2 E}{\partial z^2} = nE \tag{10.71}$$

$$\frac{\partial^2 n}{\partial \tau^2} - \frac{\partial^2 n}{\partial z^2} = \frac{\partial^2 |E|^2}{\partial z^2} \tag{10.72}$$

where E and n are proportional to $|\mathbf{E}_0|$ and n_s, respectively, and τ and z are the dimensionless time and space variables.

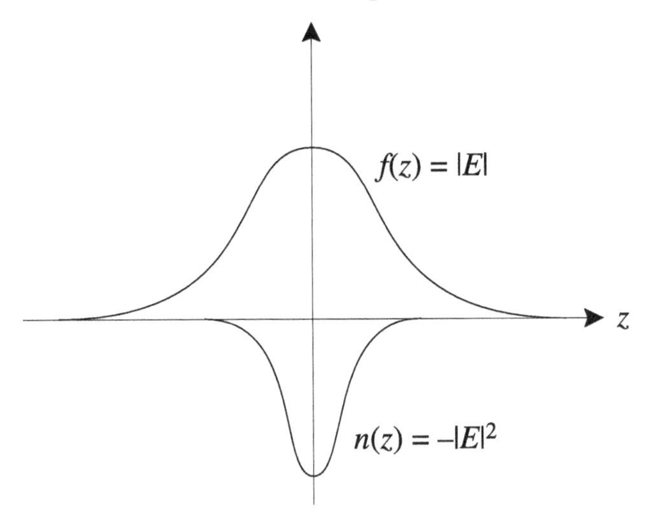

Fig. 10.13. Density depletion induced by field amplification.

Seeking a stationary solution of (10.72) we drop the first term and integrate twice with respect to z to obtain

$$n = -|E|^2 \tag{10.73}$$

where constants of integration have been set equal to zero. Substitution in (10.71) then gives the *non-linear Schrödinger equation*

$$i\frac{\partial E}{\partial \tau} + \frac{\partial^2 E}{\partial z^2} + |E|^2 E = 0 \tag{10.74}$$

Such equations occur in different contexts throughout physics and it is well-known that they have constant profile, single wave solutions known as solitary waves or *solitons*. For example, seeking a solution of the form

$$E(z,t) = e^{i\Omega\tau} f(z)$$

we find

$$f(z) = (2\Omega)^{1/2}\text{sech}(\Omega^{1/2}z)$$

This is sketched in Fig. 10.13 which also shows

$$n(z) = -2\Omega\text{sech}^2(\Omega^{1/2}z)$$

and demonstrates the effect of field concentration and density depletion that we have been discussing. Such density depletions are often referred to as *cavitons*.

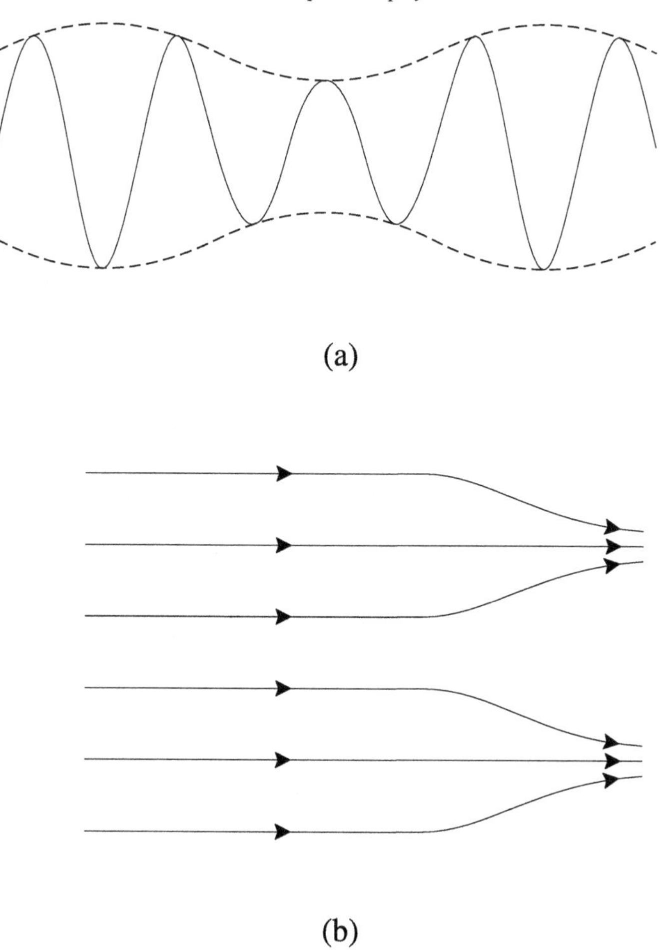

Fig. 10.14. Illustration of (a) modulation and (b) filamentation of wave due to the ponderomotive force.

Although we have found only the very simplest solution of (10.71)–(10.72), in which the wave oscillates within the static envelope $f(z)$, it is easy to see how the ponderomotive force leads to an instability. Consider the propagation of a constant amplitude wave through an almost homogeneous plasma. Any small density depletion will be matched by a corresponding amplitude increase. The ponderomotive force will then deflect plasma from this region of increased wave intensity thereby augmenting the density depletion. This is called the *modulational instability* when it refers to the modulation of wave envelope along the direction of propagation as shown in Fig. 10.14(a). Modulation of the wave profile can continue to a stage where the wave energy is confined to localized cavitons several Debye lengths in dimension. This is known as *Langmuir collapse*. The collapse in coordinate space is accompanied by so-called pumping in k space, with increasing k compensating for decreasing density in the dispersion relation $\omega^2 = \omega_p^2 + 3k^2 V_e^2$.

Although our analysis has involved only longitudinal waves, the ponderomotive effect applies equally to electromagnetic waves and can produce *filamentation* of the wave. This refers to the break-up of the wave in the direction transverse to its propagation and is illustrated in Fig. 10.14(b). It is most easily understood in terms of the refraction of the electromagnetic wave. In an inhomogeneous plasma decreasing density means increasing refractive index and consequent focusing of electromagnetic waves. The ponderomotive force then reinforces this effect by driving plasma away from the region of increased wave intensity. In this way an initially uniform beam can break up into narrow filaments. This is an important effect in laser plasma physics since it can obviously be triggered by non-uniformities in the laser beam.

10.5 Collisionless shocks

The MHD shocks discussed in Section 5.6 have widths of the order of a mean free path since collisions are responsible for the sudden change of state. In plasmas, however, shock-like changes of state are found to occur over distances much less than the mean free path. Perhaps the clearest example of this is the Earth's bow shock created by the interaction of the solar wind with the Earth's magnetic field to produce the transition from supersonic flow in the solar wind to sub-sonic flow in the magnetosheath. The shock has a thickness of about 1000 km whereas the collisional mean free path is of the order of 1 AU or 10^8 km. Clearly, collisions cannot be responsible for this change of state. Other examples arise in laboratory plasmas where changes of state occur within a few mean free paths but collisional transport is insufficient to account for this and so-called *turbulent* or *anomalous* dissipation must be involved. Any shock in which non-collisional processes play a significant role is called a *collisionless shock.*

The first important difference to note between collisional and collisionless shocks relates to the formation of the shock. In collisional shocks the wave profile results from the balance between convective and dissipative effects. The wave profile in collisionless shocks, on the other hand, is usually the result of a balance between convective and *dispersive* effects. To understand how this comes about it is useful to consider the wave in terms of its Fourier components. In the linear approximation each component propagates independently and in the absence of dispersion a wave pulse will maintain a constant profile since all components travel with the same speed ω/k. In the non-linear approximation, however, the wave pulse broadens since any pair of components $(\omega_1, \mathbf{k}_1)$ and $(\omega_2, \mathbf{k}_2)$ within the pulse, for

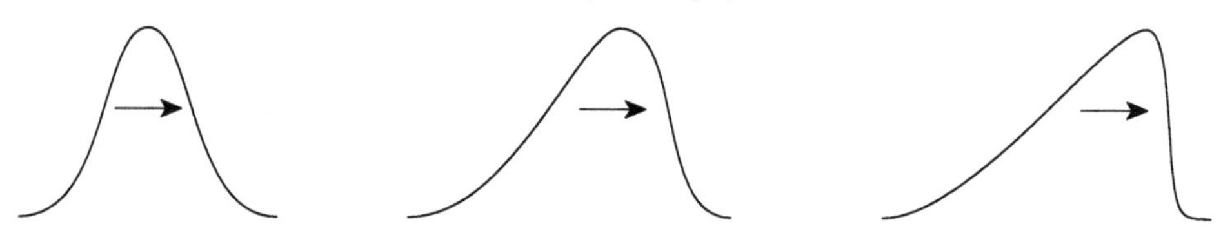

Fig. 10.15. Non-linear broadening and steepening of a wave pulse.

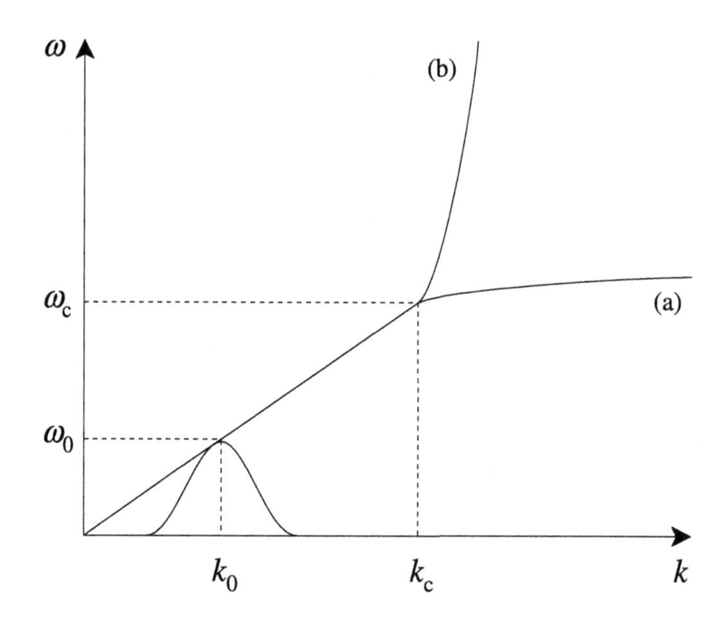

Fig. 10.16. Typical plasma dispersion curves.

which $\omega = \omega_1 \pm \omega_2$ and $\mathbf{k} = \mathbf{k}_1 \pm \mathbf{k}_2$, will be resonantly driven, leading to both longer and shorter wavelength modes. If, in addition, the wave pulse is regarded as a combination of compression and expansion waves, the effect of convection is to steepen the compression wave (the wave-front) and broaden the expansion wave so that the pulse changes its shape as illustrated in Fig. 10.15. In a collisional shock wave steepening continues until the wave-front has a sufficiently large gradient that dissipation balances convection.

Plasmas, however, are dispersive and a simple dispersion relation of the form ω/k = const. will, in general, apply only over a limited band of wavenumbers. Typical dispersion curves are shown in Fig. 10.16. Starting with a wave pulse centred around some arbitrary point (ω_0, k_0) on the straight portion of the curve, higher wavenumber modes may be generated by non-linear coupling up to the point (ω_c, k_c), where the phase velocity changes, i.e. dispersion begins. Resonant mode generation beyond this point would lead to shorter wavelength modes correspond-

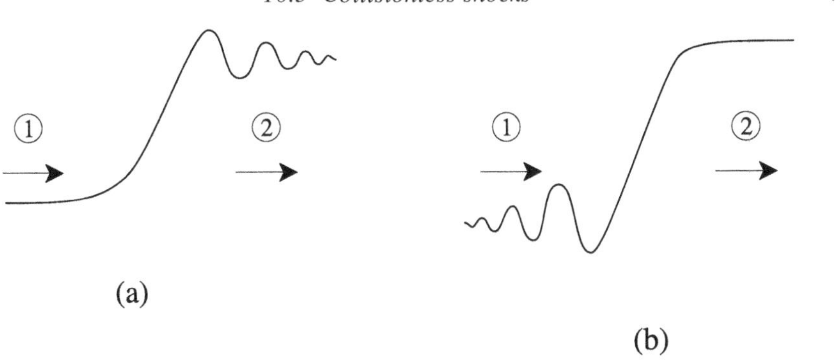

Fig. 10.17. Laminar shock profiles. In the rest frame of the shock the arrows indicate the direction of flow from (1) upstream unshocked to (2) downstream shocked plasma. Shorter wavelength waves either (a) trail behind or (b) forge ahead of the shock front.

ing to points on curves (a) or (b) but these travel at slower and faster phase speeds, respectively, and as shown in Fig. 10.17 do not remain in the shock front. Thus, steepening is limited by dispersion at a scale length $\sim k_c^{-1}$. If this is less than the scale length for the onset of dissipation then the shock is 'collisionless' in that its profile is determined by wave dispersion rather than collisional dissipation. This will obviously be the case for shocks in collisionless plasmas but may also occur in collisional plasmas.

A second important difference between collisional and collisionless shocks relates to the jump conditions across the shock. The state of a collisional plasma is determined by its density, flow velocity and temperature so that conservation of mass, momentum and energy means that the jump conditions are independent of shock structure. No such claim can be made for a collisionless shock. Even if the unshocked plasma is in an equilibrium state represented by a Maxwellian distribution there are too few collisions to re-establish a Maxwellian in the shocked plasma. So, although mass, momentum and energy must still be conserved, the final state of the plasma cannot in general be represented in terms of density, velocity and temperature alone; mathematically, the moment equations do not form a closed set. In particular, anisotropies created in the shock, in the absence of collisions, may persist into the downstream plasma. A useful aspect of this is that observation of the downstream plasma may yield information about the shock structure; for example, it may suggest which unstable waves are responsible for the turbulent dissipation.

A third point of sharp contrast shows up in shock structure. Collisions convert the upstream state to the downstream state within a collisional shock which is about a mean free path in dimension. Thus, particles from the upstream state cannot penetrate the shock without undergoing conversion to the downstream state so that the two states remain physically separated. How is this separation maintained in collisionless shocks?

One possible mechanism is to have a magnetic field perpendicular to the direction of propagation of the shock of sufficient strength that the Larmor radius $r_L \ll L_s$, where L_s is the shock thickness. If this is not the case, or if the magnetic field has a component parallel to the direction of shock propagation, fast particles can cross the wave-front where their free energy may trigger instabilities leading to turbulent dissipation. Thus, a second possibility is that the shock may have a width of the order of the mean free path for turbulent dissipation. A special case is that of low β shocks with shock velocity V_s greater than the magnetoacoustic speed; here the number of particles with speed greater than V_s will be exponentially small.

As with collisional shocks, kinetic theory is necessary for a rigorous description of shock structure though transport equations, with 'fitted' turbulent transport coefficients replacing the collisional coefficients, are frequently used. In two special cases, however, the use of a fluid description may be justified (in contrast with collisional shocks where, on account of the shock width, there is no rigorous justification):

(i) the cold plasma approximation, where thermal velocities are very much smaller than phase velocities, i.e. the thermal spread in particle velocities is unimportant and all particles of a given type experience the same local force due to the self-consistent fields.
(ii) the small Larmor radius (strong magnetic field) approximation in which particles move with the field lines.

A final comment concerns energy and entropy in collisionless shocks. Ordered energy may be in the plasma flow, the magnetic field, or coherent oscillations. Various energy conversions are possible and since we are dealing with collective interactions it is neither obvious nor necessarily true that an increase in entropy will accompany a change of state. Examples of isentropic transitions are solitons which pass through the plasma leaving the final state identical to the initial state. Alternatively, the final state may contain coherent plasma oscillations. We shall see, however, that in both cases a small amount of dissipation (collisional or turbulent) will convert these to shock transitions with increased entropy in the final state.

10.5.1 Shock classification

The most general description of a collisionless plasma is given by the Maxwell–Vlasov system of equations and for applications to collisionless shocks it is useful to make a formal separation of the dependent variables f, $\mathbf{E}$ and $\mathbf{B}$ into their average and fluctuating components. By an average quantity $\langle\phi\rangle$ we mean an ensemble average and $\delta\phi = \phi - \langle\phi\rangle$ is the fluctuation about this. In practical terms what this means is that $\langle\phi\rangle$ represents an average shock profile and $\delta\phi$ a random or turbulent variation superimposed on it. For the moment we make no assumptions about the relative magnitudes of $\langle\phi\rangle$ and $\delta\phi$.

Since $\langle\delta\phi\rangle = 0$, on taking the ensemble average the Maxwell equations being linear in f, $\mathbf{E}$ and $\mathbf{B}$ are unchanged except for the substitution of $\langle\phi\rangle$ for ϕ while the Vlasov equation for f_α ($\alpha = i, e$) becomes

$$\frac{\partial\langle f_\alpha\rangle}{\partial t} + \mathbf{v}\cdot\frac{\partial\langle f_\alpha\rangle}{\partial \mathbf{r}} + \frac{e_\alpha}{m_\alpha}[\langle\mathbf{E}\rangle + \mathbf{v}\times\langle\mathbf{B}\rangle]\cdot\frac{\partial\langle f_\alpha\rangle}{\partial\mathbf{v}} = C_\alpha \tag{10.75}$$

where

$$C_\alpha = -\frac{e_\alpha}{m_\alpha}\left\langle[\delta\mathbf{E} + \mathbf{v}\times\delta\mathbf{B}]\cdot\frac{\partial\delta f_\alpha}{\partial\mathbf{v}}\right\rangle \tag{10.76}$$

Writing the Vlasov equation in this way shows that the fluctuations act like a 'collision' term in the kinetic equation for $\langle f_\alpha\rangle$. Note, however, that C_α involves interactions between the fluctuating fields and distributions, i.e. particles. Because of this, particle momentum and energy are *not* conserved in contrast with the action of the classical collision term. Assuming no particles are created or destroyed we have

$$\int C_\alpha\,\mathrm{d}\mathbf{v} = 0$$

but

$$\sum_\alpha m_\alpha\int \mathbf{v}C_\alpha\,\mathrm{d}\mathbf{v} \neq 0 \qquad \sum_\alpha \frac{1}{2}m_\alpha\int v^2 C_\alpha\,\mathrm{d}\mathbf{v} \neq 0 \tag{10.77}$$

A further consequence of (10.77) is that C_α does not cause $\langle f_\alpha\rangle$ to relax to a Maxwellian distribution, thus producing the contrast with collisional shocks concerning the jump conditions noted in the previous section.

In principle we can now proceed to a fluid description by defining the plasma fluid variables in terms of $\langle f_\alpha\rangle$ rather than f. However, aside from the usual problem of truncating the infinite set of moment equations based on (10.75), one has the additional problem of closing the set of equations for the fluctuations $\delta\phi$ which are needed for evaluating C_α. These are obtained from the Maxwell–Vlasov equations by subtracting the ensemble averaged equations but we shall not pursue this formal approach.

Instead let us consider the separation $\phi = \langle\phi\rangle + \delta\phi$ in terms of shock classification. Shocks in which the field and plasma variables change in a coherent manner are referred to as *laminar*; any turbulence present is on a scale small enough not to destroy the coherent profile. If there is no turbulence ($C_\alpha = 0$) then there is strictly speaking no shock (unless we introduce collisional dissipation) and the solutions of the equations correspond either to solitary waves or undamped oscillations. Non-zero C_α, but such that the turbulence is weak and occurs on a scale which is small in wavelength compared with the shock thickness, gives rise to dissipation and hence to true shock solutions. Such a shock will appear laminar on a scale longer than the wavelength of the micro-turbulence. Typical profiles are shown in the shock rest frame in Fig. 10.17. Short wavelength oscillations damp out either (a) downstream or (b) upstream in accordance with the dispersion curves in Fig. 10.16. The basic procedure here is to treat C_α as a small perturbation.

As we shall see, the investigation of laminar shocks shows that they can exist only within certain parameter ranges. Beyond these ranges the fluctuations become large, C_α plays a dominant role and the shock loses its laminar profile. Such cases are referred to as *turbulent shocks*

Note, however, that there is no sharp demarcation between laminar and turbulent shocks. Rather, these are opposite ends of a spectrum embracing many possible structures. For example, if the turbulent fluctuations are small in amplitude but with a wavelength comparable with the shock thickness then this cannot be regarded as micro-turbulence and the shock is a mixture of laminar and turbulent structure. Other complications arise from the dispersive limitation of shock steepening giving rise to precursors and wakes. Also, the spread in particle velocities can lead to trapping and acceleration culminating in the emission of supra-thermal particles. All of these phenomena are discussed theoretically at various levels by Tidman and Krall (1971). Experimentally, the Earth's bow shock, which has been extensively investigated by satellite observations, is a rich source of all kinds of collisionless shock. Here we shall present only a few well-established results starting with the simplest mathematical descriptions and proceeding step by step to widen their applicability.

As noted in Section 5.6.1, shocks may also be classified by the angle θ between their direction of propagation and the magnetic field $\mathbf{B}_1$ in the unshocked plasma. Thus, shocks may be *perpendicular* ($\theta = \pi/2$), *parallel* ($\theta = 0$), or *oblique* ($0 < \theta < \pi/2$). We shall see that perpendicular shocks are in general more amenable to analysis. This is not surprising since, as already noted, a magnetic field at right angles to the flow can of itself be an effective agent for separating the upstream and downstream plasmas. On the other hand, in oblique and parallel shocks the magnetic field can act as a particle conduit between the upstream and downstream plasmas so that the physics of these shocks (and consequently the mathematics) is immediately more complex. We illustrate procedure, therefore, with perpendicular, laminar shocks.

10.5.2 Perpendicular, laminar shocks

To start with as simple a model as possible let us consider non-linear wave propagation in a cold plasma. If we put $C_\alpha = 0$ initially then we know that the definitions

$$n_\alpha = \int \langle f_\alpha \rangle \mathrm{d}\mathbf{v} \qquad \mathbf{u}_\alpha = \frac{1}{n_\alpha} \int \mathbf{v} \langle f_\alpha \rangle \mathrm{d}\mathbf{v}$$

lead to the cold plasma wave equations

$$\left. \begin{aligned} \partial n_\alpha / \partial t + \nabla \cdot (n_\alpha \mathbf{u}_\alpha) &= 0 \\ (\partial/\partial t + \mathbf{u}_\alpha \cdot \nabla)\, \mathbf{u}_\alpha &= e_\alpha (\mathbf{E} + \mathbf{u}_\alpha \times \mathbf{B})/m_\alpha \end{aligned} \right\} \tag{10.78}$$

on taking the first two moments of (10.75).

Linear wave solution

Before we seek 'shock' solutions of these equations it is of interest to identify the linear wave which, through dispersive limitation of non-linear steepening, produces the steady, finite amplitude wave. In Section 6.3.3 we showed that there were two waves which propagate perpendicular to the equilibrium magnetic field. One of these, the O mode, is a transverse electromagnetic wave which propagates at frequencies above the plasma frequency. The other wave, the X mode, has three branches and it is the lowest frequency branch ($0 < \omega < \omega_{\mathrm{LH}}$) which produces the non-linear wave we shall investigate.

Assuming $\Omega_{\mathrm{e}}^2 \ll \omega_{\mathrm{p}}^2$, i.e. $v_{\mathrm{A}}^2 \ll c^2$, from (6.36), (6.43) and (6.44) we have $\omega_{\mathrm{UH}} \approx \omega_{\mathrm{R}} \approx \omega_{\mathrm{L}} \approx \omega_{\mathrm{p}}$ and $\omega_{\mathrm{LH}} \approx |\Omega_{\mathrm{i}}\Omega_{\mathrm{e}}|^{1/2}$ so that, on using (6.29), (6.60) becomes

$$\frac{\omega^2}{k^2c^2} = \frac{\omega^2 - |\Omega_{\mathrm{i}}\Omega_{\mathrm{e}}|}{\omega^2 - \omega_{\mathrm{p}}^2}$$

The solution of this equation in the frequency range $0 < \omega < |\Omega_{\mathrm{i}}\Omega_{\mathrm{e}}|^{1/2}$ is

$$\frac{\omega}{k} \approx \frac{v_{\mathrm{A}}}{(1 + k^2c^2/\omega_{\mathrm{p}}^2)^{1/2}} \tag{10.79}$$

This is the dispersion relation of the compressional Alfvén wave which appears as the lowest branch of the X mode in Fig. 6.7. It is more commonly referred to by its finite β name as the magnetoacoustic (or magnetosonic) wave. The points to note are:

(i) for $k \to 0$, $\omega/k \approx v_{\mathrm{A}} = \text{const.}$,
(ii) dispersion becomes significant for $k \sim k_{\mathrm{c}} = \omega_{\mathrm{p}}/c$,
(iii) waves with $k > k_{\mathrm{c}}$ have phase speeds $\omega/k < v_{\mathrm{A}}$.

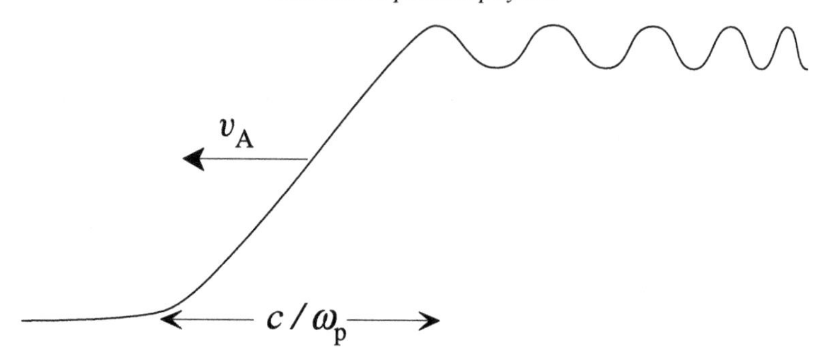

Fig. 10.18. Non-linear compressional Alfvén wave.

It follows that a finite amplitude wave generated by non-linear wave steepening and limited by dispersion would have a wave-front, of width $L_s \sim c/\omega_p$, travelling with speed v_A ahead of shorter wavelength modes as shown in Fig. 10.18.

Non-linear wave solutions

Calculations to find the structure predicted in Fig. 10.18 were first carried out by Adlam and Allen (1958), Davis, Lüst and Schlüter (1958) and, subsequently, Sagdeev (1966). It is Sagdeev's calculation, as presented by Tidman and Krall (1971), that we follow. We suppose that a steady profile has been achieved between non-linear wave steepening and dispersive limitation and we look for a time-independent, laminar solution of (10.78). Thus, we assume all variables are functions of x only and have values in the upstream region ($x \to -\infty$) given by

$$\left.\begin{aligned} n_e(-\infty) &= Zn_i(-\infty) = n_1 \\ \mathbf{u}_e(-\infty) &= \mathbf{u}_i(-\infty) = (u_1, 0, 0) \\ &\quad \mathbf{B}(-\infty) = (0, 0, B_1) \\ &\quad \mathbf{E}(-\infty) = (0, u_1 B_1, 0) \end{aligned}\right\} \tag{10.80}$$

It then follows from the Maxwell equations that

$$B_x(x) = 0, \qquad E_z(x) = 0, \qquad E_y(x) = u_1 B_1$$

Also, since the equations for B_y and $u_{\alpha z}$ decouple from the rest we may look for a solution in which these variables are zero. The remaining equations give

$$n_e u_{ex} = Zn_i u_{ix} = n_1 u_1 \tag{10.81}$$

$$u_{\alpha x} u'_{\alpha x} = \frac{e_\alpha}{m_\alpha}(E_x + u_{\alpha y} B) \qquad (\alpha = i, e) \tag{10.82}$$

$$u_{\alpha x} u'_{\alpha y} = \frac{e_\alpha}{m_\alpha}(u_1 B_1 - u_{\alpha x} B) \qquad (\alpha = i, e) \tag{10.83}$$

$$B' = \mu_0 e(n_e u_{ey} - Zn_i u_{iy}) \tag{10.84}$$

$$E'_x = \frac{e}{\varepsilon_0}(Zn_i - n_e) \tag{10.85}$$

Now assuming quasi-neutrality we put $Zn_i = n_e$ in (10.81) and (10.84) but not in (10.85) where the small inequality produces the electric field which, as we shall see later, decelerates the ion flow. The quasi-neutrality condition is $|Zn_i - n_e| \ll n_1$ which can be shown *a posteriori* to be $\Omega_e^2 \ll \omega_p^2$, as assumed in the linear wave calculation. With this assumption it follows from (10.81) and (10.83) that

$$u_{ix} = u_{ex} = u_x \tag{10.86}$$

say, and

$$u_{iy} = -\frac{Zm_e}{m_i} u_{ey} \tag{10.87}$$

Substituting these results in (10.82) and subtracting the electron equation from the ion equation then gives

$$E_x = -Bu_{ey}$$

on ignoring terms of order m_e/m_i. Likewise, multiplying (10.82) by m_α/e_α ($\alpha =$ i, e) and subtracting eliminates the electric field to give to the same order

$$\frac{m_i}{Ze} u_x u'_x = B(u_{iy} - u_{ey}) = -\frac{BB'}{\mu_0 n_e e}$$

where the second equality follows from (10.84). Substituting for $n_e u_x$ from (10.81) and (10.86) and integrating we get

$$\frac{u_x}{u_1} = 1 - \frac{1}{2M_A^2}\left[\left(\frac{B}{B_1}\right)^2 - 1\right] \tag{10.88}$$

where $M_A = u_i/v_A$ is the Mach number and $v_A = B_1/(\mu_0 n_1 m_i/Z)^{1/2}$ is the upstream Alfvén speed. This equation shows how the flow velocity decreases as the magnetic field increases.

From (10.81), (10.84) and (10.87) we get

$$u_{ey} = \frac{u_x B'}{\mu_0 e n_1 u_1} = \frac{1}{\mu_0 e n_1}\left[1 - \frac{1}{2M_A^2}\left(\frac{B^2}{B_1^2} - 1\right)\right] B' \tag{10.89}$$

where terms of order m_e/m_i have again been ignored and (10.88) has also been used. Now all variables are, at least implicitly through (10.88) and (10.89), expressed in terms of the magnetic field $B(x)$. To complete the calculation, therefore, we need an equation for $B(x)$. This is obtained from (10.84) using (10.83) to get

$$u_x \frac{\mathrm{d}}{\mathrm{d}x}(u_x B') = \mu_0 e u_x \frac{\mathrm{d}}{\mathrm{d}x}(u_x n_e u_{ey}) = -\frac{\mu_0 e^2 n_1 u_1}{m_e}(u_1 B_1 - u_x B)$$

and hence, using (10.88) and (10.89),

$$\left[1-\frac{1}{2M_A^2}\left(\frac{B^2}{B_1^2}-1\right)\right]\frac{d}{dx}\left\{\left[1-\frac{1}{2M_A^2}\left(\frac{B^2}{B_1^2}-1\right)\right]\frac{dB}{dx}\right\}$$
$$=\frac{\omega_{pe}^2}{c^2}(B-B_1)\left[1-\frac{B(B+B_1)}{2M_A^2B_1^2}\right]$$

Multiplying this equation by dB/dx gives an exact differential on the left-hand side and on integration we get

$$\frac{1}{2}\left(\frac{dB}{dx}\right)^2+\Phi(B)=0 \tag{10.90}$$

where

$$\Phi(B)=-\frac{\left\{\frac{1}{2}\left(\frac{dB_1}{dx}\right)^2+\frac{\omega_{pe}^2}{2c^2}(B-B_1)^2\left[1-\frac{(B+B_1)^2}{4M_A^2B_1^2}\right]\right\}}{\left[1-\frac{1}{2M_A^2}\left(\frac{B^2}{B_1^2}-1\right)\right]^2} \tag{10.91}$$

Although (10.90) can be formally integrated to obtain x as a function of B it is more instructive to discuss it directly since it has the form of an energy equation for a particle in a potential well; B has the role of space coordinate and x is the 'time'. When $B = B_1$, the 'potential energy' $\Phi(B) = -\frac{1}{2}(B_1')^2$ and it is easy to show that initially it decreases as B increases. Thereafter it reaches a minimum and then increases with B and we can find the range of values of B by examining the motion of the imaginary particle in the potential well. There are two possible cases depending on whether B_1' is zero or not and these are both represented in Fig. 10.19.

In case (a) ($B_1' \neq 0$) the 'particle', which starts off at the point shown with kinetic energy $\frac{1}{2}(B_1')^2$, will travel to B_{max}, at which point its kinetic energy is exhausted, and then it will roll back till it reaches B_{min}. Thereafter it will oscillate back and forth between these two points so the structure of $B(x)$ is as shown in (c), i.e. a train of finite amplitude waves of finite wavelength.

In case (b) the particle again travels from its initial point B_1, here also its minimum point, to its maximum point and back again, but because it returns to an equilibrium point it does not oscillate further. In fact, it takes an infinite 'time' (distance x) on both stages of its journey so the structure of $B(x)$ in this case is a soliton, as shown in (d). Neither of these solutions corresponds to a shock but the introduction of dissipation of some kind will convert both to laminar shock profiles. However, before we demonstrate this let us examine the soliton solution a little more closely to illustrate some of its parametric properties.

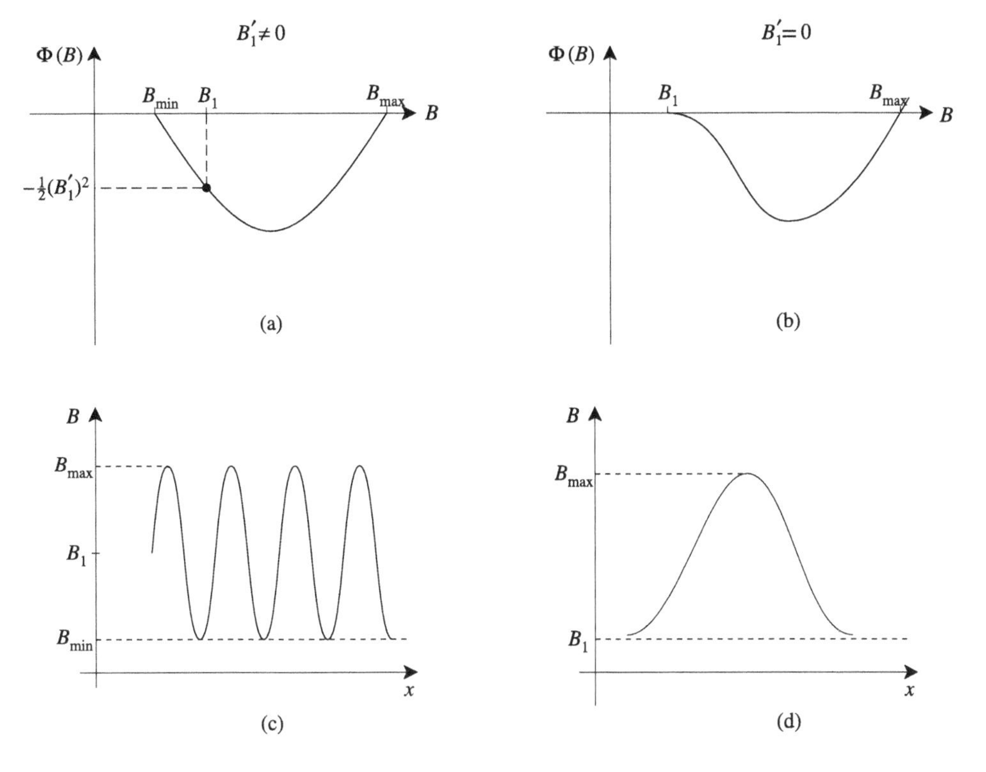

Fig. 10.19. Potential function for non-linear compressional Alfvén wave.

First of all, with $B_1' = 0$, it is easily seen from (10.91) that $\Phi'(B_1) = 0$ so that near B_1

$$\Phi(B) \approx \frac{(B - B_1)^2}{2} \Phi_1''$$

where

$$\Phi_1'' = \left[\frac{\mathrm{d}^2\Phi(B)}{\mathrm{d}B^2}\right]_{B=B_1} = -\frac{\omega_{\mathrm{pe}}^2}{c^2}\left(1 - \frac{1}{M_{\mathrm{A}}^2}\right)$$

It follows that the asymptotic solution of (10.90) is

$$B - B_1 \sim e^{\mp|\Phi_1''|^{1/2}x} \qquad (x \to \pm\infty)$$

This shows the slope predicted by the linear theory but with a modifying factor $(1 - 1/M_{\mathrm{A}}^2)^{1/2}$ which is dependent upon the amplitude of the wave. This result

$$L \sim \frac{c}{\omega_{\mathrm{pe}}} \frac{1}{(1 - 1/M_{\mathrm{A}}^2)^{1/2}} \tag{10.92}$$

for the breadth of the wave-front was obtained by Adlam and Allen (1958).

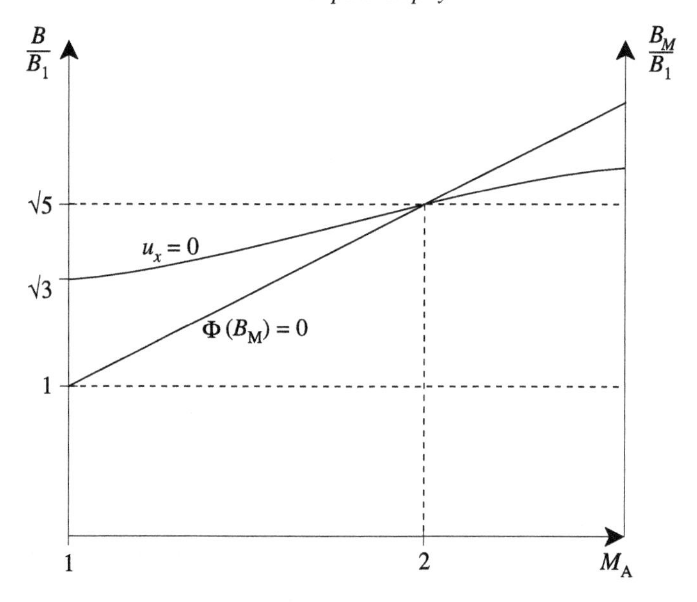

Fig. 10.20. Validity diagram for undamped non-linear wave solution.

Next, writing B_M for B_{max} and noting that $\Phi(B_M) = 0$ it follows from (10.91) that

$$B_M = B_1(2M_A - 1) \tag{10.93}$$

Also, from (10.88), (10.90) and (10.91) we see that $u_x \to 0$ and $dB/dx \to \infty$ as

$$(B/B_1) \to \sqrt{2M_A^2 + 1} \tag{10.94}$$

The solution breaks down, therefore, if $\sqrt{2M_A^2 + 1} \leq 2M_A - 1$, i.e. $M_A \geq 2$. This is illustrated graphically in Fig. 10.20; for a valid solution the $\Phi(B_M) = 0$ curve must lie below the $u_x = 0$ curve. For waves with $M_A \geq 2$ dispersive limitation of steepening fails as the limit (10.94) is approached. It is easily verified that $n_e, E_x \to \infty$ as well as dB/dx in this limit. Physically, it is clear that dissipative limitation will take over in response to these large gradients and we shall now show that the introduction of dissipation converts these isentropic, non-linear wave solutions into shocks.

Shock solutions

Dissipation is likely to arise via drift instabilities driven by the free energy in the current flowing parallel to the wave-front. Indeed, this is how one attempts to arrive at a self-consistent model of a perpendicular shock. The jump in B_z requires a

j_y which feeds energy into unstable drift waves thereby providing the turbulent dissipation and consequent increase in entropy. A proper treatment clearly requires kinetic theory but we can model this behaviour very simply by re-introducing the 'collision' term C_α in (10.75) so that first-order velocity moments of C_α now appear in the momentum equations. For simplicity, we shall introduce these only in (10.83) since it is only in the y direction that there is an appreciable drift between ions and electrons. Furthermore, we shall assume, despite the observations made in Section 10.5.1 (see (10.77)), that the net loss of momentum from particles to fields is negligible, that is

$$\sum_\alpha m_\alpha \int v_y C_\alpha \, \mathrm{d}\mathbf{v} \approx 0$$

Neither of these simplifying assumptions can be rigorously justified since unstable waves are capable of transferring momentum in all directions and from particles to waves. However, they are not unreasonable in the limit of weak turbulence and enable us to make analytic progress revealing, qualitatively at least, the effect of dissipation on the isentropic solutions.

The analysis proceeds as before but there now appears an extra term in the differential equation for $B(x)$ which becomes

$$\frac{u_x}{u_1}\frac{\mathrm{d}}{\mathrm{d}x}\left(\frac{u_x}{u_1}\frac{\mathrm{d}B}{\mathrm{d}x}\right) = \frac{\omega_{\mathrm{pe}}^2}{c^2}(B-B_1)\left[1 - \frac{B(B+B_1)}{2M_{\mathrm{A}}^2 B_1^2}\right] + \frac{\mu_0 e u_x}{u_1^2}\int v_y C_{\mathrm{e}} \, \mathrm{d}\mathbf{v} \quad (10.95)$$

Defining a (constant) 'collision' frequency ν by

$$\int v_y C_{\mathrm{e}} \, \mathrm{d}\mathbf{v} = \nu n_{\mathrm{e}} u_{\mathrm{e}y}$$

and a 'stretched' space coordinate ξ by

$$x = \xi u_x / u_1$$

(10.95) becomes, on substituting for $u_{\mathrm{e}y}$ from (10.89),

$$\frac{\mathrm{d}^2 B(\xi)}{\mathrm{d}\xi^2} = -\frac{\mathrm{d}\phi(B)}{\mathrm{d}B} - \frac{\nu}{u_1}\frac{\mathrm{d}B(\xi)}{\mathrm{d}\xi} \quad (10.96)$$

where

$$\phi(B) = \frac{\omega_{\mathrm{pe}}^2}{2c^2}(B-B_1)^2\left[\frac{(B+B_1)^2}{4M_{\mathrm{A}}^2 B_1^2} - 1\right] \quad (10.97)$$

Comparing (10.96) with

$$\ddot{x} = -\frac{\mathrm{d}V}{\mathrm{d}x} - \nu\dot{x}$$

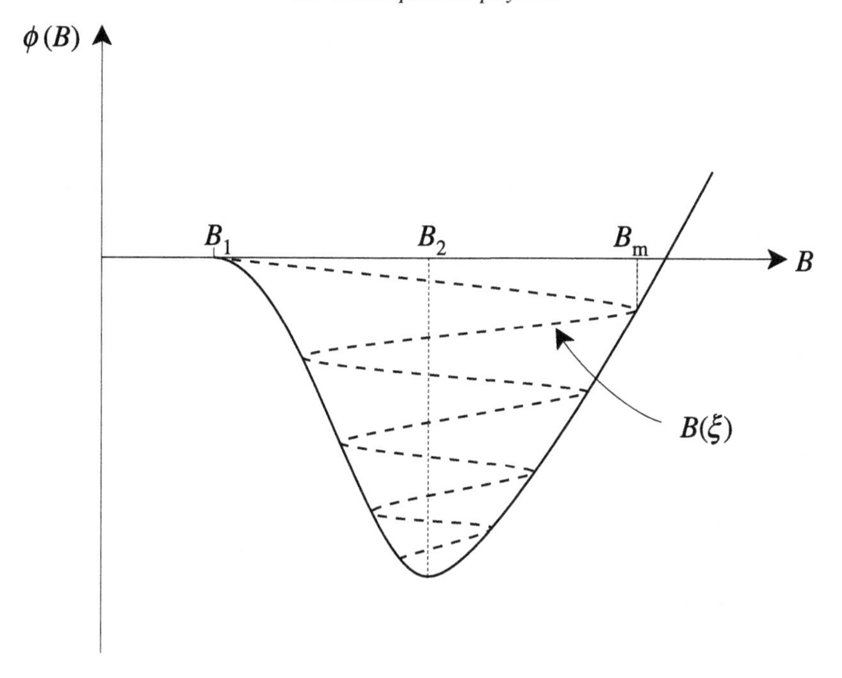

Fig. 10.21. Damped motion in a potential well.

we see that it is analogous to damped motion in a potential well. This is sketched in Fig. 10.21. The dashed lines join up the successive turning points and trace out the values of $B(\xi)$ as damping decreases the total energy of the imaginary particle in the potential well; $B_{\mathrm{m}}(< B_{\mathrm{M}})$ is now the maximum value attained by the magnetic field and B_2 is its final value. The corresponding structure of the magnetic field is shown in Fig. 10.22; the dashed line here shows the soliton solution in the absence of dissipation. There are now two scale lengths associated with the solution, c/ω_{pe} is still the width of the leading edge of the shock but this is followed by a trail of waves of decreasing amplitude over a decay length of order u_1/ν. Provided the damping is weak we may regard the structure as a train of solitons the breadth of which successively increases in accordance with (10.92).

Since $B_{\mathrm{m}} < B_{\mathrm{M}}$, wave-breaking ($u_x \to 0$) no longer occurs as $M_{\mathrm{A}} \to 2$ and the shock solution exists beyond this limit. Indeed, one might suppose that by increasing ν one could ensure a valid shock solution for arbitrary M_{A}. As $\nu \to \infty$, $B_{\mathrm{m}} \to B_2 = B_1[\sqrt{2M_{\mathrm{A}}^2 + 1/4} - 1/2] < B_1\sqrt{2M_{\mathrm{A}}^2 + 1}$, which is the value of B at which $u_x = 0$ as given by (10.94). However, this would be stretching the validity of this simple model beyond reasonable limits. Not only have we supposed that the damping is weak but we have ignored the fact that dissipation inevitably leads to plasma heating. A dissipative, *cold* plasma model is *not* self-consistent.

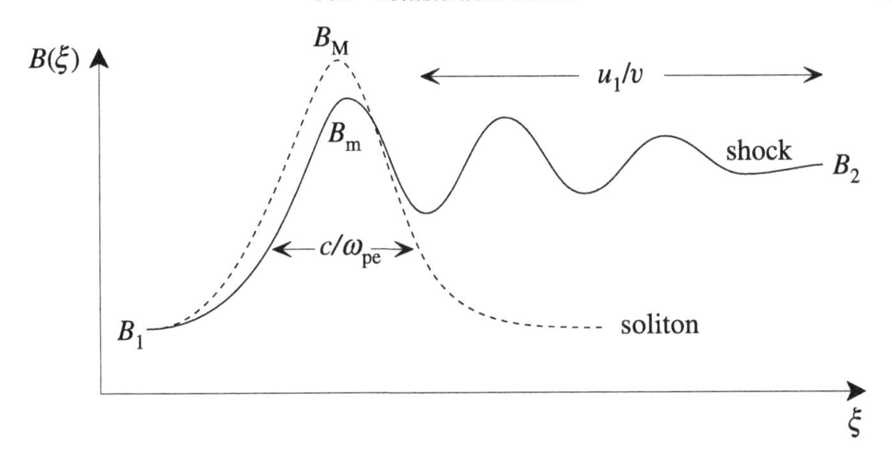

Fig. 10.22. Structure of the magnetic field in the shock solution.

Tidman and Krall (1971) discuss at some length shock solutions obtained from the warm plasma wave equations with damping and find that finite β limitations on M_A outweigh the effect of dissipation with the result that the critical Mach number, now a function of β and ν, is less than 2.

The usefulness of fluid equations for the description of finite temperature effects is limited. Any spread in particle velocities means that some ions will have sufficient kinetic energy to pass over the potential hill presented by the electric field but others will not. These slower ions are therefore reflected back upstream so that one has, in effect, two ion fluids rather than one. Furthermore, the reflected ions are turned by the Lorentz force upstream and re-enter the shock where they may again be transmitted or reflected – they bounce off the shock front, as illustrated in Fig. 10.23, until they have gained sufficient energy to pass through. Since the dynamics of the ions is dependent upon their velocity distribution a kinetic description becomes essential.

We are now ready to construct a self-consistent model of a laminar, perpendicular shock. A fluid model based on a strong magnetic field (i.e. small Larmor radius) is valid provided the scale length $L \gg r_L$. For ions this would imply a much stronger condition, $\beta_i \ll m_e/m_i$, than for electrons, $\beta_e \ll 1$, when $L = L_s \sim c/\omega_{pe}$, the width of the wave-front produced by dispersive limitation of wave steepening. Consequently, the ordering is

$$r_{L_e} \ll L_s \ll r_{L_i}$$

and the effect of the electric field, due to charge separation in the shock, is different for electrons and ions. The electrons experience an $\mathbf{E} \times \mathbf{B}$ drift in the shock front, establishing the current j_y. The ion orbits are, however, essentially straight lines

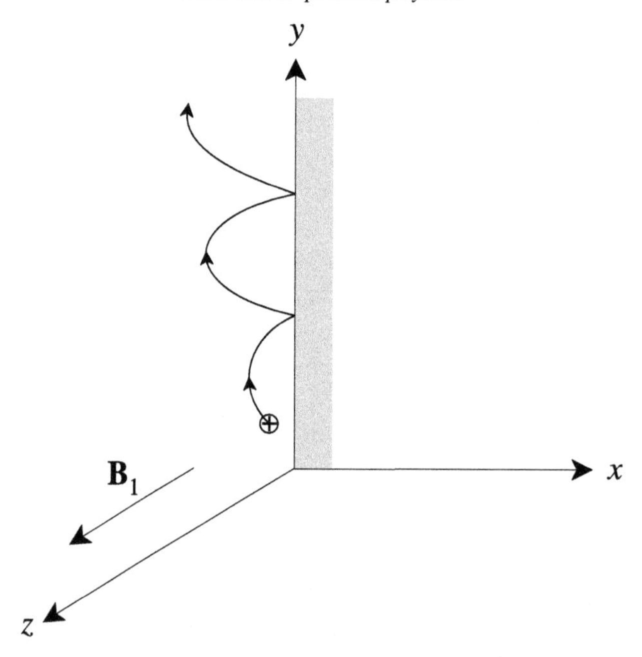

Fig. 10.23. Slow ion reflection off shock front.

so that the main effect of the electric field E_x is to slow the fluid (ion) flow. The current j_y is consistent with the increase in $B(x)$ and drives the drift instabilities which provide the dissipation to turn flow energy into thermal energy.

As we approach the critical Mach number (or increase β_i), some ions are reflected before eventually passing through the shock. These ions drag electrons with them and this plasma and the magnetic field lines that are drawn out with it form a 'foot' in the magnetic field structure in front of the main field jump, as illustrated in Fig. 10.24. This foot, which has a width $\sim v_A/\Omega_i \sim c/\omega_{pi}$, is observed as the critical Mach number is exceeded since the fraction of reflected ions becomes significant. As M_A increases, the foot dominates and $L_s \sim c/\omega_{pi}$. Dissipation in supercritical shocks is caused by ion streaming instabilities feeding on the energy in the reflected ions.

The same procedure can be used to discuss oblique (and parallel) shocks but for the reasons already mentioned, the analysis is complicated. Often transport equations with fitted 'turbulent' transport coefficients are used to obtain numerical calculations of shock structure. The main interest in this field is related to planetary and astrophysical shocks and much data has been collected about the Earth's bow shock which varies in nature between quasi-perpendicular and quasi-parallel. While this makes it a very interesting object it also makes the interpretation of data

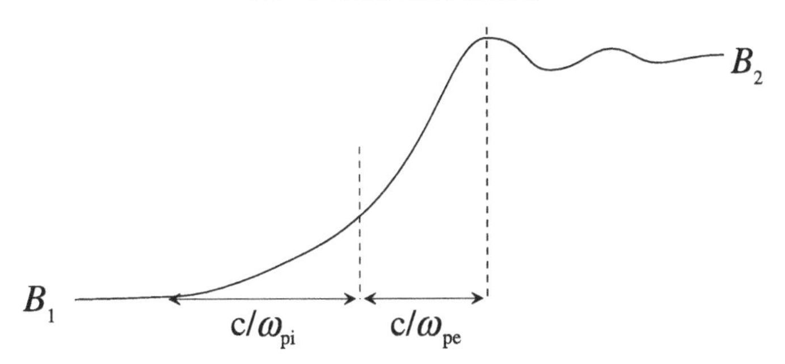

Fig. 10.24. Shock structure for supercritical shocks.

difficult. Observations, usually made by groups of satellites so that correlations may be recorded to facilitate interpretation, have not only estalished the existence of the bow shock but have also yielded information about the electron and ion foreshocks. The ion foreshock, comprising reflected ions with energies of a few keV, is the shock foot to which we have already referred. The electron foreshock consists of energetic electrons (1–2 keV) created at the quasi-perpendicular shock and travelling back into the solar wind along the field lines. A possible mechanism for this phenomenon is discussed in the following section.

10.5.3 Particle acceleration at shocks

The presence of electron and ion populations in the upstream plasma with energies far greater than the mean energy of the solar wind particles is a well-established feature of the Earth's bow shock. These particles are produced as a result of reflection and acceleration by the shock and there are any number of mechanisms which may be responsible. In the last section we noted the reflection of ions by the electric field in a perpendicular shock but when there is a component of magnetic field along the shock normal it is very easy for both ions and electrons to return into the solar wind should conditions in the bow shock propel them back along the field lines.

It is easiest to discuss this phenomenon in the de Hoffmann–Teller (HT) frame of reference introduced in Section 5.6.4. In the HT frame, by definition, the upstream velocity $\mathbf{v}_1$ and magnetic field $\mathbf{B}_1$ are parallel, a situation that is brought about by applying a Lorentz transformation from the shock frame to a frame that is moving parallel to the shock face with an appropriate velocity $\mathbf{v}_{\mathrm{HT}}$. Thus, the incident ($\mathbf{u}$) and reflected ($\mathbf{v}$) particle velocities may be resolved

$$\mathbf{u} = \mathbf{u}_{\parallel} + \mathbf{v}_{\mathrm{HT}} \tag{10.98}$$

$$\mathbf{v} = \mathbf{v}_{\parallel} + \mathbf{v}_{\mathrm{HT}} \tag{10.99}$$

where $\mathbf{u}_\parallel$ and $\mathbf{v}_\parallel$ are both guiding centre velocities along $\mathbf{B}_1$. Depending upon the reflection mechanism, the magnetic moment may or may not be conserved so we write

$$\mathbf{v}_\parallel = -\alpha \mathbf{u}_\parallel \tag{10.100}$$

where α is a positive constant.

Since there is no electric field, $\mathbf{E}_1 = -\mathbf{v}_1 \times \mathbf{B}_1 = 0$, in the upstream plasma, kinetic energy is conserved so that

$$u_\parallel^2 + u_\perp^2 = v_\parallel^2 + v_\perp^2 \tag{10.101}$$

where $u_\perp$ and $v_\perp$ are, of course, the components of the incident and reflected velocities perpendicular to $\mathbf{B}_1$, i.e. the speeds of rotation around the field lines.

From (10.98)–(10.100) it follows that

$$u^2 - v^2 = (1-\alpha^2)u_\parallel^2 + 2(1+\alpha)\mathbf{u}_\parallel \cdot (\mathbf{u} - \mathbf{u}_\parallel)$$

and hence,

$$v^2 = u^2 + (1+\alpha)^2 u_\parallel^2 - 2(1+\alpha)\mathbf{u}_\parallel \cdot \mathbf{u}$$

Thus, in the rest frame of the shock, the ratio of reflected to incident kinetic energy is

$$\frac{v^2}{u^2} = 1 + (1+\alpha)^2 \frac{u_\parallel}{u^2} - 2(1+\alpha)\frac{\mathbf{u}_\parallel \cdot \mathbf{u}}{u^2} \tag{10.102}$$

Then, if θ_{ab} is the angle between vectors $\mathbf{a}$ and $\mathbf{b}$, we have $u_\parallel \cos\theta_{Bn} = u\cos\theta_{vn}$ and $\mathbf{u}_\parallel \cdot \mathbf{u} = u_\parallel u \cos\theta_{Bv}$ so that (10.102) becomes

$$\frac{v^2}{u^2} = 1 + (1+\alpha)^2 \frac{\cos^2\theta_{vn}}{\cos^2\theta_{Bn}} - 2(1+\alpha)\frac{\cos\theta_{Bv}\cos\theta_{vn}}{\cos\theta_{Bn}}$$

Clearly, this ratio can take a wide range of values, but for quasi-perpendicular shocks $\cos\theta_{Bn} \approx 0$, making the second term dominant and leading to large increases in reflected particle energy.

Note that the reflected 'thermal' energy represented by $v_\perp^2$ is given from (10.100) and (10.101) by

$$\begin{aligned} v_\perp^2 &= (1-\alpha^2)u_\parallel^2 + u_\perp^2 \\ &\approx (1-\alpha^2)u_\parallel^2 \end{aligned}$$

in the cold solar wind approximation ($u_\perp^2 \approx 0$) showing that $\alpha \leq 1$ for physical solutions, there being no change in thermal energy for $\alpha = 1$.

Exercises

10.1 What are the small parameters that justify use of linear theory in (i) particle orbit theory, (ii) cold plasma wave theory, (iii) warm plasma wave theory, (iv) MHD, and (v) kinetic theory? Give examples of the breakdown of linear theory despite the smallness of the appropriate parameters.

10.2 Carry out the steps indicated in the text to obtain (10.6), (10.8) and (10.10).

Show that the spectral energy density $\mathcal{E}(\mathbf{k}, t)$ defined by (10.12) satisfies (10.13) and explain the physics of this equation.

10.3 With reference to Fig. 10.2, explain the relationship between the energy level of an electron and its phase space trajectory. Why is it the resonant electrons that are trapped? How does this impose a restriction on the validity of linear Landau theory and what is this restriction?

10.4 What is the essential property of the ballistic terms which gives rise to plasma echoes?

Show that if waves $\exp -i[k_1 x - \omega_1 t]$ and $\exp -i[k_2(x - l) - \omega_2 t]$ are generated at grids separated by a distance l a plasma echo may appear at a point x downstream where $x = \omega_2 l/(\omega_2 - \omega_1)$.

Note that the echo will appear only if $\omega_2 > \omega_1$. How do you explain this physically?

10.5 Starting from the coupled harmonic oscillator equations (10.28) derive the set of equations (10.32) for the amplitudes a_0, a_1, a_2. Verify (10.34) and show that it expresses conservation of energy. Why is energy conserved?

Obtain the Manley–Rowe relations (10.35).

10.6 Derive (10.39) from (10.36) by allowing for a frequency mismatch $\Delta\omega = \omega_0 - \omega_1 - \omega_2$. Show that this leads to an increase in the threshold for instability given by (10.41) and exlain why.

10.7 Obtain (10.71) and (10.72) from the one-dimensional Zakharov equations.

Show that the non-linear Schrödinger equation (10.74) has a solution of the form $E(z, t) = e^{i\Omega t} f(z)$ where $f(z) = (2\Omega)^{1/2}\text{sech}(\Omega^{1/2} z)$. Interpret this solution physically.

10.8 Carry out the steps indicated in the text to show that the magnetoacoustic wave with dispersion relation (10.79) arises on the lowest frequency branch of the X mode.

With reference to Fig. 10.18, explain the properties of a finite amplitude magnetoacoustic wave.

10.9 Explain why a laminar, perpendicular shock has a thickness which is much less than the ion Larmor radius but much greater than the electron Larmor radius. What is the significance of this for the motion of ions and electrons through such a shock? How does this lead to a self-consistent model for these shocks?

11

Aspects of inhomogeneous plasmas

11.1 Introduction

In this chapter we turn to a consideration of the physics of inhomogeneous plasmas. Since virtually all plasmas whether in the laboratory or in space are to some degree inhomogeneous, all that can be attempted within the limits of a single chapter is to outline some general points and illustrate these with particular examples. Throughout the book we have dealt in places with plasmas which were inhomogeneous in density or temperature and confined by spatially inhomogeneous magnetic fields. In the case of the Z-pinch the high degree of symmetry allowed us to find analytic solutions in studying the equilibrium. By contrast for a tokamak, even with axi-symmetry, solutions to the Grad–Shafranov equation could only be found numerically. Indeed the only general method of dealing theoretically with problems in inhomogeneous plasmas is by numerical analysis.

Nevertheless useful analytic insights may be gained in two limits. In the first, plasma properties change slowly in the sense that for an inhomogeneity scale length L and wavenumber k, $kL \gg 1$ and one can appeal to the WKBJ approximation described in Section 11.2. In this limit we shall draw on illustrations from the physics of wave propagation in inhomogeneous plasmas. If we picture a wave propagating in the direction of a density gradient, at some point on the density profile it may encounter a cut-off or a resonance. As we found in Chapter 6, propagation beyond a cut-off is not possible and the wave is reflected, whereas at a resonance, wave energy is absorbed. The WKBJ approximation breaks down in the neighbourhood of both cut-offs and resonances. We shall illustrate some aspects of this physics by means of a case history of stimulated Raman scattering, progressing from a local model for which the WKBJ approximation is a valid representation, to a global picture of the instability which can only be determined numerically.

Absorption of wave energy at a resonance is the basis of an important method of plasma heating. For example, radiofrequency heating makes a critical contribution to heating tokamak plasmas and the whole concept of inertial containment fusion is based on coupling laser energy to the target plasma. In general, radiation has to propagate to a resonance where it can be absorbed so *accessibility* is an important issue in inhomogeneous plasmas. The next stage of the heating process involves the transfer of electromagnetic energy to the plasma across the resonant region by means of *mode conversion*. Mode conversion describes the coupling of waves which individually satisfy distinct dispersion relations over a range of parameter space but which are coupled across some region. WKBJ analysis breaks down in a region of mode conversion. The second case-history we examine deals with the coupling of a longitudinal mode in the form of a Langmuir wave to a transverse electromagnetic wave in the presence of a steep density gradient.

In the second limit $kL \ll 1$. Under these conditions the change in plasma density is so steep that the inhomogeneity may sometimes be treated as a sharp boundary and jump boundary conditions applied. However, in other cases the physics of the boundary layer is important in characterizing the physics overall. Plasmas close to material boundaries often display sharp spatial variation even if relatively homogeneous outside these boundary layers. The importance of such regions was first recognized by Langmuir who showed that for plasmas in contact with a material surface, the interface between plasma and surface takes the form of a sheath several Debye lengths thick. This comes about on account of the greater mobility of electrons over ions that allows a negative potential to be established across the sheath. Most electrons are therefore reflected back into the plasma from the sheath.

11.2 WKBJ model of inhomogeneous plasma

The most widely used model for describing wave characteristics in non-uniform plasmas is the WKBJ approximation, developed independently by Wentzel, Kramers, and Brillouin to solve Schrödinger's equation for quantum mechanical barrier penetration. J recognizes the contribution of Jeffreys who had earlier developed the same approximation, albeit in a different context. The physical appeal of the WKBJ approximation is intuitive in that it is only a step beyond the familiar territory of a plane wave solution.

To keep the discussion as simple as possible consider electromagnetic wave propagation in an isotropic plasma in which the density varies spatially along Oz. For a linearly polarized transverse wave the electric field E (in the Oxy-plane) satisfies

$$\frac{\mathrm{d}^2 E}{\mathrm{d}z^2} + k^2(z)E = 0 \tag{11.1}$$

where

$$k^2(z) = \frac{\omega^2 - \omega_p^2(z)}{c^2} \tag{11.2}$$

For a homogeneous plasma the electric field satisfying (11.1) has the form $E(z,t) = A\exp[i(\phi(z) - \omega t)]$ where A is constant and $\phi(z) = kz$. For the inhomogeneous case in which the plasma density is a slowly varying function of z, we keep this form for the field and set out to find a representation for the phase $\phi(z)$, sometimes referred to as the *eikonal*. Then, dropping the time dependence,

$$\frac{\mathrm{d}E}{\mathrm{d}z} = iA\phi' e^{i\phi} \qquad \frac{\mathrm{d}^2E}{\mathrm{d}z^2} = \left\{iA\phi'' - A(\phi')^2\right\} e^{i\phi}$$

in which $\phi' = \mathrm{d}\phi/\mathrm{d}z$, $\phi'' = \mathrm{d}^2\phi/\mathrm{d}z^2$. Substituting in (11.1) gives

$$i\phi'' - (\phi')^2 + k^2(z) = 0 \tag{11.3}$$

For plasmas in which the density varies on a sufficiently long scale length the ϕ'' term may be taken to be small compared with $k^2(z)$ giving

$$\phi' = \pm k(z) \qquad \phi'' = \pm k'(z)$$

Then from (11.3)

$$\phi'(z) = \left[k^2(z) \pm ik'(z)\right]^{1/2} \simeq \pm k(z) + \frac{ik'(z)}{2k(z)}$$

$$\phi(z) \simeq \pm \int^z k(z)\mathrm{d}z + i\ \ln\sqrt{k(z)}$$

The integration constant may be set to zero by an appropriate choice for the lower limit of the integral, but is left unspecified for the time being since it does not affect the argument. Consequently the spatial dependence of the electric field takes the form

$$E(z) = \frac{A}{k^{1/2}(z)} \exp\left(\pm i \int^z k(z)\mathrm{d}z\right) \tag{11.4}$$

This WKBJ solution corresponds to right (+) and left (−) travelling waves. The wave forms in (11.4) resemble plane waves with phases expressed as an integral over the region of propagation but with an amplitude that is only weakly spatially varying. A useful rule of thumb for the validity of WKBJ solutions follows from the requirement that ϕ'' in (11.3) be much less than k^2, leading to the condition

$$\left|\frac{1}{k}\frac{\mathrm{d}k}{\mathrm{d}z}\right| \ll k \tag{11.5}$$

Clearly (11.5) is violated when $k \to 0$ (cut-off) or when $\mathrm{d}k/\mathrm{d}z \to \infty$ (resonance). When the WKBJ method fails we generally have to resort to numerical solution of the parent differential equation. For consistency we need to check how well the solutions satisfy the Maxwell equations. In the first place we find that the wave magnetic field decreases as cut-off is approached in contrast to the swelling of the electric field. Consistency provides a more precise validity condition as to what is meant by 'slowly varying'. Unless this condition is satisfied one solution generates some of the other so that the two waves in (11.4) are no longer independent and reflection occurs.

Now consider an electromagnetic wave propagating from a source at $z = 0$ into a plasma of increasing density. The electric field of this wave may be represented as

$$E_{\mathrm{i}}(z) = \frac{A}{k^{1/2}(z)} \exp\left(i \int_0^z k(z)\mathrm{d}z\right)$$

and that of the reflected wave as

$$E_{\mathrm{r}}(z) = \frac{AR}{k^{1/2}(z)} \exp\left(-i \int_0^z k(z)\mathrm{d}z\right)$$

where R denotes the reflection coefficient for wave amplitude. If we now let $z \to z_{\mathrm{c}}$, where $z = z_{\mathrm{c}}$ is a cut-off point, and equate the two fields in this region, turning a blind eye to WKBJ breakdown, it follows that

$$R = \exp\left(2i \int_0^{z_{\mathrm{c}}} k(z)\mathrm{d}z\right) \tag{11.6}$$

The integral in (11.6) is known as the *phase integral* since it measures the change in phase of the wave from source to the cut-off and back. Not surprisingly (11.6) is not a correct representation of the reflection coefficient. We shall find in the following section that it differs from the true value by a factor i.

The WKBJ method is a mathematical representation of ray-tracing and its extension to three dimensions, known as the eikonal or ray-tracing approximation, is widely used (see Weinberg (1962)). The amplitude of the wave is taken as $\mathbf{E}_0 \exp[i\phi(\mathbf{r}, t)]$, where $\phi(\mathbf{r}, t)$ satisfies $\nabla\phi = \mathbf{k}(\mathbf{r}, t)$ and $\partial\phi/\partial t = -\omega(\mathbf{r}, t)$, and the direction of energy flow is given by $\mathbf{v}_{\mathrm{g}} = \partial\omega/\partial\mathbf{k}$. If we denote the dispersion relation by

$$D(\omega, \mathbf{k}, \mathbf{r}, t) = 0 \tag{11.7}$$

we can express $\omega = \omega(\mathbf{k}, \mathbf{r}, t)$. The ray trajectories are then determined by the set of equations (see Exercise 11.3):

$$\frac{\mathrm{d}\mathbf{r}}{\mathrm{d}t} = -\frac{\partial D/\partial\mathbf{k}}{\partial D/\partial\omega} \qquad \frac{\mathrm{d}\mathbf{k}}{\mathrm{d}t} = -\frac{\partial D/\partial\mathbf{r}}{\partial D/\partial\omega} \qquad \frac{\mathrm{d}\omega}{\mathrm{d}t} = -\frac{\partial D/\partial t}{\partial D/\partial\omega} \tag{11.8}$$

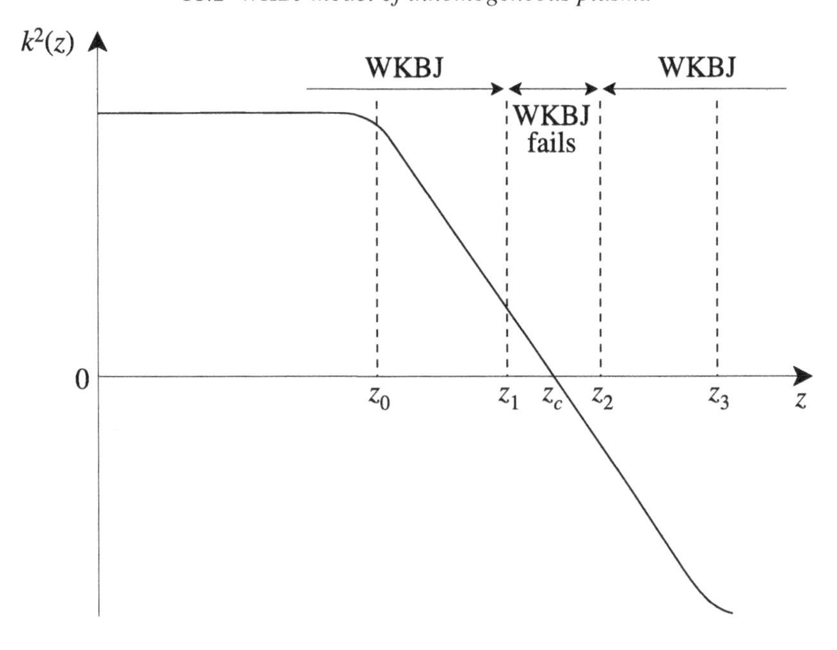

Fig. 11.1. The function $k^2(z)$ showing regions of validity of the WKBJ aapproximation.

Starting from a source at the plasma boundary that injects a given spectrum of wavenumbers, (11.8) can be integrated to find the ray characteristics. Numerical ray-tracing codes find wide application across many areas of plasma physics.

11.2.1 Behaviour near a cut-off

The next task is to see how to deal with behaviour near a cut-off. At a cut-off (or C-point) the incident wave is reflected and $k^2(z)$ changes sign from positive to negative. Figure 11.1 shows a cut-off at $z = z_c$ in a region $z_1 \leq z \leq z_2$ over which WKBJ solutions fail. For $z \leq z_1$ the solutions correspond to incident and reflected waves as we saw in the previous section. Beyond z_2, WKBJ solutions are again valid but now correspond to evanescent and amplifying waves. In the evanescent wave both electric and magnetic fields decay spatially and there is no energy flux beyond the C-point. Clearly an amplifying solution is non-physical in the absence of a source of energy. Across the region $z_0 \leq z \leq z_3$, $k^2(z)$ may be represented as

$$k^2(z) = -\left(\frac{k_c^2}{z_c}\right)(z - z_c) + O(z - z_c)^2$$

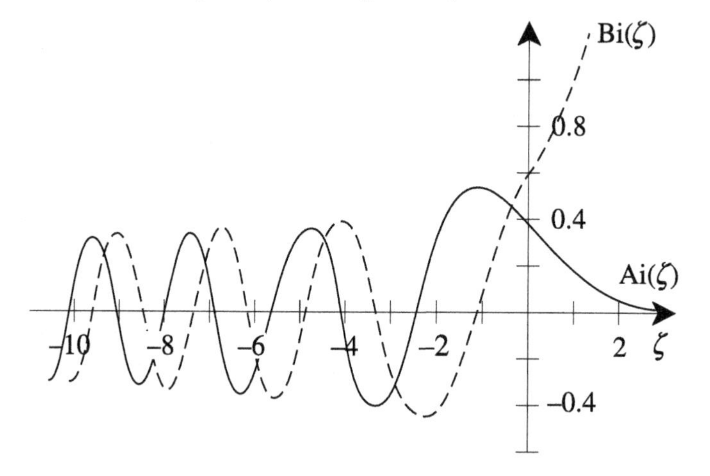

Fig. 11.2. Ai(ζ) and Bi(ζ) (dashed curve).

where k_c is a constant, not to be confused with $k^2(z_c) = 0$, and (k_c^2/z_c) is real and positive. Thus across this region (11.1) becomes

$$\frac{\mathrm{d}^2 E}{\mathrm{d}z^2} - \left(\frac{k_c^2}{z_c}\right)(z - z_c)E = 0$$

which can be cast as Stokes' equation

$$\frac{\mathrm{d}^2 E}{\mathrm{d}\zeta^2} = \zeta E \tag{11.9}$$

by making use of the transformation $\zeta = \left(k_c^2/z_c\right)^{1/3}(z - z_c)$. Stokes' equation has no singularities for finite ζ. Its solution is therefore finite and single-valued and is expressed in terms of *Airy functions* Ai(ζ), Bi(ζ) represented in Fig 11.2.

Solutions to Stokes' equation have to be considered over the complex ζ-plane. We can readily write down WKBJ solutions to Stokes' equation; disregarding constants

$$E = \zeta^{-1/4} \exp\left(\pm\frac{2}{3}\zeta^{3/2}\right) \tag{11.10}$$

A linear combination of these solutions is a good representation across the ranges indicated in Fig. 11.1. However, since these are multiply valued functions, the solution to Stokes' equation, being single-valued, cannot be represented by the *same* combination for all ζ. For example, for real positive ζ ($\arg \zeta = 0$), the WKBJ approximation to Ai(ζ) is given by

$$\mathrm{Ai}(\zeta) = A\zeta^{-1/4} \exp\left(-\frac{2}{3}\zeta^{3/2}\right) \tag{11.11}$$

On the other hand, if we keep $|\zeta|$ constant and set arg $\zeta = \pi$, this representation clearly fails. For a correct general representation we need both terms in (11.10), i.e.

$$A\zeta^{-1/4}\exp\left(-\frac{2}{3}\zeta^{3/2}\right)+B\zeta^{-1/4}\exp\left(\frac{2}{3}\zeta^{3/2}\right)$$

with the proviso that the constants A and B take on different values as ζ moves in the complex plane. The need for this was first recognized by Stokes and is known as the *Stokes phenomenon*. The exponents involved are real for arg $\zeta = 0, 2\pi/3, 4\pi/3$. For large $|\zeta|$ one contribution is exponentially large (the *dominant* term), the other exponentially small (*subdominant*). A fuller discussion of the Stokes' solutions has been given by Budden (1966).

We can now return to reconsider the reflection coefficient discussed earlier. This requires a solution of the wave equation across the entire range. The argument is quite general but for sake of reference we consider it in the context of a light wave incident from $z = 0$ on a plasma containing a region across which the density varies linearly with z. Over that part of the range beyond z_2, the solution contains only the subdominant term so that over the region $z_1 \leq z \leq z_3$ the solution is determined by $\mathrm{Ai}(\zeta)$. This solution has to fit on to the WKBJ solutions below the cut-off, i.e.

$$k^{-1/2}\left[\exp\left(i\int_0^z k(z)\mathrm{d}z\right)+R\exp\left(-i\int_0^z k(z)\mathrm{d}z\right)\right]$$

Multiplying this by the constant $\exp(-i\int_0^{z_c}k(z)\mathrm{d}z)$ and recasting in terms of ζ we find the solution valid over the range (z_0, z_3) is proportional to

$$\zeta^{-1/4}\left[\exp\left(-\frac{2}{3}\zeta^{3/2}\right)+R\exp\left(-2i\int_0^{z_c}k(z)\mathrm{d}z\right)\exp\left(\frac{2}{3}\zeta^{3/2}\right)\right]$$

This form must be identical to the approximation for $\mathrm{Ai}(\zeta)$ for arg $\zeta = \pi$ so that

$$R = i\exp\left(2i\int_0^{z_c}k(z)\mathrm{d}z\right) \tag{11.12}$$

which differs from the result (11.6) by a phase factor $\pi/2$.

In general two different representations of a function can match only over some limited range of z before phase divergence becomes serious. Implicit in the matching is the assumption that there is a region of overlap across which both the eikonal solution and the asymptotic Airy solution are each valid representations. Matching will clearly not be possible if $k^2(z)$ is significantly non-linear before valid WKBJ solutions are reached.

11.2.2 Plasma reflectometry

The reflection of a wave at cut-off lends itself to a versatile and widely used diagnostic, plasma reflectometry. Simply detecting a reflected wave is evidence that at some point in the profile the density is supercritical. The question of interest is precisely where reflection occurs. Finding the cut-off point requires a measurement of the relative phases of the incident and reflected waves. The electron density profile may be unfolded by sweeping the frequency of the incident wave. The technique has been widely applied to tokamak plasmas and has the advantage of both good spatial (≤ 1 cm) and temporal (≤ 5 μs) resolution. Reflectometry has certain features in common with interferometry, but whereas interferometry builds up an electron density profile from measurements along different chords (of a cylindrical plasma), reflectometry constructs the same profile from phase measurements at different frequencies. One of the difficulties inherent in reflectometry arises from the effect of density perturbations on propagation of the incident wave, though techniques have been developed to get round this difficulty. Different types of reflectometer are used in plasma diagnostics. In the simplest, a single frequency is used, with the frequency swept linearly across a broadband range. Strong (MHD) fluctuations in the plasma present problems should this sweep be too slow.

For simplicity we consider O-wave propagation and ignore relativistic corrections. We saw from the previous section that the phase shift of the reflected wave is given by

$$\phi = \frac{2}{c}\int_0^{z_c} (\omega^2 - \omega_{\mathrm{p}}^2(z))^{1/2}\,\mathrm{d}z - \frac{\pi}{2}$$

This phase shift may be expressed as a time delay $\tau = \mathrm{d}\phi/\mathrm{d}\omega$ and if τ is measured for frequencies within the range of interest, one may then determine the position of the cut-off from an Abel inversion:

$$z_c(\omega) = z_0 + \frac{c}{\pi}\int_0^{\omega} \frac{\tau(\omega')}{(\omega^2 - \omega'^2)^{1/2}}\,\mathrm{d}\omega' \tag{11.13}$$

From this relation, knowing the phase delay for frequencies less than ω, the position of the cut-off may be found. Since measurements cannot be carried out from $\omega = 0$, an extrapolation is needed from some level to the plasma edge, assuming some density profile.

While measurements using a single frequency spectrum are widely used in practice, this approach suffers from the limitations imposed by the length of the interval needed for the diagnostic to map the density profile. There is an intrinsic limit to the time resolution of the reflectometer since the profile is unfolded step by step. Thus the measurement of the density profile needs to be completed within an interval short enough that significant changes in plasma parameters do not occur. Density fluctuations present between the boundary of the plasma and the cut-off affect wave propagation and are the source of inaccuracies in constructing the density profile. Fluctuations with short scale length give rise to destructive interference; however, unless the plasma is in a highly turbulent state it is usually possible to unfold a density profile, albeit at the cost of reduced resolution.

11.3 Behaviour near a resonance

Resonances are less straightforward to deal with than cut-offs since the essential physics governing the resonance has to be incorporated to obtain physically meaningful results. Whereas waves undergo reflection at cut-offs, resonances are characterized by absorption of the wave energy by the plasma. As the wave approaches a resonance, $n = ck/\omega \to \infty$ so that the wave is refracted toward the resonant surface and reaches it at normal incidence. Since condition (11.5) is increasingly well satisfied, no reflection occurs. What happens to the energy carried by the wave to the resonance (or R-point)? In the strict cold plasma limit, this energy could only be stored in the form of currents, resulting in the non-physical limit of increasingly large rf power density. However in real plasmas, finite temperature ensures that the refractive index of the plasma remains finite even at a resonance. In warm plasmas, the consequent damping, however small, means that some heating takes place. Even in the cold plasma limit where there is no dissipation, a small amount of damping, ν, has to be introduced into the analysis in order to move the singularity at the resonance $(\omega_R(z) - \omega)^{-1}$ off the real axis and so determine how the solution is to be continued around the singularity. In physical terms, if one examines the transport of wave energy to the resonance, the time required to approach the R-point varies as ν^{-1}. In hot plasmas Coulomb collisions near the resonance are ineffective as an agent for energy dissipation so that an alternative means is needed. This alternative is provided by *linear mode conversion* which converts the incident wave to a warm plasma wave. Thus as well as C-points and R-points we now identify X-points, in the neighbourhood of which linear mode conversion takes place.

An issue of importance in discussing resonances in inhomogeneous plasmas is the question of their accessibility. In the radiofrequency heating of laboratory plasmas a wave is launched from an antenna configuration outside the plasma and propagates to a region within the plasma where the resonance is sited. Formally, the resonance is *accessible* if $k^2(z) > 0$ at all points on the density profile below the resonant density. However, if a C-point is present en route to the R-point then reflection there will prevent the wave from reaching the resonance. One important exception to this appears in cases where the cut-off and resonance stand back-to-back. Between cut-off and resonance the wave is evanescent ($k^2(z) < 0$). If the separation between the points is not too great some fraction of the incident wave energy can tunnel through the evanescent region beyond the C-point to reach the resonance. Such a conjunction of C- and R- points occurs not only in radio wave propagation in the ionosphere but in the resonant absorption of laser light by target plasmas.

The first analysis of the physics of a cut-off and resonance back-to-back was carried out by Budden (1966) who used a wave equation with the form

$$\frac{\mathrm{d}^2 E}{\mathrm{d}z^2} + k_0^2 \left(1 + \frac{z_c}{z}\right) E = 0 \tag{11.14}$$

This provides the simplest representation for a back-to-back cut-off and resonance, with the resonance at $z = 0$ and the cut-off at $z = -z_c$. With suitable substitutions (11.14) may be cast in standard form with solutions expressed in terms of confluent hypergeometric functions. Asymptotic solutions for large positive z may be connected to those for z large and negative, with a small amount of damping introduced to resolve the singularity present. A wave travelling to the right encounters the cut-off first and is in part reflected while some fraction tunnels through beyond the C-point. Budden found (amplitude) reflection and tunnelling (transmission) coefficients

$$|R| = 1 - e^{-2\eta} \qquad |T| = e^{-\eta} \tag{11.15}$$

where $\eta = \frac{1}{2}\pi k_0 z_c$. A wave travelling to the left meets the resonance first as shown in Fig. 11.3. The tunnelling coefficient is symmetric in the two cases but there is no reflection at the resonance for incidence from the right so that

$$|R| = 0 \qquad |T| = e^{-\eta} \tag{11.16}$$

Physically, the parameter η provides a measure of the number of vacuum wavelengths that fit between the cut-off and resonance. For right-propagating waves, $\eta \ll 1$ implies that most of the incident flux penetrates to the resonance. For $\eta \gg 1$, tunnelling is ineffective for right-propagating waves resulting in almost total reflection. It is clear from both expressions (11.15) and (11.16) that $|R|^2 + |T|^2 < 1$, i.e. energy is not conserved.

The ingredient missing from Budden's equation is mode conversion which is important in regions of the plasma over which two modes that in general satisfy distinct dispersion relations, and thus propagate as independent modes, no longer do so. Over such regions, modes can interact strongly with one another. In mode conversion a wave incident from one side of the region in question will emerge from the other side as a linear combination of two modes. Mode conversion is important in practice in the absorption of electromagnetic energy by a plasma, leading to heating.

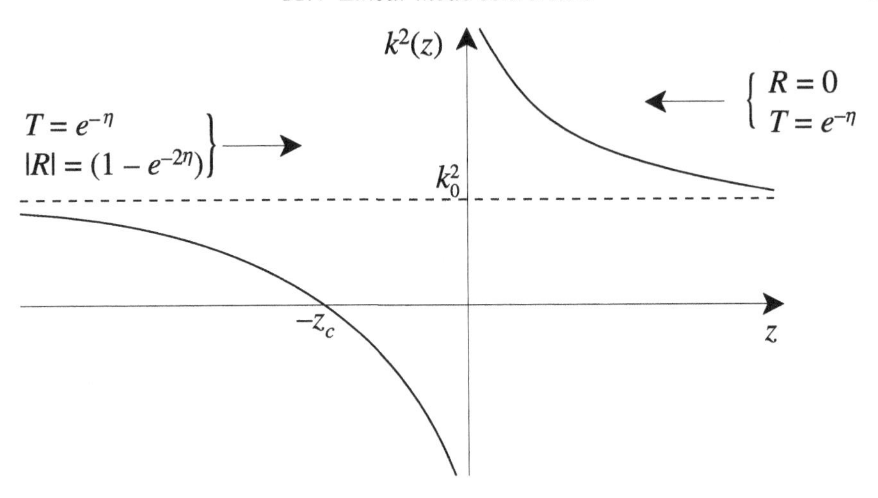

Fig. 11.3. Behaviour of $k^2(z)$ in the neighbourhood of a cut-off resonance pair.

11.4 Linear mode conversion

Mode conversion operates over regions where otherwise distinct modes have a common wavelength. Outside such a region, coupling between modes is weak and one would expect WKBJ solutions to provide a satisfactory representation of the propagation characteristics of each mode. However, within this region the WKBJ approximation breaks down and an alternative formulation is needed to characterize wave propagation. One approach extrapolates the homogeneous dispersion relation, $D(\omega, \mathbf{k}) = 0$, to represent *local* wave characteristics by means of a new relation $D(\omega, \mathbf{k}, z) = 0$. A differential equation may then be constructed from this algebraic equation by introducing a mapping $k \to -i\mathrm{d}/\mathrm{d}z$, an operation that is not in general unique.

If one considers a one-dimensional model of two modes subject to mode conversion and allows for propagation in either direction then the governing dispersion relation is fourth-order in k so that mapping in this way produces a fourth-order differential equation. One such equation constructed by Erokhin (1969)

$$y^{(\mathrm{iv})} + \lambda^2 z y'' + (\lambda^2 z + \gamma) y = 0 \tag{11.17}$$

has been widely used as a paradigm mode conversion equation. Since the coefficients are linear in z, solutions may be found in the form of a contour integral $y(z) = \int_C F(k) \exp(-ikz)\,\mathrm{d}z$. The asymptotic properties of a solution having this form are found by the method of steepest descents. The saddle points are found and the contribution to the asymptotic solution from a saddle point gives a WKBJ solution. The asymptotic solutions correspond to superpositions of the WKBJ solutions for the separate waves. However this procedure provides a solution across the region where no WKBJ solution is possible. The contour C is chosen to thread the saddle points giving the required behaviour (see Swanson (1985, 1989)). Discarding $y^{(iv)}$ in (11.17) reproduces Budden's equation. The important difference here is that the energy missing from Budden's equation is now mode-converted to the slow wave branch.

Quite apart from the non-uniqueness of the operation $k \rightarrow -i\mathrm{d}/\mathrm{d}z$, the analysis involved in such an approach is both complex and cumbersome to apply. A simpler alternative is based on the idea that is central to mode coupling, namely that just two modes are involved and the propagation of each is governed by distinct differential equations except for a limited region across which the modes (and hence the equations) are coupled. This *local* coupling provides the means by which power can flow from one mode to the other in the coupling region. One can show for example how a system of differential equations may be transformed to separate out a second-order system in the neighbourhood of a mode conversion point (see Heading (1961)). Away from the mode conversion point the solutions to this equation represent a superposition of the two modes involved. This approach has since been used and adapted by Fuchs, Ko and Bers (1981) and by Cairns and Lashmore-Davies (1983); see Cairns (1991). In fact the method was first devised in 1932, independently by Landau and by Zener, in treating the pseudo-crossing of potential energy curves in a quantum mechanical description of slow adiabatic atomic collisions; see Landau and Lifshitz (1958). Indeed the energy transmission coefficient found from the mode-coupling analysis in this section ((11.21) below) is readily recovered from the Landau–Zener transition probability after an appropriate transcription of variables.

Coupling involving two modes is illustrated in Fig. 11.4 which shows the crossing of the individual dispersion curves for the uncoupled modes $k = k_1(z)$, $k = k_2(z)$. In the neighbourhood of this curve-crossing, the dispersion relation is assumed to take the form

$$[k - k_1(z)][k - k_2(z)] = \chi \tag{11.18}$$

in which the coupling term χ is significant only in the neighbourhood of the crossing point $z = z_0$ of the (uncoupled) dispersion curves. The next step is to convert this local dispersion relation to a second-order differential equation by identifying k with $-i\mathrm{d}/\mathrm{d}z$. To get round the difficulty introduced by the non-uniqueness of this procedure Cairns and Lashmore-Davies proposed that uniqueness be ensured by a choice compatible with energy conservation. For modes with positive group velocities (as in Fig. 11.4) they introduced mode amplitudes ϕ_1 and ϕ_2 normalized

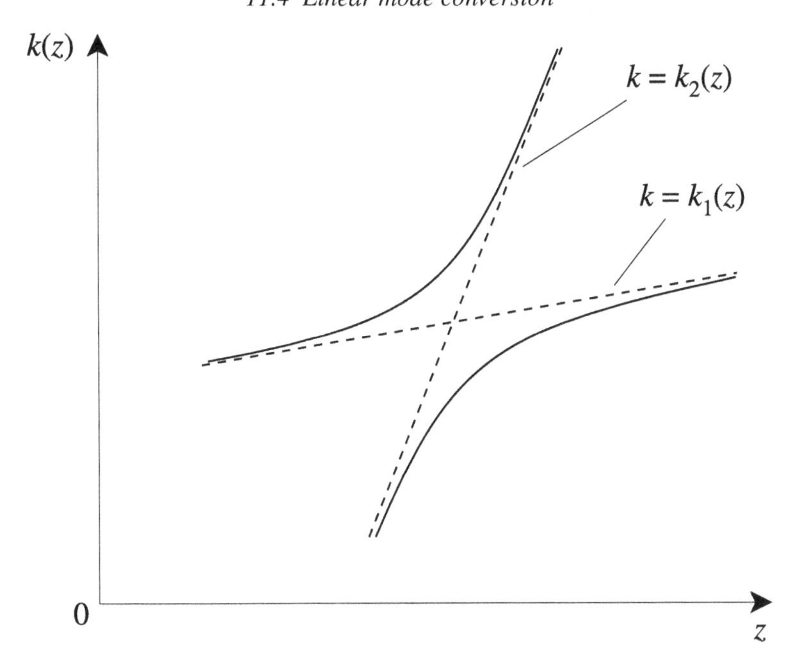

Fig. 11.4. Coupling of modes $k = k_1, k = k_2$.

so that $|\phi_1|^2$, $|\phi_2|^2$ denote energy fluxes in the respective modes. Energy (flux) conservation then requires

$$|\phi_1|^2 + |\phi_2|^2 = \text{constant} \tag{11.19}$$

A pair of differential equations satisfying (11.19) and reproducing the dispersion relation (11.18) is then

$$\left.\begin{aligned}\frac{d\phi_1}{dz} - ik_1(z)\phi_1 &= i\chi^{1/2}\phi_2 \\ \frac{d\phi_2}{dz} - ik_2(z)\phi_2 &= i\chi^{1/2}\phi_1\end{aligned}\right\} \tag{11.20}$$

Across the region of mode conversion one may assume

$$k_1(z) = a(z - z_0) \qquad k_2(z) = b(z - z_0)$$

which allows one of the amplitudes in (11.20) to be eliminated and the other determined by a second-order differential equation. This procedure leads to the Weber equation with solutions expressed in terms of parabolic cylinder functions. The asymptotics of these solutions are well known and a choice is made to represent an incoming mode with two outgoing modes, leading to expressions for the transmission and mode conversion coefficients in terms of the local dispersion relation and

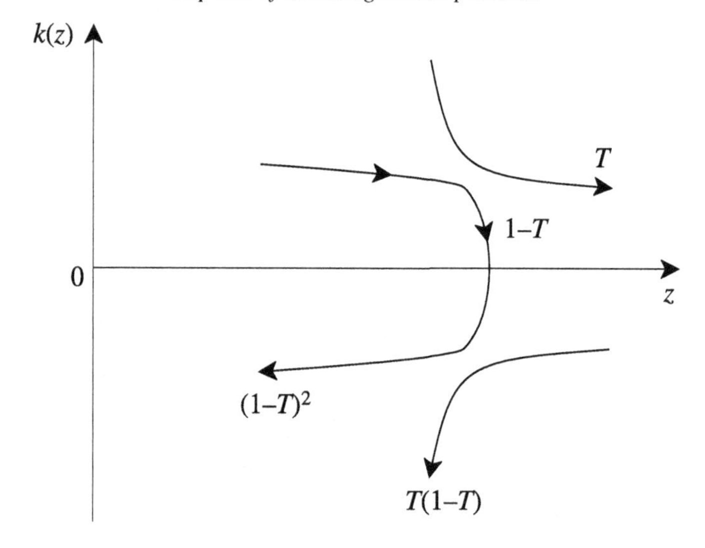

Fig. 11.5. Behaviour at a cut-off near a mode conversion region.

gradients in the region of mode conversion. Cairns and Lashmore-Davies (1983) showed that the energy transmission coefficient is given by

$$T = \exp\left[-\frac{2\pi\chi_0}{|a-b|}\right] \tag{11.21}$$

where χ_0 denotes the value of χ at $z = z_0$. The energy flux in the converted wave is $(1 - T)$ times the incident flux.

In many cases of interest one of the modes suffers a cut-off near the mode conversion point resulting in most of the energy associated with this mode being reflected and coupled to oppositely directed waves. Thus in Fig. 11.5 an incoming wave is shown approaching a C-point from the left. The fraction of energy transmitted is T while the reflected fraction $(1 - T)$ couples to a mode propagating in the opposite direction at an X-point. Here a fraction T undergoes mode conversion, giving a fraction $T(1-T)$ of the incident energy flux converted while the remainder $(1-T)^2$ is reflected. If we picture a wave approaching the X-point from the right we see that the converted mode now propagates away from the cut-off and no reflection occurs. The transmission coefficient is unchanged but the converted fraction is now $1 - T$. There is therefore an asymmetry between waves approaching from the left (resulting in reflection) and waves from the right (no reflection). These results are summarized in Table 11.1.

The appeal of this approach is that it side-steps the analytic complexity needed to deal with equations such as (11.17). On the debit side the assumption that the

Table 11.1. Transmission (T), reflection (R) and mode conversion (C) coefficients for modes incident from right (r) and left (l) of the cut-off resonance pair. Note $\eta^* = \pi \chi_0/|a - b|$.

T	R	C
$T_r = e^{-2\eta^*}$	$R_r = 0$	$C_r = 1 - e^{-2\eta^*}$
$T_l = e^{-2\eta^*}$	$R_l = (1 - e^{-2\eta^*})^2$	$C_l = e^{-2\eta^*}(1 - e^{-2\eta^*})$

coupling coefficient χ may be taken to be both spatially uniform and symmetric between the modes is unlikely to be generally valid.

11.4.1 Radiofrequency heating of tokamak plasma

One important application of mode conversion is found in radiofrequency (rf) heating of plasmas, particularly in tokamaks. We saw in Chapter 1 that the temperature corresponding to the optimum reaction rate for D–T fusion lies in the region 10–20 keV. The tokamak current heats the plasma through ohmic heating (typically to temperatures of a few keV) but as the temperature rises, ohmic heating becomes less and less effective and on its own is unable to heat the plasma to the stage where alpha particle heating can sustain fusion. Additional power is needed and this auxiliary heating in tokamaks is provided by neutral beam injection and rf heating. In schemes for rf heating, power is fed into the plasma from waveguides mounted in the wall of the torus. This power has then to be transported to the R-point deep inside the plasma so that the issue of accessibility is critical to the success of this form of heating. Various frequency ranges are used, including both ion and electron resonances. Ion cyclotron resonance heating (ICRH) operating in the range of a few tens of MHz has produced up to 16 MW of power in JET.

To illustrate the importance of accessibility of the R-point consider a simple model for electron cyclotron resonance heating (ECRH), in which an rf wave is launched in the mid-plane of the torus, from either the inside or outside edge. Typically ECRH operates at frequencies across the range 30–150 GHz. The plasma density varies approximately parabolically across the torus, while the toroidal magnetic field varies as $1/R$. Since the wavelength of the electron cyclotron mode is typically much less than the scale length of the tokamak plasma, a WKBJ representation will be valid except at a cut-off and in the neighbourhood of a resonance. Away from a resonance, wave propagation is adequately described by the cold plasma, Appleton–Hartree dispersion relation so that we may make use of the CMA diagram, introduced in Section 6.3.4. Figure 11.6 corresponds to regions 1, 3 and 5

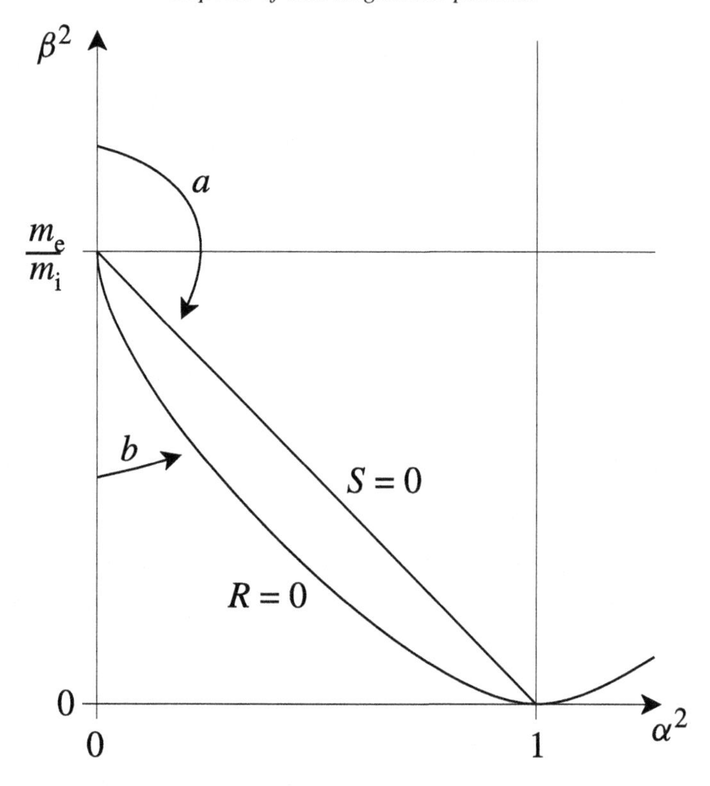

Fig. 11.6. Section of the CMA diagram showing access to a resonance via path a in contrast to path b which encounters a cut-off before the resonance.

in the CMA diagram, Fig. 6.12, and serves to highlight the role of both electron density and magnetic field in determining the propagation characteristics of these modes. A wave propagating from the wall of a tokamak into regions of higher density is represented by a point in the CMA diagram moving to the right, whereas upward movement corresponds to an increase in the magnetic field strength. A wave that follows path a in Fig. 11.6 propagates from a region of higher to one of lower magnetic field and is capable of accessing the R-point, whereas one that takes path b will fall foul of the C-point ($R = 0$) in propagating from a low density region to one of high density. Thus a wave launched from the outside wall will be reflected whereas one launched from the inside is able to reach the upper-hybrid resonance. We stress that this picture is valid strictly for a cold, loss-free plasma. For warm plasmas a crucial distinction is that electron cyclotron resonances at both the fundamental and second harmonic now become dominant. Electrons interact strongly with an RCP field whereas this is shorted out in the cold plasma limit. A concise discussion of the various schemes used in the rf-heating of plasmas may be found in Cairns (1991).

11.5 Stimulated Raman scattering

We turn next to the first of the topics chosen to illustrate aspects of the physics of inhomogeneous plasmas. Stimulated Raman scattering was introduced in Section 10.3.1 in model format. We now want to take this to a stage where it becomes possible to use the model to interpret and understand observations of SRS. In particular we shall examine the ways in which plasma inhomogeneity affects the nature of the instability. Dephasing, the fact that phase-matching conditions can be satisfied only over small regions of the plasma, is now a key concern. In addition, non-local effects, such as the reflection of the Raman decay modes as the density increases, can lead to feedback and this in turn can change the nature of the instability in important ways. In particular, a wave that is C-unstable in a homogeneous plasma may become A-unstable in the presence of a density gradient.

11.5.1 SRS in homogeneous plasmas

Before turning to the inhomogeneous case it is helpful to retrieve some Raman characteristics for a homogeneous plasma on the basis of the electron fluid equations as opposed to the model equations used in Section 10.3.1. We begin by representing the laser pump fields by

$$\mathbf{E}_0(\mathbf{r}, t) = 2\mathbf{E}_0 \cos(\mathbf{k}_0 \cdot \mathbf{r} - \omega_0 t) \qquad \mathbf{B}_0(\mathbf{r}, t) = 2\mathbf{B}_0 \cos(\mathbf{k}_0 \cdot \mathbf{r} - \omega_0 t) \quad (11.22)$$

with $\omega_0^2 = \omega_\mathrm{p}^2 + k_0^2 c^2$. Provided the laser intensity is not so high as to accelerate electrons to relativistic velocities then the electron quiver velocity in the laser electric field is given by $\mathbf{v}_0(\mathbf{r}, t) = 2\mathbf{v}_0 \sin(\mathbf{k}_0 \cdot \mathbf{r} - \omega_0 t)$. For laser intensity I_L and wavelength λ_L, the normalized quiver velocity, $v_0/c = (eE_0/m\omega_0 c) = 4.27 \times 10^{-10} I_\mathrm{L}^{1/2} (\mathrm{W\,cm^{-2}}) \lambda_\mathrm{L}(\mu\mathrm{m})$, is a critical parameter in characterizing parametric instabilities. A straightforward first-order perturbation analysis of the electron fluid equations (see Exercise 11.9) leads to a pair of coupled equations describing SRS in a homogeneous plasma:

$$\left[\frac{\partial^2}{\partial t^2} - c^2\nabla^2 + \omega_\mathrm{p}^2\right] \mathbf{v}_s = -\frac{e^2}{m\epsilon_0} n_\mathrm{e} \mathbf{v}_0 \quad (11.23)$$

$$\left[\frac{\partial^2}{\partial t^2} - 3V_\mathrm{e}^2\nabla^2 + \omega_\mathrm{p}^2\right] n_\mathrm{e} = n_0 \nabla^2(\mathbf{v}_0 \cdot \mathbf{v}_s) \quad (11.24)$$

where $\partial \mathbf{v}_s/\partial t = -e\mathbf{E}_s/m$ and subscript s denotes the scattered wave.

Fourier analysing (11.23) and (11.24) leads directly to the Raman dispersion relation

$$(\omega^2 - \omega_\mathrm{p}^2 - 3k^2 V_\mathrm{e}^2)(\omega_s^2 - \omega_\mathrm{p}^2 - k_s^2 c^2) = k^2 v_0^2 \omega_\mathrm{p}^2 \quad (11.25)$$

Physically, the laser pump in the presence of an electron density perturbation associated with a Langmuir wave is the source of a non-linear current which generates the Raman scattered light wave, given frequency and wavenumber matching. The pump and scattered light wave beat to produce a ponderomotive force proportional to $\nabla(\mathbf{E}_0 \cdot \mathbf{E}_s)$ and this in turn drives the density fluctuation. This feedback loop makes possible the development of SRS. For an instability to occur we need a pair of complex conjugate roots. Accordingly we set $\omega = \omega_{\rm L} + i\gamma$, $\omega_s = \omega_{sr} - i\gamma$, where $\omega_{\rm L}^2 = \omega_{\rm p}^2 + 3k^2V_{\rm e}^2$, $\omega_{sr}^2 = \omega_{\rm p}^2 + k_s^2c^2$, and $|\gamma| \ll \omega_{\rm p}$. Then from (11.25) the maximum homogeneous Raman growth rate γ_0 is given by

$$\gamma_0 = \frac{kv_0\omega_{\rm p}}{2(\omega_{\rm L}\omega_{sr})^{1/2}} \tag{11.26}$$

In the event that the phase matching conditions $\omega_0 = \omega + \omega_s$, $\mathbf{k}_0 = \mathbf{k} + \mathbf{k}_s$ are not exactly satisfied, mismatch will result in a growth rate $\gamma < \gamma_0$.

In practice one needs to allow for damping of the Raman decay waves. The Langmuir wave will be Landau damped but in a fluid model we can only represent this damping phenomenologically so that (11.25) is replaced by

$$(\omega^2 + 2i\omega\gamma_{\rm p} - \omega_{\rm p}^2 - 3k^2V_{\rm e}^2)(\omega_s^2 - 2i\omega_s\gamma_s - \omega_{\rm p}^2 - 3k_s^2c^2) = k^2v_0^2\omega_{\rm p}^2$$

where $\gamma_{\rm p}$, γ_s denote damping coefficients for the plasma and scattered light waves respectively. The Langmuir wave may suffer collisional as well as Landau damping so that in general $\gamma_{\rm p} = \gamma_{\rm L} + \nu_{\rm ei}/2$, where $\nu_{\rm ei}$ denotes the electron–ion collision frequency. Damping of the light wave is purely collisional with $\gamma_s = \omega_{\rm p}^2\nu_{\rm ei}/2\omega_s^2$.

Under resonance conditions

$$(\gamma + \gamma_{\rm p})(\gamma + \gamma_s) = \gamma_0^2 \tag{11.27}$$

which determines the threshold for the onset of SRS

$$\gamma_0 = (\gamma_{\rm p}\gamma_s)^{1/2} \tag{11.28}$$

and the Raman growth rate in terms of the maximum growth rate γ_0 and the wave damping coefficients, i.e.

$$\gamma = -\frac{1}{2}(\gamma_{\rm p} + \gamma_s) \pm \left[\gamma_0^2 + \frac{1}{4}(\gamma_{\rm p} - \gamma_s)^2\right]^{1/2} \tag{11.29}$$

11.5.2 SRS in inhomogeneous plasmas

In reality, plasmas in which stimulated Raman scattering occurs are inhomogeneous, often over short scale lengths. Under such conditions the extent of the region of instability is localized on account of phase matching or the finite range of the pump. Phase mismatch introduces a new loss mechanism on account of the convection of wave energy away from the localized resonances. In practice, convective loss is usually dominant over collisional and Landau damping. The question that then arises is whether SRS is an absolute or a convective instability. In other words, do the Raman daughter waves propagate away from the interaction region after reaching some maximum amplitude or do they continue to grow within the domain of instability? We shall find that convective losses suppress the absolute (temporal) growth found for the homogeneous plasma.

To keep the analysis as simple as possible we limit the plasma density inhomogeneity to one direction only, i.e. $n_0(\mathbf{r}) = n_0(x)$ and assume that the ions are stationary. Moreover we consider only the case of normal incidence so that the electric fields of both the incident and Raman scattered light waves are polarized parallel to one another. The WKBJ representation of the laser pump may then be written

$$\begin{pmatrix} \mathbf{E}_0(x,t) \\ \mathbf{B}_0(x,t) \end{pmatrix} = \frac{2}{\sqrt{k_0(x)}} \begin{pmatrix} E_0\hat{\mathbf{y}} \\ B_0\hat{\mathbf{z}} \end{pmatrix} \cos\psi \qquad \mathbf{v}_0(x,t) = \frac{2}{\sqrt{k_0(x)}} v_0\hat{\mathbf{y}} \sin\psi \tag{11.30}$$

with $\psi = \int^x k_0(x)\mathrm{d}x - \omega_0 t$ and $\omega_0^2 = \omega_\mathrm{p}^2(x) + k_0^2(x)c^2$. For convenience we suppress the WKBJ swelling factor $k_0^{-1/2}$, identifying the laser field as the local field at the Raman resonance. The first-order perturbation procedure proceeds as in the homogeneous case and now generates the set of equations (in which primes denote $\partial/\partial x$ and dots $\partial/\partial t$):

$$\left.\begin{aligned}
&\dot{n} + [n_0(x)v]' = 0 && \frac{1}{c^2}\dot{E}_s + B_s' - \mu_0 e n_0(x) v_s \\
& && \quad = \mu_0 e n v_0(x,t) \\
&\dot{v} + \frac{eE}{m} + \frac{3V_\mathrm{e}^2}{n_0(x)}n' + \nu v && \dot{v}_s + \frac{eE_s}{m} + \nu v_s = 0 \\
&\quad = -[v_0(x,t)v_s]' && \\
&E' + \frac{en}{\epsilon_0} = 0 && E_s' + \dot{B}_s = 0
\end{aligned}\right\} \tag{11.31}$$

The structure of this set of equations shows the linear characteristics of the Langmuir wave and scattered light wave on the left-hand side with the non-linear coupling through the action of the laser pump on the right. The scattered wave can propagate either backwards (stimulated Raman back-scatter, SRBS) or forwards (stimulated Raman forward-scatter, SRFS).

In general the set of equations (11.31) has to be solved numerically. To carry this through and obtain time-asymptotic solutions we first Laplace-transform (11.31). It is straightforward to eliminate v and E_s from the Laplace-transformed equations and convenient to cast the reduced set in normalized form. Times and velocities are normalized to ω_0^{-1} and c respectively and in addition $eE/mc\omega_0 \rightarrow E$, $eB_s/m\omega_0 \rightarrow B_s$ and $e^2(n_0, n)/m\epsilon_0 \rightarrow (\omega_{\rm p}^2(x), n)$. In this form the frequency matching condition now reads $\omega + \omega_s = 1$. The set of first-order equations then takes the form (see Barr, Boyd and Mackwood (1994))

$$\left.\begin{aligned} &E' + n = 0 && v_s' - \frac{p_1}{p_1 + \nu} B_s = 0 \\ &3V_{\rm e}^2 n' + [p_2^2 + 2\gamma_2 p_2 + \omega_{\rm p}^2(x)]E && B_s' - [p_1^2 + p_1\nu + \omega_{\rm p}^2(x)]v_s \\ &\quad = \omega_{\rm p}^2(x)[v_0(x)v_s]' && \quad = -v_0^*(x)n \end{aligned}\right\} \tag{11.32}$$

While the set of first-order equations is convenient for numerical integration, one can of course recover the coupled second-order equations for comparison with the homogeneous plasma equations. These follow directly from (11.32) on eliminating n and B_s to give

$$3V_{\rm e}^2 E'' + [\omega^2 + 2i\gamma_{\rm p}\omega - \omega_{\rm p}^2(x)]E = -\omega_{\rm p}^2(x)[v_0(x)v_s]' \tag{11.33}$$
$$v_s'' + [\omega_s^2 - 2i\gamma_s\omega_s - \omega_{\rm p}^2(x)]v_s = v_0^*(x)E' \tag{11.34}$$

Here $\gamma_{\rm p}(\equiv \gamma_2) = \gamma_{\rm L} + \nu/2$, $\gamma_s = \nu\omega_{\rm p}^2/2\omega_s^2$, $p_1 = i\omega_s$, $p_2 = -i\omega$. WKBJ solutions to (11.33) and (11.34) may be used provided the plasma density is only weakly inhomogeneous in the sense described in Section 11.2 and provided the Raman resonance is not in the neighbourhood of a turning point, i.e. at densities well below the quarter-critical density. Consider a Raman resonance at $x = x_{\rm r}$ where local wavenumber matching is satisfied, $k_0(x) = k(x) + k_s(x)$, with slowly varying amplitudes $a_{\rm p}$ and a_s defined such that

$$E = \frac{r a_{\rm p}(x)}{i\sqrt{k(x)}} \exp\left[i\int k(x)\mathrm{d}x\right] \qquad v_s = \frac{s a_s(x)}{\sqrt{k_s(x)}} \exp\left[-i\int k_s(x)\mathrm{d}x\right] \tag{11.35}$$

where $r = (k\omega_{\rm p}/\omega)^{1/2}$, $s = (k_s/\omega_s\omega_{\rm p})^{1/2}$ evaluated at $x = x_{\rm r}$ are constants.

The WKBJ equations reduce to (see Exercise 11.10)

$$V_{\rm p}a_{\rm p}' + (\gamma + \gamma_{\rm p})a_{\rm p} = \gamma_0 e^{-i\psi} a_s \tag{11.36}$$
$$V_s a_s' + (\gamma + \gamma_s)a_s = \gamma_0 e^{i\psi} a_{\rm p} \tag{11.37}$$

where $\psi = \int K(x)\mathrm{d}x$ and $K(x) = k_0(x) - k(x) - k_s(x)$ is the wavenumber mismatch with $K(x_{\rm r}) = 0$; $V_{\rm p}$, V_s denote the group velocities of the Langmuir and Raman scattered waves respectively. Resonant coupling is lost once a significant phase shift builds up. For a homogeneous plasma $\psi = 0$ and the growth rate of (11.29) is recovered. If we now combine the two WKBJ equations into a single second-order differential equation for (say) a_s, neglecting the spatial dependence of $V_{\rm p}$ and V_s, setting $a_s = a\exp(\int \alpha\, {\rm d}x)$ where

$$\alpha = -\frac{1}{2}[(p+\gamma_{\rm p}+iKV_{\rm p})/V_{\rm p}+(p+\gamma_s)/V_s]$$

one may show that

$$a'' - \left\{\frac{1}{4}\left[\frac{p+\gamma_{\rm p}+iKV_{\rm p}}{V_{\rm p}} - \frac{p+\gamma_s}{V_s}\right]^2 + \frac{iK'}{2} + \frac{\gamma_0^2}{V_{\rm p}V_s}\right\} a = 0 \tag{11.38}$$

For Raman back-scatter ($V_{\rm p}V_s < 0$) from a *localized* source there is a regime of convective Raman growth ($p = 0$) provided

$$\frac{\gamma_0^2}{|V_{\rm p}V_s|} \le \frac{1}{4}\left(\frac{\gamma_{\rm p}}{V_{\rm p}} + \frac{\gamma_s}{V_s}\right)^2 \tag{11.39}$$

Note that no convective growth is possible in the absence of damping. If the inequality sign in (11.39) is reversed, then stimulated Raman back-scatter (SRBS) may grow absolutely with a growth rate

$$\gamma_{\rm A} = \frac{2\sqrt{V_sV_{\rm p}}}{(V_{\rm p}+V_s)}\gamma_0 - \left(\frac{V_{\rm p}\gamma_s + V_s\gamma_{\rm p}}{V_{\rm p}+V_s}\right) \tag{11.40}$$

If we now allow a finite mismatch and consider in particular the case where this is linear we find that WKBJ theory shows that temporal growth is choked by convection so that only spatial amplification (convective gain) is possible. Expanding $K(x)$ about the Raman resonance and discarding wave damping, (11.38) becomes

$$a'' - \left\{\frac{1}{4}\left[\frac{p}{V_{\rm p}} - \frac{p}{V_s} + iK'(0)x\right]^2 + \frac{i}{2}K'(0) + \frac{\gamma_0^2}{V_{\rm p}V_s}\right\} a = 0 \tag{11.41}$$

For neither SRBS nor SRFS are solutions with $\Re p > 0$ possible so that the instability is always convective ($p = 0$). Setting $y = (K'(0))^{1/2}x$, (11.41) reduces to

$$\frac{{\rm d}^2a}{{\rm d}y^2} + \left(\frac{y^2}{4} - \frac{i}{2} \pm \lambda\right) a = 0 \tag{11.42}$$

where

$$\lambda = \frac{\gamma_0^2}{K'(0)|V_{\rm p}V_s|} \equiv \frac{1}{K'(0)L_{\rm g}^2} \tag{11.43}$$

defining a threshold scale length $L_{\rm g}$ for convective gain. In this form we see that the solution depends only on the parameter λ; $\pm\lambda$ refer to SRBS, SRFS respectively. Solutions to (11.38) may be found in terms of parabolic cylinder functions but all that is needed here is an estimate of the maximum convective amplification. The detuning of the Raman resonance condition is measured by the phase factor $\psi(x)$ and a *resonance width l* is defined by

$$\int_0^l K(x)\mathrm{d}x = \int_0^l K'(0)x\,\mathrm{d}x = \frac{\pi}{2}$$

so that

$$l = \sqrt{\frac{\pi}{K'(0)}} \tag{11.44}$$

Only if $l > L_\mathrm{g}$ will convective amplification be significant. A convective threshold is conventionally taken as that intensity which results in a gain of $\exp(2\pi)$, sometimes referred to as the *Rosenbluth gain* (Rosenbluth (1972)). Thus the gain is

$$C_G = \exp\left(\frac{2\pi\gamma_0^2}{V_\mathrm{p}V_s K'(0)}\right) \tag{11.45}$$

and the convective threshold condition is expressed as

$$\frac{\gamma_0^2}{V_\mathrm{p}V_s K'(0)} = 1 \tag{11.46}$$

For a linear density profile with $\omega_\mathrm{p}^2(x) = \omega_\mathrm{p}^2(1 + x/L)$ the threshold for SRBS is then

$$\left(\frac{v_0}{c}\right)^2 k_0 L = 1 \tag{11.47}$$

Early in the evolution of SRS, waves grow at the absolute growth rate from some initial localized noise source at the source point until such time as the waves transit the resonance. At this point the waves become aware of the finite extent of the resonance and temporal growth is saturated. Thereafter waves grow spatially (amplify) as they propagate across the resonance with both decay waves amplified in the back direction for SRBS and in the forward direction for SRFS. In this steady state the localized resonance acts as a convective amplifier.

For a given scattered frequency the resonant densities for SRBS and SRFS differ by $\Delta n \sim n(V_\mathrm{e}^2/c^2)$, with SRFS occurring at the higher density. The two resonances are not in fact independent though it has been conventional to treat them as if they were. Wave propagation allows communication between resonances at different locations, and this consideration in general results in quite distinct *global* behaviour for the instability. Attention was first drawn to the need for non-local models of parametric instabilities by Koch and Williams (1984) and later described in detail by Barr, Boyd and Coutts (1988), who solved the full system of SRS equations. The SRBS, SRFS amplifiers couple through the propagation of a plasma wave up the density profile from the SRBS resonance to that for SRFS, together with the propagation of forward scattered light from the SRFS resonance to its reflection density and then back to the SRBS resonance. The feedback loop established in this way allows amplifiers, which are convective in the absence of feedback, to grow temporally at those frequencies for which the two amplifiers are in phase.

11.5.3 Numerical solution of the SRS equations

The set of coupled equations (11.32) contains all the models that describe the time-asymptotic state of linear SRS theory and have been solved by Barr, Boyd and Mackwood (1994) for a plasma slab with appropriate boundary conditions. The density profile within the slab is arbitrary and this region is bounded by semi-infinite regions of homogeneous plasma. In these regions the four first-order equations may be solved exactly, the solutions corresponding to left and right propagating electromagnetic and Langmuir waves. In the presence of a pump field the solutions are coupled and hence are neither purely electromagnetic nor electrostatic. The set of equations (11.32) is solved in two distinct forms. Solution as an eigenvalue problem determines the complex eigenfrequencies; in this case all initial value terms are set equal to zero and the equations form a (mathematically) homogeneous system. The second form of solution determines convective gain factors at frequencies distinct from the eigenvalues and for this, initial value source terms have to be retained.

A range of physical parameters was chosen to exploit the many features of the model. A linear density ramp overdense to forward scattered light was used to allow for feedback. The parameters essential for interpreting SRS characteristics and enabling comparisons to be made with experiment include the Landau damping of the Langmuir wave, the collisional damping of both daughter waves, the convection of each wave, the degree of feedback between back- and forward-scattering amplifiers and the physical extent of the feedback loop.

Figure 11.7 shows the threshold contour $C_G = \exp 2\pi$ for Raman back-scatter. This shows two distinct regimes separated at a scattered frequency $\omega_s = 0.63\omega_0$. The approximately horizontal section of the gain contour at lower frequencies agrees well with the convective gain predicted by (11.45). At frequencies $\omega_s = 0.65\omega_0$ the steep rise in threshold corresponds to the onset of a cut-off produced by Landau damping. A rule of thumb is often applied for the onset of Landau damping when $k\lambda_{\mathrm{D}} = 0.3$ which, for the temperature used to produce these results, corresponds to $\omega_s = 0.7\omega_0$. Note that this cut-off is not predicted by the convective damping threshold (11.28) which gives zero threshold in the absence of collisions. Figure 11.8 plots the threshold predicted by (11.29). There is good agreement between the Landau cut-off seen here and the result in Fig. 11.7.

The absolute character of the instability is sustained over the whole range of emission frequencies on account of the feedback between back- and forward-

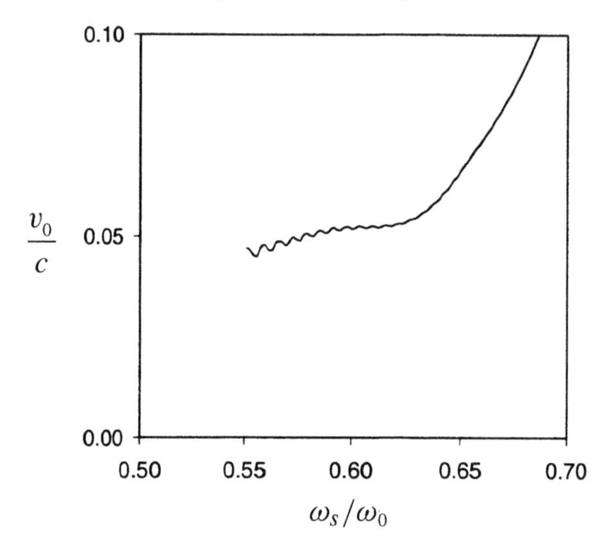

Fig. 11.7. Convective gain contour showing the threshold gain contour $C_G = \exp 2\pi$ for SRBS from a linear density ramp underdense to forward-scattered light.

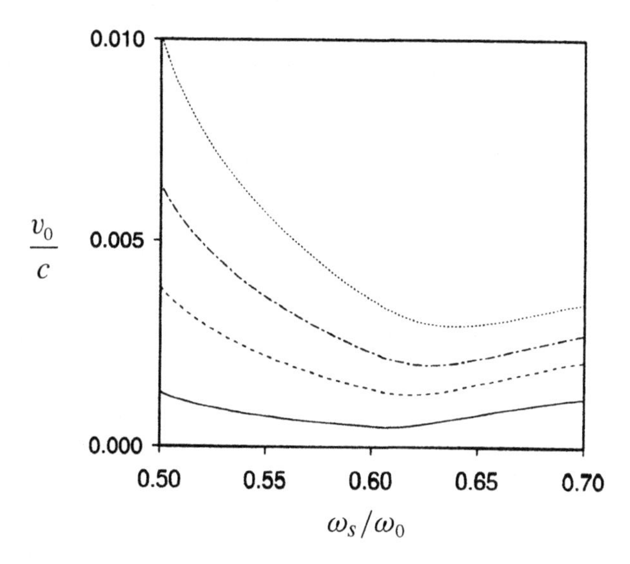

Fig. 11.8. Thresholds predicted by (11.29) for SRS from plasmas with $Z = 10$ (——), 30 (– – –), 50 (— · —) and 79 (· · · · · ·).

Raman resonances discussed above. Figure 11.9 shows absolute Raman growth rates as a function of ω_s/ω_0 for a $Z = 10$ plasma, with the corresponding thresholds plotted in Fig. 11.10 which also shows thresholds for a $Z = 79$ plasma. Raman scattering from quarter-critical density $n_c/4$ shows the most rapid growth. Figure 11.10 shows some differences between plasmas with low and high Z. For

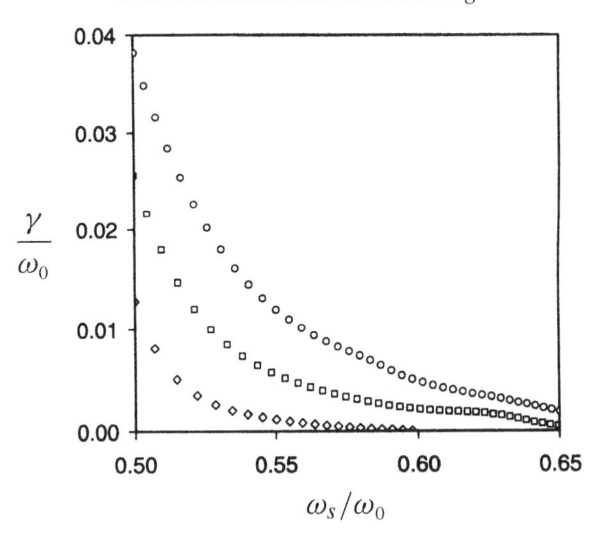

Fig. 11.9. Absolute SRS growth rates versus scattered frequency for a $Z = 10$ plasma for $v_0/c = 0.1$ (○), 0.07 (□) and 0.04 (◇).

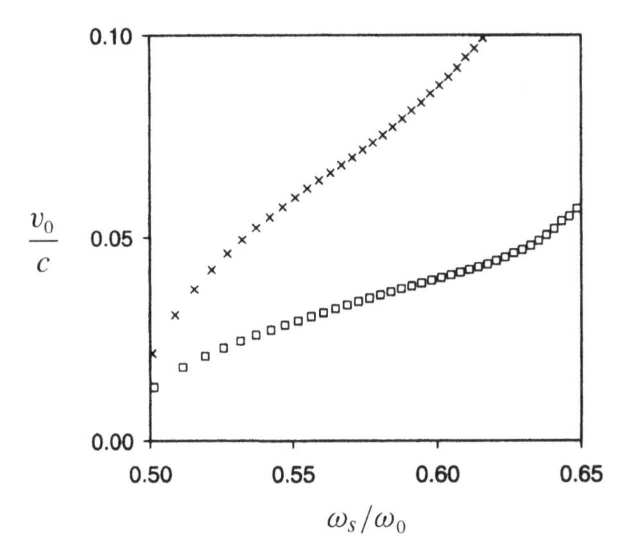

Fig. 11.10. Absolute SRS thresholds versus scattered frequency for $Z = 10$ (□) and $Z = 79$ (×) plasmas.

low Z (weak collisionality), thresholds increase only slowly with ω_s up to the onset of Landau cut-off. By contrast, absolute thresholds in the case of a strongly collisional plasma rise steeply with ω_s on account of the effects of damping on the waves within the feedback loop. This proves to be the dominant effect so that no Landau cut-off is evident for strongly collisional plasmas.

11.6 Radiation from Langmuir waves

In this section we consider a laser–plasma interaction in the presence of a density gradient so steep that WKBJ theory cannot be relied on to provide insights into the physics. The interaction in question is one in which Langmuir waves are coupled to the radiation field. Although longitudinal and transverse modes are always coupled in inhomogeneous plasmas, the coupling is generally weak for plasmas in thermal equilibrium. However, in plasmas in which a non-thermal spectrum of Langmuir waves is excited, steep density gradients may lead to significant radiation from the plasma over a narrow band around the plasma frequency.

Radiation by Langmuir waves first attracted attention in attempts at interpreting the characteristics of certain types of emission from the Sun, in particular type III solar radio noise. One of the striking characteristics of type III emission spectra is a drift in frequency over time, predominantly from high to low frequencies. Not only is the bandwidth narrow but in many instances a second harmonic band is observed. These characteristics along with the sudden onset of the emission, are consistent with a conjecture put forward by Wild (1950) to interpret type III emission. The sequence of events starts with bursts of energetic electrons produced in solar flares injected into the chromosphere, travelling outwards through the corona exciting Langmuir waves in the coronal plasma. The sudden onset of type III bursts is consistent with a threshold set by the collisional damping of Langmuir waves in the colder and denser chromospheric plasma. Likewise the bandwidth of the emission should reflect the narrow bandwidth of the Langmuir waves determined by Landau damping. Wild's original conjecture has since been supported by the direct observation of both electron beams and the Langmuir waves they generate, from satellite observations of type III bursts (see Lin *et al.* (1986)). The appearance of a second harmonic in some, though by no means all, recorded type III spectra is another signature pointing to Langmuir waves as the source of the emission. By and large more is known about the excitation phase in which suprathermal levels of Langmuir waves are generated, than the coupling phase, in which electrostatic energy is converted into radiation, with the generation of both a fundamental plasma line and its second harmonic.

In a very different corner of parameter space, high intensity laser interactions with dense plasmas afford another example of Langmuir waves coupling to the radiation field. These interactions lead to jets of energetic electrons which can excite suprathermal levels of Langmuir waves in the superdense plasma which in turn may radiate in the very steep density gradients present. The fact that this radiation is generated at the plasma frequency which, for sufficiently overdense plasmas, is far above that of the incident light suggests that plasma emission may, under suitable conditions, have potential as a source of XUV light.

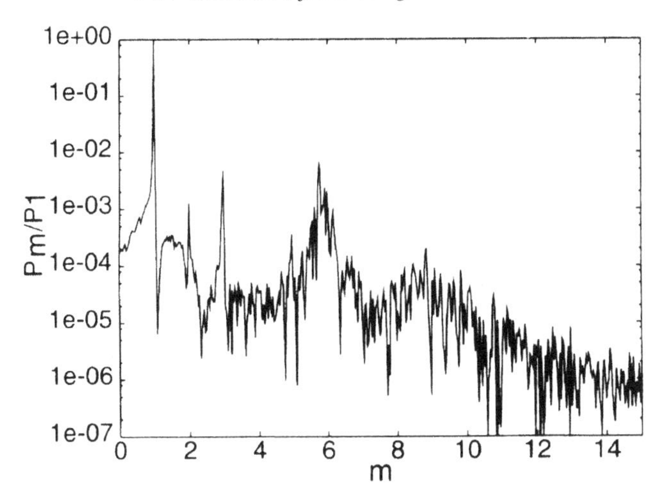

Fig. 11.11. Normalized frequency spectrum showing laser harmonics in reflection and plasma line emission for a plasma with $n/n_c = 30$ and $a_0 = 0.5$; $m = \omega/\omega_0$.

The problem of plasma emission was revisited by Boyd and Ondarza-Rovira (2000) in PIC simulations of the interaction of moderately intense laser light with slab plasmas of density up to 200 times critical density. Figure 11.11 shows the spectrum of back-reflected light from a plasma of density $n/n_c = 30$ first observed by Ondarza-Rovira (1996). In this spectrum the plasma line at $\omega/\omega_0 = (n/n_c)^{1/2} \simeq 5.5$ appears as a dominant feature against the background of harmonics of the incident laser light, comparable in intensity to the third laser harmonic. Moreover, the plasma line is in reality a broad feature in the spectrum reflecting the fact that in practice there will be some plasma emission at normalized frequencies up to $(n/n_c)^{1/2}$. However, at lower densities the intensity is correspondingly lower and is swamped by low harmonics of the incident laser light. The spectrum in Fig. 11.11 shows that as the harmonic line spectrum weakens beyond $m = 5$ the broad plasma line becomes increasingly dominant. There is no clear indication of a second harmonic of the plasma line at $2\omega_p \simeq 11\omega_0$, due in part to the fact that this is a weak feature for the parameters chosen and in part to the dominance of the broad feature that appears in the spectrum on the blue side of the plasma line centred about $\omega \simeq 9\omega_0$ with intensity about an order of magnitude below that of the plasma line. This unexpected spectral detail turns out to be a robust feature in the simulations. Figure 11.12 reproduces the reflected spectrum from a plasma of density $n/n_c = 200$, again showing a feature at $\omega \simeq 1.5\omega_p$ with intensity about an order of magnitude weaker than the plasma line. In addition a second harmonic of the plasma line appears in the spectrum with a weak third harmonic, shifted slightly to the blue. Surprisingly, additional lines

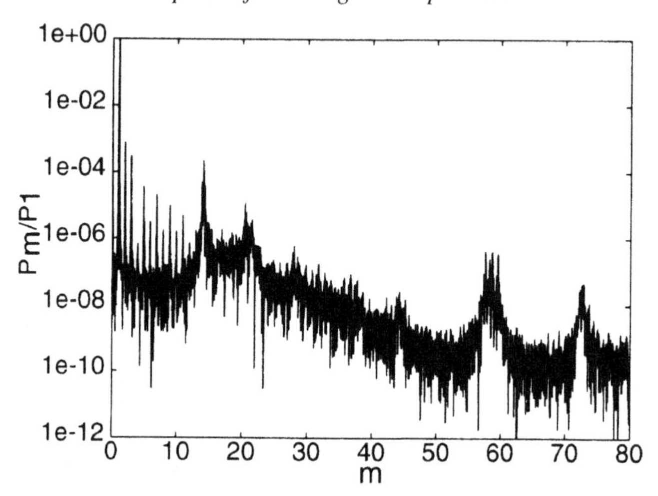

Fig. 11.12. Reflected harmonic spectrum for $n/n_c = 200$ and $a_0 = 0.5$.

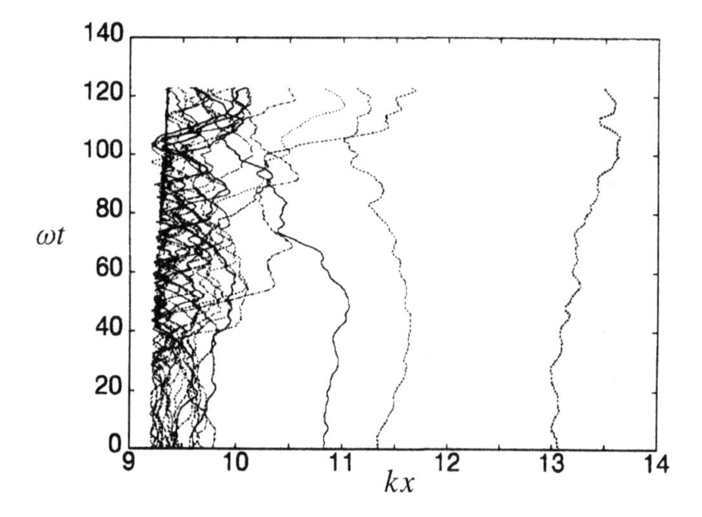

Fig. 11.13. Electron trajectories for $n/n_c = 100$ and $a_0 = 0.5$.

appear at approximately fourth and fifth harmonics of the plasma line. The PIC simulations showed clear evidence of electrons of relativistic energy from direct forward acceleration by the $\mathbf{v} \times \mathbf{B}$ force at high intensities.

Figure 11.13 shows electron trajectories with electron velocities spread over a wide band and moreover, evidence of beam-like behaviour, with electrons penetrating into the dense plasma. Across this region of penetration, Langmuir waves excited by these electrons are detected. The simulations show strongly driven Langmuir waves at the front edge of the plasma slab and this localization of the

Langmuir spectrum is important for the subsequent coupling of the plasma waves to the radiation field by means of the localized density gradient in the (perturbed) peak density region.

Plasma emission from laser-produced plasmas has been observed by Teubner *et al.* (1997) who detected a plasma line and a second harmonic although in fact the harmonic appeared at a frequency a little over 1.6 times the fundamental. They attribute the plasma line to surface emission and interpret the harmonic as due to radiation from Langmuir plasmons in the interior where the plasma density is assumed to be lower. On these grounds they expect the harmonic line to be more intense than the fundamental. However, in the simulations by Boyd and Ondarza-Rovira the opposite is the case, with the second harmonic typically between two and four orders of magnitude weaker than the fundamental. If one recalls that the line at $\sim 1.5\,\omega_p$ is between two and three orders of magnitude more intense than the second harmonic it is possible that it is this feature which has been detected by Teubner *et al.* given that the relative frequencies of the lines they observe are approximately 1:1.6.

Other aspects of the emission have been reported by Lichters, Meyer-ter-Vehn and Pukhov (1998) in simulations in which second and third harmonics of the plasma line were seen but not the fundamental. Spectra were recorded after the laser pulse had finished so that no laser harmonics are present. The absence of any effect in the spectrum in the region of the feature found by Boyd and Ondarza-Rovira suggests that the presence of the plasma line is critical for its appearance.

11.7 Effects in bounded plasmas

Boundedness and inhomogeneity in a sense go hand in hand since a bounded plasma is always inhomogeneous to some degree. Any attempt to classify modes as was done in Chapter 6 would be to miss an essential point about bounded, inhomogeneous plasmas where the boundary and the structure of the density profile are integral in determining wave characteristics. Not only do plasma boundaries modify the dispersion characteristics of modes within the plasma (the 'bulk' modes) but distinct modes may propagate along the surface of the plasma ('surface' waves) in addition to bulk modes.

11.7.1 Plasma sheaths

We look next at the inhomogeneous layer that is formed where plasma is in contact with a material boundary. Close to any wall or at any surface of contact with plasma, a *sheath* is formed across which electron and ion densities no longer balance. The reason for this stems from the very much greater mobility of electrons over ions (see Section 8.2.1). The net flux of electrons means that any surface in contact with a plasma rapidly acquires a negative charge. The potential resulting from this then acts to reduce the electron flux and enhance the flux of positive ions to the surface until a stage is reached where the two balance and the net current to the surface disappears. In this steady state the surface potential is known as the *floating potential.*

Sheath dynamics is one of the classical problems of plasma physics having been identified and largely resolved by Langmuir. It requires the solution of Poisson's equation along with the dynamical equations for both ions and electrons with suitably chosen boundary conditions. The plasma parameters are strongly inhomogeneous across the sheath and inhomogeneity, added to the non-linearity of the governing equations, means that a general solution can only be found numerically. Exceptionally, in plane geometry an approximate analytical solution is possible. The key approximation lies in the distinct assumptions made about electron and ion dynamics. For the electron fluid, the pressure gradient is dominant over the momentum term while for the ion fluid the reverse is the case, i.e. the ions are cold. With these assumptions we may integrate the respective equations of motion for the two fluids. For the electrons

$$n_{\mathrm{e}}(x) = n_0 \exp[eV(x)/k_{\mathrm{B}}T_{\mathrm{e}}] \tag{11.48}$$

where $V(x)$ is the electrostatic potential and we have made use of the boundary condition $V(x \to \infty) = 0$ and $n_{\mathrm{e}}(x \to \infty) = n_0$. Integrating the ion momentum equation and the continuity equation gives, for a hydrogen plasma,

$$n_{\mathrm{i}}(x) = n_0 \left[1 - \frac{2eV(x)}{m_{\mathrm{i}} u_{0\mathrm{i}}^2}\right]^{-1/2} \tag{11.49}$$

where $u_{0\mathrm{i}} = u_{\mathrm{i}}(x \to \infty)$. The spatial variation of electron and ion densities across the sheath is illustrated in Fig 11.14. If we now make use of (11.48) and (11.49) in Poisson's equation we find

$$\frac{\mathrm{d}^2 V(x)}{\mathrm{d}x^2} = \frac{n_0 e}{\varepsilon_0}\left[\exp\left(\frac{eV(x)}{k_{\mathrm{B}}T_{\mathrm{e}}}\right) - \left(1 - \frac{2eV(x)}{m_{\mathrm{i}} u_{0\mathrm{i}}^2}\right)^{-1/2}\right] \tag{11.50}$$

This non-linear equation is the *plasma sheath equation* in plane geometry. Before integrating (11.50) it helps to rewrite it in terms of dimensionless variables

$$\phi = -\frac{eV}{k_{\mathrm{B}}T_{\mathrm{e}}} \qquad M = \frac{u_{0\mathrm{i}}}{(k_{\mathrm{B}}T_{\mathrm{e}}/m_{\mathrm{i}})^{1/2}} \qquad \xi = \frac{x}{\lambda_{\mathrm{D}}}$$

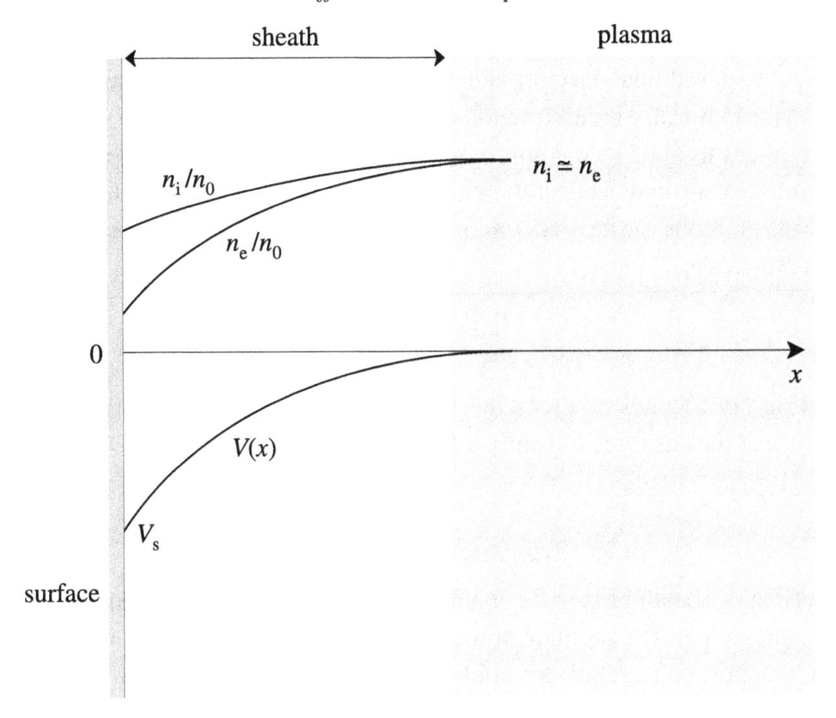

Fig. 11.14. Sheath formation at a plasma boundary.

With these definitions (11.50) reads

$$\frac{d^2\phi}{d\xi^2} = \left(1 + \frac{2\phi}{M^2}\right)^{-1/2} - e^{-\phi} \tag{11.51}$$

Multiplying by $d\phi/d\xi$ allows us to integrate (11.51) once to give

$$\frac{1}{2}\left(\frac{d\phi}{d\xi}\right)^2 = M^2\left[\left(1 + \frac{2\phi}{M^2}\right)^{1/2} - 1\right] + (e^{-\phi} - 1) \tag{11.52}$$

Note that (11.52) effectively determines $\phi(\xi)$ in terms of the 'Mach number' M. A second integration can only be done numerically unless we make an additional approximation. If we confine our attention to the plasma side of the sheath where ϕ is small we may approximate (11.52) by

$$\left(\frac{d\phi}{d\xi}\right)^2 = \left(1 - \frac{1}{M^2}\right)\phi^2 \tag{11.53}$$

which has a monotonic solution provided $M^2 > 1$. In other words

$$u_{0i} > \left(\frac{k_B T_e}{m_i}\right)^{1/2} \tag{11.54}$$

a result due to Bohm (1949) and known as the *Bohm sheath criterion.* A sheath forms at a material boundary provided the streaming velocity of ions entering the region close to a wall exceeds the ion acoustic velocity c_s.

If we return to (11.51) and approximate conditions within the sheath proper by supposing that we may neglect the electron contribution (this would only be valid for a wall or probe surface at a high enough negative potential, $|\phi| \gg 1$) then (11.51) reduces to

$$\frac{\mathrm{d}^2\phi}{\mathrm{d}\xi^2} \simeq \frac{M}{(2\phi)^{1/2}} \tag{11.55}$$

Integrating (11.55) twice across the sheath, assuming $\mathrm{d}\phi/\mathrm{d}\xi = 0$ at the edge, confirms that the sheath is typically a few Debye lengths thick.

11.7.2 Langmuir probe characteristics

This outline of sheath properties enables us to understand and interpret Langmuir probe characteristics. Langmuir probes provide reliable electron temperature and density diagnostics in relatively cool, low-density plasmas. The probe itself is a small metal electrode – cylindrical, spherical or in the shape of a disk – inserted into the plasma. The sheath that envelops the probe shields the plasma from the probe potential.

The essence of the Langmuir probe technique is to monitor the current to the probe as the probe voltage changes. Assuming that current is positive when it flows out from the probe, ion current drawn to the probe will be negative. The probe characteristic is shown in Fig. 11.15 as a current–voltage plot. For a potential more negative than the floating potential V_s the electron contribution to the currrent drops off until the probe draws only ion current given by the Bohm value, i.e.

$$I_{is} = \frac{1}{4} n_{i0} e \langle u_{0i} \rangle A = \frac{1}{2} n_{is} e A \left(\frac{2k_B T_e}{\pi m_i} \right)^{1/2} \tag{11.56}$$

Here n_{is} denotes the plasma ion density at the edge of the sheath and A the surface area of the probe. If we choose the sheath edge to be the point at which $u_{0i} \equiv 2(2k_B T_e/\pi m_i)^{1/2}$ and with a potential $V_s \simeq -k_B T_e/2e$ relative to the plasma then $n_{is} \simeq n_{es}$ so that

$$n_{is} = n_0 \exp(eV_s/k_B T_e) \simeq 0.6 n_0$$

The Bohm ion saturation current I_i^B is therefore

$$I_i^B \simeq 0.24 n_0 e A \left(\frac{k_B T_e}{m_i} \right)^{1/2} \tag{11.57}$$

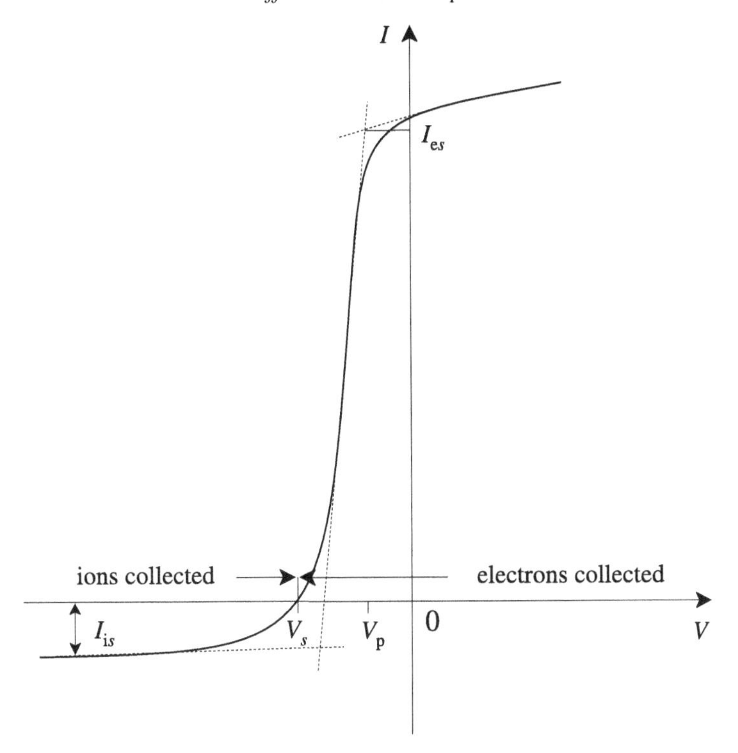

Fig. 11.15. Langmuir probe characteristic showing the variation of the probe current with probe potential.

Thus once the electron temperature is known, the Bohm saturation current determines the plasma density.

As V becomes less negative with respect to the plasma, energetic electrons from the tail of the distribution are collected by the probe until, at the floating potential V_s, electron and ion currents cancel one another. A further increase in V leads to a steep rise in electron current which eventually saturates at *space potential*, the potential of the plasma $V_p \simeq 0$.

Consider the Langmuir characteristic in the region $V_s < V < V_p$ in which electron current is drawn. Over this region the electron sheath shields the probe from electrons other than those with sufficient energy to overcome the potential barrier. The electron current to the probe is then the random current, reduced by the Boltzmann factor, i.e.

$$I_e = \frac{n_0 e A}{2} \left(\frac{2k_B T_e}{\pi m_e} \right)^{1/2} \exp\left[\frac{eV}{k_B T_e} \right] \tag{11.58}$$

so that the total current drawn to the probe is

$$I = I_e + I_i^B \tag{11.59}$$

From the slope of the characteristic we can determine the electron temperature. From (11.58), (11.59)

$$\frac{\mathrm{d}I}{\mathrm{d}V} = \frac{e}{k_{\mathrm{B}} T_{\mathrm{e}}} I_{\mathrm{e}} + \frac{\mathrm{d}I_{\mathrm{i}}}{\mathrm{d}V}$$

If we use the region of the characteristic where the slope is greatest we may disregard $\mathrm{d}I_{\mathrm{i}}/\mathrm{d}V$ so that

$$T_{\mathrm{e}} = \frac{e}{k_{\mathrm{B}}}(I - I_{\mathrm{i}}^{B}) / \frac{\mathrm{d}I}{\mathrm{d}V} \tag{11.60}$$

This can be determined directly from the characteristic by measuring its slope as shown in Fig. 11.15.

The Langmuir probe technique is a versatile diagnostic in plasmas of moderate density and temperature. If the density is high enough for electron and ion mean free paths to become comparable to probe dimensions the model is no longer valid.

Exercises

11.1 Consider a light wave propagating along Oz in a plasma of increasing density $n(z)$. By appealing to the conservation of energy flux show that $|E(z)| = E_0/\varepsilon^{1/4}$ where E_0 denotes the amplitude of the electric field in vacuo and ε is the plasma dielectric function. Compare this result with the WKBJ field amplitude. How is $|B(z)|$ related to B_0?

11.2 Show that the validity condition for WKBJ solutions to the wave equation may be expressed as

$$\frac{1}{k^2}\left|\frac{3}{4}\left(\frac{1}{n^2}\frac{\mathrm{d}n}{\mathrm{d}z}\right)^2 - \frac{1}{2n^3}\frac{\mathrm{d}^2 n}{\mathrm{d}z^2}\right| \ll 1$$

11.3 Express $\partial \mathbf{k}/\partial t = -\boldsymbol{\nabla}\omega$ as

$$\frac{\partial \omega}{\partial \mathbf{r}} + \nabla \mathbf{k} \cdot \frac{\partial \omega}{\partial \mathbf{k}} + \frac{\partial \mathbf{k}}{\partial t} = 0$$

and by considering the dispersion relation as a function of $\mathbf{k}$, $\mathbf{r}$ and t, i.e. $D\left(\omega\left(\mathbf{k}, \mathbf{r}, t\right), \mathbf{k}, \mathbf{r}, t\right) = 0$ so that

$$\frac{\partial D}{\partial \mathbf{k}} + \frac{\partial D}{\partial \omega}\frac{\partial \omega}{\partial \mathbf{k}} = 0 \qquad \frac{\partial D}{\partial \mathbf{r}} + \frac{\partial D}{\partial \omega}\frac{\partial \omega}{\partial \mathbf{r}} = 0 \qquad \frac{\partial D}{\partial t} + \frac{\partial D}{\partial \omega}\frac{\partial \omega}{\partial t} = 0$$

establish the set of equations (11.8) that determine ray trajectories.

11.4 Verify the steps leading to the reflection coefficient R given by (11.12). Note that the representation for $\mathrm{Ai}(\zeta)$ valid for $\arg \zeta = \pi$ is

$$\mathrm{Ai}(\zeta) \sim \frac{1}{2}\pi^{-1/2}\zeta^{-1/4}\left[\exp\left(-\frac{2}{3}\zeta^{3/2}\right) + i\exp\left(\frac{2}{3}\zeta^{3/2}\right)\right]$$

11.5 Consider an electromagnetic wave propagating along the density gradient in a plasma in which the electron density varies as $n(z) = n_c(z/L)$. From the analysis outlined in Section 11.2.1, the electric field is determined by the Airy function $\mathrm{Ai}(\zeta)$ shown in Fig. 11.2.

By matching this electric field to the vacuum electric field at $z = 0$, show that

$$E(\zeta) = \sqrt{2\pi}\,(k_0L)^{1/6} E_0 e^{i\phi}\,\mathrm{Ai}(\zeta)$$

where E_0 is the amplitude of the vacuum field and ϕ is a phase factor. The Airy function $\mathrm{Ai}(\zeta)$ has a maximum value 0.535 at $\zeta = -1$ corresponding to $z = L\left\{1 - (k_0L)^{-2/3}\right\}$. Show that

$$\left|\frac{E_{\max}}{E_0}\right|^2 \simeq 1.8(k_0L)^{1/3}$$

which provides an estimate of the swelling of the electric field as the wave approaches the cut-off.

Estimate the swelling of the WKBJ electric field by representing $\varepsilon_{\min}$ as the value of ε averaged over a half wavelength near cut-off. Show that $\varepsilon_{\min} \simeq (\pi c/\omega L)^{2/3}$. Hence show that $|E_{\max}/E_0|^2 \simeq 1.4(k_0L)^{1/3}$.

In contrast to the WKBJ approximation in which $k_c = 0$ it is possible to define a local incident wavenumber k_{A} from the Airy solution. Show that $k_{\mathrm{A}} \simeq 0.6k^{2/3}L^{-1/3}$.

11.6 Consider a light wave incident obliquely on a plasma slab. Take Oyz to be the plane of incidence and the angle of incidence $\theta = \cos^{-1}(\hat{\mathbf{k}} \cdot \hat{\mathbf{z}})$. In the case of oblique incidence we have to distinguish between two polarization states, S and P, in which $\mathbf{E}$ is respectively perpendicular to, and coplanar with, the plane of incidence. For an S-polarized wave the wave electric field gives rise to electron oscillations in the x-direction, along which the density is uniform. The electric field of a P-polarized wave on the other hand causes electrons to oscillate across regions of non-uniform density; in this case the wave is no longer purely electromagnetic.

For S-polarized light show that the results of Exercise 11.5 carry over after making allowance for the fact that reflection now takes place at $\varepsilon(z) = \sin^2\theta$, i.e. reflection takes place at a density below critical, where $n_e = n_c \cos^2\theta$.

By contrast, a P-polarized wave at its cut-off has its electric field aligned along the density gradient and, provided the separation between cut-off and resonance is not too large, some fraction of the incident electric field can tunnel through to the resonance and excite a Langmuir wave. Interactions between resonant electrons and the Langmuir wave excited in this way result in *resonant absorption* of the incident radiation. To estimate the electric field $\mathbf{E}_z$ along the direction of the density gradient near the critical density, first express $E_z(z) = [B(z)\sin\theta]/[c\varepsilon(z)] = E_\mathrm{d}(z)/\varepsilon(z)$. To determine the resonant field one has to find $B(L)$. A simple way of doing this uses a physical argument given by Kruer (1988), which represents $B(L)$ by its value at the turning point, $B(L\cos^2\theta)$ multiplied by a factor to allow for the exponential decay from cut-off to resonance. Show that

$$B(L\cos^2\theta) \simeq \frac{0.9E_0}{c}\left(\frac{c}{\omega L}\right)^{1/6}$$

where E_0 is the intensity of the electric field in vacuo. The decay of the field beyond cut-off may be represented by a factor $e^{-\beta}$ where $\beta = (1/c)\int_{L\cos^2\theta}^{L}\left(\omega_\mathrm{p}^2 - \omega^2\cos^2\theta\right)^{1/2}\mathrm{d}z$. Evaluate β and show that

$$E_\mathrm{d}(L) \simeq 0.9E_0\left(\frac{c}{\omega L}\right)^{1/6}\sin\theta\exp\left(\frac{-2\omega L\sin^3\theta}{3c}\right)$$

Defining $\tau = (\omega L/c)^{1/3}\sin\theta$, show that this may be written

$$E_\mathrm{d}(L) = \frac{E_0}{(2\pi\omega L/c)^{1/2}}\phi(\tau)$$

where the shape function $\phi(\tau) \simeq 2.3\tau\exp(-2\tau^3/3)$.

Note that $E_\mathrm{d}(L) \to 0$ as $\tau \to 0$, i.e. $\theta \to 0$. Similarly for large τ, $E_\mathrm{d}(L) \to 0$ corresponding to the cut-off occurring at too low a density for the wave to tunnel effectively to the critical density surface. Between these limits there is an optimum angle of incidence given by $(\omega L/c)^{1/3}\sin\theta \simeq 0.8$. The shape function found from this heuristic argument provides a good approximation to the numerical solution of the wave equation (see Denisov (1957)).

11.7 Following linear conversion to Langmuir waves the electrostatic field energy is resonantly transferred to electrons by means of Landau damping. Representing damping phenomenologically we may write

$$\varepsilon(z) = \left(1 - \frac{z}{L}\right) + \frac{i\nu}{\omega}\left(\frac{z}{L}\right)$$

Show that the energy flux that is resonantly absorbed, $I_\mathrm{RA} = f_\mathrm{RA}I_0$, is determined by

$$I_\mathrm{RA} = \frac{1}{2}\varepsilon_0\nu\int_{z_c}^{z_r}\frac{E_\mathrm{d}^2(z)}{|\varepsilon(z)|^2}\,\mathrm{d}z$$

where z_c, z_r denote C- and R-points. Show that this reduces to (Kruer (1988))

$$I_{\mathrm{RA}} = \frac{\phi^2}{4}\left(\frac{\varepsilon_0}{2}cE_0^2\right) = f_{\mathrm{RA}}\left(\frac{\varepsilon_0}{2}cE_0^2\right)$$

so that $I_{\mathrm{RA}} = \frac{1}{4}\phi^2 I_0$. Simulations of absorption of P-polarized light confirm that resonance absorption of about 40% takes place.

11.8 The frequency and wavenumber matching conditions for forward and backward Raman scattering are $\omega_0 = \omega_{\mathrm{L}} + \omega_s$, $k_0 = k_{\mathrm{L}} \pm k_s$, where the plus sign denotes SRFS and the minus, SRBS. Show that the wavenumber matching conditions may be written

$$\frac{\omega_0}{c}\left(1 - \frac{n}{n_{\mathrm{c}}}\right)^{1/2} = k_{\mathrm{L}} \pm \frac{\omega_s}{c}\left(1 - \frac{\omega_0^2}{\omega_s^2}\frac{n}{n_{\mathrm{c}}}\right)^{1/2}$$

In the limit $k_{\mathrm{L}}\lambda_{\mathrm{D}} \ll 1$ show that $k_s \simeq (\omega_0/c)(1 - 2\omega_{\mathrm{p}}/\omega_0)^{1/2}$ so that

$$\left(1 - \frac{n}{n_{\mathrm{c}}}\right)^{1/2} \simeq \frac{k_{\mathrm{L}}c}{\omega_0} \pm \left(1 - 2\sqrt{\frac{n}{n_{\mathrm{c}}}}\right)^{1/2}$$

From this it is clear that SRS matching conditions can only be satisfied for densities up to quarter-critical, $n_{\mathrm{c}}/4$. At this density $k_s \simeq 0$ so that $k_{\mathrm{L}} \sim k_0 = (\sqrt{3}\omega_0/2c)$. For densities much below $n_{\mathrm{c}}/4$, show that

$$k_{\mathrm{L}} \simeq \frac{\omega_0}{c} \mp \frac{\omega_0}{c}\left(1 - \sqrt{\frac{n}{n_{\mathrm{c}}}}\right)$$

Thus for SRBS at very low densities $k_{\mathrm{L}} = 2k_0$, while for SRFS, $k_{\mathrm{L}} = \omega_{\mathrm{p}}/c$.

11.9 Carry out a first-order perturbation analysis on the electron fluid equations to obtain the SRS coupled equations (11.23), (11.24) for a homogeneous plasma. Fourier analyse these equations to find the SRS dispersion relation (11.25).

Show that the maximum homogeneous Raman growth rate is given by (11.26). Allowing for damping of the Langmuir wave and the scattered light wave confirm that the Raman growth rate is now given by (11.29).

11.10 Recover (11.33) and (11.34), the second-order SRS equations for an inhomogeneous plasma, from (11.32).

Show that the WKBJ equations for an inhomogeneous plasma are given by (11.36) and (11.37). By combining these equations establish (11.38). Show that the convective gain is given by (11.45). In the case of a linear density profile with scale length L confirm that the SRBS threshold is determined by (11.47).

11.11 A plasma is contained in the right half-space $x > 0$, i.e. a density step from $n = 0$ to $n = n_0$ appears at $x = 0$. Consider a surface wave in which the electric field is electrostatic with the potential

$$\phi(x, y, t) = \phi_k(x) \exp i(ky - \omega t) + \text{c.c.}$$

Show that $\phi_k(x) = \phi_s \exp(-|k|x)$ where ϕ_s denotes the potential at the boundary. Applying boundary conditions at $x = 0$ show that the dispersion relation for the electrostatic surface wave is $\omega^2 = \omega_{\mathrm{p}}^2/2$.

Show that in the case of electromagnetic surface waves in the short wavelength limit, the surface wave dispersion relation becomes

$$\frac{k^2c^2}{\omega^2} = \frac{\omega_{\mathrm{p}}^2 - \omega^2}{\omega_{\mathrm{p}}^2 - 2\omega^2}$$

11.12 Tonks and Langmuir (1929) found that in a cylindrical discharge plasma the principal resonance appeared at $\omega_{\mathrm{p}}/\sqrt{2}$. Subsequently Romell (1951) observed in addition a series of weaker resonances, later studied by Dattner (1957, 1963) and others and known as *Tonks–Dattner resonances*.

Consider a slab of plasma of uniform density n_0 and thickness $2L$, unbounded in the other two dimensions. Show that the set of resonant frequencies ω_m is given by

$$\omega_m^2 = \omega_{\mathrm{p}}^2 \left[1 + 3\,(m + 1/2)^2\,\pi^2 \frac{\lambda_{\mathrm{D}}^2}{L^2}\right]$$

where λ_{D} is the Debye length. This result predicts a spacing $\left(\Delta\omega_m/\omega_{\mathrm{p}}\right)$ which is an order of magnitude too small. To fix this one has to take account of plasma inhomogeneity.

Starting from the fluid model of a warm plasma (6.86)–(6.88) with a perturbation scheme defined by

$$\left.\begin{aligned} n(x,t) &= n_0(x) + n_1(x)e^{-i\omega t} & E(x,t) &= E_0(x) + E_1(x)e^{-i\omega t} \\ v(x,t) &= \quad 0 \quad\; + v_1(x)e^{-i\omega t} & p(x,t) &= p_0(x) + p_1(x)e^{-i\omega t} \end{aligned}\right\}$$

show that, to first order, the perturbation in electron density is determined by

$$\frac{\mathrm{d}^2 n_1}{\mathrm{d}x^2} + k^2(x) n_1 = \frac{e}{3mV_{\mathrm{e}}^2}\left[\frac{\mathrm{d}}{\mathrm{d}x}(n_1 E_0) + E_1 \frac{\mathrm{d}n_0}{\mathrm{d}x}\right]$$

If we denote the scale length of the inhomogeneity by L show that the terms on the right-hand side are typically $(1/kL)$ times those on the left so that

$$\frac{d^2 n_1}{dx^2} + \frac{1}{3\lambda_{D0}^2}\left[\frac{\omega^2 - \omega_p^2(x)}{\omega_{p0}^2}\right] n_1 \simeq 0 \qquad \text{(E11.1)}$$

where ω_{p0}, λ_{D0} denote the electron plasma frequency and the Debye length at the centre of the plasma. This model shows that electron plasma waves can now propagate in the region between the boundary and a point x_c at which the plasma density becomes critical, i.e. $n(x_c) = n_c$. Since we have already assumed that the density scale length is much greater than the electron plasma wavelength we may use a WKBJ approximation to find the eigenvalues

$$\int_0^{x_c} k(x)\mathrm{d}x = \left(m + \frac{3}{4}\right)\pi \qquad m = 0, 1, 2, \ldots \qquad \text{(E11.2)}$$

Useful as this electrostatic model is, it too fails when it comes to interpreting the spectra observed experimentally. Work by Parker, Nickel and Gould (1964), in which neither the electrostatic assumption nor the WKBJ approximation was made, determined the spectrum of resonances numerically for a cylindrical plasma column and compared these results with their measured spectra. Refer to their paper for the satisfactory agreement they found between theory and experiment.

11.13 By integrating (11.55) twice across the sheath and taking $\mathrm{d}\phi/\mathrm{d}\xi = 0$ at the edge show that the sheath is typically a few Debye lengths thick.

12

The classical theory of plasmas

12.1 Introduction

In this final chapter an attempt is made to sketch the classical mathematical structure underlying the various theoretical models which have been used throughout the book. The knowledge of where a particular model fits within the overall picture helps us both to understand the relationship to other models and to appreciate its limitations. Of course, we have touched upon these relationships and limitations already so the task remaining is to construct the framework of classical plasma theory and show how it all fits together.

Since collisional kinetic theory is the most comprehensive of the models that we have discussed we could begin with it as the foundation of the structure we wish to build. Indeed, we shall demonstrate its pivotal position. This would, however, be less than satisfactory for two reasons. The first and basic objection is that, so far, we have merely assumed a physically appropriate model for collisions. We have not carried out a mathematical derivation of the collision term. In fact, enormous effort has gone into this task though we shall present only a brief resumé. In doing so, we shall show how the separation of the effects of the Coulomb force into a macroscopic component (self-consistent field) and a microscopic component (collisions) appears quite naturally in the mathematical derivation of the collisional kinetic equation. This is the second reason for starting at a more fundamental level than the collisional kinetic equation itself.

To lighten the burden of the mathematical analysis we have, wherever convenient, restricted calculations to a one-component (electron) plasma. The ions, however, are not ignored but treated as a uniform background of positive charge. Electrons interact with the ions but this appears as a 'field' rather than an interparticle interaction. Extensions of important formulae to multi-component plasmas are given at appropriate places in the text.

12.2 Dynamics of a many-body system

Going back to first principles and assuming, for simplicity, that the motion of the N electrons in a plasma obeys the laws of classical mechanics, we may write down N Newtonian equations

$$m\ddot{\mathbf{R}}_i(t) = \mathbf{F}_i(t) \qquad (i = 1, 2, \ldots, N) \tag{12.1}$$

in which m is the electron mass, $\mathbf{R}_i(t)$ is the position of the ith electron at time t and $\mathbf{F}_i(t)$is the force it experiences at that position and time. Formal solutions of these N equations are

$$\begin{aligned} \dot{\mathbf{R}}_i(t) &= \dot{\mathbf{R}}_i(0) + \frac{1}{m}\int_0^t \mathbf{F}_i(t')\,\mathrm{d}t' \\ \mathbf{R}_i(t) &= \mathbf{R}_i(0) + \dot{\mathbf{R}}_i(0)t + \frac{1}{m}\int_0^t \mathrm{d}t' \int_0^{t'} \mathbf{F}_i(t'')\,\mathrm{d}t'' \\ & \qquad (i = 1, 2, \ldots, N) \end{aligned} \tag{12.2}$$

In principle, this completely determines the motion of the plasma. In practice, it is impossible to carry out the integrations in (12.2) since $\mathbf{F}_i$ is, in general, a function of the positions and velocities of *all* the plasma particles; the N vector equations (12.1) are coupled in a very complicated way. Moreover, even if the forces were sufficiently simple that one could do the integrations, there would never be sufficient information to supply the required $2N$ initial conditions $\mathbf{R}_i(0)$, $\dot{\mathbf{R}}_i(0)$ $(i = 1, 2, \ldots, N)$.

Turning now to a description of the plasma in terms of distribution functions, the *exact N-particle distribution function* $f_N^{\text{ex}}(\mathbf{r}_1, \mathbf{v}_1, \ldots, \mathbf{r}_N, \mathbf{v}_N, t)$ is given by

$$f_N^{\text{ex}}(\mathbf{r}_1, \mathbf{v}_1, \ldots, \mathbf{r}_N, \mathbf{v}_N, t) = \prod_{i=1}^{N} \delta[\mathbf{r}_i - \mathbf{R}_i(t)]\delta[\mathbf{v}_i - \dot{\mathbf{R}}_i(t)] \tag{12.3}$$

since (12.2) formally prescribes the position and velocity of each particle. Taking the partial derivative of f_N^{ex} with respect to t and using (12.3) we get

$$\begin{aligned} \frac{\partial f_N^{\text{ex}}}{\partial t} &= \sum_{i=1}^{N}\left[\dot{\mathbf{R}}_i \cdot \frac{\partial f_N^{\text{ex}}}{\partial \mathbf{R}_i} + \ddot{\mathbf{R}}_i \cdot \frac{\partial f_N^{\text{ex}}}{\partial \dot{\mathbf{R}}_i}\right] \\ &= -\sum_{i=1}^{N}\left[\dot{\mathbf{R}}_i \cdot \frac{\partial f_N^{\text{ex}}}{\partial \mathbf{r}_i} + \ddot{\mathbf{R}}_i \cdot \frac{\partial f_N^{\text{ex}}}{\partial \mathbf{v}_i}\right] \end{aligned}$$

since $\partial\delta(x-y)/\partial y = -\partial\delta(x-y)\partial x$. However, f_N^{ex} is zero unless $\dot{\mathbf{R}}_i = \mathbf{v}_i$ and from (12.1) $\ddot{\mathbf{R}}_i = \mathbf{F}_i/m$ so this may be written

$$\frac{\partial f_N^{\text{ex}}}{\partial t} + \sum_{i=1}^{N}\left[\mathbf{v}_i \cdot \frac{\partial f_N^{\text{ex}}}{\partial \mathbf{r}_i} + \frac{\mathbf{F}_i}{m} \cdot \frac{\partial f_N^{\text{ex}}}{\partial \mathbf{v}_i}\right] = 0 \tag{12.4}$$

So far the description of plasma motion by (12.4) is equivalent to that given by (12.1). However, the lack of information concerning initial conditions leads us to seek a *statistical* interpretation of the distribution function. Since the initial conditions are unknown, probability considerations may be applied to them. Instead of the distribution function at $t = 0$ being zero everywhere except at a single point in the $6N$-dimensional *phase space* $(\mathbf{r}_1, \mathbf{v}_1, \ldots, \mathbf{r}_N, \mathbf{v}_N)$ it will be given by a smoother function $f_N(\mathbf{r}_1, \mathbf{v}_1, \ldots, \mathbf{r}_N, \mathbf{v}_N, t = 0)$ which is an *ensemble average* over the many possible (but unknown) initial starting points in phase space. As time evolves, each initial point traces out a locus in phase space, determined by the dynamics of the system, thus prescribing $f_N(t)$, which we define by the statement that $f_N(\mathbf{r}_1, \mathbf{v}_1, \ldots, \mathbf{r}_N, \mathbf{v}_N, t)\prod_{i=1}^{N} \mathrm{d}\mathbf{r}_i\, \mathrm{d}\mathbf{v}_i$ is the probability of finding the system within the volume element $\prod_{i=1}^{N} \mathrm{d}\mathbf{r}_i\, \mathrm{d}\mathbf{v}_i$ about the point $(\mathbf{r}_1, \mathbf{v}_1, \ldots, \mathbf{r}_N, \mathbf{v}_N)$ at time t. Replacing the exact distribution function f_N^{ex} by f_N in (12.4) then gives the *Liouville equation*

$$L_N f_N = 0 \tag{12.5}$$

where the *Liouville operator*

$$L_N \equiv \frac{\partial}{\partial t} + \sum_{i=1}^{N}\left[\mathbf{v}_i \cdot \frac{\partial}{\partial \mathbf{r}_i} + \frac{\mathbf{F}_i}{m} \cdot \frac{\partial}{\partial \mathbf{v}_i}\right] \tag{12.6}$$

The Liouville equation is the starting point for a statistical description of a many-body system and, for a classical system, is usually written

$$\frac{\partial f_N}{\partial t} + [f_N, H] = 0 \tag{12.7}$$

where H is the Hamiltonian of the system and the Poisson bracket

$$[f_N, H] \equiv \sum_{i=1}^{N}\left[\frac{\partial f_N}{\partial \mathbf{q}_i} \cdot \frac{\partial H}{\partial \mathbf{p}_i} - \frac{\partial f_N}{\partial \mathbf{p}_i} \cdot \frac{\partial H}{\partial \mathbf{q}_i}\right]$$

$\mathbf{q}_i$, $\mathbf{p}_i$ being the generalized coordinates and momenta. For the non-relativistic plasma that we consider

$$H = \sum_{i=1}^{N} \frac{p_i^2}{2m} + V(\mathbf{q}_1, \mathbf{q}_2, \ldots, \mathbf{q}_N, t) \tag{12.8}$$

with $\mathbf{q}_i = \mathbf{r}_i$, $\mathbf{p}_i = m\mathbf{v}_i$ and $\mathbf{F}_i = -\partial V/\partial \mathbf{r}_i$ from which the equivalence of (12.7) and (12.5) is easily demonstrated.

The aim now is to integrate (12.5) over most of the space coordinates $\mathbf{r}_i$ and velocities $\mathbf{v}_i$ to obtain equations for reduced distribution functions (containing less information) which we may hope to determine. The reduced distribution functions are defined by

$$f_s(\mathbf{r}_1, \mathbf{v}_1, \ldots, \mathbf{r}_s, \mathbf{v}_s, t) = \frac{N!}{(N-s)!} \int f_N \prod_{i=s+1}^{N} \mathrm{d}\mathbf{r}_i\, \mathrm{d}\mathbf{v}_i \tag{12.9}$$

where the normalization constant has been chosen because there are $N!/(N-s)!$ ways of choosing s electrons from a total of N. This choice introduces a second, rather more subtle, change in the nature of the distribution functions that we are seeking. The first change from f_N^{ex} to f_N acknowledged the fact that we cannot ever know the exact starting point in phase space so it makes more sense to consider a distribution of initial points $f_N(t = 0)$ leading to a corresponding distribution $f_N(t)$ at any later time. Of course, f_N is supposed to be chosen to be consistent with whatever information we do have about the plasma, but since such 'macroscopic' detail usually involves just one, or at most two, space and velocity coordinates that leaves much uncertainty about f_N. It is this indeterminate detail that we eliminate by integrating over most of the phase space coordinates. Now, by our choice of normalization constant in (12.9), we recognize that it makes no sense to talk about *specific* electrons, labelled $1, 2, \ldots, s$ being at $(\mathbf{r}_1, \mathbf{v}_1), (\mathbf{r}_2, \mathbf{v}_2), \ldots, (\mathbf{r}_s, \mathbf{v}_s)$, respectively, but only about the probability of finding (any) electrons at these coordinates since we have no way of distinguishing one electron from another. The reduced distribution functions defined by (12.9) are sometimes called *generic* distribution functions as opposed to the *specific* distribution functions which would be defined by the choice of unit normalization constant.

Integrating (12.5) over all but s spatial and velocity coordinates, assuming that f_N vanishes on the boundaries of phase space and that the only velocity-dependent forces are Lorentzian, we obtain

$$\frac{\partial f_s}{\partial t} + \sum_{i=1}^{s} \mathbf{v}_i \cdot \frac{\partial f_s}{\partial \mathbf{r}_i} + \frac{N!}{(N-s)!m} \sum_{i=1}^{s} \int \mathbf{F}_i \cdot \frac{\partial f_N}{\partial \mathbf{v}_i} \prod_{j=s+1}^{N} \mathrm{d}\mathbf{r}_j\, \mathrm{d}\mathbf{v}_j = 0 \tag{12.10}$$

Separating $\mathbf{F}_i$ into what we may call its internal component $\mathbf{F}_i^{\mathrm{int}}$ (the force due to all the other electrons) and external component $\mathbf{F}_i^{\mathrm{ext}}$ (the force due to the ions and any applied fields) it follows that

$$\frac{N!}{(N-s)!m} \int \mathbf{F}_i^{\mathrm{ext}} \cdot \frac{\partial f_N}{\partial \mathbf{v}_i} \prod_{j=s+1}^{N} \mathrm{d}\mathbf{r}_j\, \mathrm{d}\mathbf{v}_j = \frac{1}{m} \mathbf{F}_i^{\mathrm{ext}} \cdot \frac{\partial f_s}{\partial \mathbf{v}_i} \quad (i = 1, 2, \ldots, s) \tag{12.11}$$

since $\mathbf{F}_i^{\text{ext}}$ may be taken outside the integral. In the non-relativistic approximation considered here, the only internal forces are electrostatic

$$\mathbf{F}_i^{\text{int}} = \frac{e^2}{4\pi\varepsilon_0} \sum_{\substack{j=1 \\ j\neq i}}^{N} \frac{\mathbf{r}_i - \mathbf{r}_j}{|\mathbf{r}_i - \mathbf{r}_j|^3} = -\sum_{\substack{j=1 \\ j\neq i}}^{N} \frac{\partial \phi_{ij}}{\partial \mathbf{r}_i} \tag{12.12}$$

where

$$\phi_{ij} = \frac{e^2}{4\pi\varepsilon_0 |\mathbf{r}_i - \mathbf{r}_j|} \tag{12.13}$$

Hence

$$\begin{aligned} &\frac{N!}{(N-s)!m} \sum_{i=1}^{s} \int \mathbf{F}_i^{\text{int}} \cdot \frac{\partial f_N}{\partial \mathbf{v}_i} \prod_{j=s+1}^{N} \mathrm{d}\mathbf{r}_j\, \mathrm{d}\mathbf{v}_j \\ &= \frac{1}{m} \sum_{i=1}^{s} \sum_{\substack{j=1 \\ j\neq i}}^{s} \frac{\partial \phi_{ij}}{\partial \mathbf{r}_i} \cdot \frac{\partial f_s}{\partial \mathbf{v}_i} - \frac{1}{m} \sum_{i=1}^{s} \int \frac{\partial \phi_{is+1}}{\partial \mathbf{r}_i} \cdot \frac{\partial f_{s+1}}{\partial \mathbf{v}_i}\, \mathrm{d}\mathbf{r}_{s+1}\, \mathrm{d}\mathbf{v}_{s+1} \end{aligned} \tag{12.14}$$

where we have separated the sum over j in (12.12) into the first s terms and the remaining $(N - s)$ terms and then made use of the fact that all of the latter are identical. Substituting (12.11) and (12.14) into (12.10) gives the general equation for the reduced distribution functions. Using (12.6) this may be written

$$L_s f_s = \frac{1}{m} \sum_{i=1}^{s} \int \frac{\partial \phi_{is+1}}{\partial \mathbf{r}_i} \cdot \frac{\partial f_{s+1}}{\partial \mathbf{v}_i}\, \mathrm{d}\mathbf{r}_{s+1}\, \mathrm{d}\mathbf{v}_{s+1} \quad (s = 1, 2, \ldots, N-1) \tag{12.15}$$

In (12.15) L_s is the Liouville operator for s electrons and the right-hand side represents electron interactions. We can see immediately the fundamental problem with this set of equations. The equation for f_s contains f_{s+1} so that the system closes only with the Liouville equation (12.5) and no simplification has yet been achieved. This is not surprising since no approximations have been introduced thus far and the problem is therefore the one with which we started. The chain of equations represented by (12.15) is called the BBGKY hierarchy after Bogolyubov (1962), Born and Green (1949), Kirkwood (1946), and Yvon (1935).

To reduce the complexity of the theoretical description we want, in principle, some physical approximation enabling us to write f_{s+1} in terms of $f_1, f_2, \ldots, f_s$ for some small value of s, so obtaining a solvable set of equations – that is, a set of s equations, $s - 1$ of which may be used to eliminate $f_2, \ldots, f_s$ leaving a single (kinetic) equation for f_1. Even for $s = 2$ this is a formidable task. Putting $s = 1, 2$

in (12.15) the pair of equations is

$$\frac{\partial f_1}{\partial t} + \mathbf{v}_1 \cdot \frac{\partial f_1}{\partial \mathbf{r}_1} + \frac{\mathbf{F}_1^{\text{ext}}}{m} \cdot \frac{\partial f_1}{\partial \mathbf{v}_1} = \frac{1}{m} \int \frac{\partial \phi_{12}}{\partial \mathbf{r}_1} \cdot \frac{\partial f_2}{\partial \mathbf{v}_1} \, \mathrm{d}\mathbf{r}_2 \, \mathrm{d}\mathbf{v}_2 \tag{12.16}$$

$$\frac{\partial f_2}{\partial t} + \mathbf{v}_1 \cdot \frac{\partial f_2}{\partial \mathbf{r}_1} + \mathbf{v}_2 \cdot \frac{\partial f_2}{\partial \mathbf{r}_2} + \frac{\mathbf{F}_1^{\text{ext}}}{m} \cdot \frac{\partial f_2}{\partial \mathbf{v}_1} + \frac{\mathbf{F}_2^{\text{ext}}}{m} \cdot \frac{\partial f_2}{\partial \mathbf{v}_2} - \frac{1}{m} \frac{\partial \phi_{12}}{\partial \mathbf{r}_1} \cdot \left(\frac{\partial f_2}{\partial \mathbf{v}_1} - \frac{\partial f_2}{\partial \mathbf{v}_2} \right)$$
$$= \frac{1}{m} \int \left(\frac{\partial \phi_{13}}{\partial \mathbf{r}_1} \cdot \frac{\partial f_3}{\partial \mathbf{v}_1} + \frac{\partial \phi_{23}}{\partial \mathbf{r}_2} \cdot \frac{\partial f_3}{\partial \mathbf{v}_2} \right) \mathrm{d}\mathbf{r}_3 \, \mathrm{d}\mathbf{v}_3 \tag{12.17}$$

where we have used $\partial \phi_{12}/\partial \mathbf{r}_1 = -\partial \phi_{12}/\partial \mathbf{r}_2$ in (12.17).

12.2.1 Cluster expansion

As a first step towards the expression of f_3 in terms of f_1 and f_2 we introduce the *cluster expansion* by means of which we define new functions f, g and h such that

$$f_1(\mathbf{r}_1, \mathbf{v}_1, t) \equiv f(1) \tag{12.18}$$

$$f_2(\mathbf{r}_1, \mathbf{v}_1, \mathbf{r}_2, \mathbf{v}_2, t) = f(1) f(2) + g(1,2) \tag{12.19}$$

$$f_3(\mathbf{r}_1, \mathbf{v}_1, \mathbf{r}_2, \mathbf{v}_2, \mathbf{r}_3, \mathbf{v}_3, t) = f(1) f(2) f(3) + f(1) g(2,3) + f(2) g(3,1) + f(3) g(1,2) + h(1,2,3) \tag{12.20}$$

For convenience we have also simplified the notation, suppressing the time dependence in f, g, h and writing (1) for $(\mathbf{r}_1, \mathbf{v}_1)$, (2) for $(\mathbf{r}_2, \mathbf{v}_2)$, etc. The idea behind the cluster expansion is easily seen. If electrons 1 and 2 were completely independent of each other then the probability of finding 1 at $(\mathbf{r}_1, \mathbf{v}_1)$ at the same time that 2 is at $(\mathbf{r}_2, \mathbf{v}_2)$ would simply be the product $f(1) f(2)$. Thus, $g(1,2)$, being the difference between f_2 and $f(1) f(2)$, is a measure of the extent to which electrons 1 and 2 are *not* independent but *correlated* and it is called the *pair correlation function*. In a similar manner the five terms in the expansion in (12.20) represent the contributions to f_3 corresponding to all three electrons being independent of each other, each electron in turn being independent while the remaining two are correlated, and finally all three being correlated.

The cluster expansion was first introduced to deal with molecular interactions for which the range of interaction $r_c \ll \lambda_{\text{mfp}}$, the mean free path of the molecules. Thus, throughout most of its motion a molecule is unaware of the presence of other molecules and $g(1,2) \approx 0$ except when $|\mathbf{r}_1 - \mathbf{r}_2| \lesssim r_c$. Likewise, $h(1,2,3) \approx 0$ unless all three molecules are within a sphere of radius $\approx r_c$. In these circumstances we expect that binary interactions will be far more significant than three particle interactions and it can be shown that it is valid to ignore h in (12.20) thus achieving the desired truncation of the BBGKY hierarchy.

In a plasma the Coulomb interaction is anything but short range; indeed, the long range nature of electron interactions is a dominant feature of plasma dynamics. Nevertheless, the cluster expansion works for plasmas, too, but in a quite different way. First of all, it separates out the dominant long range part of electron interactions in the form of a field force. The short range interactions, the 'collisions', are then defined in terms of the pair correlation function $g(1, 2)$ and we shall see that this is again vanishingly small over most of phase space provided that the number of electrons in a Debye sphere is large. To see how this comes about let us substitute (12.18) and (12.19) in (12.16) giving

$$\frac{\partial f(1)}{\partial t} + \mathbf{v}_1 \cdot \frac{\partial f(1)}{\partial \mathbf{r}_1} + \frac{\mathbf{F}_1^{\text{ext}}}{m} \cdot \frac{\partial f(1)}{\partial \mathbf{v}_1} - \frac{1}{m}\frac{\partial f(1)}{\partial \mathbf{v}_1} \cdot \int \frac{\partial \phi_{12}}{\partial \mathbf{r}_1} f(2) \mathrm{d}\mathbf{r}_2 \, \mathrm{d}\mathbf{v}_2 = \frac{1}{m} \int \frac{\partial \phi_{12}}{\partial \mathbf{r}_1} \cdot \frac{\partial g(1, 2)}{\partial \mathbf{v}_1} \mathrm{d}\mathbf{r}_2 \, \mathrm{d}\mathbf{v}_2 \tag{12.21}$$

The first three terms in (12.21) are comparatively straightforward. The fourth term contains the average electric field experienced by one electron due to other electrons and may be written

$$-\frac{1}{m}\frac{\partial f(1)}{\partial \mathbf{v}_1} \cdot \int \frac{\partial \phi_{12}}{\partial \mathbf{r}_1} f(2) \mathrm{d}\mathbf{r}_2 \, \mathrm{d}\mathbf{v}_2 = -\frac{e\mathbf{E}}{m} \cdot \frac{\partial f(1)}{\partial \mathbf{v}_1}$$

where the field $\mathbf{E}$ is given by

$$(-e)\mathbf{E}(\mathbf{r}_1, t) = -\int \frac{\partial \phi_{12}}{\partial \mathbf{r}_1} f(2) \mathrm{d}\mathbf{r}_2 \, \mathrm{d}\mathbf{v}_2 \tag{12.22}$$

and $(-e)$ is the electronic charge. Note, however, that $\mathbf{E}$ is computed assuming that electrons are uncorrelated; the average is taken over all positions and velocities of electron 2 with no reference to any interaction between 1 and 2. This is in fact the electron contribution to the self-consistent field. Thus (12.21) may be written

$$\frac{\partial f(1)}{\partial t} + \mathbf{v}_1 \cdot \frac{\partial f(1)}{\partial \mathbf{r}_1} + \frac{\mathbf{F}}{m} \cdot \frac{\partial f(1)}{\partial \mathbf{v}_1} = \left(\frac{\partial f}{\partial t}\right)_{\mathrm{c}} \tag{12.23}$$

where $\mathbf{F}$ now includes all external and internal 'field' forces and

$$\left(\frac{\partial f}{\partial t}\right)_{\mathrm{c}} = \frac{1}{m} \int \frac{\partial \phi_{12}}{\partial \mathbf{r}_1} \cdot \frac{\partial g(1, 2)}{\partial \mathbf{v}_1} \mathrm{d}\mathbf{r}_2 \, \mathrm{d}\mathbf{v}_2 \tag{12.24}$$

is called the *collision term* or *collision integral.*

The separation of electron interactions into the self-consistent field (12.22) and the collision integral (12.24), brought about by splitting f_2 into its uncorrelated and correlated parts, is fundamental. Given that we are treating the ions as a uniform background of positive charge, any charge imbalance in the plasma must appear as an inhomogeneity in f and hence lead to a non-zero self-consistent field. There is no sense in which this term is small; indeed, we know that the tendency of a plasma to resist charge imbalance is one of its dominant features. On the other hand, as discussed in Section 7.1, it is very often an entirely satisfactory approximation to neglect the collision term (12.24) completely. This, of course, corresponds to truncation at $s = 1$ by expressing f_2 in terms of f_1, namely $f_2(1, 2) = f(1)f(2)$.

Substituting (12.19) and (12.20) in (12.17), but setting $h(1, 2, 3) = 0$ to truncate the expansion, we obtain, after some cancellation on using (12.21),

$$\begin{aligned}
&\frac{\partial g(1,2)}{\partial t} + \left(\mathbf{v}_1 \cdot \frac{\partial}{\partial \mathbf{r}_1} + \mathbf{v}_2 \cdot \frac{\partial}{\partial \mathbf{r}_2}\right) g(1,2) \\
&\quad + \left(\frac{\mathbf{F}_1^{\text{ext}}}{m} \cdot \frac{\partial}{\partial \mathbf{v}_1} + \frac{\mathbf{F}_2^{\text{ext}}}{m} \cdot \frac{\partial}{\partial \mathbf{v}_2}\right) g(1,2) \\
&\quad - \frac{1}{m}\frac{\partial \phi_{12}}{\partial \mathbf{r}_1} \cdot \left(\frac{\partial f(1)}{\partial \mathbf{v}_1} f(2) - \frac{\partial f(2)}{\partial \mathbf{v}_2} f(1) + \frac{\partial g(1,2)}{\partial \mathbf{v}_1} - \frac{\partial g(1,2)}{\partial \mathbf{v}_2}\right) \\
&\quad = \frac{1}{m}\int d\mathbf{r}_3\, d\mathbf{v}_3 \left\{\frac{\partial \phi_{13}}{\partial \mathbf{r}_1} \cdot \frac{\partial}{\partial \mathbf{v}_1}[f(1)g(2,3) + f(3)g(1,2)]\right. \\
&\quad \left. + \frac{\partial \phi_{23}}{\partial \mathbf{r}_2} \cdot \frac{\partial}{\partial \mathbf{v}_2}[f(2)g(1,3) + f(3)g(1,2)]\right\}
\end{aligned} \tag{12.25}$$

Next, consistent with the neglect of $h(1, 2, 3)$, we may drop the g terms compared with the ff terms in the last expression on the left-hand side of (12.25). Neglecting these terms is equivalent to the assumption that in the cluster expansion

$$|h| \ll |g|f \ll fff \tag{12.26}$$

and is known as the *weak coupling approximation*. This reduces our pair of equations to (12.23) and

$$\begin{aligned}
&\frac{\partial g(1,2)}{\partial t} + \left(\mathbf{v}_1 \cdot \frac{\partial}{\partial \mathbf{r}_1} + \mathbf{v}_2 \cdot \frac{\partial}{\partial \mathbf{r}_2}\right) g(1,2) \\
&\quad + \left(\frac{\mathbf{F}_1^{\text{ext}}}{m} \cdot \frac{\partial}{\partial \mathbf{v}_1} + \frac{\mathbf{F}_2^{\text{ext}}}{m} \cdot \frac{\partial}{\partial \mathbf{v}_2}\right) g(1,2) \\
&\quad - \frac{1}{m}\frac{\partial \phi_{12}}{\partial \mathbf{r}_1} \cdot \left(\frac{\partial f(1)}{\partial \mathbf{v}_1} f(2) - \frac{\partial f(2)}{\partial \mathbf{v}_2} f(1)\right) \\
&\quad = \frac{1}{m}\int d\mathbf{r}_3\, d\mathbf{v}_3 \left\{\frac{\partial \phi_{13}}{\partial \mathbf{r}_1} \cdot \frac{\partial}{\partial \mathbf{v}_1}[f(1)g(2,3) + f(3)g(1,2)]\right. \\
&\quad \left. + \frac{\partial \phi_{23}}{\partial \mathbf{r}_2} \cdot \frac{\partial}{\partial \mathbf{v}_2}[f(2)g(1,3) + f(3)g(1,2)]\right\}
\end{aligned} \tag{12.27}$$

Although we have achieved closure we are still a long way from obtaining the desired kinetic equation for f alone. The pair of simultaneous equations (12.23) and (12.27) for f and g are far too complicated for the elimination of g. Assuming *Bogolyubov's hypothesis*, which states that the time scale for changes in g is much shorter than that for f, Balescu (1960) and Lenard (1960) obtained a kinetic equation for the special case of a homogeneous plasma in the absence of external forces. The *Balescu–Lenard equation* is

$$\frac{\partial f}{\partial t} = -\frac{\pi}{m^2}\frac{\partial}{\partial \mathbf{v}_1}\cdot\int\frac{\mathbf{k}\phi^2(k)\mathrm{d}\mathbf{k}}{|\epsilon(\mathbf{k},\mathbf{k}\cdot\mathbf{v}_1)|^2}\mathbf{k}\cdot \int\left[\frac{\partial f(\mathbf{v}_2)}{\partial \mathbf{v}_2}f(\mathbf{v}_1)-\frac{\partial f(\mathbf{v}_1)}{\partial \mathbf{v}_1}f(\mathbf{v}_2)\right]\delta[\mathbf{k}\cdot(\mathbf{v}_1-\mathbf{v}_2)]\mathrm{d}\mathbf{v}_2 \tag{12.28}$$

where

$$\phi(k) = \frac{e^2}{(2\pi)^{3/2}\varepsilon_0 k^2} \tag{12.29}$$

and

$$\epsilon(\mathbf{k},\omega) = 1 + \frac{e^2}{\varepsilon_0 m k^2}\mathbf{k}\cdot\int\frac{\partial f/\partial\mathbf{v}}{(\omega-\mathbf{k}\cdot\mathbf{v})}\mathrm{d}\mathbf{v} \tag{12.30}$$

are the Fourier transforms of the Coulomb potential and the plasma dielectric function, respectively. We shall not give the derivation of (12.28) for it is not the most appropriate nor the most useful kinetic equation for most plasmas. That description is provided by the Landau kinetic equation, which we shall derive in Section 12.4, where we also discuss the Bogolyubov hypothesis.

A simpler version of (12.27) is obtained by setting the right-hand side equal to zero, the advantage of which is that the equation can then be solved for g *without* assuming plasma homogeneity. However, throwing away terms just because they are inconvenient is mathematically unconvincing, to say the least. Nevertheless, there is another special case for which (12.27) can be solved without further reduction. This is the equilibrium plasma and by solving it we shall gain insight into the significance of the terms that we wish to discard.

12.3 Equilibrium pair correlation function

In this section we consider a plasma in equilibrium and evaluate f and g. It is a well-known result of statistical mechanics that in thermodynamic equilibrium a solution of the Liouville equation (12.5) is the *Gibbs distribution function*

$$f_N = C\exp(-H/\theta) \tag{12.31}$$

where C and θ are constants and H is the Hamiltonian (12.8) expressed in terms of v_i and ϕ_{ij} as

$$H = \sum_{i=1}^{N} \left(\frac{1}{2} m v_i^2 + \sum_{\substack{j=1 \\ j \neq i}}^{N} \phi_{ij} \right)$$

It is easily verified that (12.31) satisfies (12.5) and, assuming that f_N like f_N^{ex} is normalized to unity, it follows that the constant C is given by

$$C = 1 \Big/ \int \exp(-H/\theta) \prod_{i=1}^{N} \mathrm{d}\mathbf{r}_i \, \mathrm{d}\mathbf{v}_i$$

Knowing f_N, we may in principle calculate all the reduced equilibrium distribution functions though in practice only the calculation of f_1 is simple. From (12.9) and (12.31)

$$f_1(1) = \frac{N \int \exp(-H/\theta) \prod_{i=2}^{N} \mathrm{d}\mathbf{r}_i \, \mathrm{d}\mathbf{v}_i}{\int \exp(-H/\theta) \prod_{i=1}^{N} \mathrm{d}\mathbf{r}_i \, \mathrm{d}\mathbf{v}_i} \tag{12.32}$$

The velocity integrations are separable and thus trivial. The substitutions

$$\mathbf{r}_i' = \mathbf{r}_i - \mathbf{r}_1 \qquad (i = 2, 3, \ldots, N) \tag{12.33}$$

remove $\mathbf{r}_1$ from both integrands in (12.32) with the result

$$\begin{aligned} f_1(1) &= \frac{N \exp(-m v_1^2/2\theta)}{\int \mathrm{d}\mathbf{r}_1 \, \mathrm{d}\mathbf{v}_1 \exp(-m v_1^2/2\theta)} \\ &= n \left(\frac{m}{2\pi\theta} \right)^{3/2} \exp(-m v_1^2/2\theta) \end{aligned} \tag{12.34}$$

where n is the electron number density. The constant θ may be identified with the definition of temperature T (see Section 12.5)

$$\frac{3}{2} n k_{\mathrm{B}} T = \int \frac{1}{2} m v_1^2 f_1 \, \mathrm{d}\mathbf{v}_1$$

giving

$$\theta = k_{\mathrm{B}} T$$

and the equilibrium distribution function (12.34) is thus the Maxwell distribution

$$f_1(1) = f_{\mathrm{M}}(1) \equiv n \left(\frac{m}{2\pi k_{\mathrm{B}} T} \right)^{3/2} \exp(-m v_1^2/2k_{\mathrm{B}} T) \tag{12.35}$$

Direct integration of (12.31) to obtain higher-order reduced distribution functions proves to be a formidable task requiring approximation techniques (see Montgomery and Tidman (1964)). However, the equilibrium pair correlation function may be obtained indirectly by solving the truncated equation for f_2. From (12.27), with no external forces, this is

$$\frac{\partial g(1,2)}{\partial t} + \left(\mathbf{v}_1 \cdot \frac{\partial}{\partial \mathbf{r}_1} + \mathbf{v}_2 \cdot \frac{\partial}{\partial \mathbf{r}_2}\right) g(1,2)$$
$$- \frac{1}{m}\frac{\partial \phi_{12}}{\partial \mathbf{r}_1} \cdot \left(\frac{\partial f(1)}{\partial \mathbf{v}_1} f(2) - \frac{\partial f(2)}{\partial \mathbf{v}_2} f(1)\right)$$
$$= \frac{1}{m}\int \mathrm{d}\mathbf{r}_3\, \mathrm{d}\mathbf{v}_3 \left\{\frac{\partial \phi_{13}}{\partial \mathbf{r}_1} \cdot \frac{\partial}{\partial \mathbf{v}_1}[f(1)g(2,3) + f(3)g(1,2)]\right.$$
$$\left. + \frac{\partial \phi_{23}}{\partial \mathbf{r}_2} \cdot \frac{\partial}{\partial \mathbf{v}_2}[f(2)g(1,3) + f(3)g(1,2)]\right\} \tag{12.36}$$

which is to be solved for g when $f = f_{\mathrm{M}}$. Note first that, since the velocity integrations in

$$f_2(1,2) = \frac{N(N-1)\int \exp(-H/\theta)\prod_{i=3}^{N} \mathrm{d}\mathbf{r}_i\, \mathrm{d}\mathbf{v}_i}{\int \exp(-H/\theta)\prod_{i=1}^{N} \mathrm{d}\mathbf{r}_i\, \mathrm{d}\mathbf{v}_i}$$

are separable, f_2 must be of the form

$$f_2(1,2) = f_{\mathrm{M}}(v_1) f_{\mathrm{M}}(v_2)[1 + p(r_{12})]$$

i.e.

$$g(1,2) = f_{\mathrm{M}}(v_1) f_{\mathrm{M}}(v_2) p(r_{12}) \tag{12.37}$$

where $r_{12} = |\mathbf{r}_1 - \mathbf{r}_2|$. Hence, (12.36) reduces to

$$(\mathbf{v}_1 - \mathbf{v}_2) \cdot \left[\frac{\partial p(r_{12})}{\partial \mathbf{r}_1} + \frac{1}{k_{\mathrm{B}}T}\frac{\partial \phi_{12}}{\partial \mathbf{r}_1}\right] =$$
$$- \frac{n}{k_{\mathrm{B}}T}\int \mathrm{d}\mathbf{r}_3 \left[\mathbf{v}_1 \cdot \frac{\partial \phi_{13}}{\partial \mathbf{r}_1} p(r_{23}) + \mathbf{v}_2 \cdot \frac{\partial \phi_{23}}{\partial \mathbf{r}_2} p(r_{13})\right]$$

This equation is valid for arbitrary $\mathbf{v}_1$ and $\mathbf{v}_2$, so choosing $\mathbf{v}_2 = 0$ we have

$$\frac{\partial p(r_{12})}{\partial \mathbf{r}_1} + \frac{1}{k_{\mathrm{B}}T}\frac{\partial \phi_{12}}{\partial \mathbf{r}_1} = -\frac{n}{k_{\mathrm{B}}T}\int \mathrm{d}\mathbf{r}_3\, p(r_{23})\frac{\partial \phi_{13}}{\partial \mathbf{r}_1} \tag{12.38}$$

The divergence of (12.38) with respect to $\mathbf{r}_1$, using (12.13) and

$$\nabla^2(1/r) = -4\pi\delta(\mathbf{r}) \tag{12.39}$$

then gives

$$\nabla_1^2 p(r_{12}) - \frac{e^2}{\varepsilon_0 k_{\mathrm{B}}T}\delta(\mathbf{r}_{12}) = \frac{ne^2}{\varepsilon_0 k_{\mathrm{B}}T}\int \mathrm{d}\mathbf{r}_3\, p(r_{23})\delta(\mathbf{r}_{13})$$

or

$$(\nabla_1^2 - \lambda_D^{-2})p(r_{12}) = \frac{e^2}{\varepsilon_0 k_B T}\delta(\mathbf{r}_{12}) \tag{12.40}$$

It is easily verified that the solution of (12.40) is

$$p(r_{12}) = -\frac{e^2}{4\pi\varepsilon_0 k_B T}\frac{\exp(-r_{12}/\lambda_D)}{r_{12}}$$

and hence from (12.37) the equilibrium pair correlation function is

$$g(1,2) = -f_M(v_1)f_M(v_2)\frac{\phi_{12}}{k_B T}\exp(-r_{12}/\lambda_D) \tag{12.41}$$

This result allows us to examine the assumption that $|g| \ll ff$. Clearly this is true for $\mathbf{r}_{12} > \lambda_D$ and breaks down only when electrons are sufficiently close that the potential energy of their interaction ϕ_{12} is of the order of, or greater than, their mean kinetic energy, i.e. for

$$\frac{r_{12}}{\lambda_D} \lesssim \frac{e^2}{4\pi\varepsilon_0 k_B T \lambda_D} = \frac{1}{4\pi n \lambda_D^3} \tag{12.42}$$

Provided that the number of electrons in a Debye sphere $(4\pi n\lambda_D^3/3) \gg 1$, this shows that the approximation is good even within the Debye sphere except for a tiny region at the centre given by $(r_{12}/d) < 1/4\pi(n\lambda_D^3)^{2/3}$, where $d \equiv n^{-1/3}$ is the mean distance between electrons. The chance of two (or more!) electrons being this close to each other is clearly very small and the dimensionless parameter that ensures this is the number of particles in the Debye sphere. The inverse of this number is the small parameter in the weak coupling approximation as can be seen from (12.41); generally, within the Debye sphere

$$\frac{|g|}{ff} \sim \frac{\phi}{k_B T} \sim \frac{1}{4\pi n\lambda_D^3}$$

Even more pleasing than the consistency of the result of our calculation with the assumption underlying it is the precise form of (12.41) for it demonstrates Debye shielding. This is worthy of further examination for we can see exactly where the shielding has arisen. It is clear from (12.38) that had the integral term on the right-hand side of this equation been set equal to zero we should have obtained the solution $p(r_{12}) = -\phi_{12}/k_B T$. The effect of this term, therefore, has been to replace the Coulomb potential by the shielded potential. With hindsight this is not surprising. The integral term in (12.38) has arisen directly from the integral terms in (12.36) and these are the only terms in that equation which retain any effect of the rest of the electrons (labelled 3) on the correlation between electrons 1 and 2; these terms 'sum up' such effects and describe the shielding which the other electrons provide. In the next section we use this insight to obtain an intuitive and relatively simple method for finding g in the general case.

12.4 The Landau equation

The general solution of (12.27) for g as a functional of f, as we have already observed, is no simple matter. But if we use the insight obtained in the equilibrium case and assume that the major, indeed the only important effect of the rest of the electrons on the correlation of electrons 1 and 2 is to replace the Coulomb potential by the shielded potential then we are left with an equation that is solvable. Thus, we put the right-hand side of (12.27) equal to zero and replace the Coulomb potential (12.13) by the shielded potential

$$\phi_{ij}^{\mathrm{s}} = \frac{e^2}{4\pi\varepsilon_0|\mathbf{r}_i - \mathbf{r}_j|} \exp(-|\mathbf{r}_i - \mathbf{r}_j|/\lambda_{\mathrm{D}}) \tag{12.43}$$

giving

$$\frac{\partial g}{\partial t} + \mathbf{v}_1 \cdot \frac{\partial g}{\partial \mathbf{r}_1} + \mathbf{v}_2 \cdot \frac{\partial g}{\partial \mathbf{r}_2} = \frac{1}{m}\frac{\partial \phi_{12}^{\mathrm{s}}}{\partial \mathbf{r}_1} \cdot \left[\frac{\partial f(1)}{\partial \mathbf{v}_1} f(2) - \frac{\partial f(2)}{\partial \mathbf{v}_2} f(1)\right] \tag{12.44}$$

In solving this equation we shall use Bogolyubov's hypothesis which is based on the observation that time and length scales for changes in g and f are widely separated. Since particle correlations are limited to the Debye sphere, the length scale for g is λ_{D} and the corresponding time scale is ω_{p}^{-1}, the time for an electron with mean thermal energy to cross the Debye sphere. In contrast, f relaxes to a Maxwellian under the influence of collisions on a time scale of τ_{c}, the collision time, and the corresponding length scale is the electron mean free path λ_{c}, the distance travelled by a thermal electron between collisions. Since $\omega_{\mathrm{p}}\tau_{\mathrm{c}} \gg 1$ and $\lambda_{\mathrm{D}} \ll \lambda_{\mathrm{c}}$, the assumption of Bogolyubov's hypothesis is justified.

Equation (12.44) may be solved by Green function techniques. Defining $G(\mathbf{r}_1, \mathbf{r}_1', \mathbf{r}_2, \mathbf{r}_2', t, t')$ as the solution of

$$\frac{\partial G}{\partial t} + \mathbf{v}_1 \cdot \frac{\partial G}{\partial \mathbf{r}_1} + \mathbf{v}_2 \cdot \frac{\partial G}{\partial \mathbf{r}_2} = \delta(t - t')\delta(\mathbf{r}_1 - \mathbf{r}_1')\delta(\mathbf{r}_2 - \mathbf{r}_2')$$

it is easily verified that G is given by

$$G = \Theta(t - t')\delta[\mathbf{r}_1 - \mathbf{r}_1' - \mathbf{v}_1(t - t')]\delta[\mathbf{r}_2 - \mathbf{r}_2' - \mathbf{v}_2(t - t')]$$

where $\Theta(t)$ is the step function

$$\Theta(t) = \begin{cases} 1 & t > 0 \\ 0 & t < 0 \end{cases}$$

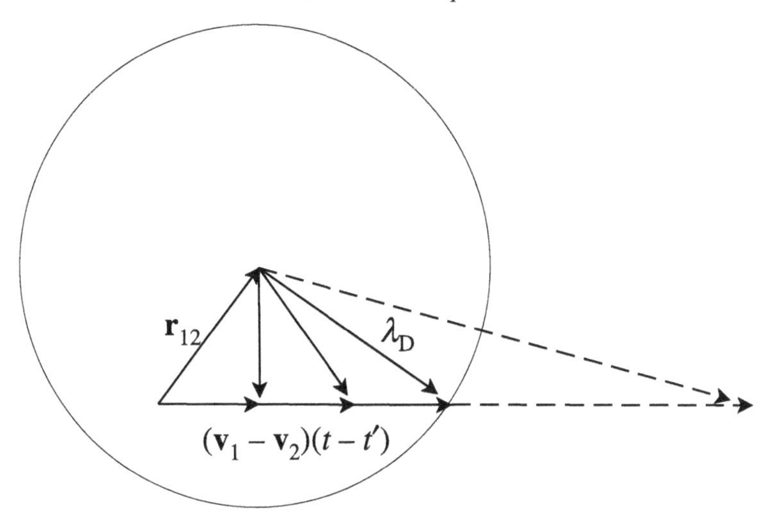

Fig. 12.1. Time variation of $|\mathbf{r}_{12} - (\mathbf{v}_1 - \mathbf{v}_2)(t - t')|$; $\phi^s \approx 0$ unless both $\mathbf{r}_{12}$ and $|\mathbf{r}_{12} - (\mathbf{v}_1 - \mathbf{v}_2)(t - t')|$ lie within a Debye sphere.

Then

$$
\begin{aligned}
g &= \frac{1}{m}\int \mathrm{d}t'\,\mathrm{d}\mathbf{r}_1'\,\mathrm{d}\mathbf{r}_2' G(\mathbf{r}_1, \mathbf{r}_1', \mathbf{r}_2, \mathbf{r}_2', t, t')\frac{\partial \phi^s(r_{12}')}{\partial \mathbf{r}_1'} \cdot \\
&\quad \left[\frac{\partial f(\mathbf{r}_1', \mathbf{v}_1, t')}{\partial \mathbf{v}_1} f(\mathbf{r}_2', \mathbf{v}_2, t') - \frac{\partial f(\mathbf{r}_2', \mathbf{v}_2, t')}{\partial \mathbf{v}_2} f(\mathbf{r}_1', \mathbf{v}_1, t')\right] \\
&= \frac{1}{m}\int_0^t \mathrm{d}t' \frac{\partial \phi^s(|\mathbf{r}_{12} - (\mathbf{v}_1 - \mathbf{v}_2)(t - t')|)}{\partial \mathbf{r}_{12}} \cdot \\
&\quad \left[\frac{\partial f(\mathbf{r}_1 - \mathbf{v}_1(t - t'), \mathbf{v}_1, t')}{\partial \mathbf{v}_1} f(\mathbf{r}_2 - \mathbf{v}_2(t - t'), \mathbf{v}_2, t')\right. \\
&\quad \left. - \frac{\partial f(\mathbf{r}_2 - \mathbf{v}_2(t - t'), \mathbf{v}_2, t')}{\partial \mathbf{v}_2} f(\mathbf{r}_1 - \mathbf{v}_1(t - t'), \mathbf{v}_1, t')\right]
\end{aligned} \tag{12.45}
$$

Substitution of this expression for g in (12.24) (with ϕ replaced by ϕ^s) does not immediately yield a tractable kinetic equation because of the arguments of f in the collision term. Indeed, one might say that it does not yet amount to a kinetic equation at all because g is not expressed as a functional (as opposed to a function) of f. However, Bogolyubov's hypothesis leads to a number of simplifications so that g does become a functional of f making (12.23) a kinetic equation.

Note first that since the arguments of the shielded potentials in (12.24) and (12.45) are r_{12} and $|\mathbf{r}_{12} - (\mathbf{v}_1 - \mathbf{v}_2)(t - t')|$ and $\phi^s(r) \approx 0$ for $r \gtrsim \lambda_D$ it follows that the time interval $(t - t') \lesssim \lambda_D/|\mathbf{v}_1 - \mathbf{v}_2| \sim \omega_p^{-1}$; this is illustrated in Fig. 12.1. Next, since we assume that f does not vary significantly on this time scale (ω_p^{-1})

nor over distances of order λ_D, the arguments of f in (12.45) may be reduced to $(\mathbf{r}_1, \mathbf{v}_1, t)$ and $(\mathbf{r}_2, \mathbf{v}_2, t)$ in which $\mathbf{r}_1 \approx \mathbf{r}_2$. Letting $\tau = t - t'$ we may now rewrite (12.45) as

$$g(1,2) = \frac{1}{m}\left(f(2)\frac{\partial f(1)}{\partial \mathbf{v}_1} - f(1)\frac{\partial f(2)}{\partial \mathbf{v}_2}\right) \cdot \int_0^\infty \frac{\partial \phi^s(|\mathbf{r}_{12} - (\mathbf{v}_1 - \mathbf{v}_2)\tau|)}{\partial \mathbf{r}_{12}} \mathrm{d}\tau \tag{12.46}$$

where we have taken the limit $t \to \infty$ in the integral, which is allowed because, as already observed, the integrand vanishes for $\tau \gtrsim \omega_p^{-1}$. This expression for g is now a functional of f and substitution in (12.24) gives

$$\left(\frac{\partial f}{\partial t}\right)_c = \frac{1}{m^2}\int \mathrm{d}\mathbf{r}_2\,\mathrm{d}\mathbf{v}_2 \frac{\partial \phi^s(r_{12})}{\partial \mathbf{r}_{12}} \cdot \frac{\partial}{\partial \mathbf{v}_1}$$
$$\int_0^\infty \mathrm{d}\tau \frac{\partial \phi^s(|\mathbf{r}_{12} - (\mathbf{v}_1 - \mathbf{v}_2)\tau|)}{\partial \mathbf{r}_{12}} \cdot \left(f(2)\frac{\partial f(1)}{\partial \mathbf{v}_1} - f(1)\frac{\partial f(2)}{\partial \mathbf{v}_2}\right) \tag{12.47}$$

Writing $\phi^s(r)$ in terms of its Fourier transform

$$\phi^s(r) = \frac{e^2}{(2\pi)^3\varepsilon_0}\int \frac{\mathrm{d}\mathbf{k}\exp(i\mathbf{k}\cdot\mathbf{r})}{k^2 + k_D^2}$$

where $k_D \equiv \lambda_D^{-1}$, (12.47) becomes

$$\left(\frac{\partial f}{\partial t}\right)_c = -\frac{e^4}{(2\pi)^6\varepsilon_0^2 m^2}\int \mathrm{d}\mathbf{r}_2\,\mathrm{d}\mathbf{v}_2\,\mathrm{d}\mathbf{k}\,\mathrm{d}\mathbf{l}\frac{\exp(i\mathbf{k}\cdot\mathbf{r}_{12})}{(k^2 + k_D^2)}\mathbf{k} \cdot \frac{\partial}{\partial \mathbf{v}_1}$$
$$\int_0^\infty \mathrm{d}\tau \frac{\exp\{i\mathbf{l}\cdot[\mathbf{r}_{12} - (\mathbf{v}_1 - \mathbf{v}_2)\tau]\}}{(l^2 + k_D^2)}\mathbf{l} \cdot \left[\frac{\partial f(1)}{\partial \mathbf{v}_1} f(2) - \frac{\partial f(2)}{\partial \mathbf{v}_2} f(1)\right] \tag{12.48}$$

To simplify this expression we integrate over $\mathbf{r}_2$ space first. In doing this we use a similar approximation to that for the time integration in (12.45) and assume f varies on a length scale much greater than λ_D; since the integrand in (12.48) contains $\phi(r_{12})$ it vanishes exponentially for $r_{12} \gg \lambda_D$ so that we may replace $\mathbf{r}_2$ by $\mathbf{r}_1$ in $f(2)$. The integration over $\mathbf{r}_2$ then involves $\exp[-i(\mathbf{k}+\mathbf{l})\cdot\mathbf{r}_2]$ only and gives $(2\pi)^3\delta(\mathbf{k}+\mathbf{l})$. The integration over $\mathbf{l}$ space is then trivial and (12.48) becomes

$$\left(\frac{\partial f}{\partial t}\right)_c = \frac{e^4}{(2\pi)^3\varepsilon_0^2 m^2}\int \frac{\mathrm{d}\mathbf{v}_2\,\mathrm{d}\mathbf{k}}{(k^2 + k_D^2)^2}\mathbf{k} \cdot \frac{\partial}{\partial \mathbf{v}_1}$$
$$\int_0^\infty \mathrm{d}\tau \exp[i\mathbf{k}\cdot(\mathbf{v}_1 - \mathbf{v}_2)\tau]\mathbf{k} \cdot \left(\frac{\partial f(\mathbf{v}_1)}{\partial \mathbf{v}_1} f(\mathbf{v}_2) - \frac{\partial f(\mathbf{v}_2)}{\partial \mathbf{v}_2} f(\mathbf{v}_1)\right) \tag{12.49}$$

The integration over τ is given by the *Plemelj formula*

$$\int_0^\infty \mathrm{d}\tau \exp[i\mathbf{k}\cdot(\mathbf{v}_1-\mathbf{v}_2)\tau] = \pi\delta[\mathbf{k}\cdot(\mathbf{v}_1-\mathbf{v}_2)] - iP\frac{1}{\mathbf{k}\cdot(\mathbf{v}_1-\mathbf{v}_2)} \tag{12.50}$$

where P stands for the principal part. However, the second term in (12.50) gives no contribution to (12.49), since it makes the integrand odd in $\mathbf{k}$ (its contribution must be zero since it is imaginary). Hence

$$\left(\frac{\partial f}{\partial t}\right)_{\mathrm{c}} = \frac{e^4}{8\pi^2\varepsilon_0^2 m^2}\int \frac{\mathrm{d}\mathbf{v}_2\,\mathrm{d}\mathbf{k}}{(k^2+k_{\mathrm{D}}^2)^2}\mathbf{k}\cdot\frac{\partial}{\partial \mathbf{v}_1}\delta[\mathbf{k}\cdot(\mathbf{v}_1-\mathbf{v}_2)]$$
$$\mathbf{k}\cdot\left[\frac{\partial f(\mathbf{v}_1)}{\partial \mathbf{v}_1}f(\mathbf{v}_2) - \frac{\partial f(\mathbf{v}_2)}{\partial \mathbf{v}_2}f(\mathbf{v}_1)\right] \tag{12.51}$$

Defining $\mathbf{g} = (\mathbf{v}_1 - \mathbf{v}_2)$, this leaves an integration over $\mathbf{k}$ space of the form

$$\int \frac{\mathbf{k}\cdot\mathbf{A}\delta(\mathbf{k}\cdot\mathbf{g})\mathbf{k}\cdot\mathbf{B}}{(k^2+k_{\mathrm{D}}^2)^2}\mathrm{d}\mathbf{k} \tag{12.52}$$

which diverges at large $\mathbf{k}$. This is a direct result of the weak coupling approximation which clearly breaks down as $r_{12}\to 0$, in fact for $r_{12} \lesssim e^2/4\pi\varepsilon_0 k_{\mathrm{B}}T$. We cut off the integration therefore at $k = 4\pi\varepsilon_0 k_{\mathrm{B}}T/e^2$ and find for (12.52)

$$\pi A_{\mathrm{i}}\left(\frac{\delta_{ij}}{g} - \frac{g_i g_j}{g^3}\right)B_j \ln\Lambda = \pi A_{\mathrm{i}}\frac{\partial^2 g}{\partial v_i\,\partial v_j}B_j\ln\Lambda \tag{12.53}$$

where $\Lambda = 4\pi n\lambda_{\mathrm{D}}^3$ and only terms in $\ln\Lambda$ have been retained.

Thus, using (12.21), the kinetic equation may finally be written

$$\frac{\partial f}{\partial t} + \mathbf{v}\cdot\frac{\partial f}{\partial \mathbf{r}} - \frac{e}{m}\mathbf{E}\cdot\frac{\partial f}{\partial \mathbf{v}} =$$
$$\frac{e^4\ln\Lambda}{8\pi\varepsilon_0^2 m^2}\frac{\partial}{\partial v_i}\int \mathrm{d}\mathbf{v}'\frac{\partial^2|\mathbf{v}-\mathbf{v}'|}{\partial v_i\,\partial v_j}\left[\frac{\partial f(\mathbf{v})}{\partial v_j}f(\mathbf{v}') - \frac{\partial f(\mathbf{v}')}{\partial v_j'}f(\mathbf{v})\right] \tag{12.54}$$

where $\mathbf{E}$ is the self-consistent field given by (12.22). This form of the collision integral for a plasma was first derived by Landau using an intuitive model for collisions, so (12.54) is known as the *Landau kinetic equation.*

At this point it is of interest to consider the relationship of the Landau equation to the Balescu–Lenard equation (12.28). A comparison of (12.28) with (12.51) shows that the dielectric function $\epsilon(\mathbf{k}, \mathbf{k}\cdot\mathbf{v}_1)$, has simply been replaced by the term $(1 + (k_{\mathrm{D}}/k)^2)$, which is the static shielding factor introduced by replacing the Coulomb potential ϕ by the shielded potential ϕ^s. In fact the main effect of the dielectric function is to produce shielding, but in the Balescu–Lenard equation it appears as a dynamic effect involving the interaction of particles and Fourier wave components through the resonant integral in (12.30). Furthermore, it can be shown, as for example in Nicolson (1983), that to the order of keeping only terms in $\ln \Lambda$ the Balescu–Lenard collision integral reduces to the Landau collision integral.

The generalization of the Landau collision integral to more than one species of particle is given by (8.36). Since the latter equation was derived from the Fokker–Planck equation this establishes an important link and justifies the applicability of the Fokker–Planck model to a plasma.

12.5 Moment equations

In Section 7.2 we showed through Jeans' theorem the formal equivalence of the collisionless (Vlasov) kinetic theory and particle orbit theory. We turn now to the relationship between collisional (Landau) kinetic theory and the MHD equations discussed in Chapters 3–5. There is no simple theorem establishing this relationship for there is not, even in a formal sense, an exact equivalence. The procedure from microscopic kinetic description to macroscopic fluid description has rather more in common with the derivation of the kinetic equation from the Liouville equation. The reduction in detail of description, in this case the removal of dependence on the velocity coordinate $\mathbf{v}$, gives rise to an infinite chain of equations – the *moment equations*. The derivation of the moment equations is the first of the three basic steps that lead to the MHD equations. The second and most formidable is the truncation and closure of the moment equations. This is where physical approximation enters and equivalence disappears; it is the subject of classical transport theory discussed in Section 12.6. The final step is the derivation of the MHD equations from the transport equations and the further physical approximations necessary for this are examined in Section 12.7.

We begin with definitions of fluid variables in which we introduce the label α (= i, e, for ions and electrons, respectively), to denote particle species. On scalars α appears as a subscript but on vectors and tensors, for which we frequently need to denote components by roman subscripts (j, k, l etc.), it is more convenient to write α as a superscript. Since $f_\alpha(\mathbf{r}, \mathbf{v}, t)\mathrm{d}\mathbf{r}\,\mathrm{d}\mathbf{v}$ is the probability at time t of finding particles of type α within a small volume element $\mathrm{d}\mathbf{r}\,\mathrm{d}\mathbf{v}$ about the point $(\mathbf{r}, \mathbf{v})$ the integral of this over all velocity space is the probability of finding particles, irrespective of velocity, within a volume $\mathrm{d}\mathbf{r}$ about $\mathbf{r}$. Hence, the (number) density, $n_\alpha(\mathbf{r}, t)$ is defined by

$$n_\alpha(\mathbf{r}, t) = \int \mathrm{d}\mathbf{v}\, f_\alpha(\mathbf{r}, \mathbf{v}, t) \tag{12.55}$$

Similarly, the flow velocity $\mathbf{u}^\alpha(\mathbf{r}, t)$ is the mean velocity of all particles of type α

in the volume $\mathrm{d}\mathbf{r}$ about $\mathbf{r}$ at time t and so is given by

$$\mathbf{u}^{\alpha}(\mathbf{r}, t) = \frac{\mathrm{d}\mathbf{r} \int \mathbf{v} f_{\alpha}(\mathbf{r}, \mathbf{v}, t)\mathrm{d}\mathbf{v}}{\mathrm{d}\mathbf{r} \int f_{\alpha}(\mathbf{r}, \mathbf{v}, t)\mathrm{d}\mathbf{v}} = \frac{1}{n_{\alpha}(\mathbf{r}, t)} \int \mathbf{v} f_{\alpha}(\mathbf{r}, \mathbf{v}, t)\mathrm{d}\mathbf{v} \tag{12.56}$$

Other fluid variables like pressure, temperature and heat flux are defined in terms of the *random* or *thermal velocities* of the particles

$$\mathbf{c}^{\alpha} = \mathbf{v} - \mathbf{u}^{\alpha} \tag{12.57}$$

Whilst the mean value of $\mathbf{c}^{\alpha}$ is clearly zero, the mean values of higher powers of its components are non-zero in general and, for example, are related to the thermal momentum flux and thermal energy flux and thereby to the pressure and heat flux. Thus, we define the elements of the pressure tensor $p^{\alpha}_{ij}(\mathbf{r}, t)$ by

$$p^{\alpha}_{ij}(\mathbf{r}, t) = m_{\alpha} \int c^{\alpha}_i c^{\alpha}_j f_{\alpha}(\mathbf{r}, \mathbf{v}, t)\mathrm{d}\mathbf{v} \tag{12.58}$$

and the heat flux tensor $q^{\alpha}_{ijk}(\mathbf{r}, t)$ by

$$q^{\alpha}_{ijk}(\mathbf{r}, t) = m_{\alpha} \int c^{\alpha}_i c^{\alpha}_j c^{\alpha}_k f_{\alpha}(\mathbf{r}, \mathbf{v}, t)\mathrm{d}\mathbf{v} \tag{12.59}$$

It is convenient to separate p^{α}_{ij} into two tensors by

$$p^{\alpha}_{ij} = p_{\alpha}\delta_{ij} + \Pi^{\alpha}_{ij} \tag{12.60}$$

where

$$\Pi^{\alpha}_{ij} = m_{\alpha} \int \left(c^{\alpha}_i c^{\alpha}_j - \frac{1}{3}|\mathbf{c}^{\alpha}|^2 \delta_{ij} \right) f_{\alpha}\,\mathrm{d}\mathbf{v} \tag{12.61}$$

is by definition a traceless tensor and

$$p_{\alpha} = \frac{m_{\alpha}}{3} \int |\mathbf{c}^{\alpha}|^2 f_{\alpha}\,\mathrm{d}\mathbf{v} \tag{12.62}$$

is, for an isotropic plasma, the scalar pressure.

It is also useful to define a kinetic temperature $T_{\alpha}(\mathbf{r}, t)$ by extrapolating the well-known result that for an ideal gas in equilibrium the thermal energy associated with each degree of freedom of the gas is $\frac{1}{2}k_{\mathrm{B}}T_{\alpha}$. Since Boltzmann's constant k_{B} is always associated with T_{α} we shall, in this chapter, absorb it into the definition of T_{α} (which is thereby measured in energy units instead of degrees). Thus, with three degrees of freedom for each of the n_{α} particles we have

$$\frac{3n_{\alpha}T_{\alpha}(\mathbf{r}, t)}{2} = \frac{m_{\alpha}}{2} \int |\mathbf{c}^{\alpha}|^2 f_{\alpha}(\mathbf{r}, \mathbf{v}, t)\mathrm{d}\mathbf{v} = \frac{3}{2}p_{\alpha} \tag{12.63}$$

Finally, the heat flux vector $\mathbf{q}^\alpha(\mathbf{r}, t)$ is the flux of thermal energy relative to the flow velocity and so is given by

$$\mathbf{q}^\alpha(\mathbf{r}, t) = \frac{m_\alpha}{2} \int |\mathbf{c}^\alpha|^2 \mathbf{c}^\alpha f_\alpha(\mathbf{r}, \mathbf{v}, t) \mathrm{d}\mathbf{v} \tag{12.64}$$

It then follows from (12.59) that

$$q_i^\alpha(\mathbf{r}, t) = \frac{1}{2} q_{ijj}^\alpha(\mathbf{r}, t) \tag{12.65}$$

These velocity integrals by which the fluid variables are defined are called 'moment integrals' of f_α. The order of a moment integral is given by the highest power of the components of $\mathbf{v}$ occurring in the integrand. Thus, the density is a zero-order moment, the flow velocity is a first-order moment, the pressure and temperature are second-order moments and the heat flux is a third-order moment.

We now generate evolution equations for these macroscopic variables by taking appropriate moments of the kinetic equation

$$\frac{\partial f_\alpha}{\partial t} + \mathbf{v} \cdot \frac{\partial f_\alpha}{\partial \mathbf{r}} + \frac{e_\alpha}{m_\alpha}(\mathbf{E} + \mathbf{v} \times \mathbf{B}) \cdot \frac{\partial f_\alpha}{\partial \mathbf{v}} = \left(\frac{\partial f_\alpha}{\partial t}\right)_\mathrm{c} \tag{12.66}$$

Since $\mathbf{r}$, $\mathbf{v}$ and t are independent variables it is clear from (12.55) that when we integrate (12.66) over velocity space (i.e. take the zero-order moment) the first term in the new equation is $\partial n_\alpha / \partial t$ so that this may be regarded as an equation for the time evolution of $n_\alpha(\mathbf{r}, t)$. It is also evident from (12.56) that the second term gives rise to $\nabla \cdot (n_\alpha \mathbf{u}^\alpha)$, that is, the evolution equation for the zero-order moment n_α contains a term involving the first-order moment $\mathbf{u}^\alpha$. We obtain an evolution equation for $\mathbf{u}^\alpha$ by taking the first-order moment of (12.66), that is we multiply by $\mathbf{v}$ and then integrate over velocity space. Now the leading term is $\partial(n_\alpha \mathbf{u}^\alpha)/\partial t$ but the second term again gives rise to a higher moment, $\int \mathbf{v}\mathbf{v} f_\alpha \, \mathrm{d}\mathbf{v}$, which by (12.58) is somehow related to the pressure tensor as well as to n_α and $\mathbf{u}^\alpha$. Clearly, this drawing down of higher moments persists no matter how many moments of the kinetic equation we take. We have removed a variable ($\mathbf{v}$) of infinite dimension but, in doing so, have generated an infinite chain of moment equations. Such mathematical transformations do not reduce the basic complexity of the description. To achieve this we need approximations based on physical assumptions. Up to this point we have merely outlined a scheme for replacing one equation in seven-dimensional space $(\mathbf{r}, \mathbf{v}, t)$ by an infinite set of equations in four-dimensional space $(\mathbf{r}, t)$ and there is, thus far, an exact equivalence between these descriptions. For the present, we shall not worry about how the real simplification is to be achieved but will go ahead and generate the first few equations in the chain.

It is useful to obtain a general moment equation by multiplying (12.66) by $\psi(\mathbf{v})$, where $\psi(\mathbf{v})$ is any polynomial function of the components of $\mathbf{v}$, and integrating over velocity space. Defining

$$\langle\psi\rangle = \frac{\int \psi(\mathbf{v}) f_\alpha \,\mathrm{d}\mathbf{v}}{\int f_\alpha \,\mathrm{d}\mathbf{v}} = \frac{1}{n_\alpha(\mathbf{r},t)} \int \psi(\mathbf{v}) f_\alpha(\mathbf{r},\mathbf{v},t)\,\mathrm{d}\mathbf{v} \tag{12.67}$$

we get

$$\frac{\partial}{\partial t}(n_\alpha\langle\psi\rangle) + \frac{\partial}{\partial \mathbf{r}}\cdot(n_\alpha\langle\mathbf{v}\psi\rangle) - \frac{n_\alpha e_\alpha \mathbf{E}}{m_\alpha}\cdot\left\langle\frac{\partial\psi}{\partial\mathbf{v}}\right\rangle - \frac{n_\alpha e_\alpha}{m_\alpha}\left\langle(\mathbf{v}\times\mathbf{B})\cdot\frac{\partial\psi}{\partial\mathbf{v}}\right\rangle = \int \psi\left(\frac{\partial f_\alpha}{\partial t}\right)_{\mathrm{c}}\mathrm{d}\mathbf{v} \tag{12.68}$$

where we have used integration by parts to obtain the third and fourth terms assuming that $f_\alpha \to 0$ faster than any inverse power of $|\mathbf{v}|$ as $|\mathbf{v}| \to \infty$, i.e. $\lim_{|\mathbf{v}|\to\infty}(\psi f_\alpha) = 0$.

The following useful relationships are easily verified using (12.57)–(12.65) and (12.67):

$$\begin{aligned}
\langle v_i v_j\rangle &= \langle(c_i^\alpha + u_i^\alpha)(c_j^\alpha + u_j^\alpha)\rangle \\
&= \langle c_i^\alpha c_j^\alpha\rangle + u_i^\alpha u_j^\alpha \\
&= \frac{T_\alpha}{m_\alpha}\delta_{ij} + \frac{\Pi_{ij}^\alpha}{m_\alpha n_\alpha} + u_i^\alpha u_j^\alpha
\end{aligned} \tag{12.69}$$

$$\begin{aligned}
\langle v_i v_j v_j\rangle &= \langle(c_i^\alpha + u_i^\alpha)(c_j^\alpha + u_j^\alpha)^2\rangle \\
&= \langle c_i^\alpha c_j^\alpha c_j^\alpha\rangle + 2\langle c_i^\alpha c_j^\alpha\rangle u_j^\alpha + u_i^\alpha\langle c_j^\alpha c_j^\alpha\rangle + u_i^\alpha u_j^\alpha u_j^\alpha \\
&= \frac{2q_i^\alpha}{m_\alpha n_\alpha} + \frac{5T_\alpha}{m_\alpha}u_i^\alpha + \frac{2\Pi_{ij}^\alpha}{m_\alpha n_\alpha}u_j^\alpha + u_i^\alpha|u_j^\alpha|^2
\end{aligned} \tag{12.70}$$

The *collisional conservation relations* are of fundamental importance in the derivation of the fluid equations. We have

$$\int \mathrm{d}\mathbf{v}\left(\frac{\partial f_\alpha}{\partial t}\right)_{\mathrm{c}} = 0 \qquad (\alpha = \mathrm{i}, \mathrm{e}) \tag{12.71}$$

$$\sum_\alpha m_\alpha \int \mathrm{d}\mathbf{v}\,\mathbf{v}\left(\frac{\partial f_\alpha}{\partial t}\right)_{\mathrm{c}} = 0 \tag{12.72}$$

$$\sum_\alpha \frac{m_\alpha}{2}\int \mathrm{d}\mathbf{v}\,v^2\left(\frac{\partial f_\alpha}{\partial t}\right)_{\mathrm{c}} = 0 \tag{12.73}$$

Thus, when $\psi(\mathbf{v}) = 1$ the collision term makes no contribution to the moment equation (12.68) for either ions or electrons. Also, when $\psi(\mathbf{v}) = m_\alpha\mathbf{v}$ or $\frac{1}{2}m_\alpha v^2$ the ion and electron collision terms are equal in magnitude but opposite in sign.

In turn we now set $\psi(\mathbf{v}) = 1, m_\alpha v_i, \frac{1}{2}m_\alpha v^2$ in (12.68) to get

$$\frac{\partial n_\alpha}{\partial t} + \frac{\partial}{\partial \mathbf{r}} \cdot (n_\alpha \mathbf{u}^\alpha) = 0 \tag{12.74}$$

$$\frac{\partial}{\partial t}(m_\alpha n_\alpha u_i^\alpha) + \frac{\partial}{\partial r_j}(m_\alpha n_\alpha u_i^\alpha u_j^\alpha + n_\alpha T_\alpha \delta_{ij} + \Pi_{ij}^\alpha) - e_\alpha n_\alpha (E_i + (\mathbf{u}^\alpha \times \mathbf{B})_i) = R_i^\alpha \tag{12.75}$$

$$n_\alpha \frac{\partial T_\alpha}{\partial t} + n_\alpha \mathbf{u}^\alpha \cdot \boldsymbol{\nabla} T_\alpha + \frac{2}{3} n_\alpha T_\alpha \boldsymbol{\nabla} \cdot \mathbf{u}^\alpha + \frac{2}{3} \Pi_{ij}^\alpha \frac{\partial u_i^\alpha}{\partial r_j} + \frac{2}{3} \boldsymbol{\nabla} \cdot \mathbf{q}^\alpha = \frac{2}{3} Q_\alpha \tag{12.76}$$

where

$$\mathbf{R}^\alpha = m_\alpha \int \mathrm{d}\mathbf{v}\, \mathbf{v} \left(\frac{\partial f_\alpha}{\partial t} \right)_\mathrm{c} \tag{12.77}$$

is the rate of transfer of momentum to species α from the other species and

$$\begin{aligned} Q_\alpha &= \frac{m_\alpha}{2} \int \mathrm{d}\mathbf{v}\, |\mathbf{v} - \mathbf{u}^\alpha|^2 \left(\frac{\partial f_\alpha}{\partial t} \right)_\mathrm{c} \\ &= \frac{m_\alpha}{2} \int \mathrm{d}\mathbf{v}\, v^2 \left(\frac{\partial f_\alpha}{\partial t} \right)_\mathrm{c} - \mathbf{u}^\alpha \cdot \mathbf{R}^\alpha \end{aligned} \tag{12.78}$$

is the rate of transfer of thermal energy. In deriving (12.78), (12.71) and (12.77) have been used. It is easily seen that consequences of the conservation relations (12.72) and (12.73) are

$$\mathbf{R}^\mathrm{e} = -\mathbf{R}^\mathrm{i} \tag{12.79}$$

and

$$Q_\mathrm{e} = -Q_\mathrm{i} - (\mathbf{u}^\mathrm{e} - \mathbf{u}^\mathrm{i}) \cdot \mathbf{R}^\mathrm{e} \tag{12.80}$$

The momentum equation (12.75) can be simplified using the continuity equation (12.74) but, with subsequent manipulations in mind, we choose not to do this. However, the energy equation (12.76) has been reduced using both (12.74) and (12.75). Because of the special role of collisions in classical (and neoclassical) transport theory the three moment equations (12.74)–(12.76) are particularly important and although, for an adequate account of transport theory, it is necessary to generate further moments these will not be displayed.

12.5.1 One-fluid variables

The fluid variables we have defined and the moment equations that have been derived for their evolution amount to a two-fluid description of the plasma. But the MHD equations, our ultimate goal, are cast in terms of a single fluid. The final exercise in this section, therefore, is to introduce one-fluid hydrodynamical and electrodynamical variables and make a (partial) transformation of the moment equations. At this stage a complete transformation to a one-fluid description would be inappropriate. The reason for this is that the exchange of thermal energy between ions and electrons is a rather inefficient process and can be as slow as the hydrodynamical changes taking place in the plasma. In transport theory the separation of fast (collisional) and slow (hydrodynamic) time scales is the key to the truncation procedure. It is essential, therefore, to retain separate evolution equations for ion and electron temperatures. The one-fluid variables are the plasma mass density

$$\rho(\mathbf{r}, t) = \sum_\alpha m_\alpha n_\alpha(\mathbf{r}, t) \tag{12.81}$$

the charge density

$$q(\mathbf{r}, t) = \sum_\alpha e_\alpha n_\alpha(\mathbf{r}, t) \tag{12.82}$$

the centre of mass (or plasma flow) velocity $\mathbf{u}(\mathbf{r}, t)$ defined by

$$\rho(\mathbf{r}, t)\mathbf{u}(\mathbf{r}, t) = \sum_\alpha m_\alpha n_\alpha(\mathbf{r}, t)\mathbf{u}^\alpha(\mathbf{r}, t) \tag{12.83}$$

and the current density

$$\mathbf{j}(\mathbf{r}, t) = \sum_\alpha e_\alpha n_\alpha(\mathbf{r}, t)\mathbf{u}^\alpha(\mathbf{r}, t) \tag{12.84}$$

With the aid of this set of equations and its inverse

$$\left.\begin{aligned}
n_\mathrm{e} &= \frac{Ze\rho - m_\mathrm{i}q}{e(m_\mathrm{i} + Zm_\mathrm{e})} \approx \frac{Z\rho}{m_\mathrm{i}} - \frac{q}{e} \approx \frac{Z\rho}{m_\mathrm{i}} \\
n_\mathrm{i} &= \frac{e\rho - m_\mathrm{e}q}{e(m_\mathrm{i} + Zm_\mathrm{e})} \approx \frac{\rho}{m_\mathrm{i}} = \frac{\rho}{m_\mathrm{i}} \\
\mathbf{u}^\mathrm{e} &= \frac{Ze\rho\mathbf{u} - m_\mathrm{i}\mathbf{j}}{Ze\rho - m_\mathrm{i}q} = \frac{Ze\rho\mathbf{u} - m_\mathrm{i}\mathbf{j}}{Ze\rho - m_\mathrm{i}q} \approx \mathbf{u} - \frac{m_\mathrm{i}}{Ze\rho}\mathbf{j} \\
\mathbf{u}^\mathrm{i} &= \frac{e\rho\mathbf{u} + m_\mathrm{e}\mathbf{j}}{e\rho + m_\mathrm{e}q} \approx \mathbf{u} = \mathbf{u}
\end{aligned}\right\} \tag{12.85}$$

we can transform the two-fluid moment equations (12.74) and (12.75) to their equivalent one-fluid moment equations for $\rho, q, \mathbf{u}$ and $\mathbf{j}$. The small electron to ion mass ratio allows us to simplify the new equations for $\mathbf{u}$ and $\mathbf{j}$ by taking the

$m_e/m_i \to 0$ limit. The first set of approximations (third column) in (12.85) corresponds to this limit. The second set of approximations (final column) is obtained in the limit $|q| \ll Ze\rho/m_i$ and allows a further simplification of these equations. This approximation is a consequence of the assumption of quasi-neutrality since what we mean by this is that the charge imbalance should be a very small fraction of the electronic or ionic charge, that is

$$|q| \ll en_e \approx Zen_i \approx Ze\rho/m_i \tag{12.86}$$

Multiplying the ion and electron versions of (12.74) by their respective masses and adding gives, on using (12.81) and (12.83),

$$\frac{\partial \rho}{\partial t} + \boldsymbol{\nabla} \cdot (\rho \mathbf{u}) = 0 \tag{12.87}$$

which is the one-fluid continuity equation. Multiplying by the respective charges and adding gives, on using (12.82) and (12.84), the equation of charge conservation

$$\frac{\partial q}{\partial t} + \boldsymbol{\nabla} \cdot \mathbf{j} = 0 \tag{12.88}$$

Both (12.87) and(12.88) are, of course, exact.

Next we add the ion and electron versions of (12.75) to obtain

$$\begin{aligned} \frac{\partial(\rho u_i)}{\partial t} + \frac{\partial}{\partial r_j}(\rho u_i u_j + P\delta_{ij} + \Pi^{\mathrm{i}}_{ij} + \Pi^{\mathrm{e}}_{ij}) & \\ - qE_i - (\mathbf{j} \times \mathbf{B})_i &= 0 \end{aligned} \tag{12.89}$$

where the total scalar pressure

$$P \approx \frac{Z\rho T_e}{m_i} + \frac{\rho T_i}{m_i} \approx n_e T_e + n_i T_i \tag{12.90}$$

and terms of order (Zm_e/m_i) and $(m_i q/Ze\rho)$ have been neglected. Likewise, multiplying (12.75) by e_α/m_α and adding we obtain in the same approximation

$$\begin{aligned} &\frac{\partial j_i}{\partial t} + \frac{\partial}{\partial r_k}\left(j_k u_i + j_i u_k - \frac{m_i}{Ze\rho} j_i j_k\right) \\ &- \frac{Ze}{m_i m_e}\frac{\partial(\rho T_e)}{\partial r_i} + \frac{\partial}{\partial r_k}\left(\frac{Ze}{m_i}\Pi^{\mathrm{i}}_{ki} - \frac{e}{m_e}\Pi^{\mathrm{e}}_{ki}\right) \\ &- \frac{Ze^2\rho}{m_i m_e}E_i - \frac{Ze^2\rho}{m_i m_e}\left[\left(\mathbf{u} - \frac{m_i}{Ze\rho}\mathbf{j}\right) \times \mathbf{B}\right]_i = -\frac{e}{m_e}R^{\mathrm{e}}_i \end{aligned} \tag{12.91}$$

This equation is the generalized Ohm's law since it relates the current to the electromagnetic fields and, under conditions yet to be discussed, reduces to the simple Ohm's law (3.35).

On using the transformation equations (12.85) (final column), (12.76) gives for the evolution of ion and electron temperatures

$$\frac{\rho}{m_{\mathrm{i}}}\frac{\partial T_{\mathrm{i}}}{\partial t} + \frac{\rho}{m_{\mathrm{i}}}\mathbf{u}\cdot\nabla T_{\mathrm{i}} + \frac{2\rho T_{\mathrm{i}}}{3m_{\mathrm{i}}}\nabla\cdot\mathbf{u} + \frac{2}{3}\Pi^{\mathrm{i}}_{ij}\frac{\partial u_i}{\partial r_j} + \frac{2}{3}\nabla\cdot\mathbf{q}^{\mathrm{i}} = \frac{2}{3}Q_{\mathrm{i}} \tag{12.92}$$

and

$$\frac{Z\rho}{m_{\mathrm{i}}}\frac{\partial T_{\mathrm{e}}}{\partial t} + \frac{Z\rho}{m_{\mathrm{i}}}\left(\mathbf{u} - \frac{m_{\mathrm{i}}}{Ze\rho}\mathbf{j}\right)\cdot\nabla T_{\mathrm{e}} + \frac{2Z\rho T_{\mathrm{e}}}{3m_{\mathrm{i}}}\nabla\cdot\left(\mathbf{u} - \frac{m_{\mathrm{i}}}{Ze\rho}\mathbf{j}\right)$$
$$+ \frac{2}{3}\Pi^{\mathrm{e}}_{ij}\frac{\partial}{\partial r_j}\left(u_i - \frac{m_{\mathrm{i}}}{Ze\rho}j_i\right) + \frac{2}{3}\nabla\cdot\mathbf{q}^{\mathrm{e}} = \frac{2}{3}Q_{\mathrm{e}} \tag{12.93}$$

Let us summarize what has been done in this section. The ion and electron kinetic equations have been replaced by an infinite series of moment equations. The leading equations in this series have been obtained explicitly and describe the evolution of the hydrodynamical (ρ, $\mathbf{u}$, T_{i}, T_{e}) and electrodynamical (q, $\mathbf{j}$) variables. However, this is *not* a closed set of equations for it contains the undetermined quantities: $\mathbf{\Pi}^{\mathrm{i}}$, $\mathbf{\Pi}^{\mathrm{e}}$, $\mathbf{q}^{\mathrm{i}}$, $\mathbf{q}^{\mathrm{e}}$, $\mathbf{R}^{\mathrm{e}}$, Q_{i} and Q_{e}. The system cannot be closed simply by generating higher moment equations because the evolution equation for a moment of order n automatically introduces a moment of order $n+1$, the evolution of which demands yet another equation in the infinite chain. Closure can be achieved only by truncating on the basis of some appropriate physical approximation. This is the subject of *classical transport theory* to which we now turn.

12.6 Classical transport theory

Plasma transport theory has much in common with the classical transport theory of neutral gases and yet it is sufficiently different that it has remained an active area of plasma research. The fullest and most rigorous account has been given by Balescu (1988) who deals in turn with the regimes of *classical* and *neoclassical transport*. Classical theory applies to plasmas in which the transport of matter, momentum and energy is governed by the interaction of the particles through binary collisions in the presence of slowly varying electric and magnetic fields; the external magnetic field is assumed to be straight, homogeneous and stationary.

Neoclassical theory applies to plasmas in curved, inhomogeneous fields. Physically, this is the only difference from classical theory but it turns out that the global geometry of the field, as opposed to its local value, plays a dominant role in the transport so that a fluid description remains valid even in the limit of very weak collisions (long mean free path). Nevertheless, even in this limit, neoclassical theory (a brief introduction to which was given in Section 8.3) remains a *collisional* transport theory because collisions, though rare, play an essential role and no account is taken of 'collective' interactions. When turbulent processes arising from wave instabilities dominate we enter the regime of *anomalous transport*.

It is impossible to give here more than a brief introduction to classical transport theory, the regime that is appropriate for the derivation of the MHD equations. Our account is not much more than a précis of Balescu's treatment. Truncation and closure of the infinite chain of moment equations is based on an analysis of the time scales that characterize the plasma dynamics. In the MHD approximation it is assumed that the hydrodynamic time scale τ_H characterizes the electric and magnetic fields $\mathbf{E}$ and $\mathbf{B}$, as well as the hydrodynamic variables $\rho, \mathbf{u}$, and T_α; in other words, field fluctuations in time and space arise from the plasma motion. One immediate consequence of this, noted in Chapter 3, is that the electric field is of order (u/c) compared with the magnetic field so that the displacement current may be neglected and the Maxwell equations become

$$\nabla \times \mathbf{E} = -\frac{\partial \mathbf{B}}{\partial t} \tag{12.94}$$

$$\nabla \times \mathbf{B} = \mu_0 \mathbf{j} \tag{12.95}$$

$$\nabla \cdot \mathbf{E} = \frac{q}{\varepsilon_0} \tag{12.96}$$

$$\nabla \cdot \mathbf{B} = 0 \tag{12.97}$$

A second consequence is that the $q\mathbf{E}$ term in (12.89) may be neglected compared with the $\mathbf{j} \times \mathbf{B}$ term since a dimensional analysis of (12.95) and (12.96) gives

$$\frac{|q\mathbf{E}|}{|\mathbf{j} \times \mathbf{B}|} \sim \frac{\varepsilon_0 \mu_0 E^2}{B^2} \sim \left(\frac{u}{c}\right)^2 \tag{12.98}$$

This has the effect of removing the charge density entirely from the hydrodynamical moment equations and makes (12.88) redundant. From (12.95) we see, in fact, that this equation now reads $\nabla \cdot \mathbf{j} \approx 0$, $\partial q/\partial t \approx 0$, these approximations being valid to order $(u/c)^2$. Thus, in the MHD approximation q is determined by (12.96) and $\mathbf{j}$ by (12.95). There are then two equations relating $\mathbf{E}$ and $\mathbf{B}$, namely (12.94) and the generalized Ohm's law (12.91), and the hydrodynamic moment equations (12.87), (12.89) (with $q = 0$), (12.92) and (12.93), for $\rho, \mathbf{u}, T_\mathrm{i}$ and T_e, respectively.

12.6.1 Closure of the moment equations

The physical approximation that enables us to achieve closure in classical transport theory is that τ_H is much longer than the collisional relaxation times τ_e and τ_i:

$$\tau_\alpha \ll \tau_\mathrm{H} \qquad (\alpha = i, e) \tag{12.99}$$

It then follows that ions and electrons will, on the short time scales τ_α, approach *local Maxwellian distributions*

$$f_{\alpha 0}(\mathbf{r}, \mathbf{v}, t) = n_\alpha(\mathbf{r}, t)\left(\frac{m_\alpha}{2\pi T_\alpha(\mathbf{r}, t)}\right)^{3/2} \exp\left\{-\frac{m_\alpha[\mathbf{v} - \mathbf{u}^\alpha(\mathbf{r}, t)]^2}{2T_\alpha(\mathbf{r}, t)}\right\} \quad (12.100)$$

These functions, when substituted in the kinetic equations, give zero contributions from the collision terms but are not solutions of the kinetic equations because n_α, $\mathbf{u}^\alpha$ and T_α are not constants but vary with $\mathbf{r}$ and t.

It should be noted that the two-fluid variables, n_α, $\mathbf{u}^\alpha$, T_α, which characterize the local Maxwellians in (12.100) vary on the slow time scale despite the presence of collision terms in the evolution equations of $\mathbf{u}^\alpha$ and T_α. That the equalization of ion and electron temperatures takes place on the slow time scale is not surprising since we have already observed that unlike particle collisions are inefficient exchangers of energy because of the disparity in the masses. The argument in the case of the flow velocities is best explained from (12.85). The ion flow velocity is approximately the plasma flow velocity and therefore slowly varying. Furthermore, the difference between ion and electron flow velocities is represented by the current $\mathbf{j}$ which, in the MHD approximation, is defined by (12.95) in terms of the magnetic field and, therefore, it also changes on the hydrodynamic time scale. Finally, as discussed in the next section, short time scale variation arising from collision terms is merely transient.

The distribution function is now written as

$$f_\alpha(\mathbf{r}, \mathbf{v}, t) = f_{\alpha 0}[1 + \chi_\alpha(\tilde{\mathbf{c}}; \mathbf{r}, t)] \quad (12.101)$$

where χ_α measures the deviation of f_α from $f_{\alpha 0}$ and for convenience we have introduced the dimensionless velocity†

$$\tilde{\mathbf{c}} = \left(\frac{m_\alpha}{T_\alpha(\mathbf{r}, t)}\right)^{1/2} [\mathbf{v} - \mathbf{u}^\alpha(\mathbf{r}, t)] \quad (12.102)$$

Standard procedure in transport theory is then to expand χ_α in a series of orthogonal polynomials in $\tilde{\mathbf{c}}$ with coefficients that are functions of $\mathbf{r}$ and t related to moments of f_α. There are two desirable features of any chosen procedure. One is to find an expansion which is rapidly convergent so that a truncated set of equations (for the truncated set of moments) determines transport properties to the desired degree of approximation. The second requirement is to choose a method of expansion that is physically transparent. The singular achievement of Balescu's approach is the combination of both these features.

The tensorial Hermite polynomials are a natural choice for the expansion of χ_α since they are orthogonal with respect to the weight function $f_{\alpha 0}$. This choice

† Clearly, $\tilde{\mathbf{c}}$ should have the label α but this is suppressed to avoid cluttering the notation.

was used by Grad (1949) who developed the moment method of solution of the kinetic equations. The method suffers a serious disadvantage, however, due to the lack of a one-to-one correspondence between successive terms in the expansion and the physical moments. The reason for this is that the expansion is in terms of *reducible* Hermitian tensors so that there are, for example, contributions to the vector moment from the contracted (reduced) third-order tensor. This also causes slow convergence in this method. Balescu overcomes this disadvantage by first writing χ_α in a representation which exhibits explicitly the possible anisotropies of the distribution function, that is

$$\chi_\alpha(\tilde{\mathbf{c}}; \mathbf{r}, t) = A_\alpha(\tilde{\mathbf{c}}; \mathbf{r}, t) + \tilde{c}_k B_k^\alpha(\tilde{\mathbf{c}}; \mathbf{r}, t) + \left(\tilde{c}_k\tilde{c}_l - \frac{1}{3}\tilde{c}^2\delta_{kl}\right) C_{kl}^\alpha(\tilde{\mathbf{c}}; \mathbf{r}, t) + \cdots \tag{12.103}$$

where A_α is a scalar function, B_k^α is a vector function and C_{kl}^α is a symmetric traceless second-order tensor. Further terms (which we shall not need) involve successively higher-order, *irreducible* tensors and tensor functions. Balescu then expands the functions A_α, B_k^α, C_{kl}^α in series of irreducible tensorial Hermite polynomials which maintain the Cartesian representations of the vector $\tilde{\mathbf{c}}$. This has an advantage over the Chapman–Enskog method which takes (12.103) as a basic ansatz but writes $\tilde{\mathbf{c}}$ in spherical polar coordinates and expands χ_α in an infinite series of Laguerre–Sonine polynomials and spherical harmonics.

The chief advantage of the representation of χ_α in (12.103), however, lies in the relationship between the terms in the mathematical series and the physical quantities. In a transport theory that is linear in χ_α it is clear that the heat flux $\mathbf{q}^\alpha$, for example, being a vector moment of f_α is completely determined by the vector part of χ_α, i.e. by B_k^α. Likewise, Π_{kl}^α is completely determined by C_{kl}^α. Furthermore, choosing the exact values of $n_\alpha(\mathbf{r}, t)$, $\mathbf{u}^\alpha(\mathbf{r}, t)$ and $T_\alpha(\mathbf{r}, t)$, as defined by (12.55), (12.56) and (12.63), to define the local equilibrium distribution (12.100) puts the following constraints on χ_α:

$$\left.\begin{aligned} \int \mathrm{d}\tilde{\mathbf{c}}\, f_{\alpha 0}\chi_\alpha &= 0 \\ \int \mathrm{d}\tilde{\mathbf{c}}\, f_{\alpha 0}\tilde{\mathbf{c}}\chi_\alpha &= 0 \\ \int \mathrm{d}\tilde{\mathbf{c}}\, f_{\alpha 0}\tilde{c}^2\chi_\alpha &= 0 \end{aligned}\right\} \tag{12.104}$$

It follows that in linear theory only the vector and traceless second-order tensor terms in (12.103) are required. Balescu shows, moreover, that the deeper physical significance of this lies in the fact that only these terms contribute to entropy production in the linear regime. Thus, (12.103) is replaced by

$$\chi_\alpha(\tilde{\mathbf{c}}; \mathbf{r}, t) = \tilde{c}_k B_k^\alpha(\tilde{\mathbf{c}}; \mathbf{r}, t) + \left(\tilde{c}_k\tilde{c}_l - \frac{1}{3}\tilde{c}^2\delta_{kl}\right) C_{kl}^\alpha(\tilde{\mathbf{c}}; \mathbf{r}, t) \tag{12.105}$$

The expansions for B_k^α and C_{kl}^α may be truncated at various levels with each successive level adding three vector plus five (independent) tensor components so that allowing for the five moments corresponding to n_α, $\mathbf{u}^\alpha$ and T_α the choices are for $5, 13, 21, 29, \ldots$ moment approximations. Balescu shows that whereas there is a considerable difference between the 13 moment (13M) approximation and the 21M approximation no significant improvement is provided by the 29M approximation. There is not the space here to present the mathematical analysis of even the 13M approximation. Rather we shall describe the procedure and present the main results.

12.6.2 Derivation of the transport equations

Balescu's procedure leads to a natural classification of the moments which arises mathematically but is rooted in the physics. In the two-fluid description of the plasma the first class contains the variables n_α, $\mathbf{u}^\alpha$ and T_α which Balescu calls the *plasmadynamical moments*. The physical significance of these moments is that they determine the local equilibrium distribution. All the other moments in the two-fluid description (the complementary set) are called the *non-plasmadynamical moments* and have the complementary property that they are identically zero in the local equilibrium state; this follows from the orthogonality properties of the Hermite polynomials. However, we shall not be directly concerned with the non-plasmadynamical moments because our goal is the one-fluid MHD description for which we make the transformations

$$\begin{pmatrix} n_e \\ n_i \end{pmatrix} \to \begin{pmatrix} \rho &=& m_e n_e + m_i n_i \\ q &=& -e n_e + Z e n_i \end{pmatrix} \tag{12.106}$$

$$\begin{pmatrix} \mathbf{u}^e \\ \mathbf{u}^i \end{pmatrix} \to \begin{pmatrix} \rho\mathbf{u} &=& m_e n_e \mathbf{u}^e + m_i n_i \mathbf{u}^i \\ \mathbf{j} &=& -e n_e \mathbf{u}^e + Z e n_i \mathbf{u}^i \end{pmatrix} \tag{12.107}$$

In this description the classification is based on properties of the evolution equations by means of which moments are first separated into those which are *hydrodynamical* and all others which are *non-hydrodynamical*. The class of hydrodynamical moments includes ρ, $\mathbf{u}$, T_e, T_i whose evolution equations contain no collision term or, in the case of the temperatures, one which is negligibly small. The charge density q shares this property and could be added to this class but, as we have seen already, it has no role in the MHD approximation and may be left out of the discussion. The other electrodynamical moment $\mathbf{j}$ belongs to the non-hydrodynamical class because its evolution equation (12.91) contains a significant collision term. This is also true of the electron and ion fluxes defined by

$$\boldsymbol{\Gamma}^\alpha = n_\alpha \mathbf{u}^\alpha \qquad (\alpha = e, i) \tag{12.108}$$

which are sometimes (e.g. in neoclassical theory) used instead of $\mathbf{j}$. These moments share this property with the pressure tensors and heat fluxes. Another shared property of these moments is that the evolution equations for $\mathbf{\Gamma}^\alpha$, $\mathbf{j}$, Π^α_{rs} and $\mathbf{q}_\alpha$ all contain a source term which is a function of the hydrodynamical moments. Since these are the only non-hydrodynamical moments which have this property and, furthermore, these are the only non-hydrodynamical moments appearing in the evolution equations for hydrodynamical moments, it is clear that they play a special role and the class of non-hydrodynamical moments is, therefore, further subdivided into these *privileged non-hydrodynamical moments* and all others which are *non-privileged*.

The physical significance of this subdivision, as Balescu explains, is the one-to-one correspondence between the solutions of the evolution equations for the privileged moments and the transport equations of non-equilibrium thermodynamics. The latter relate *thermodynamic fluxes*, J_n, and *forces*, X_n, by a set of *transport equations*

$$J_n = \sum_m L_{nm} X_m$$

in which the elements of the *transport matrix*, L_{nm}, are the *transport coefficients*, having the Onsager symmetry $L_{nm} = L_{mn}$. Thus, the privileged non-hydrodynamical moments may be identified with the thermodynamic fluxes and the source terms (involving the hydrodynamical moments) with the thermodynamic forces. On the other hand, the solutions for the non-privileged moments have a quite different structure and their main role is to improve the accuracy of the transport coefficients as one goes to higher moment approximations.

This classification of the moments is highly significant also for the next stage of classical transport theory. Although closure of the set of moment equations is achieved by adopting, say, the 21M approximation, there remains the task of solving a sub-set of these equations so that most of the moments may be eliminated leaving a much smaller set of equations for the physically most interesting variables. Specifically, the goal is to find expressions for the privileged non-hydrodynamical moments in terms of the hydrodynamical moments which can then be substituted in the evolution equations of the latter thereby yielding the set of MHD equations.

A subsidiary but necessary task is to express the collision terms as functions of the moments. This is accomplished by first substituting (12.105) in (12.101) and then (12.101) in (12.77), (12.78) and the other collision terms arising in other moment equations; the Landau collision integral (12.54) is used for $(\partial f_\alpha/\partial t)_\mathrm{c}$. The collision terms that we shall need explicitly are given by

$$\mathbf{R}^\mathrm{e} = \frac{m_\mathrm{e}}{\tau_\mathrm{e}}\left(\frac{\mathbf{j}}{e} + \frac{3\mathbf{q}^\mathrm{e}}{5T_\mathrm{e}} + \cdots\right) = -\mathbf{R}^\mathrm{i} \tag{12.109}$$

and

$$
\begin{aligned}
Q_{\mathrm{i}} &= \frac{3Zm_{\mathrm{e}}n_{\mathrm{i}}}{m_{\mathrm{i}}\tau_{\mathrm{e}}}(T_{\mathrm{e}} - T_{\mathrm{i}}) \\
&= -Q_{\mathrm{e}} - (\mathbf{u}^{\mathrm{e}} - \mathbf{u}^{\mathrm{i}}) \cdot \mathbf{R}^{\mathrm{e}} \\
&= -Q_{\mathrm{e}} + \frac{1}{en_{\mathrm{e}}}\mathbf{j} \cdot \mathbf{R}^{\mathrm{e}}
\end{aligned} \tag{12.110}
$$

where we have used (12.85). In the expression for $\mathbf{R}^{\mathrm{e}}$ successive terms arise from progressively higher moment aproximations; the $\mathbf{j}$ term is there in the 5M approximation, the $\mathbf{q}$ term in the 13M approximation, etc. In principle, we are discussing the 21M approximation and so the next term in the bracket is included in Balescu's calculations. This can only be expressed in terms of a collision coefficient and the next Hermite coefficient in the expansion of B_k^{e} which we have not defined. It turns out, however, that we shall not need it explicitly for reasons that will become apparent and so we do not write it out in (12.109). The point to notice about Q_{i} is the presence of the factor $m_{\mathrm{e}}/m_{\mathrm{i}}$ confirming that transfer of thermal energy from electrons to ions is a very slow process.

For what comes next, it should be noted that the collision terms in general are all proportional to χ_α, since $f_{\alpha 0}$ is Maxwellian and $\chi_\alpha = 0$ in (12.101) would make all collision terms vanish. In linear transport theory, solution of the moment equations is carried out to first order in the small parameter $\epsilon = \tau_\alpha/\tau_{\mathrm{H}}$. If all moments are written in dimensionless variables then all terms have dimensions of an inverse time which is either τ_{H} or τ_α. A self-consistent linear transport theory is then obtained by assuming that χ_α is at most of first order in the small parameter $\epsilon = \tau_\alpha/\tau_{\mathrm{H}}$. The hydrodynamical moments are taken to be of order zero with non-hydrodynamical moments and collision terms at most of order one. No assumption is made about Larmor frequencies which enter through the Lorentz force, so $\Omega_\alpha \tau_\alpha$ is treated as zero order.

Now if we examine the evolution equations (12.87), (12.89), (12.92) and (12.93) for the hydrodynamical moments, we find that these contain either no collision terms or, in the case of T_{e} and T_{i}, collision terms which are either of order $m_{\mathrm{e}}/m_{\mathrm{i}}$ or ϵ^2 and therefore negligible. The hydrodynamical moments consequently change only on the slow time scale, as anticipated earlier.

The equations for the privileged non-hydrodynamical moments, on the other hand, contain both slow and fast time scale terms with the leading order terms showing that fast time scale collision terms are balanced by slow time scale terms consisting of gradients of hydrodynamical moments and electromagnetic field terms. In (12.91), the only example of such an equation that we have explicitly presented, the leading terms are the third, fifth, sixth and seventh so that the *lin-*

earized, generalized Ohm's law is

$$\frac{Ze\rho}{m_\mathrm{i}}(\mathbf{E}+\mathbf{u}\times\mathbf{B})+\frac{Z}{m_\mathrm{i}}\nabla(\rho T_\mathrm{e})-\mathbf{j}\times\mathbf{B}=\mathbf{R}^\mathrm{e} \tag{12.111}$$

It could be argued that the $\partial\mathbf{j}/\partial t$ term should also be retained since, through the collision term, **j** can vary on the fast time scale. This is quite true but Balescu shows that fast time scale changes cause only transient behaviour which disappears after a few collision times so that asymptotic results can be obtained by simply dropping the time derivatives and solving the resulting algebraic equations. This transient behaviour is not without physical interest for it is in the transition from transient to asymptotic state that the system becomes Markovian. For a few collision times the plasma maintains a dependence on its (unknown) initial state and its dynamical behaviour is a function of the whole history of its motion up to that point. But for $t \gg \tau_\alpha$ collisions have wiped out all knowledge of the initial state and any dependence on past history extends no further back than about $7\tau_\alpha$ which is negligible in a consistent linear theory. There is a striking analogy here with the derivation of the kinetic equation from the BBGKY hierarchy, with non-hydrodynamic moments playing the role of the correlation function and hydrodynamic moments the role of the one-particle distribution function; of course the collision time τ_α has changed its role from being the slow time scale (compared with τ_c the duration of a collision) to being the fast time scale compared with τ_H.

Balescu's transport equations expressing the privileged non-hydrodynamical vector moments in terms of hydrodynamical moments and the electric and magnetic fields are

$$j_k = \sigma_{kl}\hat{E}_l - \alpha_{kl}\partial T_\mathrm{e}/\partial r_l \tag{12.112}$$

$$q_k^\mathrm{e} = \alpha_{kl}T_\mathrm{e}\hat{E}_l - \kappa_{kl}^\mathrm{e}\partial T_\mathrm{e}/\partial r_l \tag{12.113}$$

$$q_k^\mathrm{i} = -\kappa_{kl}^\mathrm{i}\partial T_\mathrm{i}/\partial r_l \tag{12.114}$$

$$\tag{12.115}$$

where the effective electric field

$$\hat{\mathbf{E}} = \mathbf{E}+\mathbf{u}\times\mathbf{B}+\frac{1}{en_\mathrm{e}}\nabla(n_\mathrm{e}T_\mathrm{e}) \tag{12.116}$$

includes a thermoelectric component. The transport coefficients are second-order tensor quantities representing electrical conductivity, σ_{kl}, the thermoelectric coefficients, α_{kl}, and thermal conductivity, κ_{kl}^α. They have just three independent components and relative to Cartesian axes with $\mathbf{B} = B\hat{\mathbf{z}}$, take the form

$$\mathbf{L} = \begin{pmatrix} L_\perp & L_\wedge & 0 \\ -L_\wedge & L_\perp & 0 \\ 0 & 0 & L_\parallel \end{pmatrix} \tag{12.117}$$

where $L_\parallel$, $L_\perp$ are the coefficients parallel and perpendicular to $\mathbf{B}$ and $L_\wedge$ is the only non-zero, off-diagonal element. This representation expresses the invariance of these transport coefficients under rotations of the axes in the plane perpendicular to $\mathbf{B}$. The traceless pressure tensors, Π^α_{kl}, are already of second-order and so they introduce a fourth-order viscosity tensor, μ_{klmn}, through the relationship

$$\Pi^\alpha_{kl} = -\mu^\alpha_{klmn} \nu_{mn} \tag{12.118}$$

where

$$\nu_{mn} = \left(\frac{\partial u_n}{\partial r_m} + \frac{\partial u_m}{\partial r_n} - \frac{2}{3} \delta_{mn} \nabla \cdot \mathbf{u} \right) \tag{12.119}$$

However, the same symmetries that reduce the second-order tensors to just three independent elements mean that μ^α_{klmn} has only five independent elements which we label $\mu^\alpha_\parallel, \mu^\alpha_1, \mu^\alpha_2, \mu^\alpha_3, \mu^\alpha_4$. Balescu shows that (12.118) may then be written

$$\left.\begin{aligned}
\Pi^\alpha_{xx} &= -\tfrac{1}{2}(\mu^\alpha_\parallel + \mu^\alpha_4)\nu_{xx} - \tfrac{1}{2}(\mu^\alpha_\parallel - \mu^\alpha_4)\nu_{yy} + \mu^\alpha_3 \nu_{xy} \\
\Pi^\alpha_{yy} &= -\tfrac{1}{2}(\mu^\alpha_\parallel - \mu^\alpha_4)\nu_{xx} - \tfrac{1}{2}(\mu^\alpha_\parallel + \mu^\alpha_4)\nu_{yy} - \mu^\alpha_3 \nu_{xy} \\
\Pi^\alpha_{xy} &= -\tfrac{1}{2}\mu^\alpha_3 \nu_{xx} + \tfrac{1}{2}\mu^\alpha_3 \nu_{yy} - \mu^\alpha_4 \nu_{xy} = \Pi^\alpha_{yx} \\
\Pi^\alpha_{xz} &= -\mu^\alpha_2 \nu_{xz} + \mu^\alpha_1 \nu_{yz} = \Pi^\alpha_{zx} \\
\Pi^\alpha_{yz} &= -\mu^\alpha_1 \nu_{xz} - \mu^\alpha_2 \nu_{yz} = \Pi^\alpha_{zy} \\
\Pi^\alpha_{zz} &= -\mu^\alpha_\parallel \nu_{zz}
\end{aligned}\right\} \tag{12.120}$$

A further reduction arises from the relationships

$$\mu^\alpha_4(\Omega_\alpha \tau_\alpha) = \mu^\alpha_2(2\Omega_\alpha \tau_\alpha), \qquad \mu^\alpha_3(\Omega_\alpha \tau_\alpha) = \mu^\alpha_1(2\Omega_\alpha \tau_\alpha)$$

discovered by Braginsky (1965) so that effectively we can again consider just three coefficients.

12.6.3 Classical transport coefficients

A number of well-known effects arise from the non-diagonal elements of the transport coefficients. For example, a non-zero E_x gives rise to a component of current in the y direction, $j_y = \sigma_{yx} E_x$; this is the *Hall current*. Similarly, a temperature gradient in the x direction produces a heat flux in the y direction, $q^\alpha_y = -\kappa^\alpha_{yx} \partial T_\alpha / \partial x$, known as the *Righi–Leduc effect*. The *Nernst effect* is the heat flux $q^{\mathrm{e}}_y = \alpha_{yx} T_{\mathrm{e}} E_x$ produced by an electric field in the x direction, and the *Ettinghausen effect*, produced by the same coefficient, is the electric current, $j_y = -\alpha_{yx} \partial T_{\mathrm{e}} / \partial x$, due to an electron temperature gradient in the x direction.

The transport coefficients are, of course, dependent on the plasma parameters as well as the magnetic field. They can be written in terms of dimensionless functions of the atomic number Z and $X_\alpha = \Omega_\alpha \tau_\alpha$ as follows:

$$\left.\begin{aligned} \sigma_{kl} &= \left(\frac{e^2 n_e}{m_e}\right) \tau_e \tilde{\sigma}_{kl}(Z, X_e) \\ \alpha_{kl} &= \left(\frac{5}{2}\right)^{1/2} \left(\frac{e n_e}{m_e}\right) \tau_e \tilde{\alpha}_{kl}(Z, X_e) \\ \kappa^\alpha_{kl} &= \left(\frac{5 n_\alpha T_\alpha}{2 m_\alpha}\right) \tau_\alpha \tilde{\kappa}^\alpha_{kl}(Z, X_\alpha) \\ \mu^\alpha_p &= n_\alpha T_\alpha \tau_\alpha \tilde{\mu}^\alpha_p(Z, X_\alpha) \quad (p = \|, 1, 2, 3, 4) \end{aligned}\right\} \tag{12.121}$$

with

$$\left.\begin{aligned} \tau_e &= \frac{6\sqrt{2}\pi^{3/2}\varepsilon_0^2 m_e^{1/2} T_e^{3/2}}{Z^2 e^4 n_i \ln \Lambda} \\ \tau_i &= \frac{6\sqrt{2}\pi^{3/2}\varepsilon_0^2 m_i^{1/2} T_i^{3/2}}{Z^4 e^4 n_i \ln \Lambda} \end{aligned}\right\} \tag{12.122}$$

In (12.121) the dimensionless transport coefficients (denoted by the tilde) are tabulated functions of Z and X_α the precise details of which depend upon the number of moments taken in the approximation.

In the limit of zero magnetic field, $X_\alpha \to 0$, we find

$$L_\perp = L_\|, \qquad L_\wedge = 0 \tag{12.123}$$

where L stands for σ, α or κ^α, and

$$\mu^\alpha_2 = \mu^\alpha_4 = \mu^\alpha_\|, \qquad \mu^\alpha_1 = \mu^\alpha_3 = 0 \tag{12.124}$$

Thus, (12.112)–(12.118) reduce to

$$\mathbf{j} = \sigma_\| \hat{\mathbf{E}} - \alpha_\| \nabla T_e \tag{12.125}$$

$$\mathbf{q}^e = \alpha_\| T_e \hat{\mathbf{E}} - \kappa^e_\| \nabla T_e \tag{12.126}$$

$$\mathbf{q}^i = -\kappa^i_\| \nabla T_i \tag{12.127}$$

$$\Pi^\alpha_{kl} = -\mu^\alpha_\| v_{kl} \tag{12.128}$$

We see that all the transport coefficients have become scalars whose values are given by the parallel coefficients in this limit. The ion heat flux is simply given by Fourier's law (12.127). The asymmetry between this equation and (12.126) for the electron heat flux, as between (12.113) and (12.114), arises from the mass ratio. Only ion–ion collisions are effective for ion transport, ion–electron collision terms being of order m_e/m_i in comparison, so there is a decoupling of the ions.

The electron fluxes, on the other hand, are subject to both electric and hydrodynamic forces. If the latter are absent ($\nabla T_e = 0$) we retrieve the simple *Ohm's law*

$$\mathbf{j} = \sigma_\| \mathbf{E} \tag{12.129}$$

and also find a heat flux driven by the electric field

$$\mathbf{q}^{\mathrm{e}} = \alpha_{\|} T_{\mathrm{e}} \mathbf{E} \tag{12.130}$$

which is related to the *Peltier effect* in thermocouples.

Fourier's law for electrons

$$\mathbf{q}^{\mathrm{e}} = -\kappa_{\|}^{\mathrm{e}} \boldsymbol{\nabla} T_{\mathrm{e}} \tag{12.131}$$

is recovered when $\hat{\mathbf{E}} = 0$, i.e. when the electric force on the electrons balances the electron pressure gradient, $en_{\mathrm{e}}\mathbf{E} = -\boldsymbol{\nabla} p_{\mathrm{e}}$. In this case there is still an electric current produced by the electron temperature gradient

$$\mathbf{j} = -\alpha_{\|} \boldsymbol{\nabla} T_{\mathrm{e}}$$

This is the *thermoelectric effect.*

It is obvious from (12.121) that for $\mathbf{B} = 0,\ X_{\alpha} = 0$ and so all of the transport coefficients are proportional to a collision time,

$$L_{\|} \propto \tau_{\alpha} \tag{12.132}$$

The implication of this is that collisions impede transport parallel to the magnetic field, as one would expect since it is only collisions that interrupt the flow of matter, momentum and energy in response to the thermodynamic forces. In this case the dependence on density and temperature can be simply expressed because the dimensionless functions $\tilde{L}_{\|}$ are functions of Z only. From (12.121) and(12.122), using $Z_{\mathrm{e}} = -1,\ Z_{\mathrm{i}} = Z$, and writing

$$A = \frac{6\sqrt{2}\pi^{3/2}\varepsilon_0^2}{\ln \Lambda} \tag{12.133}$$

we find

$$\sigma_{\|} = \frac{AT_{\mathrm{e}}^{3/2}}{e^2 m_{\mathrm{e}}^{1/2}} \frac{\tilde{\sigma}_{\|}(Z)}{Z} \tag{12.134}$$

$$\alpha_{\|} = \sqrt{\frac{5}{2}} \frac{AT_{\mathrm{e}}^{3/2}}{e^3 m_{\mathrm{e}}^{1/2}} \frac{\tilde{\alpha}_{\|}(Z)}{Z} \tag{12.135}$$

$$\kappa_{\|}^{\alpha} = \frac{5AT_{\alpha}^{5/2}}{2e^4 m_{\alpha}^{1/2}} \frac{\tilde{\kappa}_{\|}^{\alpha}(Z)}{|ZZ_{\alpha}^3|} \tag{12.136}$$

$$\mu_{\|}^{\alpha} = \frac{Am_{\alpha}^{1/2} T_{\alpha}^{5/2}}{e^4} \frac{\tilde{\mu}_{\|}^{\alpha}(Z)}{|ZZ_{\alpha}^3|} \tag{12.137}$$

Ignoring the weak density and temperature dependence contained in A through $\ln \Lambda$ we see that all coefficients increase with temperature, with the thermal conductivities and viscosities showing a particularly strong dependence, and all coefficients are independent of density. This latter feature is typical of the weak coupling approximation applied to the Landau kinetic equation and is true for the transport coefficients of a dilute neutral gas as well.

The electrical conductivity and thermoelectric coefficient are, of course, determined by the electrons because of their much smaller mass and this has already been taken into account by ignoring terms of order m_e/m_i. But now from (12.136) and (12.137) we can compare electron and ion thermal conductivities and viscosities. We see that

$$\frac{\kappa_{\parallel}^{i}}{\kappa_{\parallel}^{e}} = \left(\frac{m_e}{m_i}\right)^{1/2}\left(\frac{T_i}{T_e}\right)^{5/2}\frac{1}{Z^3}\frac{\tilde{\kappa}_{\parallel}^{i}(Z)}{\tilde{\kappa}_{\parallel}^{e}(Z)} \tag{12.138}$$

and

$$\frac{\mu_{\parallel}^{i}}{\mu_{\parallel}^{e}} = \left(\frac{m_i}{m_e}\right)^{1/2}\left(\frac{T_i}{T_e}\right)^{5/2}\frac{1}{Z^3}\frac{\tilde{\mu}_{\parallel}^{i}(Z)}{\tilde{\mu}_{\parallel}^{e}(Z)} \tag{12.139}$$

in which the mass ratio dependence indicates the predominance of electron thermal conductivity and ion viscosity. Thus, parallel to the magnetic field, energy is mainly transported by the electrons and momentum by the ions though it should be noted that for high Z or $T_e \gg T_i$ electron viscosity could become significant.

All of this discussion of the parallel transport coefficients holds good when $\mathbf{B} \neq 0$. This is because the equations for the parallel moments decouple from those for the perpendicular moments and are independent of the magnetic field. In other words, if we were to plot the variation with $X_\alpha = \Omega_\alpha \tau_\alpha$ of the dimensionless transport coefficients, the parallel coefficients $\tilde{L}_{\parallel}(Z)$, being functions of Z only, would be horizontal straight lines, as in Fig. 12.2. This is not the case for the perpendicular coefficients, of course. From (12.123) and (12.124) we see that some of these, $\tilde{\sigma}_{\perp}, \tilde{\alpha}_{\perp}, \tilde{\kappa}_{\perp}^{\alpha}, \tilde{\mu}_2^{\alpha}$ and $\tilde{\mu}_4^{\alpha}$, start at $X_\alpha = 0$ with values equal to the corresponding parallel coefficient while the others are all zero at $X_\alpha = 0$. Representing all the initially non-zero set by the label $\tilde{L}_{\perp}$ and the others by $\tilde{L}_{\wedge}$ we find that for fixed Z, $\tilde{L}_{\perp}$ decreases in magnitude as X_α increases, while $\tilde{L}_{\wedge}$ first increases in magnitude and then decreases. This is shown schematically in Fig. 12.2 where the decrease in $\tilde{L}_{\perp}$ is presented as monotonic. This is true for all except $\tilde{\alpha}_{\perp}$ which is actually negative at $X_\alpha = 0$ and first decreases in magnitude, passes through zero, and then asymptotically approaches zero as $X_\alpha \to \infty$. Although the figure is only schematic it should be noted that all the $\tilde{L}_{\perp} \sim X_\alpha^{-2}$ as $X_\alpha \to \infty$ whereas, with the single exception of $\tilde{\alpha}_{\wedge}$, all the $|\tilde{L}_{\wedge}| \sim |X_\alpha|^{-1}$; $\tilde{\alpha}_{\wedge} \sim X_e^{-3}$ decays fastest of all.

The asymptotic behaviour of the perpendicular transport coefficients

$$\tilde{L}_{\perp} \sim \frac{1}{(\Omega_\alpha \tau_\alpha)^2} \tag{12.140}$$

means that classical transport across magnetic field lines decreases with the square

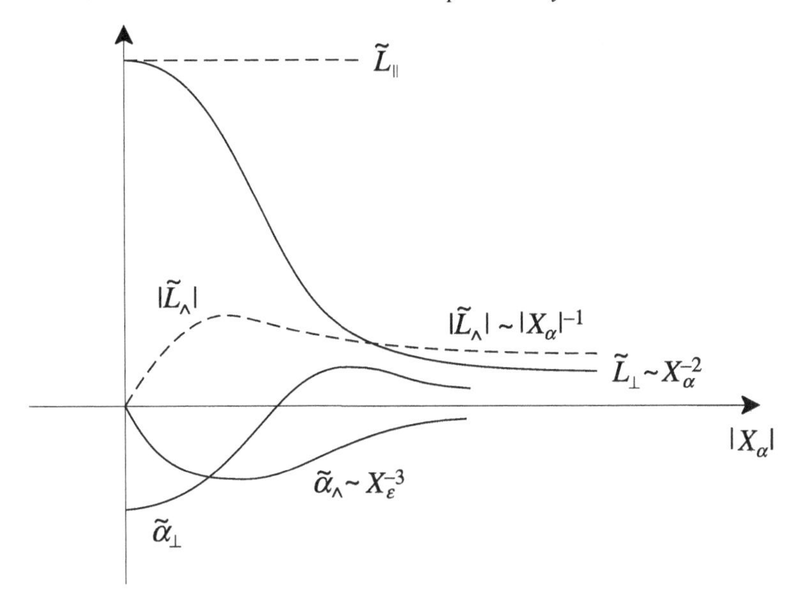

Fig. 12.2. Schematic illustration of variation of transport coefficients with $|X_\alpha|$.

of the magnetic field. Unfortunately, this turns out not to be true of toroidal confinement devices where other factors, notably the geometry of the field, come into play. Transport is neoclassical rather than classical and the perpendicular coefficients vary more like B^{-1}.

The dependence on collision time is also very interesting and shows a sharp contrast with the parallel coefficients. From (12.121) and (12.140) we see that the perpendicular coefficients vary like τ_α^{-1} as $X_\alpha \to \infty$, that is they increase with the collision frequency $\nu_\alpha = \tau_\alpha^{-1}$. Thus, while collisions impede parallel transport they increase perpendicular transport. The reason for this is easily understood and was discussed in some detail in Section 8.2.1. In a strong magnetic field particles are restricted to Larmor orbits in the plane perpendicular to **B** so transport is impeded by the field. But collisions disrupt this ordered motion and allow particles to slip across field lines thereby enhancing perpendicular transport. Note that the contrasting dependence on collision frequency of parallel and perpendicular coefficients means that in the presence of a strong field, as the collision frequency is increased, effective transport of matter, momentum and energy is transferred progressively from the parallel to the perpendicular direction.

The dependence on density and temperature shows a marked difference as well. From (12.121) and (12.140) it is easily seen that all perpendicular coefficients increase asymptotically with the square of density and decrease with temperature; $\sigma_\perp$ and $\alpha_\perp$ are proportional to $T_e^{-3/2}$ while $\kappa_\perp^\alpha$ and μ_2^α decrease like $T_\alpha^{-1/2}$. Yet another striking result arises from a comparison of ion and electron thermal conductivities. We find

$$\frac{\kappa_\perp^{\mathrm{i}}}{\kappa_\perp^{\mathrm{e}}} \sim Z^2 \left(\frac{m_\mathrm{i}}{m_\mathrm{e}}\right)^{1/2} \left(\frac{T_\mathrm{e}}{T_\mathrm{i}}\right)^{1/2} \tag{12.141}$$

Comparing this with (12.138) we see that each of these factors now favours ion transport over electron transport instead of the other way around. Hence, conclusions in the parallel direction are reversed in the perpendicular direction where energy transport is dominated by the ions.

On the other hand, comparison of ion and electron viscosities strongly reinforces the conclusions based on (12.139). For the perpendicular viscosities we get

$$\frac{\mu_2^{\mathrm{i}}}{\mu_2^{\mathrm{e}}} \sim \frac{1}{Z} \left(\frac{m_\mathrm{i}}{m_\mathrm{e}}\right)^{3/2} \left(\frac{T_\mathrm{e}}{T_\mathrm{i}}\right)^{1/2} \tag{12.142}$$

The stronger dependence on mass ratio and the weaker dependence on Z and temperature ratio means that ions dominate momentum transport even more emphatically in the perpendicular direction than in the parallel direction.

Finally, turning to the non-diagonal transport coefficients we see that, with the exception of $\tilde{\alpha}_\wedge \sim X_\mathrm{e}^{-3}$, all have the asymptotic form

$$\tilde{L}_\wedge \to -\frac{1}{X_\alpha}, \qquad |X_\alpha| \to \infty \tag{12.143}$$

Not only are these coefficients larger by a factor $|X_\alpha|$ than the perpendicular coefficients in this limit but they are also independent of any collision coefficients arising from the collision term. When we substitute (12.143) in (12.121) we find

$$\left.\begin{array}{lcl} \sigma_\wedge & \to & n_\mathrm{e}/eB \\ \kappa_\wedge^{\mathrm{e}} & \to & 5n_\mathrm{e}T_\mathrm{e}/2eB \\ \kappa_\wedge^{\mathrm{i}} & \to & -5n_\mathrm{i}T_\mathrm{i}/2ZeB \\ \mu_1^{\mathrm{e}} & \to & n_\mathrm{e}T_\mathrm{e}m_\mathrm{e}/eB \\ \mu_1^{\mathrm{i}} & \to & -n_\mathrm{i}T_\mathrm{i}m_\mathrm{i}/ZeB \end{array}\right\} \tag{12.144}$$

showing no dependence on collisions at all, an effect also noted in Section 8.2.1. Thus, in the strong **B** limit, the fluxes associated with these coefficients are *non-dissipative*, arising entirely from the anisotropy introduced into plasma motion by the magnetic field. Balescu shows, in fact, that in general the non-diagonal fluxes make no contribution to entropy production and it is for this reason that there is no contradiction implied by the negative coefficients $\kappa_\wedge^{\mathrm{i}}$ and μ_1^{i}.

Balescu compares his results with other treatments of transport theory. The most important of these is that by Braginsky (1965) and there is remarkably good agreement between Braginsky's results and those of Balescu. A particularly important feature of Balescu's presentation which we have scarcely touched upon is the role of entropy. Balescu shows how the second law of thermodynamics underlies the classification of the moments and the interpretation of the results, many of which are quite general and independent of the level of truncation of the moment approximation.

12.7 MHD equations

We are now ready to take the third and final step in the progression from kinetic equations to MHD equations. The major change from microscopic to macroscopic description has already been established in Section 12.6.2 where we obtained a closed set of equations entirely in terms of macroscopic variables. But substitution of the transport equations (12.112)–(12.118) and the collision moments into the hydrodynamic evolution equations (12.87), (12.89), (12.92) and (12.93), yields a set of equations that is, for most practical purposes, far too complicated. Further simplification is therefore essential.

To see how this may be done it is useful to summarize the full, closed set of dissipative MHD equations. Starting with the hydrodynamic evolution equations we have the continuity equation (12.87)

$$\frac{\partial \rho}{\partial t} = -\nabla \cdot (\rho \mathbf{u}) \tag{12.145}$$

The momentum balance equation (12.89), with $q\mathbf{E}$ neglected, is

$$\frac{\partial (\rho \mathbf{u})}{\partial t} = -\nabla \cdot (\rho \mathbf{u}\mathbf{u} + P\mathbf{I}) + \mathbf{j} \times \mathbf{B} - \nabla \cdot (\mathbf{\Pi}^{\mathrm{e}} + \mathbf{\Pi}^{\mathrm{i}}) \tag{12.146}$$

The electron and ion temperature equations are obtained from (12.92) and (12.93). Using (12.109) and (12.110) for the collision terms we have

$$\begin{aligned}\frac{\partial T_{\mathrm{e}}}{\partial t} &= -\hat{\mathbf{u}} \cdot \nabla T_{\mathrm{e}} - \frac{2}{3} T_{\mathrm{e}} \nabla \cdot \hat{\mathbf{u}} - \frac{2}{3n_{\mathrm{e}}} \mathbf{\Pi}^{\mathrm{e}} : \nabla \mathbf{u} \\ &\quad - \frac{2}{3n_{\mathrm{e}}} \nabla \cdot \mathbf{q}^{\mathrm{e}} + \frac{2}{3en_{\mathrm{e}}^2} \mathbf{j} \cdot \mathbf{R}^{e} - \frac{2m_{\mathrm{e}}}{\tau_{\mathrm{e}} m_{\mathrm{i}}} (T_{\mathrm{e}} - T_{\mathrm{i}})\end{aligned} \tag{12.147}$$

$$\begin{aligned}\frac{\partial T_{\mathrm{i}}}{\partial t} &= -\mathbf{u} \cdot \nabla T_{\mathrm{i}} - \frac{2}{3} T_{\mathrm{i}} \nabla \cdot \mathbf{u} - \frac{2}{3n_{\mathrm{i}}} \mathbf{\Pi}^{\mathrm{i}} : \nabla \mathbf{u} \\ &\quad - \frac{2}{3n_{\mathrm{i}}} \nabla \cdot \mathbf{q}^{\mathrm{i}} + \frac{2Zm_{\mathrm{e}}}{\tau_{\mathrm{e}} m_{\mathrm{i}}} (T_{\mathrm{e}} - T_{\mathrm{i}})\end{aligned} \tag{12.148}$$

where

$$\hat{\mathbf{u}} = \mathbf{u} - \frac{m_{\mathrm{i}}}{Ze\rho} \mathbf{j} = \mathbf{u} - \frac{1}{en_{\mathrm{e}}} \mathbf{j} \tag{12.149}$$

Note that in the $\mathbf{\Pi}^{\mathrm{e}}$ term in (12.147) we have dropped the $\mathbf{j}$ component because the product $\mathbf{\Pi}^{\mathrm{e}} \cdot \mathbf{j}$ is quadratic in small quantities.

Using the approximation (12.90) together with (12.145) and (12.146) we can combine (12.147) and (12.148) to obtain a single equation for the total plasma pressure P

$$\begin{aligned}\frac{\partial P}{\partial t} &= -\mathbf{u}\cdot\nabla P - \frac{5}{3}P\nabla\cdot\mathbf{u} + \frac{5}{3e}\mathbf{j}\cdot\nabla T_{\mathrm{e}} - \frac{2}{3en_{\mathrm{e}}}\mathbf{j}\cdot\nabla(n_{\mathrm{e}}T_{\mathrm{e}}) \\ &\quad - \frac{2}{3}(\mathbf{\Pi}^{\mathrm{e}} + \mathbf{\Pi}^{\mathrm{i}}) : \nabla\mathbf{u} - \frac{2}{3}\nabla\cdot(\mathbf{q}^{\mathrm{e}} + \mathbf{q}^{\mathrm{i}}) + \frac{2}{3en_{\mathrm{e}}}\mathbf{j}\cdot\mathbf{R}^{\mathrm{e}}\end{aligned} \tag{12.150}$$

The collision term $\mathbf{R}^{\mathrm{e}}$ and the privileged non-hydrodynamical moments are given by (12.109) and the transport equations (12.112)–(12.114) and (12.120), where the effective electric field $\hat{\mathbf{E}}$ is given by (12.116) and the transport coefficients by (12.121). The final group of equations closing the set are the reduced Maxwell equations (12.94)–(12.97).

This is the complete set of *dissipative MHD equations* but, as already remarked, some further reduction is desirable for practical purposes. To make contact with the MHD equations in Chapters 3–5, we rewrite the hydrodynamic equations (12.145)–(12.150) as follows:

$$\begin{aligned}\frac{\partial\rho}{\partial t} &= -\nabla\cdot(\rho\mathbf{u}) \\ \frac{\partial(\rho\mathbf{u})}{\partial t} &= -\nabla\cdot(\rho\mathbf{u}\mathbf{u} + P\mathbf{I}) + \mathbf{j}\times\mathbf{B} + [DISS] \\ \frac{\partial T_\alpha}{\partial t} &= -\mathbf{u}\cdot\nabla T_\alpha - \frac{2}{3}T_\alpha\nabla\cdot\mathbf{u} + [DISS] \\ \frac{\partial P}{\partial t} &= -\mathbf{u}\cdot\nabla P - \frac{5}{3}P\nabla\cdot\mathbf{u} + [DISS]\end{aligned}$$

where only the non-dissipative terms are shown explicitly and in each equation $[DISS]$ represents all the dissipative terms, i.e. all the terms involving collisions. Note that in most cases this is via the transport coefficients†.

Now recalling that the assumption underlying our approximation scheme is that hydrodynamical moments are of order zero in $\epsilon = \tau_\alpha/\tau_{\mathrm{H}}$ while non-hydrodynamical moments are of order one, it is easily seen that dissipative terms are of order ϵ times the non-dissipative terms. The kind of further simplification we might look for, therefore, would be to drop some or all of the dissipative terms. However, this needs to be done with caution. The mathematical procedure used in the derivation of the transport equations was a precise and correct linearization in ϵ. Dropping dissipative terms which are of higher differential order than the non-dissipative terms, on the other hand, may change completely the nature of

† In the following discussion we ignore the complication that in certain limits some of the transport coefficients are independent of τ_α.

the solutions. Put the other way round, the predictions of non-dissipative theory may be very significantly changed by the inclusion of even the smallest amount of dissipation. If dissipation is to be correctly treated as a small perturbation we ought to use singular perturbation theory. We shall not pursue this but merely remind the reader of what was said in Section 4.1. Narrow regions of very steep gradients (sheaths) arise so that we should not expect a reduced set of equations to be universally valid even for a given range of physical parameters.

12.7.1 Resistive MHD

The first simplification that we consider is the neglect of all dissipative terms except electrical resistivity. The justification† for picking out this particular dissipative term lies in the moment approximation scheme. Putting $\chi_\alpha = 0$ in (12.101) eliminates all the non-hydrodynamical moments except $\mathbf{j}$ which survives because $\mathbf{u}_\mathrm{i} \neq \mathbf{u}_\mathrm{e}$. Resistive MHD is therefore equivalent to the 5M approximation.

Setting

$$\mathbf{q}^\alpha = 0 \qquad \boldsymbol{\Pi}^\alpha = 0 \tag{12.151}$$

in (12.113), (12.114) and (12.120) gives

$$\alpha_{kl} = 0 \qquad \kappa^\alpha_{kl} = 0 \qquad \mu^\alpha_p = 0 \tag{12.152}$$

for all k, l, and p. Referring back to the discussion of the transport coefficients in the previous section we see that (12.152) can be achieved by a combination of the $|\Omega_\alpha \tau_\alpha| \to \infty$ limit, to make non-parallel coefficients vanish, and the $T_\alpha \to 0$ limit (see (12.135)–(12.137)) to make parallel coefficients vanish. Since $\tau_\alpha \propto T_\alpha^{3/2}/n_\alpha$ (see (12.122)) we are really talking about low pressure plasmas in strong magnetic fields which are to be found in diffuse space plasmas such as the solar wind or planetary atmospheres but not in fusion devices. Alternatively, from (12.121) we see that (12.152) can be satisfied in the $|\Omega_\alpha|\tau_\alpha \to 0$ limit, i.e. in a very strongly collision-dominated plasma. Neither space nor fusion plasmas satisfy this condition.

Applying (12.151) to (12.109) gives

$$\mathbf{R}^\mathrm{e} = \frac{m_\mathrm{e}}{e\tau_\mathrm{e}}\mathbf{j} \tag{12.153}$$

and substituting this in (12.111) we get

$$\frac{m_\mathrm{e}}{\tau_\mathrm{e}}\mathbf{j} + e\mathbf{j} \times \mathbf{B} = e^2 n_\mathrm{e}(\mathbf{E} + \mathbf{u} \times \mathbf{B}) + e\boldsymbol{\nabla}(n_\mathrm{e} T_\mathrm{e}) \tag{12.154}$$

† In view of the following discussion about the vanishing of the other coefficients, the retention of σ should be regarded as an investigation of the effect of keeping one dissipative term.

or, on solving for **j**,

$$\mathbf{j} = \boldsymbol{\sigma} \cdot \left[\mathbf{E} + \mathbf{u} \times \mathbf{B} + \frac{1}{en_e} \boldsymbol{\nabla}(n_e T_e) \right] \tag{12.155}$$

where

$$\boldsymbol{\sigma} = \frac{e^2 n_e}{m_e} \tau_e \begin{pmatrix} \dfrac{1}{1+X_e^2} & \dfrac{-X_e}{1+X_e^2} & 0 \\ \dfrac{X_e}{1+X_e^2} & \dfrac{1}{1+X_e^2} & 0 \\ 0 & 0 & 1 \end{pmatrix}$$

is identified as the electrical conductivity by comparison of (12.155) with (12.112).

Further simplification of Ohm's law is obtained by assuming that the ion Larmor radius r_{L_i} is much smaller than the hydrodynamic length scale L_H

$$r_{L_i} \ll L_H \tag{12.156}$$

This is, in fact, the drift approximation which is good for space and fusion plasmas. Then comparing the last two terms in (12.155) we find

$$\frac{|\boldsymbol{\nabla}(n_e T_e)|}{en_e |\mathbf{u} \times \mathbf{B}|} \sim \frac{T_e}{eBuL_H} \sim \left(\frac{r_{L_i}}{L_H}\right) \left(\frac{c_s}{u}\right) \left(\frac{T_e}{T_i}\right)^{1/2} \tag{12.157}$$

where $c_s = (T_e/m_i)^{1/2}$ is the ion acoustic speed. Assuming $u \sim c_s$ and $T_e \sim T_i$ we can then drop the $\boldsymbol{\nabla}(n_e T_e)$ term in Ohm's law (12.155).

The final step in the simplification of Ohm's law is to assume that $|\mathbf{u}^e - \mathbf{u}^i| \ll \mathbf{u}$ or

$$|\mathbf{j}| \ll en_e |\mathbf{u}| \tag{12.158}$$

which seems reasonable in view of the fact that n_e and **u** are hydrodynamical moments of order zero while **j** is a privileged non-hydrodynamical moment of order one in our approximation scheme. Here again (12.158) should be used with caution. It clearly is not valid in static equilibrium problems where $\mathbf{u} = 0$ but $\mathbf{j} \neq 0$. Adoption of (12.158) means that the $\mathbf{j} \times \mathbf{B}$ term in Ohm's law can also be neglected so that with (12.156) and (12.158) we get from (12.154)

$$\mathbf{j} = \frac{e^2 n_e \tau_e}{m_e} (\mathbf{E} + \mathbf{u} \times \mathbf{B}) = \sigma_\parallel (\mathbf{E} + \mathbf{u} \times \mathbf{B}) \tag{12.159}$$

which is the well-known simple form of Ohm's law for a conductor moving with velocity **u**.

The main point to note about (12.159) is that the conductivity tensor has been reduced to a scalar. For strong magnetic fields this grossly over-estimates the current perpendicular to the magnetic field for if we were to use (12.155) with

$\nabla(n_e T_e)$ neglected we know from our discussion of transport coefficients in the previous section that $\sigma_\perp, \sigma_\wedge \to 0$ as $X_\alpha \to \infty$. There is, therefore, no consistent derivation of (12.159) and it must, at best, be regarded as a model equation adopted for mathematical simplicity.

The derivation of the MHD equations is completed by applying (12.151), (12.153), and (12.158) to (12.150) giving

$$\frac{\partial P}{\partial t} + \mathbf{u} \cdot \nabla P = -\frac{5P}{3}\nabla \cdot \mathbf{u} + \frac{2(\nabla \times \mathbf{B})^2}{3\sigma_\parallel \mu_0^2} \tag{12.160}$$

where (12.95) has been used to substitute for $\mathbf{j}$ in the last term. Note that T_e no longer appears in (12.160) and, since neither T_e nor T_i appear except implicitly through P in the momentum equation (12.146), we can use (12.160) instead of both (12.147) and (12.148).

In summarizing the set of resistive MHD equations it is useful to eliminate some of the variables. Thus the current density $\mathbf{j}$ is determined by (12.95) and the electric field $\mathbf{E}$ by (12.159). Then using (12.145), (12.95) and (12.151) the momentum balance equation (12.146) becomes

$$\rho \frac{D\mathbf{u}}{Dt} = -\nabla P - \frac{1}{\mu_0}\mathbf{B} \times (\nabla \times \mathbf{B}) \tag{12.161}$$

Finally, substituting (12.95) in (12.159) and (12.159) in (12.94) gives, in view of (12.97),

$$\frac{\partial \mathbf{B}}{\partial t} = \nabla \times (\mathbf{u} \times \mathbf{B}) + \frac{1}{\mu_0 \sigma_\parallel}\nabla^2 \mathbf{B} \tag{12.162}$$

The set has been reduced to (12.145) and (12.160)–(12.162) for the three hydrodynamical variables ρ, $\mathbf{u}$ and P and the magnetic field $\mathbf{B}$. It is easily seen that these correspond to the equations in Table 3.1 for $\gamma = 5/3$, in accordance with our assumption that the plasma is a perfect gas having three degrees of freedom.

Exercises

12.1 Show that the Liouville operator L_N is invariant under interchanges of like particles and hence deduce that the generic N-particle distribution function obeys the Liouville equation (12.5).

12.2 In the case of no external forces, compare the order of magnitude of the terms in (12.23). Taking ω_p^{-1} and λ_D as characteristic time and length scales show that the collision term (12.24) is of order g/ff compared with all the other terms.

12.3 By Fourier transform, or otherwise, solve (12.40) and hence obtain the equilibium pair correlation function (12.41).

12.4 Obtain the general moment equation (12.68) from the kinetic equation (12.66) and show that the first three moments are given by (12.74)–(12.76). Explain physically why there is no term containing **B** in (12.76).

12.5 Invert the set of equations (12.81)–(12.84) defining the one-fluid variables to obtain the set (12.85).

12.6 Carry out the steps indicated in the text to transform the two-fluid moment equations (12.74) and (12.75) to the equivalent set of equations (12.87)–(12.91) in terms of one-fluid variables.

12.7 Under what circumstances might electron viscosity become significant? How can this be explained physically?

12.8 Invert (12.154) to obtain (12.155). What is the physical significance of the gradient term and how do you explain it?

Appendix 1

Numerical values of physical constants and plasma parameters

Physical constants		
Velocity of light	c	$2.998 \times 10^{8}\ \mathrm{m\,s^{-1}}$
Electron charge	e	$1.602 \times 10^{-19}\ \mathrm{C}$
Electron mass	m_e	$9.109 \times 10^{-31}\ \mathrm{kg}$
Proton mass	m_p	$1.673 \times 10^{-27}\ \mathrm{kg}$
Proton/electron mass ratio	$m_\mathrm{p}/m_\mathrm{e}$	1.836×10^{3}
Classical electron radius	r_e	$2.818 \times 10^{-15}\ \mathrm{m}$
Thomson scattering cross-section	σ_T	$6.652 \times 10^{-29}\ \mathrm{m^2}$
Planck's constant	h	$6.626 \times 10^{-34}\ \mathrm{J\,s}$
Electron volt	eV	$1.602 \times 10^{-19}\ \mathrm{J}$
Ionization potential of hydrogen		$2.180 \times 10^{-18}\ \mathrm{J}$
Boltzmann's constant	k_B	$1.381 \times 10^{-23}\ \mathrm{J\,K^{-1}}$
Vacuum permittivity	ε_0	$8.854 \times 10^{-12}\ \mathrm{F\,m^{-1}}$
Vacuum permeability	μ_0	$1.257 \times 10^{-6}\ \mathrm{H\,m^{-1}}$
Gravitational constant	G	$6.673 \times 10^{-11}\ \mathrm{m^3\,kg^{-1}\,s^{-2}}$
Solar mass	$M_\odot$	$1.989 \times 10^{30}\ \mathrm{kg}$
Solar radius	$R_\odot$	$6.960 \times 10^{8}\ \mathrm{m}$
Sun–Earth distance	1 AU	$1.496 \times 10^{11}\ \mathrm{m}$
Earth's gravitational acceleration	g	$9.807\ \mathrm{m\,s^{-2}}$
Temperature equivalent to 1 eV		$1.160 \times 10^{4}\ \mathrm{K}$
Plasma parameters		
Electron plasma frequency	$\left(n_\mathrm{e}e^2/m_\mathrm{e}\varepsilon_0\right)^{1/2}$	$56.4\,n_\mathrm{e}^{1/2}\ \mathrm{rad\,s^{-1}}$
Ion plasma frequency	$\left(n_\mathrm{i}(Ze)^2/m_\mathrm{i}\varepsilon_0\right)^{1/2}$	$1.32\,Z(n_\mathrm{i}/A)^{1/2}\ \mathrm{rad\,s^{-1}}$
Electron cyclotron frequency	eB/m_e	$1.76 \times 10^{11}\,B\ \mathrm{rad\,s^{-1}}$
Ion cyclotron frequency	ZeB/m_i	$9.58 \times 10^{7}\,(ZB/A)\ \mathrm{rad\,s^{-1}}$
Debye length	$\left(\varepsilon_0 k_\mathrm{B}T_\mathrm{e}/n_\mathrm{e}e^2\right)^{1/2}$	$69.1\,(T_\mathrm{e}/n_\mathrm{e})^{1/2}\ \mathrm{m}$

Electron Larmor radius	$m_e v_\perp / eB$	$5.68 \times 10^{-12}\ v_\perp B^{-1}$ m
Ion Larmor radius	$m_i v_\perp / ZeB$	$1.04 \times 10^{-8}\ v_\perp (A/ZB)$ m
Electron thermal speed	$(k_B T_e / m_e)^{1/2}$	$3.89 \times 10^3\ T_e^{1/2}$ m s^{-1}
Ion thermal speed	$(k_B T_i / m_i)^{1/2}$	$90.9\ (T_i/A)^{1/2}$ m s^{-1}

In the above formulae densities are expressed in m^{-3}, B in teslas, T_e, T_i in K and $v_\perp$ (m s^{-1}) is the component of particle velocity perpendicular to **B**.

Appendix 2

List of symbols

The following is a list of symbols which appear in the text beyond the location of their definition or first introduction and indicates the page on which this may be found. Some of these symbols also appear with subscript e(i) indicating electron (ion) equivalents. Likewise, symbols with subscript $\|$ ($\perp$) indicate the component parallel (perpendicular) to the magnetic field.

Roman alphabet

Symbol	Description	Page
$\mathbf{A}$	magnetic vector potential	32
A	atomic number	9
a	minor radius of torus	95
a_s	$(m_s/2k_\mathrm{B}T_s)^{1/2}$	313
$\mathbf{B}$	magnetic field intensity	13
$\tilde{\mathbf{B}}$	vacuum magnetic field	75
$\mathbf{B}_\mathrm{ind}$	magnetic field induced by magnetization current	16
B	magnetic field magnitude	11
B_p	poloidal magnetic field	33
B_t	toroidal magnetic field	33
$\mathbf{b}$	unit vector in direction of $\mathbf{B}$	43
b	impact parameter	309
b_0	impact parameter for $\pi/2$ scattering	10
C_v	specific heat at constant volume	184
C_α	'collision term' for wave–particle interactions	409
$\mathbf{c}^\alpha$	random or thermal velocity for species α	481
c	speed of light	18
c_s	ion acoustic speed	65
D	component of cold plasma dielectric tensor	205

D	diffusion coefficient	299
$D(k, p)$	plasma dielectric function	258
D_a	ambipolar diffusion coefficient	301
$\mathbf{E}$	electric field	13
$\hat{\mathbf{E}}$	effective electric field	494
$\mathbf{E}_s$	scattered electric field	355
$\mathbf{E}_T$	electric field of transverse wave	395
E	energy	3
e	magnitude of electronic charge	7
e_j	charge on particle j	13
$\mathcal{E}$	internal energy per unit mass	55
$\mathcal{E}(\mathbf{k}, t)$	spectral energy density of electrostatic field	380
$\mathbf{F}$	force per unit mass	52
$\mathbf{F}_i(t)$	force on electron i at time t	465
F	field reversal parameter	144
F	fast wave	218
$F_0(u)$	one-dimensional distribution function	258
$f(1)$	one-particle distribution function	469
$f(\mathbf{r}, \mathbf{v}, t)$	particle distribution function	252
$f_K(\mathbf{r}, \mathbf{v}, t)$	Klimontovich distribution function	252
f_N	N-particle distribution function	466
f_s	distribution function of scattering particles	309
f_s	generic s-particle distribution function	467
$f_{\alpha 0}$	local Maxwellian distribution for species α	489
$G(\mathbf{v})$	Rosenbluth potential	312
$G(\omega b_0/v)$	Gaunt factor	336
$\mathbf{g}$	relative velocity of colliding particles	309
g	plasma parameter	8
g	$1 - \mathbf{n}\cdot\boldsymbol{\beta}$	326
$g(1, 2)$	pair correlation function	469
$\bar{g}(\omega, T)$	Maxwellian-averaged Gaunt factor	338
$H(\mathbf{v})$	Rosenbluth potential	312
$\mathbf{I}$	unit dyadic	78
I	total plasma current	87
I	internal energy	183
I_p	poloidal current	98
$I_\omega(\mathbf{s})$	radiation intensity	330
J	longitudinal invariant	28
$\mathbf{j}$	current density	13

$\mathbf{j}_G$	current density due to grad B drift	23
$\mathbf{j}_g$	guiding centre current density	70
$\mathbf{j}_M$	magnetization current density	16
$\mathbf{j}_P$	polarization current density	37
$\mathbf{j}_{p(t)}$	poloidal (toroidal) current density	99
K	magnetic helicity	103
K	common coupling coefficient	391
$\mathbf{k}_L$	wavenumber of Langmuir wave	394
$\mathbf{k}_S$	wavenumber of ion acoustic wave	394
$\mathbf{k}_s$	wavenumber of scattered wave	356
$\mathbf{k}_T$	wavenumber of transverse wave	394
k_B	Boltzmann's constant	3
k_c	critical wavenumber	240
L	scale length	12
L	component of cold plasma dielectric tensor	205
L	left circularly polarized wave	210
L	Langmuir wave	394
L_H	hydrodynamic length scale	49
L_s	shock thickness	408
$L_\parallel, L_\perp, L_\wedge$	transport coefficients	495
$\tilde{L}_\parallel$	dimensionless $L_\parallel$	497
$\tilde{L}_\perp, \tilde{L}_\wedge$	dimensionless $L_\perp, L_\wedge$	498
$\ln \zeta$	Euler's constant	338
$\ln \Lambda$	Coulomb logarithm	10
$\mathbf{M}$	magnetization per unit volume due to Larmor motion	16
M	Mach number	187
M_A	Alfvén Mach number	174
$M_\odot$	solar mass	170
m	particle mass	14
m_j	mass of particle j	13
m_s	mass of scattering particle	309
N	total particle number density	57
$\mathbf{n}$	unit vector normal to surface	74
$\mathbf{n}$	dimensionless wave propagation vector	203
$\hat{\mathbf{n}}$	unit vector in direction of shock propagation	184
n	particle number density	3
n	refractive index	331
n	quantum state	339
n_b	number density of electron beam	241

O	ordinary wave	214
P	pressure	54
P	component of cold plasma dielectric tensor	205
P	power radiated by accelerated charge	326
P	total scalar pressure	486
P_l	power in harmonic line l	345
$\mathcal{P}$	wave polarization	199
$\boldsymbol{\mathcal{P}}$	polarization dyadic	355
Q	heat energy per unit mass	55
Q	$1 + k^2c^2/\omega_{pe}^2$	232
Q_α	rate of thermal energy transfer to species α	484
$\mathbf{q}$	heat flux	297
$\mathbf{q}^\alpha$	heat flux for species α	482
q	charge density	13
q	safety factor	34
q_T	test particle charge	287
$\mathbf{R}^\alpha$	rate of momentum transfer to species α	484
R	hydrodynamic Reynolds number	61
R	pressure ratio across shock	185
R	component of cold plasma dielectric tensor	205
R	right circularly polarized wave	210
R	$\lvert\mathbf{r} - \mathbf{r}_0\rvert$	325
R	reflection coefficient	428
R_E	Earth's radius	7
R_M	magnetic Reynolds number	61
R_y	Rydberg constant	340
R_0	tokamak major radius	33
R_0	gas constant	58
$R_\odot$	solar radius	173
$\mathbf{r}_g$	guiding centre position vector	15
$\mathbf{r}_j$	position vector of particle j	13
r	tokamak minor radius	33
r	density ratio across shock	185
r_A	Alfvén radius	172
r_e	classical electron radius	328
r_L	Larmor radius	15
$\mathbf{S}$	Poynting vector	200
S	Lundquist number	141
S	entropy	184

S	component of cold plasma dielectric tensor	205
S	slow wave	218
S	sound wave	394
$S(\mathbf{k}, \omega)$	form factor	358
$\mathbf{T}$	Maxwell stress tensor	78
T	temperature	3
T	characteristic time scale	12
T	transverse wave	395
T	tunnelling (transmission) coefficient	434
T'	scattered transverse wave	394
T_α	temperature of species α (in energy units)	481
t'	retarded time	325
t_{F}	time scale for magnetic field changes	29
U	total energy density	78
$\mathbf{u}$	velocity of fluid element	49
$\mathbf{V}(t)$	test particle velocity	313
V	warm plasma thermal speed	227
V	wave phase velocity	269
$V(x)$	electrostatic potential in plasma sheath	454
V_{b}	electron beam thermal speed	273
$V_{p(s)}$	group velocity of Langmuir (Raman scattered) wave	444
V_{th}	thermal speed	10
$\mathbf{v}_{\mathrm{B}}$	combined grad B and curvature drift velocity	22
$\mathbf{v}_{\mathrm{b}}$	flow velocity of electron beam	241
$\mathbf{v}_{\mathrm{c}}$	curvature drift velocity	21
$\mathbf{v}_{\mathrm{E}}$	$\mathbf{E} \times \mathbf{B}$ drift velocity	17
$\mathbf{v}_{\mathrm{G}}$	grad B drift velocity	20
$\mathbf{v}_{\mathrm{g}}$	guiding centre velocity	43
$\mathbf{v}_{\mathrm{P}}$	polarization drift velocity	37
$\mathbf{v}_{\mathrm{p}}$	poloidal velocity	34
$\mathbf{v}_s$	velocity of scattering particle	309
v_{A}	Alfvén speed	130
v_{d}	electron–ion drift velocity	243
v_{g}	group velocity	201
v_{p}	phase velocity	72
v_0	quiver velocity	441
W	particle kinetic energy	14
W	total potential energy	120
$W_{\mathrm{p(s,v)}}$	plasma (surface, vacuum) potential energy	122

X	extraordinary wave	214
X_α	$\Omega_\alpha \tau_\alpha$	498
Z	charge state	9
$Z(\zeta)$	plasma dispersion function	265
z	atomic number of scattered particle	310
z_c	cut-off point in WKBJ model	428
z_s	atomic number of scattering particle	310

Greek alphabet

α	Clebsch variable	32
α^2	ω_p^2/ω^2	222
α_{kl}	thermoelectric coefficient	494
α_ω	absorption coefficient	331
$\boldsymbol{\beta}$	$\mathbf{v}/c$	47
β	Clebsch variable	32
β	ratio of plasma to magnetic pressure	61
β^2	$\lvert\Omega_i\Omega_e\rvert/\omega^2$	222
β_e	$\lvert\Omega_e\rvert/\omega$	222
β_i	Ω_i/ω	222
$\beta_{p(t)}$	poloidal (toroidal) β	89
$\boldsymbol{\Gamma}$	particle flux	299
Γ	$z^2e^2 \ln\Lambda/4\pi\varepsilon_0^2 m^2$	311
γ	$(1 - v^2/c^2)^{-1/2}$	38
γ	ratio of specific heats	57
γ	instability growth rate	121
γ	$\alpha^2/\beta^2 = c^2/v_A^2$	222
γ	Landau damping decrement	261
Δ'	discontinuity in magnetic field gradient	152
ϵ	inverse aspect ratio	33
ϵ	ratio of collision time to hydrodynamic time	493
ϵ_ω	emission coefficient	331
$\boldsymbol{\varepsilon}$	cold plasma dielectric tensor	203
ε_0	vacuum permittivity	8
η	plasma resistivity	140
θ	pitch angle	27
θ	poloidal angle	33
θ	wave propagation angle relative to $\mathbf{B}$	205
θ	scattering angle	309
θ_{res}	resonant angle	207

ι	rotational transform	88
$\boldsymbol{\kappa}$	curvature of magnetic field	122
κ	coefficient of thermal conductivity	56
κ_{kl}^{α}	element of thermal conductivity tensor for species α	494
λ	$k^2 k_B T / m\Omega_e^2$	280
λ_c	collisional mean free path	63
λ_D	Debye length	8
$\boldsymbol{\mu}_B$	magnetic moment	16
μ	coefficient of viscosity	54
μ_0	vacuum magnetic permeability	13
μ_{klmn}^{α}	fourth-order viscosity tensor for species α	495
$\mu_{1,2,3,4,\parallel}^{\alpha}$	independent elements of viscosity tensor	495
ν	kinematic viscosity	192
ν_c	collision frequency	10
ν_{ei}	electron–ion collision frequency	10
$\boldsymbol{\xi}(\mathrm{r}, t)$	displacement vector	109
Π_{ij}^{α}	traceless pressure tensor for species α	481
ρ	mass density	51
$\boldsymbol{\sigma}$	electrical conductivity tensor	203
σ	collisional cross-section	3
σ	electrical conductivity	56
$\sigma(\lvert\mathbf{g}\rvert, \theta)$	differential scattering cross-section	309
σ_{kl}	element of electrical conductivity tensor	494
τ	$t - x/c$	38
τ	optical depth	332
τ_A	Alfvén transit time	140
τ_B	bounce time	306
τ_c	collision time	62
$\tau_{e(i)}$	electron (ion) collision time	62
τ_G	gravitational time scale	155
τ_H	hydrodynamic time	58
τ_L	Larmor period	32
τ_P	precession time	32
τ_R	resistive diffusion time	140
$\tau_\parallel$	longitudinal invariant transit time	29
Φ	magnetic flux	31
Φ_{ij}	element of stress tensor	52
ϕ	magnetic scalar potential	113
$\phi(x)$	error function	314

$\phi(z)$	wave phase or eikonal	427
ϕ_{ij}	electrostatic potential between electrons i and j	468
χ_α	susceptibility of species α	362
χ_α	normalized deviation from local Maxwellian	489
$\Psi(x)$	$[\phi(x) - x\phi'(x)]/2x^2$	314
ψ	flux function	96
Ω	Larmor frequency	15
Ω	solid angle	326
Ω	relativistic Larmor frequency (Ω_0/γ)	344
ω_{L}	L-mode cut-off frequency	210
ω_{L}	Langmuir wave frequency	394
ω_{LH}	lower hybrid resonance frequency	209
ω_{p}	plasma frequency	8
ω_{R}	R-mode cut-off frequency	210
ω_{s}	ion acoustic wave frequency	394
ω_{T}	transverse electromagnetic pump wave frequency	394
ω_{UH}	upper hybrid resonance frequency	209

References

Adlam, J. and Allen, J. (1958). *Phil. Mag.* **3**, 448.
Alfvén, H. (1940). *Ark. f. Mat. Astr. o Fysik* **27A** No. 22.
Alfvén, H. (1951). *Cosmical Electrodynamics* (Oxford University Press).
Alfvén, H. and Fälthammer, C.-G. (1963). *Cosmical Electrodynamics*; 2nd edn. (Clarendon Press, Oxford).
Alfvén, H. and Herlofson, N. (1950). *Phys. Rev.* **78**, 616.
Allan, W. and Sanderson, J.J. (1974). *Plasma Phys.* **16**, 753.
Allis, W.P., Buchsbaum, S.J. and Bers, A. (1963). *Waves in Anisotropic Plasmas* (MIT Press, Cambridge, Mass.).
Armstrong, R.J. *et al.* (1981). *Phys. Lett.* **85A**, 281.
Backus, G.E. (1958). *Ann. Phys.* **4**, 372.
Baldis, H.A. *et al.* (1996). *Phys. Rev. Lett.* **77**, 2957.
Baldis, H.A., Villeneuve, D.M. and Walsh, C.J. (1986). *Canadian J. Phys.* **64**, 961.
Balescu, R. (1960). *Phys. Fluids* **3**, 52.
Balescu, R. (1982). *J. Plasma Phys.* **27**, 553.
Balescu, R. (1988). *Transport Processes in Plasmas* (North-Holland, Amsterdam).
Barr, H.C., Boyd, T.J.M. and Coutts, G.A. (1988). *Phys. Rev. Lett.* **60**, 1950.
Barr, H.C., Boyd, T.J.M. and Mackwood, A.P. (1994). *Phys. Plasmas* **1**, 993.
Batchelor, G.K. (1967). *An Introduction to Fluid Dynamics* (Cambridge University Press, Cambridge).
Bateman, G. (1978). *MHD Instabilities* (MIT Press, Cambridge, Mass.).
Beckers, J.M. and Schröter, E.H. (1968). *Solar Phys.* **4**, 192.
Bekefi, G. (1966). *Radiation Processes in Plasma* (Wiley, New York).
Bernstein, I.B. (1958). *Phys. Rev.* **109**, 10.
Bethe, H.A. (1939). *Phys. Rev.* **55**, 434.
Bethe, H.A. and Critchfield, C.L. (1938). *Phys. Rev.* **54**, 248.
Bhatnagar, P.L., Gross, E.P. and Krook, M. (1954). *Phys. Rev.* **94**, 511.
Biermann, L. (1950). *Z. Naturforsch.* **5a**, 65.
Biskamp, D. (1986). *Phys. Fluids* **29**, 1520.
Biskamp, D. (1993). *Nonlinear Magnetohydrodynamics* (Cambridge University Press, Cambridge).
Bodin, H.A.B. and Newton, A.A. (1980). *Nucl. Fusion* **20**, 1255.
Bogolyubov, N.N. (1962). Problems of a dynamical theory in statistical physics, in *Studies in Statistical Mechanics*, Vol.1, ed. J. deBoer and G.E. Uhlenbeck (North-Holland, Amsterdam).

Bohm, D. (1949). Qualitative description of the arc plasma in a magnetic field, in *The Characteristics of Electrical Discharges in Magnetic Fields*, ed. A. Guthrie and R.K. Wakerling (McGraw-Hill, New York).
Book, D.L. (1987). *NRL Plasma Formulary* (Naval Research Laboratory, Washington, DC).
Born, M. and Green, H.S. (1949). *A General Kinetic Theory of Liquids* (Cambridge University Press, London).
Bowles, K.L. (1958). *Phys. Rev. Lett.* **1**, 454.
Boyd, T.J.M. (1966). *The effect of Coulomb collisions on light scattering by plasmas* UKAEA Culham Laboratory Report CLM-R52 (HMSO).
Boyd, T.J.M., Gardner, G.A. and Gardner, L.R.T. (1973). *Nucl. Fusion* **13**, 764.
Boyd, T.J.M. and Ondarza-Rovira, R. (2000). *Phys. Rev. Lett.* **85**, 1440.
Boyd, T.J.M. and Sanderson, J.J. (1969). *Plasma Dynamics* (Thomas Nelson, London).
Braginsky, S.I. (1965). Transport processes in a plasma, in *Reviews of Plasma Physics*, Vol. 1, ed. M.A. Leontovich (Consultants Bureau, New York).
Braginsky, S.I. (1991). *Geophys. Astrophys. Fluid Dyn.* **60**, 89.
Briggs, R.J. (1964). *Electron Stream Interaction with Plasmas* (MIT Press, Cambridge, Mass.).
Bryant, D.A. (1993). Space plasma physics, Part I, in *Plasma Phys.*, ed. R.O. Dendy (Cambridge University Press, Cambridge).
Budden, K.G. (1966). *Radio Waves in the Ionosphere* (Cambridge University Press, Cambridge).
Bunemann, O. (1959). *Phys. Rev.* **115**, 503.
Cairns, R.A. (1991). *Radiofrequency Heating of Plasmas* (Adam Hilger, Bristol).
Cairns, R.A. and Lashmore-Davies, C.N. (1983). *Phys. Fluids* **26**, 1268.
Chen, F.F. (1984). *Introduction to Plasma Physics*, 2nd edn. (Plenum, London).
Chen, F.F. (2001). *Phys. Plasmas* **8**, 3029.
Chen, S.-Y., Maksimchuk, A. and Umstadter, D. (1998). *Nature* **396**, 653.
Cheng, C.Z. and Knorr, G. (1976). *J. Comp. Phys.* **22**, 330.
Chew, G., Goldberger, M. and Low, F. (1956). *Proc. Roy. Soc.* **A236**, 112.
Cowley, S.W.H. (1995). *Phys. World* **8**, 46.
Cowling, T.G. (1976). *Magnetohydrodynamics* (Adam Hilger, London).
Dattner, A. (1957). *Ericsson Technics* **13**, 309.
Dattner, A. (1963). *Ericsson Technics* **19**, 3.
Davidson, R.C. (1972). *Methods in Nonlinear Plasma Physics* (Academic Press, London).
Davis, L., Lüst, R. and Schlüter, A. (1958). *Naturforsch.* **13a**, 916.
Dawson, J.M. (1962). *Phys. Fluids* **5**, 445.
Dawson, J.M. and Oberman, C. (1962). *Phys. Fluids* **5**, 517.
Denisov, N.G. (1957). *Sov. Phys. JETP* **4**, 544.
DeSilva, A.W., Evans, D.E. and Forrest, M.J. (1964). *Nature* **203**, 1321.
DiMarco, J.N. (1983). *Mirror-Based and Field-Reversed Approaches to Magnetic Fusion*, Vol. 2, 681, ed. R.F. Post *et al.* (International School of Plasma Physics, Varenna).
Dougherty, J.P. and Farley, D.F. (1960). *Proc. Roy. Soc.* **A259**, 79.
Dougherty, J.P., Barron, D.W. and Farley, D.F. (1961). *Proc. Roy. Soc.* **A263**, 238.
Drummond, W.E. *et al.* (1970). *Phys. Fluids* **13**, 2422.
Dum, C.T. (1978). *Phys. Fluids* **21**, 956.
Dungey, J.W. (1953). *Phil. Mag.* **44**, 725.
Dungey, J.W. (1961). *Phys. Rev. Lett.* **6**, 47.
Dyson, A. (1998). University of Essex Ph.D. thesis (unpublished).
Elwert, G. (1948). *Z. Naturforsch.* **A3**, 477.

Ericson, W.B. and Bazar, J. (1960). *Phys. Fluids* **3**, 631.
Erokhin, N.S. (1969). *Ukr. Fiz. Zh.* **14**, 2055.
Evans, D.E. (1970). *Plasma Phys.* **12**, 573.
Evans, D.E. and Carolan, P. (1970). *Phys. Rev. Lett.* **25**, 1605.
Fejer, J.A. (1960). *Canadian J. Phys.* **38**, 1114.
Fejer, J.A. (1961). *Canadian J. Phys.* **39**, 716.
Feldman, W.C. *et al.* (1983). *J. Geophys. Res.* **88**, 96.
Fermi, E. (1949). *Phys. Rev.* **75**, 1169.
Fidone, I. and Granata, G. (1994). *Phys. Plasmas* **1**, 1231.
Field, G.B. (1995). The origin of astrophysical magnetic fields, Internat. Conf. on Plasma Physics ICPP 1994, *AIP Conference Proceedings* **345**, 417.
Forrest, M.G., Carolan, P.G. and Peacock, N.J. (1978). *Nature* **271**, 718.
Freidberg, J.P. (1987). *Ideal Magnetohydrodynamics* (Plenum Press, New York).
Fuchs, V., Ko, K. and Bers, A. (1981). *Phys. Fluids* **24**, 1251.
Furth, H.P., Killeen, J. and Rosenbluth, M.N. (1963). *Phys. Fluids* **6**, 459.
Galanti, M. and Peacock, N.J. (1975). *J. Phys. B* **8**, 2427.
Gary, S.P. (1970). *J. Plasma Phys.* **4**, 753.
Gary, S.P. (1993). *Theory of Space Plasma Microinstabilities* (Cambridge University Press, Cambridge).
Gasiorowicz, S., Neuman, M. and Riddell, R.J. (1956). *Phys. Rev.* **101**, 922.
Gentle, K.W. and Lohr, J. (1973). *Phys. Fluids* **16**, 1464.
Ginzburg, V.L. (1969). *Elementary Processes for Cosmic Ray Astrophysics* (Gordon and Breach, New York).
Giovanelli, R.G. (1947). *Mon. Not. Roy. Astron. Soc.* **107**, 338.
Glatzmaier, G.A. and Roberts, P.H. (1995). *Nature* **377**, 203.
Glenzer, S.H. *et al.* (1999). *Phys. Plasmas* **6**, 2117.
Goldstein, H. (1959). *Classical Mechanics* (Addison-Wesley, Reading, Mass.).
Gordon, W.E. (1958). *Proc. IRE* **46**, 1824.
Grad, H. (1949). *Commun. Pure Appl. Math.* **2**, 325.
Grad, H. (1967). *Magneto-Fluid and Plasma Dynamics*, Vol. 18, ed. H. Grad (American Mathematical Society, Providence).
Griem, H.R. (1964). *Plasma Spectroscopy* (McGraw-Hill, New York).
Hagfors, T. (1961). *J. Geophys. Res.* **66**, 1699.
Heading, J. (1961). *J. Res. Natl. Bur. Stand.* **D65**, 595.
Herzenberg, A. (1958). *Phil. Trans. Roy. Soc.* **A250**, 543.
Hinton, F.L. (1983). Collisional transport in plasma, in *Basic Plasma Physics I* (North-Holland, Amsterdam).
Hirshfield, J.L., Baldwin, D.E. and Brown, S.C. (1961). *Phys. Fluids* **4**, 198.
Hughes, D.W. (1991). *Advances in Solar System Magnetohydrodynamics*, ed. E.R. Priest and A.W. Hood (Cambridge University Press, Cambridge) 77.
Hutchinson, I.H. (1988). *Principles of Plasma Diagnostics* (Cambridge University Press, Cambridge).
Innes, D.E. *et al.* (1997). *Nature* **386**, 811.
Jackson, J.D. (1975). *Classical Electrodynamics* (Wiley, New York).
Jeffrey, A. (1966). *Magnetohydrodynamics* (Oliver and Boyd, Edinburgh).
Johnson, J.L. *et al.* (1958). *UN Geneva Conf.* **31**, 198.
Jones, R.D. (1980). *Los Alamos National Laboratory Report* LA-UR-80-2857.
Jorna, S. and Wood, L. (1987). *Phys. Rev.* **A36**, 397.
Kaufman, A.N. (1960). Plasma transport theory, in *The Theory of Neutral and Ionized Gases*, ed. C. De Witt and J.F. Detoeuf (John Wiley, New York).

Kibble, T.W.B. (1964). *Phys. Rev.* **138**, B740.
Kiepenheuer, K.O. (1950). *Phys. Rev.* **79**, 738.
Kirkwood, J.G. (1946). *J. Chem. Phys.* **14**, 180.
Kivelson, M.G. and Dubois, D.F. (1964). *Phys. Fluids* **7**, 1578.
Koch, P. and Williams, E.A. (1984). *Phys. Fluids* **27**, 2346.
Kogan, V.I. (1961). *Plasma Physics and the Problem of Controlled Thermonuclear Reactions* **1**, 153.
Kramers, H.A. (1923). *Phil. Mag.* **46**, 836.
Kruer, W.L. (1988). *The Physics of Laser Plasma Interactions* (Addison-Wesley, Redwood City).
Kruskal, M.D. (1962). *J. Math. Phys.* **3**, 806.
Kruskal, M.D. and Schwarzschild, M. (1954). *Proc. Roy. Soc.* **A223**, 348.
Kulsrud, R.M. (1983). MHD description of plasma, in *Basic Plasma Physics I*, ed. A.A. Galeev and R.N. Sudan (North Holland, Amsterdam).
Labaune, C. *et al.* (1995). *Phys. Rev. Lett.* **75**, 248.
Labaune, C. *et al.* (1996). *Phys. Rev. Lett.* **76**, 3727.
Landau, L.D. (1946). *J. Phys. (U.S.S.R.)* **10**, 25.
Landau, L.D. and Lifshitz, E.M. (1958). *Quantum Mechanics* (Pergamon Press, Oxford).
Landau, L.D. and Lifshitz, E.M. (1960). *Electrodynamics of Continuous Media* (Pergamon Press, Oxford).
Landau, L.D. and Lifshitz, E.M. (1962). *The Classical Theory of Fields*, 2nd edn. (Pergamon Press, Oxford).
Langdon, A.B. (1980). *Phys. Rev. Lett.* **44**, 575.
Lawson, J.D. (1957). *Proc. Phys. Soc.* **B70**, 6.
Lehnert, B. (1964). *Dynamics of Charged Particles* (North-Holland, Amsterdam).
Lenard, A. (1960). *Ann. Phys. NY* **10**, 390.
Lichters, R., Meyer-ter-Vehn, J. and Pukhov, A. (1998). Superstrong fields in plasmas, ed. M. Lontano *et al. AIP Conf. Proc.* **426**, 41.
Lifshitz, E.M. and Pitaevskii, L.P. (1981). *Physical Kinetics* (Pergamon Press, Oxford).
Lighthill, M.J. (1956). Viscosity effects in sound waves of finite amplitude, in *Surveys in Mechanics*, ed. G.K. Batchelor and R.M. Davies (Cambridge University Press, London).
Lin, R.P. *et al.* (1986). *Astrophys. J.* **308**, 954.
Littlejohn, R.G. (1979). *J. Math. Phys.* **20**, 2445.
Littlejohn, R.G. (1981). *Phys. Fluids* **24**, 1730.
McGowan, A.D. and Sanderson, J.J. (1992). *J. Plasma Phys.* **47**, 373.
Malmberg, J.H. and Wharton, C.B. (1964). *Phys. Rev. Lett.* **13**, 184.
Malmberg, J.H. and Wharton, C.B. (1966). *Phys. Rev. Lett.* **17**, 175.
Mirnov, S.V. and Semenov, I.B. (1971). *Sov. At. Energy* **30**, 22.
Miyamoto, K. (1989). *Plasma Physics for Nuclear Fusion* (MIT Press, Cambridge, Mass.).
Moffat, H.K. (1993). *Phys. World* **6**, 38.
Montgomery, D.C. and Tidman, D.A. (1963). *Plasma Kinetic Theory* (McGraw-Hill, New York).
Moorcroft, D.R. (1963). *J. Geophys. Res.* **68**, 4870.
Morozov, A.I. and Solov'ev, L.S. (1966). *Reviews of Plasma Physics*, Vol. 2, 201 ed. M.A. Leontovich (Consultants Bureau, New York).
Mostovych, A.N. and DeSilva, A.W. (1984). *Phys. Rev. Lett.* **53**, 1563.
Nicholson, D.R. (1983). *Introduction to Plasma Theory* (John Wiley and Sons, Chichester).

Northrop, T.G. (1961). *Ann. Phys. (N.Y.)* **15**, 209.
Northrop, T.G. (1963). *The Adiabatic Motion of Charged Particles* (Interscience, New York).
Northrop, T.G. and Teller, E. (1960). *Phys. Rev.* **117**, 215.
Ondarza-Rovira, R. (1996). University of Essex Ph.D. thesis (unpublished).
O'Neil, T.M., Winfrey, J.H. and Malmberg, J.H. (1971). *Phys. Fluids* **14**, 1204.
Oppenheimer, J.R. (1970). *Lectures on Electrodynamics* (Gordon and Breach, New York).
Parker, E.N. (1955). *Astrophys. J.* **121**, 491.
Parker, E.N. (1958). *Astrophys. J.* **128**, 664.
Parker, E.N. (1963). *Astrophys. J. Suppl.* Ser. **8**, 177.
Parker, E.N. (1979). *Cosmical Magnetic Fields* (Clarendon Press, Oxford).
Parker, J.V., Nickel, J.C. and Gould, R.W. (1964). *Phys. Fluids* **7**, 1489.
Parks, G.K. (1991). *Physics of Space Plasmas* (Addison-Wesley, New York).
Peacock, N.J. *et al.* (1969). *Nature* **224**, 488.
Pechacek, R.E. and Trivelpiece, A.W. (1967). *Phys. Fluids* **10**, 1688.
Penrose, O. (1960). *Phys. Fluids* **3**, 258.
Petschek, H.E. (1964). *AAS/NASA Symposium on the Physics of Solar Flares*, 425, ed. W.N. Hess (NASA, Washington, DC).
Potter, D.E. (1973). *Computational Physics* (John Wiley and Sons, London).
Press, W.H., Flannery, B.P., Teukolsky, S.A. and Vetterling, W.T. (1989). *Numerical Recipes* (Cambridge University Press, Cambridge).
Priest, E.R. (1987). *Solar Magnetohydrodynamics* (D. Reidel, Dordrecht).
Priest, E.R. and Forbes, T.G. (1986). *J. Geophys. Res.* **91**, 5579.
Priest, E.R. and Sanderson, J.J. (1972). *Plasma Phys.* **14**, 951.
Romell, D. (1951). *Nature* **167**, 243.
Rosenbluth, M.N. (1972). *Phys. Rev. Lett.* **29**, 565.
Rosenbluth, M.N., MacDonald, W.M. and Judd, D.L. (1957). *Phys. Rev.* **107**, 1.
Sagdeev, R.Z. (1966). Co-operative phenomena and shock waves in collisionless plasmas, in *Reviews of Plasma Physics*, Vol. 4, ed. M.A. Leontovich (Consultants Bureau, New York).
Salpeter, E.E. (1952). *Phys. Rev.* **88**, 547.
Salpeter, E.E. (1960). *Phys. Rev.* **120**, 1528.
Salpeter, E.E. (1961). *Phys. Rev.* **122**, 1663.
Sanderson, J.J. and Priest, E.R. (1972). *Plasma Phys.* **14**, 959.
Schamel, H. *et al.* (1989). *Phys. Fluids* **B1**, 1.
Schott, G.A. (1912). *Electromagnetic Radiation* (Cambridge University Press, Cambridge).
Shafranov, V.D. (1970). *Sov. Phys. Tech. Phys.* **15**, 175.
Sheffield, J. (1975). *Plasma Scattering of Electromagnetic Radiation* (Academic Press, New York).
Shklovsky, I.S. (1953). *Astron. J. USSR* **30**, 15.
Shklovsky, I.S. (1960). *Cosmic Radio Waves* (Harvard University Press, Cambridge, Mass.).
Smith, M.F. and Rodgers, D.J. (1991). *J. Geophys. Res.* **96**, 11617.
Solov'ev, L.S. (1968). *Sov. Phys. JETP* **26**, 400.
Sonnerup, B.U. (1970). *J. Plasma Phys.* **4**, 161.
Steenbeck, M., Krause, F. and Rädler, K.H. (1966). *Z. Naturforsch.* **21a**, 369.
Stix, T.H. (1992). *Waves in Plasmas* (American Institute of Physics, New York).
Stringer, T.E. (1963). *Plasma Phys. (J. Nucl. Energy*, Part C) **5**, 89.
Sturrock, P.A. (1958). *Phys. Rev.* **112**, 1488.

Sturrock, P.A. (1994). *Plasma Physics* (Cambridge University Press, Cambridge).
Swanson, D.G. (1985). *Phys. Fluids* **28**, 2645.
Swanson, D.G. (1989). *Plasma Waves* (Academic Press, New York).
Sweet, P.A. (1958). *Nuovo Cimento Suppl.* **8**, (10), 188.
Takeda, T., *et al.* (1972). *Phys. Fluids* **15**, 2193.
Taylor, J.B. (1974). *Phys. Rev. Lett.* **33**, 1139.
Teubner, U. *et al.* (1997). *Opt. Commun.* **144**, 217.
Tidman, D.A. and Krall, N.A. (1971). *Shock Waves in Collisionless Plasma* (Wiley-Interscience, New York).
Tonks, L. and Langmuir, I. (1929). *Phys. Rev.* **33**, 195.
Trubnikov, B.A. (1958). *Sov. Phys. Doklady* **3**, 136.
Twiss, R.Q. (1950). *Phys. Rev.* **80**, 767.
Twiss, R.Q. (1952). *Phys. Rev.* **88**, 1392.
Walén, C. (1946). *Arkiv. f. Mat.* **A30** No. 15; *ibid.,* **A33**, No. 18.
Weber, E.J. and Davis, L. (1967). *Astrophys. J.* **148**, 217.
Weinberg, S. (1962). *Phys. Rev.* **126**, 1899.
Weizsäcker, C.F. von (1938). *Phys. Z.* **39**, 633.
Wesson, J.A. (1981). MHD stability theory, in *Plasma Physics and Nuclear Fusion Research*, ed. R. Gill (Academic Press, London).
Wesson, J.A. (1987). *Tokamaks* (Clarendon Press, Oxford).
White, R.B. (1983). Resistive instabilities and field line reconnection, in *Basic Plasma Physics I*, ed. A.A. Galeev and R.N. Sudan (North-Holland, Amsterdam).
Wilcox, J.M., Boley, F.I. and DeSilva, A.W. (1960). *Phys. Fluids* **3**, 15.
Wild, J.P. (1950). *Australian J. Sci. Res.* **A3**, 541.
Woltjer, L. (1958). *Proc. Nat. Acad. Sci.* **44**, 489; *ibid.* 833.
Wong, A.Y. and Baker, D.R. (1969). *Phys. Rev.* **188**, 326.
Wong, A.Y., Motley, R.W. and D'Angelo, N. (1964). *Phys. Rev.* **133**, A436.
Yvon, J. (1935). *La théorie des fluids et l'equation d'état* (Hermann, Paris).
Zakharov, V.E. (1972). *Sov. Phys. JETP* **35**, 908.
Zhang, Y.Q., DeSilva, A.W. and Mostovych, A.N. (1989). *Phys. Rev. Lett.* **62**, 1848.

Index

Abel inversion, 432
acoustic wave, 132, 211
adiabatic gas law, 61, 176, 228
adiabatic invariance, 24–36
adiabatic invariants, 12, 25, 28, 32
  first, *see* magnetic moment
  relativistic, 38
  second, 25, 28–31, 36, 68
  third, 25, 31, 32
Airy functions, 430
Airy wavenumber, 459
Alfvén critical point, 174, 195
Alfvén Mach number, 147, 174
Alfvén radius, 172
Alfvén wave, 197
  compressional, 122, 132, 211, 215, 223, 235
  shear, 122, 131, 211, 233, 235, 237
alpha-effect, 165–167
amplifying waves, 276, 429
Ampère's law, 58
anisotropic plasmas, 67
anisotropy parameter, 321
anomalous dissipation, 405
anomalous transport, 268, 488
anti-dynamo theorem, 163
Appleton–Hartree theory, 197
arc discharge
  characteristics of, 141
  plasma parameters in, 10
argument principle, 269

Balescu–Lenard equation, 472, 479
ballistic term, 379, 388
banana orbit, 35, 46, 305
BBGKY hierarchy, 468, 469, 494
Bennett relation, 87
Bernstein modes, 277, 286
  electron, 278–283
    Landau damping of, 281
  ion, 293
Biermann's battery mechanism, 163, 194
black body intensity, 333
  Rayleigh–Jeans form, 333
Bogolyubov hypothesis, 472, 476, 477
Bohm ion saturation current, 456
Bohm sheath criterion, 456
Boltzmann collision integral, 307
Boltzmann distribution, 7, 293
Boltzmann equation, 254
Boltzmann's H-theorem, 313
bounce frequency, 382
bounce time, 45, 306
boundary conditions, 49
  electromagnetic, 74
boundary sheath, 8
bow shock, 177, 197, 244, 318, 405, 410, 421
  electron foreshock, 421
  ion foreshock, 421
  thickness, 405
bremsstrahlung, 3
  as plasma diagnostic, 343
  emission coefficient, 337, 339
  plasma corrections to, 342
  power loss, 3
  power radiated, 335
Budden equation, 434
buoyancy force, 106

carbon cycle, 2
cavitons, 403
characteristic time, 12
charge conservation equation, 486
charge density, 13, 485
charge separation, 17, 23
chromosphere, 6, 148
classical thermodynamics
  relations of, 55
Clebsch variables, 32, 104, 134
cluster expansion, 469, 471
CMA diagram, 221, 224
cold plasma, 198
  approximation, 72, 408
  dielectric tensor, 203
  dispersion relations for oblique propagation, 222–227
  general dispersion relation, 205
    properties of, 217
  wave equations, 72, 411

collision broadening, 347
collision coefficients
  for energy exchange, 317
collision frequency, 10, 254, 296, 315, 499
  effective, 306
  electron–electron, 316
  electron–ion, 316
  in collisionless shock, 417
  ion, 300
  ion–electron, 316
  ion–ion, 316
collision integral, 470
  Balescu–Lenard, 480
  Landau, 480
collision term, 297, 470
  BGK model, 297, 307
  Fokker–Planck, 307–312
collision time, 62, 476
  effective, 306
  electron, 62
  electron–electron, 317
  electron–ion, 10
  ion, 62
  ion–ion, 317
collisional cross-section, 3
collisional dissipation, 407
collisional relaxation, 317, 319
collisionless approximation, 73
collisionless MHD, 69–70
  equations, 69, 71
collisionless shocks, 181, 405
  classification, 408
  Earth's bow shock, *see* bow shock
  energy, 408
  entropy, 408
  foot in magnetic field structure, 420
  laminar, 409
  oblique, 410, 420
  parallel, 410, 420
  perpendicular, 410
    self-consistent model of, 419
  quasi-parallel, 420
  quasi-perpendicular, 420, 422
  structure, 407
  supercritical, 420
  turbulent, 410
collisions, 8, 10, 470
  Coulomb, 308
  electron–ion, 3
compression wave, 406
Compton field, 47
conductivity
  electrical, *see* electrical conductivity
  thermal, *see* thermal conductivity
confinement
  inertial, 5
  magnetic, 84
  toroidal, 304
confinement time, 5
conservation equations, 182
conservation relations, 78
  collisional, 483, 484
containment time, 5
continuity equation, 78, 486
  mass, 52
convective derivative, 50, 392
convective gain, *see* Raman gain, convective
correspondence principle, 340
Couette flow, 133
Coulomb logarithm, 10, 311
Coulomb potential, 472, 475
critical density, 342, 396
critical surface, 395
current
  Chapman–Ferraro, 177
current density, 13, 485
  magnetization, 23, 70
  plasma, 22
  polarization, 37
current filamentation, 155
current sheet, 141, 146, 178
  model, 147
curvature drift, 22, 25
cut-offs, 207, 210, 219
  back-to-back with resonance, 433
  wave behaviour near, 429
cyclotron frequency, 12, 15
cyclotron resonance, 47

de Broglie wavelength, 335
de Hoffmann–Teller frame of reference, 189, 421
Debye length, 8, 262
Debye shielding, 475
  potential, 288
Debye sphere, 8, 10, 256, 307, 475, 476
degrees of freedom
  in MHD shock, 181
  number of, 57
differential scattering cross-section, 309
diffusion
  ambipolar, 300
    coefficient of, 301
  coefficient, 299
    in Fokker–Planck equation, 312
    in neoclassical transport, 306
  in magnetic field, 301
  random walk model of, 304
  time, 82
direct drive, 6
dispersion curves
  crossing of for uncoupled modes, 436
dispersion relation, 200
  Appleton–Hartree, 225, 230
  Bernstein modes, 280
  compressional Alfvén wave, 411
  electron cyclotron wave, 212
  electron plasma waves, 260, 400
  extraordinary mode, 226
  ideal MHD waves, 131
  ion acoustic wave, 230, 267

dispersion relation (*cont.*)
ion cyclotron wave, 212, 224, 293
oblique propagation in warm plasma, 250
longitudinal waves
beam-carrying cold plasma, 239
warm plasma, 229
low frequency warm plasma waves, 232
stimulated Raman scattering
in homogeneous plasma, 441
transverse waves, 210
two-stream instability, 241
whistler wave, 212
oblique propagation, 236
displacement current, 23, 58, 236
distribution function, 252, 312
N-particle, 466
bi-Maxwellian, 302, 319
bump-on-tail, 380
collisional relaxation of, 318
D-shaped, 179
equilibrium, 283, 284, 473
exact N-particle, 465
generic, 467
Gibbs, 472
ion, 179
isotropic two-temperature, 318
Klimontovich, 252
Maxwellian, 260, 289, 473
local, 59, 297, 489
relativistic, 349
non-Maxwellian, 318
self-similar, 318
specific, 467
test particle, 313
Doppler broadening, *see* electron cyclotron radiation, Doppler broadening
Doppler shift
of light frequency, 40
double adiabatic approximation, 67
double adiabatic theory, 70
dressed test particle, *see* Thomson scattering, coherent, dressed test particle approach
drift approximation, 504
drift instabilities, 283, 416, 420
drift velocity, 43
$\mathbf{E} \times \mathbf{B}$, 17, 18, 43, 285, 419
curvature, 21, 43
due to grad B and curvature, 22, 34, 45
due to non-electromagnetic force, 18
grad B, 19, 20, 43, 283
in plane-polarized wave, 40
ion–electron, 285
macroscopic, 284
polarization, 37
drift waves, 417
dynamical friction, 313
coefficient of in Fokker–Planck equation, 312

Earth's core
characteristics of, 141
Earth's ring current, 45
echo wave, 388
efficiency factor, 3
eikonal, 427
approximation, 428
Einstein relation, 299, 301
electric field
effective, 65, 494
Feynman representation of, 325
electrical conductivity, 298, 321, 494, 498, 504
scalar
condition for, 65
tensor, 203, 504
electron acoustic mode, 292
electron cyclotron emission
as diagnostic for electron temperature, 347
electron cyclotron radiation, 344
Doppler broadening, 347
emission lines, 344
Doppler shift, 345
relativistic frequency shift, 345
harmonic lines, 346
plasma cyclotron emissivity, 346
shape function, 347
electron cyclotron wave, 215, 237
electron flux, 491
electron line density, 87
electron runaway, 314
electron trapping, 33
electrostatic force, 58
energy exchange
ion–electron, 62
energy flux, 79, 200
energy principle, 111, 119–123
entropy, 388, 490, 500
fluctuations, 369
of perfect gas, 184
equation of motion
of fluid element, 52–55
equations of state, 55
internal energy, 57
temperature, 57
Erokhin mode conversion equation, 435
Ettinghausen effect, 495
evanescent waves, 429
evolution equations, 482
electron temperature, 487
ion temperature, 487
expansion wave, 406
extraordinary mode, 214, 411

Fermi acceleration, 36
field reversal parameter, 144
filamentation, 405
finite element method, 123
finite transit time effect, 357
floating potential, 454

flow velocity, 49, 70, 480
  electron, 489
  ion, 489
  plasma, 485, 489
fluid element, 49, 268
  dimension of, 63
flux conservation, 81
flux function, 96, 98
flux invariant, *see* adiabatic invariants, third
flux transfer events, 179
flux tube, 80
  length of, 68
  model, 105
Fokker–Planck equation, 307–313, 318, 322, 480
force-free fields, 102–105, 144
Fourier law, 297
  electron, 497
  ion, 496
free–bound transitions, 339
free–free absorption, 341
  coefficient, 341
frequency mismatch, 393
frictional coefficient, 314
frozen flux theorem, 79, 80
fusion plasmas, 67, 141, 504

gas constant, 58
Gaunt factor, 335, 336
  Born approximation for, 339
  Maxwellian-averaged, 338, 339
geomagnetic field, 44, 169, 177, 197
Grad–Shafranov equation, 98–100, 134
  numerical solution of, 100
gravitational drift, 19, 136
gravitational field
  solar, 106
gravitational force
  equivalent, 125
group velocity, 201
guiding centre, 15, 29, 31–34, 41
  approximation, 19, 41
  current, 70
  drift, 22, 23, 42, 93
  drift orbit of, 25
  model, 12
  motion of, 16, 19, 22, 41
  plasma model, 69, 70
  precession of, 32
  velocity of, 17, 43
gyro-frequency, 15
gyro-magnetic flux, 303
gyro-phase, 15
gyro-radius, *see* Larmor radius

Hall current, 66, 495
Hall effect, 65
Hartmann number, 133
heat conduction, 56
  coefficient of, 56
heat energy, 56
heat flux, 299, 491, 495, 497
  electron, 496
  ion, 496
  tensor, 481
  vector, 482
Helmholtz equation, 104
hohlraum, 6
Hugoniot relation
  hydrodynamic, 184, 190
  hydromagnetic, 184, 195

ideal MHD, 60, 77
  approximations, 62, 67
  equations, 61, 77
  stability, 70
impact parameter, 310
indirect drive, 6
induced scattering, 389
induction equation, 59, 166, 193
  for variable resistivity, 149
inertial confinement fusion, 5
  plasma parameters in, 10
instabilities
  absolute, 244–248, 276
  ballooning, 119, 129
  beam–plasma, 241, 272
    growth rate, 273
    kinetic, 273
  bump-on-tail, 268, 273
    growth rate, 273
  Buneman, 243, 274
  convective, 244–247, 276
  current-driven, 122
  disruptive, 160
  drift wave, 285
  fixed boundary, 122
  fluid, 268
  flute
    growth rate, 136
  free boundary, 122
  gravitational interchange, 155
  in beam–plasma systems, 238
  interchange, 123, 124
  ion acoustic, 285
    in current-carrying plasma, 274
  kink, 118, 122
    stability condition, 118
  magnetic buoyancy, 136
  MHD, 121
  Mirnov, 160
  modulational, 402, 404
  non-linear saturation of, 377
  parametric, 392–397
    parametric decay, 395
    stimulated Brillouin, *see* stimulated Brillouin scattering
    stimulated Raman, *see* stimulated Raman scattering
    two plasmon decay, 396
  pressure-driven, 122

Rayleigh–Taylor, 124
MHD analogue of, 124
resistive, 148
characteristics of, 156
rippling mode, 156
sausage, 115, 128
marginal stability condition for, 116
sawtooth, 160
streaming, 244
tearing mode, 146, 148, 151, 161, 194
growth rate, 154
two-stream, 240, 314
interactions
collective, 8, 408
collisional, 253
Coulomb, 309, 470
electron, 468, 470
laser–plasma, 392, 394
resonant, 262, 274, 282
three-wave, 392
instability threshold condition, 393
wave–particle, 198, 409
wave–wave, 389, 392
internal energy, 56, 67
of fluid element, 55
interstellar plasma, 11
inverse aspect ratio, 33, 99, 305
inverse bremsstrahlung, 341
inverse Faraday effect, 47
ion acoustic wave, 230, 236, 237, 394, 397
Landau damping of, 265
ion cyclotron wave, 235, 237
first, 235
second, 235
ion flux, 491
ion plasma wave, 236
ionosphere, 197, 224
plasma parameters in, 10
isentropic transitions, 408
isobaric surface, 84

Jeans theorem, 255, 283
Joule heating, 56

kinematic dynamo problem, 163
kinetic energy
average, 16
conservation of, 14
flow, 56
particle, 14
plasma, 16
kinetic equation, 253, 479
collisional, 254, 289, 296
collisionless, 253
Landau, 472, 479, 498
kinetic temperature, 481
Klimontovich distribution function, *see* distribution function, Klimontovich
Kruskal–Shafranov stability criterion, 118

Landau damping, 227, 256, 274
decrement, 261, 382
electron, 265, 267, 275
experimental verification, 263
ion, 265, 267
non-linear, 383
physical origin, 262
Landau equation, *see* kinetic equation, Landau
Landau growth, 262, 274
Landau theory
non-linear, 377, 388
Langmuir collapse, 404
Langmuir probe characteristics, 456
Langmuir waves, 229, 263, 394
Landau damping of, 289, 295
radiation from, 450
combination line, 451
in laser-produced plasmas, 453
Larmor formula, 328, 335
Larmor frequency, 15, 18
relativistic, 38
Larmor motion, 16, 19
Larmor orbit, 18, 19, 281
Larmor period, 16
Larmor precession, 25
Larmor radius, 12, 15, 21, 302
electron, 282
instantaneous, 18
laser-produced plasmas, 394
channel formation in, *see* ponderomotive force, channel formation
harmonic emission from, 451
plasma line emission, 451
resonance absorption in, *see* resonant absorption
stimulated Raman scattering in, *see* stimulated Raman scattering, in inhomogeneous plasmas
Lawson criterion, 3–5, 88
Leibnitz theorem, 50, 79
lighthouse effect, 351
linear mode conversion, 435
mode conversion coefficient, 437, 438
reflection coefficient, 438
transmission coefficient, 438
Liouville equation, 466, 468
Liouville operator, 466, 468
Liénard–Wiechert potentials, 325
longitudinal invariant, *see* adiabatic invariants, second
longitudinal invariant surface, 32
Lorentz equation
non-relativistic, 13
relativistic, 38
Lorentz gas model, 297
Lorentzian line shape, 369
loss cone, 28
lower hybrid wave, 215
Lundquist number, 141, 146, 193

Mach number, 413
critical, 419, 420
magnetic bottle, 4, 27, 28

magnetic buoyancy, 106
magnetic containment fusion, 5
magnetic convection equation, 68
magnetic curvature, 128
magnetic energy, 104
magnetic field
  bumpy, 44
  force-free, 99
  generation, 162
  geomagnetic tail, *see* magnetotail, magnetic field in
  induced, 16
  interplanetary, 195
  measurement by Thomson scattering, *see* Thomson scattering, coherent, effect of magnetic field
  reconnection, 146
  reversal, 144
  solar, 106
magnetic field lines
  ergodic, 94
  pitch of, 88
magnetic flux
  poloidal, 95
  toroidal, 96
magnetic helicity, 102, 103, 143, 166
magnetic islands, 142, 158
magnetic knot, 105
magnetic mirror, 25, 26, 33, 36, 305
magnetic modulation of scattered light spectrum, 360
magnetic moment, 16, 25, 68
  invariance of, 25, 34, 36, 44, 68
magnetic pressure, 83
magnetic reconnection, 142, 148, 177–179
  driven, 145
  spontaneous, 145
magnetic relaxation, 142
magnetic shear, 22, 125, 129, 155, 158
magnetic surface, 80
magnetic tension, 83
magnetic trap, 31
magnetic well, 93, 129
magnetization, 16, 23
magneto-ionic theory, 224
magnetoacoustic wave, 233, 411
  fast, 132
  slow, 132
magnetohydrostatic condition, 106
magnetopause, 31, 177–179
magnetosheath, 177–179, 405
magnetosphere, 7, 146, 169, 177–179, 197
magnetotail, 7, 177
  magnetic field in, 44
Manley–Rowe relations, 392
Markov system, 494
Markovian, 307
mass conservation, 52
mass density, 485
Maxwell equations, 13, 23, 182, 255, 409
  reduced, 488
Maxwell stress tensor, *see* stress tensor, Maxwell
mean free path, 469
  collisional, 302, 304, 405
  electron, 10, 63, 476
  electron–ion, 176
  ion, 63
  turbulent dissipation, 408
MHD approximation, 58, 63, 488, 489
  non-relativistic, 63
MHD equations, 48, 480, 488
  dissipative, 501
MHD equilibria
  solar, 105
MHD instabilities, *see* instabilities, MHD
micro-instabilities, 262, 268
minimum B stability condition, 129
Mirnov oscillations, 158
mirror ratio, 27
mirror trap, 27–30
mobility
  electron, 300
  ion, 300
mode conversion, *see* linear mode conversion
mode rational surface, 157
moment equations, 480–487
  general moment equation, 483
moment integrals of distribution function, 482
  electrodynamical, 491
  hydrodynamical, 491–494
  non-hydrodynamical, 491–494
  non-plasmadynamical, 491
  plasmadynamical, 491
momentum equation
  parallel, 70
  perpendicular, 70

Navier–Stokes equation, 55
neoclassical theory, 491
neoclassical transport, 304, 487
Nernst effect, 495
neutral sheet, 44, 142, 146
non-linear Schrödinger equation, 403, 423
number density, 7, 480
Nyquist diagrams, 270

Ohm's law, 59, 64, 496, 504
  generalized, 64, 73, 486, 494
    in MHD approximation, 65
    in warm plasma wave equations, 73, 231
omega-effect, 165–167
Onsager symmetry, 492
optical depth, *see* radiative transport equation, optical depth
optically thick plasma, 333
optically thin plasma, 332
ordinary mode, 214, 411

pair correlation function, 469, 470
  equilibrium, 475
parameter space, 10, 218, 219
parametric amplification, 376
parametric waves, 377

particle acceleration, 35
  at shocks, 421
particle drifts, 22
particle energy, 32
particle orbits
  in torus, 93
particle trapping, 382–386, 397
  in beam–plasma instability, 384
passing particles, 33–34
path variables, 56
Peltier effect, 497
Penrose criterion, 272
perfect gas, 57
Petschek model, 147–148, 193
  maximum reconnection rate in, 147
phase integral, 428
phase mixing, 388
phase shift, 432
phase space, 466
phase velocity, 201
photosphere, 6, 7, 106, 107
pinch parameter, 144
pitch angle, 27, 36, 45
plasma beta, 83, 86, 88, 105
  poloidal, 89
  toroidal, 89
plasma bremsstrahlung, 334–339
  spectrum
    classical, 336
    quantum mechanical, 338
plasma cavities, 402
plasma characteristics, 7
plasma containment, 4, 93, 108, 161
plasma diamagnetism, 16, 23
plasma dielectric function, 258, 276, 472, 479
plasma dispersion function, 265
plasma echoes, 388, 423
plasma frequency, 8
  electron, 8, 254, 258
  ion, 9
plasma oscillations, 8, 23, 260
  coherent, 408
  electron, 197, 229
  longitudinal, 209
plasma parameter, 8, 10
plasma reflectometry, 432
plasma resistivity, 140
  variable, 149
plasma sheath, 453
  equation, 454
plasma sheet, 31, 44, 244
plasma transport, 158, 296
plasma wave equations, 49
plasma waves
  electron, 227, 397
  ion, 230
Plemelj formula, 479
Poisson equation, 378
polarization
  circular, 199
  elliptical, 199
  extraordinary mode, 217
  linear, 199
polarization drift, 37
polarization dyadic, 355
ponderomotive force, 40, 398, 401, 404
  channel formation, 41
power radiated by accelerated charge, 326
Poynting flux, 357
Poynting theorem, 200
Poynting vector, 200
  for radiation field, 326
pressure
  fluid in motion, 54
  scalar, 481, 486
  tensor, 481, 491
    traceless, 481, 495
  thermodynamic, 54
prominences, 108
proton–proton cycle, 2
pseudo-resonance, 235, 236

quasi-linear theory, 377–382, 388
quasi-neutrality condition, 57, 64, 413, 486
quiver motion
  electron, 40
quiver velocity, 441

radial force balance, 85
radial pressure balance, 101
radiation
  absorption coefficient, 331
  broadening, 347
  emission coefficient, 331
  flux, 330
    spectral density of, 330
  frequency spectrum from accelerated charge, 328
  from Langmuir waves
    combination line, *see* Langmuir waves, radiation from, combination line
    harmonic emission, *see* laser-produced plasmas, harmonic emission from
    plasma line, *see* laser-produced plasmas, plasma line emission
  intensity, 330
  pressure, 355
  temperature, 334
radiation transport in plasma, 330
radiative recombination, 339
radiative transport equation, 332
  optical depth, 332
  source function, 332
radio waves, 197
radiofrequency heating
  ion cyclotron resonance, 439
  of tokamak plasma, 439
Raman decay waves, 442
Raman gain, convective, 445

Raman resonance, 444
  detuning condition, 446
  width, 446
Raman scattered light wave, 442
Raman threshold
  convective, 446
Rankine–Hugoniot equations, 184
ratio of specific heats, 57
ray path, 331
ray-tracing approximation, 428
Rayleigh-Jeans limit, *see* black body intensity, Rayleigh-Jeans limit
recombination radiation, 339
  emission coefficient, 340
reflected ions, 419
reflection coefficient, 431
reflectometry, *see* plasma reflectometry
resistive MHD, 59, 503
  approximations, 60, 66
  equations, 60, 505
resistivity, *see* plasma resistivity
resonance, 207
  accessibility, 433
  conditions, 389
    induced scattering, 389
    linear Landau damping, 389
    particle trapping, 389
    wave–wave interactions, 389
  electron cyclotron, 208, 212, 226, 236
  heating
    electron cyclotron, 439
  ion cyclotron, 208, 212, 223, 235, 236
  lower hybrid, 209, 215
  principal, 207, 210, 219
  upper hybrid, 209, 216
  wave behaviour near, 433
resonant absorption, 460
  shape function, 460
  tunnelling, 459
resonant angle, 207
resonant electrons, 382
resonant surface, 128, 129, 152, 156
resonant triad, 389
response function, 277
retarded potentials, 325
retarded time, 325
reversed field pinch, 99, 143
Reynolds number, 192
  hydrodynamic, 61
  magnetic, 61, 140, 193
Righi–Leduc effect, 495
Rosenbluth convective gain threshold, 446
Rosenbluth potentials, 311
rotating star, 194
rotational transform, 88, 94

safety factor, 34, 88, 118, 130, 160, 305
sawtooth oscillations, 159
scale length, 12
  diffusion in magnetic field, 302
  dispersive steepening, 407
scale length (*cont.*)
  sunspot, 105
  wave-front, 412, 415, 418
  wavetrain decay, 418
screw pinch, 87, 99
self-consistent field, 12, 470, 479
Shafranov shift, 101, 134
shape function
  electron cyclotron radiation, 347
  Salpeter, 364
  synchrotron radiation, 352
sheath thickness, 301, 463
shock waves, 285
  astrophysical, 420
  collisional, 181, 406
  collisionless, *see* collisionless shocks
  fast, 190
  formation of, 405
  jump conditions, 184, 186, 407
  low β, 408
  MHD, 179
  planetary, 420
  slow, 190
  structure, 181, 190, 407
  switch-off, 190
  switch-on, 187, 190
  thickness, 180, 192
skin current, 74
small Larmor radius approximation, 408
solar corona, 6, 169
  characteristics of, 141
  plasma parameters in, 10
solar flares, 28, 108, 142, 146, 177
solar wind, 7, 31, 169, 405, 421
  characteristics of, 169
  plasma parameters in, 10
solitons, 403, 408, 414, 418
sound waves
  propagation of, 70
space charge, 300
space plasmas, 67, 141, 503, 504
space potential, 457
spectral density of radiation flux, *see* radiation, flux, spectral density of
spectral energy density of electrostatic field, 380
stability
  radial, 5, 98, 99
  toroidal, 5, 98
state variables, 55
static equilibrium
  configurations, 82
    cylindrical, 85
    toroidal, 89
  equations, 82
stimulated Brillouin scattering, 397
stimulated Raman scattering, 397, 441
  back-scatter, 445
  equations
    numerical solution of, 447
  forward-scatter, 445
  growth rate, 442

stimulated Raman scattering (*cont.*)
in homogeneous plasmas, 441
maximum growth rate, 442
in inhomogeneous plasmas, 442
absolute growth rate, 445
convective growth, 445
non-local behaviour, 446
wave damping coefficients, 442
Stokes equation, 430
Stokes phenomenon, 431
stop-band, 210
stress tensor, 52
equilibrium, 67
isotropic, 53
Maxwell, 78
non-isotropic, 54
total, 82
viscous, 54
Sun
chromosphere, *see* chromosphere
convection zone, 6
corona, *see* solar corona
magnetic field of, *see* magnetic field, solar
photosphere, *see* photosphere
radiation zone, 6
sunspots, 105, 108, 134
characteristics of, 141
diffusion time, 141
supra-thermal particles, 410
surface charge, 74
surface current, 74
surface wave, 462
Suydam criterion, 130, 136
Sweet–Parker model, 146
reconnection rate, 193
synchrotron radiation
by ultra-relativistic electrons, 351
emission intensity, 353
emission spectrum, 352
emission spectrum from Crab nebula, 354
lighthouse effect, 351
spectral index, 353
spectral width, 352
from hot plasmas, 348
emission coefficient, 349
harmonic structure, 350
relativistic broadening, 351

Taylor hypothesis, 143
temperature equilibration, 62
test particle, 287, 288, 313
model, 313, 315
thermal conductivity, 299, 494
coefficient of, 66, 297
electron, 498, 500
ion, 498, 500
thermal velocity, 481
thermodynamic equilibrium
local, 334
thermodynamic fluxes, 492
thermodynamic forces, 492
thermodynamics
first law of, 55
second law of, 184, 186, 501
thermoelectric coefficients, 494, 498
thermoelectric effect, 497
thermonuclear fusion, 2, 6
controlled, 2, 3
thermonuclear power, 3
theta-pinch, 85, 93
Thomson cross-section, 328
Thomson scattering
coherent, 361
dressed test particle approach, 361
effect of Coulomb collisions, 369
effect of impurity ions, 366
effect of magnetic field, 360
experimental verification, 365
form factor, 362
plasma diagnostic, 365
scattering parameter, 364
spectral power density, 362
electron temperature diagnostic, 358
incoherent, 355
form factor, 358
scattered power, 357
relativistic non-linear, 374
time scales
convective, 140
gravitational, 155
hydrodynamic, 62, 488
resistive diffusion, 140
temperature equilibration, 317
tokamak, 4, 99, 119
banana orbit, *see* banana orbit
characteristics of, 141
disruptions, 142
equilibria, 100, 101
fields, 5
instabilities, 157
Mirnov oscillations, *see* Mirnov oscillations
particle orbits in, 33
passing particles, *see* passing particles
plasma parameters in, 10
sawtooth oscillations, *see* sawtooth oscillations
trapped particles, *see* trapped particles
Tonks–Dattner resonances, 462
toroidal force balance, 89, 92, 101
transit time, 29
Alfvén, 140
proton, 31
transport coefficients, 55, 297, 492–500, 503
anomalous, 318
turbulent, 420
transport equations, 190, 420, 492
transport matrix, 492
transport theory, 296
classical, 487
transverse waves, 210
trapped particles, 33–35, 305
Trubnikov function, 350

turbulent dissipation, 407
type III solar radio noise, 450

upper hybrid frequency, 280

Van Allen radiation belts, 30–31
virial theorem, 84, 103
viscosity
  bulk, 54
  coefficient of, 66, 300
  electron, 498, 500
  ion, 498, 500
  kinematic, 61
  shear
    coefficient of, 54
  stress, 54
  tensor, 495
Vlasov equation, 254, 289, 296, 377, 409
  electron, 256
  equivalence to particle orbit theory, 254
  numerical solution of, 293

warm plasma
  general dispersion relation, 230
  theory, 227
  wave equations, 72, 73, 398
  waves, 227
wave dispersion, 200, 406
  normal, 201
wave normal surface, 198, 218
  destructive transition, 222
  intact transition, 222
  re-shaping transition, 221
wave packet, 200
wave-front, 406
wavenumber mismatch, 394, 397
weak coupling approximation, 256, 471, 475, 479, 498
whistler wave, 215
winding numbers, 95
WKBJ approximation, 426
  electric field swelling, 428
  equations for stimulated Raman scattering, 444
  solutions to Stokes equation, 430
    dominant term, 431
    subdominant term, 431
  validity of solutions, 427
Woltjer theorem, 143

Z-pinch, 4, 86, 90, 108, 115, 133
Zakharov equations, 397–402
  first, 400
  second, 401

Printed in the United States
By Bookmasters